Second Edition

Developments in Pressure-Sensitive Products

T0132594

Second Edition

Developments in Pressure-Sensitive Products

Edited by
István Benedek

CRC Press
Taylor & Francis Group
Boca Raton London New York

CRC Press is an imprint of the
Taylor & Francis Group, an **informa** business

A TAYLOR & FRANCIS BOOK

CRC Press
Taylor & Francis Group
6000 Broken Sound Parkway NW, Suite 300
Boca Raton, FL 33487-2742

First issued in paperback 2019

ISBN-13: 978-1-57444-542-8 (hbk)
ISBN-13: 978-0-367-39183-6 (pbk)

Library of Congress Card Number 2005043937

Library of Congress Cataloging-in-Publication Data

Developments in pressure-sensitive products / edited by Istvan Benedek.-- 2nd ed.
    p. cm.
  Includes bibliographical references and index.
  ISBN 1-57444-542-1
  1. Pressure-sensitive adhesives. 2. Adhesives. I. Benedek, Istvan, 1941-

TP971.D48 2005
668'.3--dc22   2005043937

Visit the Taylor & Francis Web site at
http://www.taylorandfrancis.com

and the CRC Press Web site at
http://www.crcpress.com

Preface to the Second Edition

Over the short time since the first edition of this book, advances in macromolecular science and engineering have led to new pressure-sensitive products and applications by using multiple ways to design them. Practice has confirmed that my attempt to integrate the various pressure-sensitive products — that may differ in their manufacturing technology, but not in their use — in a sole category of self-adhesives was correct and necessary. Pressure-sensitive adhesives (PSAs) represent only a part of the science and technology of pressure-sensitives. The whole domain includes them as well as other engineering solutions. There are few books addressing pressure-sensitive adhesives and this one is unique in discussing the competitive technologies also.

This is the first comprehensive book bridging the gap between the fundamental concepts of pressure sensitivity and its application. As stated in the previous edition, this is the first attempt to examine comparatively the various products with or without adhesive that have pressure-sensitive behavior during their application. Therefore, this book simultaneously addresses the pressure-sensitive adhesives, the products based on pressure-sensitive adhesives, and the adhesiveless products which work like pressure-sensitive adhesives (due to their various chemical or physical characteristics). The scope of this investigation is to establish their end-use parameters that must be generally valid. Although the competition in application practice of adhesive- and plastics-based pressure-sensitive products is a reality, because of their composite structure, their mechanism of functioning can vary. Therefore, a strong emphasis is placed on establishing a clear understanding of the complex interaction between the fundamentals of pressure-sensitive adhesion and the manufacture and application technology of self-adhesive products.

Economic considerations forced the development of pressure-sensitive products without adhesives. These products have a growing industrial application especially in the packaging industry and are based on self-adhesive polymer films which replace adhesive-based products for several applications where removability is required. Unlike classical pressure-sensitive adhesives that are viscoelastomers, such products are based on plastomers. They are developed from common plastics and have special macromolecular characteristics. On the other hand, certain classic plastomers without self-adhesivity can be transformed into pressure-sensitive adhesives using simultaneous plasticizing and crosslinking. The classic approach to manufacture pressure-sensitive adhesives uses hydrophobic elastomers that are tackified (and crosslinked). The new technology tackifies hydrophilic plastomers. Thus pressure-sensitive hydrogels are manufactured, which at a first glance, do not confirm the common theory of pressure sensitivity, but work like pressure-sensitives. Such developments confirm the conviction of the author about the necessity and opportunity of a global examination and explanation of pressure-sensitive products.

Pressure-sensitive adhesives were described in the previous works of the author: *Pressure-Sensitive-Adhesives Technology* (1997, Dekker), *Pressure Sensitive Formulation* (2000, VSP) and *Pressure-Sensitive Adhesives and Applications* (2004, Dekker). Pressure-sensitive products were discussed in the first edition of this book: *Developments in Pressure-Sensitive Products* (1999, Dekker). Recent advances in both domains (that is, pressure-sensitive adhesives and pressure-sensitive products) make possible the formulation of general principles concerning pressure-sensitive behavior. Therefore, in this book a new chapter is included that discusses the "Molecular Fundamentals of Pressure-Sensitive Adhesion" (see Chapter 4). This is a phenomenological analysis of the relationship between pressure-sensitive adhesion and the characteristics of molecular mobility and cohesive strength of PSA polymers. In this way, our work becomes the first global theoretical approach to pressure-sensitives.

Manufacture- and application-related developments made it necessary to insert a new chapter describing the features of "Crosslinking of Solvent-Based Acrylics" (see Chapter 6) that constitute the most important class of special products; this chapter also includes radiation-curing technology. A special segment of the book is dedicated to application-related formulations. The quite new class of pressure-sensitive hydrogels and their special high-level end-use is investigated in Chapter 9, "Molecular Design of Hydrophilic Pressure-Sensitive Adhesives for Medical Applications."

Although this book is based on its first edition and includes a large amount of the published and unpublished data arising from the fundamental and application-related research work of the author, the rapid expansion and diversification of pressure-sensitive products and the development of macromolecular science and technology imposed a cooperation with scientists and engineers, specialists in their domains. The author is very grateful for their precious contribution.

István Benedek

Preface to the First Edition

Over the last two decades, the manufacture and application fields of pressure-sensitive products have developed from a largely empirical body of accumulated practical knowledge, to an increasingly sophisticated science, utilizing the most advanced techniques of physics, chemistry, and engineering. Pressure-sensitive labels, tapes, protective films, seals, business forms, etc., are used in medicine, pharmaceutical applications, electronic circuits, assembly of machine parts, as well as in product promotion, coding, and packaging.

Due to the wide utility and consumer acceptance of these products, a high level of basic research and product development has evolved over the last few years and is continuing to grow. Pressure sensitivity, the main performance characteristic of these various products, possesses a common scientific basis in macromolecular chemistry and physics. The ways and means, however, to achieve it differ, and include the technology of adhesives and plastics as well. In the final use, no one asks about the construction of the product whether it is pressure-sensitive, or with or without adhesive. The technical solution should work. This book is an attempt to integrate the different technologies to give the same result, a pressure-sensitive product.

This is a book about pressure-sensitive products. This work is intended as a companion volume to the book I wrote earlier with L.J. Heymans, *Pressure-Sensitive Adhesives Technology* (Dekker, 1997). My main aim in writing this work was to bridge the gap between theory and practice, between engineering of plastics, the technology of adhesives, and the conversion and integration of the practical aspects including the engineering fundamentals. This work is a guide to the entire field of pressure-sensitive products, with or without adhesives, and discusses the engineering steps (of paper, plastics, adhesives, and other materials) required for their manufacture.

This monograph covers a broad spectrum of knowledge, and is designed for production and manufacturing managers, production engineers, material scientists, chemists, new product specialists, and other technologists involved in the efficient producing or use of pressure-sensitive products and in new process and product developments. I discuss the whole complex of buildup, manufacture, testing, and application of pressure-sensitive products. The focus of the description of the technology is on specific examples of application rather on theory. The basic principles of this technical domain, however, are always presented, and the book summarizes our present understanding of the construction and functioning of pressure-sensitive products.

It is not the aim of this book to give a detailed discussion of the science of adhesives or plastics, nor does it constitute a practical vade mecum. It is an attempt to integrate technical domains which are belonging together, as an aid for those involved in the understanding, design, and use of pressure-sensitive products.

István Benedek

Contributors

István Benedek
Pressure Sensitive Consulting
Wuppertal, Germany

Gary W. Cleary
Corium International
Redwood City, California
USA

Zbigniew Czech
Szczecin University of Technology
Polymer Institute
Poland

Mikhail M. Feldstein
Topchiev Institute for
    Petrochemical Synthesis
Moscow, Russia

Nicolai A. Platé
Topchiev Institute for
    Petrochemical Synthesis
Moscow, Russia

Table of Contents

1 Introduction

István Benedek

Pressure-sensitive tapes were first used about 150 yr ago. Pressure-sensitive labels came to the market 90 yr later. About 10 yr after that pressure-sensitive protective films were manufactured. Pressure-sensitive products (PSPs) such as pressure-sensitive adhesive (PSA) coated web have been defined by the special nature of this adhesive although the definition of PSA is not completely clear. The German technical term *Haftkleber* (i.e., adhesive which adheres) supposes that it is possible to differentiate between adhesion and building up of an adhesive bond. In English the term pressure-sensitive adhesives (i.e., adhesives that bond when pressure is applied) admits pressure as an indispensable condition for their function. In reality, as known from loop tack measurements and touch blow labeling, almost no pressure is required for label application but high pressures are needed for protective films in coil coating. *Autocollants*, the French name does not define the application conditions. It refers only to the bonding behavior. The common characteristic of these products is ensured by their special viscoelastic behavior, manifested as permanent cold flow, where the chemistry of the adhesive plays only a secondary role [1]. The development of PSPs without a coated PSA layer (in the classical sense known from the converting industry) makes the definition of this product group more difficult. Adhesive-free PSPs have been developed some decades ago. In this case adhesive-free means that the self-adhesive component is not coated on the product surface. It is included in the carrier, that is, the carrier *per se* is pressure sensitive. In some cases, an adhesive-free composition is used and pressure sensitivity is provided by physical treatment of the carrier surface or application conditions (temperature, pressure). According to the definition given in Ref. [2], PSAs are adhesives "which in dry form are aggressively and permanently tacky at room temperature . . . and adhere without the need of more than finger or hand pressure, require no activation by water, solvent, or heat."

Most PSPs do not meet these requirements. However, they manifest self-adhesivity and, under well-defined conditions, can be applied like a PSA-coated classical PSP, that is, like an adhesive acting via viscous flow and debonding like a viscoelastic compound. Such behavior is achieved by a complex buildup and reciprocal interaction of the product components and in some cases by special application or deapplication conditions. Obviously, a physically treated hot laminating plastomer film applied under pressure or a warm laminating film based on a partially viscoelastic olefin copolymer and applied under pressure cannot have the same chemical basis as a PSA-coated product or a plastic-carrier material that includes PSA. As mentioned, their application conditions are quite different also. However, all these products work as viscoelastic bonding elements and are used in application domains of classical, PSA-coated products. Therefore, they can be considered as PSPs. Plastic processing specialists possess the know-how of the manufacture of plastic-based PSPs. Specialists in converting/coating are skilled in the design and testing of PSPs. They control the market also. Therefore, from an economic point of view, both domains belong together.

PSAs have been used for about a century for medical tapes and dressing. Natural adhesives mixed with natural resins, waxes, and fillers were applied as the first PSPs in the form of medical plasters [3]. In 1845, Horace H. Day prepared and patented a plaster composed of a mixture of natural rubber and tackifier resin coated on a cloth [4]. According to Refs. [5,6], the first tape patented by Paul Beiersdorf was a zinc oxide/rubber-based plaster. At the end of the

19th century, masking tapes and cellophane tapes were the early nonmedical PSPs. For such products, natural rubber was preferred as the raw material. In this period of time, PSAs were used for plasters, labels, and tapes [7]. Industrial tapes were introduced in the market in the 1920s and 1930s and self-adhesive labels in 1935–36 [8,9]. In the 1930s, Stanton Avery developed Kum-Kleen labels [10]. In 1955, the firm Sassions of York was licensed by Avery to produce PSA labels in Europe [11].

Tapes are self-adhesive, self-wound web-like materials used (mainly) as a continuous web. Generally, tapes are produced by coating of a nonadhesive web with PSA, but some tapes have a self-adhesive carrier material. The pressure-sensitive layer is protected by the back side of the carrier material. The permanent web-like character of tapes allows the use of higher forces in their application. The lower converting degree of common tapes permits their design mainly for their adhesive properties. In this case, the adhesive performances need not be balanced. Therefore, theoretically, tapes may also be formulated as PSA-free products.

Labels are self-adhesive, laminated carrier materials. Generally, they possess continuous web-like character only during their manufacture. Labels can also be produced as separate items. Quite unlike tapes, labels are used as discrete objects with a well-defined geometry. Because of their adhesive- and surface characteristics, the self-adhesive layer of labels must be protected with a supplemental solid-state abhesive material (release liner). The first release material was wax, as used by Stanton Avery [12]. Silicone release coatings have been in the market since the mid-1950s [13,14]. Labels preserve their laminate character until their application. Because of their discontinuous character, limited contact surface, high application speed, and low application pressure, labels have to exhibit well-balanced adhesive characteristics. Therefore, most of them are manufactured in the classical way, by coating a nonadhesive carrier material (face stock) with a PSA.

Protective films are removable, self-adhesive webs based on a carrier material that possesses built-in or built-on self-adhesive properties. The role of protective films is to protect a product by adhering to it, covering its surface with a mechanically resistant supplemental layer. This is a time-limited function, that is, the bond should be removable, allowing the protective sheet to be separated from the protected surface. Protective films are packaging materials, not so much in a legislative sense as in a functional one. Unlike classical packaging materials, where functionality concerns protection of the product during transport and storage and aesthetic, marketing-related design characteristics are determinant, protective films are technological components of a product, attached in many cases to the raw product, and passing through the entire manufacturing process up to the finished product, that is, undergoing the working steps of fabrication. Such products are applied by lamination/delamination of large surfaces. Therefore, the resultant bonding/debonding forces are much higher than those applied for labels or tapes. In contrast, protective films have to be removable. Therefore, the instantaneous adhesive performance of protective films plays only a secondary role. It is evident that in this case adhesive-free constructions may be equivalent to PSA coated products.

The development of different product classes has been conditioned by the development of the raw materials and of the coating and application technology. Pressure-sensitive labels, stickers, and other products have seen considerable development over the years with the appearance of new materials and combinations of materials, as well as new processing technologies. By the end of the 1920s, acrylics (ACs) had been synthesized. Acrylics possess adequate die-cutting properties and do not manifest migration (they contain no low molecular products). Their introduction made possible the development of label manufacture. Acrylics display resistance to aging and plasticizers, which allowed their use on transparent carrier materials. Tapes and protective films on polyvinylchloride (PVC) and polyolefin carriers have been produced. Envelopes, wall covering, forms etc. have been manufactured. The nonirritating behavior of acrylics permits their use for medical tapes. Fixing, transfer, carpet, or electrical insulating tapes are also made with acrylics.

The development of thermoplastic elastomers allowed the use of less expensive coating equipment via hot-melts. Styrene–olefin block copolymers have been developed in 1965 [15].

Hydrocarbon-based resins have been introduced as tackifier around 1935 [16]. Polyvinyl acetate emulsions have been produced since 1940 [17]. Vinyl acetate copolymer dispersions together with water-based acrylics allowed the development of water-based technology. In 1960, ethylene–vinyl acetate copolymers were introduced in the market. The first class of raw materials versatile enough to be used as hot-melt, nonadhesive, and adhesive carriers appeared. Owing to this development, new ways were opened to manufacture products having pressure-sensitive properties without coating and without PSAs. Some years before, the plasticizing technology of PVC had made possible the manufacture of plastomers, elastomers, and adhesives on the basis of the same raw material. Parallel advances were made in the field of plastomers and elastomers. The development of cling films, tackified carrier films, sheet-like laminating adhesives, etc. confirmed the possibility of manufacturing self-adhesive products without PSA. The nonPSA-related PSP technology gained an important market segment (especially in the field of protective films and tapes). In contrast, the curable prepolymer technology advanced from the experimental to the industrial domain. In this situation, the PSP technology became a complex field of industrial procedures from the adhesive manufacture (synthesis and formulation) to coating and from the plastomer/elastomer manufacture to processing. The development of plastomer-based pressure-sensitive hydrogels opened new ways for the formulation of hydrophilic PSPs [18]. Some of the most important features of PSAs, their chemistry, formulation, coating technology, and test methods are described in companion volumes of the author [19–22]. The goal of this book is to discuss the PSPs technology as a whole, relating adhesive and adhesiveless manufacture to their common background, macromolecular science.

REFERENCES

1. I. Benedek, *Eur. Adhes. Sealants*, (2), 25, 1996.
2. J.H.S. Chang, 0,179,628/Merck & Co., Inc., Rahway, NJ, USA, EP 1984.
3. J. Andres, *Allg. Papier Rundschau*, (16), 444, 1986.
4. R. Jordan, *Adhäsion*, (1/2), 17, 1987.
5. *Coating*, (11), 46, 1990.
6. *Coating*, (11), 307, 1990.
7. R. Gutte, *Reichold Albert Nachrichten*, (4), 26, 1970.
8. P. Foreman and P. Mudge, *EVA-Based Waterborne Pressure-Sensitive Adhesives*, in Proceedings of Tech 12, Technical Seminar Proceedings, Itasca, IL, USA, May 3–5, 1989, p. 203.
9. M. Fairley, *Labels Label. Int.*, (5/6), 76, 1997.
10. *Etiketten-Labels*, (3), 8, 1995.
11. *Der Siebdruck*, (3), 70, 1986.
12. C.M. Brooke, *Finat News*, (3), 34, 1987.
13. H. Brus, C. Weitemeyer, and J. Jachmann, *Finat News*, (3), 84, 1987.
14. *Adhes. Age*, (8), 28, 1986.
15. D.J. St. Clair and J.T. Harlan, *Adhes. Age*, (12), 39, 1975.
16. D.R. Tucker, *Adhäsion*, (7), 248, 1971.
17. F.M. Rosenbaum, *Adhes. Age*, (6), 32, 1972.
18. A.A. Chalykh, A.E. Chalykh, M.B. Novikov, and M.M. Feldstein, *J. Adhes.*, 78, 667, 2002.
19. I. Benedek and L.J. Heymans, *Pressure-Sensitive Adhesives Technology*, Marcel Dekker, Inc., New York, 1997.
20. I. Benedek, *Pressure-Sensitive Formulation*, VSP, Utrecht, 2000.
21. I. Benedek, *Development and Manufacture of Pressure-Sensitive Products*, Marcel Dekker, Inc., New York, 1999.
22. I. Benedek, *Pressure-Sensitive Adhesives and Applications*, Marcel Dekker, Inc., New York, 2004.

2 Buildup and Classification of Pressure-Sensitive Products

István Benedek

CONTENTS

Pressure-sensitive products (PSPs) may have a simple or sophisticated construction, depending on their end-use. More or less expensive products can be used in the same application field; therefore, the buildup of PSPs differ according to their product class and special use. For a better understanding of their function, we first consider the buildup of the PSPs.

I. CONSTRUCTION OF PRESSURE-SENSITIVE PRODUCTS

Generally, PSPs are sheet-like constructions which exhibit self-adhesion. In principle, such products include a component that ensures the required mechanical properties and a component that provides adhesivity.

A. GENERAL BUILDUP OF PRESSURE-SENSITIVE PRODUCTS

Supposing that, in principle, a pressure-sensitive laminate is built up from a carrier material, an adhesive, and a release liner, such product may be defined as shown in Figure 2.1. The complex,

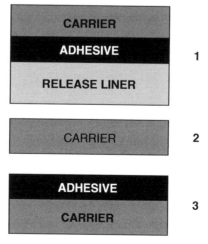

FIGURE 2.1 Buildup of the main PSPs. (1) Label; (2) extruded protective film; (3) tape.

multilayer structure of labels containing separate, solid-state carrier and release components bonded by an adhesive can be simplified for tapes. Tapes and adhesive-coated protective films possess only one solid state-carrier component, and it is coated with adhesive. The new generation of protective films is built up from an (adhesive) carrier without a pressure-sensitive adhesive (PSA) layer (Figure 2.1).

As can be seen in Figure 2.1, theoretically such protective films are the simplest PSPs, built up as one-component pressure-sensitive (self-adhesive) carrier material. Tapes have to be more aggressive; therefore, generally, they need a PSA layer coated on a nonadhesive carrier material. Labels with balanced adhesive performances and high speed machine application require a separate release liner. Because of the discontinuous (nonweb-like) character of labels their handling and automatic application require a continuous, supplemental carrier material, that is, a release liner. The liner allows the labels to be processed as a continuous web and protects their adhesive layer. It can be concluded that PSPs, generally, have either a carrier and a PSA layer, or a pressure-sensitive carrier. Other constructions are known also.

Some PSAs may be used as PSP *per se*, without a carrier material. In other cases, such as decalcomania, transferable letters, etc., the PSA layer also plays the role of information carrier. However, because of its discontinuity and for mechanical resistance it also requires a release liner.

The main characteristics of PSPs are their pressure-sensitive bonding and debonding. Such performances have to be ensured by one of the product components or by the assembly as a whole. In classical PSA-coated products adhesivity was given by the PSA. As discussed earlier, generally, application of PSPs requires a solid-state carrier material too. The simplest classical PSP can be designed as an adhesive coated carrier material. Such product has to adhere on the substrate surface. Theoretically, the PSA is the bonding component of the PSP; the carrier should only allow its application.

Depending on its end-use, the nature of the solid-state carrier material and the character of the bond of a PSP can be quite different. Therefore, supposing the classical construction of a label as shown in the Figure 2.1, the nature and geometry of both components (solid-state carrier material and PSA) vary. Because of different manufacturing possibilities the construction of PSPs is more sophisticated. The particular buildup is a function of required performance and manufacturing procedure.

For instance, certain tapes have to be primed to ensure good anchorage of the adhesive on the carrier. For some tapes a release layer should be coated on the back side of the carrier. Such a layer is not necessary if the material of the carrier exhibits abhesive properties, like certain nonpolar plastic films. For special tapes a separate release film should be interlaminated (Figure 2.2).

B. PARTICULAR BUILDUP OF PRESSURE-SENSITIVE PRODUCTS

The buildup of labels has been described in a detailed manner in the companion volumes [1,2]. Historically, the production of the PSPs started with the manufacture of tapes, that is, with the production of monowebs coated with PSA. The development of removable PSAs allowed the manufacture of removable labels, tapes and later the manufacture of large surface PSPs displaying permanent adhesion during processing of the laminate and removability after. Such products are protective films. Generally, the adhesive layer of tapes and protective films is protected by self-laminating (self-winding) only. Such products are manufactured and used as constructions having only one solid-state carrier material, that is, as monowebs. Labels, tapes, and protective films are generally designed as products having a continuous solid-state carrier coated with a continuous adhesive.

Special tapes exist that have an adhesive with a carrier-like character. Such behavior can be achieved by crosslinking, foaming, filling, or reinforcing the adhesive layer (Figure 2.3). The adhesive can contain a metallic network to ensure electrical conductivity [3].The adhesive (if coated on a carrier) can be discontinuous also. Producers are looking to produce their own base

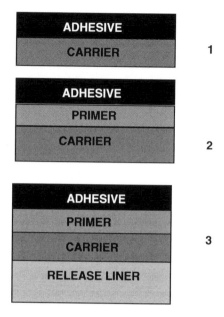

FIGURE 2.2 Buildup of tapes. (1) Adhesive-coated carrier; (2) adhesive-coated primed carrier; (3) adhesive-, primer-, and release-coated carrier.

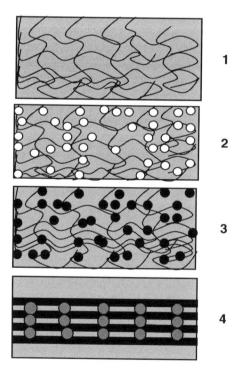

FIGURE 2.3 Carrier-less pressure-sensitive constructions. (1) Crosslinked adhesive; (2) foamed and crosslinked adhesive; (3) filled and crosslinked adhesive; (4) reinforced adhesive.

label stock or special stripe-, patch- or spot-coated adhesive construction and liner-less labels, labels having adhesive-free zones, form and text combinations [4]. Figure 2.4 presents the main pressure-sensitive constructions.

1. Monoweb Constructions

Classical monoweb constructions (tapes, protective films, etc.) possess a PSA adhesive layer coated on a nonadhesive carrier material. As discussed later (see Chapter 8), the adhesive layer can be applied using various coating techniques such as those used in the converting (coating or printing) industry or in plastics manufacture (extrusion). The development of macromolecular chemistry and extrusion technology allowed the manufacture of carrier materials with built-in pressure-sensitivity, that is, PSPs constructed like an uncoated monoweb but behaving when applied, like a coated web (see also Chapter 5).

a. Uncoated Monoweb

PSPs which are built up like an uncoated monoweb are composed of a carrier material only. This carrier must have a special chemical nature or undergo special physical treatment to allow self-adhesion under special application conditions (pressure, temperature, and surface treatment). Generally, such conformable, auto-adhesive monowebs are plastic films (see Chapter 11). According to Djordjevic [5], "films are planar forms of plastics, thick enough to be self-supporting but thin enough to be flexed, folded, or creased without cracking." The upper dimensional limit for a film is difficult to define and is situated between 70 and 150 μm depending on the polymer used as raw material. Uncoated monowebs used as PSPs can have a homogeneous or heterogeneous structure. The whole carrier can be auto-adhesive (e.g., ethylene–vinyl acetate (EVAc) copolymers or very low density polyolefin based films) or it may possess an adhesive layer defined by manufacture (coextrusion) or by diffusion of a self-adhesive, built-in component (see Figure 2.5).

The manufacture of an uncoated self-supporting adhesive material is a complex procedure. One possibility is the production of a self-supporting film. It is known that the adhesive bond is the result of chemical attraction, as well as physical anchorage. Both require contact surface and interpenetration, that is, flow. Cold flow depends on chemical basis, products geometry, and application conditions. Unfortunately, mechanically resistant, self-supporting products exhibit only a limited flow (see Chapter 3, Section II). Owing to its plasticizing ability, polyvinylchloride (PVC) is an ideal material to achieve self-adhesive performance and to conserve mechanical strength. Decorative

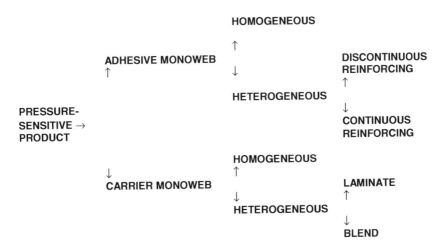

FIGURE 2.4 The main pressure-sensitive constructions.

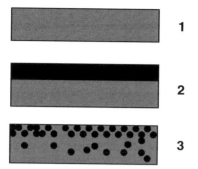

FIGURE 2.5 Buildup of self-adhesive PSPs. (1) Homogeneous adhesive carrier; (2) heterogeneous partially adhesive carrier; (3) heterogeneous adhesive carrier.

decals (e.g., adhesive films based on PVC with a very high plasticizer level) possess a monoweb construction too. Depending on its formulation and softness, PVC can be used as self-adhesive medical tape (carrier) or as an adhesive coating [6]. "Hardening" of an adhesive can also lead to a carrier-like product. Sealing tapes (without carrier) based on tackified butyl rubber, are applied with an extruder. Sheet-like hot-melts are used for thermal lamination of various web-like materials. EVAc, and ethylene–propylene copolymers, copolyesters, copolyamides, vinylchloride copolymers, and thermoplastic polyurethanes (TPUs) have been developed for such applications.

Transfer tapes are another class of uncoated and self-supporting monowebs (see Chapter 11). Such carrier-less tapes are prefabricated glues having more than sufficient dimensional stability to permit high speed lamination. In actuality, such products are adhesive layers that have a higher mechanical resistance that allows their transfer during application, but does not allow their manufacture, storage, and handling (see Chapters 8 and 11). Therefore, before use they must have a laminate structure. An adhesive material in film form (without surface coating) can display many advantages during its application. Additional benefits are: cleanliness, controlled, uniform thickness compared with a liquid system, and positionability. Such carrier-less glues are also used for nonpressure-sensitive applications also. For instance, a modified heat-curable epoxy adhesive-based film (25 μm) has been manufactured as a carrier-less adhesive sheet for printed circuits [7]. In order to achieve better mechanical performances, the adhesive can be reinforced. As an example, for liquid crystal-based thermometers the (repositionable) acrylic (AC) adhesive is a 2 mil fiber reinforced free film [8].

Acrylic foam has been developed as a carrier-free adhesive construction for mounting tapes [9]. Such products are transfer tapes. As mentioned earlier, transfer tapes have a temporary solid-state component that forms the release liner, supporting the adhesive core (Figure 2.6).

The adhesive core may be a continuous, homogenous, or a semicontinuous heterogeneous adhesive layer. The heterogeneity of the latter may be due to included solid, liquid, or gaseous particles (holes), that is, the adhesive layer can be a foam also [10]. Such structural adhesives are used in carrier-free adhesive tapes for bonding dissimilar substrates [11]. Structural tape combines the properties of classical tapes and those of PSA [12,13]. It should be mentioned that virtually carrier-less sealing tapes based on butyl rubber have been used in the automotive industry also. Such "carrier-free" tapes are reinforced with a metallic wire included in the elastomer.

Transfer tapes without carrier can have a core of pure adhesive, or include a reinforcing material. The reinforcing component can be a continuous web (e.g., a network) or a discontinuous filler-like material. For instance, thick PSA tapes (0.2 to 1.0 mm) have been prepared by ultraviolet light (UV)-initiated photopolymerization of acrylics. In this case, the PSA matrix can contain glass microbubbles [14]. The filled layers can be laminated together with the unfilled layers. Such acrylic formulations can be polymerized as a thick layer (up to about 60 mils) or the thick layer may be

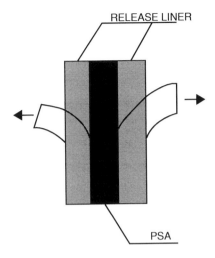

FIGURE 2.6 Buildup of a transfer tape.

composed of a plurality of layers, each separately photopolymerized (Figure 2.7). The thickness of the layer is a main factor, a thicker layer requires a greater degree of exposure. Thick multilayered PSAs or PSPs are made in this way. If the thick layer is sandwiched between two thinner layers, it may be considered a carrier, although it has pressure-sensitive properties also. The support or carrier layer is of 25–45 mils thick, it conforms well to substrates that are themselves not flat. The thick layer may include a filler, such as glass microbubbles as disclosed in Ref. [15]. The thinner layers are about 1–5 mils thick. Thick, triple-layered adhesive tape can also be manufactured, with fumed silica as filler material in the center layer. The viscoelastic properties of the layers are regulated using different photoinitiators and crosslinking monomer concentrations. A plastomer (polyvinyl acetate) can be used as filler for the "carrier" layer too (see Table 2.1). Such a carrier may be nonadhesive but for certain applications the carrier itself possesses some pressure-sensitivity.

A carrier-less, self-sustaining pressure-sensitive film can be manufactured by laminating together an adhesive-based film and a rubber-based film. Such a tape is produced by bonding a 5 mil acrylic adhesive-based film with a 5 mil rubber compounds-based film (application temperature 180°C) [16]. The adhesive strength of this type of tapes was found to have decreased 10–45% after 1 year. The use of EVAc copolymers for carrier-less self-adhesive films (tapes) applied for bonding roofing insulation and laminating dissimilar materials was proposed by Li [17]. Such formulations contain EVAc–polyolefin blends, rosin, waxes, and antioxidants. They are processed as

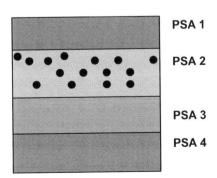

FIGURE 2.7 Multilayer pressure-sensitive composite manufactured by photopolymerization.

TABLE 2.1
Fillers for the Pressure-Sensitive Layer

Product	Filler		Reference
	Nature	Function	
Transfer tape	Glass micro-bubbles	Mechanical reinforcing	[14,15]
	Plastomer particles	Mechanical reinforcing	[15]
	Plastic scrim	Mechanical reinforcing	[153]
Removable tape	Expanded polymer particles	Stiffening and contact surface reduction	[38]
	Elastic polymer particles	Contact surface reduction	[79]
Thermometer label	Continuous fibers	Mechanical reinforcing	[8]
Medical tape	Inorganic filler	Reinforcing, crosslinking	[28]
	Crosslinked polysiloxane	Reinforcing	—
	Water	Electrical conductivity	[81]
	Air	Porosity	[30]
	Antimicrobial agent	Medical	[81]
Ironing labels	Plastic powder	Sealability	[129]

hot-melt and cast as 1–4 mm films. Soft solid PSA has been prepared from high carbon number fatty acid metal soaps containing 20–200% tackifier [18].

In some cases, the self-supporting adhesive layer is really a reinforced one. Such constructions are called tape prepreg because the fiber-based reinforcing matrix is impregnated with the adhesive. In such tapes without backing strength can be enhanced with a tissue-like scrim (cellulose or polyamide (PA)) included and coextensive with the adhesive layer. Such tapes behave like classic, adhesive-coated, carrier-based products. Foam-like carrier-less tapes can be considered as a development of tape prepregs.

Generally, constructions having porous carrier materials in which the adhesive can penetrate into the carrier, that is, it can impregnate, can be considered always as partially impregnated, partially carrier-less, and partially adhesiveless tapes. Such PSPs have layers with both carrier and adhesive characteristics. In reality, transfer tapes with a carrier need the carrier as technological aid only. During manufacture they are carrier-based; during application they are carrier-less. Monoweb labels without built-in carrier have also been manufactured [19].

In some cases, the tape is an adhesive monoweb, but behaves before or after use like a nonadhesive product. Such behavior is due to its superficial crosslinking. According to Meinel [20], in this way a carrier-less PSP can be obtained (the adhesive layer possesses adequate strength to permit it to be used without a carrier material), that has better conformability. The product exhibits both adhesive and adhesion-free surfaces. The tack-free surface is achieved by superficial crosslinking of the adhesive. This crosslinked portion of the adhesive has greater tensile strength and less extensibility. In some cases, the polymer can be crosslinked only along the edges of the adhesive layer. The tack-free edge prevents oozing and dirtiness. The superficially crosslinked adhesive layer may be stretched to fracture the crosslinked portions, exposing the tacky core in order to bond it. The surface of the adhesive becomes virtually nontacky under light pressure, but becomes tacky when the product is pressed against the adherent surface (Figure 2.8).

Such monoweb tapes are mostly special or experimental products. It should be emphasized that only the production of superficially soft plastic film carrier materials and the development of thermoplasts having elastic (rubber-like) and viscoelastic properties allowed the manufacture of noncoated PSPs.

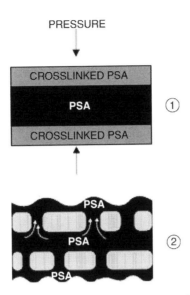

FIGURE 2.8 Buildup of a virtually adhesive-free tape. (1) Stored tape; (2) applied tape.

b. Coated Monoweb

Coated monowebs are PSPs based on a solid carrier material and an adhesive (Figure 2.9). For most applications, PSPs must have special adhesive characteristics. As discussed later (see Chapter 5), because of the broad range of raw materials available for PSAs and the sophisticated adhesive coating technologies now in use, the adhesive properties of a PSP can be easily regulated by coating of a solid-state carrier material with a low viscosity PSA. Therefore, special requirements for aggressive PSPs or removable products can be fulfilled only by coating.

Because of the balanced character of their adhesive performances (see Chapter 7) and their low pressure, high speed application technology (automatic labelling), it is very difficult to manufacture labels without a separate release liner, that is, labels with a monoweb structure. However, special monoweb labels (roll labels) have also been developed. Liner-less labels are supplied as a continuous tape-like monoweb material. A special coating on the top surface of the label prevents blocking of the adhesive layer [21]. John Waddington in the UK launched monoweb in the 1980s; its application equipment is more expensive and its printing is more complex than those of the other types of label manufacture [22].

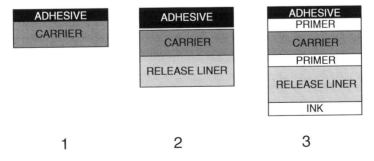

FIGURE 2.9 Schematic buildup of a coated monoweb. (1) Adhesive-coated carrier; (2) adhesive- and release-coated carrier; (3) primed, printed, adhesive- and release-coated carrier.

Tapes were the first PSPs to have a coated monoweb construction. Theoretically, pressure-sensitive tapes have the same construction as wet adhesive tapes. Wet tapes are not adhesive before humidification, that is, they do not need an abhesive layer. For PSA tapes based on polymeric films, the release is usually the surface of the polymeric film remote from the adhesive, thus enabling the tape to be conveniently stored in the form of a spirally wound roll. The development of such products has been enhanced by the use of PVC polymerized in emulsion (EPVC). The surface of films based on EPVC is coated with a thin layer of emulsifier. This layer acts as an abhesive substance and allows low resistance unwinding of the rolls. In practice, for pressure-sensitive tapes with high tack adhesive a coated release layer is required. This is not necessary if the carrier itself is abhesive, like certain nonpolar plastic films or those containing slip or other abhesive additives. It is evident that the abhesivity of the carrier depends on the nature and geometry of the coated adhesive also. As stated by Kuminski and Penn [23], different unwinding behavior is to be expected for a tape based on a hard acrylic or a soft hot-melt PSA (HMPSA). There is a trend in the market to provide customers with printed packaging tapes. The text or graphics are imprinted on the nonadhesive side of the carrier material before the tape is made. The problem with HMPSA is that a release coating has already been applied to the backing material before it is imprinted. Such coating is necessary because hot-melts do not release easily from the nonadhesive side of the carrier material. In some cases, a separate release film (paper) should be interlaminated. Double side-coated tapes are products of this type (Figure 2.10).

At the beginning of their introduction in the market, protective films have been tape-like products with a special removable adhesive. Masking tapes are really narrow web, masking (protective) films. Such products can also be manufactured as coated monoweb (Figure 2.11). The buildup of special tapes is more complex. Although most tapes are monowebs, laminated and double-laminated constructions are also manufactured (see in a later chapter). Transfer PSAs (so called adhesives from the reel) coated temporarily on a siliconized release liner (double-side siliconized) can be considered as adhesive-coated monowebs. In this case, their solid-state carrier material is the release liner. Transfer printing materials like Letraset can be considered carrierless or temporary laminates having (after application) a monoweb character. In such cases, the printing ink includes a pressure-sensitive component and (sometimes) a release component also (Figure 2.12). It may also contain a vinyl polymer with a high tensile strength. This type of product is applied under pressure.

The number of coated layers on a monoweb depends on its end-use requirements. For instance, a laminated PSA tape having good chemical, heat, and water resistance that is to be used for soldering portions of a printed circuit board is manufactured using a craft paper carrier material

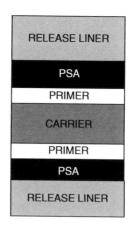

FIGURE 2.10 Buildup of a double sided-coated tape.

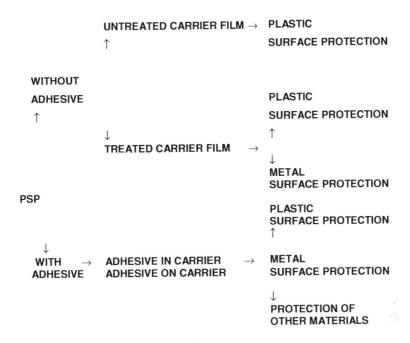

FIGURE 2.11 Buildup and end-use of protective films.

impregnated with latex, treated with a primer or corona discharge, and coated one side with a rubber-resin adhesive, and on the other side with a polyethylene (PE) film and release [24]. Special coated monowebs can have quite different construction characteristics (Table 2.2). Continuous or discontinuous carrier materials or adhesive layers having a special geometry and multilayer composite buildup can be used.

The discontinuity of the carrier and adhesive may serve to fulfill special mass or heat transfer requirements or to allow regulation of the adhesive performance. Some monoweb tapes include a partially discontinuous carrier material, a discontinuous adhesive, or both. For instance, a carrier for tapes for low temperature application has transverse holes [25]. A porous carrier may also be required [26]. As a carrier for medical tapes, porous PE has been applied in order to allow the diffusion of humidity and blood [27]. In this case, the adhesive should be porous also. Air- and moisture-permeable nonwovens (polyester nonwoven, embossed nonwoven, etc.) or air-permeable tissue are used for medical tapes for prolonged application of a medical device on the human skin. Such products are developed for pharmaceuticals companies, ostonomy appliances,

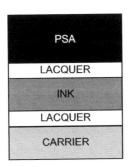

FIGURE 2.12 Buildup of a decalcomania.

TABLE 2.2
Constructive Characteristics of Special PSPs

Characteristic	Product Component		Manufacture Method
	Carrier	Adhesive	
Porosity	×		Textile, nonwoven
			Plastic foam
			Perforation
		×	Discontinuous coating
			Foaming
			Shrinkage
Special geometry	×		Special cross section
			Discontinuous web
		×	Profilated coating
			Special filler
			Special carrier
Multilayer structure	×		Coextrusion
			Lamination
			Noncompatible blend
			Lacquering
			Superficial crosslinking
		×	Crosslinking
			Noncompatible blend
			Multilayer coating
			Primer

diagnostic apparatus, surgical grounding pads, transdermal drug delivery systems, and wound care products. They minimize skin irritation due to their ability to transmit air and moisture through the adhesive system. The air porosity rate has to meet a given value. A fluid-permeable adhesive useful for attaching transdermal therapeutic devices to human skin for periods up to 24 h is based on an acrylic, urethane, or elastomer PSA mixed with a crosslinked polysiloxane. The therapeutic agent passes through the adhesive into the skin. Uninterrupted liquid flow through the adhesive has to occur over a prolonged period (6–36 h) and at a constant rate [28].

Polyvinyl ether is recommended for medical tapes alone or with polybutylene with titan dioxide as filler. Such compounds are coated warm (like a transfer mass without solvent) on the carrier web. Thus a porous adhesive is achieved by cooling [29]. A porous adhesive can also be manufactured making a latex foam [30]. Tissue tapes are used as temporary holders for the back section of the seat upholstery [31]. Such carrier materials may include a textile network. In this case, the fibers carry the forces and the matrix acts to stabilize the filaments and to introduce the forces into the fiber.

The carrier can have an asymmetric construction (geometry, surface quality, etc.) also. A patterned PSA transfer tape may possess an asymmetric carrier material, which has one surface with a series of recesses and another that is smooth [32]. In some cases, the cross section of the carrier differs from a usual parallelpiped. Masking tapes have special carrier constructions to allow conformability. According to Lipson [33], a masking tape carrier has a stiffened wedge-shaped, adhesiveless, longitudinal section (the thickest away from the tape centerline) extending from one edge, with a pleated structure to conform to small radii.

There are various possibilities to reduce contact surface between the adhesive and the substrate. Figure 2.13 presents the main ways in which an adhesive surface with reduced contact area can be achieved.

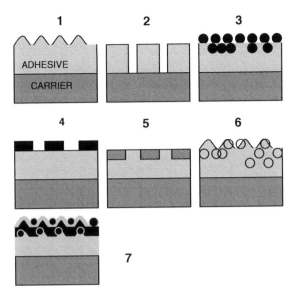

FIGURE 2.13 Adhesive surface with reduced contact area. (1) Contoured surface; (2) patterned surface; (3) filled surface; (4) release patterned surface; (5) partially crosslinked surface; (6) foamed surface; (7) textile inlay.

Peel resistance and removability can be controlled by regulating the ratio between adhesion surface and application surface [34] (see also Chapter 7). For a removable product their ratio should be smaller than one. In order to achieve such conditions the adhesive layer should be discontinuous or should also have a virtually discontinuous surface. Pattern or strip coating allows the use of nonremovable raw materials for removable applications.

Natural rubber (NR) latex as adhesive layer with a profiled surface can be used on paper to give removable protective web which is adequate for acrylic items [35]. A pressure-sensitive removable adhesive tape useful for paper products comprises two outer discontinuous layers (based on tackified styrene–isoprene–styrene (SIS) block copolymer) and a discontinuous middle release layer [36]. The tape can be manufactured with less adhesive than conventional tapes and its top side can be printed and perforated. Adhesive tapes with a narrower adhesive coated width than that of the substrate, and one edge with an adhesive-free strip have easy untie property and are useful for bundling electronic parts, building materials, vegetables, and other commodities [37].

Removable pressure-sensitive tapes containing resilient polymeric microspheres (20–66%), hollow thermoplastic expanded polymer (acrylonitrile–vinylidene copolymer) spheres having a diameter of 10–125 μm, a density of 0.01–0.04 g/cm^3, and shell thickness of 0.02 μm in an iso-octyl acrylate–acrylic acid copolymer, have been prepared [38]. The particles are completely surrounded by the adhesive, having a thickness of at least 20 μm. When the adhesive is permanently bonded to the backing and the exposed surface has an irregular contour, a removable and repositionable product is obtained, when the PSA forms a continuous matrix, that is strippable bonded to the backing, having a thickness of more than 1 mm. A foam-like transfer tape or foam tape is manufactured. The (40 μm) cellulose acetate-based tape could be removed from paper without being delaminated.

An embossed coating cylinder can coat a PSA in pattern on a paper face stock material. A discontinuous adhesive layer can be achieved by spraying also. For instance, polyurethane (PUR) foam alone or with film is used for medical tapes, for fractures. For this product, the adhesive is sprayed achieving a 25–75% adhesive-free back-side [39]. The adhesive layer can have a rough surface also, with channels to enclose air flow between the adherent and the sheet. The surface roughness of the sheet has a $50-1000$ μm width and a $10-1000$ μm height of convexes [40]. An adhesive tape may have an adhesive only along the lengthwise edges (to allow easy perforation) [41]. A removable display poster is manufactured by coating of distinct adhesive and nonadhesive strips on the carrier [42]. The adhesive strips are situated on the same plane, the plane being elevated with respect to the product surface. A multilayer adhesive construction can be manufactured also. Here, the PSA coated carrier material functions as a transfer base for a heat-activatable adhesive. After application of the heat-sensitive adhesive the PSA-coated tape can be peeled off [43].

c. Comparison of Uncoated/Coated Monoweb Constructions

As a result of the continuous development of the PSAs technology the adhesive coated monoweb constructions fulfill the most sophisticated end-use requirements. Such products can be used for almost all PSP classes. Therefore, it was mainly economic considerations that led to the development of uncoated self-adhesive PSPs. It is evident, that a coated fluid pressure-sensitive layer having a thickness of minimum $1.0-1.5$ μm allows better conformability for better contact with a solid surface than a self-supporting plastomer (or plastomer/viscoelastomer) with a minimum thickness of $7-25$ μm (see Chapter 3, Section I). The more conformable carrier-less (transfer) tapes are also manufactured by coating technology. Coating allows the combination of adhesive and nonadhesive areas on the same surface, this is not possible for extruded PSPs.

In some cases, a coated PSP can have an enclosed adhesive layer also. For certain products, an adhesive modification of the bulky carrier material is required in order to improve the anchorage of the adhesive coating, without a supplemental primer coating. In such cases, the improved self-adhesivity of the top layer requires the use of an adhesive repellent (release) layer also. Biaxially oriented multilayer polypropylene films for adhesive coating have been manufactured by mixing the polypropylene (PP) with particular resins to improve their adhesion to the adhesive coating. To prepare an adhesive tape which can easily be drawn from a roll without requiring an additional coating on the reverse side, at least two different layers having different compositions are coextruded with the thin back layer containing an antiadhesive component [44]. In some cases a primer is necessary for the anchorage of the adhesive (see Chapter 5). Theoretically, primer thicknesses approaching a single molecular layer are required, but in reality a 0.5 to 1.0 μm layer is coated [45]. According to Ref. [46], tapes are generally 80 μm thick constructions, with $1-1.5$ μm accounted for the primer. Polymer analogous reactions can be used also to modify the carrier surface before coating. In certain applications the PSA is coated first on a flexible PUR, acrylic (AC), or other foam, and then is laminated on the carrier surface. The opposite surface of the foam may also be provided with a PSA layer. The polyolefin surface of the carrier can be polarized by graft polymerization using an electron beam (EB) with a higher than 0.05 Mrad dosage level. A special composite PSA construction is made with the adhesive on the lateral portions of the carrier only and with a longitudinal split in the carrier [47]. Such sophisticated constructions can be manufactured using the coating and laminating technology only.

It should be emphasized that generally noncoated monowebs allow lower bond strength than coated products. Because of the bulk monolayer nature of the adhesive and its limited flow in such constructions the nature of the adhesive break can be easily controlled.

d. Comparison of Coated Monoweb Constructions

The coated monoweb constructions can have many different application conditions and end-uses. Within the same class of labels, tapes, or protective films the end-use requirements may also differ. Therefore, coated monoweb constructions are various also. The main representatives are described in Chapters 8 and 11.

2. Multiweb Constructions-Laminates

Among the PSPs, multiweb constructions were first introduced as labels and special tapes. Labels, generally, need a separate release liner. Their face stock and release liner are laminated together by PSA to form a multiweb construction. Tapes include double-sided products and transfer tapes which need a separate solid-state release or carrier layer. As seen from the Table 2.3, a multiweb construction can include multiple solid-state components (face stock and release liner) or multiple adhesive components. For instance, double-sided mounting tapes may have different adhesives on each side of the carrier [48] (see Chapters 8 and 11). Masking tapes may also have a special construction in order to allow conformability. According to Djordjevic [5], a masking tape has a removable release liner. Abrasion-resistant automotive decoration films also have a separate release liner [49]. Such products are made on PUR or poly(EVAc)–PVC (EVAc/PVC) basis. The carrier is coated with a PUR primer. Self-adhesive wall covering consists of a layer of fabric having a visible surface, a barrier paper that has one surface fixed to another fabric layer, a PSA coated on the barrier paper, and a release paper [50]. The release liner itself may also have a multilayer construction. Different degrees of release are achieved for a double-sided liner and a release paper coated on one side with PE and using the same release component.

Multiweb products are not new. They were developed in the packaging materials industry. The introduction of new face stock materials, especially of film-like face stock materials, required the improvement of certain carrier characteristics, such as dimensional stability, cuttability, and temperature resistance. This has been possible partially due to the development of laminated face stock materials. Such multiweb constructions have been produced by bonding the solid-state components via adhesive coating or extrusion. Theoretically, such adhesive coated laminates differ from PSPs with respect to the nature of the adhesive and the character of the adhesive bond. They are manufactured using a permanent adhesive. Generally, both temporary and permanent laminates are used in buildup of PSPs. Permanent laminates serve as carrier materials; temporary laminates as PSPs (Figure 2.14).

Multiweb constructions can include solid state components and adhesives that differ in their chemical nature and buildup (e.g., paper, plastic films, fabrics, nonwovens, foam etc). For instance, double-sided coated tapes can have a PSA on one side and a reaction mixture of uncrosslinked PUR on the other side. A PVC tape applied on a PUR foam reacts with the adherent and adheres to it chemically [51]. Business forms include permanent and temporary laminates also (see Chapter 11). A thermal tag is produced as a reinforced laminate construction for subsequent conversion by label producers. Taking an airline luggage as an example, a central plastic layer gives its strengths, the self-adhesive layer allows wrapping around luggage handles, and the upper layer is a top coated thermal paper onto which the data are printed as barcode [52].

TABLE 2.3
Multiweb PSPs

Product	Multiweb Status During Manufacture	During Application	Solid-State	Coated
Label	Yes	No	Carrier, release liner overlaminating film	Adhesive, primer release, ink lacquer
Double-sided coated tape	Yes	No	Carrier, release liner	Multiple adhesive primer
Transfer tape	Yes	No	Multiple release liner	Adhesive
Form	Yes	Yes	Multiple carrier multiple liner	Multiple adhesive, primer, multiple release

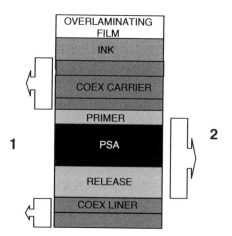

FIGURE 2.14 Laminates used for PSPs. (1) Permanent laminates; (2) temporary laminates.

The PSP laminate manufactured in order to protect the adhesive coated surface of a tape or a label is a temporary construction. In such a product, the release liner may protect the adhesive or the solvent incorporated in the adhesive layer also. Another role of the separate release liner is that of a continuum passing through the coating, converting, and labeling machines carrying the discontinuous label. Decals or labels applied per hand do not need this function. In some cases, such as the manufacture of transfer tapes, the use of a solid-state component is a technological modality only to buildup and apply the product. A patent [53] describes the photopolymerization (UV) of acrylic monomers directly on the carrier to manufacture a tape. This patent uses a temporary carrier, an endless belt, which does not become incorporated in the final product. The tapes have PSA on both sides of the carrier. In this case, the laminate is an auxiliary construction built up in order to allow the construction of the final laminate from the label or tape and the substrate. Labels and certain tapes are built up during manufacturing and application as temporary laminate and are used as permanent laminates.

Theoretically, labels are built up by coating a solid-state face stock material with a PSA, and covering the adhesive surface with an abhesive one. In the praxis transfer coating is common, that is, the release liner is coated with the adhesive and laminated together with the face stock material. The result is a temporary sandwich. During conversion, the continuous web-like face stock material is die-cut, and during labeling the discontinuous face stock material is transferred from the release liner to the substrate and the temporary laminate is destroyed. In the case of double-sided coated tapes, first, one of the release liners is taken away, then the other.

Forms are labels that have a multiweb, multilaminate structure, where continuous and discontinuous carrier materials and PSAs with different adhesive performances are laminated together, in order to allow a time- and stress-dependent controlled delamination (Figure 2.15). Their delamination is carried out in several steps during manufacture and application.

Some PSPs are constructed with both types of laminates, temporary and permanent. Security labels may have a complex buildup. Such labels are used for hangtags, that are printed with advertising text, product information, and a barcode and include an electronic security element [54]. Some forms also include both temporary and permanent laminates.

C. COMPONENTS OF PRESSURE-SENSITIVE PRODUCTS

As discussed earlier, during their application PSPs have to be laminated. Therefore, they need a carrier component. In order to bond they have to display adhesivity; therefore, they contain an adhesive component also. Theoretically such products possess at least a carrier and an adhesive component.

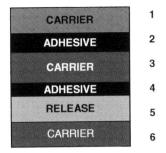

FIGURE 2.15 Buildup of a form. (1) Carrier; (2) adhesive; (3) carrier; (4) adhesive; (5) release; (6) carrier.

The technical requirements concerning the components of a pressure-sensitive laminate have changed in parallel with the development of PSPs. Table 2.4 lists the requirements for the components of a pressure-sensitive laminate.

It is evident, that in the first period of development of pressure-sensitive labels and tapes the main quality of the adhesive was its adhesion force, that is, peel resistance. The carrier used for labels had to allow their conversion of printing and cutting. Therefore, its mechanical properties, dimensional stability, and surface quality played an important role. For tapes the main requirement was mechanical resistance. Productivity increase forced the improvement of converting performances. Some performance characteristics are needed during processing of the temporary laminate; others are required after its application. Therefore, the components of the PSPs play a special role in their design and manufacture.

1. Carrier Material

Most PSPs are web-like constructions laminated together during their manufacture or application. Exceptions include some special cases where the adhesive layer itself carries information or plays

TABLE 2.4
Requirements for PSP Components

Laminate Components	Technical Requirements				
	Initial		Actual		
	Label	Tape	Label	Tape	Protective Film
Adhesive	Peel level	Peel level	Peel level, substrate peel control, time dependent peel control, removability	Peel level, substrate peel control time dependent peel control, removability	Peel level, substrate removability
Carrier	Printability, mechanical resistance, dimensional stability, die-cuttability	Mechanical resistance	High quality printability, mechanical resistance, dimensional stability	Printability, mechanical resistance, deformability	Printability, mechanical resistance, deformability
Release	Removability	Removability	Controlled release	Release	Release

an aesthetic or protective role and the carrier is the laminate component with a packaging and protection function, that is, the component with special aesthetic and mechanical characteristics. Generally, the carrier acts as a face stock material or as release liner. In some special cases it also plays the role of the adhesive. Table 2.5 summarizes the role of the carrier material in PSPs.

a. The Role of the Carrier Material

The carrier should have adequate mechanical characteristics that satisfy the end-use requirements of the product. In practice, such requirements may vary considerably. Dimensional stability of the carrier material is required for labels and to some extent for certain tapes (e.g., packaging and mounting tapes). For other tapes (e.g., masking or wire wound tapes, hygienic products, deep-drawable protective films), carrier deformability is needed [55]. Tamper evident labels should display excellent conformability and destructibility (low internal strength) [56]. Except for labels and some common packaging tapes, application requires regulating the dimensions and dimensional stability of the carrier material (i.e., control of its plasticity, elasticity, elongation, shrinkage etc.). If dimension stability is required, ideally it should not depend on the product geometry. Low gauge carrier materials have to display the same characteristics as thick or reinforced products. Isotropy is another desired property (see Chapter 3, Section II).

Cuttability of the carrier material is required for labels during their manufacture. For tapes it is needed during their lamination also. For protective films it is necessary after their application in laminated status (e.g., laser cuttability). For all these products cuttability is also required in the converting conversion phase of manufacture.

Generally, the carrier plays the role of a face stock material. It is the most important solid-state part of the product, functioning as a packaging, protecting, and information carrying component. This is a general role of the carrier material in labels, tapes, and protective films, and there are only a few web-like PSPs without a solid-state face stock material (e.g., transferable letters, decalcomania, transfer tapes, etc.) [57].

According to the end-use of the PSPs, quite different performance characteristics are required for the carrier used as face stock material (i.e., adhesive-coated permanent top layer) or release (see also Chapter 8). Therefore, the nature of the carrier materials also differs. According to Ref. [58], a tape carrier can be any reasonably thin, flat, and flexible material. Thus it may be woven, non-woven, metallic, electrically-resistant, natural or synthetic, tear-resistant or fragile, water-resistant or water-soluble (see also Chapter 8). For instance, carrier materials used for tamper evident labels should display low mechanical resistance. Fragile pressure-sensitive materials are either very thin or have low internal strength [59]. They are coated with a very aggressive PSA. Slits or perforations improve the tamper evident performances. For such products, the mechanical properties of liner

TABLE 2.5
The Role of the PSP Carrier Material

Function	Product		
	Label	Tape	Protective Film
Mechanical strength	x	X	x
Protection	—	x	X
Information carrier	X	x	x
Dosage	x	x	x
Adhesion	—	x	X

Note: X, primary function; x, secondary function.

complement the low mechanical properties of the face stock material during manufacture and application.

The carrier geometry also influences end-use performance. For instance, masking tapes can have special carrier constructions to allow conformability. A masking tape carrier has a stiffened, wedge shaped, adhesiveless longitudinal section (the thickest away from the tape centerline) extending from one edge, with a pleated structure, to conform to small radii. An adhesive carrier material can be perforated along the centerline of the long direction; it is suitable for tying together printer paper [40]. The carrier for tapes can be reinforced; a special buildup (folding) of the reinforcing fibers ensures extensibility [60]. A study in the early 1990s found that reinforced tapes made up 4% of European packaging tape production [61]. Fiber reinforced tapes include biaxially oriented polypropylene (BOPP), release, adhesive, filaments, and a hot-melt adhesive. For a pressure-sensitive label having a wrinkle-resistant, lustrous, opaque facing layer for application to collapsible wall containers (squeeze bottles), the carrier has a thermoplastic core layer, with upper and lower surfaces and voids, a void-free thermoplastic skin layer fixed to the upper surface and optionally to the lower surface of the core layer, and discrete areas of PSA. As the core layer a blend of isotactic PP/polybutylene terephtalate, with PP as skin layer, and circular dots of HMPSA as the adhesive [62]. Woven plastics have been proposed as carrier material for double layered PSA tape [63]. Paper is suggested as extensible carrier for medical tapes [64]. Polymer impregnated paper can also be used as the carrier for tapes. For automotive masking tapes, pleated paper is applied [65]. Woven fiberglass scrim has also been recommended as carrier material for tapes.

The face stock material can repel, absorb or, contain chemicals, such as solid-state water-absorbent particles. For such application it serves a dosage and storage function. The carrier materials can also act as protective surfaces. Stain-resistant, nonpaper face materials help with the staining problem. For temperature-resistant tapes, PE, polyester, and Teflon are suggested. Modified poly(ethyl methacrylate) can be used as a protective film for various substrates such as PVC, acrylonitrile–butadiene–styrene and polystyrene (PS), wood, paper or metal [66]. It is weathering-resistant and UV-absorbent.

It is evident that independently from the special end-use requirements the main surface-related performance required of a face stock carrier material is anchorage of the adhesive. The role of the carrier as the basis for adhesive anchorage is illustrated by the data of Haddock [67]. The same medical adhesive coated on cloth backing gives an average adhesive transfer index of 1.5 in comparison with a vinyl carrier displaying a value of 3.5.

As discussed earlier, labels need a separate solid-state abhesive component (release liner). In this case, the abhesive layer is coated on a carrier material. The need for a continuous, solid-state, separate abhesive laminate component for labels arises from their aggressive adhesivity and discontinuous character. Such a liner can also act as an information carrier. A large variety of products, for example, common paper, satinated craft papers, clay coated papers, polymer coated papers, and plastic films are used as raw materials for liners [68] (see Chapter 8). For tapes and protective films the backside of the self-adhesive (coated or uncoated) materials can act as liner. This function may be ensured by a supplemental release coating or by the choice of an adequate carrier material with low level of adhesivity (e.g., nonpolar polyolefins or EPVC). The abhesive layer can be coated on the backside of the carrier material, or included in it, generally as a separate layer. For instance, for a biaxially oriented multilayer PP film used as carrier or liner for tapes, the second layer of the film which faces the adhesive has a thickness of less than one-third of the total thickness of the adhesive tape and contains an antiadhesive substance [69]. Another construction includes the release substance randomly distributed in the carrier material. Abhesive substances can also be built into the carrier material. For instance, a self-adhesive tape for thermal insulation used to secure a tight connection between heat exchange members (metal foils) and PUR foam-based thermal insulation materials, is prepared by applying an adhesive that is able to adhere to refrigerated surfaces on an olefin copolymer carrier material (containing fatty acid amides or silicone oils as release agents or lubricants) that is bonded to the PUR [70].

b. The Carrier as Adhesive

The first application domain of adhesive-free laminates was the lamination of packaging films. This development continued in other areas and led to PSPs without a coated adhesive layer. Hot laminated corona treated common plastic films and special plastic films have been introduced on the market. These films do not contain special viscous components (tackifiers) or viscoelastic components (thermoplastic elastomers or low molecular weight (MW) elastomers) and are applied at a temperature above or near the softening range of the products (see also Chapters 8 and 11). Later plastic films having viscoelastic properties (e.g., very low density polyolefins, copolymers of ethylene with polar vinyl or acrylic monomers) or containing viscoelastic additives (e.g., polybutylene copolymers, thermoplastic elastomers (TPEs)) were developed for use at lower or room temperature [71]. It is evident, that classical plastic films without an adhesive layer cannot possess the tack of PSA. Plastics can flow, but such flow is a time (stress)- or temperature-related performance (see high pressure hot laminating in Chapter 11, Section IV). It should be mentioned that high application pressure, temperature, or special surface treatment alone are not sufficient to allow hot laminating of plastic films. A special chemical or macromolecular composition that ensures an improved flow is also required. That means, that really adhesive-free hot laminating protective films are products with a special carrier film composition.

The need for an "adhesive" carrier film having improved adhesive anchorage became evident during the development of polyolefin carrier materials. In this case, an adhesive level is required that ensures the anchorage of the adhesive on the carrier. Monolayer and multilayer PP carrier materials with imbedded special resins have been developed for this purpose.

c. Other Functions of the Carrier

The carrier can function as a storage or dosage system. Its surface characteristics can be modified by built-in additives. In some cases, the carrier contains special ingredients for certain purposes. For instance, cover tapes for bathroom use are made with a bacteriostatic agent (which inhibits bacterial growth) incorporated in the vinyl formulation [72]. Tackifier resin can be included in the carrier also. A butadiene–styrene elastomer is mixed with a colophonium-based tackifier and coated (impregnated) on a paper carrier. This carrier is coated with a PSA on NR. Special paper is required for direct thermal coating. Common organic type direct thermally printable papers include a colourless leuco dye and acidic colour developer [73]. These are coated and held onto the surface of the paper with a water soluble binder. During printing the two components melt together and react chemically to form the colour. Inorganic fillers such as calcium carbonate, chalk or clay are used to limit this image to the heated area. The problem of image stability (protection again chemicals) is solved by applying a topcoat, as a transparent film forming layer. It is also possible to apply a barrier coat to the underside of the paper to prevent the migration of adhesive or plasticizer from the opposite side.

2. Coating Components

The classical way to produce PSPs uses coating as the main manufacture step. A carrier material is coated with a PSA. The manufacturing process may include multiple coating, where different surface coated layers are built up, that have various (e.g., adhesive, abhesive, machining, aesthetic) functions.

a. Adhesive Components

Generally, pressure-sensitive adhesivity is provided by a PSA component built into the product. In some special cases, (e.g., protective or decorative films) pressure sensitivity can be a characteristic of the carrier material or may be the result of special physical treatment and of the application conditions (pressure and temperature).

In the first stage of PSP development, PSAs were used to impart pressure sensitivity. Therefore, the adhesive was the most important component of PSPs. Classical self-adhesive products are pressure-sensitive. Really good PSAs do not need a measurable application pressure. In blow touch labeling operations, labels fly through the air and land without pressure on the substrate. This is possible because of their high instantaneous adhesivity called tack (see Chapter 7). Tack is the result of molecular mobility, and is based on a complex chemical formulation (see Chapter 8). The PSA can be coated on or built into the bulk carrier material. Coated adhesives are such components that are deposited on the surface of a carrier material. As discussed later (Chapter 8), there are various coating methods, depending on the adhesive, the carrier and end-use requirements. Built-in adhesives are chemical compounds that can be mixed in with the components of the carrier material and processed as a homogeneous or heterogeneous (i.e., laminate) composite. Such adhesives are processed together with the main thermoplastic component of the carrier material and, therefore, in this case no supplemental adhesive coating is necessary.

A number of raw materials display (alone or in a mixture with viscous components) pressure-sensitive properties. In the classical recipe, PSAs are (at least) two-component formulations composed of elastic and a viscous component. Later, one-component viscoelastic raw materials were developed (see Chapter 5). The bonding function of the adhesive is supplemented with other functions. In some cases, the adhesive has to display the role of a porous carrier material too. For instance, one such adhesive is manufactured from a thermoplastically deformable composition that is calendered using a release-coated embossed cylinder that perforates the adhesive layer [25]. Fluid-permeable adhesive is required for transdermal applications; such an adhesive must not irritate the skin [74]. The irritation caused by the removal of the tape was overcome by including certain amine salts in the adhesive [75,76]. The role of the adhesive as storage place for different end-use components may be various. Postapplication crosslinking is proposed to improve shear. This is achieved by storing latent crosslinking agents in the adhesive [77]. An antimicrobial agent such as iodine can be incorporated in the skin adhesive as a complex also [78]. As filler to reduce the contact surface and decrease buildup, elastic polymer particles having a diameter of 0.5–300 μm can be included in the adhesive [79]. To improve their detackifying effect, the particles may include an ionic low tack monomer [80]. The adhesive used for bioelectrodes applied to the skin contains water, which affords electrical conductivity. Electrical performances can be improved by adding electrolytes to the water [81].

The adhesive can also have a multilayer structure. Such structure may serve as a modality to control creep compliance according to Gobran [82], with a plurality of superimposed adhesive layers having different gradients of shear creep compliance. Multilayer heterogeneous adhesive is obtained by UV crosslinking of 100% solid acrylics [83]. The radiation is partially absorbed, partially transmitted, and partially reflected. The maximum radiation level is experienced at the top of the adhesive layer, therefore, the crosslinking degree is maximum on this side. Therefore, direct and transfer coating give different adhesive characteristics. A uniform, isotropic adhesive layer can be achieved only by using a UV-transparent face stock (backing) and double-sided irradiation. As discussed earlier, a thick, triple layered adhesive tape (having filler material in the center layer) can be manufactured by photopolymerization also. This is really a carrier-less tape that has the same chemical composition for the "carrier" (i.e., self-supporting and adhesive) middle layer and the outer adhesive layers. The second adhesive layer has the same composition as the first one [18]. A tape made according to a patent [19] by superficial crosslinking of the adhesive also has a heterogeneous multilayer adhesive. The formulation of PSAs was described in detail in Ref. [84].

b. Abhesive Components

Because of the high tack of many PSAs or PSPs, an abhesive layer should be used as a protective or intermediate component. In the first period of PSA, development paper was used as carrier material.

Because of its texture and polarity most PSAs give a permanent bond with paper. Therefore, it was necessary to coat the surface of paper with an abhesive coating to buildup a protecting material (release liner). For labels, the adhesive properties of the liner have to be exactly controlled. For tapes and protective films, such performances play only a secondary role. In the actual stage of development the precise regulation of the release force is made possible by use of a separate, abhesive-coated, solid-state laminate component only. This is an expensive and environmentally inadequate technical solution. In addition, it does not satisfy some special requirements (resistance to abrasion, lubrication, etc.) for tapes and protective materials.

c. Printing Components

To improve the adhesive performances of the PSA layer or the surface characteristics of the carrier material, other nonadhesive or abhesive coatings can also be applied. These include primers, printing inks, antistatic agents, and other materials (see Chapter 5).

II. CLASSES OF PSPs

The major product areas in flexible coatings market are plastic coatings, PSPs, coated laminated film composites, and magnetic media. A general classification of the laminates would be possible on the basis of their carrier materials: paper, film, and laminates with other carrier materials. According to Merrettig [85], the most important PSPs based on films are: packaging tapes (based mainly on "hard" films); office and decorative tapes (based mainly on "hard" films); insulation tapes (based on soft films); corrosion protection tapes (based on "hard" and "soft" films), and decoration films (on printed "hard" film, laminated on paper).

The most important paper- and textile-based PSPs are [85]: labels; crepe tapes, and textile tapes (for technical and medical use).

Depending on their bonding characteristics, laminates may be permanent, removable, or wet removable. PSPs can be divided into classes according to their end-use. Some of these classes contain well-known products having a broad range of representatives (e.g., labels, tapes, etc.); others are special or one-of-a-kind products. Because of the continuous expansion of their application field, the number of special products is increasing.

Labels and tapes are the main PSPs (Table 2.6). Both have been developed by replacing classical wet adhesives with PSPs. PSAs have been used since the late 1800s for medical tapes

TABLE 2.6
The Main PSPs

Product	Buildup			
	Monoweb	Multiweb	Adhesiveless	With Adhesive
Label	—	×	—	×
Tape	×	×	×	×
Protective film	×	×	×	×
Decalcomania	—	×	—	×
Form	—	×	—	×
Decor film	—	×	×	×
Envelope	×	×	—	×
Separation film	×	—	×	×
Wall covering	×	×	—	×

Note: —, not used; ×, used.

and dressing. Industrial tapes were introduced on the market in the 1920s and 1930s, self-adhesive labels in 1935 [86]. Another product class including protective, cover, and separation films has been developed to replace carrier-free protective coatings. A comparison of the requirements for these product classes shows that labels and tapes have similarities with respect to the nature of the adhesive, and protective films and tapes exhibit common features concerning the nature of the carrier. In 1983 about 11% (w/w) of European adhesives were used for the primary pressure-sensitive applications (tapes, labels, and postage) and the annual growth (3.2%) of this segment has been higher than the mean value for the adhesives industry (2.5%) [87]. According to Ref. [88] in the mid 1980s, the most obvious applications for traditional pressure-sensitive polymers were labels, tapes, and decals. A decade later 55% of PSAs have been used for tapes, 34% for labels, and 11% for specialties. Now, about 60% of PSAs are coated on tapes.

A. Labels

In 1987, labels have been the most important PSPs in Europe, their proportion in the global pressure-sensitive market was estimated at 72% and it was forecast, that by 1990 it would be 76% [89]. As coated carrier material the label/tape ratio attained 7/1. Label production continued to increase and 7 mrd m^2 were produced each year [90]. It was forecast, that self-adhesive labels will cover about 60% of the label market and glue applied market share will continuously decrease [91]. Actually self-adhesive labels are used for more than 80% of the labeled products.

The definition of this successful product is difficult. Is it a packaging material? Packaging materials have to fulfill the following main functions [92]: They have to (1) contain, (2) preserve, and (3) present.

Containing is a mechanical function. Labels (except for some special dermal dosage products) do not fulfill this requirement. Although there are some antitheft or closing labels, the function to preserve is not a general requirement for labels. Their principal function is presentation. As stated by Fust [93], the main function of labels is to carry an image such as information for a special product. According to Pommice et al. [94], packaging is used not only as a protection for the products, but also as a support for advertisements and sales aid. This is also true for labels. The computer label business is one of the fastest growing segments of the label market [95]. Variable data bar-coded labels (routing labels for mailing, inventory control labels, document labels, individual part and product labels, supermarket shelf labels, and health sector labels) are the most important sector of nonimpact printing. Stanton Avery developed Kum-Kleen labels in the 1930s [96]; these were the first pressure-sensitive labels. Other types of labels have been used for almost 100 years. The growth in the use of PSA labels has been a result of substitution.

Labels are discontinuous items that serve as carriers of information and can be applied on different substrates. Their geometry, material, buildup, processing, and application technology can vary widely. Materials weighing up to 350 g/m^2 are used. Fan-fold material can be processed. Single labels, strip labels, and label strips can be printed. Labels between 30.2 and 164 mm wide cover almost all areas of application. In the 1970s, the major types of labels were wet adhesive labels, pressure-sensitive labels, gummed labels, heat adhesion labels, and shrink labels.

In Europe the traditional labeling techniques are [97] roll feed labeling (wrap around labeling), cut-and-stack labeling, and self-adhesive labeling. The types of labels now produced include wet glue applied, pressure-sensitive, gummed, heat seal, shrink sleeve, and in-mold. Pressure-sensitive labels have a higher manufacturing cost than wet adhesive labels, but their application is easier. Wet adhesive labels are suggested for application fields with very high speed, but low quality requirements (metal cans, glass and plastic bottles, etc.) [98]. Shrink labels use a shrinkable 40–70 μm film. In-mold labeling inserts the label in the mold prior to molding [99]. The insertion of chip cards is also recommended [100]. Heat adhesion labels and papers have been used in the packaging industry. Some heat adhesion labels adhere immediately and others have a delayed adhesive affect. The advantages of the heat-activatable or heat applied labels are that they can be applied at high speed, they require no preliminary operations before application, they are rapidly activated, their use requires no solvent (no pollution) and no humidity, there is no chemical yellowing, they give good permanent adhesion, they need short sealing times, and they exhibit no edge splitting. On much the same principle hot stamps, that is, labels without a solid-state carrier. Dormann [101] classified pressure-sensitive labels as cold sealing agents. Generally, sealing agents can be divided into hot seal and cold seal systems. Hot seal systems include hot seal dispersions, hot seal varnishes, hot seal films, and extrusion coatings. Cold seal systems include PSAs and cold seal dispersions.

Certain label functions such as removability can be provided only by pressure-sensitive labels. As discussed earlier, labels are not packaging materials *per se*, but they are used together with packaging materials, to which they are applied to present the product, to advertise, to serve as decoration and to provide information about the product [102].

1. Buildup and Requirements for Labels

Labels are temporary laminates composed of a face stock material with an adhesive pressure-sensitive surface that is protected by a separate abhesive component, the release liner. According to Havercroft [103], such products can be considered a special case of soft laminates that allow reciprocal mobility (shrinkage and elongation) of the solid-state components. The nature, geometry, and buildup of the carrier material; the nature and geometry of the adhesive; the nature and geometry of the release liner; and the buildup of the finished product may vary greatly. Labels can be round, square, rectangular or elliptical, or irregular in shape. Their purpose may be to impart essential information, that is, company address, product description or weight statement. In addition, depending on the product, they may give instructions and warning statements. Base labels are almost without exception printed in one colour and coated with permanent or removable adhesive. Decorative labels are printed in many colours, on various materials. The larger ones are coated with a so-called semipermanent adhesive, with low initial tack.

Labels may have a very complex construction. A label construction contains seven layers as follows: release liner, release layer, adhesive, primer, face stock, ink, and top coat [104]. Multilayer self-adhesive labels comprise a carrier label containing silicones, an adhesive layer, a printed message, a carrier layer (e.g., polyester) a release layer (e.g., silicones), a second adhesive layer, a second carrier layer (e.g., PE) and a printed message. Labels may have more than two adhesive-carrier layer-message layer units [105]. A German label manufacturer [106], has succeeded in placing original perfume oils in a special two layer label structure (scent labels). The pharmaceutical industry requires the use of so-called wrap around labels with a flag, that is, an overlapping part. Such multilayer labels may display supplemental information, constant message on one side, and variable message on the other side. In contrast, the flag can be used for the bill [107]. Decalcomanias can be built up as labels also. Decalcomanias are manufactured using screen printing. First a clear carrier lacquer is printed; this layer provides the mechanical resistance of the product. The image is coated (mirror printing) on this layer with screen or offset printing. The next layer is transparent and is followed by a screen printed PSA [108]. In an other procedure, the PSA is coated on release paper and the PSA coated liner is laminated together with a printed film [109].

a. Face Stock Material

Because of the different substrates used and the various requirements concerning the aesthetic, mechanical, and chemical characteristics of the label material, various raw materials have been developed as face stock for labels. Fabric, PVC, cellulose acetate, and polyester films have been suggested for labels [110] (see also Chapter 6). Material combinations are also used. For instance, a PE/paper laminate with the PE layer enclosed between the paper layers is recommended for a special application where the product should be partially transparent after delamination [111].

Concerning the face stock, changes in label usage have to be taken into account. The development of reel labels and of nonpaper labels is faster than the average growth [112]. Label form and function can also change. For instance, some special pharmaceutical labels may have take-off parts to allow multiple information transfer; parts of the label remain on the drug or on the patient [113], that is, such label works like a business form. Advances in document security and identification affect the label market (identification systems, document security, anticounterfeiting, corporate fingerprinting, copyable labels, etc.) and label construction. According to Thorne [65], labels are used mainly to provide information on packs (83%), information on cases (71%), for design purposes (32%), for off-pack promotions (20%), to carry bar codes (20%), and for security (15%). Unlike consumer goods, the promotional function does not have to convince the purchaser of the need to buy. Now, bar code labels cover more than 30% of label market; security labels attain 20% of the label end use.

It is also possible to classify labels according to type of carrier and application domain. For instance, according to Waeyenbergh [97], self-adhesive labels printed on oriented PP are used for: metal cans (preserves, paints and drinks); glass bottles (champagne, wine and cosmetics); cardboard containers (foodstuffs); flexible packaging (food and nonfood items); technical uses (labels, stickers and rating plates), and special domains (pocket labels).

b. Adhesives for Labels

Because of the broad range of bonding forces required in label application, the differences in types of bonding, debonding speed, and bond break nature; and various possible types of coating a number of different adhesives are used for labels. The main PSAs applied in this field are acrylics and rubber-resin formulations. First rubber-resin PSAs based on natural raw materials (NR and rosin derivatives) were used to produce PSPs. As adhesives for labels they possess the advantage of having well-balanced adhesive properties, easy regulating of adhesive performances and low costs. Unfortunately, they have a limited resistance to aging. Later acrylics and carboxylated styrene–butadiene dispersions and HMPSAs based on styrene–butadiene block copolymer (SBC) were introduced as competitors to acrylics. For special applications a broad range of commercial and experimental low volume products are available (see Chapters 8 and 11). Acrylics are applied as solvent-based, water-based, or hot-melt formulations for labels having various buildup and end-use. Labeling is the most important field of water-based pressure-sensitive adhesives (WBPSA) [114]. Removable price labels used almost exclusively a paper carrier coated with rubber-resin adhesive [115]. Two decades ago in Europe about 30–40% the amount of HMPSA were used for labels [116]. First SBCs, later acrylate-based HMPSAs were introduced. The manufacture of UV-crosslinkable (100% solids) acrylates is more expensive than that of acrylic dispersions. Such products are an alternative to solvent-based adhesives, where without too much capital investment a common hot-melt coating line can be equipped with UV curing lamps [83].

Although EVAc-based adhesives are produced as hot-melt, solvent-based or water-based formulations, because of their unbalanced adhesive performance they have been used more in other than web coating applications [117]. Water-based EVAc copolymers have been tested for labels, as a less expensive alternative to acrylics. The adhesives for labels are described in detail in Ref. [84].

c. Release Liner for Labels

The release liner for labels is a separate carrier-based component of the laminate. It has to display abhesive characteristics. These are (generally) given by a coated abhesive layer. Therefore, the construction of the release liner is similar to the face stock buildup. A solid carrier bears a coating layer. Various carrier and coating materials and various coating techniques are used for the manufacture of the release liner.

2. Principle of Functioning of Labels

Labels are sheet-like pressure-sensitive items that carry an information, made to be applied (laminated) on a solid-state adherend surface. In comparison with other PSPs, such as tapes or protective films which are manufactured and applied as a continuous web, labels constitute the temporary component of a laminated web, applied as discrete discontinuous items. The main requirement towards labels are the following: flexibility, aesthetic value, low cost, quality, ease of application, speed of application, strength, and moisture resistance [96]. It has been stated [96], that of the technical criteria, ease and speed of application have been the most difficult to fulfill.

It should be mentioned that the technical functions of labels have also changed, with the development of the labeled products. As an example, price labels in addition to communicating such information as product name, weight, composition, expiration date, and manufacturer's name, must be able to capture the potential buyer's attention. For such effects labels had to become more and more brilliant and bigger. They had to replace direct conversion of the packaging material. Also, the direct printing of bottles is limited by the quality of the image [99]; here labeling is recommended.

3. Special Characteristics of Labels

As mentioned earlier, like other PSPs, labels are manufactured as a continuous web, but applied as a discontinuous item, that is, discrete parts of this web. Therefore, their construction must allow them to be separated from the rest of the laminate and applied on the adherend surface. To allow them to be separated labels have to be die-cut and must have an abhesive layer on the backside. They must possess sufficient mechanical stability and adhesivity to be applied. As information carriers, labels have to fulfill aesthetic, coating, and mechanical demands.

4. Classes of Labels

The diversity of labels is astonishing. According to Ref. [104], 30 years ago more than 100 paper label laminates have been produced. In 1993, the main application domains of labels were computers, foods, cosmetics, pharmaceuticals, and miscellaneous markets. Pressure-sensitive booklets, coupons, and piggy-backs can also be considered labels [118]. There are various principles used to categorize labels according to their buildup, dimensions, labeling methods, end-uses and so on. Principally, they can be grouped as general labels and special labels. Within each class there are a wide variety of products. The adequacy of a label is judged for a particular use by the customer. According to Ref. [96], the marketing department has a major say in the type of label to be used.

a. General Labels

Labels are manufactured as web-like products, but they can be applied in other forms. One main classification of labels concerning their application technology divides them in roll and sheet labels. Roll labels are those manufactured in roll form and applied from the roll, where their continuous web-like backing allows them to be handled as a continuous material after their confectioning (die-cutting) as separate item. Sheet labels are those manufactured in roll form, and cut in discontinuous sheet-like finite product (as finite laminate) that cannot be applied with a common labeling gun. According to Ref. [119], roll labels constituted 66% of the labels produced in 1988 in the US. The proportion of roll labels is growing [89]. According to a market survey in 1990 [120], about 95% of the label printers in Europe were reel label printers; only 54% also had printed sheet-like products. Although home–office printing and desk top printing slightly increased the amount of sheet-printed labels, actually reel labels cover more than 95% of the printed labels.

Using another general classification based on the nature of the carrier material, labels may be considered as paper or film labels. Labels are also classified as permanent and temporary labels, according to the character of the adhesive bond. Repositionable labels are a special class of removable labels, which stick to different surfaces, but remove cleanly and can be reapplied. The final adhesion builds up over a few hours. The major production of labels (75–80%) uses permanent paper label stock [121].

b. Special Labels

Special labels are manufactured according to one of the general label classes discussed earlier, but they have to satisfy certain special end-use criteria (e.g., water resistance or water solubility, mechanical resistance or loss of mechanical strength, etc.). The special requirements for different labels and their end-use characteristics are discussed in Chapter 11). Special labels can also be classified according to the type of adhesive, as permanent, repositionable/semipermanent, or removable. The hot-melt-coated products are classified according to the labelstock manufacture into computer, thermo, price marking, freezer, film, airline luggage, office and retail labels, and shipping documents. Special labels can be classified according to their processibility, application field, function, and so on. The main class of postprocessible labels includes computer labels and copyable labels.

Generally, writability and printability are common features of labels. Postprintability using computer, that is, digital printing is a performance characteristic of a separate label class called computer labels. Other special labels like table and copy labels may also belong to this product group. For table labels, printability and lay flat are required. Such labels are computer-imprintable. Computer-imprintable labels may be transferable or nontransferable and based on paper, Tyvek, PVC, cardboard, aluminum, or other carrier material (see Chapters 8 and 11). There are different digital postprinting methods; therefore, the construction of computer labels can also vary. Laser printable labels are a special class of computer-printable labels. Computer labels can be classified as label sheets, endless labels, and folded labels (with pinholes for transport) [122]. They are manufactured as printed or bianco, nonprinted labels. Their printing is carried out using dot matrix, ink jet, laser or thermo-transfer printers, or with copying machine. In 1995 in Europe, about 40% of labels were computer labels; now they cover more than 50% of the market. According to their end-use, they are applied in addressing, marking, organizing, logistics, and other fields.

The original requirements for computer labels included machining and printing quality, application-related properties, and environmental performance. Starting from these requirements the most important development concerns printing quality; modern labels must accept 600 dpi (dots per inch). Release properties have also changed. Some years ago, slow-running machines allowed very slight release forces of 0.1 N/25 mm. Actually at high speed a better mutual anchorage of the laminate components is required, so the release force has increased to more than 0.2 N/25 mm.

Copy labels have to allow light to penetrate through the face stock in order to achieve a copy of the image on the liner. Drop on demand printing devices for product coding are able to print logos, barcodes, auto-dating, numbering, graphic program, and have a printing speed of up to 5 m/sec [123]. Suppliers for labeling technology offer software for label printing, label printer, printing and labeling systems, and materials. For instance, a supplier offers 25 printing systems and 15 types of printing–labeling units, which use more than 100 standard materials [124].

Medical labels can be defined as sterilization labels also [22]. Like medical tapes and bioelectrodes they have a special conformable, removable, porous, skin tolerant adhesive, with well-defined water solubility and electrical properties, and a special, porous carrier material (see Chapters 8, 9 and 11) or a carrier-less construction.

Tamper evident labels are special permanent labels with a sophisticated construction that does not allow debonding. Tamper evident cast films ensure that the printed label will fracture if remova is attempted. For this purpose low strength carrier materials are combined with high strength adhesives and optically working printed elements [125]. Their design shows similarities to that of special closure tapes. Special labeling machines allow the labeling of pharmaceutical products with labels and with a closure for originality [126]. The optical display of temperature sensitivity is applied in temperature labels also. Common end-use areas are airline tags, caution labels, automotive labels, police forces, and gaming machines [127].

Temperature labels are used as thermometer. They possess a conformable, heat- or cold-, water- and oil-resistant carrier material and a special temperature indicating component. Such products can be applied between 37.8–260°C [128].

There are quite different application fields where labels having water resistance or water solubility are used (see Chapter 11). Solubility or dispersibility in alkaline or neutral aqueous solutions at hot or room temperature and resistance to humid atmospheres or immersion in water are desired. Applicability on wet or condensation covered surface or frozen surfaces is related to water solubility of the PSAs. Although they are more expensive than classical wet glues in such applications, the industry appreciates the fact that PSA labels can be applied at high speed from roll stock, elimination of the need for cut label inventories and glue applicators.

Some special labels are applied as PSPs, but use other bonding mechanisms (e.g., iron-on labels) or are applied as alternatives to pressure-sensitive labels (e.g., in-mold labels). For iron-on labels, used mostly for marking textile products, the final bonding is achieved via molten plastomer (PE, polyamide, etc.) embedded in the PSA [129]. In-mold labeling (IML) allows the manufacture of the plastic item (via injection molding) and its labeling in the mold with a special label placed in the mold. Form fill and seal cups can also be labeled using this procedure [130]. Classical in-mold labels are not pressure-sensitive. They are coated with a heat-activatable (210–230°C, for injection molding, 100–110°C for deep drawing). Now, film insert molding can add a soft touch soft feel surface, using a flexible TPU film. One of the strengths of the elastic TPUs is that, in many cases, they do not have to be formed prior to injection molding [131]. New in-mold labels that work like self-adhesive films have also been developed [132].

IML is a special case of injection molding with in-mold parts. In-mold fixing techniques [133] include (1) insert molding, (2) outsert molding, (3) back melting of textiles, (4) removable tool technology, (5) in-mold decoration, and (6) IML. In these cases, part of the finished product is fixed in or out of the mold by the molten polymer. In-mold decoration and IML are used to fix a nonfunctional part of the item. The principle of the method is simple. The label is placed in the mold, and its back side is fixed with the molten polymer. For this fixing no PSA is needed. The handling of the label is difficult. Special apparatus is used to place the labels in the mold. Stacker systems have been developed to feed the label into an in-mold production machine [134]. IML is used for injection molded items, but it can also be applied with thermoformed plastics too [135]. No supplemental postlabeling or postprinting of the finished item is necessary. The label is made from the same polymer as the labeled item, and both are recycled together.

Taking into account the increasing number of requirements concerning environmental considerations, labels could also be possibly classified as repulpable and non repulpable (see Table 2.7). There are many applications where recyclability is desirable (e.g., address and franking labels, magazine supplements, etc.) [136].

It is evident that the existent range of labels — printed circuit board labels, library labels, warehouse labels, floor labels, retro-reflective labels, magnetic labels, asset labels, tamper evident labels, decorative labels, multifunctional labels, bar code labels, imprinted labels, scented labels, rub-off labels, labels with a no label look, aluminum labels, harsh environment labels, ceramic labels, titanium labels, UV protective labels, textile labels, laboratory labels, etc. — will broaden, as the result of new label applications and new engineering concepts [137]. Taking into account the growing importance of hydrophilic, hydrogel-based pressure-sensitive systems, Chapter 9 discusses this product class.

B. TAPES

PSA tapes have been used for more than half a century for a variety of marking, holding, protecting, sealing, and masking purposes. Industrial tapes were introduced in the 1920s and 1930s

TABLE 2.7
Classification of Labels

Classification Criterion	Product Classes			
	Main			Special
Form	Sheet	Reel	—	—
Carrier	Paper	Film	Others	—
Adhesion	Permanent	Removable	Repositionable	Permanent, destructible; Permanent, nondestructible Dry removable, wet removable
Recycling	Repulpable	Nonrepulpable	—	—
Web number	Monoweb	Multiweb	—	—
Printability	Classical	Digital	—	Thermal, laser, ink jet, dot matrix
Image	Label look	No-label look	—	—
Labeling	Manual	Automatic	—	—
Application field	General	Special	—	Medical; security; office; food; logistic; computer

followed by self-adhesive labels in 1935 [86]. Single-coated PSA tapes were used in the automotive industry in the 1930s [138]. Self-adhesive packaging tapes replaced gummed tapes on paper basis [139]. According to Ref. [140], the main tape markets in the US have been packaging tapes (mostly HMPSA coated), electrical tapes (solvent-based), industrial tapes (HMPSA- and solvent-based), health care tapes (HMPSA and WBPSA), masking tapes (water-based and solvent-based), and consumer tapes (HMPSA, solvent-based, and water-based).

Tapes are continuous web-like PSPs applied in continuous form. Generally, their role is to assure the bonding, fastening or assembling of adherend components due to their mechanical characteristics. PSA tape is defined as a PSA-coated substrate in roll form, wound on a core, and at least 0.305 m (12 in.) in length [141]. PSTC developed a guide of pressure sensitive-tapes [141]. The guide covers their significance, standard width, labelling, test, units of measurement, and tolerances. The advantages of tapes as an adhesive system are discussed by Bennett et al. [142]. They are positionable, have a controlled coating weight, allow automatic use by having die-cut parts, distribute stress, and exhibit a low level of cold flow. Tapes constitute the largest group of PSPs [143,144].

1. Buildup and Requirements for Tapes

Generally, tapes are PSPs that have a solid-state carrier with a coated (built-in) pressure-sensitive layer. Some tapes also have a release layer. Tapes were the first PSPs to be produced with a coated monoweb construction. Theoretically, pressure-sensitive tapes have the same construction as wet adhesive tapes. For PSA tapes based on polymeric film carrier, the release surface is usually the surface of the polymeric film remote from the adhesive, which allows the tape to be conveniently stored in the form of a spirally wound roll. In practice, pressure-sensitive tapes that have a high tack adhesive (or coating weight), for example, double-sided tapes or transfer tapes, need to have a separate solid-state release component.

The development of the carrier materials and adhesive components made it possible to manufacture special tapes as a continuous web, which facilitates applications, that are not used for bonding in a classical sense.

Double-sided mounting tapes may have a primer coating on the back-side to allow the anchorage of a postfoamed layer [86]. They must have a separate release-coated liner. A tape with a good adhesion on *in situ* foamed PUR includes a PVC carrier coated with a crosslinkable PUR adhesive, which cures in use.

Masking tapes may have special carrier constructions to give them conformability. According to Ref. [33], a masking tape carrier has a stiffened, wedge-shaped, adhesiveless longitudinal section extending from one edge, with pleated structure, to conform to small radii. This region is formed of a material different of that of the main tape. The nonadhesive face of the tape is heat-reflective. Conformability is also required for medical tapes. A pleated carrier (which gives extensibility) is needed for some mounting tapes too. Medical tapes require the adhesive, carrier, and the tape construction as a whole to be porous. Electrical conductivity or nonconductivity may be required for insulating, packaging and medical tapes (see Chapter 11). Temperature-dependent adhesion and temperature-independent bonding are needed for certain tapes. The temperature at which a PSA tape is laminated is critical [145]. Unwinding resistance is also important for tapes. Shrinkage and elongation of the tape during unwinding and deapplication; peel force, fresh and aged; weathering at low temperature; peel from back side, shear resistance at various temperatures are also measured (see Chapter 11). Water resistance or water transmission and oil resistance are required. Tear and dart drop resistance are needed. Some standards for tapes specify requirements with respect to unwinding resistance and unwinding behavior, water permeation, and water vapour permeation.

Electrical tapes have to possess high dielectric strength and good thermal dissipation properties. They are used for taping generator motor and coils and transformer applications, where the tape serves as over-wrap, layer insulation or connection, and lead-in tape. The carrier is woven glass cloth impregnated with a high temperature resistant polyester resin or impregnated woven polyester glass cloth. This last variance is used where conformability is required. For some applications transfer tapes are die-cut [146]. In other cases (e.g., special packaging tapes and masking tapes), an easy tear carrier or high strength carrier (closure tapes) is required. Carrier-less mounting tapes may have monolayer or multilayer, filled or unfilled, foamed or unfoamed construction.

a. Carrier for Tapes

Different web-like materials can be used as carrier for tapes. Films, woven and nonwoven textiles, foams and combinations of these materials are applied (see also Chapter 8). Woven fiberglass can be used for special tapes. PUR-coated PVC can also be used [147]. PUR foams were the first foam-like carriers [148]. Acetate films are used as carrier material for self-adhesive tamper-evident products. A 50 μm cellulose acetate film has been modified to be a brittle film, with a low tear strength, but it has to be die-cut [149]. Extensible, deformable paper is required for medical tapes [150]. Well-defined mechanical, electrical, and thermal properties are necessary for insulation tapes [151]. Elongation, shear and, peel resistance are also needed. In some cases, shear resistance should be measured after solvent exposure of the PSP. Thermal resistance and low temperature conformability and flexibility are required. Bacteriostatic performance may also be necessary. Cover tapes for bathroom are made on special PVC basis, with a bacteriostat (which inhibits bacterial growth) incorporated in the vinyl formulation. They are embossed to produce a secure nonslip surface. The tapes are coated with a repositionable adhesive which builds up adhesion from 26.7 oz/in to 36.2 oz/in [152]. To enhance immediate adhesion to rough and uneven surfaces a resilient foam backing can be applied [153]. The use of PURs with an extremely high level of elongation at break will ensure a sufficiently high impact strength.

b. Adhesives for Tapes

In the first stage of development, rubber-resin-based formulations were used as solvent-based adhesives for tapes. Later synthetic rubbers were also introduced. The sealing tape industry had relied heavily on the HMPSA based on SBCs because they offer a good balance of shear, tack, and peel resistance at reasonable price. Water-based acrylics replaced NR and block copolymers for many general applications. Acrylic emulsions represent an attractive alternative to HMPSA. Waterborne acrylic PSAs do not need tackifiers, exhibit better shear and adequate low temperature behavior, which are important in the manufacture of mounting tapes [154]. For special uses, cross-linkable, water soluble, electrically conductive PSAs have been developed (see also Chapter 8).

Acrylic adhesives are used for general purpose and special tapes. They are applied as solvent- or water-based or hot-melt formulations for tapes of various buildups and end-uses. Solvent-based acrylics are used for special medical, insulation, and mounting applications. They allow easy regulation of conformability, porosity, and removability. Hot-melt acrylics are expensive products, and are used mostly for special medical and sanitary tapes. They are suggested for mounting tapes also [153]. Radiation cured 100% acrylics are recommended for transfer tapes and medical application. The development of such crosslinkable and highly filled acrylics allowed the design of carrier-less tapes having a foam-like character [74,155–157]. Glass microbubbles have been incorporated to enhance immediate adhesion to rough and uneven surfaces [156]. Such tapes are prepared by polymerizing *in situ* with UV radiation. Water-based acrylics are preferred for general packaging tapes; they are also formulated as removable PSAs (e.g., application tapes).

The first rubber-resin PSAs were used for tapes with a paper carrier. Because of the relatively simple regulation of the adhesive properties for formulations based on NR or blends of NR with synthetic elastomers (via crosslinkers and active fillers), recipes were developed for almost every application field (packaging, mounting, and medical tapes) for permanent or removable adhesives. Masticated, calendered compositions have been coated without solvent. The introduction of the meltable SBC allowed the coating of rubber-resin formulations as HMPSA. Solvent-based and hot-melt rubber-resin formulations are the most common raw materials for inexpensive tape applications. Water-based, carboxylated rubber dispersions replaced acrylics for some packaging applications as a less expensive raw material. According to Wabro et al. [153], for certain tapes rubber-resin formulations cannot be replaced with other raw materials.

c. Release Liner for Tapes

Tapes generally do not need a separate solid-state abhesive component. Plastic carriers used for tapes may or may not have a coated or built-in release layer depending on the type of carrier and adhesive (see Chapter 8). Double-sided coated and transfer tapes require a separate solid-state release liner based on a common uncoated or coated plastic or paper carrier. Release liners used for double-sided tapes have special requirements [153]. Such materials have to display adequate unwinding performances, controlled adhesion to the tape, dimensional stability, tear resistance, weather and environmental stability, and confectionability/cuttability.

2. Principle of Functioning of Tapes

Tapes are manufactured as a continuous web and applied as web-like or discontinuous items (portions of this web) by laminating on a solid-state surface.

3. Special Characteristics of Tapes

Because of the need to be applied as a continuous web and due to their general use as a bonding element of multiple adherends, tapes generally have to possess a mechanically resistant carrier material with a PSA layer. The mechanical characteristics of the carrier material and the adhesive characteristics of the built-in or coated adhesive differ according to the end-use of the product. Unlike labels, tapes are not built up as laminates before their use (except for their self-wound character or some special tapes). Because they are unwound at high speed, noise reduction is very important for tapes. For certain tapes the use of a tape applicator is required.

4. Classes of Tapes

Adhesive tapes are manufactured for packing, masking, office uses, protection, marking of pipes and cables, fixing of carpets, drug delivery, floor marking, etc. Such product can be printed in one or many colours or blank. Tapes may be divided into groups according their buildup and end-use (see Table 2.8). End uses include packaging, mounting, construction tapes, medical, decoration, and others. HMPSAs can be incorporated into film tapes, mounting tapes, textile tapes, and insulation tapes. Like labels a primary classification of tapes is based on their general or special characteristics.

a. General Tapes

According to their carrier material, tapes may be divided into paper-based and film-based tapes. Some tapes are manufactured with no carrier. Tapes with carrier are classified in single-sided coated and double-sided coated products. Depending on their adhesive characteristics, tapes may be grouped into permanent and removable products. According to their end-use there is a broad range of possible classifications. Packaging, masking, protecting, marking, closure, fixing, mounting, and insulating tapes are some examples [149].

Paper-based tapes are manufactured using paper as carrier material. They are adhesive-coated PSPs that have various paper qualities and adhesive coatings as depending on their end-use. Extensible, pleated, and conformable paper is required as carrier for special tapes [149]. Double-sided coated and transfer tapes can have a paper carrier or release liner also (see Chapters 8 and 11).

Film-based tapes are manufactured using plastic films as carrier material. These products can have a coated adhesive, a built-in adhesive or a virtually adhesive-free construction. The self-adhesivity and conformability of certain plastic films and the nonpolar abhesive surface of some plastic carrier films allow the design of adhesive-free and release layer-free tape constructions.

Carrier-less tapes do not have a solid state carrier material after their application. These products are the so-called transfer tapes, or tapes from the reel. Transfer tapes have a PSA layer inserted between two release liners. They are manufactured as a continuous web supported by a

TABLE 2.8
Classification of Tapes

Classification Criterion	Major Product Classes	Special Product Classes
Form	Continuous; discontinuous	—
Carrier	Paper; film; other	Textiles; metals; laminates
Adhesion	Permanent; removable, repositionable	Permanent destructible; permanent nondestructible Dry removable, wet removable
Buildup	With carrier; carrier-less With adhesive; adhesiveless	Transfer —
Recycling	Repulpable; nonrepulpable	—
Web number	Monoweb; multiweb	One-side coated Double-sided coated
Printability	Classical; digital	Thermal; laser; ink jet; dot matrix
Application method	Manual; automatic	—
Application field	General; special	Packaging; mounting; assembling; office; closure; masking; medical; logistic; splicing; electrical

solid-state carrier. This allows them to be processed and applied as a common tape, but it serves as a temporary aid only. The adhesive layer of the tape is detached from the carrier during application. Such tapes without carrier can be used on multidirectionally deformed substrates having a various shape, up to 150°C because the temperature resistance is given by the adhesive only [153].

Single-sided coated tapes represent the classical construction of tapes, having a solid-state carrier material coated one side with a PSA. Products displaying pressure sensitivity on both sides of the carrier material are double-sided coated. Such tapes may have another coated layer on the back-side also, for example, a primer. A different release degree is achieved for a double-sided coated release liner using the same release component but one-sided PE-coated release paper. The release paper assists in the application of the tape without damaging the coating which is facing the release paper. These tapes are normally used as the bonding agent when combining two materials. The tape is pulled of its roll, and the exposed PSA is placed against the first material. Then the release paper is pulled from the tape and the exposed PSA bonds with the second material. The converting industry uses a large amount of two-sided coated PSA tapes like flying splicing tapes.

The principle of adhesive detachment from the inserted solid-state release liner is used for transfer tapes. Both surfaces of the carrier may have low adhesion coatings, one of which is more effective than the other. When the tape is used as a transfer tape, when unwound the adhesive layer remains wholly adhered to the higher adhesion surface from which it can be subsequently removed.

Permanent tapes give permanent adhesive bonding. The main representatives of this class are the packaging tapes. High strength carrier materials and aggressive permanent PSAs are recommended for these products.

Certain tapes ensure a removable adhesive bonding which is required for closure, medical, masking tapes, among others (see Chapters 9 and 11). Like labels, tapes can also be repositionable. For instance, cover tapes for bathroom are repositionable [72]. Stone impact-resistant automotive decor tapes have to be repositionable [49]. Certain closure tapes, for example, diaper closure tapes have to be removable [158]. Storable crosslinkable adhesives allowing a built-in controlled removability have been developed. Postapplication crosslinking is proposed to improve shear [155]. Such postcrosslinking is achieved by using free radical [77] or photoinitiated reactions [156]. Post-crosslinking leads to adhesion-less surfaces, that is, easier delamination. According to Bedoni and Caprioglio [61] self-adhesive tapes produced in Europe can be divided into the following classes: packaging (64%); masking (12%); protective (5%); double-sided (5%); dielectric (3%) and stationary, and others (7%). Unfortunately, the evaluation criteria of this tape classification are not known and it is difficult to delimit exactly the difference between masking and protective tapes (and films), stationary tapes and labels, dielectric (insulating) and protective, products etc. Packaging tapes are still the main grade produced, and the productivity developments in the last decade relate mostly to these products. The production speed of such tapes increased from about 250 m/min (1985) to about 600 m/min (1995) and the coating machine width increased from ca. 1400 mm in 1985 to about 2000 mm in 1995. Improvement of the average coating speed is less impressive from about 100 m/min to about 300 m/min. The average coating machine width increased from 1300 mm to 1500 mm. A new coating machine for tapes having a width of 1200 to 2400 mm is running with HMPSA, on a 25 μm BOPP film as carrier material, with a speed of 400 to 600 m/min.

According to Becker [164] technical tapes can be divided according their end-use into: closing; bonding; reinforcing; marking, and protection tapes. Closing, bonding, and reinforcing are the main functions of packaging tapes.

In this domain water-resistant tapes form a special category. It is obvious that tapes could also be classified according to their carrier material. For the same carrier class, writable and printable tapes are considered special products [159]. Table 2.9 summarizes the main tape categories according to their construction.

TABLE 2.9
The Main Tape Categories According to their Construction

Type of Web	Tape Buildup				Components	Tape Grade Application Field
	Number of Carrier Layers		Number of Adhesive Layers			
	One	Many	One	Many		
Monoweb	Yes	—	Yes	—	Adhesive primer 1 carrier primer 2 release ink	One-side-coated tape; Packaging, office use, closure, insulating
	Yes	—	—	—	carrier ink	Adhesiveless tape; Masking, insulating
Multiweb	—	Yes	Yes	—	Adhesive primer 1 carrier 1 release primer 2 carrier 2 ink	One-side-coated tape; Insulating, medical assembling, closure
	—	Yes	Yes	—	Adhesive release 1 primer 1 carrier primer 2 release 2	Carrier-less transfer tape; Mounting Assembling, insulating
	—	Yes	—	Yes	Adhesive 1 adhesive 2 carrier 1 carrier 2 release 1 release 2	Double side-coated tape; Splicing, mounting medical

As can be seen from Table 2.9 there are tapes based on a solid-state carrier and carrier-less tapes. Carrier-less tapes may be virtually or actually carrier-free. Tapes that are virtually carrier-less possess a reinforcing layer (network) embedded in the adhesive mass; this layer ensures the required mechanical strength of the tape.

According to their active (adhesive) surfaces, tapes can be classified as one- or double-sided coated. As seen from Table 2.10, double-sided coated tapes are manufactured with or without carrier. Tapes with carrier can be manufactured with a paper, textile, or plastic carrier. The textile carrier used may be woven or nonwoven. Textile carriers ensure nonextensibility and good anchorage [153]. A plastic carrier may be a film or a foam. A film carrier acts as a barrier against chemicals (plasticizers, surfactants, antioxidants, etc.). A foam carrier ensures conformability on uneven surfaces and equalizing of the thermal dilatation coefficients.

b. Special Tapes

At the beginning of their use, tapes were applied mostly as fastening and bonding elements. Packaging tapes still play this role. The development of the new carrier materials made possible

TABLE 2.10
Special Tapes

Product	Buildup		Manufacture	Application
Double-sided adhesive tape	Double side-coated		Coating, laminating	Splicing, mounting, insulation, assembling
	Self-adhesive	Blend, coex	Extrusion	Masking, insulation,
	Carrier-less adhesive	Foamed, filled, reinforced, and crosslinked	Coating, extrusion	Mounting, insulation, medical
Foam tape	Foamed carrier	Elastomer foam, plastomer foam	Foaming, extrusion	Mounting, insulation, assembling, medical
	Foamed adhesive	Acrylic, rubber	Foaming, coating, extrusion	Mounting, insulation, assembling, medical
Filled tape	Solid-state filler in adhesive	Continuous fiber discontinuous fiber, powder-like filler	Calendering, extrusion, coating	Packaging, medical, electrical
	Liquid filler in adhesive	Electrolyte, antimicrobial agent	Coating	Medical
	Air in adhesive	Foam tapes	Foaming, extrusion	Medical
	Solid-state filler in carrier	Reinforcing fiber powder-like filler	Extrusion, calendering	Packaging, medical

the manufacture of special tapes. These products are used as sealants, mounting elements, etc. Some of them are described subsequently.

Textile-based tapes are applied on various substrates. Insulation, medical applications, and packaging are the main end-use domains. As stated earlier, electrical tapes have to possess high dielectric strength and good thermal dissipation properties. The carrier for such tapes is woven glass cloth or woven polyester glass cloth impregnated with a high temperature resistant polyester. In medical domain, the textile-based carrier acts as a network that allows the diffusion (of air, pharmaceutical agents, vital liquids, etc.). Like high strength wet tapes, pressure-sensitive packaging tapes may also have a reinforcing textile carrier.

The plastic web of tapes may be a foam-like product or a combination of film and foam. Double-sided coated tapes using neoprene, PVC, PUR, PE as carrier have been developed (see Chapter 6). Carrier-less foam tapes are also manufactured. Glass microbubble filled mounting tapes imitate the conformability given by cellular (air-filled) structures. Table 2.11 lists the main application domains of special tapes.

As indicated in Table 2.11 such tapes possess two active surfaces; therefore, they can bond in plane as common packaging tapes, as well as between two surfaces. In their early development stage double-sided tapes were manufactured with a carrier. Later carrier-less tapes were produced. In principle, such tapes may bond multilaterally, like an overall adhesive profile. As an example, the tapes used for mounting of rubber profiles in cars have a special form (see Chapter 11).

Insulations tapes have to assure thermal, electrical, sound, vibration, humidity, chemical, or other type of insulation. Depending on their application field they are made with various carrier materials that have different thermal, electrical, and mechanical properties.

Medical tapes are used as fixing or dosage elements. According to Ref. [160], the earliest plaster-like, adhesive medical products were crude mixtures of masticated NR tackified with rosin derivatives and turpentine, and filled with zinc oxide as pigment. Cloth served as carrier material. Later, extensible, deformable paper was used. Now, plastic films and plastic-based

TABLE 2.11
Main Application Domains of Special Tapes

Tape	Main Characteristics	Application Domain
Foam tape	Conformability Damping Multiple adhesive surface (bonding in space)	Mounting Vibration and thermal insulating Sealing Assembling
Carrier-less transfer tape	Conformability Damping Multiple, variable adhesive surface (bonding in space)	Mounting Vibration and thermal insulating Sealing Assembling
Double side-coated tape	High shear resistance Instantaneous bonding Multiple adhesive surface (laminating)	Splicing Mounting
Self-adhesive tape	Conformability Deformability Removability	Masking Insulating
Discontinuous tape	Automatic applicability	Mounting Assembling
Removable tape	Removability Repositionability	Mounting
Dosage tape	Storage and dosage Conformability Removability	Medical mounting Anticorrosive protection

textile-materials are used. As adhesive, a broad range of macromolecular compounds (NR and synthetic rubbers, PURs, acrylics, special vinyl polymers) are tested. The special requirements for different tapes and their end-use characteristics are discussed in Chapter 11.

Seal tapes are used as insulating and mounting elements, having sealant-like functions. The three general classes of sealants are: hardening, plastic, and elastic sealants [161]. Elastic sealants behave like rubber bands. They deform if a force is exerted upon them, but return to their original state if this force is removed. Sealing tapes possess intermediate characteristics: they are elastoplastic and plastoelastic, and their ability to absorb stresses is about 8–15 and 15–20%, respectively. The edge joint of insulating glass is presently manufactured worldwide predominantly using sealants and sealing tapes. During the lifetime of a glazing unit the edge joint is subjected to many different types of stresses. After installation mechanical stresses result from the wind pressure and suction, flexion of the insulating glass pane due to temperature and pressure fluctuations etc. Chemical and physical stresses appear also [161].

The insulating glass sealant or tape has to satisfy a number of requirements, the most important being (1) optimal adhesion to glass and metals, (2) aging resistance, (3) high flexibility (even at low temperatures), and (4) low water permeability.

C. PROTECTIVE FILMS

Protective films form a special class of PSPs, manufactured as self-wound webs and applied as the laminating component for both web-like or discrete products. They can be used to protect finished surfaces from damage during manufacture, shipping and handling, and to mask areas of a surface from exposure during spraying operations by painting. Such products can be formed using a stencil and undergo the same processing steps as the protected item. Taking into account their protective function, they could be considered as packaging material, but actually they cannot be defined as packaging material because of their aesthetics. According to Djordjevic [5], packaging is a process with a functional definition, playing both a protective and cosmetic role. It should be mentioned that masking tapes may be considered as a special class (narrow web) of protective films.

Table 2.12 summarizes the main types of products used for surface protection. As shown in this table, protective materials can be carrier-less (coatings) or based on a solid-state carrier (protective webs). Coatings display an adhesive contact with the surface to be protected. Protective webs may (e.g., protective films and separation films) or may not (e.g., packaging films and cover films) exhibit an adhesive contact with the surface to be protected. This "adhesive" contact may be the result of an interaction between an adhesive and a substrate (e.g., protective film) or between two adhesiveless surfaces (e.g., cover film). For certain protective webs in adhesive contact with the surface to be protected, this contact is the result of lamination (e.g., laminating and separation films); in other cases (e.g., cover films) no lamination is carried out during application and the contact between protected surface and protective web is not uniform.

The nature of the bond (temporary or permanent) may also be different. Common lacquers buildup a permanent coating with the protected surface. The bond between laminating films used for surface protection is permanent too. Other protective materials (the majority of film-like products) are removable.

Table 2.13 lists the main removable products used for surface protection. As can be seen, both carrier-less coatings and carrier-based materials are used for surface protection. Carrier-less coatings were the first to be produced. Protective films were developed as a replacement for masking paper and peelable varnishes (see also Chapter 11).

Peelable varnishes have to fulfill the following requirements [162]: coatability as a monolayer lacquer, weatherability, water and temperature resistance, chemical resistance, nonflammability, transparency, film-like rigidity and flexibility, and low but sufficient adhesion. According to Zorll [162], such formulations occupy an intermediate position between primers and varnishes, concerning their adhesion and cohesion properties. Primers have to display an excellent anchorage on the surface to be coated but medium cohesion. Varnishes have to possess higher cohesion and mechanical resistance. The mechanical resistance and flexibility of removable protective

---

TABLE 2.12
The Main Products Used for Surface Protection

Protective Materials	Characteristics				
	Components			Application Method	
	Carrier	Adhesive	Release	Lamination	Wrapping
Packaging material	×	—	—	—	×
Protective coating	—	—	—	—	—
Protective web					
Cover films	×	—	—	—	×
Separation films	×	× /—	—	×	—
Overlaminating films	×	× /—	× /—	×	—
Protective films	×	× /—	× /—	×	—
Masking tapes	×	×	× /—	×	—
Protective labels	×	×	×	×	—

Note: ×, with; —, without.

TABLE 2.13
The Main Removable Products for Surface Protection

Product	Chemical Basis	Thickness (μm)	Application Method	Deapplication Method
Peelable varnish	Wax, low molecular EVAc, acrylics, polyvinyl butyral	5–10	Spray coating, dipping	Mechanical peel off; washing with water (alkali), or solvents
Peelable low strength films	PVC plastisols, crosslinked polybutadiene, polybutylene/resin, PP/resin, (VC/AC) copolymer	10–70	Spray coating dipping	Mechanical peel off; washing with water (alkali)
Protective papers	Paper	30–100	Lamination	Delamination
Protective films	PE, PP, PET, PVC	12–120	Lamination	Delamination
Separation films	PE, PP, PVC	25–50	Lamination	Delamination

lacquers has to be as high as for varnishes, but their adhesion should be lower (their adhesion is about 1 N/mm^2).

Coating films based on EVAc copolymers and waxes have been used [163,164]. Polyvinylbutyral has been applied for removable coatings [165]. Low molecular weight acrylic polymers have been developed as temporary automotive coatings. They have to exhibit weathering resistance, surface protection, and slight removability with alkali solutions after a long period of application or storage. Low molecular solvent-based acrylics have been marketed [162]. Such coatings have to be coatable as monolayer lacquer and should exhibit weather-resistance, chemical-resistance, nonflammability, mechanical strength, and flexibility and adequate adhesion. For temporary protection of automobiles during transport and manufacture, copolymers of C$_{3-4}$ alkyl methacrylate and methacrylic acid have been proposed [166]. Such coatings sprayed on the surface to be protected, give a 6–10 μm film after drying in air (10–15 min), which is removable with a 10% NaOH solution. Crosslinked elastomers have also been used as protective coating for metals. A patent describes the use of a polyisocyanate-cured butadiene polymer [167]. Later, protective papers replaced carrier-less protective coatings. Water-removable protective papers have also been manufactured [168].

1. Buildup and Requirements for Protective Films

Protective films are web-like, self-adhesive, removable materials, that are used in an intimate contact with the surface to be protected, and require a self-supporting carrier in order to display the mechanical resistance necessary for the protective function and for the removability. The web-like attribute refers to the solid-state character of the product, which is manufactured as a solid-state continuum (reel). Such products were introduced in the 1960s for the protection of coated coils. As stated in Ref. [169], both cold and hot systems are known. Cold systems use an adhesive-coated film, whereas hot systems apply a meltable film that bonds with the coating layer of the coil (at high temperatures) to produce the desired protection. In contrast, this bond should allow removability after the coil is processed. It should be mentioned that laminating of coils has been used as a coating method for many years. Such (adhesive) lamination of PVC and polyvinylidene fluoride films has been carried out at 180–200°C.

It is evident, that the simplest way (at least theoretically) to buildup a protective film, is to manufacture an uncoated plastic film, that possesses the mechanical strength of a carrier film, and has a built-in adhesivity (see also Chapter 8).

The monoweb as a challenge for label producers, and a reality for tape manufacturers, is undergoing further development for protective films. To understand the direction of this development, let us have a look at the requirements for protective films (Table 2.14). As shown in the Table 2.14 conformability, mechanical strength, balanced adhesion to the protected surface, and removability are the main application criteria for protective films. A number of protective materials fulfill these requirements at least partially. As discussed earlier, there are many types of materials for surface protection. Some, like packaging materials, are self-supporting, mechanically resistant carrier materials that do not need to be in intimate contact with the product to be supported; thus their removability does not affect the product. Other products, like varnishes or lacquers, develop an intimate contact with the surface to be protected. Generally, these are permanent coatings, that do not need a self-supporting carrier. In some cases, even permanent coatings may display sheet-like character (laminates). Some packaging materials (e.g., cover films) auto-adhere to the protected product too, although they need no laminating to perform their function.

As shown in Table 2.14 protective materials differ according to their self-supporting or coating-like character (with carrier or without), their contact with the product to be protected (adhesion or overlapping only), and their removability from the product (with or without bond breaking). Protective films include a broad range of products that exhibit pressure-sensitive behavior, although they are built up quite differently.

From the history of protective films, it is well known that they appeared on the market as improved variants of PSA-coated protective papers. Paper masking tapes were introduced in the 1920s in the automobile industry [153]. Later, paper was replaced with the more conformable plastic film. During the 1950s, Japanese PVC-based products were the sole materials used for protective film. PVC has since been replaced with polyolefins. Taking account of this development from the technical point of view, it was easier to manufacture the protective film in a two-step process, in which common plastic films (mainly polyolefins) were off-line coated with a PSA. Therefore, classical protective films are pressure-sensitive self-adhesive films. They are manufactured by a film conversion process.

Adhesive-free laminating films are not new; they have been used for many years [170]. They have been applied on different substrates. A thermoplastic heat-activated adhesive film has been designed for laminating calendered soft PVC foils to PE foam, on aluminum, fiberboard, and woven fabric. Bonding has been achieved with conventional flame bonding machines, at speeds

TABLE 2.14
Requirements for Protective Films

Application Requirements	End-Use Requirements	Performance Given By Carrier	Adhesive
Laminating ability	—	×	×
Cuttability	×	×	—
Mechanical resistance	×	×	—
Machining performances	×	×	—
	Deformability	×	—
	Adhesive bond stability	×	×
	Aging, stability and weatherability	×	×
	Delaminating ability	×	×

Note: —, not required; ×, required.

of 10–20 m/min. Coating paper with film forming (tack-free) EVAc copolymers has been carried out to manufacture packaging materials with barrier and sealing properties [171].

In the PSA label and tape industry, the face stock materials used by converters were common materials developed for the packaging industry (PVC, cellulose derivatives, and later polyolefines). The main requirements for these packaging materials include: mechanical strength, flexibility, printability, and the lack of self-adhesivity (blocking). Only the later development of plastic films for a quite new packaging technology (shrink and cling films), in parallel with the development of tackifiers for hot-melt formulations, allowed the manufacture of plastic films having a self-adhesive surface. It should be pointed out that, unlike that of classical PSAs, which are fluid (cold flow) at normal temperature, the adhesivity, that is, the laminating ability of the new self-adhesive films (SAFs) can be used only under well-defined temperature and pressure conditions; that is, these materials are adhesive only under exactly defined laminating conditions. Moreover, their removability also depends on the laminating conditions. Another disadvantage of the uncoated protective films, is their substrate-dependent adhesivity, that is, they do not have (actually) a general usability for quite different application fields. It is evident, that manufacture of such films requires the skill of specialists from the plastic film production (extrusion) field.

As can be seen from Table 2.14, within the range of protective materials, protective films form a separate class, characterized as a laminatable monoweb construction. From the range of monoweb products, cover films (like protecting films) are functional, technological packaging materials, but they are used without laminating [172]. The application of the tape to the product to be protected also differs from the use of protective film. It is a low pressure, discontinuous process. Other products like carrier-less protective coatings are not pressure-sensitive. They are tack-free polymer layers, like varnishes, having a good adhesion on the coated surface; their removability is mainly chemically controlled.

a. Carrier for Protective Films

The carrier in adhesive-coated protective films is, generally, a common polymer used in the manufacture of packaging films. PVC, polyolefins, and polyester are the most commonly applied materials (see Chapter 8). For self-adhesive protective films tack-free elastomers (polyolefins) or viscoelastic compositions (based on EVAc, ethylene-butyl acrylate copolymers or tackifyed plastomers) are suggested. Adhesive-coated protective films possess thicknesses of 40–120 μm. Generally, it is not possible to coat all thicknesses with every adhesive. Adhesives giving higher peel strength need higher gauge films. It is recommended that films possess enough adhesion to avoid film detachment but not high enough to produce a deposit.

b. Adhesives for Protective Films

Depending on their construction, adhesive-coated or self-adhesive, protective films may or may not contain an adhesive. Self-adhesive protective films may also include tackifying components (e.g., polyisobutylene copolymers). For coated protective films crosslinked adhesives are, generally, used to achieve removability. Various raw materials, elastomers (natural and synthetic rubber) and viscoelastomers (acrylics and PURs) have been suggested for PSA formulations for protective films (see Chapter 8).

Although harder than rubber-resin based compositions, acrylic adhesives are used as solutions or dispersions for protective films. Generally, such formulations are not tackified. Solvent-based acrylics are crosslinked recipes. Acrylic dispersions can be applied in certain cases as non-crosslinked formulations also. Formulations based on rubber-resin adhesives (solutions) are the most important adhesives for protective films. Because of their softness and easy control of removability (via mastication, crosslinking, and tackifying) for certain uses they can not be replaced with other adhesives (see also Chapter 8).

c. Release Liner for Protective Films

Generally, protective films do not need a separate release liner. In most cases, they need no release coating. For products with a higher coating weight or tacky adhesive, special nonsilicone-based release coatings are applied (see Chapter 8).

2. Principle of Functioning of Protective Films

As mentioned earlier, protective films adhere to the surface to be protected. This is necessary in order to ensure a monoblock character of the product, in order to leave its working (technological) performances intact. This is a main requirement towards protective films, because, unlike classical packaging materials, they are used at the start and not at the end a technological procedure, that is, they are applied before the product to be protected undergoes its technological cycle. The contact with and adhesion to the surface to be protected is achieved by laminating the web with the product. Thus, the definition of protective films may be that they are self-supporting, removable adhesive webs, that buildup a laminate with the product to be protected. As discussed earlier, the main characteristics of a protective film are its self-supporting character and its surface adhesivity.

The self-supporting character is given by the use of a solid-state carrier material (film, paper, etc.). This material has to display the mechanical resistance required during application of the web (bonding and debonding) and during the life of the laminate (protected material/protective web).

Application and deapplication of the protective web refers to its lamination on the surface to be protected, and its separation from that surface. As schematically shown in Figure 2.16, during lamination and delamination the film is stressed by a tension; thus its tensile strength is a very important property.

Quite different from the laminating of other PSPs (labels, tapes, etc.), where low laminating pressure, high laminating speed, and (generally) normal temperature are used, the lamination of

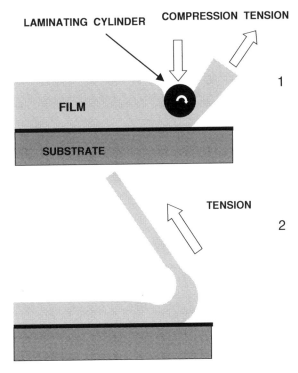

FIGURE 2.16 (1) Lamination and (2) delamination of a protective film.

protective films may require a high laminating pressure and temperature (Table 2.15). Their debonding (delaminating) speed is higher than that of other products. Therefore, the mechanical resistance requirements for the carrier material during application/deapplication of protective films are relatively high in comparison with those of labels and tapes. In this operation, the mechanical resistance has to ensure the dimensional stability of the film, necessary to allow standard debonding (peel) values (see Removability in Chapter 7, Section I).

Once applied, the film protects the product against mechanical damage during different working/logistic steps. Depending on the nature of the product, the nature of the applied stresses will also differ. The most common working operations are cutting, drawing, and punching. Therefore, the film itself has to resist, shear, tensile, and compress stresses (see Chapter 3).

On the contrary, like web-less, unsupport protective coatings, the protective web has to conform to the surface profile of the product. Unlike these products, which are liquid during application, and possess an "unlimited" flow, and as a consequence a perfect surface conformance, the solid-state carrier film (protective film) has to allow flow in the solid-state. As discussed later in more detail (see Chapter 8), this requirement limits the choice (formulating freedom) of the self-supporting material. In conclusion, the choice of a solid-state, self-supporting carrier, as base material for protective films is influenced by the application conditions (laminating/delaminating), by the end-use (product working conditions), and by economic considerations.

The application conditions depend on the self-adhesive character of the film. Its influence is very complex. Laminating and delaminating require high mechanical resistance (solid state-like behavior). Bond formation is a function of deformability, or plasticity (liquid-like behavior) of the product.

a. The Self-Adhesive Character

Protective films are laminated on the product to be protected. Unlike the manufacture of classical composite structures this lamination does not occur simultaneously with the manufacturing of the adhesive film (i.e., converting), but after it, like labeling, or the application of a pressure-sensitive tape. Protective films are manufactured as preformed materials, and undergo laminating after an indefinite storage time. Therefore, they have to possess a permanent adhesivity, given by the cold flow of macromolecular compounds above the T_g (see Chapter 3).

Protective films (like tapes) are monowebs, that is, pressure-sensitive laminates, where the PSA layer, is protected by the back side of the web (carrier material). In some cases, because of the relatively high tack of the PSA layer, protective films need a release layer on their back side. Generally, the release layer of protective films is built up on nonsilicone basis; therefore, it displays a relatively low abhesivity. This is acceptable because of the low tack and peel of the adhesive layer. In contrast, it is necessary because of the increased mechanical requirements for this release coating too. Economic considerations also play a role in the choice of the release layer (see Chapter 8).

In principle, the self-adhesive character of a film may be achieved for the whole material (bulk) or for a built-in or built-on layer of it; that is, the carrier itself can possess pressure-sensitive

TABLE 2.15
Laminating Conditions for Protective Films

Protective Film	Protected Product	Laminating Conditions Temperature (°C)	Pressure
Hot laminating film	Coil	230	High
Adhesive-coated film	Coil	Room temperature	High
Self-adhesive EVAc film	Cast plastic plate	Room temperature	Medium
Self-adhesive EVAc film	Extruded plastic plate	50–70	Medium

properties, or it can be coated with a layer of pressure-sensitive material. Independently from the construction of the self-adhesive layer, this should display removability.

b. Removability of Protective Films

The general principles of removability are discussed in Chapter 7. In the case of protective films the protected items are (in comparison with the other components of the laminate), generally, mechanically resistant, bulky materials (e.g., profiles, coils, and plates); thus, it can be assumed, that their internal cohesion (R_s) is much higher than the resistance of the adhesive joint (R_{aj}):

$$R_s \gg R_{aj} \tag{2.1}$$

In working with protective film coated surfaces, their force of adhesion should ensure a non-destructible, permanent bonding (as in the case of permanent adhesives). On the other hand, delaminating of the protective film after its use requires a very low peel force level, theoretically as low as possible. That means, that the real adhesion force is a compromise between the protection-related high peel and the low peel required for easy delamination.

Generally, relatively thin plastic films are used as common carrier materials for protective films. Their mechanical resistance is the result of their formulation and geometrics. It is given by the empirical experience of film and foil manufacturers with common packaging materials. Therefore, in the first instance, it may be supposed, that manufacture of protective films requires knowledge in the manufacture of removable film coatings and in the mechanics of adhesive joints for plastics.

3. Special Characteristics of Protective Films

Because of their laminating, end-use, and delaminating conditions protective films have to display some special characteristics. These are given by an adequate physical and chemical basis (see Chapters 3–6).

The performances of protective films should allow their practical use which includes their application on the product to be protected, performance of their protecting function, however, the protected item needs it, and their debonding or separation behavior. Both the application and deapplication (separation) of the protective film depend on its adhesive properties. Therefore, protective films have to display special adhesive characteristics manifested as an unbalanced low tack and high cohesion adhesion. They have to be conformable in order to allow lamination on difficult adherend contures. Protective films have to display plasticity in order to deform in parallel with the protected, processed adherend. They have to exhibit permanent adhesion during end-use and removable low force adhesion during deapplication.

4. Classes of Protective Films

On the basis of their buildup, protective films can be classified as adhesiveless and adhesive-coated films. The adhesiveless products are also called self-adhesive films. The SAFs may work physically or chemically, depending on their principle of functioning, that is, their buildup. The SAFs include different chemical compounds as pressure-sensitive component. According to the nature of the adhesive polymer, the main types of SAFs based on: polyolefins; EVAc copolymers; ethylene-butyl acrylate copolymers; polyisobutylene copolymers, and other chemical compounds.

Protective films with adhesive can be classified according to adhesive nature, type of carrier film, adhesive properties, bonding/debonding nature, and application domain (Table 2.16).

The self-adhesive protective films can be classified according to the main application fields (and conditions) into SAFs for hot laminating and SAFs for warm or room temperature laminating. The self-adhesive films for hot laminating can be used for metal, plastics, and paper surfaces.

TABLE 2.16
Classification of Protective Films

Classification Criterion	Product Classes			
	Main			Special
Adhesive presence				
	Adhesiveless	Built-in adhesive	Coated adhesive	—
Nature	Acrylic	Rubber-resin		—
	Crosslinked/ uncrosslinked, curing agent	Primered/ unprimered		
	Solvent-based/ water-based			
Aggressivity	Low	Medium	High	Easy peel; cleavage peel
Coating weight	Low	Medium	High	—
Recyclability	Yes	No	—	—
Carrier nature	Polyolefin LDPE, LDPE-HDPE, LDPE-PP-LLDPE	PVC	PET	Other materials
Buildup	Homogeneous Clear/pigmented	Heterogeneous Blend/coex	—	Colored
Thickness	Low (25–40 μm)	Medium (40–80 μm)	High (80–120 μm)	—
Release presence	With release liner	Without release liner	—	—
	With release layer	Without release layer	—	—
Nature	Silicone	Carbamate	Fluoro-polymers	Other materials Water-based/ solvent-based
Laminating temperature	Room temperature laminating	High temperature laminating	—	—
Pressure	Low	High	—	—
Method	Manual	Automatic	—	—
Processibility	Nonprocessible	Processible	—	Deep drawable Laser-cuttable
Weatherability	UV-resistant	NonUV-resistant	—	—
Protected surface				
Nature	Metal	Plastic	Textile	Copper
Roughness	Rough	Polished	Glossy	Soft PVC; enameled metal
Cleanliness	Clean	Oiled	—	—
Delaminating	Manual	Automatic	—	—

Hot laminating SAFs can be divided into films for uncoated and those for coated adherend surfaces. The films for coated surfaces can be applied on coated lacquered or coated laminated adherends.

Adhesive-coated protective films may be used for metal or nonmetal (glass, plastic, stone, etc.) surfaces. They can be classified according to their function (storage or processing protection), their place of application (face or back-side protection), to the permanent (mirror tape, label over-laminating film, etc.) or temporary (masking films) character of the bond. Nonmetal surfaces may be film-like ones, plates, profiles, and textiles (e.g., carpet).

D. OTHER PRODUCTS

According to Ref. [153], only about 5% of PSAs is used for the manufacture of labels and tapes; the main portion is applied in the paper processing industry (27.5%) and building (28.7%). However, PSPs other than labels, tapes, and protective films having a lower production volume have also achieved economic importance. Business forms, decals, decorative films, and plotter films are the main representatives of these specials products.

Business forms are complex, multilayer laminated sheet-like products designed the carry information to be distributed to different end-users at different times in different forms. Form-label combinations, cut sheet and laser imprintable forms, security forms have been developed. A business form with removable label, with tape pieces adhesively mounted on the paper substrate, is described in Ref. [173]. As usual, the business form has die-cut label areas that can be removed from it. This laminate is made of paper, printed continuously in a press. Pieces of adhesive tape are applied at spaced points of the printed paper. The formed binary areas having a greater thickness than the carrier include the tape in relief to the substrate. One of the layers is die-cut.

Plotter films are pressure-sensitive films designed to be used in graphics as signs, letters, etc. [159]. They are cut from special PSA-coated plastic films, using computer programs that allow the design and mounting of letters or text fragments having dimensions of 10–1000 mm [174]. Stickies are carrier free PSPs designed to transfer (usually) graphical information. They use the solid-state carrier material as release liner [175]. According to Refs. [176,177], decor films are soft PVC films coated with PSA, and printed on the backside with a special design. A detailed description of special PSPs is given in Chapter 11.

REFERENCES

1. I. Benedek and L.J. Heymans, *Pressure Sensitive Adhesives Technology*, Marcel Dekker, Inc., New York, 1997, Chap. 9.
2. I. Benedek, *Pressure Sensitive Adhesives and Applications*, Marcel Dekker, Inc., New York, 1997, chap. 9.
3. T.J. Kilduff and A.M. Biggar, U.S. Patent, 3,355,545, *Coating*, (7), 210, 1969.
4. *Etiketten-Labels*, (3), 10, 1995.
5. D. Djordjevic, *Tailoring Films by the Coextrusion Casting and Coating Process*, Speciality Plastics Conference '87, Polyethylene and Copolymer Resin and Packaging Markets, Dec. 1, 1987, Maack Business Service, Zürich, Switzerland.
6. M. von Bittera, D. Schäpel, U. von Gizycki, and R. Rupp (Bayer AG, Leverkusen, Germany), EP 0,147,588 B1/1985.
7. *Coating*, (12), 390, 1969.
8. *Adhes. Age*, (9), 30, 1986.
9. *Adhäsion*, (6), 14, 1985.
10. F. Altenfeld and D. Breker, *Semi Structural Bonding with High Performance Pressure Sensitive Tapes*, 3M Deutschland, 2 ed., Neuss, 1993, p. 278.
11. *Eur. Adhes Sealants*, (6), 37, 1995.
12. G. Bennett, P.L. Geiß, J. Klingen, and T. Neeb, *Adhäsion*, (7–8), 19, 1996.
13. *Etiketten-Labels*, (3), 32, 1995.
14. U.S. Patent, 4223067, in D.K. Fisher and B.J. Briddell (Adco Product Inc., Michigan Center, MI, USA), EP 0,426,198 A2/1991.
15. R. Mast, J.D. Muchin, and J.A. Neri (Lee Pharmaceuticals, Ascutek Adhesive Specialities), EP 262,271/1988, *CAS*, 19, 3, 1988.

16. J. Suchy, J. Hezina, and J. Matejka, Czech Patent, 247,802/1987, *CAS*, 19, 4, 1988.

17. Xinhua Li, Fuaming Zuanly Sheqing Kongkai Shuomingshu, Chinese Patent, 86,102,255/1987, in *CAS*, 19, 4, 1988.

18. D.K. Fisher and B.J. Briddell (Adco Product Inc., Michigan Center, MI, USA), EP 0,426,198 A2/1991.

19. F.C. Larimore and R.A. Sinclair (Minnesota Mining and Manuf. Co., St. Paul, MN, USA), EP 0,197,662A1/1986.

20. G. Meinel, *Papier und Kunststoff Verarbeiter*, (19), 26, 1985.

21. United Barcode Industries, Denmark, *Drucker Zubehör*, Technical Booklet, 1996.

22. C.M. Brooke, *Finat News*, (3), 34, 1987.

23. F.M. Kuminski and Th.D. Penn, *Plast. Process.*, XLII (2), 25, 1972.

24. R. Higginson (DRG, UK), PCT/WO 88,03,477/1988, *CAS, Colloids* (*Macromol. Aspects*), 22, 6, 1988.

25. Mystik Tape Inc., IL, U.S. Patent, 3,161,533, in *Adhäsion*, (6), 277, 1966.

26. Johnson & Johnson, New Jersey, U.S. Patent, 3,161,554, *Adhäsion*, (6), 277, 1966.

27. K.G. Lohmann, Neuwied, Germany, U.S. Patent, 3,086,531, *Coating*, (6), 185, 1969.

28. J.R. Pennace, C. Ciuchta, D. Constantin, and T. Loftus (Flexcon Co. Inc., Spencer, MA, USA), WO 8,703,477A/1987.

29. *Coating*, (6), 185, 1969.

30. Johnson & Johnson, British Patent, 799,424, *Coating*, (6), 185, 1969.

31. *Adhes. Age*, (4), 6, 1983.

32. D.C. Stillwater, D.C. Koskenmaki, and M.H. Mazurek (Minnesota Mining and Manuf. Co., St. Paul, MN, USA), U.S. Patent, 5,344,681/1995, *Adhes. Age*, (5), 14, 1996.

33. R.B. Lipson (Kwik Paint Products), U.S. Patent, 5,468,533, *Adhes. Age*, (5), 12, 1996.

34. U.S. Patent, 3,364,063, in J.C. Pasquali, EP 0,122,847/1984.

35. Rohm & Haas Co., Philadelphia, PA, USA, U.S. Patent, 3,152,921, *Adhäsion*, (3), 80, 1966.

36. K. Palli, M. Tirkkonnen, and T. Valonen (Yhtyneet Paperitehtaat Oy), Finnish Patent, 74723/1987, *CAS, Adhesives*, 14, 4, 1988, 108:222786.

37. M. Satsuma (Nitto Electric Ind. Co.), Japanese Patent, 6,348,381/1988, in *CAS, Adhesives*, 14, 4, 1988.

38. W.K. Darwell, P.R. Konsti, J. Klingen, and K.W. Kreckel (Minnesota Mining and Manuf. Co., St. Paul, MN, USA), EP 257,984/1986.

39. A.D. Little, Inc., U.S. Patent, 3,039,893, *Coating*, (6), 185, 1969.

40. Rikkidain K.K., Japanese Patent, 07,278,508/1995.

41. H. Nakahata, Japanese Patent, 07,316,510/1995, *Adhes. Age*, (5), 11, 1996.

42. M. Hasegawa, Tokyo, U.S. Patent, 4,460,634/1984.

43. British Patent, 1,102,244, *Adhäsion*, (10), 303, 1972.

44. G. Crass, A. Bursch, and P. Hammerschmidt (Hoechst AG, Frankfurt am Main, Germany), EP 4,673,611/1987.

45. Th.J. Bonk and J.Th. Simpson (Minnesota Mining and Manuf. Co., St. Paul, MN, USA), EP 0,120,708/1987.

46. J.M. McClintock (Morgan Adhesives Co., Stow, OH, USA), U.S. Patent, 4,345,678/1986, *Adhes. Age*, (5), 26, 1987.

47. *Coating*, (11), 89, 1984.

48. *Adhäsion*, (11), 37, 1994.

49. N.W. Malek (Beiersdorf AG, Hamburg, Germany), EP 0,095,093/1983.

50. Minnesota Mining and Manuf. Co., St. Paul, MN, USA, German Patent, 1,486,514, *Coating*, (12), 363, 1973.

51. E. Borregard, U.S. Patent, 3,454,458, *Coating*, (12), 353, 1970.

52. *Conver. Today*, (7/8), 6, 1990.

53. Belgian Patent, 675,420, in D.K. Fisher and B.J. Briddell (Adco Product Inc., Michigan Center, MI, USA), EP 0,426,198A2/1991.

54. *Packlabel News*, Packlabel Europe '97, Ausgabe 2.

55. *Coating*, (3), 65, 1974.

56. *Adhes. Age*, (3), 8, 1987.

57. *Adhes. Age*, (8), 8, 1983.
58. J.H.S. Chang (Merck & Co., Inc., Rashway, NJ, USA), EP 0,179,628/1984.
59. D. Lacave, *Labels Labelling*, (3/4), 54, 1994.
60. I.P. Rothernberg (Stik-Trim Industries Inc., New York), U.S. Patent, 4,650,704/1987.
61. D. Bedoni and G. Caprioglio, *Modern Equipment for Label and Tape Converting*, in 19. Münchener Klebstoff u. Veredelungsseminar, Munich, Germany, 1994, p. 37.
62. G.L. Duncan (Mobil Oil Co.), U.S. Patent, 4,720,416/1988.
63. Johnson & Johnson, U.S. Patent, 3,483,018, *Coating*, (1), 28, 1971.
64. Universum Verpackungs GmbH, Rodenkirchen, Germany, German Patent, 12,977,990, *Coating*, (1), 9, 1990.
65. P. Thorne, *Finat News*, (3), 47, 1988.
66. Minnesota Mining and Manuf. Co., St. Paul, MN, USA, DBP 1277483, *Coating*, (7), 210, 1969.
67. T.H. Haddock (Johnson & Johnson Products Inc., New Brunswick, NJ), EP 0,130,080 B1/1985.
68. *Papier und Kunststoff Verarbeiter*, (9), 57, 1988.
69. G. Crass and A. Bursch (Hoechst AG, Frankfurt am Main, Germany), U.S. Patent, 4,673,611/1987.
70. G. Camerini (Coverplast Italiana SpA), EP 248,771/1986.
71. I. Benedek, *Eur. Adhes. Sealants*, (2), 25, 1996.
72. *Adhes. Age*, (10), 125, 1986.
73. E. Park, *Paper Technol.*, (8), 15, 1989.
74. U.S. Patent, 33,21,451, in S.E. Krampe and C.L. Moore (Minnesota Mining and Manuf. Co., St. Paul, MN, USA), EP 0,202,831 A2/1986.
75. U.S. Patent, 4,260,659, in S.E. Krampe and C.L. Moore (Minnesota Mining and Manuf. Co., St. Paul, MN, USA), EP 0,202,831 A2/1986.
76. U.S. Patent, 4,374,883, in S.E. Krampe and C.L. Moore (Minnesota Mining and Manuf. Co., St. Paul, MN, USA), EP 0,202,831 A2/1986.
77. R.R. Charbonneau and G.L. Groff (Minnesota Mining and Manuf. Co., St. Paul, MN, USA), EP 0,106,559 B1/1986.
78. H. Miyasaka, Y. Kitazaki, T. Matsuda, and J. Kobayashi (Nichiban Co. Ltd., Tokyo, Japan), Offenlegungsschrift, DE 3,544,868 A1/1985.
79. Japanese Patent, 2736/1975, in H. Miyasaka, Y. Kitazaki, T. Matsuda, and J. Kobayashi (Nichiban Co. Ltd., Tokyo, Japan), Offenlegungsschrift, DE 3,544,868 A1/1985.
80. U.S. Patent, 3,691,140/1972, in H. Miyasaka,Y. Kitazaki, T. Matsuda, and J. Kobayashi (Nichiban Co. Ltd., Tokyo, Japan), Offenlegungsschrift, DE 3,544,868 A1/1985.
81. C. Engel, U.S. Patent, 514,950/1983, in F.C. Larimore and R.A. Sinclair (Minnesota Mining and Manuf. Co., St. Paul, MN, USA), EP 0,197,662A1/1986.
82. P. Gobran, U.S. Patent, 4,260,659, in J.N. Kellen and C.W. Taylor (Minnesota Mining and Manuf. Co., St. Paul, MN, USA), EP 0246 A2/1987.
83. G. Auchter, J. Barwich, G. Rehmer, and H. Jäger, *Adhes. Age*, (7), 20, 1994.
84. I. Benedek, *Pressure-Sensitive Formulation*, VSP, Utrecht, 2000.
85. Joachim Merretig, *Coating*, (3), 50, 1974.
86. P. Foreman and P. Mudge, *EVA-Based Waterborne Pressure-Sensitive Adhesives*, in Proceedings of Tech 12, Technical Seminar Proceedings, Itasca, IL, USA, May 3–5, 1989, p. 203.
87. W.M. Stratton, *Adhes. Age*, (6), 21, 1985.
88. *Finat News*, (3), 31, 1994.
89. *Druck Print*, (10), 32, 1987.
90. *Etiketten-Label*, (5), 25, 1995.
91. M. Fairley, *Labels Labelling Int.*, (5/6), 76, 1997.
92. G. Bulian, *Extrusion Coating and Adhesive Laminating: Two Techniques for the Converter*, Polyethylene '93, Maack Business Service, Conference, Zürich, Switzerland, 4/6, 1993.
93. K. Fust, *Coating*, (2), 65, 1988.
94. J.C. Pommice, J. Poustis, and F. Lalanne, *Paper Technol.*, (8), 22, 1989.
95. A. Prittie, *Finat News*, (3), 35, 1988.
96. *Etiketten-Labels*, (3), 8, 1995.
97. L. Waeyenbergh, 19. Münchener Klebstoff u. Veredelungsseminar, München, Germany, 1994, p. 138.
98. *Etiketten-Labels*, (3), 9, 1995.

99. H.J. Teichmann, *Papier und Kunststoff Verarbeiter*, (11), 10, 1994.
100. W. Keller, *Kunststoffe Synthetics*, (9), 32, 1995.
101. K. Dormann, *Coating*, (6), 150, 1984.
102. K.W. Holstein, *Neue Verpackung*, (4), 59, 1991.
103. W.E. Havercroft, *Paper, Film and Foil Converter*, (10), 52, 1973.
104. *Paper, Film and Foil Converter*, (5), 45, 1969.
105. U.P. Seidl, (Schreiner Etiketten und Selbstklebetechnik GmbH und Co.), Ger. Offen, DE 3625904/ 1988, *CAS, Siloxanes and Silicones*, 16, 3, 1988.
106. *Pack Report*, (3), 26, 1997.
107. *Etikettierkompetenz für die Pharma Industrie*, Technical Booklet, Pago AG, Buchs, Switzerland 10, 2001.
108. H. Hadert, *Coating*, (1), 11, 1969.
109. Johnson & Johnson, Canadian Patent, 583,367, in H. Hadert, *Coating*, (1), 11, 1969.
110. *Coating*, (3), 68, 1974.
111. *Verpackungs-Rundschau*, (9), 994, 1983.
112. *Etiketten-Labels*, (3), 23, 1995.
113. *Neue Verpackung*, (1), 156, 1991.
114. G. Bonneau and M. Baumassy, *New Tackifying Dispersions for Water-Based PSA for Labels*, 19. Münchener Klebstoff u. Veredelungsseminar, Munich, Germany, 1994, p. 82.
115. H. Mueller and J. Türk (BASF AG, Ludwigshafen, Germany), EP 0118726/1984.
116. R. Schieber, *Adhäsion*, (5), 21, 1982.
117. I. Benedek, *Adhäsion*, (12), 17, 1987.
118. *Papier und Kunststoff Verarbeiter*, (6), 18, 1995.
119. *Adhäsion*, (11), 9, 1988.
120. Cham Tenero, *Label Printers*, 1990.
121. L. Heymans, *Developments in PSA for Labels and Tapes*, Second European Tape & Label Conference, Brussels, Belgium, 1993, p. 115.
122. *Etiketten-Labels*, (1), 17, 1995.
123. *Verpackungsberater*, (5), 20, 1996.
124. *Systemlösungen, komplett aus einer Hand, Bluhm Systeme*, Technical Booklet, Bluhm Systeme GmbH, Unkel/Rhein, Germany, 2001.
125. *Package Print Design Int.*, (1/2), 8, 1997.
126. *Verpackungs-Berater*, (3), 18, 1997.
127. Mat Bateson, *Finat News*, (3), 29, 1989.
128. *Prodoc*, (2), 3, 1992.
129. *Etiketten-Labels*, (5), 136, 1995.
130. *Allg. Papier Rundschau*, 22, 787, 1986.
131. *Brit. Plast.*, (10), 47, 2001.
132. *Etiketten-Labels*, (1), 10, 1996.
133. W. Keller, *Kunststoffe Synthetics*, (9), 32, 1995.
134. *Finat News*, (4), 10, 1996.
135. C. Stöver, *Kunststoffe*, 84(10), 1426, 1994.
136. *Paper Eur.*, (9), 11, 1993.
137. *Label Buyer International*, Spring '97, p. 20, 26.
138. F. Altenfeld and D. Breker, *Semi-Structural Bonding with High Performance Pressure Sensitive Tapes*, 3M Deutschland, 2nd ed., Neuss, 1993, p. 278.
139. *Klebeband Forum*, Hoechst Films, Hoechst AG, Wiesbaden, Germany, October 1989, No. 27.
140. Second European Tape & Label Conference, April 28, 1993, Brussels, Belgium, *Coating*, (8), 277, 1993.
141. *Adhes. Age*, (7), 42, 1994.
142. G. Bennett, P.L. Geiß, J. Klingen, and Th. Neeb, *Adhäsion*, (7–8), 19, 1996.
143. *Coating*, (11), 46, 1990.
144. H. Kniese, *Coating*, (11), 394, 1990.
145. L.A. Sobieski and Th.A. Tangney, *Adhes. Age*, (12), 26, 1988.
146. *Adhäsion*, (1/2), 29, 1987.
147. *Kunststoff J.*, (9), 92, 1986.

148. *Convert. Today*, (11), 9, 1991.

149. Johnson & Johnson, U.S. Patent, 3,403,018, *Coating*, (1), 24, 1969.

150. *Adhäsion*, (6), 14, 1985.

151. *Adhes. Age*, (10), 125, 1986.

152. *Adhes. Age*, (9), 8, 1986.

153. K. Wabro, R. Milker, and G. Krüger, *Haftklebstoffe und Haftklebebänder*, Astorplast GmbH, Alfdorf, Germany, 1994, p. 48.

154. F.M. Kuminski and T.D. Penn, *Resin Review*, 17(2), 25, 1990.

155. U.S. Patent, 4,181,752, in R.R. Charbonneau and G.L. Groff (Minnesota Mining and Manuf. Co., St. Paul, MN, USA), EP 0,106,559B1/1984.

156. U.S. Patent, 2,925,174, in R.R. Charbonneau and G.L. Groff (Minnesota Mining and Manuf. Co., St. Paul, MN, USA), EP 0,106,559B1/1984.

157. U.S. Patent, 4,286,047, in R.R. Charbonneau and G.L. Groff (Minnesota Mining and Manuf. Co., St. Paul, MN, USA), EP 0,106,559B1/1984.

158. C. Parodi, S. Giordano, A. Riva, and L. Vitalini, Styrene–butadiene block copolymers in hot-melt adhesives for sanitary application, in 19. Münchener Klebstoff und Veredelungsseminar, Munich, Germany, 1994, p. 119.

159. H. Becker, *Adhäsion*, (3), 79, 1971.

160. S.E. Krampe and C.L. Moore (Minnesota Mining and Manuf. Co., St. Paul, MN, USA), EP 02,02,831A2/1986.

161. A. Wolf, *Kaut. Gummi, Kunststoffe*, 41(2), 173, 1988.

162. U. Zorll, *Adhäsion*, (9), 237, 1975.

163. *Coating*, (6), 188, 1969.

164. *Adhäsion*, (3), 83, 1974.

165. *Coating*, (8), 200, 1984.

166. O. Cucu, C. Ilie, N. Moga, and M. Popescu, Romanian Patent 92,466/1987, *CAS, Coating Inks Related Products*, 11, 6, 1988.

167. Hitco, S.A., French Patent, 2,045,508, *Coating*, (11), 336, 1972.

168. P. Gleichenhagen and I. Wesselkamp (Beiersdorf AG, Hamburg, Germany), EP 0,058,382B1/1982.

169. Horst Bucholz, *Oberflächentechnik*, (10), 484, 1973.

170. Thermoplastic adhesive film, *Adhes. Age*, (12), 8, 1986.

171. C. Bohlmann, *Papier und Kunststoff Verarbeiter*, (9), 12, 1982.

172. I. Benedek, E. Frank, and G. Nicolaus (Poli-Film Verwaltungs GmbH, Wipperfürth, Germany), DE 4,433,626 A1/1994.

173. R.C. Lomeli and G.E. Stewart (Trade Printers), U.S. Patent, 4,379,573/1983.

174. *Der Siebdruck*, (8), 14, 1986.

175. (Minnesota Mining and Manuf. Co., St. Paul, MN, USA), DBP 1,277,483, *Coating*, (7), 210, 1969.

176. BASF, Tl-2.2-21 d/November 1979, Teil 3, B.3, Selbstklebende Dekorationsfolien.

177. G. Fuchs, *Adhäsion*, (3), 24, 1982.

3 Physical Basis of Pressure-Sensitive Products

István Benedek

CONTENTS

Pressure-sensitive products (PSPs) are constructed in various ways depending on their application. Their components are manufactured from various raw materials. However, the astonishingly broad range of materials and structures leads to a common characteristic: pressure sensitivity. Why? In order to answer this question, let us make a short examination of the theoretical (physical and chemical) basis of PSPs.

PSPs are viscoelastic materials. As discussed in Chapter 2, such materials are built up as various constructions. The viscoelasticity and flow properties of PSPs play a key role in their application. Their performance results from their intrinsic characteristics and from the interaction of their components. Their mechanical performance characteristics allow their converting and end-use. Such performance characteristics depend on their rheology. Other physical properties influence the performance characteristics of PSPs also. For certain PSPs, the polarity and the electrical properties of the carrier material affect coatability and processibility. Therefore, in this chapter we deal with the rheology, mechanical characteristics, and electrical properties of PSPs.

I. RHEOLOGY OF PSPs

The rheology of pressure-sensitive adhesives (PSAs) is described in detail in Refs. [1,2]. As discussed earlier (see Chapter 2), a nonPSA-related technology has also been developed to manufacture PSPs. The first PSPs were manufactured from masticated rubber and filler. Later, the elastomer was transformed by formulation in a viscoelastic material used as coated PSA. The newest PSPs are adhesiveless films made of plastomer (Figure 3.1).

The first carrier materials were nonconformable, nondeformable, and nonadhesive. For such products, pressure sensitivity and self adhesivity were given by a special coated layer. Plastics developed more recently as carrier materials are conformable, deformable, and may exhibit adhesivity. The main components of the classical PSP construction are plastomers (carrier) and modified elastomers (PSA). Among the rubbery products, thermoplastic elastomers (TPEs) can be processed like plastomers. The construction of the new PSPs includes a single viscoelastic material processed as plastomer. It can be concluded that elastomers and plastomers are the base components of PSPs. Both must be modified to achieve adequate viscoelastic properties.

Classical (nonadhesive) plastic carrier materials exhibit viscoelastic properties that are reflected in their mechanical performance during product manufacture and application. Their conformability and deformability do not allow pressure-sensitive adhesion but influence bonding and debonding (stress transfer). PSAs display the viscoelasticity necessary to bond and debond. "One-component" PSPs must also exhibit viscoelastic characteristics that allow them to carry out their mechanical (carrier) and adhesive function. Pressure-sensitive adhesives exhibit cold flow. It is well known that plastics also display cold flow, but to a different degree. In their common use, plastics are virtually cold flow free. PSAs must exhibit a real cold flow. The development of macromolecular chemistry allowed the synthesis of new materials (e.g., ethylene–vinyl acetate [EVAc] copolymers, ethylene–butyl acrylate (EBA) copolymers, etc.) having a PSA-like rheology and the processibility of plastics. Embedded in the carrier film, such products can impart pressure-sensitive properties.

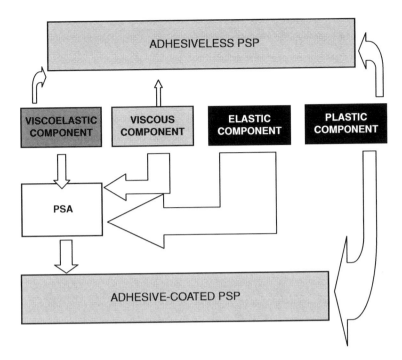

FIGURE 3.1 Raw material basis and manufacture of PSPs.

PSAs are based on elastic substances. Rubbery elasticity allows high elastic deformations. Film-like carrier materials are based on plastics. Their elastic deformation is limited; high force levels or the use of thin gauge materials are required to produce elastic or reversible deformations in them. PSAs undergo continuous self-deformation. Although there are considerable differences between the behavior of PSAs and that of plastomers under stress, it is possible to manufacture plastic carrier materials that will deform like PSAs at a low force level by bonding and debonding. How? What is the theoretical background for the design of such products? Rheological considerations help to answer this question.

For classical adhesives, their chemical nature plays an important role. The nature of the substrate is less important. For PSAs, their physical status is more important, and the surfaces of the carrier material and the substrate determine the type of bond. Because of the mutual influence of the solid-state components and the adhesive, both influence the peel value and the bonding and debonding characteristics too. For many classical applications, the dimensions and physical/mechanical characteristics of the solid-state components of the joint are less important for the bond. For thin plastic film-based PSPs, the deformation of the carrier material (its rheology) and consequently its mechanical characteristics play a considerable role. Both the carrier material and the adhesive suffer deformation under stress. Nonreinforced plastics possess mechanical resistance and deformability of the same order of magnitude as the crosslinked adhesive layer [3]. Because of the mutual deformability of the adhesive and of the carrier material and their same order of magnitude, it is difficult to quantify their participation in the stress transfer (Figure 3.2).

On the basis of the industrial requirements, the following questions can be formulated:

1. Is it possible to achieve pressure-sensitive behavior for carrier-like materials also?
2. If it is possible to achieve "plastomer-like" PSPs or "carrier-like" PSAs, is their mechanisms of functioning the same?
3. For "full-plastic" PSPs, what are the roles (from rheological point of view) of the carrier material and that of the adhesive?

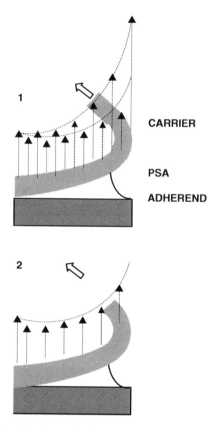

FIGURE 3.2 Mutual deformation of the adhesive and carrier during debonding of the PSP. (1) Plastic carrier; (2) paper carrier.

The manufacturer of PSPs can be (and for certain products must be) the manufacturer of the plastic carrier material also. Therefore, he is involved in the regulation of polymer rheology during processing at high temperatures and that of the film-like material at room temperature. The rheology of the molten plastic material is decisive in carrier manufacture. The rheology of the molten or diluted adhesive is important for its coating. The rheology of the softened plastic film is determinative for its application. Therefore, these aspects should be discussed separately.

A. RHEOLOGY OF CARRIER MATERIAL

Some carrier materials and all PSAs are viscoelastic materials. They differ only in the ratio of elastic to plastic components and their degree of linearity. A purely elastic (Hookean) material is characterized by a linear relationship between stress and strain, whereas a purely viscous fluid exhibits a stress directly proportional to the strain rate but independent of strain itself. According to Gerace [4], viscoelasticity means that the bond strength is not constant for varying strain rates and temperatures. Lines of equivalent adhesion can be drawn over all combinations of rate and temperature. Any incompletely elastic body subjected to cyclic strains will respond with an out-of-phase induced stress. If the body is linearly viscoelastic, a simple shift in the phasing of the induced stress with respect to the cyclic strain will occur. Nonlinear viscoelastic behavior is more complex [5]. Pressure sensitivity supposes nonlinear viscoelastic behavior. Pressure sensitivity requires fluidity for low stress rate, instantaneous adhesion, and solid-state behavior for high-speed debonding stress.

The experiments in which bodies are subjected to periodic oscillatory stresses due to sinusoidal strains (γ), the stress and the strain rates are never in phase or $\pi/2$ out-of-phase together, but somewhere in between. Such materials when strained store a part of the energy and dissipate some of it as heat. This behavior is described as viscoelastic. A viscoleastic body subjected to sinusoidal strain:

$$\gamma = \gamma_0 \sin \bar{\omega} t \tag{3.1}$$

where γ_0 is the maximum strain amplitude and $\bar{\omega}/2\pi$ is the frequency, will exhibit a sinusoidal stress (δ) of identical frequency but out-of-phase with the strain (linear viscoelasticity) [5]. In this case, the stress can be expressed as follows:

$$\sigma = \gamma_0 [G'(\omega) \sin \omega t + G''(\omega) \cos \omega t] \tag{3.2}$$

where G' is the storage modulus, a measure of the energy stored and recovered in cyclic deformation, and G'' is the loss modulus. The energy loss per cycle hysteresis is given by:

$$\pi (\gamma_0)^2 G'' \tag{3.3}$$

The rheological behavior of solid-state materials is characterized by the modulus; the behavior of liquids is described by the viscosity. A classical carrier material is a solid-state component. A classical PSA is liquid. Theoretically, a classical plastic material is a plastomer that does not respond elastically when stressed. Theoretically, rubber (which is the main classical PSA component) is a "pure" elastomer without viscous flow. In reality, ordered structures in both classes of materials allow elastic behavior and excessive stresses cause remnant deformation. Theoretically, the flow properties of the solid-state carrier material and those of a permanently liquid PSA should be discussed separately. In reality, the viscous flow of plastics is temperature dependent like the viscoelastic properties of PSA. The rheology of both materials is a function of time and temperature. The dynamic modulus and the viscosity depend on the temperature and stress rate. The influence of temperature on the viscosity is characterized by the shift factor (a_T):

$$a_T = \frac{\eta_0(T)}{\eta_0(T_0)} \tag{3.4}$$

where T_0 is the reference temperature and η_0 is the zero viscosity, that is, the viscosity that does not depend on the deformation rate (in the linear region, at small shear rates). For amorphous polymers such as PSAs, the dependence of the shift factor on the temperature is characterized by the Williams–Landel–Ferry (WLF) equation including the material correlated parameters c_1 and c_2:

$$\ln a_T = -\frac{c_1(T - T_0)}{c_2 + (T - T_0)} \tag{3.5}$$

For partially crystalline polymers (e.g., common carrier materials), the Arrhenius equation is valid:

$$\ln a_T = \frac{E_0}{R(1/T - 1/T_0)} \tag{3.6}$$

where E_0 is the activation energy of the flow related to the material [6]. PSAs are generally amorphous polymers; common plastics are partially crystalline. It is evident that the hot laminating conditions and the bonding behavior of polyisobutylene-based self-adhesive films (SAFs) differ from the laminating conditions of a partially crystalline polyvinyl-acetate (PVAc). For instance, for a (PVAc) film between -60 and $+68°C$ at a frequency of 1000 Hz, there is a change in the dynamic modulus of elasticity of 10^2 dyn/cm^2 [7]. At lower frequencies, the difference is more pronounced.

1. Solid-State Rheology

As a function of the end-use of PSPs, a variety of carrier materials have been introduced (see Chapter 2 and Chapter 8). Almost all carrier materials are based on macromolecular compounds. Paper and plastics are the main representatives. Paper has a relatively small influence on the rheology of the PSP. It is a polymeric material with a high glass transition temperature (T_g) and crosslinked structure, that is low deformability. According to Dunckley [8], for common grades of paper, paper strain appears at 1500 g/in. force during debonding of the pressure-sensitive laminate. The actual forces in the adhesive are higher because they are partially absorbed by the deformation of the paper (Figure 3.2). In practice, it may be supposed that the paper's deformation during the manufacture or application of PSPs is relatively low when compared with the deformation of other product components. As known, the influence of temperature on the rheology of a macromolecular compound can be quantified by the value of the glass transition temperature. Above this temperature, the macromolecules have an unordered, free-flowing status and, therefore, they can form bonds. Above T_g, the polymer is expanded to the extent that molecular motion is possible. The fluidity of a polymer depends on the position of its T_g, relative to the application temperature. The higher the T_g, the higher the fluidity. Carrier materials for PSPs are mainly thermoplasts. Thermoplasts possess a relatively high T_g (Table 3.1). Their end-use temperature is situated below their glass transition temperature. They are processed at temperatures above their T_g.

Elastomers (which are the main components of classical PSAs) display a much lower T_g than plastomers. They do not melt above this temperature; they are transformed in rubbery elastic materials. They keep this rubbery elastic status due to their crosslinked structure up to their thermal destruction [26]. TPEs have a relatively low-glass transition temperature like elastomers; above this temperature they are also transformed into rubbery elastic materials. This status is, however, less stable than for rubbery materials. As temperature and stresses increase, their stability decreases and a second T_g is displayed. At this temperature, the TPEs melt, and above it, they are plastic and can be processed like elastomers (Figure 3.3).

Plastomers are the common carrier materials for PSPs. Elastomers can also be used as carriers, and (reinforced) viscoelastomers play the role of carrier for special PSPs (e.g., transfer tapes). TPEs have been developed for uses other than PSA. Their self-adhesivity is low. To be used as PSAs, such materials require tackifying like natural rubber. However, their example shows that it is possible to manufacture materials having plastomer-like processibility (as films), which exhibit the properties of elastomers or viscoelastomers. Such performance characteristics have been achieved by development of a broad range of TPEs and pseudo-TPEs (see Chapter 5).

In contrast, macromolecular compounds that are viscoelastic manifest viscous flow and elasticity. The importance of the viscous or elastic part of their behavior can be characterized by the values of their viscosity or elastic modulus. These parameters are not material characteristics only; they depend on the temperature and time, that is on the strain rate applied. Paper has a relatively high modulus, that is high dimensional stability (at a given atmospheric humidity). Therefore, its influence on the rheology of PSPs is manifested more through its interaction with other laminate components; this interaction is regulated primarily by its surface characteristics.

Plastic films exhibit a different behavior. They display cold flow as a result of their relatively low T_g and modulus. The modulus is a function of chemical composition and macromolecular characteristics (see also Chapter 5). The nature of the comonomers, their sequence distribution, and the molecular weight (MW) of the polymer influence the modulus. The nature of the comonomers affects functionality and reactivity of the macromolecules. Functionality refers to the chemistry of the polymer side chain, whereas reactivity refers to the crosslinkability. Inter- or

TABLE 3.1
Glass Transition Temperature of Main PSPs Components

	Component		
Carrier Component	**Adhesive Component**	T_g (°C)	**Reference**
—	Polybutadiene	−85	[9]
Polyethylene	—	−80 to −90	—
—	Natural rubber	−56 to −64	—
Ethylene–propylene copolymer	—	−51 to −59	[10]
—	Cohesive acrylate	−55	[11]
		−53	[12]
—	Carboxylated styrene–butadiene rubber	−55	[13]
—	Neoprene	−44	[14]
—	Acrylate	−43	[15]
—	Tacky acrylate	−42	[16]
—	Vinyl acetate–acrylate	−40	[17]
—	Isoprene–styrene, block copolymer, electron beam-cured	−28	[18]
—	Acrylate	−20	[19]
—	Vinyl acetate–ethylene, self-crosslinking	−15	[20]
—	Acrylate, thermally crosslinkable	−7	[21]
—	Vinyl acetate–dibutyl maleate	−5 to +10	—
—	Very hard acrylate	8	[22]
Polypropylene	—	−11	[10]
Polypropylene, cast	—	−3 to +23	[23]
Vinyl acetate–ethylene copolymer	—	−10 to +15	—
Polyvinyl chloride, plasticized (30–45%)	—	−40 to 0	—
Vinyl acetate, homopolymer plasticized (40% plasticizer)	—	5	[24]
Polyvinyl chloride, plasticized (10–30%)	—	0 to +40	—
Vinyl acetate, homopolymer	—	40	[24]
Polyvinyl chloride	—	75 to 105	[25]
Polyimide	—	310 to 365	[26]

intrachain interactions due to the functionality or crosslinking due to chain reactivity may reinforce the polymer. The reinforcing effect of the sequence distribution is well known for block copolymers and is taken advantage of in the production of hot-melt pressure-sensitive adhesives (HMPSA). The effect of the MW on the cold flow of polymers having the same chemical composition is illustrated by migration, penetration, and blocking. Linear low density polyethylene (LLDPE) contains low MW species and requires more antiblock additive to avoid blocking [27].

The glass transition temperature also has a special influence on aging. A glassy polymer may undergo aging. The temperature range in which this phenomenon occurs has been located by Struik [28] between T_g and the first secondary transition T_β. The data of Struik were obtained from creep measurements. The kinetics of aging has been characterized by the shift rate (M) related to creep (cold flow) at different times according to the following correlation:

$$M = -\frac{\partial \log a}{\partial \log t} \tag{3.7}$$

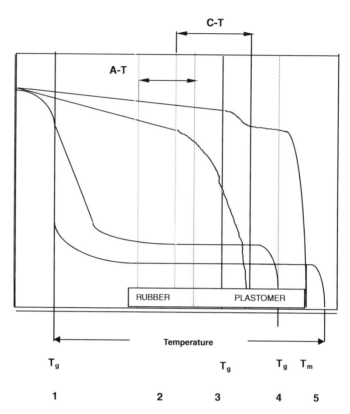

FIGURE 3.3 Phase transformations of the PSP raw materials. Application domains of rubber, TPE, amorphous plastomer (A-T), and partially crystalline plastomer (C-T). (1) T_g of rubber; (2) room temperature; (3) T_g of amorphous plastomer; (4) T_g of segmented copolymer; (5) melting point of crystalline domains.

where t denotes the aging time and a the horizontal shift factor required to superpose two creep curves related to different aging times. According to Bauwens [29], for the polymers studied, the shift rate is found to reach a constant value, near unity, over a more or less wide temperature range. That means that an increase of creep time by a factor of ten has the same effect on the creep compliance as an increase of the aging time by a factor of ten. As mentioned earlier, pressure sensitivity requires flow for instantaneous bonding. Let us see what cold flow means for a plastomer, an elastomer, and a viscoelastomer.

2. Cold Flow and Self-Adhesion

Pressure sensitivity is the result of cold flow. Cold flow (F_c) considered as a deformation (ε) of the material under a constant load (σ_0) during a given period of time (t) is expressed by the following correlation [30]:

$$F_c = \left(\frac{1}{\sigma_0}\right) \cdot \varepsilon(t) \tag{3.8}$$

Cold flow is a phenomenon which allows bonding, and the dissipation of energy by debonding (W_{F_c}), which can be described as a function of the deformation rate (v_{F_c}) and modulus (E) [30]:

$$W_{F_c} = \frac{\sigma_0^2}{2E} + \sigma_0^2 \int_0^t V_{F_c}(t)\, dt \tag{3.9}$$

As stated by Köhler [31], the most important property of a PSA is its permanent liquid status, that is cold flow. This allows instantaneous adhesion. Different adhesion theories attempt to explain why one material adheres to another. The best known are the mechanical theory and the adsorbtion theory. According to the mechanical theory, the adhesive flows and fills microcavities on the surface of the substrate. In general, mechanical anchoring is the prime factor in bonding porous substrates (paper, nonwovens, etc.). It is evident that for this theory the role of adhesive flow is important, but the adsorbtion theory also supposes intimate contact between adhesive and adherend. The adsorbtion theory states that adhesion results from molecular contact between two materials and the surface forces that develop. Continuous contact is established between an adhesive and the substrate by wetting. Hot-melts, solvent- or dispersion-based adhesives, contact adhesives, and PSAs work physically [32]. It is evident that (generally) the magnitude of the cold flow differs for plastics and PSAs. Intermediate values are achieved for SAFs with embedded viscous component and for warm or hot laminated PSPs (see also Chapter 8).

For adhesive-coated PSPs, the cold flow of the adhesive depends on its coating weight and viscosity. Therefore, for tropically resistant tapes a low coating weight is recommended. For these tapes, filled adhesive with higher viscosity is used [33]. Special fillers or multilayer structures allow the "combination" of adhesive and carrier-like behavior for transfer tapes. Thick PSA tapes (0.2–1.0 mm) have been prepared by ultraviolet (UV) photopolymerization of acrylics (ACs). The filled layer can be laminated together with the unfilled one. The PSA matrix may contain glass microbubbles [34]. The formulation can be polymerized as a thick layer (up to 60 mil), which may be made up of a plurality of layers, each separately photopolymerized. The middle layer having a higher degree of polymerization can be considered the carrier. Such structure with a plurality of superimposed adhesive layers having different gradients of shear creep compliance may serve as modality to control the creep compliance, according to Gobran [35].

In contrast, as is known from the plasticizing of polyvinylchloride (PVC), special plastomers can be formulated as adhesive, and copolymerization may lead to materials having the performance characteristics of plastomers (carrier), elastomers (adhesive component), or viscoelastical compounds (PSA). A vinyl acetate (VAc) content of 32% in EVAc copolymers gives a partially crystalline polymer; at 40% VAc content, a completely amorphous polymer is synthesized [36]. Polymers with a level of 15% VAc exhibit polyethylene (PE)-like properties. Polymers having 15–30% VAc exhibit PVC-like characteristics; polymers with more than 30% VAc are elastomers. Copolymerization of ethylene with acrylic acid (AA) softens the polymer also. Kirchner [37] showed that the high modulus (of about $1600\,kg/cm^2$, ASTM 638-58) of high density polyethylene (HDPE) decreases to $430\,kg/cm^2$ for an ethylene-acrylic acid (EAA) copolymer having 19% polar comonomer. Figure 3.4 illustrates schematically that "softening" of the plastomer carrier material and "hardening" of the adhesive lead to the same result, that is, pressure sensitivity.

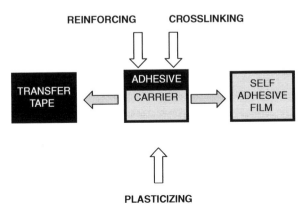

FIGURE 3.4 Technical possibilities to obtain carrier-less and adhesiveless PSPs.

The self-deformation (cold flow) of many polymeric materials allows an intimate contact to be established between the polymer and the substrate. Adhesion is possible if the flow of the materials allows enough surface contact and the chain segments can interpenetrate [38]. This supposes an increased mobility of the chains. Therefore, crystalline or crosslinked materials are not adhesive above their T_g. As found by the study of mutual adhesion, tack and green strength of ethylene–propylene–diene multipolymer (EPDM) rubber and autoadhesion depend on the (mutual) diffusion of polymer chains, their flexibility, and polarity [39]. Studying the contact adhesion of rubber to glass, Carre and Roberts [40] stated that because of energy loss at the border of the contact surfaces, the contact buildup energy is lower than the debonding energy. On wet surfaces, capillary water improves the contact surface, that is, increases the energy of contact buildup and decreases the debonding energy.

External application conditions may allow better cold flow. Prinz [41] demonstrated that applying a pressure of more than 130 bar to a polyethylene film with aluminum at room temperature (for some seconds) increases the effective contact surface; cold flow of the polymer occurs and self-adhesion appears. By pressing crosslinked butadiene–styrene copolymer rubber between aluminum plates, a laminate was produced that was used as an adhesive joint model by Carre and Schultz [42]. As stated by Prinz [43], the actual contact surface between two films is very low (for a polyethylene monocrystal it is about 1–3%). Therefore, the possibility of building up self-adhesion by mutual contact is also low. Corona treatment produces a double layer of electrical charges, which causes a better contact between the surfaces, that is, increases the effective contact surface. This increase in the contact surface results in better adhesion. The effect of improving the instantaneous contact surface and adhesion by corona treatment is used for self-adhesive (adhesive-less) films such as hot laminated polyolefins or poly(EVAc) films and cold laminatable polyolefin-based cover films [44] (see also Chapter 8 and Chapter 11).

Cold flow is time- and temperature dependent. Such behavior is illustrated by silicone rubbers. Spontaneous adhesion of slight crosslinked silicone rubbers on polar substrates builds up linearly over time in the presence of ammonia and sufficiently high humidity and temperature. Reactive groups in the polymer react with the substrate [45]. Cold flow as surface contact buildup phenomenon and as a component of self-adhesion plays a special role in the manufacture and application of adhesive-free (plastic-based) PSPs (protective films, ornamental labels, etc.). Because it allows the bulk deformation of the carrier material and of the pressure-sensitive laminate during application (e.g., tape or protective film lamination) and end-use (e.g., processing of protective film or adherend laminates), cold flow plays a very important role for all PSPs that include a plastic carrier material (see also Chapter 7). Cold flow is also important for cold seal adhesives (for their both adhesion and blocking behavior), which are usually mixtures of natural rubber and fillers tackified with VAc copolymers [46].

There are various theories to explain the adhesion between polymers [47]. New attempts to clarify the role of molecular interdiffusion in self-adhesion of polymer melts draw attention to the effect of labile bonds on the strength of adhesion and of plastic yielding in semicrystalline polymers [48]. Macromolecular diffusion influences the autoadhesion of rubbers. The self-diffusion coefficient of polymers gives information about autoadhesion. The buildup of adhesion is generally proportional to self-diffusion coefficient (one exception being polybutadiene) [49]. Autoadhesion increases with the temperature. It is known from the praxis of film tests, which measure the coefficient of friction (COF), that a film with low COF at $23°C$ often displays slip or stick problems at $50°C$.

The role of the softness of the carrier material in auto-adhesion can be illustrated by cold seal adhesives, where the maximum coating weight (without blocking) recommended for paper carrier is about 6 g/m^2. The same parameter for film carrier is only 3 g/m^2 [50]. Here, the slow diffusion of the low tack adhesive is facilitated by the pressure used and by the form of the pressing tool. The adhesion of styrene-butadiene rubber (SBR) on a glass surface is virtually independent of the degree of crosslinking. However, after the application pressure, the adhesion was found to be related to the degree of crosslinking [51]. The surface characteristics of the adherend determine the softness of the adhesive to be used; for a vinyl adherend, an adhesive with a glass transition temperature of $-10°C$; for fiberboard, $-3°C$; and for steel, $+7°C$ have been suggested [52].

Self-adhesion of plastic films is used industrially for plastic film manufacture by means of heat laminating. Thick bioriented polypropylene (PP) films cannot be produced economically using the common stretching process. They are used by combining several thin films to form a multilayer film by means of heat laminating [53].

In industrial practice, cold flow and autoadhesion of plastic films may lead to blocking. According to Schwab [54], blocking is an undesirable adhesive bonding of different surfaces. However, sometimes such adhesion is desired. It is known that the low level of crystallinity of very low density polyethylene provides an intrinsic cling to extruded films that are used in coextruded structures for industrial stretch wrapping [55]. The examination of the cold flow of common and modified (more viscous) plastomers reveals that although their cold flow is limited (when compared with the creep of PSA), it can be improved by use of the proper laminating conditions.

3. Elasticity

Elasticity is related to intramolecular conformational changes. Such changes may occur in elastomers and in the amorphous portion of partially crystalline plastomers. The main problem has to do with how different elastomer structures contribute to such mobility and whether intermolecular interactions are also possible. Common rubber is a crosslinked material and may also be crystalline. It can also be artificially crosslinked. The thermomechanics of the crosslinked networks is considered from the point of view of interchain entropy and energy contributions to the free energy of deformation and the temperature coefficient of the unperturbed chain dimensions. The thermodynamic quantity of primary interest is the energy contribution exhibited by a polymer network. The classical Gaussian theory of rubber elasticity predicts the intrachain entropy and energy contributions at simple deformations of the network and their independence of the deformation (at small and moderate deformations). Analysis of the entropy and energy effects resulting from the simple extension of the stress-softened networks filled with different fillers show that in many cases the energy and entropy contribution are dependent on the concentration of the fillers, which contradicts the classical theory of rubber elasticity. Earlier molecular theories of rubber elasticity supposed that the elasticity of a polymer network is exclusively entropic. The classical theory of rubber elasticity takes into account the change of conformational energy by incorporating the front factor (f) into the equation of state for simple elongation and compression [54]:

$$f = \nu kTL_i^{-1} \frac{\langle r^2 \rangle_i}{\langle r^2 \rangle_0} (\lambda - \lambda^{-2}) \qquad (3.10)$$

where ν is the number of elastically active chains in the network, $\langle r^2 \rangle_i$ is the mean square end-to-end distance of a network in undistorted state, $\langle r^2 \rangle_0$ is the mean square end-to-end distance of the corresponding free chains, k is a constant, T is the temperature, and L_i the length of the undistorted sample. From the Equation (3.10) one may derive Equation (3.11), which relates the intramolecular energy changes to the mean square end-to-end distance of unperturbed chains:

$$\left(\frac{\Delta U}{W} \right)_{V,T} = Td \ln \langle r^2 \rangle_0 \, dT \qquad (3.11)$$

where U is the internal energy and W is the deformational work.

4. Ordered/Reinforced Systems

Similarities and differences exist between the thermomechanical and thermodynamical behavior of chemically crosslinked polymer networks, filled networks, rubber-like thermoplastics, and crystalline networks. Depending on the composition and dimensions of the reinforcing network, plastomer- or elastomer-like behavior may appear. The proportion of such "interpenetrating" networks in a material also varies. For instance, polyethylene contains 85–95% crystallinity, EVAc 55–65%, and rubber 0–10%.

The physical reason for the appearance of rubber-like elasticity in styrene block copolymers (SBCs) is connected to immiscibility and microphase separation, which lead to domain structure with domain size on the order of 100 Å. These rigid domains act as crosslinks. Thermomechanical studies of styrene–butadiene–styrene (SBS) and styrene–isoprene–styrene (SIS) block copolymers with a hard block content of less than 40% show that the energy contributions accompanying uniaxial extension are independent on the hard block content and degree of deformation [54]. The energy contributions for diene blocks coincide well with the data on chemically crosslinked diene networks. The energy contributions for polyisoprene and polybutadiene are in good agreement with the results for common chemically crosslinked networks. This seems to indicate that there are no considerable intermolecular changes in the rubbery matrix even at very large deformations. The hard block content also has no influence on the energy contribution. The thermomechanical behavior of the segmented polymers with low MW soft blocks and large content of hard blocks is determined not only by intrachain conformational changes but also by intermolecular changes, in both the soft and hard blocks.

The block lengths of segmented PVR and polyesters are much shorter than those with polystyrene (PS)–polydiene block copolymers, which tends to limit their extensibility (see also Chapter 5). Their stress softening is accompanied, as a rule, by a considerable residual deformation. Such residual deformation is a consequence of plastic deformation of rigid domains and orientation of the domains in stretching direction. Therefore, the energy changes in samples containing less than 50% hard block differ from those with 50% or more. Polymers with low hard block level behave similarly to typical SBS. Experimental evidence indicates that the free energy of the strained rubber-like block copolymer with a relatively high hard component content cannot originate only within the chains of the network. The interchain effects play a considerable role.

Although it is widely accepted that the free energy of the uniaxial deformation of the two-phase crystalline networks is purely intrachain, calorimetric measurements of Godovsky [56] showed that the thermodynamics of the deformation of these networks is controlled by interchain changes in the amorphous region. Crystalline polymers are two- phase systems also consisting of both crystalline and amorphous regions and, therefore, above their T_g they can be considered as networks in which crystallites are the solid filler and act as multifunctional crosslinks. The elastic properties of such systems are related to the conformational changes in the amorphous region. According to this approach, the free energy of deformation of two-phase crystalline polymers above T_g is also purely intrachain. According to the thermomechanics of solids, work (W) is a parabolic function of strain (ε) and heat (Q) is a linear function of strain [38]:

$$W = \frac{E\varepsilon^2}{2} \tag{3.12}$$

$$Q = \beta T E \varepsilon \tag{3.13}$$

Therefore, the heat or work ratio is a hyperbolic function of strain:

$$\frac{Q}{W} = \frac{2\beta T}{\varepsilon} \tag{3.14}$$

The presence of crystallites in these networks prevents the amorphous chains from deforming due to conformational rearrangements. The deformation is accompanied by a volume change, the volume increasing under extension. For undrawn crystalline polymers, the thermomechanical properties cannot be related to the conformational changes of the extended tie molecules in the noncrystalline region. Cold drawing of crystalline polymers leads to a change in the sign of the thermal effect accompanying the reversible stretching of a drawn sample, that is, reversible stretching of drawn polymers is accompanied by evolution of heat. This is a consequence of the stretching of highly oriented tie molecules in amorphous regions like that postulated for crystalline networks. The drawn polymers are able to deform reversibly at 1–30% and such deformations can be related to the intramolecular conformational changes [56].

As discussed earlier, the rheological behavior of segmented or ordered polymers (e.g., TPE, partially crosslinked, or crystalline polymers) is approximated in many cases by the behavior of filled systems, supposing that the hard segments or crosslinking points behave like filler particles imbedded in the polymer matrix. In the presence of reinforcing fillers, the elasticity modulus (E) of the elastomers at small and moderate deformation increases in the first approximation according to the Guth–Smallwood [57] equation:

$$E = E_0(1 + 2.5\phi + 14.1\phi^2) \tag{3.15}$$

where E is the modulus of the filled material, E_0 is the modulus of the polymer, and ϕ is the volume fraction of filler. Thus, for filled elastomers, the mechanical work of deformation, the change of entropy, and internal energy should include the parameter ϕ. The dependence of the energy contribution on the amount of filler and its reinforcing ability demonstrates that in filled elastomers the energy contribution seems to lose its obvious meaning as only a measure of the intrachain effects.

5. Molten-State Rheology

The flow properties of the warm and molten polymers depend on their structure and on the forces acting on them. Mobile, fluid-like polymers are the result of the destruction of ordered or segregated structures to achieve chain mobility.

6. Destruction of Ordered Structures

The flow of the polymer during processing, its extrudability and deformability in molten state, which allows it to form a tubular or sheet-like film is very important for plastic film manufacturers. The polymers used as raw material for film manufacture are non-Newtonian fluids; their processing viscosity depends on the temperature and pressure and on the deformation rate. The relationships between the processing conditions and flow properties and between the flow properties and the chemical or macromolecular parameters are very important for the film manufacturer [58]. For viscoelastical liquids, the rate of flow along the wall depends on the shear stress on a potential function [59]. The shear rate during processing of raw materials is quite different for calendering (e.g., $10–10^2$ sec^{-1}) and extrusion ($10^2–10^3$ sec^{-1}); therefore, the residual tension in the materials is also different [60]. In comparison, in the slot-die and film forming of hot-melts, shear values of $10^2–10^5$ sec^{-1} are common [61]. According to Ref. [62], a shear rate of 1000 sec^{-1} is used in hot-melt coating.

Both the plastomers used as carrier material and the elastomers used as carrier and adhesive generally have a low-glass transition temperature. The excellent mechanical characteristics of plastomers are due to the presence of ordered structures. The plastomers used as carrier material possess a multiphase structure given by the presence of crystalline, crosslinked, or filled domains. Such domains disappear above the melting point of the crystallites only. Thermoplastic, segmented elastomers display multiphase portions due to the segregation of their constituents. Such ordered domains can have an amorphous or crystalline structure. Quite different bonding forces (covalent, Van der Waals, dipolar, hydrogen, etc.) can hold them together. In comparison, the mechanical characteristics of paper are due to the presence of such ordered structure imposed by hydrogen bonds. In the crystalline region, the modulus of paper (E_c) is unaffected by moisture, therefore:

$$E_c = K_c n_c^{1/3} \qquad (3.16)$$

where K is the average value of the force constant and n is the number of hydrogen bonds effective in taking up strain under tension at a given moisture content [63]. The mechanical characteristics of such segregated systems depend on the degree of segregation, that is, the degree to which a two-phase construction is buildup. Strain may induce ordered structures in elastomers or plastomers. For instance, strain-induced crystallization appears in polyurethane (PUR) elastomers also. Such crystallization is enhanced by structural crosslinking bimodality [64]. The melting point of the crystalline domains is an important characteristic of such materials. Above their processing temperatures, plastomers exhibit viscous flow and elastomers exhibit viscoelastic flow. Studying the order–disorder transition in SBCs in comparison with low density polyethylene (LDPE), Han and Kim [65] evaluated the ratio of loss modulus (G') and storage modulus (G'') of SIS block copolymers over a broad range of temperature (140–240°C). The G'/G'' ratio of such copolymers is a function of the temperature (as a result of the order–disorder transitions). Above 165°C, the specific volume of SBS polymers increases due to the change in the domain structures [66]. For LDPE there is no such dependence.

The flow properties of a molten polymer are a function of its chemical characteristics. Chemical composition, MW, molecular weight distribution (MWD), and branching are their main regulating parameters. These parameters can change the viscosity of the molten polymer and also (because of the non-Newtonian character of its flow) the dependence of the viscosity on the shear rate. Their influence is clearly illustrated by the processing of polyolefins.

LDPE is highly branched, has a broad MWD, and exhibits medium mechanical properties and excellent processability. The extensive branching of LDPE promotes easy extrusion and bubble stability by blown film manufacture (see also Chapter 8). In the early 1970s, a LLDPE was introduced, which had no long branching, had a narrow MWD, and gave better mechanical performance than LDPE [67]. However, its narrow MWD reduced the ease of extrusion and bubble stability. There are fundamental rheological differences between LDPE and LLDPE. LLDPE is less viscous at low shear rates and more viscous at higher shear rates. LDPE hardens as film is drawn off the die surface. The lower viscosity LLDPE does not strain harden and has low-bubble stability and little tolerance to high-speed air flows [68].

Melt index is a measure of viscosity. High melt index polymers have a lower viscosity, lower MW and better flow than those of lower melt index. Lower melt index grades are required for melt strength and toughness, that is, blown film. In spite of its universal use, melt index is a poor measure of processibility and molecular structure, being a single-point measurement at low shear rate. MWD has to be considered together with melt index for polymer characterization.

7. Polymer Blends

The mechanical properties of polymers are basically determined by their mutual solubility. Processibility (melt viscosity) depends on the compatibility of the polymer blend also. Interchain interactions responsible for the compatibility in polymer blends tend to reduce the entanglement but increase the friction between dissimilar chains. This friction appears to arise from the local reduction of chain convolution due to segmental alignment, with the latter arising from increased interchain interaction. The entanglement probability and the friction coefficient between dissimilar chains correlate with the strength of specific interactions. The free volume tends to be linearly additive but may deviate influenced by segmental conformation and packing rather than specific interactions. For instance, the number of branches and branch MW influences both the solid-state morphology of SIS block copolymers and their processing [69]. The reduced entanglement and the free volume additivity tend to reduce the melt viscosity, whereas the increase in friction tends to increase it. The former two effects are often stronger than the latter, resulting in the reduction of melt viscosity of compatible polymer blends [70].

B. RHEOLOGY OF ADHESIVES

The flow properties of the coated adhesive influence the adhesive and end-use performance of the PSP. The flow properties of the adhesive during processing influence its coatability. The rheology of uncoated PSAs was discussed in detail in Refs. [1,2]. This chapter summarizes the special features of the rheology of coated adhesives.

1. Rheology of Coated Adhesive

PSAs are macromolecular compounds whose viscoelastic nature is manifested through viscous flow and elasticity and characterized by modulus and glass transition temperature.

2. Rheological Parameters

The dilatation coefficient increases above the T_g because of the increase of the free volume [71] (The role of the free volume in pressure sensitivity will be discussed in detail in Chapter 4). Adhesive bonding exists only above the glass transition temperature [72]. For a given temperature of application there is an optimum T_g. It means that for PSAs (used generally at room temperature), a T_g value of less than $-15°C$ is required, but values between -40 and $-60°C$ are preferred. As can be seen from Table 3.1, the raw materials for the PSAs and for certain carrier films possess low T_g values. External and internal plasticizing (copolymerization) allow the decrease of the T_g; cross-linking increases it. From the point of view of the chain mobility, pressure sensitivity can be achieved with very different raw materials. Although the T_g is a material constant, its value depends on the purity of the material. As shown by Burfield [72], there is a difference (0.7– 0.9° K) between the glass transition temperature of *cis*-polyisoprene and natural rubber. As listed by Druschke [73], there are "hardening" (e.g., AA, acrylonitrile, methyl methacrylate, styrene, VAc, methyl acrylate, etc.) and "softening" (ethyl acrylate, vinyl isobutylether, butyl acrylate, isobutylene, ethyl hexyl acrylate, isoprene, butadiene, etc.) monomers. Their use as comonomer influences the T_g, and T_g influences the modulus of elasticity. The T_g showing the temperature domain where chain mobility exists and the modulus indicating the level and nature of material resistance (deformation and change of molecular structures) against external stress are necessary to characterize the polymer. The Dahlquist's criterion of tack establishes that the elasticity modulus of the PSA should be of the order of 10^5 Pa. Feldstein et al. [74,75] appreciate the physical meaning of the Dahlquist's phenomenological criterion at a most fundamental, molecular level, as the ratio between cohesive interaction energy and free volume within pressure-sensitive polymers (see Chapter 4). The key result is that the free volume and cohesive strength of polymer need to be of appropriate magnitude and in a strictly specified ratio to each other for adhesion to appear. The maximum in debonding force corresponds to the minimum value of the ΔCpT_g product. The ΔCpT_g product, which can be defined as a measure of heat that has to be expended to provide the polymer transition from the glassy to the viscoelastic state and to impart translational mobility to the polymeric segments, is constituted from contributions of free volume and the energy of cohesion. Hardness and stiffness of polymers increase with the T_g [76].

Adhesion is related to the glass transition temperature of the polymer. The adhesive fracture energy exhibits a maximum in the temperature range above the glass transition region. In this temperature range, the mechanical behavior is determined by intermolecular interactions, which form entanglements. A network of temporary crosslinks is formed. High tack values require good deformability of the polymer, that is, a sufficiently low modulus, which means that the material must have an entanglement network with long chain molecules between two entanglements. For instance, according to Ref. [77], a radiation curable base composition for PSA has to possess a T_g lower than 20°C and a controlled crosslinked network. The higher chain flexibility and higher free volume of rubber provide a lower T_g for rubber and a lower modulus [78]. The polymers for PSA must have a T_g 30–70°C below the application temperature [79]. According to Zawilinski [80], room temperature PSAs should possess a T_g between −15 and +5°C. The best SBR lattices usable for PSA exhibit a T_g between −60 and −35°C. PSAs can be used for tapes, veneers, and wallpapers [81]. The T_gs of such PSAs should be within the range of −35 to −25°C. Copolymers of maleic anhydride with AAs and vinylacetate have been synthesized for PSAs having glass transition temperatures between −45 and −65°C [82]. The regulation of the T_g can be achieved by mixing base viscoelastic components that have different T_g values. For HMPSAs used for surgical tapes, a mixture of AA polymers with low T_g (−80 to −10°C) and high T_g (10–40°C) is recommended [83]. Elastomers for sealants have a T_g range of −46 to −50°C [84]. The role of the glass transition temperature as a versatility index of raw materials usable for PSAs is illustrated by amorphous propylene copolymers, where long chain comonomers give lower T_g and can be used as one component HMPSA also. Amorphous polypropylene (APP) obtained as by-product has a glass transition temperature of −14°C. The polymer produced by direct synthesis exhibits a T_g of −12°C [85]. A propylene–butene APP copolymer displays a T_g of −16°C. Even detackifying additives must have a low T_g. The composition of the PSA polymer, according to one patent [86], includes a crosslinking agent that is also used for both the matrix polymer and the polymer particles (suspension polymerized) in the matrix, as elastic detackifier. The T_g of the particle should be lower than 10°C.

As discussed in Refs. [1,2] the T_g of the PSA is the resultant of the components of the recipe. The elastomer or viscoelastic compound and the tackifier resin are the main components of the formulation. Common resins have a T_g situated above room temperature. In certain cases, the introduction of a glassy polymer in an elastomer can lead to a product with a broad glass transition [87]. In such cases, the rigid polymer component restricts the segmental mobility of the elastomer. Besides the broadening of the tan δ peak, a reduction of the peak value is also readily observable. The T_g allows a forecast for possible tackifier loading [4]. Therefore, certain formulations (e.g., HMPSA) need a liquid plasticizing component also. According to Hughes and Looney [88], fully saturated petroleum resins can be used having a well defined M_n and a T_g of less than 45°C for an HMPSA formulation based on TPEs, without plasticizer oils. It is known that HMPSA formulations incorporate a high level of plasticizer, usually a naphthenic oil or a liquid resin. The use of plasticizers results in a number of disadvantages including long-term degradation of the adhesive bond. Resins synthesized from the C_5 feedstock with a defined diolefin/monoolefin ratio and a M_w of 800–960 and M_n of 500–600 (M_w/M_n ratio of at least 1.3) display a T_g of about 20°C and need no oils for formulation with saturated midblock TPEs. Oils can be replaced with other low T_g components also. Low MW polyisoprenes (35,000–80,000) having a low T_g (−65 to −72°C) have been used as viscosity regulators for HMPSA [89]. Used as replacement for plasticizers, they improve the migration resistance and low temperature adhesion. The elastomer sequence of a rubbery block copolymers must be designed to adequately supress crystallinity without, at the same time, rising the T_g. A saturated olefin rubber block has to be obtained with the lowest possible T_g and the best rubber characteristics. The rubbery elasticity is maximized when the rubbery segment is completely amorphous. In commercial styrene–ethylene–butylene–styrene (SEBS) block copolymers, the best rubbery performances are obtained with a rubber block having a T_g of about −50°C and no detectable crystallinity [90].

The flow properties of the adhesive are determinant for its debonding also. It has been shown that for many applications (e.g., labels, tapes, protective films, etc.), PSPs should be removable (see also Chapter 7). Removability requires a breakable bond at the adhesive or substrate interface. Bond breaking is an energetic phenomenon. For removability, the whole debonding energy should be absorbed by the adhesive itself in such a manner that no failure occurs in the adhesive mass. In this case, the energy is used only for the viscous flow and elastic motion of the macromolecules. Therefore, a special balance of the plastic or elastic behavior of the PSA is required to allow energy-absorbing flow. The major danger is, if the viscous flow is too pronounced, bonds will fail within the adhesive mass. To avoid this danger, the cohesion of the adhesive should be improved in parallel with the reduction of its adhesion to the substrate.

Rheology is a function of the chemical composition and structure of macromolecular compounds. This is illustrated by the investigations of Class and Chu [91], who used low molecular polymers of styrene, t-butyl styrene, and polyvinyl cyclohexane as tackifiers for natural rubber and styrene–butadiene copolymers. As stated, for optimal rheological behavior the components of the mixture must be compatible. Compatibility depends on their MW and structure. The investigation of the T_g and polystyrene content of TPEs shows that independently of the polystyrene content (15, 19, and 23%), tack and adhesion properties are similar. Cohesion properties and hot-melt viscosity differ due to the pronounced dilution. Tack and adhesion properties were found to differ mainly as a function of the midblock phase. The glass transition temperature and the correlation of adhesive properties as a function of plateau modulus allow the prediction of adhesive properties independently of the polymer [92]. The glass transition temperature provides information about the viscoelastic properties, and the value of the modulus quantify them. The value of the plateau modulus (G_n^0) defined as the storage modulus at the minimum tan δ value of a linear polymer is an important parameter. Its value depends on the chemical and macromolecular characteristics of the material.

For instance, natural rubber latex and natural rubber (even when ground or crushed) exhibit higher plateau moduli ($> 10^6$ dyn/cm^2) than carboxylated styrene–butadiene rubber (CSBR) (10^5–10^6 dyn/cm^2) [93]. Degradation of natural rubber does not change the tan δ peak (glass transition temperature). In CSBR, only a high styrene content (46%) gives a high plateau modulus, but such polymer does not respond well to the addition of resin. Lower styrene content gives a smaller modulus at high temperature.

For a given tensile stress (σ) or shear stress (τ), the interdependence between the deformation force ε and the deformation γ is affected by the value of the tension modulus (E) or shear modulus (G):

$$\sigma = \varepsilon E \tag{3.17}$$

and

$$\tau = \gamma G \tag{3.18}$$

Plotting the modulus as a function of the temperature reveals a dependence that has the form of the curves in Figure 3.3, where T_g for an amorphous polymer appears as the transition point from the glassy state to the rubbery elastic state. As can be seen from the figure, for partially crystalline polymers at higher temperatures above the T_g, the crystalline portions of the polymer structure may coexist with the amorphous rubbery parts, reducing the mobility of the chain segments. This is the case for certain PSPs made as self-adhesive carrier (e.g., EVAc copolymers or ethylene-butyl acrylate copolymers embedded in a crystallizable plastomer matrix) or based on crystallizable raw materials (e.g., chloroprene, propylene copolymers). It is evident that due to the reduced polymer mobility the tack of such compositions is lower. In contrast, if the MW is lower than the value required for rubbery performance, the diagram shows no rubbery-elastic plateau. This is the case of the "postfinished" pressure-sensitive raw materials. As known, the modulus is not a pure material constant, it depends on time–temperature. As discussed in Ref. [94], the frequency above which a macromolecular compound becomes a rubber-like material, decreases with MW. In contrast, low MW polymer fractions and (generally) a broad MWD are also required to ensure pressure sensitivity. Low MW fractions provide the fast relaxation times necessary to achieve a large real contact area within a short contact time.

Mechanically resistant plastomers or elastomers need a high modulus value; easy bonding adhesives require a low modulus value. The development of adhesive-free PSPs made softer carrier materials necessary. Making the carrier film thinner decreases its stiffness. There are special tape applications where the conformability of the finished product requires a softness of the same order of magnitude for the carrier and the adhesive. For conformable medical tapes and labels, the films should have a tensile modulus of less than about 4×10^5 psi (in accordance with ASTM D-638 and D-882) [95]. Conformability of PSAs used for medical tapes is measured as creep compliance with a creep compliance rheometer. Acceptable values lie between 1.2×10^{-5} and about 2.3×10^{-5} cm^2/dyn [96]. (It has been found that the higher the creep compliance the greater the adhesive residue left on the skin after removal.) Adhesive polymers used for surgical adhesive tapes often exhibited a dynamic modulus too low for outstanding wear performance. A low storage and loss modulus result in a soft adhesive with adhesive transfer. An adhesive suitable for the use on human skin should display a storage modulus of 1.0–2.0 N/cm^2, a dynamic loss modulus of 0.6–0.9 N/cm^2, and a modulus ratio (tan δ) of 0.4–0.6 as determined at an oscillation frequency sweep of 1 rad/sec at 25% strain rate at body temperature (36°C). In contrast, crosslinking of the adhesive (necessary in certain applications or for certain coating methods) hardens the adhesive. It can be concluded that there are simultaneous material and product developments that lead to products that barely fall within the definition of carrier or adhesive, that is, with a well-defined elasticity or plasticity.

According to the Newtonian correlation, the dependence between shear stress τ and the viscosity η as a function of the rate of deformation (dγ/dt) is given as follows:

$$\tau = \frac{\eta \, d\gamma}{dt} \tag{3.19}$$

The macromolecular compounds undergo relaxation after a force has acted on them. Such compounds are viscoelastic; therefore, the force that produces permanent deformation is transformed into viscous energy according to the equation:

$$\frac{d\gamma}{dt} = \frac{1}{G}\frac{d\tau}{dt} + \frac{\tau}{\eta} = 0 \tag{3.20}$$

As discussed earlier [see Equation (3.1) and Equation (3.2)], in the practice, there is a shift between the active force and the deformation produced due to the viscoelasticity of the material. Therefore, the dynamic modulus is the sum of a synchronous (storage) modulus G'' and a delayed (loss) modulus G'. Their ratio (G''/G') is the loss angle, that is, tan δ. The storage modulus is proportional to the average energy storage in a cycle of deformation; the loss modulus is proportional to the average dissipation of energy in a cycle of deformation; and tan δ represents the overall behavior of the material with respect to elasticity and plasticity (dissipation). Segregated, ordered structures may exhibit many tan δ peaks. Block copolymers or noncompatible mixtures display two tan δ peaks. The relationship of tan δ to temperature, the tan δ minimum temperature ($T_{\delta\text{min}}$) and the tan δ peak temperature ($T_{\delta\text{max}}$), describe the rheological behavior of the formulation. The loss tangent vs. temperature plot characterizes the cohesive strength. The lower the tan δ value, the greater the cohesive strength. For tackified formulations, it is admitted that for a given softening point and resin concentration the higher the increase of the tan δ peak temperature the better the compatibility [97]. G' (the shear modulus) and G' (the loss modulus) are the most useful parameters in PSA analysis. G' allows a direct measurement of the tack strength, and G'' provides direct measurement of the peel strength. In such investigations, the time–temperature superpositioning is used. The underlying basis for time–temperature superpositioning are (1) that the processes involved in molecular relaxation or rearrangements in viscoelastic materials occur at accelerated rates at higher temperatures and (2) there is a direct equivalency between time (the frequency of the measurement) and temperature [98]. Earlier studies [99,100] have shown that the modulus G' value at low frequencies can be related to the wetting and creep behavior (bonding) of the adhesive and the modulus at high frequencies (100 rad/sec) can be related to the peel or quick stick (debonding).

The storage modulus can be used as a measure of the crosslinking density. According to Auchter et al. [101], at 50°C the noncrosslinked soft AC polymer has a storage modulus of about 10^3 Pa, and this value may be increased up to 10^4 Pa by increasing the irradiation level. Such characterization of the flow properties has usually been made by tensile strength measurements [102]. The cohesive (tensile) strength of an uncrosslinked AC adhesive is 2500 kPa; that of the crosslinked adhesive can attain 6000 kPa (depending on the crosslinking conditions). The T_g value can also be used to characterize such crosslinked compositions. According to Donker et al. [103], the T_g of electron beam-cured PSA formulations should be situated at about 30–70°C below their temperature of use. The cohesive strength is proportional to the concentration of the hard monomer (giving a polymer with T_g higher than -25°C).

Admitting that the storage modulus value at low frequency (less than 10^{-2} rad/sec) is related to wetting and creep (cold flow), whereas the value at high frequencies may be related to peel or quick stick properties (i.e., high-speed debonding evaluated by tack or peel evaluation), both values are necessary for characterizing an adhesive. It is assumed that the application of the adhesive (tack test) would occur at an angular frequency near 0.1 rad/sec. It is further assumed that the peel test would approximate the higher frequency of 10 rad/sec. The G' range of 50,000–200,000 Pa is accepted as an ideal PSA performance [98].

For block copolymers, the mobility of the midblock can be evaluated by dynamic mechanical analysis (DMA). It is related to the softness or viscous flow, that is, energy loss modulus [104]. In contrast, the melting disappearance of the end block domains (i.e., processibility and temperature resistance) is indicated by $T_{\delta min}$ value and the crossover temperature of the loss and storage modulus (T_{cross}). Between $T_{\delta min}$ and T_{cross}, the end domains soften and dissappear. Most tackifiers used for HMPSAs are aliphatic, midblock compatible resins that do not influence the styrenic domain. Generally a mixture of resins is used for PSAs formulated for tapes. One of the resins is midblock compatible; the other is compatible with the end blocks. If resins are used that are partially compatible with the styrene domains (e.g., rosin esters, a modified terpene resin, or a modified aliphatic resin), both middle block mobility increase and end block softening are achieved. Such resin may have an aromatically modified aliphatic hydrocarbon as basis with a higher MW. In a standard HMPSA formulation, such tackifier give similar $T_{\delta max}$ and loss modulus values but a lower T_{cross}, that is, a lower softening temperature for block domains. According to Donker et al. [103], for optimal PSA packaging tape formulation $T_{\delta max}$ values of -6–0°C and loss modulus values at tan δ values of 72 and 92 kPa are recommended.

Acrylic block copolymers have a soft and hard AA phase also. They have a star-shaped, radial structure. The storage modulus has a plateau value of $10^{9.5}$ dyn/cm^2. The soft phase has a T_g of -45°C; the hard phase, a T_g of 105°C [104]. As known, because of their more ordered structure, the radial copolymers (SBCs also) generally show a longer linear portion of the rubbery plateau and lower values of tan δ [105]. The two tan δ peaks characteristic of block copolymers are more extreme (-87 and $+85$°C) for a radial SBC than for the linear one (-82 and $+75$°C). The butadiene used in these types of block copolymers normally has a loss tangent peak at -90°C, the styrene, one at $+100$°C.

Tack needs low modulus for bonding but high modulus at the strain rates and elongations which occur during bond breaking. At high frequencies, during debonding the resin stiffens the system. The T_g increases and the elastic modulus also increases. The lower the modulus at low frequency, and the higher it is at high frequency, the better the adhesive. The adhesive strengths increases [106]. Energy absorbtion produces the cold flow of the adhesive necessary for bonding. It means that a high loss modulus (peak) at the practical frequency of bonding helps to attain adequate bonding. In contrast, at the higher debonding frequency the elastic behavior of the adhesive (storage modulus) is required. An index of the energy absorbtion is the value of the loss modulus. Cold flow (loss modulus) should be maximum during bonding and minimum during debonding (except for removable formulations). Taking into account the different stress frequencies caused by bonding and debonding, dynamic mechanical investigations into the modulus value should be carried out at different frequencies.

In an experiment concerning the PSAs for diaper tapes, the storage modulus ratios of different formulations were determined at high frequency (100 rad/sec) and low frequency (0.1 rad/sec). A proportionality has been found between the G' ratios at different frequencies and quick stick values (e.g., a $G'_{100/0.1}$ value of 60.9 corresponds to a quick stick value of 24.1, and a $G_{100/0.1}$ of 5.5 gives a quick stick value of 15.3). Increasing storage modulus ratios lead to higher quick stick values. In a similar manner, decreasing tan δ ratios (at different frequencies) increases the quick stick values. The compliance (J) given by the correlation [107]:

$$J = \frac{\gamma}{\tau} \qquad (3.21)$$

where τ is the constant stress for formulations based on radial block copolymers and is lower than 0.2 mi/Pa.

As mentioned with respect to the elevated temperature application of SAFs, the surface temperature of the adherend is another factor to be taken into account. Most general purpose adhesives are formulated to have tack at room temperature. If the adherend temperature is lower than room temperature, a higher degree of adhesive cold flow is required to provide proper wet-out. Sometimes products are labeled at room temperature but subjected to lower temperatures later in their life cycle. Deep freeze labels should display the same viscoelastic properties at 40°C as at +20°C [8]. It is recommended that such labels have a lower storage modulus value than common labels (10^4 Pa). The peel of an untackified AC PSA at 0°C is about 210% lower than the peel value at 23°C. For tackified formulations, peel reduction at 0°C may attain 300% [108]. Owing to the increase of the modulus of the adhesive at low temperatures, a loss of wetting and a jerky debonding are observed with certain formulations.

Tests with different stress frequencies at different temperatures (-50 to $+16$°C) have been carried out by Debier and de Keyzer [107]. Modulus curves plotted at different frequencies and temperatures are superimposed to yield master curves. The logarithm of the dynamic modulus (on the ordinate, in dyn/cm^2) is plotted against that of the reduced frequency, log ω (rad/sec). Generally the plot corresponding to test (room) temperatures is used.

As discussed earlier, the main part of PSPs are composites with a segregated structure. Their rheology is the result of molecular and macroscopic interactions. TPEs exhibit a segregated structure on the molecular scale. Their properties depend on their morphology, which is a function of the relative concentration of the components. The morphology of a SBC polymer in the solid-state depends on its polystyrene content. At low polystyrene content (less than 20%), spheres of polystyrene are dispersed in a matrix of elastomer, whereas at higher polystyrene content (20–30%), the spheres begin to interconnect with each other and develop a cylindrical structure. Above 30–35% a morphology of alternating lamellae is observed. If the polystyrene content is further increased, a phase inversion occurs, and the elastomer becomes the discontinuous phase, in a continuum of polystyrene. It is evident that a polymer that has globular polyisoprene "particles" in a continuous polystyrene matrix (i.e., an inverse construction) will display quite different rheological and mechanical properties [109]. The plateau modulus (G_n^0) of a linear polymer is related to the MW between entanglements (M_e), density (ρ), and temperature (T) [110–113]:

$$(G_n^0) = \frac{\rho}{M_e} RT \qquad (3.22)$$

where R is the universal gas constant. The gas constant appears in the correlation due to the hypothesis that the ideally rubber behaves like an ideal gaseous conformations network. In such an ideal "gas," the pressure is given as a function of the number of phantom chains (N), the temperature (T), and the volume (V):

$$p = \frac{NkT}{V} \tag{3.23}$$

where k is the constant of Boltzmann. Because the number of phantom chains depends on density and MW, the equation for the nominal force (f) for such system can be written as a function of the deformation (D) as follows [114]:

$$f = \frac{\rho RT}{M_e} D \tag{3.24}$$

Replacing the force with the modulus, Equation (3.24) becomes Equation (3.22). Two theories have been proposed to predict the value of plateau modulus of styrenic block copolymers; both assume that the plateau modulus depends only on the styrene content. According to the filler theory, the polystyrene domains work as a filler dispersed in the continuous polydiene matrix. In this case, the plateau modulus is given by the equation [see correlation (3.15) also]:

$$G_n^0 = \frac{\rho}{M_e} RT(1 + 2.5\phi + 14.1\phi^2) \tag{3.25}$$

where ϕ is the volume fraction of the filler. In this case the segmented, two-phase block copolymer is considered as a gaseous filled network, where the modulus is given by the combination of correlations derived from the physics of gases and the hydrodynamics of filled liquids (given subsequently). The critical molecular weight M_c, which is normally twice the value of M_e, is extremely dependent on the MW and styrene content according to Bishop and Davison [115]. At high polystyrene content, the morphology no longer resembles that of a spherical filler, and the above equation is no more valid. According to the structure theory of Lewis and Nielsen [116], for phase inverted systems, the shear moduli of the filled material (G) and the unfilled matrix, that is, rubber, (G_1) are related to the fraction of the filler ϕ_2 by the following correlation:

$$\frac{G}{G_1} = \frac{1 + AB\phi_2}{1 - BU\phi_2} \tag{3.26}$$

where the constant A takes into account the geometry of the filler and the Poisson ratio of the matrix; B is a function of the moduli, and A, and U is a function of ϕ_2 and ϕ_m the maximum packing fraction of filler. The modulus of a thermoplastic rubber between these two extreme cases, where the rubber or the polystyrene acts as matrix, is given by a logarithmic rule of mixtures:

$$\log G = \phi_U \log M_U + \phi_L \log M_L \tag{3.27}$$

where M_U and M_L are the upper (rubber dispersed in polystyrene) and lower (polystyrene dispersed in rubber) modulus, respectively. According to the filler theory, the modulus of a diluted polymer (e.g., thermoplastic rubber in resin and oil) is given by:

$$GN^0 = \phi_2^2 \frac{\rho}{M_e} RT(1 + 2.5c + 14.1c^2) \tag{3.28}$$

where ϕ_2 is the fraction of the polymer in the polydiene phase, c is the fraction of the polystyrene block in the entire composition, and ρ is the density of the adhesive. The tackifier resin should be compatible with the elastomer midblock. If the resin is not fully compatible with the elastomer mid-block or its concentration is too high, it associates with the polystyrene domain or forms a separate phase, that is, it does not contribute to the tack. The same can occur by high oil content. Because of the differencies between resin and oil, different exponents are used in the equation of the plateau modulus. For instance, for Cariflex TR-1107-Foral 85 blend the exponent is 1.86 and for Cariflex 1107-Sfellflex 451 blend the exponent is 2.22.

3. Rheology of Reinforced Systems

The adhesive itself or the adhesive layer or both can have a composite structure. As discussed earlier, segregated, crosslinked polymers (mainly elastomers) are used as base materials for PSAs, carrier materials, and SAFs. Dispersed adhesive systems contain dispersants. Adhesives and carrier materials can include fillers. The rheology of such ordered systems on macromolecular or product scale is discussed in terms of a filled system.

For carrier-less transfer tapes, the use of fillers or filler-like materials (e.g., voids) is a practical method for strengthening according to Equation (3.15). The influence of the filler nature and con-centration on the flow properties of such products has been studied in a detailed manner. Mixtures of glass spheres ranging in size from 10 to 200 μm and suspended in a viscous medium have been investigated [117]. The volume fraction dependence of the viscosity could be fit into an equation developed by using ϕ as a fitting parameter [118]:

$$\eta_r = \left(\frac{1-\phi}{\phi_m}\right)^{-2}$$

(3.29)

where η_r is the relative viscosity, ϕ is the volume fraction, and ϕ_m is the volume fraction at maximum packing (i.e., the volume fraction at which the viscosity becomes infinite). This expression is related to the Krieger–Dougherty equation for the relative viscosity [119]:

$$\eta_d = \frac{\eta}{\eta_s}\left(\frac{1-\phi}{\phi_m}\right)^{-[\eta]\phi_m}$$

(3.30)

where η_d is the viscosity of the dispersion, η_s is the viscosity of the solvent, and η is the intrinsic viscosity. The displacement force (f_d) in a filled adhesive layer is given by the correlation:

$$f_d = 1.7 \times 6\pi\eta\sigma\Phi^2$$

(3.31)

where Φ is the sphere radius, σ is the shear rate, and η is the viscosity.

4. Rheology-Adhesive Characteristics

As discussed earlier, cold flow and elasticity are the main phenomena influencing the rheological behavior of a PSP. They also influence the adhesive performance characteristics. Toyama and Ito [120] used the data of creep testing (shear resistance) to calculate viscosity according to the relation:

$$\eta = \frac{Fh}{A}\frac{1}{dx/dt}$$

(3.32)

where F is the load, h is the thickness of the samples, A is the contact area, dx/dt is the slope of the displacement vs. time, and η is the steady-state shear viscosity. As seen from the aforementioned correlation, sample thickness (coating weight) is a determinant parameter of shear resistance. For a non-Newtonian fluid, the previously mentioned dependence becomes more complex as a function of the time to failure (t_0), the initial overlap (H_0), the reference viscosity at a shear rate of 1 (η_0), and the slope of a plot of log η vs. log shear rate [121]:

$$t_0 = \frac{(1-n)H_0^{(2-n)/(1-n)}(w\eta_0)^{1/(1-n)}}{(2-n)(Fg)^{1/(1-n)}h_a} \tag{3.33}$$

where w is the width and h_a is the thickness of the adhesive, F is the load, and g is the gravitational factor. As seen from the aforementioned correlation, holding power is a viscosity effect. According to Woo [122], the shift factors used to plot shear measurements at different temperature on a master curve were almost identical to the shift factors used for modulus curves [correlation (3.5)]. Thus, the time to fail in shear can be predicted from viscoelastic functions. As known, the viscosity depends on the loss modulus (viscous flow is the phenomenon that "loses" energy). It also depends on the shear rate (frequency):

$$\eta' = \frac{G''}{\omega} \tag{3.34}$$

At low frequencies, η' and the steady-state viscosity are the same. A static shear test is based on low frequency deformation; therefore, creep may be regulated by the viscosity.

As stated by Krecenski et al. [121], the peel resistance is described by some authors as an elongational flow phenomenon but is viewed by others as shear. Kaelble and Reylek [123,124] found a relationship between a_T (the shift factor from the WLF equation), time (t or $1/\omega$), rate of peel (v_p) and the cleavage stress concentration factor β:

$$\frac{a_T}{t} = \beta v_p a_T \tag{3.35}$$

where:

$$\beta = \left(\frac{9G(\omega)}{8E_c h_c^3 h_a}\right)^{1/4} \tag{3.36}$$

In the previously mentioned correlation, E_c is the Young modulus of the carrier, $G(\omega)$ is the shear modulus of the PSA at frequency ω; h_c is the half thickness of the flexible carrier, and h_a is the thickness of the adhesive. Peel data at different rates and temperatures have also been reduced to master curves. The interdependence between the rheology of the PSA and its adhesive characteristics was discussed in detail in Ref. [125].

5. Rheology during Processing

For the manufacturer of PSPs, the rheology of the product components before and during their use (production) also is important. It may be different from rheology of the final product. The manufacture of plastic carrier materials and of certain adhesives (hot-melts and heat-curable compositions) entails thermal processing, where elevated temperatures (and pressure) allow the required flow and forming properties. For other (coated) adhesives, fluidity is achieved via dispersing by manufacture of polymer solutions or dispersions. A controlled rheology during mixing of the components (formulation) and converting (coating) of the formulated (plastic or adhesive) components is necessary. Formulation rheology was discussed in detail in Ref. [126].

Generally, the main components of the formulation of solvent-based or water-based adhesives do not influence the processing rheology. This depends more on the choice of the liquid vehicle and dispersing additives.

In contrast, it has been shown that the coating rheology is strongly influenced by the molecular characteristics. The coalescence of aqueous PSAs depends on their MW. For instance, particle coalescence in poly(n-butyl methacrylate) latex films is attributed to interdiffusion of particles. Increasing the MW of the polymer decreases the rate of interdiffusion [127]. The HMPSA formulation can influence both the processing rheology and the rheology of the coated adhesive. The styrene end block of SBCs provides the cohesion and mechanical properties and influences the processing conditions. Its hardness is reduced by oils that are compatible with the styrene domains. Unfortunately, oils reduce cohesion and the shear adhesion failure temperature (SAFT) values are mostly influenced by the oil level (see also Chapter 8).

Coating rheology is characterized by wettability. According to Rantz [128], wettability is the extent to which the liquid spreads. Wettability displays a special role for coating systems having a low viscosity (solvent-based, water-based, or radiation-curable compositions) or for coating on difficult substrates (transfer coating and coating on special plastics). The coating rheology of the printing inks is also important. It includes the viscosity, flow limit, thixotropy, and tack of the ink [129]. These parameters influence the transport of the ink on the machine and its transfer on the web to be coated.

6. Energetical Aspects

As stated by Zosel [130], the separation energy of adhesive bonds can be considered as the sum of the thermodynamic adhesion work (depending on surface characteristics) and of a dimension-less term, taking into account the viscoelastic properties of the components. At high strain, the area limited by the stress–strain curve and the strain axis, illustrates the energetical changes in the deformation process.

A cyclic stress–strain curve for rubber compounds exhibits a hysteresis loop (Figure 3.5). The area of the loop can be considered as indicative of the energy loss E_L per cycle (hysteresis), the area under the return curve (E_D) as indicative of the energy elastically stored in the system due to the input strain energy E_τ:

$$E_\tau = E_L + E_D \tag{3.37}$$

The change in internal energy during deformation, determined by measurement of the deformational work and the heat developed, follows the first law of thermodynamics. At small strains,

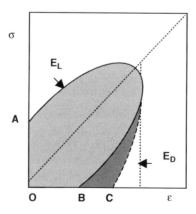

FIGURE 3.5 Hysteresis by bonding and debonding. See text for explanation.

elastomers should react with a positive heat flow above the glass transition point, but at higher strains there is a superimposed negative current that occurs as a result of the decrease of the entropy of stretched polymer chains [131]. This behavior is important for plasticization of rubber and cuttability of rubber-based coatings. The dissipation energy of polymers is involved in peel energy measurement [132].

Generally for elastomers, the dependence of the deformation (ε) on the increasing working force (tension σ) and the decreasing working force is not reversible. If viscous flow occurs, remanent deformation appears. Writing the mathematical correlation for the deformation and its disparition (tensioning and contraction) as the sum of the work of deformation (W_d) and loss of deformation (W_r), their difference (ΔW) gives the remanent deformation due to the viscous components:

$$\Delta W = W_d + W_r \tag{3.38}$$

Illustrating this phenomenon as a hysteresis plot where the tension varies among O, A, and B (Figure 3.5), the corresponding integrals of the positive and negative work of deformation can be written in the following form:

$$\Delta W = \int_0^{\sigma_A} \sigma_A \, d\varepsilon + \int_0^{\sigma_B} \sigma_B \, d\varepsilon \tag{3.39}$$

It is known that viscoelastic compounds exhibit viscous flow permanently. For a pressure-sensitive adhesive, the viscous deformation due to debonding is very pronounced; therefore,

$$\Delta W \gg 0 \tag{3.40}$$

In a first approximation, admitting that for a removable adhesive, the elastic and plastic deformation occur in the adhesive and do not affect the adherend (adhesions break), the starting point of the deformation can be fixed at the adhesive's limit, on the surface of the adherend (Figure 3.6). For a permanent adhesive where debonding destroys at least the surface layer of the adherend also, the starting point of the deformation is situated in the adherend. Both the adherend and the PSA suffer elastic and plastic deformation. For a removable PSA, low forces are generally sufficient to

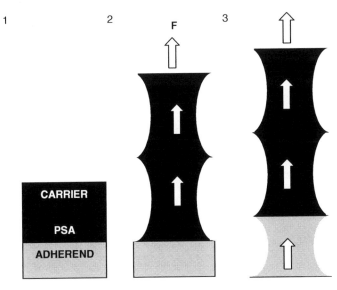

FIGURE 3.6 Strain in removable and permanent PSPs. (1) Nonstressed PSP; (2) removable PSP; (3) permanent PSP.

produce debonding. Such level of forces causes measurable deformation of the adhesive only. It can be supposed that in this case only the PSA suffers plastical deformation. Theoretically, for such an adhesive, it would be possible to minimize the distance OB (Figure 3.5) as a function of the stress rate and relaxation time of the adhesive. For a permanent adhesive, where remanent deformation includes the partial destruction of the carrier also (distance BC on Figure 3.5), this is not possible.

From an energetical point of view, the difference between a permanent and a removable adhesive is given by the ratio of the areas in which remanent deformation is produced with carrier destruction (OAC) or adhesive flow (OAB) (Figure 3.5). If this is true, the deformation work which causes a remanent deformation is the sum of the deformation work that leads to a residual deformation in the adherend (ΔW_{rad}) and the deformation work of the adhesive (ΔW_{psa}):

$$\Delta W = \Delta W_{rad} + \Delta W_{psa} \tag{3.41}$$

$$\Delta W_{rad} + \Delta W_{psa} = \int_0^{\sigma_C} \sigma_A \, d\varepsilon + \int_0^{\sigma_B} \sigma_B \, d\varepsilon \tag{3.42}$$

Admitting that the debonding energy is absorbed by the remanent deformation, for a removable adhesive, the remanent deformability must be much higher, that is:

$$\Delta W_{psa} \gg \Delta W_{rad} \tag{3.43}$$

During deformation, the energy of deformation is transformed partially in heat. The decohesion energy (fracture toughness) is deduced from energy balance considerations and is taken to be the only parameter governing the fracture process. For filled PSAs (e.g., transfer tapes) where friction causes supplementary heat, such energy losses are higher. The work of adhesion between two surfaces having surface energies of γ_a and γ_b is given by the correlation [133,134]:

$$\gamma_a + \gamma_b = \gamma_{ab} + W_{ab} \tag{3.44}$$

This work of adhesion is transformed into heat.

C. RHEOLOGY OF ABHESIVE

As a function of the product's construction and application requirements, different macro- or micro-molecular compounds are used as abhesive materials (see Chapter 5 and Chapter 8). The main products are silicones. Unlike adhesives or plastic carrier materials, in which bonding or debonding rheology plays the most important role, for abhesives the coating or processing rheology is more important. For the manufacturer of PSPs, the flow properties of the liquid silicone formulation are more important.

D. PRODUCT RHEOLOGY

Because of their various application fields, PSPs have to display different flow properties. Therefore, the special aspects of their rheology are discussed separately for each of the main product classes.

1. Rheology of Labels

Labels are complex products whose various viscoelastic components are built-in as thin layers. The introduction of polymeric films as face stock material increased the number of label components with pronounced cold flow. The rheological properties of labels are manifested during their conversion (cutting, printing), application (labeling), and end-use (storage, deapplication). Classical (paper-based permanent) labels are products that suffer deformation as a result of the cold flow of the adhesive. Because of the supposed high dimensional stability (i.e., nondeformability) of its solid-state components, the small dimensions of the product, and the low forces during application, the rheology of the label is that of the coated PSA. Cold flow-related phenomena like bleeding, migration, and cuttability are mostly attributable to the adhesive.

2. Rheology of Tapes

Tapes are PSPs for which the role of the carrier material as a functional component during application is more important than it is for labels. This is due to the web-like application of tapes where mainly a nonisotropic mechanical character of the web is required. For such applications, except for some (dry or wet) removable tapes, the rheology of the adhesive layer is less important. From the classical point of view, tapes are such products where product deformation is mainly the result of carrier deformability.

3. Rheology of Protective and Special Films

The industrial production of protective films is based mainly on empirical knowledge. However, future development requires the design of the products to have an engineering basis. This will become possible by studying the physical or chemical basis of the finished product and product components.

As discussed previously, a protective film forms a laminate together with the product to be protected. Laminating of the self-supporting film occurs via a pressure-sensitive surface layer. In the classical construction of protective films (and labels or tapes), the PSA is a separate component of the product coated on the nonadhesive solid-state carrier web (see Chapter 2). PSAs are permanently liquid systems that undergo cold flow. In the case of protective films, at least one of the solid-state components of the final laminated construction (protective film or protected item), namely, the web-like, deformable plastic carrier also exhibits cold flow. Cold flow is a result of the viscoelastic behavior of macromolecular compounds. Some of the polymers used as carrier materials for protective films (e.g., polyethylene) have low T_g values (Table 3.2). Others (e.g., polyester, polyvinyl chloride, etc.) possess higher T_g values, but the interval between their application temperature and T_g is very narrow. Protective films are applied on plastic surfaces too. As can be seen from Table 3.3, certain plastic substrates to be protected also exhibit low T_g. Such adherends can also exhibit pronounced cold flow.

TABLE 3.2
Glass Transition Temperature of Plastics Used as Substrate

Material	T_g (°C)	Reference
PVAc	29	[135]
Polystyrene	100	[135]
Polyvinylchloride	−40 to +105	[23]
Polymethyl methacrylate	105	[125]
Polycarbonate	150	[23]
Polyacrylate	190	[23]
Lacquers and varnishes	−40 to +350	—
Printing inks	−60 to +220	—

TABLE 3.3
End-Use Conditions for Protective Films

Protective Film	Protected Product	End-Use Conditions	
		Laminating	Processing
Hot laminating film	Lacquered coil	120–130°C, pressure	Room temperature (°C)
Adhesive-coated film	Lacquered coil	Room temperature, pressure	25–100
Adhesive-coated film	Cast PMMA plate	Room temperature, pressure	25–70
Adhesive-coated film	Extruded PMMA film	50–70°C, pressure	25–40
Self-adhesive EVAc film	Extruded PC plate	Room temperature, pressure	150
Self-adhesive EVAc film	Extruded PMMA plate	50–90°C, pressure	25–40

Therefore, cold flow may occur for all components of the laminate, and interpenetration of the solid-state and liquid-state laminate components is possible. Because the mechanical performance characteristics of the components of a protective film laminate are of the same order of magnitude (Figure 3.7), such products can be considered PSPs where the deformation of the finished product is the result of the deformation of all its components.

In such an interaction, the viscoelastic properties of the carrier material influence the rheology of the product. Actually, the majority of protective webs are manufactured using thin plastic films as carrier material. These are common plastics from the range of materials used as packaging films. In 1974, Toyama and Ito [120] described PSAs as being coated "onto rigid (relative to the adhesive) backings without the use of solvents or heat." For plastic-based protective films, this statement is no longer valid. Their cold flow and adhesion are improved by modern application conditions. Table 3.3 lists the application conditions for some protective films.

As illustrated by the data in Table 3.3, in certain applications protective films are laminated on the plastic products to be protected at temperatures higher than room temperature (up to 120–130°C) or are processed with the protected item at elevated temperatures (see also Chapter 11). In some cases, such application temperatures result from the manufacturing technology of the products (e.g., plates of polymethyl methacrylate or polycarbonate). In other applications, the high temperature is the processing domain for the protected product (e.g., deep drawing,

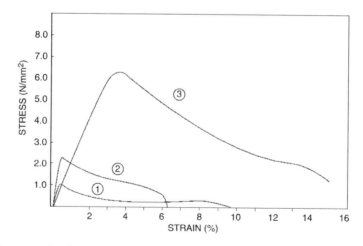

FIGURE 3.7 Stress–strain plot for PSP components. Curve 1, PSA; curve 2, crosslinked PSA; curve 3, plastomer carrier.

cutting, etc.). In certain uses, only the surface layer of the protected item is polymer based. Uncrosslinked or crosslinked polymers (varnishes and lacquers, inks deposited on metal surface) are protected and undergo thermal treatment. When the working temperature is much higher than the T_g of the protective film components, and (for a short period of time) it can attain the softening or melting range of the base polymer (e.g., hot laminating films), a pronounced viscous flow of the carrier material can occur.

As discussed subsequently (see Chapter 7), protective films possess an unbalanced adhesive character. Because of the relatively low tack of the self-adhesive layer and its low deformability, a high laminating pressure is sometimes necessary to improve the intimate contact between the protective film and the product to be protected. The use of high temperature softening (or partial melting) of the carrier film to improve its adherence (surface penetration) allowed the development of so-called "hot laminating films" without a self-adhesive layer. Here, the polarization of the plastic surface and its functionalization by corona treatment (see Chapter 8) provides the adhesive force necessary for the temporary bonding of the protective film to the adherend surface to be protected. The initial requirement to resist elevated temperatures (imposed by the manufacture of the products to be protected) was used later as a high temperature adhesion correction (tack replacement). Classical plastic films without a chemical adhesive layer cannot possess the tack of PSAs. As discussed earlier, such products exhibit flow, but it is time (stress) or temperature related (see high pressure hot laminating). It should be mentioned that high pressure, high temperature, and corona treatments are not sufficient to bond hot laminating protective films. A special chemical composition allowing improved flow is also necessary (see also Chapter 5). This means that adhesive-free hot laminating protective films are special products with special carrier film compositions.

If a special carrier composition can be used to "simplify" the construction of the protective films (and the formulation works under elevated temperature), the next step of development would be the formulation of carrier films displaying both carrier and adhesive properties at normal application temperatures. In this case, the rheology of the protective film is more complex because the product should exhibit a more pronounced viscous flow without the loss of mechanical characteristics. It is known that both surface and bulk properties of the carrier material influence the viscoelastic properties of the adhesive layer and vice versa [1,2]. The adhesive properties related to bonding on the substrate (tack and peel) depend on the anchorage of the adhesive on the carrier. The anchorage is a function of the chemical affinity and physical structure of the surface. In contrast, migration from the surface into the carrier (or from the bulk plastic material to the surface) will also influence the anchorage and adhesive properties too. As known, slip agents from polyolefins, plasticizers and surfactants from polyvinylchloride, or stabilizers from the plastics will dilute the adhesive, altering its properties. As a secondary effect the film may harden.

Debonding force is transmitted to the adhesive by the carrier. Generally (for slight removable protective films) the value of the debonding force, which causes bond failure at the adhesive or substrate interface, is lower than the tensile strength of the plastic film. As known, the mechanical resistance of the film is the result of its own internal characteristics and geometry, which depends on its formulation. Reducing the thickness of the film increases its deformability (Figure 3.8).

It is well known that for high peel force, permanent (film) labels tested in laboratory, results are denatured by the deformation of the face stock material. Its elongation and relaxation absorb energy and allow the relaxation of the PSA too; therefore, peeling off will be delayed in time (the strain diminution and strain rate will be reduced). Protective films display low peel values (see Chapter 7). In contrast, their mechanical characteristics and thickness are also lower (Table 3.4 and Table 3.5). Thus, elongation of the film carrier may strongly decrease the measured peel value even when the same adhesive, coating weight, substrate, and laminating conditions are used (Figure 3.9).

As illustrated in Figure 3.9, there is a critical carrier film tensile strength. Below this limit (of about 6–8 N/10 mm), the adhesion values are degraded (reduced) due to the pronounced

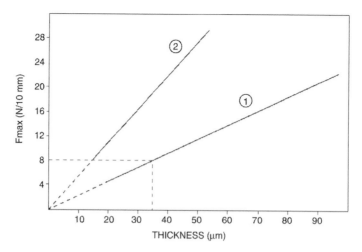

FIGURE 3.8 Dependence of the carrier film deformability on its thickness. (1), Monolayer LDPE film; (2), coextruded LDPE/LLDPE film.

deformation of the carrier. The same adhesive and coating weight lead to much lower peel values (0.06–0.10 N/10 mm). For common "nondeformable" carrier materials, values of 0.6–0.8 N/10 mm have been obtained [136]. This statement is of great importance in manufacture and test of film-based special PSPs (see Chapter 7 and Chapter 8). It demonstrates that for PSPs that work in the creep domain, debonding may be the result of the viscous flow of both components (carrier and adhesive); carrier deformation may produce debonding. The time of maximum deformation does not coincide with the time of maximum load as it does for elastic materials. Instead, maximum deformation always occurs after the maximum load. This delay occurs because the carrier continues to respond by creep. Therefore, carrier design according to the criteria of classical mechanics (i.e., tensile strength) is not more valid.

TABLE 3.4
Mechanical Characteristics of Carrier Materials Used for Adhesive-Coated Polyolefine Protective Films

Carrier Thickness (μm)	Tensile Strength (N/10 mm)				Elongation (%)				Comment
	At Maximum Load		At Break		At Maximum Load		At Break		
	MD	CD	MD	CD	MD	CD	MD	CD	
45	13.6	8.2	13.3	8.1	250	600	259	604	Clear
47	11.4	8.7	11.1	8.6	241	643	244	647	Clear
45	11.9	8.5	11.7	8.3	351	627	353	626	Clear, printed
46	14.7	6.9	14.4	6.7	167	660	171	687	Clear
48	13.4	9.5	13.3	9.3	264	674	265	676	Clear
74	19.3	10.9	19.1	10.4	258	596	261	608	Clear
78	18.6	10.4	18.3	10.4	248	455	248	437	Black/white, printed
80	18.7	9.3	18.3	9.0	239	435	242	440	Black/white, coextruded, printed
111	19.2	15.5	18.9	15.4	352	503	354	504	Black

Note: MD, machine direction; CD, cross direction.

TABLE 3.5
Mechanical Characteristics of Carrier Materials Used for Common Protective Films[a]

Tensile Strength (N/10 mm)		Elongation at Break (%)		Modulus (N/mm²)	
MD	CD	MD	CD	MD	CD
19	11	171	458	19	18
19	11	153	421	20	19
21	12	151	452	21	17
19[b]	15[b]	271[b]	511[b]	13[b]	13[b]

[a]Polyethylene film with a thickness of 50 μm except as noted.

[b]Film thickness 80 μm.

From the theoretical point of view, the aforementioned statement imposes the detailed study of joints based on PSAs and deformable (elasto-plastic film) carrier materials or adherends. For a given macroscopic energy release rate, the plastically deforming adherends do not allow high levels of opening stresses to be developed within the adhesive. Toughness of joints may not be the same for substrates that remain elastic as opposed to cases where they deform plastically. As stated recently by Yang and Thouless [137], for small values of the adherend thickness, the assumption of bending dominated crack propagation becomes invalid when the opening forces become significant. Kendall [138], compares the adhesion force for different types of adhesive joints and debonding tests. Peel test with flexible carrier materials at large angle, with rigid carrier (disc sample), with semiflexible carrier, lap joint failure, and small angle peel were compared, and shows that stiffening adhesive joints usually leads to more strengths. As demonstrated by Kim and Aravas [139], plastic yielding of the carrier will not occur as long as it is sufficient thick, so that

$$\sqrt{\frac{6E^*FR}{H}} < \tau \qquad (3.45)$$

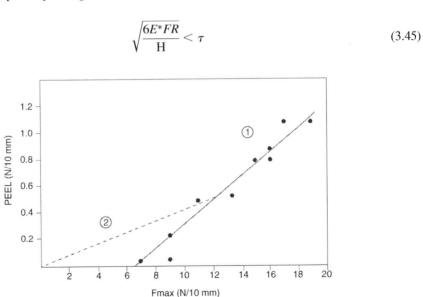

FIGURE 3.9 Denaturation of the peel values by carrier deformation. Dependence of the peel on stainless steel as a function of the tensile strength (F_{max}) of the plastic carrier material. (1) Measured peel values; (2) theoretical dependence of the peel on carrier strength. An LDPE carrier and solvent-based AC PSA were used.

where F is the peel force per unit width, τ is the yield stress of the carrier in shear and

$$E^* = \frac{E}{(1 - v^2)} \qquad (3.46)$$

where E is the Young modulus and v is the Poisson's ratio. As illustrated in Ref. [140], the modulus values of certain carrier films are of the same order of magnitude as those of the adhesives. Carrier deformation can also lead to changes in peel angle. According to Kawashita et al. [141], with an increasing peel angle a decrease in fracture toughness is predicted. Recent investigations of Chung et al. [142] demonstrate that carrier flexibility affects cavitation and fibrillation too. Unfortunately, there are only insufficient experimental data concerning the mutual interaction of PSA and flexible carrier. According to Ref. [143], during the DMA of a PSA tape, two maxima are observed corresponding to the adhesive and to the carrier.

In choosing a face stock material for labels, its dimensional stability is a decisive factor. The material's mechanical properties are reported as stress–strain plot. A criterion for dimensional stability is the value of the force required for a given minimum (5%) deformation (see Table 3.6). For other products, other end-uses require other force or deformation values as given in Table 3.6.

For protective films, which exhibit a more pronounced plastic deformation of the carrier film during debonding, it would be more practical to examine the value of the plastic deformation as a criterion of applicability instead of the development of the force (Table 3.7).

4. General Considerations

A comparison of the adhesive characteristics of the main classes of PSPs (labels, tapes, and protective films) is schematically presented in Figure 3.10. As Figure 3.10 (positions 1 and 4) illustrates, labels and tapes have balanced adhesive properties. Their tack allows instantaneous adhesion without pressure, their peel resistance ensures the formation of a bond, and their cohesion provides a permanent, mechanically resistant bond. It is evident that for common tapes this balance is shifted towards higher cohesion. Because of their reduced tack, protective films have to be applied under pressure. A new class of pressure-sensitive materials is presented by Figure 3.10, position 4. These are the so-called "postfinished" products which need a chemical or physical treatment to complete their synthesis in order to achieve balanced adhesive properties. This class of products includes

---

TABLE 3.6
Deformability of Carrier Material for Selected PSPs

PSP	Tensile Stress at a Given Deformation				Unit	Method	Reference
	5%	10%	30%	50%			
Harness wrap tape	—	1.5	4.9	7.4	MPa	ASTM-419	[144]
Easy tear tape	—	3.0–6.5	6.5–13.0	13.0–15.5	MPa	ASTM D-1000	[145]
Flame-resistant tape	—	1.5–5.4	5.0–12.3	8.5–14.0	MPa	ASTM-412	[146]
Closure tape	—	>70.0	—	—	N/15 mm	DIN53455	—
Protective film	8(5)	—	—	—	N/10 mm	—	—
Label	19(18)	—	—	—	N/10 mm	—	—
Insulating tape	11(6)	—	—	—	N/10 mm		—
Insulating tape	91	—	—	—	N/10 mm	—	—
Application tape	16(16)	—	—	—	N/ mm^2	—	—
Protective film	13(13)	—	—	—	N/10 mm	—	—

Note: For tensile stress at 5% deformation values given are MD and CD.

TABLE 3.7
Dependence of Peel Resistance on the Carrier Deformation

Carrier Elongation (%)				Peel Resistance[a] (N/25 mm)	
Nominal		Actual			
MD	CD	MD	CD	Nominal	Actual
220	660	—	—	2.75	—
—	—	680	820	—	0.35
230	600	—	—	3.00	—
—	—	530	800	—	0.38
250	550	—	—	5.00	—
—	—	750	850	—	1.50
300	530	—	—	7.50	—
—	—	730	820	—	2.25

[a]The 180° peel resistance of a crosslinked AC adhesive (coated on polyolefin carrier) on stainless steel. Dwell time 1 day.

mainly low MW 100% solids, coatable as warm melts. As stated in Refs. [101,147,148], even a high MW linear, highly viscous 100% polyacrylate, processible as hot melt, is not of high enough MW to display balanced pressure-sensitive properties. Pressure sensitivity (sufficient shear resistance) is achieved only by crosslinking. The polymer itself is a highly viscous fluid at room temperature (before crosslinking); it can be processed at 120–140°C. Therefore, it should be crosslinked. Special AC raw materials for HMPSAs have been developed that allow postcuring. In this way, polymacromerization competes with the synthesis of macromolecular compounds having segregated structures (see Chapter 4). The result of both procedures is a new class of "internal" composites having better end-use performance characteristics but a more complex rheology.

Pressure sensitivity (Π) is the resultant of two phenomena, viscoelastic bonding (B_{ve}) and viscoelastic debonding (D_{ve}):

$$\Pi = B_{ve} + D_{ve} \qquad (3.47)$$

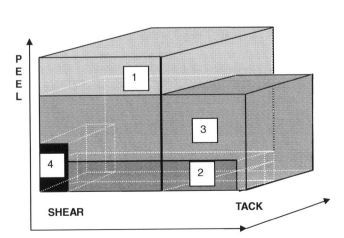

FIGURE 3.10 Schematic presentation of the adhesive properties of the major PSPs. (1) Label; (2) protective film; (3) tape; (4) postfinished PSA.

During bonding and debonding, relaxation occurs [see Equation (3.9) and Equation (3.20)]; therefore, pressure sensitivity will be the sum of deformation and relaxation by bonding (index b) and debonding (index d):

$$\Pi = \left[\left(\frac{1}{G}\right)\left(\frac{\mathrm{d}\tau}{\mathrm{d}t}\right) + \frac{\tau}{\eta}\right]_{b} + \left[\left(\frac{1}{G}\right)\left(\frac{\mathrm{d}\tau}{\mathrm{d}t}\right) + \frac{\tau}{\eta}\right]_{d} \tag{3.48}$$

Writing $1/G$ as J, $\mathrm{d}\tau/\mathrm{d}t$ as τ^0, and τ/η as f, the Equation. (3.47) becomes:

$$\Pi = (J\tau^0 + f)_b + (J\tau^0 + f)_d \tag{3.49}$$

For a pressure-sensitive construction based on carrier and adhesive, the previously mentioned correlation can be written as the sum of the bonding and debonding components for the carrier (index c) and adhesive (index a):

$$\Pi = [(J\tau^0 + f)_a + (J\tau^0 + f)_c]_b + [(J\tau^0 + f)_a + (J\tau^0 + f)_c]_d \tag{3.50}$$

Taking into account the different "material characteristics" of the carrier and adhesive for the bonding and debonding and grouping the terms for the carrier and those for the adhesive, Equation (3.46) becomes:

$$\Pi = (J\tau_b^0 + f_b + J\tau_d^0 + f_d)_a + (J\tau_b^0 + f_b + J\tau_d^0 + f_d)_c \tag{3.51}$$

Bonding and debonding occur at different stress frequencies. Because of the dependence of the modulus on the shear rate, correlation (3.49) cannot usually be simplified. The forces acting on the adhesive are different from those acting on the carrier. Laminating and debonding forces are transmitted to the adhesive by means of the carrier. Force transfer occurs at the same time as carrier deformation and relaxation. According to Equation (3.49), generally the pressure-sensitive behavior of a PSP is the sum of the bonding and debonding behavior of the components. For an one-component PSP based on self-adhesive carrier (e.g., protective film on EVAc basis), the above correlation can be simplified and written:

$$\Pi_{SAF} = (J\tau_b^0 + f_b + J\tau_d^0 + f_d)_c \tag{3.52}$$

Taking into account that for such product, laminating occurs at higher temperature than delaminating and under pressure, and delaminating is carried out with a much higher stress rate at much lower temperatures than laminating; in Equation (3.50) the terms related to debonding become more important. For such a SAF, pressure sensitivity should be designed for delamination and relaxation:

$$\Pi_{SAF} = (J\tau_d^0 + f_d)_c \tag{3.53}$$

Most protective films are coated with a hard, crosslinked adhesive and are applied under pressure (and in certain cases with the application of heat). Therefore, Equation (3.51) is valid for such products also. For a carrier-less transfer tape, the Equation (3.49) becomes:

$$\Pi = (J\tau_b^0 + f_b + J\tau_d^0 + f_d)_a \tag{3.54}$$

For such a filled, crosslinked adhesive mass, pressure sensitivity should be designed for bonding:

$$\Pi = (J\tau_b^0 + f_b)_a \tag{3.55}$$

Taking into account the differences in the deformability of paper and plastic materials during conversion, application, and deapplication, for practical use it would be more convenient to write the general form of the Equation (3.49) as follows:

$$\Pi = (J\tau_b^0 + f_b + J\tau_d^0 + f_d)_a + C(J\tau_b^0 + f_b + J\tau_d^0 + f_d)_c \tag{3.56}$$

where C is a constant taking into account the deformability of the carrier material during debonding and having values between one (soft plastic carrier) and zero (nondeformable carrier).

The relaxation related to bonding and debonding is a complex process. Generally, the debonding energy of a composite is described as the reversible energy of adhesion or cohesion and microscopical loss factor related to viscoelastic energy losses and a molecular loss factor related to the degree of polymer crosslinking [149]. Relaxation phenomena on a molecular scale are very complex and depend on the buildup of the macromolecules. It has been stated that a relatively simple elastomer such as cis-polyisoprene displays three different types of relaxation processes. Near T_g, the relaxation is related to crystallization and Van der Waals bonds in the polymeric network. At higher temperatures, relaxation is related to the interactions between the segments which produce physical contacts of an oversegmental microstructure [150]. After studying relaxation transitions in polybutadiene and poly(butadiene–methylstyrene), Bartenev and Tulinova [151] stated that there existed 17 transitions. As discussed in Ref. [152], in branched polymers of polybutadiene, the maximum of relaxation time increases with the MW of the branches. As discussed earlier, commercial AC adhesives have a very broad spectrum of relaxation times; low MW fractions provide the fast relaxation times necessary to achieve a large real contact area within a short contact time. According to Novikov et al. [153], in pressure-sensitive hydrogels based on high MW polyvinyl pyrrolidone and short chain poly ethylene glycol, two different and rapidly relaxing networks determine the pressure-sensitive adhesion and the mechanical properties (see Chapter 4).

E. ADVANCES IN ADHESIVE RHEOLOGY

As summarized earlier, the glass transition temperature (T_g) and the modulus (G) are the main rheological parameters of PSAs. The temperature domain that allows pressure-sensitive behavior (i.e., viscoelasticity) for a macromolecular compound is defined by the T_g, the extent of this behavior is quantified by the modulus. Principally, the polymer used as PSA should be above its glass transition temperature, and in order to allow a rapid wetting of the adherend, its hardness (defined by the modulus) must be adequate. After the first empirical stage of development of PSAs, the T_g and the modulus were accepted as main scientific criteria to design polymers for PSAs.

Dahlquist proposed its famous criterion which correlates pressure sensitivity with a critical value of the modulus. The Dahlquist's criterion of tack establishes that the elasticity modulus of PSA should be of the order of 10^5 Pa. In practice, instead of the modulus (hardness), the creep compliance (softness) was used as evaluation criterion. Conformability of PSAs is measured as creep compliance.

Unfortunately, both of these rheological criteria, glass temperature and modulus, give only an imperfect definition of pressure sensitivity. Developments in ethylene (polar) copolymers, in TPEs, and in special branched macromolecular compounds allowed the synthesis of viscoelastic materials with apparently too high T_g but practical usability as PSA. Olefin-based monomers with neo structure (vinyl neodecanoate and vinyl neododecanoate) became commercially available [154]. Because of their particularly branched structure, the homopolymers of these monomers exhibit unusually high entanglement molecular weight (M_e), and thus, they lower the plateau modulus (see Equation 3.22). Since PSA bonding typically occurs at the plateau region of the viscoelastic curve, the bonding is facilitated by lowering the G' at this region. PSA debonding, however, happens at much higher frequency, typically between 100 and 12,000 rad/sec. The loss modulus at this frequency is in the transition region of the viscoelastic curve and its value is highly dependent on the T_g of the polymer. Increasing the T_g results in higher G'' is this region, the overall peel and tack properties will be improved.

Developments in mechanics and DMA showed that the modulus is not a material constant, it is time–temperature dependent. Therefore, in the investigation of polymers, the notion of the Young's modulus of the classical mechanics was replaced with the dynamical modulus, having its frequency-dependent, energy-depositing, and energy-losing parts (G' and G''). Increasing the modulus G' results in adhesives with higher adhesion and lower tack. Adhesives with lower creep compliance (higher G') will have better cohesive strength (creep resistance). However, if the modulus is too high, the adhesive may not rapidly wet the substrate. It is evident that in this situation, the quantification of the modulus as a restrictive criterion of pressure sensitivity (e.g., the Dahlquist's criterion) becomes more difficult. Because of the quite different bonding and debonding frequencies, the modulus values have to be multiple. As discussed by Lakrout and Creton [155], the creep compliance itself depends on the coating weight too. For very large values of a/h (contact radius/adhesive thickness), the compliance of the adhesive layer is inversely proportional to the bulk modulus of the material.

In this situation, the use of the DMA, that is, the measurement of the modulus values as a function of the frequency, seems to be a better variance for the choice of an adequate polymer for PSA and allows the definition of so-called application windows. Unfortunately, in this experimental domain, there are also numerous discrepancies. Bamborough [156] stated that for water-based adhesive systems, the loss tangent peak temperature (i.e., the interdependence of storage modulus and loss tangent peak temperature) is not a reliable predictor of PSA performance.

In contrast, although the new pressure-sensitive hydrogels obey the Dahlquist's criterion of tack (elasticity modulus of the order of 10^5 Pa, related to the ultimate tensile modulus at polymer fracture), their composition-regulated behavior is more complex and their dynamic modulus components deviate from the common materials [74] (see Chapter 4). Such deviations of the expected rheological behavior, in sense of the "classical" PSA-theory, impose the need not only to quantify the rheological behavior of PSAs but to explain it. It must be correlated with the contact and fracture mechanics (i.e., the "visible" macro-microscopic investigations based on classical mechanics) and with the macromolecular fundamentals of the adhesive system.

Classical PSAs display instantaneous bonding and various types of debonding that differ in their mechanism and the force level required. In the practice, bonding is tested by debonding; debonding is quantified by the force required, that is, the energy required to destroy the joint. Like bonding, macroscopic debonding is also the sum of microscopic steps. Ad absurdum the measuring scale of the debonding forces must be reduced to that of the macromolecules in order to correlate the measurable macroscopic stress with the internal cohesion of the macromolecules. Over the last years, an appreciable progress has been achieved in the quantitative description of the micromechanics of PSA debonding [157,158]. These works consider the nucleation of cavities within the PSA polymer and the extension of fibrils as the major factors leading to the dissipation of applied energy. The resistance in PSA layers appears to be dominated by the cavitation behavior of the adhesive. This observation is apparent from stress-separation tests, where the strongly rate-sensitive peak stress (cavitation stress, σ_c) dictates the area under the stress separation curve. The cavitation behavior of PSA can be understood in terms of the well-established mechanics model [159]. For a material obeying the simple kinetic theory of rubber elasticity, the critical pressure at which a cavity will grow without limit (tear to for an internal crack or fibrillate) has been shown to be proportional to the initial elastic modulus of the material. The cavitation stress obtained from stress-separation tests is plotted as a function of strain rate and exhibits the same trend as the frequency-dependent shear storage modulus (G'). These data indicate that the cavitation stress in PSAs is indeed proportional to the elastic modulus. The recent advances in PSA-related contact mechanics were discussed in detail in Ref. [160]. In principle, the failure of a single fibril may occur by distanglement of the polymer chain or fracture of the polymer chain. According to Feldstein et al. [161], the force (P) required to debond the adhesive obeys the well-known Kaelble equation presented in the form:

$$P = bl\left(\frac{aN\tau D}{12RT}\right)\sigma_f^2 \qquad (3.57)$$

where σ_f is the critical tensile stress of the polymer at fibril failure, D is the self-diffusion coefficient of polymer segment, b is the width and l is the thickness of the adhesive layer, a is a size of diffusing polymer segment, N is the Avogadro number, and τ is a segmental relaxation time. As can be seen from Equation 3.37, the cavitation–fibrillation theory correlates macroscopic debonding with microscopic one (fibril failure) and with the modulus. Moreover, in Equation 3.37, the molecular parameters like self-diffusion coefficient and segmental relaxation time, correlating the polymer structure with pressure sensitivity also appear. Such parameters depend on the free volume and internal cohesion. The key result is that the free volume and cohesive strength of polymer need to be of appropriate magnitude and in a strictly specified ratio to each other for adhesion to appear. A similar conclusion was previously derived from the finding that the maximum in debonding force corresponds to the minimum value of the ΔCpT_g (see Chapter 4). The ΔCpT_g product, defined as a measure of heat that has to be expended to provide the polymer transition from the glassy to the viscoelastic state and to impart translational mobility to the polymeric segments, is in turn constituted from contributions of free volume and the energy of cohesion [162]. Thus, the applicability of the modified Kaelble equation in couple with the earlier established relevance of the ΔCpT_g quantity as a predictive criterion of adhesive behavior for various polymers argues in favor of the interplay of free volume against cohesive energy in polymers as a general determinant of pressure-sensitive adhesion. This new theory covering the fundamentals of pressure sensitivity is described in detail in Chapter 4.

It should be accentuated that the apparition of the relaxation parameters in this theory is necessary because pressure-sensitive bonding depends on relaxation too. The real area of contact of an elastic film on a rough surface depends on two parameters: the elastic modulus at the bonding frequency (G') and the relaxation properties of the polymer which govern the change of the real area of contact with time. Relaxation as phenomenon includes time and energy related aspects. PSAs are viscoelastic materials. Contacts to viscoelastic materials are poorly understood. There is very little quantitative data and, until quite recently, there were no theoretical models that include both viscoelastic response and adhesion. As discussed in Ref. [163], Johnson and Unertl have shown that crack tip and long range creep phenomena generally occur on much different time scales. Depending on the characteristic relaxation time of the viscoelastic material and the experimental measurement time, both can be important. Hui et al. [164] have put forward a model that accounts for linear viscoelastic response at all length and time scales. Contacts involving viscoelastic materials, are more difficult to analyze because not all of the applied energy instantaneously reaches the crack tip. Elastic materials have no energy dissipation in bulk; in viscoelastic materials, energy dissipation occurs and such a phenomenon may be (like their deformation) linear or not. The first PSAs were rubber-based compositions, elastic materials "denatured" to viscoelastics using viscous compounds. In some cases, such materials may display linear elastic deformation. The discovery of pressure-sensitive hydrogels based on plastomers plasticized and crosslinked simultaneously with viscous compounds forced a new approach for pressure sensitivity, since new PSAs will not be necessary linear elastic materials. From this point of view, the Dahlquist's criterion of tack can be considered as a simplification; pressure-sensitive adhesion is a time and polymer structure related process (see Chapter 4).

Fibrillation as debonding-related phenomenon is not "new"; it can be observed in the industrial practice during peeling off of removable PSAs ("legging") or tack measurement by rolling cylinder (or ball). Fibrillation during tack measurement with indenter-based methods allows a simple correlation of the modulus with the measured debonding force. Unfortunately, this method also possesses certain disadvantages. Some of them are correlated with the bonding phenomenon. Wetting of PSA depends on a large number of parameters such as the roughness of the adhesive layer, flexibility of the film face stock, the rate at which the air entrapped between the adhesive and the substrate can diffuse out of the system, and the rheological characteristics of the adhesive. The debonding mechanism is also influenced by roughness. As known from the industrial practice, quite different tack values may be obtained using various tack test methods (e.g., probe tack, loop tack, or rolling ball). The discrepancies between the experimental data depend on the adhesive nature; pronounced deviations (in detriment of probe tack) are obtained for "hard" adhesives, like SBCs, or dispersed adhesive systems. Unfortunately, there is no systematic study concerning the correlation of various tack test methods with the type of the adhesive system tested. As stated in Ref. [165], more sophisticated models are needed for fibrillation experiments with soft viscoelastic materials. In this case, cavities start to appear well below the maximum stress, their growth rate is not linear, and the cavitation stress depends on surface roughness.

II. MECHANICAL PROPERTIES OF PSPs

The main PSPs are web-like, supported or self-supporting materials. Their support has to resist stresses during manufacture. During their application, the support also has a mechanical function. The mechanical properties of the carrier material influence bonding and debonding. Therefore, for the design and manufacture of PSPs, it is necessary to evaluate the mechanical performances of the carrier, the adhesive, and the finished product.

According to Hensen [166], for a polymer used as the raw material for a plastic film carrier, the following characteristics are the most important: mechanical properties (resistance, elongation, and shrinkage); chemical resistance; permeability (to oil, gas, and water); sealability; surface quality; stiffness; thermal resistance; adhesion; coefficient of friction (COF); deep drawability; and melting temperature. From these characteristics, the resultant mechanical properties (resistance to different stresses, elongation, and stiffness), the properties related to dimensional stability (shrinkage and deep drawability), the performances related to surface quality (COF and sealability), and the performances related to thermal resistance (melting temperature, sealability, and warm deep drawability) have a special importance for carrier materials used for PSPs. As a function of their various application fields and manufacture technology, materials with different mechanical properties and dimensions are processed as PSP components under various stress conditions. Electronic converting machines for labels cut, die-cut, punch, and perforate the paper at running speeds of 200 m/min. Carrier materials as different as $15-20$ μm thin films and cardboard of 300 g/m^2 are used [167]. Generally, the main problem in designing structural bondings with elastic adhesives is the insufficient knowledge about their mechanical behavior. Therefore, it is necessary to examine the mechanical properties of the adhesive layer also.

A. RHEOLOGICAL BACKGROUND OF THE MECHANICAL PROPERTIES OF PSPs

The main PSPs are web-like products based on a solid-state carrier or combination of solid-state carrier materials (laminate) that exhibit pressure-sensitive properties. During the manufacture of the PSPs, the solid-state carrier material has to withstand mechanical stresses due to the tensions in the continuous web running at high speed on machine parts having different relative velocities. Stresses are encountered during conversion and application of the material also (see Chapter 8 and Chapter 11). Tapes are wound with variable forces, protective films are laminated with high pressure. Labels are applied with high speed as discontinuous items, die-cut from a high-speed web that has to support (as matrix) the same stresses as the laminated web. The liner plays a weighty role in the functionality and cost of most PSPs. Tear strength, dimensional stability, lay flatness, and surface characteristics are the most important features of the release liner [168]. Tear strength is important for label converting and dispensing. Good caliper control and liner hardness (lack of compressibility) are required. Dimensional stability, the ability to maintain the original dimensions when exposed to high temperature and stresses, is important for print-to-print and print-to-die registration. If the liner stretches under heat and tension, graphics will be distorted. Liner stretch can affect label dispensing also. Lay flat is very important for sheet labels or pin-perforated and folded labels (see Chapter 10). Labels that are butt cut, laser printed, and fan-folded must lie flat to get proper feeding and stacking performance. In contrast, for the application of PSPs on adherents having a complex contour or rough surface, softness and conformability are required. The carrier of tamper-evident labels requires excellent conformability and destructibility [169]. Medical tapes require conformability also [170]. In certain cases, for better conformability a carrier-less PSP is applied. For such products, the adhesive layer possesses adequate strength to permit it to be used without a carrier material (see Chapter 11). Tensile strength, stiffness, tenacity, and Elmendorf tear resistance are important features of sheet labels. The peel resistance results on both adhesion and dissipation energy. Dissipation energy is due to deformation of the plastic carrier, supposing that the adhesion is high enough to result in a deformation during the test [171] [see Equation (3.49)].

1. Structural Background of Mechanical Properties

The mechanical properties depend on the chemical basis and rheological behavior of the macromolecular compounds. Glass transition temperature, viscoelastic response, and yield behavior of crosslinked systems are explained by extending the statistical mechanical theory of physical aging, taking into account the transition of the WLF-dependence to an Arrhenius temperature dependence of the relaxation time in the vicinity of T_g [172] [see Equation (3.5) and Equation (3.6)]. For classical PSPs, the mechanical properties of the nonadhesive carrier material and of the adhesive are of different order of magnitude. The carrier is designed to resist stresses without deformations; the adhesive is designed to allow high deformations without being destroyed. For plastic carrier materials, the carrier is considered in a first approximation as nondeformable. Such behavior arises from the nature of plastics. Polymers suffer deformation under stress. The creep modulus E_c is given as a function of the stress (σ, constant in time) and the time-dependent deformation ($\varepsilon_{(t)}$) by the correlation:

$$E_{c(t)} = \frac{\sigma}{\varepsilon_{(t)}} \qquad (3.58)$$

At very low elongation ($<1°\%$), a spontaneous reversible elastic deformation exists and Hooke's law is valid. The time-dependent viscoelastic deformation takes place between 0.5% and 1% elongation. For this domain, the laws of linear viscoelasticity are valid [173]. At higher deformation, a time-dependent viscous deformation takes place. Rubber-like materials allow higher elastic deformation.

The mechanical properties of carrier materials are regulated by the choice of raw materials and by film processing conditions. The choice of the film manufacturing procedure (see Chapter 8) and the postextrusion processing affects the film quality. For blown film, the mechanical characteristics of the film can be regulated by bubble form [174]. By regulating the blow-up ratio (BUR), freezing height, or melt temperature one can control the film thickness and orientation. The mechanical, thermal, and optical properties of the film are functions of the orientation of the molecules. Orientation occurs during processing and is fixed through freezing. The mechanical properties of plastic films depend on their mono or coextruded buildup also. As stated by Acierno et al. [175], for semi-compatible or almost compatible polymers (e.g., LDPE/LLDPE), the mechanical properties of coextruded films are close to those of films made from blends. In the case of polymers giving rise to incompatible blends, deep minima are observed in many graphs of mechanical performance depending on the composition. In such cases, the mechanical properties of coextruded films are expected to be better than those of films made from blends.

The mechanical properties of the adhesive or of a self-adhesive carrier film can be regulated by the choice of raw materials and by their processing. PSAs are based on rubber-like, viscoelastic, and viscous products. The mechanical behavior of elastomers is determined by their multiphase network structure. Composite structures allowing controlled mechanical performance can be achieved by synthesis, formulation, or processing of the plastic or elastomer PSP components.

2. Molecular Order-Mechanical Characteristics

As discussed earlier, the buildup of an ordered structure in the adhesive can change its mechanical performance. Ordered structures are formed via crystallization, association, crosslinking, and bulk fillers (Figure 3.11). The resulting mechanical properties depend on the chain mobility in the "new" construction. This behavior is illustrated by Sun and Mark [176] for polyisobutylene. Viscoelastic polyisobutylene was transformed in an elastomer by crosslinking; the mechanical properties of the material were improved. Tensile strength and elongation increased. Later, by generating *in situ* reinforcement of this elastomer with inorganic filler particles, other mechanical properties, for example, stiffness, were improved. The effect of crosslinking and filling on the mechanical performance is complex and depends on the fine structural changes produced. The different methods of building an ordered structure are not interchangeable. Increasing the level of a filler (which does not allow high crosslinking of rubber), decreases the modulus [177]. The hardness of PVRs containing acrylonitrile copolymers dispersed (before polymerization) in the oligoether component of a PVR is higher than that of a filled (inorganic filler) PVR [178]. The existence of a network (composite structure) is the base condition for excellent mechanical properties.

The mechanical resistance ("green strength") of unvulcanized natural rubber is given by the bound cocoons linked together via stress-induced crystal lamellae, providing a stable network structure [179]. For synthetic rubber, the type of crosslinking influences the mobility of the chain segments, that is, the storage modulus. Below the secondary transition temperature for a crosslinked plastic material, the deformation mechanism that governs the yield does not depend on the degree of crosslinking. In this region, the mechanism of deformation can be regarded as a dislocation glide mode [180]. Above the material's T_g, the degree of crosslinking comes into play

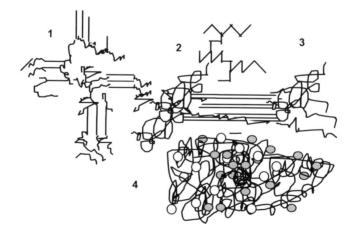

FIGURE 3.11 Network structures in base PSP polymers. (1) Molecular association; (2) polymer crosslink; (3) crystalline order; (4) filled system.

with respect to yield. Chain length distribution affects the tearing energy of crosslinked networks. Bimodal networks showed a higher tear energy than monomodal constructions [181]. Uncrosslinked and unentangled molecules do not contribute to the mechanical properties of the polymer. It has been shown that a styrene–isoprene diblock has a tensile strength of 20 psi, whereas a fully coupled SIS triblock exhibits 300 psi tensile strength [182]. A shear holding time of 387 days has been obtained for a pure triblock polymer when compared with 86 days for a commercial one.

The mechanical properties are strongly influenced by the reinforcing effect of fillers. Numerous mechanisms have been discussed in relation to their effect. The most significant ones concern the contributions to increased strength, increased stiffness, and increased hysteresis [183].

For filled materials, the nature of filler and its concentration affect the type of structure formed and the resultant mechanical characteristics. Adding EPDM to SBS to improve its ozone resistance, at low filler content (10–20 phr), the particles are dispersed randomly in the polymer matrix; at higher level (20–40 phr) a quasinetwork is built up [184]. In many cases, the energy and entropy changes related to the thermomechanical behavior of filled elastomers depend on the concentration of the fillers [56]. The geometry of the filler particles play an important role also.

It has been reported that carboxylated diene-based rubbers become strong without reinforcing fillers or covalent crosslinks when the carboxylic acid is neutralized. Their Young's modulus levels off as zinc oxide concentration increases; the Young's modulus of specimens containing carbon black is much higher than those without carbon black [185]. The effect of the carbon black or zinc oxide loading on the Young's modulus is much higher than the theoretical values calculated using the Guth–Gold equation. This suggests that the effect of surface characteristics is much stronger. The idea of the hydrodynamic or strain amplification effect (which takes into account the concentration of the particle only) is based on the concept that the local strain between the filler particles is amplified because the particles can be considered rigid (compared with the rubbery matrix) [186]. Such effect is similar to the enhancement of the viscosity of a fluid due to the incorporation of rigid fillers and has been described by Einstein [187] according to the equation:

$$\mu = \mu_0(1 + 2.5c) \tag{3.59}$$

where μ is the viscosity of the filled liquid, μ_0 that of unfilled liquid, and c the volume fraction of filler.

According to the paradox of Slayter [188], in a composite structure, the mechanical characteristics of the higher strength component are improved. For reinforcing with short fibers according to the Halpin-Tsai [189] equation, the modulus of the composite (E_C) is a function of the modulus of the matrix (E_M), the volume fraction of the filler (ϕ_F), and the coefficients A and B:

$$\frac{E_C}{E_M} = \frac{1 + AB\phi_F}{1 - B\phi_F} \tag{3.60}$$

where A is a function of the length L and diameter d of the fibers,

$$A = \frac{2L}{d} \tag{3.61}$$

and B is a function of the moduli of elasticity of the composite (E_C), fiber (E_F), and matrix (E_M):

$$B = \frac{E_F/E_M - 1}{E_F/E_M + A} \tag{3.62}$$

It should be mentioned that the shape of the filler particles has been taken into account by Guth also with the aspect ratio (f) in the equation for the magnification factor of aspheric fillers [186]:

$$X = 1 + 0.67f\phi + 1.62f^2\phi^2 \tag{3.63}$$

Generally for monodirectional constructions, the modulus and the strength of the composite can be calculated from the data of the components. According to the paradox of Griffith [190], a fiber-like material possesses a much higher strength than the same material in other form. Such fillers are used to improve the shear resistance of PSAs. The length of the filler particles is taken into account by the paradox of tensioning or stressed length. It may be supposed that this paradox influences the effect of the thickness of adhesive and of the primer also. According to Klein [191], adhesion decreases if primer thickness increases.

The thermomechanical behavior of segmented block copolymers having a low concentration of soft segments is determined by intermolecular interactions in the soft and hard blocks. First it was supposed that the free energy of monoaxial deformation (elongation) of crystalline two-phase networks influences the internal entropy of the polymer chains. Later, it was demonstrated that the deformation produces internal changes in the chains of the amorphous phase [56]. It is supposed that in segmented PVR elastomers, during application of a stress the soft polymer matrix is sheared between the the small nondeformable "plates" of the hard segments [192], that is, like an adhesive between the solid-state components of the laminate. The mechanical performances of elastic foams as function of their blowing degree have been described by considering such systems as two-phase (elastomer phase and cavity phase) systems [193]. As shown in Ref. [194], in dispersed PSAs, even though emulsion particles are internally crosslinked (after drying), the particles only interact via entangled dangling ends at the interface between particles, and the adhesive essentially behaves as a filled polymer melt.

3. Mechanical Properties-Adhesive Performance

The performance of TPEs illustrates that the segregated and ordered structure is a condition sine qua non for adequate mechanical properties. Star-branched polyamides (PAs) (nylon 6) retained part of their mechanical integrity even after melting, indicating the presence of enhanced entanglements [195]. Bocaccio [196] synthesized a sequenced styrene–isoprene copolymer by depolymerizing natural rubber (MW 50,000) and grafting styrene onto the normal rubber (NR) chains producing a TPE. Because of the only slight segregation in such compounds, the mechanical characteristics of the polymers are not as good as those of SIS or SBS. In contrast, the bond strengths of HMPSAs formulated with such compounds are comparable to those based on commercial SBCs, illustrating the complex nature of the dependence between the mechanical and adhesive properties. Evaluating the adhesive properties of a PSA tape with the rolling adhesive moment tester, Yoshiaki and Kentaro [197] demonstrated that the mechanical properties of the AC adhesive and those of the rubber-based adhesive each had different effects on adhesion. Of block copolymers with different rubbery blocks, but similar block MWs and about 30% styrene, the SEBS had the highest modulus [185]. The rubbery elasticity of the polymer is maximized when the rubbery polymer segment is completely amorphous and has a low T_g. In commercial SEBS copolymers, the best rubbery characteristics are achieved with rubbery blocks having a T_g of $-50°C$. The mechanical performance should be examined as a whole. Macromolecular compounds of quite different chemical composition or structure can display similarity of some performances, but differences concerning their global behavior. As an example, thermoplastic polyurethanes (TPUs) synthesized with hydroxy-terminated polybutadiene and methylenedi-isocyanate (MDI) possess the same modulus values as polymers manufactured with poly(oxypropylene–glycol) or polybutadiene and MDI. However, their tensile strength and elongation at break are different [198].

B. Main Mechanical Performances

The main mechanical performances allow the transport, processing, application and deapplication of web-like PSPs. Generally these characteristics are given by the carrier material, but the buildup of the finished product also affects the mechanical characteristics of PSPs. The interface has a significant effect on the structure and properties of the boundary layers, resulting in a change in the packaging density of the macromolecular chains, limitation of their conformational status, decrease in segmental mobility, and inhibition of relaxation processes near the solid surface [199]. A significant change occurs in the mechanical properties of a polymeric material on a support or in the presence of a solid-state surface of a filler in thin layers, when the thickness of the layers becomes comparable to that of the boundary layer. Such effects are "translated" into the adhesive properties of primed PSAs or into the flow properties of filled, carrier-less transfer tapes. There is a gradient of the segmental mobility and mechanical properties near the phase boundary in moving away from the solid surface. The value of the modulus of the coated material depends on its distance from the solid surface and the nature (high or low energy) of the solid surface. There could also be a shielding influence which decreases the distance through which deformation energy is transmitted from outside to the boundary layer and the more distant layers of the macromolecular coating [200]. The most important mechanical characteristics are tensile strength, tear resistance, and stiffness.

1. Tensile Strength

Tensile strength, tension yield, and elongation are evaluated as stress-elongation diagram. If stretched in a stress or strain machine, once a certain degree of elongation is achieved, both elastic and plastic materials tear apart. The force required to rupture the film is expressed as a function of the cross section of the material and is referred to as tensile strength. Direct comparisons cannot be made between the tensile strength values as they are measured with various elongations at rupture. Comparisons can be made using the moduli, that is, the forces at specific elongation, reduced to the cross section of the sample.

For polymers, the mechanical properties strongly depend on the chemical composition, macromolecular features, and processing conditions. For polyolefin films, the MW and its distribution, crystallinity, the form and order of crystalline parts, chain branching, unsaturation, and polar groups have a weighty influence on the mechanical performances. The mechanical properties of plastics are also function of the structural ordering achieved by postextrusion processing, that is, orientation [201]. The tensile strength of plastic carrier films depends on their raw materials and manufacturing technology (see Chapter 8). The MW and its distribution affect mainly the tensile strength, brittleness, and tear resistance [202]. Crystallinity has a greater influence on the softening point, elastic and flexural moduli, cold flow, and hardness. The texture of the film is given by crystallinity and its buildup conditions. As stated in Ref. [203], the modulus of polyethylene increases with the degree of orientation (achieving values of $10-60 \, kN/mm^2$), but it depends on the MWD. The tensile strength for PE homopolymer increases with its density and generally does not depend on the melt index or MW. For LDPE or LLDPE mixtures processed as blown film, higher LLDPE concentrations increase the degree of orientation of the film and the stiffness. The stiffness of a film containing more than 50% LLDPE is greater than that of the LLDPE [204]. An increase in the MW of a 300 μm PVAc film from 200×10^3 to 2000×10^3 increases its tensile strength from 7 to $50 \, kg/cm^2$ [205].

Fillers modify the tensile strength and tear resistance of a polymer. As an example, a PUR without filler exhibits an elongation of 360%, but this value decreases to 100% with a filler [206]. Pigment particles alter the characteristics of the film in two ways. If they have a higher tensile modulus, compared with the polymer, then they will increase the yield stress of the composite, and this will favor brittle fracture. They also flow within the film and act as stress concentrators reducing the film's overall impact resistance. As is known, the resistance of materials can be examined by the buildup of flow regions in amorphous thermoplasts, by the formation of microcrazes in partially crystallized polymers, and by debonding of adhesive or cohesive joints in fiber-reinforced plastics [207]. The study of the long- range order in craze microstructure of deformed polystyrene–

polybutadiene block copolymers showed that crazes grew parallel to the ordered structures where possible (as in a filled construction) but crossed them when the mismatch between the growth direction and the orientation exceeded a critical value [208].

Tensile strength plays a different role in the evaluation of the quality of carrier material used for different product classes. For labels, adequate tensile strengths provide the dimensional stability required mostly during conversion. Tear strength and tensile strength of the face material for labels depend on conversion and application methods used [209]. The layout should be in the long grain direction to provide maximum conformability. Tear strength values of 1.6–2.7 in the cross direction (CD) (across the grain) and 2–4 in grain direction are recommended for paper labels [210]. The characteristics of paper that should be taken into account for its use as release liner [211] are weight, thickness, density, transparency, stiffness (cross direction [CD]/machine direction [MD]), tear resistance, and tensile strength.

For tapes, tensile strength ensures performance of the bonding and assembling function. For protective films, tensile strength should achieve the minimum level necessary for web processing, lamination, and delamination. For special products like transfer printing elements (e.g., Letraset), a vinyl polymer with high tensile strength ensures the dimensional stability of the letters during high pressure application [212]. Table 3.8 lists some important mechanical characteristics of a number of common PSPs. It can be seen that the tensile strength values required for common packaging tapes are much higher than those for other product classes. The tensile strength values for other special (nonpackaging) tapes are given in Table 3.9.

The mechanical resistance of PSPs depends on the characteristics of the component materials and the product geometry. Protective films are based on thin carrier materials (35–120 μm), mostly polyolefins; therefore, their down gauging can lead to pronounced elongation of the carrier material

TABLE 3.8
Mechanical Characteristics of Selected PSPs

Product	Carrier Material	Carrier Thickness (μm)	Mechanical Characteristics MD/CD Tensile Strength	Elongation (%)
Automotive protective film	Polyolefin	50	27/13 N/10 mm	550/570
Automotive masking film	Polyethylene	100	19/14 N/10 mm	350/340
Thermoforming protective film	Polyolefin	70	30/27 N/10 mm	680/650
Overlaminating film	Polypropylene	25	8 kg/25 mm	125
Overlaminating film	Polyester	25	10 kg/25 mm	125
Packaging tape	PVC	37	150/70 N/mm^2	—
Packaging tape for carton sealing	PVC	36	7 kg/10 mm	70
Sealing tape for heavy packaging	PVC	70	12 kg/10 mm	70
Packaging tape	Polypropylene, reinforced	50	35 kg/10 mm	—
Palettizing tape	Polypropylene	30	6 kg/10 mm	90
Packaging tape	Polypropylene, oriented	40	250/30 N/mm^2	—
Masking tape	Crepe paper	85–90[a]	3.5 kg/10 mm	—
Label	Polypropylene	50	100/140 N/mm^2	100/40

[a]Weight, g/m^2

TABLE 3.9
Mechanical Characteristics of Carrier Materials for Special Tapes[a]

Product	Force at Break (N/10 mm)	Elongation (%)	Maximum Force (N/10 mm)	Elongation at Maximum Force (%)
Insulating tape	11	252	14	242
Insulating taping tape	209	26	115	26
Building masking tape	14 (11)	138	19 (19)	14 (11)
Masking tape	11 (8)	374	13 (12)	366 (13)
Application tape	29 (20)	340 (887)	—	—
Application tape	27 (19)	292 (814)	—	—
Application tape	21 (14)	250 (625)	—	—
Medical tape	25 (21)	250 (350)	—	—

[a]Some values are given in machine and cross directions, MD (CD).

during processing and lamination. Such elongation and the relaxation (energy absorbtion) related to the deformation may decrease the value of the peel force (Figure 3.9). The use of a coextruded film increases the tensile strength of the film as well as the critical (minimum) thickness (Figure 3.8).

It should be emphasized that the value of the tensile strength alone does not allow the characterization of the dimensional stability of the carrier material for different product classes. The elongation must also be known. As is known, elastomers display excellent tensile strength, a property associated with rubbery elasticity. Elastical carrier materials do not conform (permanently) to the product surface and may cause delamination. Therefore, such products (i.e., LLDPE alone) are not suggested for PSPs having large contact surfaces (e.g., protective films). They also cause difficulties in cutting and die-cutting (see Chapter 10, Section II). Other polymers such as HDPE or polyethylene terephtalate (PET), although they have excellent tensile strength (Table 3.10), are not sufficiently conformable to afford good contact.

A comparable elongation of the carrier material can be achieved with an appropriate formulation. Extensibility may be given by tensioned multilayer carrier materials calendered together [221] or by the use of a filled polypropylene [222]. For many applications, the direction of extensibility is very important. For medical tapes, cross-direction elasticity may be given by a special nonwoven material [223].

For most PSPs, the carrier material is a common film manufactured for other applications also. Theoretically, it would be possible to design it specially for PSP applications knowing the peel value that appears during delamination. Knowing that the mechanical resistance of the carrier material must be higher than that of the adhesive bond, it would be (at least theoretically) possible to design and manufacture tailored carrier materials. As known from the production of labels, for common, permanent paper labels, a peel value higher than 22–25 N/25 mm causes the paper to tear during label removal. Taking into account a similar dependence between the tensile resistance of the carrier material and the bond strength necessary for a tamper-evident product, films with sufficiently low mechanical resistance can be manufactured for such products. For certain other products, estimation of the required peel value (and of the tensile strength related to it) is more difficult. Some diaper tapes, medical tapes, and protective films have a so-called cleavage or breaking peel value that is higher than the usual so-called continuous peel value (see Chapter 7). For these products, the carrier has to be designed for the maximum tensile strength required. Diaper closure tapes are used as refastenable closure systems for disposable diapers, incontinence garments, and similar items. [105]. Two- and three-tape systems are known. The two-tape system comprises a release tape

TABLE 3.10
The Main Mechanical Characteristics of Common Plastic Carrier Materials

Material	Tensile Strength (N/mm^2)		Elongation (%)		Elastic Modulus	Reference
	MD	CD	MD	CD	MD (N/mm^2)	
HDPE	20–30	—	100–1000	—	600–1400	[213]
		—	220–800	—	—	[214]
LDPE	8–10	—	300–1000	—	150–500	[213]
Biaxially oriented	12–13	—	—	—	—	[215]
MDPE	14–25	—	225–500	—	28–35	[202]
Biaxially oriented and crosslinked	56–91	—		—	—	[214]
LLDPE	—	—	300–1000	—	400–800	[213]
PP	21–70	—	200–500	—	630–840	[202]
			200–600	—	—	[214]
Biaxially oriented	120–180	300–400	100–200	20–50	—	[214]
	140–200	280–200	—	—	2000–2500	[216]
Calandered	>25	25	≥200	≥200	—	[217]
EVAc	10–20	—	600–900	—	130–700	[213]
HPVC	50–75	—	10–50	—	2900–3500	[213]
Oriented	150–170	45–70	—		4000	[216]
	130	50	120	80		[218]
VAc	30–40	—	6–10	—	—	[219]
Plasticized	10	—	100–1000	—	—	[219]
SPVC	10–25	—	17–400	—	—	[213]
PC	60–65	—	80–120	—	2100–2400	[213]
PET	47	—	25	—	2800–3100	[213]
Oriented	20	23	110	90	50–130	[215,220]
Chemically treated	450	550	125	80	—	220
Filled	450	560	120	90	—	214
Cellulose acetate	37–98	—	—	—	25–45	215

and a fastening tape. The fastening tape includes a carrier material such as paper, polyester, or polypropylene. The preferred material is polypropylene (50–150 μm) with a finely embossed pattern on each side. Such systems allow reliable closure and reclosure with a debonding force that depends on the debonding rate. To test a special, UV- crosslinked adhesive tape [224], first the force required to start the breaking of the bond (initial breakaway peel) was measured, and then, the force needed to continue the breaking of the bond (initial continuing peel) was measured. The initial continuing peel was found to be about 30% lower than breakaway peel.

The mechanical performances of the adhesives and adhesive raw materials are generally measured in order to compare different base materials (Table 3.11). According to a BASF publication [225], such tests "allow a comparison of the order of magnitude" of the mechanical characteristics only. The measured property values depend on the chemical composition, rheology, and geometry of the samples. However, such properties are useful for comparison because of the use of standard test methods and their universal character. They allow a comparative test of different components of the pressure-sensitive laminate. For instance, a comparison of the tensile strength of carboxylated rubber (CSBR) and AC PSA shows that CSBR exhibits better tensile strength than ACs, and there is a good agreement between its tensile strength at 300% elongation and its shear resistance. The addition of SBCs to natural rubber improves its mechanical properties.

TABLE 3.11
Mechanical Characteristics of PSA Raw Materials

Adhesive Raw Material	Tensile Strength (N/mm^2)	Elongation (%)	Reference
Soft acrylate	0.01	>3800	[16]
Hard acrylate	0.03	2500	[15]
Hard acrylate	0.15	3000	[226]
Hard acrylate	0.20	>3800	[11]
Hard acrylate	0.25	>2000	[19]
Hard acrylate	0.27	2000	[12]
Hard acrylate	0.50	1500	[227]
Hard acrylate	1.00	1340	[228]
Hard acrylate	1.50	300	[229]
Hard acrylate	2.70	1100	[21]
Hard acrylate	7.50	550	[230]
EVAc, segmented copolymer	11.00	1400	[231]
SBS, linear block copolymer	15.00	950	[105]
SBS, radial, block copolymer	22.00	620	[105]

Increasing the thermoplastic rubber level in a natural rubber-thermoplastic rubber mixture (0–10 phr) provides a better yield (15–200 kg/cm^2). As can be seen from Table 3.11, the tensile strength of some segmented elastomers attains that of soft, plasticized plastomers.

The strength and stiffness of the block copolymer depend on the chemical nature of the segments and their mutual interaction. Block copolymers having the same styrene content, and about the same block MW but different types of midblocks, display different mechanical properties [232]. The SEBS polymer has the highest strength and modulus. As discussed earlier, the stress–strain behavior of such rubbery polymers can be adequately modeled by the Mooney–Rivlin equation as an elastical network (considering the trapped entanglements as finite crosslink junctions) or using the Guth–Gold equation relating the stress level (modulus) to the hydrodynamic effects of the styrene blocks considered as filler. The mechanical properties of the block copolymers depend on their buildup also. The tensile strength of a radial SBC is higher (22 MPa) than for a linear SBC (15 MPa), and the 300% modulus is also higher (5 vs. 2.5). Elongation at break is smaller (620 vs. 950%) [105]. The star-shaped (radial) SBS copolymer Solprene 416x with more styrene has a lower MW but higher tensile strength (200/90 kp/cm^2). The copolymer with lower styrene content shows better peel and tack values.

Tensile stress and elongation play an important role for sealants and sealing tapes. Elastomers used for sealants must exhibit good elastic recovery; low temperature ($-30°C$) flexibility; resistance to oil, ozone, and aging; elongation (>200%); and tensile strength (>10 MPa) [84]. Tensile strength, stiffness, density, flexural modulus, flammability, and impact resistance are the most important properties for foams [233]. The mechanical properties of finished products are also tested. The modulus, maximum elongation, and tensile strength are determined for sealants and for carrier-less transfer tapes used as sealants.

The plastic anisotropy ratio (R) is used to study the effect of applying a tensile force to plastic films by considering the change in original width (w_0) and thickness (h_0) resulting from the stress [234]:

$$R = \frac{\ln(w_f/w_0)}{\ln(h_f/h_0)} \tag{3.64}$$

where w_f and h_f are the final width and thickness of the specimen. For most films (especially for mono-axially oriented materials), the R value varies with direction in the material. This variation ΔR is called planar anisotropy. Dimensional stability is strongly affected by R. As known, shrinkage and die-cuttability are functions of R (see Chapter 10, Section II). Considering the planar anisotropy as a sum of the dimensional changes in different directions (angles), the component for normal aniso-tropy R' is defined as:

$$R' = 0.25(R_0 + 2R_{45} + R_{90}) \tag{3.65}$$

According to Ref. [234], if R' is greater than 1, the material would resist thinning, which would lead to better drawability. Drawability is a characteristic known to be required for deep drawable protective films (see Chapter 11).

Tensile strength is related to the shrinkage of plastics also. For heat-shrinkable tapes, the nominal value of the shrinkage stress is given. For instance, bilayer heat-shrinkable insulating tapes for anticorrosion protection of petroleum and gas pipelines have been manufactured from photochemically cured LDP with an EVAc sublayer [235]. The liner dimensions of the two-layer insulating tape decreased 10–50% depending on the degree of curing (application tempera-ture 180°C). The application conditions and physicomechanical properties of the coating were determined for heat shrinkable tapes with 5% shrinkage, at a curing degree of 30% and shrinkage test tension of 0.07 MPa [235].

2. Tear Strength Properties

The tear resistance of carrier materials used as face stock plays a pronounced role in their coating and conversion and end-use. Web coating and laminating tensions, splitting stresses induced by cutting and die-cutting, and forces acting on the matrix after die-cutting are very important.

3. Surface Tear Strength

Paper tear is a well-known criterion in evaluating the quality of permanent paper labels. Generally, if the adhesion between PSA and substrate is better than the inherent cohesion of paper during peel off, the debonding forces destroy the paper face stock material. In some cases, only an apparent tear destruction occurs. In these cases, the cohesion of the surface layer of paper (top coat) and its adhesion to the paper fibers are not sufficient. Clay-coated papers tear if the superficial strength of the paper is not sufficient [236]. This phenomenon is also common in offset printing, where it is mostly due to the printing ink and printing conditions [237]. It also occurs for top-coated and uncoated paper. Paper humidity (which reduces the mechanical resistance of paper) can also influ-ence paper surface tear strength especially for clay-coated papers. Therefore, for nonhomogeneous carrier materials (e.g., paper or top-coated or laminated films), both surface tear resistance and bulk tear resistance should be taken into account.

4. Bulk Tear Strength

Fragile pressure-sensitive materials are either very thin or have low internal strength. They are coated with a very aggressive PSA [238]. Various fragile materials can be used such as acetate for envelopes, PVC for warning labels, on electronic equipment, and paper for sterile needle car-tridges. For common permanent label, the peel resistance of the adhesive must be higher than the resistance of the carrier. The resultant peel of TPE-based removable formulations should be 2.5–40 oz, with an initial peel of 1.0–16 oz for 1×6 in. strips (on stainless steel). Values above this can result in a paper tear [239].

For plastic carrier materials, tenacity and tear resistance have to be examined together. These properties depend on the orientation of the molecules. Among the machine parameters, the neck influences the orientation [240]. Elmendorf tear resistance decreases with neck length for bimodal HDPE. At high neck length, the MD values are higher [241]. The orientation of the molecules is also a function of blow-up ratio (BUR). The greater the constriction, the higher the real BUR. The bimodal grades have excellent dart drop impact (DDI) values at high BUR. Tear strength is a function of the density also. As crystallinity decreases, the toughness of the film increases. Elmendorf tear strength increases with the melt flow index (MFI) and decreases with the density. Flexural crack resistance of C_8 LLDPE decreases with density. The isotropy of the tear strength (MD/CD) depends on the polymer and its processing conditions. Because of the tendency of polypropylene to undergo orientation during processing, its tear strength in the MD is very low [202].

Tear strength plays an important role for many tapes. Easy tear, breakable tapes are tapes that are easily torn or split during handling. Easy tear breakable hank tapes are used in the same types of finished products as easy tear paper tapes. The tape is hand applied and used for hanking and spot tape when fast and easy removal is required. This tape possesses low elongation and easy break force. Different materials differ in their ability to spread the strain away from the strained zones, thus avoiding the formation of local regions of high strain, which lead to necking and fracture. The strain hardening ability of the material is approximated by the Ludwick–Holloman equation [234]:

$$\sigma = K\varepsilon^n \tag{3.66}$$

where K is a constant and n is the strain hardening exponent. A high n value is desirable for stretch forming and is required by many tape applications and for deep drawable protective films. As discussed earlier, strain hardening also plays an important role in the mechanical strength of various polymers used as raw material for PSAs, (by debonding in the near-tip region, by fibrillation, etc.).

5. Impact Properties

For special tape applications and for security protective films, the impact properties of the carrier play an important role. Abrasion-resistant automotive decorative films that resist the impact of stones have been developed [242]. These products are made on a PUR or EVAc or PVC basis. They must be bubble-free laminates, which are heat deformable, and repositionable. Thick films (80–900 μm) are used as the carrier.

Among the classical polyolefins, LDPE possesses the highest impact strength and lowest modulus [202]. Bimodal PE grades have excellent DDI values at high BUR. Elastic polyolefins and polyolefin copolymers display adequate impact properties. The impact resistance of C_8 LLDPEs is better, increasing with MFI and decreasing with density. EVAc copolymers (with 16–30% VAc) have better impact strength. The use of PVRs with an extremely high level of elongation at break ensure a sufficiently high impact strength, which is particularly important in the manufacture of sealing and mounting tapes. For foams, especially "supersoft" foams, a low compression set and rapid recovery from static deformation are required [243].

6. Stiffness

The stiffness of a carrier (expressed as its modulus) is very important for flatness and machinability. Wrinkle buildup depends on the stiffness of the paper. Wrinkle buildup during printing is mainly due to overdrying of the paper carrier [244]. Its humidity content may decrease to 1–2%, changing its dimensions. Stiffness of the paper is a function of its geometry and grammage, construction, and humidity balance. For lower weight papers (40–80 g/m^2), the buildup of wrinkles is more accentuated. The wrinkles have a shorter wavelength than those formed in papers having a weight of 100 g/m^2. Concerning paper buildup, the choice of paper fiber is strongly influenced by sulfite pulps. Sulfite pulps are used for the manufacture of papers for soft flexible packaging. Machine finish and supercalendering also affect paper stiffness. High stiffness can be obtained by using higher grammage or a more voluminous paper grade. The grade of paper is usually given as substance (mass per unit area of sheet). For labels, paper of at least 70 g/m^2 is used; lower substances make it difficult to provide the necessary strength to enable the paper to be peeled from the substrate. There is no specific functional upper limit of substance for the coating carrier, but paper would not usually have a substance greater than 100 g/m^2 and films not more than 150 g/m^2.

The stiffness of paper and board is a function of the modulus of elasticity and the third power of thickness. Special microspheres used as filler in paper improve the paper modulus if added as a filled middle layer [245]. There is a trend toward the use of high filler loads in fine papers, providing both economic and technical advantages. When the common filler content of a paper increases to 20%, the thickness of paper decreases by 5% and its stiffness by 15%. With the addition of 0.6% special, expandable filler to the paper layer, its original stiffness is regained.

Tenacity and tear resistance of plastics are examined together and depend mainly on the orientation of the macromolecules. For LDPE, the maximum possible stiffness is about 450 N/m^2 [246]. TPU films have moduli of 10–650 MPa [247].

Tape and protective film application depend on the conformability of the film. Stiffness is the main characteristic used for comparison of different carrier materials for tapes [248]. For labels, lay flat in printing is complemented by on-pack wrinkle-free squeezability. The conformability of different carrier materials depends on their stiffness. Stiffness is characterized by the modulus value (Table 3.12). Conformable synthetic films should have a tensile modulus of less than 4,000,000 psi [95] as measured in accordance with ASTM D-638 and D-882, preferably less than about 300,000 psi. Conformable fabric backings should have a tensile modulus of 4,000,000 psi. In comparison, for sealants, modulus values of 0.4–0.6 N/mm^2 are required [249]. Typical examples of conformable carrier materials used for medical labels and tapes include nonwoven fabric, woven fabric, and medium to low tensile modulus plastic films (PE, PVC, PUR, low modulus PET, and ethyl cellulose) [95]. PET is considered as a nonconformable carrier material.

As shown by Table 3.12 VAc polymers exhibit a broad range of modulus values. The high values are comparable with the data for common carrier materials; the low ones attain the range of adhesive raw materials. It can be also observed that postprocessing (orientation) has a pronounced influence on the modulus value of films.

As discussed in Ref. [250], stiffness influences cuttability also. According to Ref. [251], the stiffness of the main carrier materials used for tapes decreases as follows:

$$PET \cong HPVC > OPP \gg HDPE \gg LDPE \qquad (3.67)$$

Die-cuttability of the same materials decreases in a similar sequence:

$$PET > HPVC > OPP \gg HDPE \gg LDPE \qquad (3.68)$$

The stiffness of polyethylene increases with its density. Density depends on crystallinity, which is a function of chain length and branching. The crystallinity-related properties decrease with increasing polar monomer content. Stiffness decreases as the crystallinity decreases. Bimodality has no effect on stiffness. For terpolymers of ethylene with maleic anhydride as polar component, the addition of LDPE in the terpolymer increases the peel strength because it modifies the stiffness of the film. The higher the density of PE, the more it increases adhesion.

Stiffness also depends on molecular orientation. A special PE having high MFI, low processing temperature, and high degree of orientation gives high stiffness films. A modulus of 400 N/mm can be achieved for a density of 0.930 g/cm^3 [261]. Most label films are oriented. Orientation gives a 50–100 % increase in rigidity [202] (see also Chapter 8). Polypropylene tends to orient during processing. Its stiffness is 3–4 times greater than that of polyethylene [262]. The blow manufacturing

TABLE 3.12
Modulus of PSP Components

Material	Modulus (N/mm²)				
	Carrier Component[a]	Adhesive Component[a]	MD	CD	Reference
PP homopolymer, broad MWD, cast film	×	—	800	—	[252]
PP homopolymer, narrow MWD, cast film	×	—	700	—	[252]
Blend of PP homopolymer, broad MWD, low crystallinity	×	—	670	—	[252]
PP random copolymer, low comonomer content	×	—	950	—	[252]
PP random copolymer, low comonomer content	×	—	800	—	[252]
PP random copolymer, low comonomer content	×	—	500	—	[252]
PP cast film	×	—	670–1400	—	[23]
PP bioriented, blown film	×	—	2500	—	[216]
PP bioriented, cast film	×	—	2000	—	[216]
PP mono-oriented	×	—	200	—	[216]
Polyethylene (HDPE)	×	—	40–150	—	[253]
Polyethylene (LDPE)	×	—	40–150	—	[253]
Polyester	×	—	38–56	—	[215]
Polyester, oriented	×	—	700	1000	[254]
Polyester, oriented	×	—	2200	2200	[255]
Polyvinyl chloride, plasticized	×	×	67	61	—
Polyamide (PA6)	×		150–320	—	[253]
Polyamide (PA66)	×		40–150	—	[253]
PUR	×		170	—	[204]
Polystyrene	×		300–360	—	[253]
Fabric for medical use	×		35–40	—	[244]
EVAc, with 50–70% VAc			20–160	—	[256]
EVAc, segmented copolymer	—	×	28	—	[231]
Natural rubber	—	×	0.9	—	[257]
			1–5	—	[78]
Natural rubber, cured	—	×	0.9	—	[258]
Butadiene–styrene copolymer	—	×	7–11	—	[259]
Butadiene–styrene, radial block copolymer	—	×	5	—	—
Butadiene–styrene, linear block copolymer	—	×	25	—	[105]
EAA copolymer (hot-melt)	—	×	22	—	[260]
Isoprene–styrene block copolymer	—	×	1.8–7.5	—	—

[a] ×, present; —, absent.

process provides high MD stiffness and tensile strength for ease of register control, stripping, and dispensing. Balanced orientation makes good die cutting possible during high-speed conversion. Films made using a MD orientation exhibit outstanding stiffness in the MD and flexibility in the CD.

Stiffness is a function of crosslinking also. For crosslinked structures, the Young's modulus is given by the correlation [263]:

$$E = \frac{CT\delta_r}{M}$$

(3.69)

where δ_r is the crosslinking density, C a constant, M the molecular weight, and T the temperature. In contrast, plasticizers added, to enable the polymer chains to slide over each other, cause a decrease in the modulus (see also Chapter 5). The influence of the degree of crosslinking on the modulus and stiffness can be observed in the practice of protective film application. Because of the differences in the "density" of the crosslinks produced by various crosslinking agents (e.g., isocyanates or aziridine derivatives) and the differences in the mobility of the chain segments between crosslinking points, different degrees of softness of the cured adhesive are achieved, (see also Chapter 6). Therefore, rubber-based adhesive-coated protective films conform better than cured acrylates, and aziridine-cured formulations are softer than isocyanate-based formulations.

Active fillers serve to stiffen the carrier material or adhesive [87]. Modulus improvement by fillers has been evaluated by a number of authors [264]. The best known equations of Guth and Smallwood were discussed earlier. In a first approximation, tackifier resins were considered as fillers. Such "fillers" possess a much higher modulus than rubber. As stated by Druschke [73] for blends of natural rubber and tackifier resin, the modulus ranges from about 10^{-1} to 2–5×10^{-1} N/mm^2 at room temperature. The modulus of a tackifier resin is about 1. Like a filler, the resin level influences the final modulus value.

Flexibility and extensibility of the carrier material influence its energy-absorbtion properties, which help to determine the nature of the debonding, that is, removability (see Chapter 7). Generally the amount of energy (U) that can be stored elastically in a carrier deformed by flexion can be written as a function of the stress (σ) and deformation (ε) as [265]:

$$U = \int \sigma\, d\varepsilon$$

(3.70)

For unidirectionally (e.g., film-like) stressed materials under well-defined conditions, the amount of stored energy can be expressed as a function of the modulus of elasticity (E):

$$U = 0.5E\varepsilon^2$$

(3.71)

As can be seen from the aforementioned relation, the deformability of the material is more important with respect to energy storage than its modulus. Elastic, extensible, low modulus carrier materials can store too much energy, during application (e.g., lamination), which causes non-uniform debonding and adhesive deposits. Such materials (e.g., LLDPE or some EVAc copolymers) are not recommended as carriers for protective films (see Chapter 8). It is evident that energy storage may be advantageous for products with high cleavage peel (zip peel) (see Chapter 7).

C. REGULATING THE MECHANICAL PROPERTIES OF PSPS

To fulfill the various requirements for different product classes and applications, there is a need to regulate the mechanical properties. This can be accomplished by controlling the formulation of the product components (carrier, liner, and adhesive), the choice of product geometry, or the choice of product buildup.

1. Regulating Mechanical Properties by Formulation

Theoretically, the simplest way to regulate the mechanical properties of carrier materials is to control their chemical or macromolecular formulation (see Chapter 7, Section II). In reality, the range of commercially available raw materials is limited. In contrast, the technology of web-like materials allows the manufacture of various products based on the same raw materials, which display quite different mechanical characteristics.

2. Regulating Mechanical Properties by Geometry

The mechanical properties of PSPs are the result of intrinsic chemical or structural characteristics and product geometry. Generally, product geometry includes the in-plane dimensions of the carrier and its thickness. For labels, the product contour (in plane) can influence the die cutting properties. For tapes, the cross-sectional contour affects the mechanical characteristics. Special operations such as folding, pleating, punching, and perforating modify mechanical performance also (see Chapter 10).

3. Regulating Mechanical Properties by Manufacture

As discussed earlier, the mechanical characteristics of a carrier material depend on its chemistry and processing technology. During processing orientation, tensioning, or relaxation of the material may occur. Such phenomena are partially inherent characteristics of the web buildup and manipulating technology being affected by the rate differences among the moving parts of the equipment, or they are brought about in a separate technological operation. Tensions are due to thermal effects also. Differential cooling and crystallization may cause tensions. The regulation of mechanical characteristics is discussed in Chapter 8.

D. The Influence of the Mechanical Properties on Other Performance Characteristics of PSPs

Chapter 7 discusses the influence of the mechanical properties of the PSP components on the adhesive performance. The mechanical properties also influence the conversion and end-use performance characteristics of the PSPs. During convertion, the components of the PSP or the finished product undergo physical, chemical, and mechanical transformations. Conversion is achieved by using high-speed, high-productivity machines, where the machinability of the product or product components plays a special role. The dimensionally stable, web-like behavior of PSPs during conversion is the result of tailored mechanical characteristics. Laminating and delaminating ability are also given by the mechanical properties. The transformability of the web in discontinuous PSPs (e.g., labels or special tapes) is ensured by well-defined mechanical features. The influence of the mechanical characteristics on the conversion performance characteristics is described in Chapter 10.

During their end-use, PSPs have to support increased stresses. For web-like products, unwinding, lamination under pressure, and delamination for the labeling with discontinuous items (separating the label from the reinforcing continuous liner) involve stresses. It is evident that the mechanical properties of the carrier material and the adhesive and those of the finished product play an important role in these operations (see Chapter 11).

III. OTHER PHYSICAL CHARACTERISTICS OF PSPs

Among the physical characteristics influencing the manufacture and the end-use properties of the PSPs, polarity and electrical characteristics play an important role.

A. ELECTRICAL CHARACTERISTICS OF PSPs

In the manufacture of the PSPs, coating is one of the main operations. Various components (adhesive, abhesive, primers, printing inks, etc.) are coated on solid-state carrier materials. Their anchorage depends on the polarity of the surface. The polarity of the surface can be improved by physical or chemical treatments (see Chapter 11), which are related to the improvement of the electrical characteristics. The electrical characteristics influence the processibility of the laminate components and the application of the finished product. According to Ref. [266], wetting out of plastic films depends on the temperature, vapor saturation of the atmosphere, and electric charges on the surface. Pitzler [267] studied the transfer coating of water-based and solvent-based PVRs and ACs on siliconized liner in order to determine the limits of reusability of the temporary carrier. He disclosed a discrepancy between the laboratory and industrial results. In practice, the reusability of the liner is limited. After a number of uses (five to eight, depending on the nature of the coated liquid), the release force increases. In laboratory test, there is also an increase of the release (peel off) force (for this increase, the polar part of the surface energy is responsible) but at such a low level that it does not affect the reusability (more than 20 times) of the liner. According to Pitzler, under industrial high-speed manufacturing conditions, the accumulation of electrical charges on the silicone surface can lead to sparks that penetrate through the abhesive layer and cause its destruction.

1. Antistatic Performance-Surface Resistivity

Static electricity can be defined as an excess or deficiency of electrons on a surface. It occurs when two nonconductive bodies rub or slide together or separate from one another. It is a triboelectric effect. Many factors influence the polarity and size of the charge: cleanness, pressure of contact, surface area, and speed of rubbing and separating. As is known from the use of protective films, their delamination from plastic plates produces statical electricity. Therefore, before processing, the plate should be given an antistatic treatment [268].

A second source of static electricity is an electrostatic field formed by charged bodies when they are near each other or near noncharged bodies. This field will induce a charge on a nearby nonconductive object. This voltage can be discharged when the nonconductive body carrying the charge comes in contact with another body at a sufficiently different potential. This discharge may be in the form of an arc or spark. Electrical charges on the surface facilitate deposition of suspended particles from the air. To avoid this phenomenon, the surface conductivity has to attain a well-defined value. To avoid deposition of dirt (dust) from the air on a polypropylene surface, it should possess a surface resistance of 10^{11} Ω [269]. The prevention of static charge is a requirement for packaging materials for electronic parts [270,271].

The surface resistance R depend on the specific electrical conductivity ϑ as follows [270]:

$$R = \left(\frac{1}{\vartheta}\right)\left(\frac{1}{h}\right)\left(\frac{a}{b}\right) \tag{3.72}$$

where h is the layer thickness, a and b are the sides of the sample quadrate. Reported as specific surface resistance for a quadratic surface where a and b have the same dimensions:

$$R = \frac{1}{\vartheta}h \tag{3.73}$$

In practice, ohms per square (quadratic surface resistance) is used. For example, a polyalkoxythiophene is having a quadratic resistance of 4×10^4. The common use of antistatic agents includes the domain of 10^6–10^{11} Ω/square. For the evaluation of antistatic agents, the maximum surface charge (U) as a function of negative ($U_{(-)}$) and positive ($U_{(+)}$) charges [268]:

$$U = \left(\frac{1}{2}\right)\left[(U_{(+)})^2 - (U_{(-)})^2\right] \tag{3.74}$$

is used together with the charge decay time τ as a function of the decay time of positive ($\tau_{(+)}$) and negative ($\tau_{(-)}$) charges,

$$U = \left\{\left(\frac{1}{2}\right)\left[(\tau_{(+)})^2 - (\tau_{(-)})^2\right]\right\}^{1/2} \tag{3.75}$$

As a function of the alkylrest and alkoxymethyl group from their constitution, the surface resistance of 1-alkyl-3-alkoxymethylimidazolium chloride-treated LDPE varies between 4×10^6 and 2.5×10^8 Ω, and the charge decay time is situated between 0 and 0.24 sec. An excellent antistatic effect is given for a surface resistivity of less than 10^9 Ω and a shelf life of zero [272,273].

To accomodate the slitting and winding of plastic carrier materials, antistatic films must have a resistance of 10^{11}–10^{13} Ω/square or less. The relationship between surface resistance and charge decay time (in seconds) allows the evaluation of the antistatic behavior. A charge decay time of 10–60 sec corresponds to a surface resistance of 10^{11}–10^{12} Ω and a medium antistatic behavior.

Antistatic agents consist of ionic or hydrophyllic materials that prevent charge accumulation and are either applied to the surface of or compounded with the polymeric raw material (see Chapter 8). To provide the required resistivities, conductive materials can be also incorporated. Reproducible antistatic protection in PE is more difficult to achieve than in PVC because of the semicrystalline nature of PE. A better solution is to incorporate conductive fillers and crosslink the product [274].

2. Electrical Conductivity

Electrically conductive plastics are used in antistatic products, self-heating plastics, shielding materials (radio-frequency and electromagnetic insulation), and electromagnetic radiation-absorbing materials. Electroconductive adhesives are also well known [275]. Anisotropically, electroconductive adhesives are made by dispersing electroconductive particles in an electrically insulating matrix.

The most important parameter characterizing an ionic conducting polymer is the temperature dependence of its ionic conductivity. Plastics can be formulated to fit different levels of conductivity. For the electronic industry, electrostatic discharge protection and electromagnetic radio-frequency interference (EMI/RFI) protection are very important.

Modification of thermoplastics by the addition of a conductive material to the resin matrix results in a conductive product that can be used for protection against electrostatic discharge (see Chapter 5 and Chapter 8). The surface resistance of the adhesion layer for a cover tape used in the electronic industry is 10^{13} Ohm/square or less [276]. Common PE insulating tape possesses an electrical resistivity higher than 10^{14} Ω cm; a PVC tape has a surface resistivity of more than 10^{12} Ohm cm [202].

Special applications require adhesives and carrier materials with excellent electrical resistivity. APP-based HMPSAs have been known for many years [277]. A patent [278] discloses a PSA formulation based on APP, SBCs, and tackifier. The formulation discussed by Wakabayashi and Sugii [278] is based on polypropylene, SBC, tackifier, and wax and displays special electrical properties as a very low dissipation factor (less than 0.01 at 1 kHz) and high volume resistivity of 1×10^{14} Ω cm. Silicone polymers have excellent electrical properties such as arc resistance, high electrical strength, low loss factor, and resistance to current leakage. These characteristics are very important for electrical insulating tapes [279].

Some applications require electrically conductive films. Such films possess a dielectrical constant of 200,000 and electrical resistance of 10^3-10^8 Ω cm [280]. Conductive fillers as carbon black, carbon fibers, metal powders, flakes, fibers, and metallized glass spheres and fibers are used to achieve such conductivity. Common plastics contain carbon black to improve their electrical conductivity. Such thermoplasts work as electromagnetical shield up to 30 dB [281]. The structure of carbon black and the polymer morphology influence the conductivity of carbon black filled plastics. The electrical conductivity of such materials has been explained as a system based on elastomer and carbon black connected in parallel [282]. The electrical characteristics of such systems depend on their degree of mechanical stretching [283]. Different techniques of compounding conductive powders with polymers (mixing the polymer in the plastic or molten state, polymerization or polycondenzation of monomer–oligomer mixtures, mixing of powdered polymer with conductive powder, suspension mixing, aqueous mixing, etc.) have been developed [284]. Much work has been done in connection with the conductivity of these filled systems. An excellent review of the possible conduction mechanisms (classical, tunnel, etc.) is given by Miyasaka [285].

Very important roles are played by the polarizability of polymer surfaces and by the methods used for polarizability. The majority of SAFs need polarization to achieve instantaneous peel. Polarization methods are discussed in Chapter 8.

B. Optical Characteristics of PSPs

Some PSPs are information carrier. For these products, optical properties are are basic performance characteristics. These characteristics are discussed in Chapters 8 through 11.

REFERENCES

1. I. Benedek and L.J. Heymans, *Pressure-Sensitive Adhesives Technology*, Marcel Dekker, New York, 1997, chap 2.
2. I. Benedek, *Pressure-Sensitive Adhesives and Applications*, Marcel Dekker, New York, 2004, chap. 2.
3. G. Fauner, *Klebgerechte Gestaltung und Festigkeitsverhalten von Kunststoffklebungen*, Swiss Bonding, 88, May 4, 1988, Rapperswil, Switzerland.
4. M. Gerace, *Adhes. Age*, (8), 84, 1983.
5. M. Gerspacher, C.P.O. Farrel, and H.H. Yang, *Elastomerics*, (11), 23, 1990.
6. *Kaut. Gummi, Kunststoffe*, (6), 556, 1988.
7. H. Gramberg, *Adhäsion*, (3), 97, 1966.
8. P. Dunckley, *Adhäsion*, (11), 19, 1989.
9. *Adhes. Age*, (4), 118, 1974.
10. S.N. Gan, D.R. Burfield, and K. Soga, *Macromolecules*, 18, 2684, 1985.
11. Dilexo AK 691, Produktinformation, Condea GmbH, Moers, Germany.
12. Acronal 80D, Technische Information, BASF, Ludwigshafen, Germany.
13. A. Midgley, *Adhes. Age*, (9), 18, 1986.
14. *Kaut. Gummi, Kunststoffe*, 41(1), 34, 1998.
15. Dilexo AK 694, Produktinformation, Condea GmbH, Moers, Germany.
16. Dilexo AK 680, Produktinformation, Condea GmbH, Moers, Germany.
17. *Adhes. Age*, (6), 21, 1985.
18. E.E. Ewins, Jr. and J.R. Erickson, *Tappi J.*, (6), 155, 1988.
19. Acronal V303, Technische Information, 08.1985, BASF, Ludwigshafen, Germany.
20. Mowilith DM 154, Merkblatt, 03.1985, Hoechst AG, Frankfurt, Germany.
21. Acronal LA449S, Technische Information, 02.1994, BASF, Ludwigshafen, Germany.
22. Acronal 330D, Technische Information, 03.1984, BASF, Ludwigshafen, Germany.

23. Cast film new grades for low temperature storage, Technical Information, Himont Italia, Centro Ricerche "G. Natta," 1994.
24. *Adhäsion*, 10(3), 20, 1968.
25. *Coating*, (1), 12, 1984.
26. H.G. Drössler, *Kaut. Gummi, Kunststoffe*, (1), 28, 1987.
27. W.J. Busby, Polyethylene '93, Session VI, 3-3, Maack Business Services, Zürich, Switzerland, p. 3.
28. L.C. Struik, *Physical Ageing in Amorphous Polymers and Other Materials*, Elsevier, Amsterdam, 1978, pp. 19–23, 75–78, 42–47.
29. J.C. Bauwens, *Plastics and Rubber Processing and Applications*, 7, 143, 1987.
30. J.C. Pommier, J. Poustis, and J.J. Azens, *Allg. Papier Rundschau*, 31, 848, 1988.
31. R. Köhler, *Adhäsion*, (2), 41, 1972.
32. *Coating*, (12), 432, 1994.
33. B.B. Blackford, British Patent 8,864,365, *Coating*, (6), 185, 1969.
34. U.S. Patent, 4,223,067, in D.K. Fisher and B.J. Briddell, EP 0,426,198 A2, Adco Product Inc., Michigan Center, MI, USA, 1991.
35. E. Gobran, U.S. Patent, 4,260,659, in J.N. Kellen and Ch.W. Taylor, EP 0246352 A2, Minnesota Mining and Manuf. Co., St. Paul, MN, USA, 1987.
36. *Adhäsion*, (3), 83, 1974.
37. C. Kirchner, *Adhäsion*, (10), 398, 1969.
38. *Coating*, (3), 65, 1974.
39. A.K. Bhowmick, P.P. De, and A.K. Bhattacharyya, *Polym. Eng. Sci.*, 27(15), 1195, 1987.
40. A. Carre and A.D. Roberts, *J. Chim., Phys., Physico-Chim. Biol.*, 84(2), 252, 1987.
41. E. Prinz, *Coating*, (10), 271, 1979.
42. A. Carre and J. Schultz, *J. Adhes.*, 18, 207, 1985.
43. E. Prinz, *Coating*, (10), 271, 1978.
44. I. Benedek, E. Frank, and G. Nicolaus, DE 4433626 A1, Poli-Film Verwaltungs GmbH, Wipperfürth, Germany, 1994.
45. H.W. Kammer, *Acta Polymerica*, 34, 112, 1983.
46. A. Lamping and W. Tellenbach, Swiss Patent, 664,971, Corti A.G., 1988, *CAS*, 22(4), 109: 130446e, 1988.
47. A.N. Gent and P. Vondracek, *J. Appl. Polymer Sci.*, 27, 4357, 1982.
48. A.N. Gent, *Peel Strength of Model Pressure-Sensitive Adhesives*, in Proceedings of PTSC XVII Technical Seminar, May 4, 1986, Woodfield Shaumburg, IL, USA.
49. C.M. Roland and G.G. Boehm, *Macromolecules*, 18, 1310, 1985.
50. L. Placzek, *Coating*, (3), 94, 1987.
51. M.E.R. Shanahan, P. Schreck, and J. Schultz, *C.R. Acad Sci., Ser, 2*, 306(19), 1325, 1988.
52. P.P. Hoenisch and F.T. Sanderson, *Aqueous Acrylic Adhesives for Industrial Laminating*, in Proceedings of Adhesives '85, Conference Papers, Sept. 10–12, 1985, Atlanta, GE, USA.
53. *Papier und Kunststoff Verarbeiter*, (2), 18, 1996.
54. U. Schwab, *Coating*, (5), 171, 1996.
55. A. Barbero and A. Amico, *A New Performance ULDPE/VLDPE from High Pressure Technology — Potential Applications*, in Proceedings of Polyethylene '93, Oct. 4, 1993, Maack Business Services, Zürich, Switzerland.
56. Yu.K. Godovski, *Progr. Colloid Polym. Sci.*, 75, 70, 1987.
57. E. Guth, *J. Appl. Phys.*, 16, 20, 1945.
58. D. Djordjevic, *Tailoring Films by the Coextrusion Casting and Coating Process*, in Proceedings of Speciality Plastics Conference '87, Polyethylene and Copolymer Resin and Packaging Markets, Dec. 1, 1987, Maack Business Services, Zürich, Switzerland.
59. H.C. Lau and W.R. Schowater, *J. Rheol.*, 30, 193, 1986.
60. J. Metall, *Kaut. Gummi, Kunststoffe*, (3), 228, 1987.
61. J. Pietschman, *Adhäsion*, (5), 21, 1982.
62. *HMPSA for Tape Application*, Adhesives Update, Exxon Chemical, Winter 94/95, Brussels, Belgium.
63. L. Salmén, *Tappi J.*, (12), 190, 1988.
64. C.C. Sun and J.E. Mark, *J. Polym. Sci. Polym. Phys.*, 25(10), 2073, 1987.

65. C. Han and J. Kim, *J. Polym. Sci., Polym. Phys.*, 25(8), 1741, 1987.
66. *Kaut. Gummi, Kunststoffe*, 40(10), 981, 1987.
67. W.W. Bode, *Tappi J.*, (6), 133, 1988.
68. S. Wu, *J. Polym. Sci., Part B, Polym., Phys.*, 25(12), 2511, 1987.
69. D.B. Alward, D.J. Kinning, E.L. Thomas, and L.J. Fetters, *Macromolecules*, 19, 215, 1986.
70. S.M. Collo and G. Garrabe, *Plast. Mod. Elastomers*, 38(8), 134, 1986.
71. *Coating*, (7), 187, 1984.
72. D.R. Burfield, *Polym. Commun.*, 29(1), 19, 1988.
73. W. Druschke, AFERA, Conference Paper, Oct., 1986, Edinburgh, UK, p. 13.
74. M.M. Feldstein, N.A. Platé, A.E. Chalykh, and G.W. Cleary, *General Approach to the Molecular Design of Hydrophylic Pressure-Sensitive Adhesives*, in Proceedings of 25th Annual Meeting of Adh. Soc., and the Second World Congress on Adhesion and Related Phenomena, February 10–14, 2002, Orlando, FL, USA, p. 292.
75. M.M. Feldstein, *Polymer*, 43, 7719, 2001.
76. H.G. Bubam and H. Ullrich, *Allg. Papier Rundschau*, (5), 108, 1986.
77. R. Hinterwaldner, *Adhäsion*, (10), 26, 1985.
78. U. Eisele, *Kaut. Gummi, Kunststoffe*, 40(6), 539, 1987.
79. Man Chium Chang, Chung Ling Mao, and R.R. Vargas, Canadian Patent, 1,225,792, 1987.
80. A. Zawilinski, *Adhes. Age*, (9), 29, 1984.
81. P.R. Mudge, EP 022,541 A2, National Starch and Chem. Co., Bridgewater, NJ, USA, 1987.
82. U.S. Patent, 1,645,063, National Starch and Chem. Co., New York, NY, USA, *Coating*, (7), 184, 1974.
83. R.L. Sun and J.F. Kennedy, U.S. Patent, 4,762,888, Johnson & Johnson Products, Inc., USA, *CAS, Hot Melt Adhesives*, 26, 1, 1988.
84. D. Hübsch, *Kaut. Gummi, Kunststoffe*, (3), 239, 1988.
85. B.W. Foster, *A New Generation of Polyolefin Based Hot Melt Adhesives*, in Proceedings of Tappi Hot Melt Symposium 1972 June 7–10, 1987, Monterey, CA, USA.
86. U.S. Patent, 3,691,140, in H. Miyasaka, Y. Kitazaki, T. Matsuda, and J. Kobayashi, Offenlegungsschrift, DE 3,544,868 A1, Nichiban Co., Ltd., Tokio, Japan, 1985.
87. M. Akay, S.N. Rollins, and E. Riordan, *Polymer*, 29(1), 37, 1988.
88. V.L. Hughes and R.W. Looney, EP 0,131,460, Exxon Research and Engineering Co., Florham Park, NJ, USA, 1985.
89. T.R. Mecker, *Low Molecular Weight Isoprene Based Polymers-Modifiers for Hot Melts*, in Proceedings of TAPPI Hot Melt Symposium, 1984, *Coating*, (11), 310, 1984.
90. *Kaut. Gummi, Kunststoffe*, 37(4), 285, 1984.
91. J.B. Class and S.G. Chu, *Org. Coat. Appl. Polym. Sci. Proceed.*, (48), 126, 1983.
92. M. Dupont and N. Keyzer, *Evaluation of Adhesive Properties of Hot-Melt-Based on Thermoplastic Elastomers with Different Polystyrene Contents*, in Proceedings of PTSC XVII Technical Seminar, May 4, 1986, Woodfield Shaumburg, IL, USA.
93. K.F. Foley and S.G. Chu, *Adhes. Age*, (9), 24, 1986.
94. H. Lakrout, C. Creton, D. Ahn, and K. Shull, *Adhesion of Monodisperse Acrylic Polymer Melts to Solid Surfaces*, in Proceedings of the 23th Annual Meeting of the Adhesion Society, Myrtle Beach, SC, Feb. 20–23, 2000, p. 46.
95. U.S. Patent, 3,321,451, in S.E. Krampe and Ch.L. Moore, EP 0,202,831 A2, Minnesota Mining and Manuf. Co., St. Paul, MN, USA, 1986.
96. T. Huang Haddock, EP 0,130,080 B1, Johnson & Johnson Products Inc., New Brunswick, NJ, USA, 1985.
97. U.S. Patent, 4,223,067, in R.R. Charbonneau and G.L. Groff, EP 0,106,559 B1, Minnesota Mining and Manuf. Co., St. Paul, MN, USA, 1984.
98. S.R. Aubuchon, R. Ulbrich, and S.B. Wadud, *Characterization of Pressure-Sensitive Adhesives and Thermosets by Controlled Stress Rheology and Thermal Analysis*, in Proceedings of 24th Annual Meeting of Adh. Soc., Feb. 25–28, 2001, Williamsburg, VA, USA, p. 427.
99. E.J. Chang, *Adhesion*, 34, 189, 1991.
100. H. Yang, *J. Appl. Polym. Sci.*, 55, 645, 1995.
101. G. Auchter, J. Barwich, G. Rehmer, and H. Jäger, *Adhes. Age*, (7), 20, 1994.

102. Y. Sasaaki, D.L. Holguin, and R. Van Ham, EP 0,252,717 A2, Avery Int. Co., Pasadena, CA, USA, 1988.

103. C. Donker, R. Luth, and K.van Rijn, *Hercules MBG 208 Hydrocarbon Resin: A New Resin for Hot-Melt Pressure-Sensitive Tapes*, in Proceedings of 19th Munich Adhesive and Finishing Seminar, 1994, Munich, Germany, p. 64.

104. P.A. Mancinelli, *New Developments in Acrylic Hot-Melt Pressure-Sensitive Adhesive Technology*, in Proceedings of Tech 12, Advances in Pressure Sensitive Tape Technology, Technical Seminar Proceedings, May, 1989, Itasca, IL, USA, p. 161.

105. C. Parodi, S. Giordano, A. Riva, and L. Vitalini, *Styrene–Butadiene Block Copolymers in Hot Melt Adhesives for Sanitary Application*, in Proceedings of 19th Munich Adhesive and Finishing Seminar, 1994, Munich, Germany, p. 119.

106. L. Jacob, *New Development of Tackifiers for SBS Copolymers*, in Proceedings of 19th Munich Adhesive and Converting Seminar, 1994, Munich, Germany, p. 107.

107. E. Debier and N. de Keyzer, *The Use of Rheometry to Evaluate the Compatibility of Thermoplastic Rubber Based Adhesives*, in Proceedings of 16th Munich Adhesive and Finishing Seminar, Oct. 28, 1991, Munich, Germany.

108. G. Bonneau and M. Baumassy, *New Tackifying Dispersions for Water-Based PSA for Labels*, in Proceedings of 19th Munich Adhesive and Finishing Seminar, Oct. 28, 1994, Munich, Germany, p. 82.

109. F. Buehler and W. Gronski, *Makromol. Chem.*, 188(12), 2995, 1987.

110. L.E. Nielsen, Mechanical properties of polymers and composites, in E. Debier and N. de Keyzer, Eds., *The Use of Rheometry to Evaluate the Compatibility of Thermoplastic Rubber-Based Adhesives*, Proceedings of 16th Munich Adhesive and Finishing Seminar, Oct. 28, 1991, Munich, Germany, Vols. 1 and 2, Marcel Dekker, New York, 1974.

111. J.D. Ferry, Viscoelastic properties of polymers, in E. Debier and N. de Keyzer, Eds., *The Use of Rheometry to Evaluate the Compatibility of Thermoplastic Rubber-Based Adhesives*, Proceedings of 16th Munich Adhesive and Finishing Seminar, Oct. 28, 1991, Munich, Germany, 3rd Edition, John Wiley, New York, 1980.

112. J. Kim, C.D. Han, and S.G. Chu, *J. Polymer Sci.*, B26, 677, 1988.

113. D.W. van Krevelen and P.J. Hoftijzer, Properties of polymers, correlation with chemical structure, in E. Debier and N. de Keyzer, Eds., *The Use of Rheometry to Evaluate the Compatibility of Thermoplastic Rubber-Based Adhesives*, Proceedings of 16th Munich Adhesive and Finishing Seminar, Oct. 28, 1991, Munich, Germany, Elsevier Publishing Co., Amsterdam, 1972.

114. H.G. Kilian, *Kaut. Gummi, Kunststoffe*, 39(8), 689, 1986.

115. E.T. Bishop and S. Davison, *J. Polymer Sci.*, C26, 59, 1969.

116. T.B. Lewis and L.E. Nielsen, *J. Appl. Polymer. Sci.*, 14, 1449, 1970.

117. F. Buckmann and F. Bakker, *Eur. Coat. J.*, (12), 922, 1995.

118. D. Quemada, *Lecture Notes in Physics Stability of Thermodynamic Systems*, Springer, Berlin, 1982, p. 210, in F. Buckmann and F. Bakker, *Eur. Coat. J.*, (12), 922, 1995.

119. I.M. Krieger and T. Dougherty, *J. Trans. Soc. Rheol.*, (3), 137, 1959.

120. M. Toyama and T. Ito, *Pressure Sensitive Adhesives, Polymer Plastics Technology & Engineering*, Vol. 2, Marcel Dekker, New York, 1974, p. 161.

121. M.A. Krecenski, J.F. Johnson, and S.C. Temin, *JMS-Rev. Macromol. Chem. Phys.*, C26(1), 143, 1986.

122. L. Woo, *Study on Adhesive Performance by Dynamic Mechanical Techniques*, in Proceedings of National Meeting of the American Chemical Society, Seattle, Washington, USA, March, 1983, in M.A. Krecenski, J.F. Johnson, and S.C. Temin, *JMS-Rev. Macromol. Chem. Phys.*, C26(1), 143, 1986.

123. D.H. Kaelble and R.S. Reylek, *J. Adhes.*, (1), 124, 1969.

124. D.H. Kaelble, *Physical Chemistry of Adhesion*, Wiley-Interscience, New York, 1961, in M.A. Krecenski, J.F. Johnson, and S.C. Temin, *JMS-Rev. Macromol. Chem. Phys.*, C26(1), 143, 1986.

125. I. Benedek, *Pressure-Sensitive Adhesives and Applications*, Marcel Dekker, New York, 2004, chap. 6.

126. I. Benedek, *Pressure-Sensitive Formulation*, VSP, Utrecht, 2000, chap. 2 and 3.

127. K. Hahn, G. Ley, and R. Oberthur, *Colloid Polym Sci.*, 266(7), 631, 1988.

128. L.E. Rantz, *Adhes. Age*, (5), 15, 1987.

129. R. Dehnen, *Der Siebdruck*, (3), 48, 1986.

130. A. Zosel, *Colloid Polym. Sci.*, 263, 541, 1985.

131. G. Göritz, *Kaut. Gummi, Kunststststoffe*, 40(10), 966, 1987.
132. M. Bowtell, *Adhes. Age*, (7), 38, 1987.
133. J. Wiedemeyer, *VDI Berichte*, Nr. 600, 74, 1987.
134. J. Wiedemeyer, *VDI Berichte*, Nr. 600, 3, 171, 1987.
135. *Adhäsion*, (4), 118, 1984.
136. I. Benedek, *Bond Failure in Pressure-Sensitive Removable Thin Plastic Film Laminates*, in Proceedings of the 22nd Annual Meeting of the Adh. Soc., Feb. 21–24, 1999, Panama City Beach, FL, USA, p. 418.
137. Q.D. Yang and M.D. Thouless, *A Fracture Analysis of Symmetrical 90° Peel-Joints Failure with Extensive Plastic Deformation*, in Proceedings of the 23th Annual Meeting of the Adh. Soc., Feb. 20–23, 2000, Myrtle Beach, SC, p. 6.
138. K. Kendall, *Molecular Adhesion and Elastic Deformations*, in Proceedings of 24th Annual Meeting of Adh. Soc., Feb. 25, 2001, Williamsburg, VA, USA, p. 11.
139. K.S. Kim and N. Aravas, *Int. J. Solid Structure*, 24, 417, 1988.
140. K. Fukuzawa and T. Uekita, *New Methods of Evaluation for Pressure-Sensitive Adhesives*, in Proceedings of the 22nd Annual Meeting of the Adh. Soc., Feb. 21–24, 1999, Panama City Beach, FL, USA, p. 78.
141. L.F. Kawashita, D.L. Moore, and J.G. Williams, *Peel Tests: Measuring the Energy Dissipation in Plastic Bending*, in Proceedings of the 27th Annual Meeting of the Adh. Soc., Feb. 15–18, 2004, Wilmington, NC, USA, p. 281.
142. J.Y. Chung, A. Ghatak, and M.K. Chaudhury, *Critical Normal Stress on the Surface of Thin Plastic Films During Peeling*, in Proceedings of the 27th Annual Meeting of the Adh. Soc., February 15–18, 2004, Wilmington, NC, USA, p. 182.
143. N. Willenbacher, *High Resolution Dynamic Mechanical Analysis (DMA) on Thin Films*, in Proceedings of 24th Annual Meeting of Adh. Soc., Feb. 25–28, 2001, Williamsburg, VA, USA, p. 263.
144. Packard Electric, Engineering Specification, ESM-4037,1992, USA.
145. Packard Electric, Engineering Specification, ESM-2359,1992, USA.
146. Packard Electric, Engineering Specification, ESM-2147,1992, USA.
147. C. Harder, *Acrylic Hot-Melts Recent Chemical and Technological Developments for an Ecologically Beneficial Production of Adhesive Tapes State and Prospects*, in Proceedings of European Tape and Label Conference, Apr. 28–30, 1993, Brussels, Belgium.
148. A.I. Everaerts, P.A. Stark, and K. Zieminsky, *Physical Crosslinking of Acrylic Hot-Melt Pressure-Sensitive Adhesives Using Acid–Base Interaction*, in Proceedings of the 28th Annual Meeting of the Adh. Soc., Feb. 13–16, 2005, Mobile, AL, USA, p. 44.
149. A. Carre and J. Scultz, *J. Adhes.*, 17, 135, 1984.
150. *Kaut. Gummi, Kunststoffe*, 39(10), 1004, 1986.
151. G.M. Bartenev and V.V. Tulinova, *Vysokomol. Soed.*, A29, 1055, 1987.
152. P.M. Toporowski and J. Roovers, *J. Polym Sci., Polym. Chem.*, 24, 3009, 1986.
153. M.B. Novikov, A. Roos, C. Creton, and M.M. Feldstein, *Polymer*, 44(12), 3559, 2003.
154. H.W. Yang, *Effect of Polymer Structural Parameters on PSAs*, in Proceedings of the 22nd Annual Meeting of the Adh. Soc., Feb. 21–24, 1999, Panama City Beach, FL, USA, p. 63.
155. H. Lakrout and C. Creton, *Probe Tack Tests of PSAs with Flat and Spherical Punches*, in Proceedings of the 22nd Annual Meeting of the Adh. Soc., Feb. 21, 1999, Panama City Beach, FL, USA, p. 313.
156. D.W. Bamborough, *Water-Based Adhesive Dispersions for Labels; Prediction of Adhesive Performance Using Dynamical Mechanical Analysis*, in Proceedings of 19th Munich Adhesive and Finishing Seminar, 1994, Munich, Germany, p. 96.
157. A.J. Crosby, K.R. Shull, H. Lakrout, and C. Creton, *J. Appl. Phys.*, 88(5), 2956, 2000.
158. C. Creton and H. Lakrout, *J. Polym. Sci., Polym. Phys. Ed.*, 38, 965, 2000.
159. A.N. Gent and C. Wang, *J. Mat. Sci.*, 26, 3392, 1991.
160. I. Benedek, *Pressure-Sensitive Adhesives and Applications*, Marcel Dekker, New York, 2004, chap. 3, sec. 3.
161. M.M. Feldstein, A.E. Chalykh, R.Sh. Vartapetian, S.V. Kotomin, D.F. Baraimov, T.A. Borodulina, A.A. Chalykh, and D. Geschke, *Molecular Insight into Rheological and Diffusion Determinants of Pressure-Sensitive Adhesion*, in Proceedings of the 23th Annual Meeting of the Adh. Soc., Feb. 20–23, 2000, Myrtle Beach, SC, USA, p. 54.

162. N. Tanaka, *Polymer*, 19, 770, 1978.
163. M. Giri, D. Bousfield, and W.N. Unertl, *Adhesion Hysteresis in Visco-Elastic Contacts*, in Proceedings of 24th Annual Meeting of Adh. Soc., Feb. 25–28, 2001, Williamsburg, VA, USA, p. 15.
164. C.K. Hui, J.M. Baney, and E.J. Kramer, *Langmuir*, (14), 6570, 1988.
165. A. Chiche and C. Creton, *Cavitation in a Soft Adhesive*, in Proceedings of the 27th Annual Meeting of the Adh. Soc., Feb. 15–18, 2004, Wilmington, NC, USA, p. 296.
166. F. Hensen, *Papier und Kunststoff Verarbeiter*, (11), 32, 1988.
167. *Etiketten-Labels*, (5), 91, 1995.
168. J.R. DeFife, *Labels and Labelling*, (3/4), 14, 1994.
169. *Adhes. Age*, (8), 8, 1983.
170. F.C. Larimore and R.A. Sinclair, EP 0,197,662 A1, Minnesota Mining and Manuf. Co., St. Paul, MN, USA, 1986.
171. J.F. Kuik, *Papier und Kunststoff Verarbeiter*, (10), 26, 1990.
172. *ACS Symp. Ser.*, 367 (Cross-Linked Polym.), 124, (1988), in *CAS, Crosslinking Reactions*, 19, 5, 1988.
173. *Plastverarbeiter*, 37(4), 48, 1986.
174. Lupolen B 581, Technical Booklet, d/12.92, BASF Kunststoffe, Ludwigshafen, Germany, pp. 12,13, 33.
175. D.F. Acierno, F.P. La Mantia, and G. Titomanlio, *Acta Polymerica*, (11), 697, 1986.
176. C.C. Sun and J.E. Mark, *Am. Chem. Soc. Div. Polym. Chem. Prepr.*, 27, 230, 1986.
177. K.M. Davis, R. Lonnet, and C.R. Stone, *Müanyag, Gummi*, 22, 233, 1985.
178. V.A. Novak, O.Yu. Krasnova, and R.A. Gommen, *Plast. Massy*, (3), 32, 1987.
179. W.J. McGill et al. A theory of green strength in natural rubber, *Kaut. Gummi, Kunststoffe*, (19), 962, 1987.
180. C.G. Coulon and B. Escaig, *Polymer*, 29(5), 808, 1988.
181. L.C. Yanio and F.N. Kelley, *Rubber Chem. Technol.*, (1), 78, 1987.
182. K.E. Johnsen, *Adhes. Age*, (11), 29, 1985.
183. J.B. Donnet, M.J. Wang, E. Papirer, and A. Vidal, *Kaut. Gummi, Kunststoffe*, 39(6), 510, 1986.
184. W. Xianglong et al., *Kaut. Gummi, Kunststoffe*, 39(12), 1216, 1986.
185. K. Sato, *Rubber Chem. Technol.*, 56, 942, 1987.
186. E.A. Meinecke and M.I. Taftaf, *Rubber Chem. Technol.*, 61, 534, 1987.
187. A. Einstein, *Ann. Phys.*, (Leipzig), 19, 289, 1906, in E.A. Meinecke and M.I. Taftaf, *Rubber Chem. Technol.*, 61, 534, 1987.
188. G. Slayter, *Two Phase Materials, Sci. Am.*, 206(1), 124, 1962, in D.W. van Krevelen, *Kaut. Gummi, Kunststoffe*, 37(4), 295, 1984.
189. J.C. Halpin, *J. Composite Mater.*, 3, 732, 1969, in D.W. van Krevelen, *Kaut. Gummi, Kunststoffe*, 37(4), 295, 1984.
190. A.A. Griffith, *Phil. Trans. Roy. Soc. (London)*, A221, 163, 1920, in D.W. van Krevelen, *Kaut. Gummi, Kunststoffe*, 37(4), 295, 1984.
191. H. Klein, *Coating*, (12), 430, 1986.
192. A.J. Owen, *Colloid Polymer Sci.*, 263, 991, 1985.
193. I. Skofic and D. Susteric, Polym. Jugoslav Casop. Plast Gumu., *Kaut. Gummi, Kunststoffe*, 41(8), 833, 1988.
194. J.J. Elmendorp, D.J. Anderson, and H. De Koning, *Dynamic Wetting Effects in Pressure-Sensitive Adhesives*, in Proceedings of 24th Annual Meeting of Adh. Soc., Feb. 25–28, 2001, Williamsburg, VA, USA, p. 267.
195. L.J. Mathias and M. Allison, *ACS Symp. Ser.*, 367, 66, 1988, *CAS, Crosslinking Reactions*, 21, 3, 1988.
196. G. Bocaccio, *Caoutch. Plast.*, 62(653), 83, 1985.
197. U. Yoshiaki and Y. Kentaro, *J. Adhes.*, 25(1), 45, 1988.
198. T. Yukinari, *Kaut. Gummi, Kunststoffe*, 39(6), 545, 1986.
199. R.A Veselovsky, V.I. Pavlov, and T.P. Muravskaya, Proceedings of the Sixth All-Union Conference on the Mechanics of Polymer and Composite Materials, Riga, *Mekhanika Kompozitnyh Materialov*, (2), 225, 1987.
200. Yu.S. Lipatov, N.P. Pasechnik, and V.F. Babich, *Dokl., Akad., Nauk, SSSR*, 239(2), 371, (1978), in R.A. Veselovsky, V.I. Pavlov, and T.P. Muravskaya, in Proceedings of the Sixth All-Union

Conference on the Mechanics of Polymer and Composite Materials, Riga, *Mekhanika Kompozitnyh Materialov*, (2), 225, 1987.

201. *Plastverarbeiter*, 37(10), 207, 1986.
202. P. Kriston, *Müanyag Fóliák*, Müszaki Könyvkiadó, Budapest, 1976, p. 45.
203. *Kaut. Gummi, Kunststoffe*, 40(10), 977, 1987.
204. L. Bayer, in Proceedings of Polyurethanes World Congress, Sept. 29–Oct. 2, 1987, p. 373.
205. G. Menges, H.J. Roskothen, and H. Adamczak, *Kunststoffe*, 62, 12, 1972.
206. Ch.S. Henkee and E.L. Thomas, Proc., Annu., Meet., Electron Microsc. Soc. Am., 45th, 524, 1987, in *CAS, Colloids*, 21, 3, 1988.
207. J. Hansmann, *Adhäsion*, (9), 251, 1976.
208. A.W. Norman, *Adhes. Age*, (4), 35, 1974.
209. G. Greiner, *Packing & Transport*, (6), 28, 1983.
210. Letraset Ltd., London, U.S. Patent, 1,545,568, *Coating*, (6), 71, 1970.
211. H. Schreiner, *Allg. Papier Rundschau*, 16,172, 1986.
212. Com-Tech Inc., U.S. Patent, 3,361,609, *Coating*, (7), 274, 1969.
213. *Kunststoff Produkte*, BASF, Ludwigshafen, Germany, 1985, p. 53.
214. *Adhäsion*, (1), 15, 1974.
215. *Adhäsion*, (5), 208, 1967.
216. P. Hammerschmidt, *Allg. Papier Rundschau*, (7), 223, 1986.
217. *Pentaform TZ* 887/53, Kalandrierte PP Folie, Technisches Datenblatt Klöckner Pentaplast, Montabaur, Germany.
218. Klebeband Träger und Abdeckfolien, Hoechst Folien, Ausgabe 07/92, Datenblatt, Wiesbaden, Germany.
219. *Coating*, (3), 65, 1974.
220. Nu Roll, Film di Poliestere, Technical Booklet, 1997, Isea Film SPA, Genova, Italy,
221. Minnesota Mining and Manuf. Co., St. Paul, MN, USA, British Patent, 1,102,296, 1987.
222. Beiersdorf A.G. Hamburg, Germany, DBP 1667940, *Coating*, (12), 363, 1973.
223. D.K. Fisher and B.J. Briddell, EP 0,426,198 A2, Adco Product Inc., Michigan Center, MI, USA, 1991.
224. *Kaut. Gummi, Kunststoffe*, 37(4), 285, 1984.
225. Acronal 79 D, Technische Information, 01.1982, BASF, Ludwigshafen, Germany.
226. Acronal 81 D, Technische Information, 10.1983, BASF, Ludwigshafen, Germany.
227. Acronal 85 D, Technische Information, 01.1982, BASF, Ludwigshafen, Germany.
228. Acronal 50 D, Technische Information, 03.1984, BASF, Ludwigshafen, Germany.
229. Acronal A120, Technische Information, 12.1988, BASF, Ludwigshafen, Germany.
230. Acronal 330 D, Technische Information, 03.1984, BASF, Ludwigshafen, Germany.
231. R. Koch and C.L. Gueris, *Allg. Papier Rundschau*, 16, 460, 1986.
232. L.D. Jurrens and O.L. Mars, *Adhes. Age*, (8), 31, 1974.
233. S.B. Driscoll, L.N. Venkateshwaran, C.J. Rosis, and L.C. Whitney, *Soc. Plast. Eng. Annu., Tech. Conf., Tech. Pap.*, 1, 450, 1985, in *Kaut. Gummi, Kunststoffe*, 39(2), 161, 1986.
234. Y.W. Lee, N.N.S. Chen, and P.I.F. Niem, *Rubber Processing and Applications*, 7(4), 222, 1987.
235. V.M. Ryabov, O.I. Chernikov, and M.F. Nosova, *Plast. Massy*, (7), 58, 1988.
236. *Coating*, (1), 35, 1988.
237. W. Walenski, *Offset Technik*, 10, 32, 1988.
238. D. Lacave, *Labels and Labelling*, (3/4), 54, 1994.
239. I.J. Davis, U.S. Patent, 4,728,572, National Starch, Bridgewater, USA, 1988.
240. E.B. Parker, Polyethylene '93, Session III, Oct. 4, 1993, Maack Business Service, Zürich, Switzerland, p. 6.
241. R. Nurse, *HDPE Applications, PE Developments*, Polyethylene '93, Session 3, Oct. 4, 1993, Maack Business Service, Zürich, Switzerland, p. 1.
242. N.W. Malek, EP 0,095,093, Beiersdorf AG, Hamburg, Germany, 1983.
243. S.L. Hager, D.A. Sullivan, B.D. Harper, and L.F. Lawler, *J. Cellular Plastics*, 22(11), 512, 1986.
244. *Druckwelt*, (12), 34, 1988.
245. Ö. Söderberg, *Paper Technology*, (8), 17, 1989.
246. *Papier und Kunststoff Verarbeiter*, (11), 28, 1986.

247. *Neue Verpackung*, (5), 64, 1991.
248. P. Hammerschmidt, *Coating*, (4), 193, 1986.
249. A. Wolf, *Kaut. Gummi, Kunststoffe*, 41(2), 173, 1988.
250. I. Benedek and L.J. Heymans, *Pressure-Sensitive Adhesives Technology*, Marcel Dekker Inc., New York, Hong Kong, Basel, 1996, chap. 7.
251. *Adhäsion*, (9), 19, 1984.
252. W. Schoene, Speciality Plastics Conference '89, Dec.4, 1989, Maack Business Service, Zürich, Switzerland, p. 47.
253. M. Heinze, *Kunststoffe*, 83(8), 630, 1993.
254. R. Kasoff, *Paper, Film and Foil Converter*, (9), 85, 1989.
255. PET Clearyl, Data Sheet, Clearyl® Products, Germany, 1997.
256. T. Matsumoto, K. Nakmae, and J. Chosokabe, *J. Adh. Soc. Japan*, 11(5), 249, 1975.
257. A. Zosel, *Adhäsion*, (3), 16, 1986.
258. D.K. Das, R.N. Datta, and D.K. Bash, *Kaut. Gummi, Kunststoffe*, (1), 60, 1988.
259. *Kaut. Gummi, Kunststoffe*, 40(1), 40, 1987.
260. Primacor Hotmelt Adhesive Resin, 5980, Hot Melt Adhesives, Aug. 1992, Technical Information, Dow Europe, Horgen, Switzerland.
261. HamLet, Polyethylene '93, Session III Oct. 4, 1993, Maack Business Service, Zürich, Switzerland, p. 56.
262. *Polyethylenes, Copolymers and Blends for Extrusion and Coextrusion Coating Markets*, Polyethylene '89, 1989, Maacks Business Service, Zürich, Switzerland, p. 107.
263. I.C. Petrea, *Structura Polimerilor*, Ed. Didactică şi Pedagogică, Bucureşti, 1971, p. 75.
264. *Adhäsion*, (7), 242, 1972.
265. N.E. Jung, *Automobiltechnische Zeitschrift*, 90(3), 99, 1988.
266. H. Kamusewitz, *Die thermodynamische Interpretation der Adhäsion unter besonderer Beachtung der Folgerungen aus der Theorie von Girifalco und Good*, Dissertation, Halle, 1988, in G. Pitzler, *Coating*, (6), 218, 1996.
267. G. Pitzler, *Coating*, (6), 218, 1996.
268. Plexiglas GS, XT, Röhm, Technische Information, Produkt-beschreibung, Ke. Nr. 212-1, November, 1993, Darmstadt, Germany.
269. W. Kahle, *Pack Report*, (6), 87, 1991.
270. K.H. Kochem, H.U. ter Meer, and H. Millauer, *Kunststoffe*, 82(7), 575, 1992.
271. F. Shikata, Japanese Patent 63,105,187, Teijin Ltd., 1988, *CAS, Adhesives*, 22, 7, 1988, 109:130717.
272. J. Pernak, A. Skrzypczak, E. Gorna, and A. Pasternak, *Kunststoffe*, 77(5), 517, 1987.
273. R.C. Rombouts, Speciality Plastics Conference, Maack Business Services, Dec. 1–4, 1987, Zürich, Switzerland, p. 401.
274. M. Narkis, *Modern Plastics International*, (6), 28, 1983.
275. T. Kawagushi, T. Nogami, and K. Nei, U.S. Patent, 4,624,801, Shin-Etsu Polymer Co., Tokyo, Japan, 1986.
276. S. Maeda and T. Miyamoto, U.S. Patent, 5,456,765, Sumitomo Bakelite Co., Ltd., Tokyo, 1994, *Adhes. Age*, (5), 15, 1995.
277. T. Wakabayashi and S. Sugii, EP 0285430 B1, Minnesota Mining and Manuf. Co., St. Paul, MN, USA, 1988.
278. U.S. Patent, 3,686,107, in T. Wakabayashi and S. Sugii, EP 0285430 B1, Minnesota Mining and Manuf. Co., St. Paul, MN, USA, 1988.
279. L.A. Sobieski and Th.J. Tangney, *Adhes. Age*, (12), 23, 1988.
280. *Adhäsion*, (5), 211, 1967.
281. D.M. Bigg, *J. Rheol.*, 28, 501, 1984.
282. K. Baricza, *Müanyag, Gummi*, 24(11), 329, 1987.
283. D.L. Parris, L.C. Burton, and M.G. Siswanto, *Rubber Chem. Technol.*, 60(4), 705, 1987.
284. T. Slupkowski, *Internat. Polymer Sci. Technol.*, 13(6), 80, 1986.
285. K. Miyasaka, *Internat. Polymer Sci. Technol.*, 13(6), 41, 1986.

4 Molecular Fundamentals of Pressure-Sensitive Adhesion

Mikhail M. Feldstein

CONTENTS

As discussed in Ref. [1], pressure-sensitive adhesives (PSAs) are viscoelastic materials with flow properties playing a key role in the bond forming; their elasticity plays a key role in the storage of energy (i.e., the debonding process). The balance of these properties governs their time-dependent repositionability and bonding strength (i.e., their removability). Classical pressure-sensitive products (PSPs) based on adhesives were designed to bond under very light pressure over a short-time; generally they display high tack. Later, practical requirements required the design of PSPs with various levels of dwell time, at a given temperature and application pressure. High tack (adhesive bond-forming property) is necessary but insufficient condition for pressure-sensitive adhesion (see Chapter 3 and Chapter 7). Pressure-sensitive adhesion is a complex and multiform phenomenon that includes the tack as a necessary component. For proper adhesion, the liquid-like fluidity of adhesive material is needed to wet a surface of substrate under applied compressive force and facilitate the formation of proper adhesive contact. However, the elasticity of the adhesive is also of particular importance, which is a characteristic feature of solid materials. The elasticity is required to provide maximum expenditure of energy in the course of adhesive bond failure. This liquid–solid duality in the properties of PSAs is difficult to realize in practice when new PSAs are under development.

Over past years, various theories have been proposed in order to explain the driving forces and mechanisms of adhesion. The majority of them are the mechanical interlocking, diffusion, adsorption, and electronic theories of adhesion, which have been reviewed in Ref. [2]. More recently it has become generally accepted that, while the adsorption theory has the widest applicability, each of the others may also be appropriate in certain circumstances. Nevertheless, neither of these theories may be regarded as allowing the most universal description of the pressure-sensitive adhesion. As described earlier (Chapter 3, Section I.D) after a first empirical period of development pressure-sensitive adhesion was correlated with the rheology of polymers and their mechanical characteristics. At the same time, it has been well recognized that for rational design of new materials with tailored performance properties the structure–property approach can be highly productive. Actually, as all the properties of materials are often the complex function of their molecular structures, the examination of the interrelationships between different properties of some typical material, coupled with direct studies of its molecular structure, makes feasible the molecular design of other materials with similar performance.

Because the adhesion is a macroscopic property, which involves numerous processes at a molecular level, eliciting of molecular structures underlying the adhesive behavior of materials represents a great challenge. Resolution of this problem is also complicated by the fact that existing PSAs, as a rule, are multicomponent systems, based on elastomers, which are widely varied in their chemical structures (see Chapter 5). For this reason, elaboration of a model PSA becomes the problem of paramount importance.

I. POLY(*N*-VINYL PYRROLIDONE) [PVP] — POLY(ETHYLENE GLYCOL) [PEG] BLENDS AS MODEL PSAs

High molecular weight (MW), glassy poly(*N*-vinyl pyrrolidone) (PVP, $M_w = 1,000,000$; $M_n = 360,000$ g/mol) is easily soluble in low molecular weight, liquid poly(ethylene glycol) (PEG, $M_w = 400$ g/mol) [3], giving single-phase homogeneous blends [4]. Miscibility of polymers results most frequently from a specific favorable interaction between macromolecules in blends [5] (see also Chapter 5). However, from the structures shown in Figure 4.1, both PVP and PEG contain only electron-donating groups in their backbones but no complementary proton-donating groups, and they are therefore expected to be immiscible. And really, at ambient temperature PVP is immiscible with high molecular weight fractions of PEG ($M_w > 600$ g/mol) [6] and this behavior implies the contribution of proton-donating hydroxyl groups at the ends of PEG short chains into the PVP–PEG compatibility.

As per the findings of FTIR spectroscopy [7], the PVP solubility in liquid short-chain PEG is due to H-bonding between terminal hydroxyl groups of PEG and carbonyl C=O groups in PVP repeat units. Schematic view of proposed structure of the PVP–PEG complex is presented in Figure 4.2. Because every PEG chain bears two reactive terminal OH-groups, the PEG acts as H-bonding crosslinker of longer PVP macromolecules.

In Figure 4.3, 180° Peel adhesion (*P*) is plotted against the composition of PVP–PEG blends and the content of water absorbed as vapor from surrounding atmosphere [8]. Although neither PVP nor PEG-overloaded blends demonstrate any pressure-sensitive adhesion, high adhesion appears in a very narrow range of PEG content (in the vicinity of 36 wt.% PEG). Absorbed water affects the adhesion in a complex manner. Dry blends possess no appreciable adhesion. For the blends containing less than 36% of PEG the water enhances adhesion, whereas the blends overloaded with PEG (45 wt.% and more) follow inverse pattern and the water sorption inhibits their adhesion. Both PEG and water are good plasticizers of glassy PVP. They decrease the glass transition temperature of PVP ($T_g = 175°C$), although the plasticizing effect of PEG is much stronger. The blends containing less than 36% of PEG have higher T_g and plasticizing effect of water promotes the adhesion. The PVP blend with 36% PEG is in viscoelastic state at ambient temperature, and the adhesion passes through a maximum at 12% level of hydration. The blends with 39% of PEG and higher are too fluid-like and the higher the water sorption, the worse the adhesion.

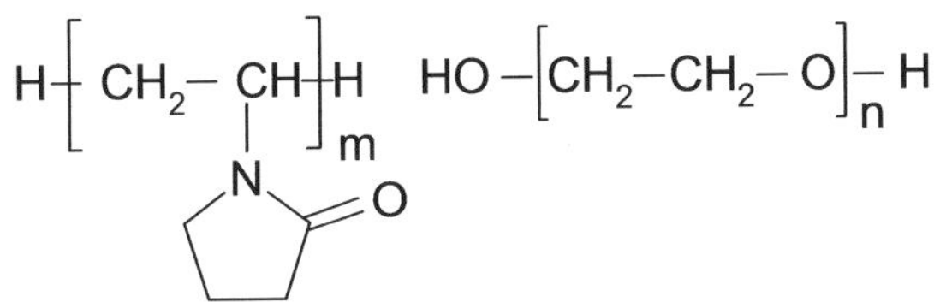

$$H \left[CH_2 - \underset{\underset{O}{\overset{|}{\underset{N}{\bigcirc}}}}{CH} \right]_m H \qquad HO - \left[CH_2 - CH_2 - O \right]_n H$$

FIGURE 4.1 Chemical structure of PVP (left) and PEG (right). $m \approx 1000$, $n = 9\text{--}10$.

The water acts here rather as a solvent that causes the swelling of PVP–PEG adhesive and dilutes the entanglement structure of the blend giving a lower modulus. With the increase of the content of both plasticizers (PEG and water), the mode of adhesive joint failure changes from adhesive to cohesive. The maximum peel strength at 36% PEG corresponds to a transition point [8].

The peel force behavior shown in Figure 4.3 makes the PVP–PEG system a very convenient model from which we are able to elicit the molecular structure responsible for pressure-sensitive adhesion. With this purpose, we merely have to compare the structures and properties of adhesive and nonadhesive PVP–PEG blends. The question is however pertinent: taking into account the rather untypical chemical composition of PVP–PEG blends that has nothing to do with conventional PSAs, which are mainly formulated with hydrophobic rubbers, is the PVP–PEG system an adequate model?

We consider that common properties found for the PVP–PEG complex and for conventional adhesives might be of particular importance for their adhesive behavior. Consequently, in order to elucidate general criteria for pressure-sensitive adhesion we need to compare the properties of those PVP–PEG blends, which provide best adhesion, with the properties of conventional PSAs, and find any similarities. At the same time, if common features in the behavior of PVP–PEG and conventional PSAs are of particular importance for our comprehension of the necessary conditions for pressure-sensitive adhesion, any distinctions might be due to the contribution of the network of H-bonds, which is only typical for PVP–PEG adhesive hydrogels.

II. PRESSURE-SENSITIVE ADHESION AS MATERIAL PROPERTY

As discussed earlier (Chapter 3, Section I.D) pressure sensitivity was correlated with the T_g and modulus of the base polymer. On the contrary, recent advances in contact mechanics allowed

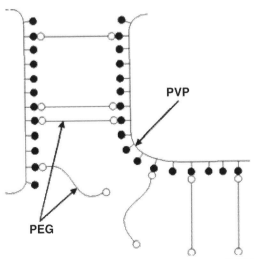

FIGURE 4.2 Simplified scheme of proposed molecular structure of PVP–PEG H-bonded complex.

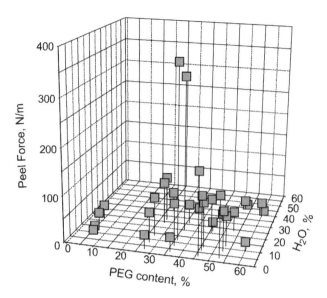

FIGURE 4.3 Peel adhesion of PVP–PEG blends as a function of PEG concentration and content of sorbed water (% of water, sorbed per 100% PVP + PEG).

the correlation of the modulus with the mechanical properties of the adhesive joint. In the following section the interrelationship between adhesion and mechanical properties of PSAs will be described.

A. INTERRELATIONSHIP BETWEEN ADHESION AND MECHANICAL PROPERTIES OF PSAs

Among numerous properties of PSAs on a macroscopic scale, the mechanical properties are of greater importance for adhesive performance (see Chapter 3, Section II). Actually, let us recall that the very name of PSAs implies the importance of their rheological behavior under a compressive force for their adhesive performance. Moreover, entire process of pressure-sensitive adhesion represents the respond of adhesive material to mechanical stress, be it shear stress in the course of adhesive bonding, or tensile stress under debonding.

Following Fukuzawa [9], as an adhesive tape being peeled away from a rigid substrate with a peeling force P acting at the angle θ to the plane of the substrate, the peel force relate to the work of deformation of adhesive layer and flexible backing film by the following equation:

$$P = \frac{b}{1 - \cos\theta}\left[W_a + l\int_0^{\varepsilon_b}\sigma\,d\varepsilon + W_b\right] \qquad (4.1)$$

where b and l are tape width and adhesive layer thickness, W_a is the work of adhesion that relates to the energy of interaction at adhesive–substrate interphase, W_b is the work of bending of adhesive tape backing, σ and ε are debonding stress and the strain of adhesive, respectively. The value of ε_b denotes the maximum adhesive deformation at the break of adhesive joint. If the tape is peeled away from a flexible substrate, the fourth term should be additionally included in the Equation (4.1), describing the work of deformation of the substrate. Similar correlations were proposed by several authors, for example, Bikermann [10], Kaelble [11], Piau et al. [12], Kinloch et al. [13], etc.

The law stated by the Equation (4.1) holds for any types of PSAs [14], and the behavior of PVP–PEG adhesive is not exceptional to this rule. Figure 4.4 shows how the debonding process

FIGURE 4.4 Front view of adhesive joint failure of the PVP blend with 36 wt.% of PEG in the course of the peel test. Intact adhesive layer of 0.25 mm in thickness is seen as a light band at the border between the backing film and the PE substrate [8].

looks in the plane of separation of the PVP–PEG (36 wt.%) adhesive and polyethylene (PE) substrate (peel test).

The fracture mechanics of adhesive debonding of PVP–PEG PSA during peel test involves dramatic stretching and fibrillation of adhesive layer (see Chapter 3, Section I.D). The length of extended fibrils is 10–20 times greater than the thickness of intact adhesive layer. The fibrils are located throughout the entire width of adhesive film at nearly equal intervals, implying that the mechanism of fibril nucleation is not random and that the adhesive material is spatially arranged into a three-dimensional network. The entire layer of the adhesive is thus subjected to elongational flow in fibrils, providing resistance to detaching stress and energy dissipation. The failure occurs in the region that is closer to substrate surface than to the backing film. This means that the locus of failure is cohesive. The viscoelastic deformation of the adhesive in extension is a major energy-consuming mechanism for all the PSAs as well as the adhesive fibrillation and cohesive mode of failure (see Chapter 3, Section I.B).

More comprehensive data on the micromechanics of PSA cavitation and fibrillation can be obtained with Probe Tack test. Indeed, as Lacrout et al. [15] have shown, the parallel geometry of flat-ended probe is better adapted to examine the details of the debonding mechanisms of soft deformable PSAs than the geometry provided by peel testing.

According to Ref. [15], the microscopic mechanisms involved in the detachment of PSA film from a flat probe can be commonly divided into four parts:

1. Homogeneous deformation before σ_{max}.
2. Cavitation (in rapid lateral growth of the cavities during the steep decrease of nominal σ, then slow growth of these cavities in the direction parallel to the tensile direction, and finally the bulk of the adhesive film or at the probe/film interface) around σ_{max}. These cavities are seen in microphotographs in Figure 4.5 as light spots.
3. Rapid lateral growth of the cavities during the steep decrease of nominal σ, then, if there is a plateau in the stress–strain curve, slow growth of these cavities in the direction parallel to the tensile direction.
4. Elongation of the walls between cavities (fibrillation).

These fibrils eventually either break cohesively or detach from the probe surface causing complete debonding. From the data shown in Figure 4.5, the behavior of PVP–PEG model PSA follows

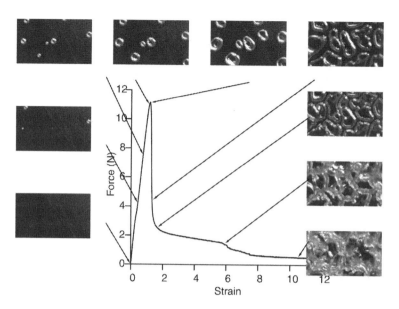

FIGURE 4.5 Direct observation of the debonding mechanisms and force vs. strain curve in the course of probe tack test for the PVP blend with 36 wt.% PEG at a debonding rate of 1 μm/s. Each image corresponds to a specific point in the stress–strain curve [16].

this general description fairly reasonably and we obtained adhesive debonding from the probe in all cases except liquid-like PEG-overloaded compositions, tested at low debonding velocities.

Let us now consider the relationship between peel adhesion and the work of viscoelastic deformation of adhesive material (Equation (4.1)). Because the major mode of the deformation of the adhesive under high angle peeling is in extension (Figure 4.4) and the contribution of shear is negligible [17], it is logical to compare the behaviors of peel force and the work of deformation of PSA under its uniaxial drawing up to fracture for the PVP–PEG model adhesives of various compositions, which provide different levels of the adhesion. This relationship is illustrated in Figure 4.6.

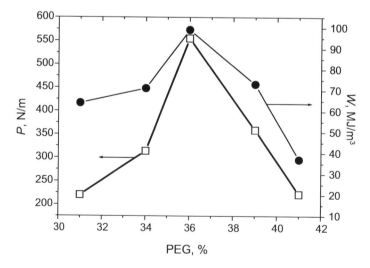

FIGURE 4.6 Effects of PVP–PEG composition on 180° peel force, *P*, and the work of viscoelastic deformation of the adhesive film up to break, *W*, under uniaxial extension. The peel and drawing rates are 20 mm/min.

Evident correlation between peel adhesion and the work of viscoelastic deformation of the adhesive, shown in Figure 4.6, signifies the controlling contribution of the process of viscoelastic deformation to adhesive performance.

By replotting the values of peel force vs. the work of deformation to break for the PVP–PEG model PSAs, we obtain an insightful and illustrative relationship presented in Figure 4.7. The analysis of this relationship allows us to gain an insight into the factors governing the PSA behavior at a most fundamental, molecular level.

B. Toward Molecular Nature of Pressure-Sensitive Adhesion

First of all, the data in Figure 4.7 establish a demarcation line between the adhesive and nonadhesive PVP–PEG blends. Both PEG-overloaded (41% PEG) and underloaded (31% PEG) blends demonstrate the same, comparatively moderate adhesive capability, but only the latter blend belongs to the class of PSAs, whereas the former one is in essence a tacky liquid. To be PSA, a tacky material should dissipate appreciable amount of energy in the course of debonding, and the value of the work of viscoelastic deformation to break the tacky film under its uniaxial extension may be taken as a measure of the dissipated energy (60 MJ/m^3 and greater, see Chapter 3, Section I.B).

Second, the linear relationship between the peel force and the work of deformation in Figure 4.7 has, most likely, a general character, spanning not only hydrophilic PVP–PEG, but and hydrophobic rubber-based PSAs. Indeed, the values of peel adhesion and deformation work for traditional PSA, based on styrene–isoprene–styrene (SIS) triblock copolymer, are aligned with those for PVP–PEG adhesives.

The linear relationship in Figure 4.7 can be described by the following equation:

$$P = kbl \int_0^{\varepsilon_b} \frac{\sigma}{2} \, d\varepsilon \qquad (4.2)$$

where k is a constant that takes into account the contributions of the first and the third terms in the Equation (4.1) and other parameters have the same meaning as in the Fukuzawa Equation. Assuming that the deformation of adhesive film in the course of both debonding and uniaxial extension

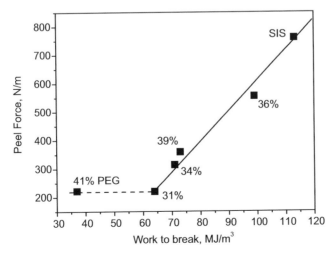

FIGURE 4.7 The contribution of the work of viscoelastic deformation of the PVP–PEG model adhesives and SIS-based PSA (DURO-TAK 34-4230, National Starch & Chem. Corp.) into their peel adhesion toward PET substrate. The contents of PEG in the blends with PVP (in wt.%) are also indicated.

follows the linear elastic law, the Equation (4.2) can be written as:

$$P = kbl\frac{\sigma_b}{2}\varepsilon_b \tag{4.3}$$

For the PVP–PEG (36%) adhesive, whose behavior deviates from ideal linear elasticity significantly [18], the approximation of the work of viscoelastic deformation of adhesive film under its uniaxial stretching up to break with the $(\sigma_b\varepsilon_b)/2$ quantity has been shown to be adequate with an accuracy of about 20% [8].

Expression of maximum elongation at the break of adhesive material, ε_b, through the ultimate tensile stress, σ_b, and tensile modulus, E, leads to the following equation:

$$P = k\frac{bl\sigma_b^2}{4E} \tag{4.4}$$

Equation (4.4) is similar to the Kaelble Equation (4.5), derived for conventional PSAs [17]:

$$P = \frac{bl\sigma_f^2}{4E} \tag{4.5}$$

where σ_f is a critical value of ultimate stress at the fracture of adhesive joint, E is the tensile modulus of adhesive material. The constant $k = 1$ if the contributions of the W_a and W_b terms in Equation (4.1) are lacking. The implication of this finding is that the Kaelble Equation (4.5) holds for any types of PSAs, including the hydrophilic PVP–PEG.

Equation (4.5) can be easily modified to express the peel resistance by debonding, P, as an explicit function of the relaxation time, τ, and the self-diffusion coefficient, D, of a PSA polymer. Indeed, let us assume in the first approximation that an adhesive can be characterized with a single apparent relaxation time, τ, and a microviscosity (or monomer–monomer friction coefficient of polymer chain), η. Taking into account that $E = 3\eta/\tau$, we obtain:

$$P = \frac{bl\tau\sigma_f^2}{12\eta} \tag{4.6}$$

In turn, the value of microviscosity can be further expressed through the self-diffusion coefficient of polymer segment (D) using the De Gennes equation [19]:

$$D = \frac{kT}{\eta aN} \tag{4.7}$$

where N is a number of monomer units of size a in a segment of polymer chain. Substitution of obtained values into the Equation (4.6) yields:

$$P = bl\frac{aN_AD\tau}{12RT}\sigma_f^2 \tag{4.8}$$

where a is now the size of polymer chain segment, N_A is Avogadro's number, R is the universal gas constant, and T is temperature.

Equation (4.8) is illustrative but holds only in linear elastic region of the deformation process and ignores the existence of the spectrum of relaxation times and thus provides a very crude model, which is applicable only for qualitative estimations. Nevertheless it predicts correctly the significance of diffusion and relaxation processes for the adhesive behavior of polymers. It is worthy of note, that Equation (4.6) has been derived on the basis of the analysis of deformation contribution to peel adhesion without resorting to so-called a diffusion theory of adhesion [2,20]. According to Equation (4.8), the pressure-sensitive adhesion requires a coupling of high molecular mobility, embedded by the self-diffusion coefficient of adhesive polymer segment, D, with long-term relaxation processes outlined by great values of the relaxation times, τ, and a high cohesive strength of adhesive polymer, expressed in the terms of critical tensile stress at the point of fracture of stretched adhesive under cohesive mode of debonding, σ_f.

In turn, a fundamental quantity that underlies a high value of the self-diffusion coefficient at molecular level is a fraction of free volume, f_v [21]:

$$D = A \exp\left(\frac{-B}{f_v}\right) \qquad (4.9)$$

where A and B are constants. In this way, specific feature of PSAs is that they should combine a high energy of cohesive interaction with a large free volume. Most commonly, strong cohesive interaction between macromolecules causes a drastic decrease in the free volume, and this explains why the pressure-sensitive adhesion is comparatively rare phenomenon. In the PVP–PEG system these apparently conflicting properties are nevertheless conciliated due to the location of reactive hydroxyl groups at the opposite ends of the PEG chains and as a result of their appreciable length and flexibility (see scheme in Figure 4.2).

The combination of high cohesion energy with large free volume, featured for PSAs, is also embedded in such fundamental characteristics of adhesive material as specific values of solubility parameters, defined as the cohesive energy density or the ratio of the energy of cohesion to the total volume, or the glass transition temperatures, T_g. The latter value relates to the energy of cohesion and free volume by the equation [22]:

$$T_g = 0.445 \frac{z\langle D_o \rangle}{R} \qquad (4.10)$$

where z is the coordination number, a value that is inversely proportional to the free volume, and $\langle D_o \rangle$ is the total interaction energy of atoms forming a polymer segment.

C. Physical Significance of the Dahlquist Criterion of Tack

While the modified Kaelble Equation (4.8) and such fundamental characteristics of materials as the glass transition temperature, the solubility parameter (cohesive energy density) or the change in heat capacity at T_g (ΔC_p) have embedded the combination of the contributions of high cohesion energy and large free volume, the question arises of how can we measure these two contributions separately? The tensile test data allow us to estimate the cohesion in terms of the ultimate tensile stress at the break, σ_b, which is a direct measure of integral cohesive strength of adhesive material. The free volume in the PVP–PEG blends has been measured with positron annihilation lifetime spectroscopy (PALS) [23]. Figure 4.8 compares the behaviors of free volume and maximum elongation at the break of adhesive film as the functions of the composition of PVP–PEG blends.

Plasticization of PVP with liquid PEG causes the increase in both fractional free volume and maximum elongation at break. Consequently the ε_b value, which characterizes a process of elongational flow, can be taken as an indirect measure of the free volume.

With the rise in PEG concentration, the cohesive strength of PVP–PEG adhesives, embedded by the σ_b quantity, grows until 36% PEG concentration is achieved and then decreases (Figure 4.9). This implies a twofold role of PEG in the blends. In PVP-reached blends, where H-bonded complex is forming, PEG acts simultaneously both as plasticizer, decreasing cohesive strength, and as cross-linker (cohesive strength enhancer). At PEG concentrations more than 36%, when PVP–PEG

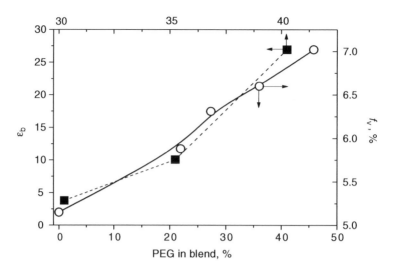

FIGURE 4.8 Fractional free volume, f_v, and maximum elongation of adhesive film at the break of uniaxially stretched material, ε_b, as the functions of PVP–PEG composition.

complex is fully formed, the former process dominates and the PEG acts as plasticizer, decreasing cohesive strength and increasing free volume.

The free volume, described by the ε_b quantity, increases with PEG concentration (Figure 4.8 and Figure 4.9). The product $\sigma_b\varepsilon_b$, which approximates the work of viscoelastic deformation to break, passes through a maximum at 36% PEG content, in full accordance with the data shown in Figure 4.6. The $\sigma_b\varepsilon_b$ product is a measure of energy dissipated during a process of deformation of PSA material, and it is therefore of no surprise that the maximum of the $\sigma_b\varepsilon_b$ product corresponds to the maximum adhesion.

The σ_b/ε_b ratio can be interpreted physically as an average modulus of the adhesive at the moment of fracture. As the data in Figure 4.9 indicate, the maximum energy that has to be expended in order to draw and break a unit volume of the PVP–PEG adhesive ($\sim$40 – 90 MJ/m^3) corresponds to the apparent tensile modulus $\sigma_b/\varepsilon_b \approx 2$–$9 \times 10^5$ Pa. Let us recall the well-known Dahlquist's criterion of tack,

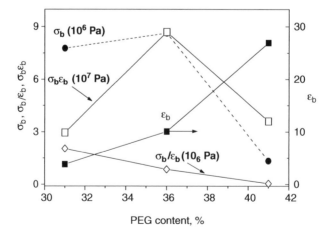

FIGURE 4.9 Relationship of ultimate tensile stress (σ_b) and maximum elongation at break (ε_b), their product and ratio to the content of PEG-400 in the blends with PVP [8].

defining the elasticity modulus of various PSAs in the order of 10^5 Pa [24] (see also Chapter 3, Section I.D; the modulus values for various materials used for PSA laminates are given in Ref. [25]). Although the chemical composition and structure of PVP–PEG blends are absolutely dissimilar compared with those for conventional PSAs, nevertheless the behavior of PVP–PEG H-bonded network complex near the fracture of adhesive bond obeys also the Dahlquist's criterion of tack, and this fact allow us to appreciate the physical meaning of this phenomenological criterion. At a most fundamental, molecular level, the Dahlquist's criterion of tack specifies the ratio between cohesive interaction energy and free volume within PSA polymers.

D. PVP–PEG BLENDS VS. HYDROPHOBIC ADHESIVES

While the Dahlquist's criterion holds for PVP–PEG adhesive blends in the region of large deformations, it still appears to be invalid under small strains [16,18]. Figure 4.10 shows a typical example of the DMA data obtained at a frequency of 1 Hz for the PVP–PEG blend containing 36 wt.% PEG.

At a first glance, the PVP–PEG blend displays elastic behavior represented by the storage modulus G', and viscous dissipation represented by the loss modulus G'', which is typical of the high molecular-weight polymers used as PSAs, with a glassy region, a transition region, and a higher-temperature plateau region (see Chapter 3, Section I.A). However, a close observation of the G' and G'' curves reveals a range of peculiarities that is only observed for PVP–PEG PSA.

Typically, at a deformation rate of 1 Hz and at ambient temperature, the value of G' for a PSA lies in the range 0.01–0.1 MPa [26–30]. The value of G' for the 36% PEG is more in the range 3–5 MPa, a value clearly incompatible with the well-known Dahlquist's criterion for tackiness [24], which specifies that an adhesive loses its tack if its elastic modulus at 1 Hz is higher than about 0.1 MPa. As a result, it is impossible to predict even qualitatively the adhesive properties for PVP–PEG blends from G' and G'' data in stark contrast with conventional PSA. Also, typical PSAs are used in a temperature range corresponding to the beginning of the high-temperature rubbery plateau or the end of the transition region [31,32], while the 36% PEG blend has been found to be tacky in a region corresponding to the center of the transition region [8,16].

The storage modulus, G', measures the elasticity of the adhesive. The loss modulus, G'', is on the contrary, associated with energy dissipation during deformation. The G''/G' ratio, defined as

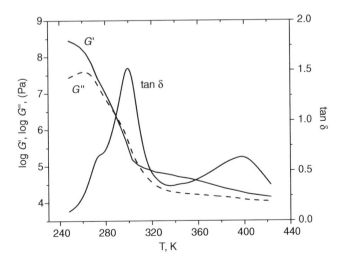

FIGURE 4.10 Master curves of the dynamic shear moduli, G' and G'', and tan δ as a function of temperature for adhesive PVP blend with 36 wt.% PEG-400, containing 11% of sorbed water. $\varpi_{\text{ref}} = 1$ Hz [18]. The amplitude of deformation was chosen to be in the linear regime of the elastic modulus G' (from 0.1 to 1%).

the loss tangent, tan δ, is the balance of viscous/elastic behavior. Tan $\delta = 1$ is a limiting value. Above this value the adhesive is generally considered as a viscoelastic fluid while below this value, the adhesive can be considered as a viscoelastic solid (see Chapter 3, Section I.B). Typical values of tan δ for conventional PSAs range from 0.1 (SIS block copolymer) to 0.7 (acrylic PSA). In contrast to this behavior of classical rubber-based PSAs, typical G' values for swollen bioadhesive hydrogels based on chemically crosslinked PVP have been reported to range from 3×10^3 to 8×10^4 Pa, while relevant G'' values vary between 1×10^2 and 1×10^4 depending on the degree of swelling.

As shown in Figure 4.10, the value of tan δ is about 1.25 for the PVP–PEG adhesive hydrogel at a temperature 293°K. This value is obviously atypical with respect to the classical PSAs. The found value indicates that, based on the tan δ data the PVP–PEG adhesive is much more dissipative than classical PSAs and behaves rather like a liquid, whereas based on the G' data it is stiffer and provides appreciable cohesion, in comparison with conventional PSA. Although coupling the liquid-like fluidity with cohesive toughness is a characteristic feature of all the PSAs, in the PVP–PEG system this property has been found to result from the molecular mechanism of PVP–PEG interaction [33,34].

Another distinguished feature of the PVP–PEG adhesive is also a multimodality of the tan δ curve (Figure 4.10), which shows a shoulder at 273°K (tan $\delta = 0.67$), a peak at 298°K (tan $\delta = 1.57$), and a maximum at 403°K (tan $\delta = 0.65$). This character results from the peculiar phase behavior of the PVP–PEG blends [34].

The PVP blend with 36% PEG exhibits the best adhesion in the vicinity of the tan δ peak (298°K). This behavior is absolutely unique among the PSAs described to date. To gain an insight into the nature of this phenomenon, we have to take into consideration the mechanism of PVP–PEG interaction.

E. Relation of Adhesion to Phase State and Molecular Structure of PVP–PEG Complex

Following the earlier presented data, a specific feature of the PVP–PEG adhesives is a very strong decoupling between the small-strain and the large-strain properties of the adhesive indicative of a pronounced deviation from rubber elasticity in the behavior of the blends. This deviation, also seen on tensile tests, is attributed to the peculiar phase behavior of the blends. The latter, in turn, affects the adhesive properties.

Figure 4.11 compares the adhesive behavior with the phase-state of PVP–PEG blends. The PVP–PEG system is plausibly among the first and most illustrative examples of miscible single-phase polymer blends, which reveal two distinct relaxation transitions. These transitions are seen on DSC scans as heat capacity jumps resembling glass transitions, and corresponding glass transition temperatures demonstrate coherent compositional behavior. The behavior of the upper T_g in PVP–PEG blends has been found to obey the rule of homogeneous PVP–PEG mixing or glassy PVP dissolution in liquid PEG due to PEG chains H-bonding to PVP through one terminal group only [34]. This uncrosslinked and labile PVP–PEG complex requires a small amount of heat for dissociation and is comparatively unstable. In contrast, the lower T_g is due to the formation of hydrogen bonded PVP–PEG network complex (gel), which behaves like a new chemical entity. In this stoichiometric complex, nearly 20% of PVP repeat units have been shown to be crosslinked by PEG terminal OH-groups (via H-bonding). The lifetime of crosslinked PVP–PEG complex is much longer due to multiple hydrogen bonds involved in its formation [34]. This two T_g behavior is typical of proteins, where lower T_g relates to glass transition and upper T_g corresponds to denaturation.

The data in Figure 4.11 allow us to understand why the composition region wherein the adhesion appears in PVP–PEG blends is so narrow and centered at the blend containing 36% PEG. Adhesion of PVP–PEG model system is due to H-bond crosslinking of PVP macromolecules

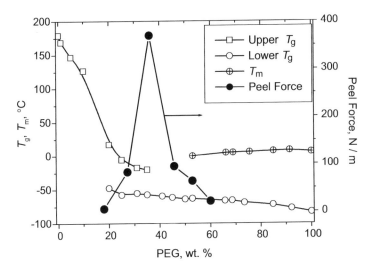

FIGURE 4.11 Peel adhesion and phase state of PVP–PEG systems. T_g is glass transition temperature and T_m is melting temperature of PEG in blends with PVP.

through PEG short-chains. However, the crosslinked complex (the lower T_g-phase) cannot be formed in PVP-reached blends where upper T_g is above room temperature. The pressure-sensitive adhesion is known to occur in polymers, which are in viscoelastic state (30–50°C above T_g). The PVP blend with 36% PEG matches this requirement and demonstrates high adhesion. However, as PEG content becomes higher 36%, the adhesion drops dramatically due to vanishing the upper T_g-phase that is much more solid-like compared with a liquid-like lower T_g-phase. Thus, the liquid–solid dualism of PSAs is embedded in the PVP–PEG model adhesive by the coexistence of the upper and lower T_g-phases. Finally, as PEG content is higher than 45%, free PEG appears in the blends, which is uninvolved into H-bonding network complex and capable of crystallizing at low temperatures (Figure 4.11). This unbound PEG acts now as a solvent, diluting the blend and causing the swelling of PVP–PEG network complex (gel). Adhesion in PEG-overloaded phase disappears due to the lack in solid-like, the upper T_g-phase.

Mechanical and adhesive properties examined with tensile strength, probe tack, and peel tests show also a pronounced decoupling between the small-strain and large-strain behaviors [8,16,18]. Such decoupling is due to the existence of two types of network in PVP–PEG blends. The first network is mainly provided by physical entanglements of longer PVP chains and contributes mainly to the material behavior under small deformations. The second network is associated with much shorter and flexible chains of PEG, and is formed by H-bonding of two terminal hydroxyl groups of the PEG to complementary carbonyl groups in monomer units of PVP. The contribution of H-bond network affects appreciably both the adhesive and mechanical properties of PVP–PEG blends. The effects of debonding and deformation rates on the mechanism of adhesive joint failure and deformation provide one of the most illustrative examples of these structure–property relationships. Probe tack tests have shown [16], for the blend containing 36% PEG, a clear transition from debonding without fibrils to debonding with extensive fibril formation; this occurs without any change in σ_{max} with the rise in probe velocity. This result implies that an adhesive could be designed to have a high adhesion energy at low debonding rates but a good release at high debonding rates. Adhesion tests performed with the probe method also show an uncharacteristically high sensitivity to the velocity of removal of the probe, with a sharp transition from detachment by fibril formation at low probe velocity to brittle fracture at high probe velocity. This transition occurs in a very narrow velocity range suggesting the existence of a very well-defined relaxation time in the polymer network. This well-defined relaxation time may be related to the breakdown of the H-bonded network structure formed by the interaction between OH terminal groups of the PEG with the carbonyl groups of the pyrrolidone repeat units ([16], see Figure 4.2).

From Figure 4.12, another particular feature of the hydrogen bonded PVP–PEG network is the existence of a well-defined time for its structural rearrangement, which in turn depends upon the life-time of H-bonds under applied mechanical stress. Under slow drawing rates, the PVP–PEG hydrogel behaves as a ductile, uncrosslinked elastomer, whereas at faster extension rates it deforms and breaks as a tight, cured rubber. The transition from the ductile to the tight stretching mode occurs in a narrow range of deformation rates (20–50 mm/min). Similarly, the type of extension changes from ductile to tight with a very small decrease in the concentration of both plasticizers in blend, PEG (between 36 and 34%) and water (between 6.5 and 3%) [18].

Figure 4.13 displays the peculiarities of the tensile stress–strain behavior of PVP–PEG H-bonded network complex in comparison with that of two typical PSA polymers: highly tacky, uncrosslinked low molecular weight PIB, and elastic SIS block copolymer, which is crosslinked physically through glassy polystyrene (PS) domains. The latter is filled with relevant tackifier to yield a blend coupling high tack with perfect elasticity and adhesion. Comparison of the data presented in Figure 4.12 and Figure 4.13 clearly shows the fact that the 10-fold increase in drawing rate affects the ductile–tight behavior of the H-bonded PVP–PEG adhesive to a much greater extent than in conventional hydrophobic adhesives. In other words, the ductile–tight transition in the SIS and PIB-based adhesives is much wider than that in PVP–PEG blends. Thus, the sharp transition between the ductile and tight deformation modes with the change of extension rate is featured only for H-bonded PVP–PEG system and untypical of SIS and PIB based adhesives. Taking into account the transient character and fast reformation of the H-bonded network, we believe that under slow extension the intermolecular H-bonds in PVP–PEG blends have time to rupture and reform anew at another place during deformation and do not contribute appreciably to the resistance to strain until the onset of a critical, strain hardening region, where the final rupture of H-bonded crosslinks between the PVP chains occurs. In contrast, at higher extension rates the H-bonds have insufficient time for rearrangement at new places and behave like pseudo crosslinks, which have to be ruptured in order to deform the polymer. The narrow transition from ductile to tight stretching with the rise of drawing rate in Figure 4.12 corresponds to a well-defined rate of the rearrangement of H-bonded network under drawing the PVP–PEG adhesive. Assuming that breakup and reformation of hydrogen bonds forming PVP–PEG

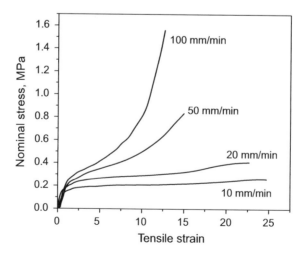

FIGURE 4.12 Stress–strain curves to break the PVP–PEG hydrogel, containing 36 wt.% of PEG-400 and 8–9% of water, at drawing rates ranging from 10 to 100 mm/min [18].

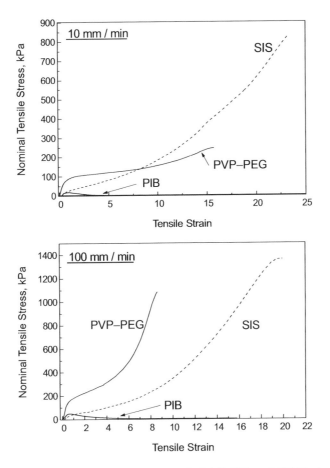

FIGURE 4.13 Stress–strain curves to break the PVP–PEG (36%), PIB, and SIS block copolymer (DURO-TAK) adhesives at extension rates of 10 and 100 mm/min [18].

network can occur below the critical deformation rate of 0.05 s^{-1} we can identify the characteristic time for this process to occur at about 20 sec [18].

III. PRESSURE-SENSITIVE ADHESION AS A PROCESS

Judging by the Dahlquist's criterion of tack, small strain behavior of PVP–PEG adhesives is untypical of conventional PSAs and the PVP–PEG system therefore cannot be regarded as belonging to the class of PSAs. At the same time, large strain and adhesive behaviors of the PVP–PEG blend matches the properties of the PSAs fairly well. This conclusion implies that our interpretation of the phenomenon of pressure-sensitive adhesion as the property of a material solely is not quite adequate for diverse PSAs. For improved insight into the nature of pressure-sensitive adhesion we also have to consider the pressure-sensitive adhesion as a process of continuing rearrangements of the structure and properties of adhesive material, which are caused by applied stress and result in appreciable deformation of adhesive film. In other words, the term "the process of pressure-sensitive adhesion" refers to the development of adhesive behavior in time, in the course of adhesive material straining under applied bonding and debonding stresses. With this purpose we have to examine the change of the properties of adhesive material over entire course of its lifetime and service, from adhesive bonding up to the failure of adhesive joint.

The lifetime of PSAs can be divided into three consequent stages. The Squeeze–Recoil test provides an adequate characterization of all three stages (Figure 4.14, [35]). At the first stage (I), adhesive film is compressed between a bottom immovable plate and upper cylindrical rod under a fixed squeezing force applied to the rod. The rod displacement (gap between the plates, $h(t)$), is measured as a function of time. As the compression proceeds, material is squeezed from a gap between the upper and the bottom plates and the deformation of tested material is registered in the terms of Δh. This stage of testing (squeezing flow) imitates a process of adhesive bond formation under fixed compressive force.

Upon the removal of squeezing force, the creep recovery occurs and the sample returns more or less to its initial shape, lifting the upper rod. This stage of strain relaxation (II) is termed as squeeze–recoil and imitates the stage of the service of adhesive film. Finally, as a detaching force is applied to the upper rod (III), debonding process proceeds and the strength of adhesive joint can be evaluated in terms of durability. The Squeeze–Recoil test allows simultaneous measuring of a range of important rheological characteristics of adhesive material.

A. RHEOLOGICAL PROPERTIES OF PVP–PEG PSA UNDER CONDITIONS MODELING ADHESIVE BONDING

To gain additional insight into the nature of tack in PVP–PEG blends, we have to consider the mechanism of PVP–PEG deformation under conditions resembling the formation of the adhesive bond. Because the PSAs form adhesive bonds under a fixed light pressure, it is reasonable to employ a simple and illustrative Squeeze–Flow test for this purpose.

A particular feature of the PVP–PEG blends is that the sample thickness never achieves zero thickness under applied compressive force as would a Newtonian fluid, but tends toward a limiting value, h_∞, which corresponds to yield stress of the fluid. In this way, the squeeze–flow analysis is a simple way to evaluate the yield stress defined as a critical value of shear stress at which the material ceases to flow under a fixed compressive force. The value of yield stress is a measure of integral cohesive strength of adhesive.

As no squeezing force is applied, both the shear stress and shear rate are zero, and the apparent shear viscosity is utterly high (Figure 4.15), that is typical of polymers possessing yield stress. However, under a fixed compressive force the shear stress and shear rates grow almost instantaneously. Because the compressive force has been kept constant, the shear stress reduces gradually, following the pattern given by a decreasing gap height (h, Figure 4.14). The shear rate reduces as well, while the apparent viscosity increases, tending toward infinity at considerably

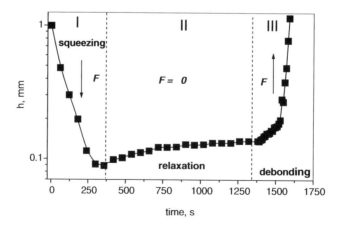

FIGURE 4.14 Typical protocol of Squeeze–Recoil testing of PVP–PEG (36%) adhesive film at 40°C. Compressive and debonding forces are 0.2 N [35].

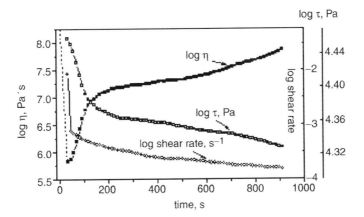

FIGURE 4.15 Dynamics of shear stress, shear rate, and apparent shear viscosity behaviors for PVP–PEG adhesive blend over the time of its squeezing under compressive force of 1 N [36].

longer times when yield stress is achieved. The significance of the data in Figure 4.15 is that the squeeze–flow of the adhesive material is only appreciable at the early stages of the compression process (within the first 3 min under a compressive stress of 35 kPa).

It is well recognized that the phenomenon of tackiness features a high shear flow of an adhesive material under a compressive force. In the process of adhesive bonding a high liquid-like fluidity of the adhesive is required to wet a surface of substrate over few seconds while compressive force is applied. However, in order to form a tough adhesive bond, the shear flow of adhesive material should be short-term. As seen from the flow curves in Figure 4.15, for the PVP–PEG adhesive the shear rate is only high during first few seconds upon the application of compressive force, and the less the compressive force, the shorter the period of flow. This means that for proper pressure-sensitive adhesion the light compressive forces are preferred.

The adhesive PVP blend with 36 wt.% PEG has been established to flow like a viscoplastic (yield stress) liquid with a power law index, n, of about 0.12. While Newtonian liquids are reported to possess the n index close to 1, the power law liquids typically have n values ranging between 0.6 and 0.3. The value $n = 0.12$ found in the present study signifies that the PVP–PEG hydrogel deviates drastically even from the behavior of the majority of power law liquids. The implication of this fact is that the elastic contributions to the squeeze–flow of PVP–PEG adhesive hydrogel can never be ignored.

The study of yield stress as a function of PVP–PEG composition, content of sorbed water, molecular weight of PVP, and temperature shows that the occurrence of a yield stress in the blends results from a noncovalent crosslinking of PVP macromolecules through short PEG chains by means of H-bonding. (Crosslinking is described in a detailed manner in Chapter 5 and Chapter 6). Indeed, only crosslinked or highly structured polymers (e.g., liquid crystalline ones) possess yield stress. In order to characterize the respective contributions of H-bonded and entanglement networks to the flow curve in Figure 4.15, the dependence of the yield stress on the molecular weight of the PVP chains has to be considered (Figure 4.16).

As follows from the results of the uniaxial extension of PVP–PEG blends under tensile stress, the average molecular weight between neighboring entanglements of PVP chains is about 220,000 g/mol for adhesive PVP blends with 36% of PEG-400 [37]. Consequently, the contribution of the PVP chains entanglement network can be essentially reduced or even eliminated by using a lower molecular weight PVP polymer ($M_w = 2000$–8000 g/mol).

The occurrence of a clearly pronounced yield stress for the blends of higher and lower molecular weight PVP confirms the important role of their network structure (Figure 4.16). The strength of the network provided by the high molecular weight PVP is approximately 60 times greater than that

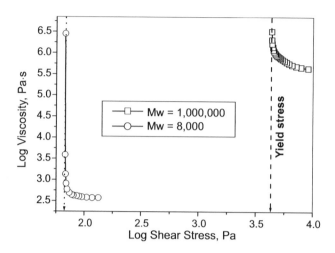

FIGURE 4.16 Effect of the molecular weight of PVP on flow curves in PVP–PEG (36%) hydrogel for PVP $M_w = 1 \times 10^6$ and $2-8 \times 10^3$ g/mol.

of lower molecular weight fraction. Accordingly, the steady-state viscosity within the Newtonian region is about 1200 times higher for the blend of high molecular weight PVP with PEG. The blends of higher molecular weight PVP dissipate more mechanical energy for the deformation of macromolecules, thus stabilizing the network formed by the PVP–PEG H-bonds. The fact that both PVP–PEG blends exhibit yield stress indicates that the yield stress is a property of the H-bonded network.

If adhesive tape is used in order to adhere any device to a vertical surface of a substrate, yield stress serves like a barrier that prevents shear flow of adhesive polymer and detaching of attached burden under stress imposed by its own weight. Flow curves in Figure 4.17, measured with parallel-plate shear plastometer, illustrate this process. Under shear stress of 1200 Pa, which is lower than yield stress of PVP–PEG adhesive, no shear flow occurs upon an initial section of the shear curve referring to elastic response of adhesive material to applied load. If the shear stress provided by attached weight is above a value of yield stress, which has been found to be around 1500 Pa,

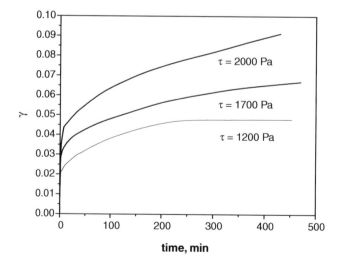

FIGURE 4.17 Shear strain of the PVP–PEG adhesive under fixed shear stress as a function of time.

adhesive material reveals unrecoverable plastic deformation and is unable to hold attached weight in its position.

B. RELAXATION CRITERIA FOR PRESSURE-SENSITIVE ADHESION

By comparison of the adhesive and relaxation behaviors of different PSAs at the stage of strain relaxation upon the removal of bonding force (region II in Figure 4.14), the relaxation criteria for pressure-sensitive adhesion can be stated [38,39]. Relaxation behavior of the range of conventional hydrophobic PSAs and hydrophilic PVP–PEG adhesive has been shown to reveal two values of retardation time: the shorter retardation time, τ_1, of 10–50 sec and the longer time, τ_2, of 300–750 sec. These times can be associated, respectively, with small- and large-scale processes. The former process is thought to characterize mainly an elastic recovery of the conformations of polymer chains. The latter process relates to the large-scale recovery of the structure of adhesive polymers and involves translational movement of polymer segments. It takes appreciable time and provides a most energy-dissipating mechanism of strain recovery due to self-diffusion.

What retardation time is of most importance for the adhesive behavior? To appreciate the significance of the shorter and longer retardation times for pressure-sensitive adhesion, Figure 4.18 compares retardation times and peel adhesion of PVP–PEG adhesive with those featured for hydrophobic PSAs.

As follows from the data in Figure 4.18, the greater adhesion is associated with the values of longer retardation time ranged from 300 to 750 sec. This finding can be considered as the first relaxation criterion for pressure-sensitive adhesion. The second criterion takes into account the moduli corresponding to the retardation times. For proper adhesion, the relaxation modulus, G_2, relating to the longer retardation time, is to be preferably higher than the modulus, G_1, corresponding to the shorter retardation times.

Because the G_2 and G_1 values are the measures of energy dissipated, respectively, for predominantly diffusive and elastic mechanisms of squeeze recoil, and the amount of the energy dissipated in the course of debonding process is the measure of adhesion, this requirement signifies prevailing importance of the diffusion mechanism for the pressure-sensitive adhesion. The third criterion reads that optimum adhesion can be achieved as the absolute values of the G_2 and G_1 moduli range between 1.0–2.5 and 0.30–1.50 MPa, respectively. It is evident that further work is needed

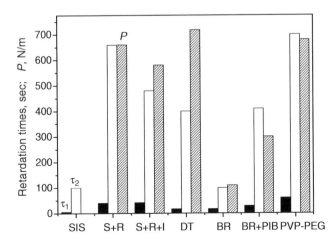

FIGURE 4.18 Retardation times, τ, and peel adhesion, P, found for hydrophilic PVP–PEG and hydrophobic adhesives based on SIS triblockcopolymer, SIS mixed with tackifier regalite (R) and plasticizer isolene (I), DURO-TAK 34-4230 PSA (DT), butyl rubber (BR) and its blend with low molecular weight polyisobutylene (PIB).

to demonstrate whether the established values of retardation times and relevant moduli are also typical of entire variety of PSAs, which are currently available.

Relaxation properties of PSAs characterize the pressure-sensitive adhesion as a process. As in the course of deformation, accompanying adhesive joint failure, the role of large-scale processes progressively increases in time, this means that the contribution of relaxation is enhanced at the later stages of debonding process under detaching force.

C. BEHAVIOR OF PSA IN THE COURSE OF THE DEBONDING PROCESS

The third and final stage of the process of pressure-sensitive adhesion, debonding under detaching force (III, Figure 4.14), is of crucial importance for adhesive performance because its mechanism governs the strength of adhesive joint. As peeling force is applied under high angle, the adhesive material is subjected to large tensile deformation causing the change in the structure of adhesive layer. This most long-term stage of the process of pressure-sensitive adhesion consists in turn of few subsequent stages and can be therefore considered as a process within the process.

Figure 4.19 illustrates schematically our idealized view of the mechanism of PVP–PEG hydrogel deformation in the process of uniaxial drawing.

Initial isotropic structure is formed due to combination of two interpenetrating networks: PVP chain entanglements and H-bond network through PEG chains. We suppose that the PVP chain entanglements contribute mainly to the small-strain behavior (region I, Figure 4.19), whereas the PVP–PEG H-bond crosslinking affects predominantly the resistance to elongational flow and hydrogel ductility within a steady–state region II. The transition between regions I and II in the stress-strain curves is associated with the orientation of long PVP chains along the direction of stretching, and with the onset of elongational flow.

The strain hardening effect and a transition from the region II to region III (Figure 4.19) is thought to be associated with maximum stretching of PVP chains under tensile stress and with onset of the fibrillation of the adhesive film (observed both in the course of probe tack [16] and peel [8] testing). Finally, within the region IV (Figure 4.19) the fibrils of PVP–PEG hydrogel elongate to a limiting degree and fail cohesively. Because the failure occurs always through the rupture of loosest bonds, and H-bonds are much looser than the covalent bonds, we consider that rather the pullout of long PVP chains from the stretched fibrils underlies the mechanism of the break of PVP–PEG film under uniaxial extension than the rupture of covalent bonds in the PVP backbones. In this way, the ultimate stress is supposed to be proportional to the total energy of ruptured H-bonds, which held together the stretched PVP chains within the fibrils (Figure 4.19).

It is obvious that different structure of polymer at the stages I, II, III, and IV of debonding process (Figure 4.19) results in different adhesive and mechanical properties. As the debonding

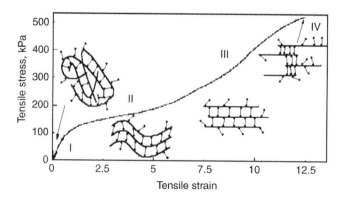

FIGURE 4.19 Schematic presentation of idealized mechanism of structural rearrangements within PVP–PEG adhesive blend in the course of uniaxial drawing to break [18].

occurs in time under fixed detaching tensile force, the time, t^*, required to rupture the adhesive bond characterizes the durability of the joint. The durability of adhesive joint is a fundamental quantity that characterizes the pressure-sensitive adhesion as the process. Two-stage kinetics of adhesive debonding is shown in Figure 4.20 for the PVP–PEG blends of different compositions. Similar two-stage debonding curves have been reported for conventional, hydrophobic PSAs [40]. The second stage of faster elongation and fracture of adhesive film follows the first, the slower stage of the debonding process. The first and slower stage is supposed to involve the orientation of polymer chains under applied detaching stress. The second, much faster stage, is the elongation flow of polymer chains in fibrils until their cohesive fracture occurs. The moment of fibril formation corresponds to the edge between the slower and faster debonding stages. All the blends in Figure 4.20 demonstrate comparatively sharp transition between two debonding stages, whereas for the composition that possesses the best adhesion (36% PEG) a wider transition period is typical. The durability of the PVP–PEG (36%) adhesive is more than three orders of magnitude higher compared with all others blends examined (Figure 4.20). In the Squeeze–Recoil test we did not observe a complete fracture of polymer fibrils, but only their maximum elongation under plates separation, limited by the maximum gap value of the instrument cell. A fairly reasonable correlation has been established between the results of the peeling and Squeeze–Recoil tests (Figure 4.21). With the rise in PEG concentration, adhesive durability passes through a maximum at 36% PEG.

As has been shown by Vinogradov et al. [42], at cohesive fracture the long-term durability of adhesive joints, t^*, relates to the debonding stress value, σ, temperature, and the molecular weight of a polymer, M_w, by equation:

$$t^* = B_1(\sigma)^{-m} \exp\frac{U}{RT}M_w^\alpha \qquad (4.11)$$

where U is an activation energy of fracture process, B_1, m, and α are constants. The B_1 and m values have been found to be highly sensitive to the mode of fracture. They decrease by several orders of magnitude as the fracture type changes from cohesive to adhesive [42].

Figure 4.22 indicates the applicability of the Vinogradov Equation (4.11) for description of fracture mechanics under debonding of the PVP–PEG adhesive. For the PVP–PEG blend with 36 wt.% of PEG-400, following parameters of the Equation (4.11) have been evaluated: $U = 70.1$ kJ/mol (at debonding stress of 33 kPa), $m = -0.43$ and $B_1 = 20.00$ (at temperature of 60°C).

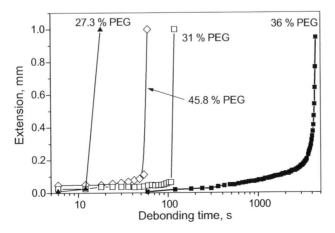

FIGURE 4.20 Effect of PVP–PEG composition on the kinetics of adhesive debonding under detaching force of 0.92 N [41].

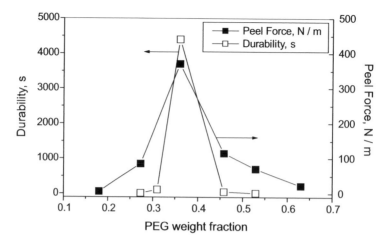

FIGURE 4.21 Correlation between peel adhesion and durability of PVP–PEG blends.

As is obvious from the data in Figure 4.22, temperature dependence of the logarithm of durability follows fairly reasonable the Arrhenius relationship, allowing evaluation of the activation energy of adhesive debonding process for PVP–PEG adhesives. The latter value is plotted vs. the composition of PVP–PEG adhesives in Figure 4.23 along with the durability of adhesive joint and the activation energy for PEG self-diffusion, determined by a Pulsed-Field Gradient NMR method as is described in Refs. [43,44].

As the data in Figure 4.23 have shown, the activation energy for adhesive bond failure follows the pattern provided by the compositional profile of the self-diffusion activation energy, climbing sharply with the decrease of PEG concentration below 36%, when maximum adhesion has been achieved. For the blend with maximum adhesion, the activation energy for adhesive debonding is 10–15 kJ/mol higher than the energy for self-diffusion. In this way, the translational mobility measured in the terms of self-diffusion takes an appreciable part of the activation energy for adhesive debonding in the blends, which exhibit a miscellaneous (adhesive–cohesive) mechanism

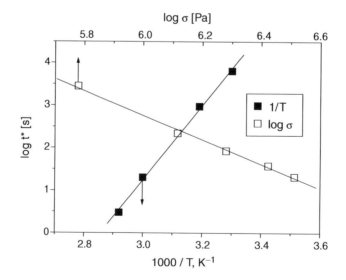

FIGURE 4.22 The Vinogradov plot of the durability of PVP–PEG adhesive bond under detaching force vs. debonding stress and reciprocal of temperature [41].

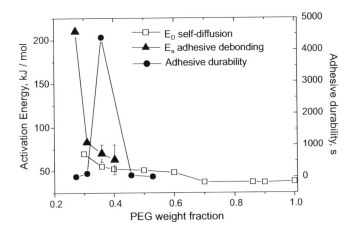

FIGURE 4.23 Relation of adhesive joint durability, activation energy for adhesive debonding and that for PEG self-diffusion to the composition of PVP–PEG blends [41].

of adhesive joint failure [8]. In the blends overloaded with PVP, wherein the upper glass transition temperature is above room temperature (see Section II.E, Figure 4.11), the activation energy of debonding process achieves a value of 210 kJ/mol. Within this composition range the lack in molecular mobility makes impossible the cohesive fracture of adhesive joint and the fracture proceeds through predominantly adhesive mechanism.

The activation energy for self-diffusion, E_D, is a function of the product of the cohesive energy density (CED) and the volume of a mole of cylindrical cavities required for diffusion of a polymer chain segment of diameter d over a jump length λ [45]:

$$E_D = \frac{\pi d^2}{4}\lambda\langle\text{CED}\rangle = (1 - f_v)E_c \qquad (4.12)$$

where E_c is the cohesive energy and f_v is the fractional free volume in polymer.

Thus, a phenomenological analysis of the relationship between pressure-sensitive adhesion and the characteristics of molecular mobility and cohesive strength of PSA polymer, outlined by the Equation (4.8), Equation (4.9), and Equation (4.12), highlights the energy of favorable intermolecular interactions (cohesion) and free volume as the major molecular determinants of the adhesive behavior of polymers.

In this chapter the attempt has been made to define the molecular fundamentals of pressure-sensitive adhesion based on the study of the structure and properties of PSAs. In Chapter 9 we will try to resolve a reverse task and use the approach described above for development of new PSAs with tailored performance properties.

REFERENCES

1. I. Benedek, *Pressure-Sensitive Adhesives and Applications*, Marcel Dekker, New York, 2004, chap. 2.
2. A.J. Kinlock, *Adhesion and Adhesives. Science and Technology*, Chapman and Hall, London, 1987, chap. 3.
3. V. Bühler, *Kollidon®: Polyvinylpyrrolidone for the Pharmaceutical Industry*, BASF, Ludwigshafen, Germany, 1996, p. 20.
4. D.F. Bairamov, A.E. Chalykh, M.M. Feldstein, R.A. Siegel, and N.A. Platé, *J. Appl. Polym. Sci.*, 85, 1128, 2002.

5. P.C. Painter and M.M. Coleman, Formulation, in *Polymer Blends*, D.R. Paul and C.B. Bucknall, Eds., Vol. 1, John Wiley & Sons, New York, 2000, chap. 4.

6. D.F. Bairamov, A.E. Chalykh, M.M. Feldstein, and R.A. Siegel, *Macromol. Chem. Phys.*, 203(18), 2674, 2002.

7. M.M. Feldstein, T.L. Lebedeva, G.A. Shandryuk, S.V. Kotomin, S.A. Kuptsov, V.E. Igonin, T.E. Grokhovskaya, and V.G. Kulichikhin, *Polym. Sci.*, 41(A), 1316, 1999.

8. A.A. Chalykh, A.E. Chalykh, M.B. Novikov, and M.M. Feldstein, *J. Adhesion*, 78, 667, 2002.

9. K. Fukuzawa, in *Advances in Pressure Sensitive Adhesive Technology*, D. Satas, Ed., Satas & Associates, Warwick, RI, USA, 1998, chap. 3.

10. J.J. Bikermann, *Trans. Soc. Rheol.*, 2, 9, 1957.

11. D.H. Kaelble, *Trans. Soc. Rheol.*, 3, 161, 1960.

12. J.M. Piau, G. Ravilly, and C. Verdier, *Experimental and Theoretical Investigations of Model Adhesives Peeling*, Proceedings of the 24th Annual Meeting Adhesion Society, February 25, 2001, Williamsburg, VA, USA, p. 287.

13. A.J. Kinloch, B.R.K. Black, H. Hadavinia, M. Paraschi, and J.G. Williams, Proceedings of the 24th Annual Meeting Adhesion Society, February 25, 2001, Williamsburg, VA, USA, p. 44.

14. I. Benedek, *Pressure-Sensitive Adhesives and Applications*, Marcel Dekker, New York, 2004, chap. 6, sec. 1.2.

15. H. Lacrout, P. Sergot, and C. Creton, *J. Adhesion*, 69, 307, 1999.

16. A. Roos, C. Creton, M.B. Novikov, and M.M. Feldstein, *J. Polym. Sci. Polym. Phys.*, 40, 2395, 2002.

17. D.H. Kaelble, in *Handbook of Pressure Sensitive Adhesive Technology*, 3rd edn, D. Satas, Ed., Satas & Associates, Warwick, RI, USA, 1999, chap. 6.

18. M.B. Novikov, A. Roos, C. Creton, and M.M. Feldstein, *Polymer*, 44(12), 3559, 2003.

19. P.G. De Gennes, *J. Chem. Phys.*, 55, 572, 1971.

20. S.S. Woyutskii, *Autohesion and Adhesion of High Polymers*, Wiley Interscience, New York, 1963.

21. H. Fujita, *Fortsch. Hochpolym. Forsch.*, Bd. 3, 1, 1969.

22. A.A. Askadskii and Y.I. Matveev, *Chemical Structure and Physical Properties of Polymers*, Chemistry, Moscow, 1983, pp. 24–48.

23. Y. Li, R. Zhang, H. Chen, J. Zhang, R. Suzuki, T. Ohdaira, M.M. Feldstein, and Y.C. Jean, *Biomacromolecules*, 4, 1856, 2003.

24. C.A. Dahlquist, in *Treatise on Adhesion and Adhesives*, R.L. Patrick, Ed., Vol. 2, Marcel Dekker, New York, 1969, p. 219.

25. I. Benedek, *Pressure-Sensitive Adhesives and Applications*, Marcel Dekker, New York, 2004, chap. 3, sec. 2.3.

26. C. Creton and L. Leibler, *J. Polym. Sci., Polym. Phys. Ed.*, 34, 545, 1996.

27. C. Donker, R. Luth, and K. van Rijn, *Hercules MBG 208 Hydrocarbon Resin: A New Resin for Hot-Melt Pressure-Sensitive Tapes*, 19th Munich Adhesive and Converting Seminar, 1994, Munich, Germany, p. 64.

28. E.J. Chang, *J. Adhesion*, 34, 189, 1991.

29. A. Zosel, *J. Adhesion*, 30, 135, 1989.

30. D.P. Bamborough and P.H. Dunckley, *Adhes. Age*, 11, 20, 1990.

31. M. F. Tse and L. Jacob, *J. Adhesion*, 56, 79, 1996.

32. E. J. Chang, *J. Adhesion*, 60, 233, 1997.

33. M.M. Feldstein, S.A. Kuptsov, G.A. Shandryuk, and N.A. Platé, *Polymer*, 42(3), 981, 2001.

34. M.M. Feldstein, A. Roos, C. Chevallier, C. Creton, and E.D. Dormidontova, *Polymer*, 44(6), 1819, 2003.

35. S.V. Kotomin, T.A. Borodulina, M.M. Feldstein, and V.G. Kulichikhin, Proceedings of the XIIIth Intern. Congress on Rheology, Cambridge, U.K., 4, 44, 2000.

36. T.A. Borodulina, M.M. Feldstein, S.V. Kotomin, V.G. Kulichikhin, and G.W. Cleary, Proceedings of the 24th Annual Meeting Adhesion Society, Williamsburg, VA, USA, 2001, p. 147.

37. C. Creton, A. Roos, M.B. Novikov, and M.M. Feldstein, Proceedings of the 26th Annual Meeting Adhesion Society, 34, 2003.

38. M.B. Novikov, T.A. Borodulina, S.V. Kotomin, V.G. Kulichikhin, and M.M. Feldstein, Proceedings of the 26th Annual Meeting Adhesion Society, 2003, pp. 402.

39. M.B. Novikov, T.A. Borodulina, S.V. Kotomin, V.G. Kulichikhin, and M.M. Feldstein, *J. Adhesion*, 81(1), 7–107, 2005.

40. A. Zosel, Proceedings of the 5th European Conference on Adhesion (EURADH'2000), Lyon, France, 2000, p. 149.
41. M.M. Feldstein, T.A. Borodulina, R.Sh. Vartapetian, S.V. Kotomin, V.G. Kulichikhin, D. Geschke, and A.E. Chalykh, Proceedings of the 24th Annual Meeting Adhesion Society, Williamsburg, VA, USA, 2001, p. 137.
42. G.V. Vinogradov, A.I. Elkin, and S.E. Sosin, *Polymer*, 19, 1458, 1978.
43. R.S. Vartapetian, E.V. Khozina, J. Kärger, D. Geschke, F. Rittig, M.M. Feldstein, and A.E. Chalykh, *Colloid Polym. Sci.*, 279(6), 532, 2001.
44. R.Sh. Vartapetian, E.V. Khozina, J. Karger, D. Geschke, F. Rittig, M.M. Feldstein, and A.E. Chalykh, *Macromol. Chem. Phys.*, 202(12), 2648, 2001.
45. V. Stannet, in *Diffusion in Polymers*, J. Crank and G.S. Park, Eds., Academic Press, New York, 1968, p. 68.

5 Chemical Basis of Pressure-Sensitive Products

István Benedek

CONTENTS

As discussed in the previous chapter, pressure-sensitivity supposes the existence of pronounced flow during bonding and elastic, solid-state-like behavior during debonding. Elasticity requires a reversibly deformable network. The existence of a network does not allow viscous flow. This dilemma can be avoided by synthesizing and formulating viscoelastic compounds that have partially segregated structures. Therefore, the chemical basis of pressure-sensitive products (PSPs) includes raw materials with composite structures. As mentioned earlier, intramolecular or intermolecular segregation is achieved with a crosslinked network, crystalline structure, or reinforcement with fillers. The nature of segregated structures determines their mobility and elasticity. A simplified approach explains the functioning of such networks comparing them with filled liquids (see Chapter 3). It should be emphasized that this approach assumes that the filler particles are more mechanically resistant than the liquid and that they do not interact with the matrix. This is not always true. A second hypothesis concerns the elastomer. Rubber is considered an ideal network for use as a base elastomer. But rubber-based formulations used for PSPs are tackified and crosslinked; their networks cannot be ideally elastic. Nevertheless, such formulations work well in practice.

"Diluting" the segregated structure with viscous components also allows plastomers to experience pronounced flow. Thus, the first requirement for pressure-sensitivity is fulfilled. The second requirement is the ability to undergo large elastic deformation. How can that be achieved? To answer this question, the chemical basis of the PSPs must be examined. Except for some special materials (e.g., metals, ceramics, glass) used as carrier, the main raw materials of PSPs are organic compounds, mostly polymers. To understand how PSP components work, let us make a short evaluation of their macromolecular basis and chemical composition. Because of the quite different role of the solid-state carrier material and the liquid adhesive (at least for the classical PSP construction), the raw materials of these components are also different. Therefore, the carrier material and the other components are discussed separately here. The chemical basis of pressure-sensitive adhesives (PSAs) is described in detail in [1–3]. These adhesives are often considered the components responsible for pressure-sensitivity, which is true for labels. For other products, the PSA-carrier assembly or (in extreme cases) the carrier itself is the component that displays pressure-sensitivity. In some cases such behavior can be compared with the "true" pressure-sensitivity of the PSA (see also Chapter 7 and Chapter 11). For certain products the adhesive properties result from the application conditions only. A detailed examination of the common macromolecular basis of all components of the pressure-sensitive assembly is necessary for its understanding, design, and manufacture. Recent advances concerning the molecular fundamentals of pressure-sensitive adhesion are discussed in a separate chapter (see Chapter 4).

I. MACROMOLECULAR BASIS AND CHEMICAL COMPOSITION OF CARRIER MATERIAL

The carrier materials used for PSPs have to display special mechanical characteristics (see Chapter 3). Therefore, they are manufactured on the basis of high molecular weight polymers. At first, paper was used as the carrier; later organic or inorganic nonpaper materials were used. Paper and the main polymeric carrier materials have chemical compositions and macromolecular structures that do not allow enhanced chain mobility at the surface, that is, surface adhesivity. Paper is a unique carrier material based on natural raw material (cellulose derivative) and has its own chemistry and technology. It is not the aim of this book to discuss them. It should be only mentioned that paper contains other inorganic or organic macromolecular compounds usually as a top coat. The surface layer of top-coated papers is based on about 90% white pigment and 10% binder [4]. The mean coating weight of the top coat is about 15 g/m^2, that is, the layer has an average thickness of 12 μm. Caseine, starch, polyvinylalcohol, acrylics (AC) or styrene–butadiene latex is used as binder. The printing and adhesion properties of the top coat depends on its chemical composition.

The development of plastics allowed the manufacture of "full plastic" carrier materials with better mechanical properties. Both paper and common plastic films are nonadhesive carrier materials. Their behavior is due to their partially ordered structure. Although ordered and cross-linked, natural rubber (NR) exhibits slight self-adhesion. Its self-adhesivity can be improved by transforming (by tackifying) the crosslinked elastic structure in a mobile construction that exhibits elastic deformation and viscous flow (due to enhanced chain entanglement and molecular motion). As discussed in Chapter 3, the buildup characteristics of the crosslinked network regulate its mobility. The "conformational gas" of ideal rubber with long elastic bridges between the crosslink points allows a high degree of deformation. Crystalline, ordered structures generally do not allow rubber-like high elastic deformations. Their flow occurs in their amorphous domain. Therefore, the deformational behavior of plastics is different from that of rubber. Rubber can act as an adhesive; a plastomer cannot (at room temperature). For classic PSP constructions the plastomer is the non-adhesive carrier, and rubber is the adhesive component.

However, reducing crystallinity and enhancing chain mobility in the amorphous region can lead to more elasticity and more plasticity for such materials also. The empirics of tackification of natural rubber using liquid resins and the plasticization of cellulose derivatives or polyvinyl chloride (PVC) demonstrated the possibility of transforming an elastomer or a plastomer into a visco-elastomer (Figure 5.1). Internal plasticizing competes with external plasticizing. The synthesis of vinyl chloride copolymers, ethylene–vinyl acetate (EVAc) copolymers, propylene copolymers, and branched polyethylenes (PEs) led to macromolecular compounds with greater chain mobility.

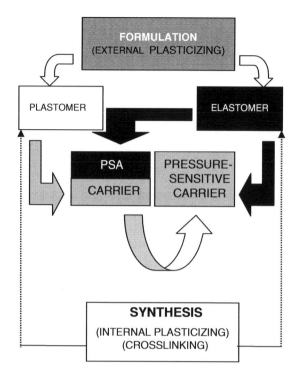

FIGURE 5.1 Chemical methods to obtain PSPs.

Such compounds exhibit viscoelastic flow and are self-adhesive. It should be emphasized that there is a big difference between PSAs and plastomers. The main characteristic of PSAs is their flow; their mechanical strength is generally not sufficient for them to be self-supporting. In some cases their cohesion is not sufficient to ensure acceptable shear resistance. The main characteristic of plastomers is their excellent mechanical resistance, their flow is weaker and allows self-adhesion only with high temperature or pressure lamination. Recent developments in hydrogels demonstrate the possibility to achieve pressure-sensitive viscoelastic structures by simultaneous crosslinking or plasticizing of a plastomer (see Chapter 4).

Various industrial methods can be used to manufacture bulky self-adhesive products (Table 5.1). Synthesis, formulation and application conditions together ensure their pressure-sensitive adhesivity.

The inverse possibility to stiffen the rubbery network has also been studied. The crosslinking of natural rubber or of viscoelastic compounds such as acrylates made it possible to increase the modulus and to develop carrier-less transfer tapes. On the molecular scale, the buildup of segregated structures in "liquid" macromolecular compounds allowed temperature-dependent molecular stiffening, the development of thermoplastic elastomers (TPEs). Thermoplastic elastomers exhibit rubber-like elasticity at room temperature but behave like thermoplastic resins above their melting point. Such compounds are free from curing agents, accelerators, and reinforcing materials, which reduces the amount of mixing required as well as the amount of materials.

A. GENERAL CONSIDERATIONS

The mechanical characteristics and viscoelastic behavior of PSPs depend on their chemical basis (chemical composition and macromolecular characteristics). As is known from the work with PSPs, the molecular weight of the adhesive components, coating weight, carrier thickness, debonding angle, and force influence the peel (debonding) resistance [5]. Molecular weight,

TABLE 5.1
The Main Manufacture and Application Parameters for Pressure-Sensitive Behavior

Product Buildup		Application	
Base Component	Additional Component/Parameter	Product Mechanism	Working Mechanism
Plastomer	Physical treatment laminating conditions	Hot laminating PE film	Enhanced surface polarity and flow
Plastomer	Viscous additives	Plasticized PVC protective film	Enhanced conformability (lower T_g)
Plastomer	Viscoelastic additives	Polyethylene film tackified with isobutylene copolymer	Surface layer of tacky, viscoelastic compound, due to incompatibility
Plastomer	High molecular weight viscoelastic additives; laminating conditions	Polyethylene film tackified with ethylene–butyl acrylate copolymer	Surface layer of tacky, viscoelastic high polymer
Plastomer	Elastical, polar, low T_g polymer as additive; physical treatment	Polyolefin blends with EVAc copolymers	Bulky viscoelastical properties of compatible blends
Plastomer	Physical treatment; laminating conditions	Very low density polyethylene	High conformability due to amorphous structure
Elastomer	Tackification	PSA	Viscoelastic compound
Viscoelastomer	—	PSA	Viscoelastic compound

molecular weight distribution (MWD), even shape and structure (short chain branching, frequency, distribution and length, long chain branching, and unsaturation) affects the viscoelastic behavior of macromolecular compounds.

1. Parameters of Molecular Construction

The main parameters that can be manipulated in the polymerization process are the choice of monomers, molecular weight, MWD, short-chain branching, long-chain branching, and unsaturation. For instance, common Ziegler–Natta (ZN) catalysts lead to PP with M_w/M_n ratios of 5–6, and special catalysts allow M_w/M_n ratios of 3–4; using the technology of controlled rheology (oxidation in extruder) M_w/M_n ratios of 2–3 can be obtained [6]. The parameters allow the buildup of multiphase, ordered structures which are characteristic of rubbery and plastomeric materials also.

a. Monomers

The most common carrier materials are based on polyolefins. These are generally nonpolar, chemically inert, partially crystalline plastomers. Among the structural characteristics due to the monomers, crystallinity is most important for mechanical properties and surface adhesivity. To improve their surface wetting properties and elasticity, copolymers of polar monomers (vinyl acetate (VAc), acrylics and maleic derivatives) have been developed. Copolymerization of ethylene with polar vinyl and acrylic derivatives gives rise to polar copolymers, whose properties differ greatly from those of low density polyethylene (LDPE) [7]. The term ethylene-copolymers is restricted to polymers containing more than 50% ethylene. Polymer analogous reactions have also been used to obtain new copolymers. One such method is grafting.

Copolymers of ethylene with vinyl acetate and butyl acrylate and acid–ethylene copolymers have been manufactured to achieve self-adhesion (i.e., low temperature sealability). The comonomer provides polarity in the chain and decreases the crystallinity of the material. Both favor the wettability of the substrate. At the same time its melting point decreases [8]. Butyl acrylate decreases crystallinity of polyethylene, increases the tack and provides good mechanical properties. Maleic anhydride increases the adhesion on polar surfaces and allows initiation of covalent chemical bonding with some polymers. Terpolymers of ethylene acrylic ester and maleic anhydride have also been synthesized. Ethylene–acrylic acid (EAA), ethylene–maleic anhydride (EMAA), ethylene–butyl acrylate (EBA), and ethylene–methyl acrylate copolymers have been produced [9]. Because of their lower crystallinity these are very low modulus polymers (the value of the flexural modulus of such polymers is situated at 8, respectively 18 MBA). Consequently, they also have a greater degree of tack. The presence of functional groups in ethylene copolymers cause them to be more chemically reactive than LDPE. EVAc copolymers have a processing temperature limitation of approximately 230°C. At higher temperatures, acetic acid can split off from the functional group. EAA copolymers can be processed at 330°C [10].

The introduction of a functional monomer into conventional elastomers is a common strategy for altering physical characteristics or designing the macromolecules to fulfill a particular function. The carboxylic acid group is reactive with many agents such as amine and epoxy compounds, which can be used for covalent crosslinks.

b. Molecular Weight and Molecular Weight Distribution

Higher molecular weight polymers allow the manufacture of stronger films at a penalty in processability. As known, the tensile strength (S_T) depends on the number average degree of polymerization (DP_n) according to the correlation [11]:

$$S_T = S_{T\infty} - C/DP_n \qquad (5.1)$$

where C is a polymer constant. Unfortunately, with high molecular weights, excessive pressures and extrudate melt fracture becomes a problem [12]. Low molecular weight polymers allow better hot seal performances. As stated in [13], a hot seal PSA formulation may include polyethylene (10–50% w/w) with a (low) molecular weight of 1000–10,000.

Blown film equipment designed optimally for processing of one kind of polyethylene is generally not suitable for processing other types of polymers. A polymer with a broader MWD will be more processible. On the contrary, plate-out and blocking occurs with molecular weight species. Linear low density polyethylene (LLDPE) sometimes has more low molecular weight species and requires more antiblock additive. Molecular weight and MWD influence melt strength and processibility [14]. To avoid blocking with blown high density polyethylene (HDPE), the best results are obtained by using high blow-up ratios, and high frost kine heights; therefore, very broad and preferably bimodal grades are needed [15]. For flat films (monofilaments and tapes), HDPE with a narrow molecular weight distribution is used [16]. Good mechanical properties during and after stretching are the main performances required for these products. For blown film manufacture medium MWD is recommended, but for higher melt strength very large MWD grades may be needed too. The blown film is applied for tapes also. Here mechanical strength and elongation are required. On the contrary, broad MWD and low MW (high melt flow index) can result in plate-out and blocking of the film and poorer mechanical properties. Bimodal MWD ensures a combination of excellent mechanical properties and good processibility. Bimodal HDPE can be blended with polypropylene (PP) also.

When more distinctly, different molecular weights are present, the material is described as bimodal. Such materials combine optimum processibility and excellent mechanical properties. The broad molecular weight HDPEs have low melt memory, so the die must be designed to allow time for any melt strain to disappear [15] (see also Chapter 8). The MWD has no effect on the stiffness, which depends only on density [14]. At low film thicknesses and low blow-up ratio, the broad MWD increases the dart drop impact of the film (see also Chapter 3, Section II). For special mechanical characteristics ultra high molecular weight PE is used [17].

The degree of polymerization and molecular weight of the plastomers or elastomers used as carrier materials and the additives used to transform them into self-adhesive carriers play an important role in the melting point depression and spherulite growth rate. The melting point depression (ΔT_m) of a crystallizable polymer blended with an amorphous polymer in a compatible mixture according to the Flory–Huggins theory [18,19] can be written as a function of the heat of fusion (ΔH^0), molar volumes of the repeat units of the two polymers (V_1 and V_2), degree of polymerization (m_1, m_2), and volume fractions (Φ_1, Φ_2) of the polymers:

$$-\lceil \Delta H^0 V_1 / R V_2 (1/T_m - 1/T_m^0) + \ln \Phi_2 / m_2 + (1/m_1 - 1/m_2)\Phi_1 \rfloor = \chi_{12}\Phi_1^2 \qquad (5.2)$$

where χ_{12} is the interaction parameter of the components of the mixture. As can be seen from the earlier correlation the molecular weight (degree of polymerization) strongly influences the melting point depression. Melting point depression possesses a special importance for hot laminating films, self-adhesive films (SAFs) based on EVAc–PE mixtures or PIB–PE mixtures, and resin mixtures (see also Chapter 8). On the contrary, the volume fraction of the polymers Φ (which depends on their structure, i.e., branching) and the diffusional processes (which depend on the thickness of the macromolecular layer, b_0) influence the spherulite growth rate (G) of the crystallizable polymer in a mixture according to the correlation [20]:

$$\lg G - \lg \Phi_2 + U^*/2.3\,R(T_c - T_\infty) - 0.2\,T_m \ln \Phi_2/2.3\,\Delta T = \lg G_0 - k_g/2.3\,T_c\Delta T_f \qquad (5.3)$$

where G_0 is a pre-exponential factor, ΔT the undercooling, $U^*/R(T_c - T_\infty)$ is a term for the contribution of diffusional processes, f a correction factor for the heat of fusion (ΔH_0) and k_g is the nucleation factor, which depends on the thickness of the layer:

$$k_g = f(b_0) \qquad (5.4)$$

Spherulite growth influences the diffusion of viscous components in self-adhesive films and the decrease of bond performance in semi pressure-sensitive-adhesives (see Chapter 11).

c. Branching

Long chain branching provides shear sensitivity, that is, high processibility [21]. Low density polyethylene is highly branched, gives much entanglement, possesses broad MWD and displays poor mechanical properties. In the early 1970s, linear low density PE was introduced, with no long branching, little entanglement, narrow MWD, and improved mechanical properties [22]. The entanglement of the short-chain branches, (e.g., linear low density polyethylene [LLDPE] with octene) provides excellent stretchability and flexibility [23].

2. Segregation

As discussed in Chapter 3, segregated multiphase structures allows the combination of viscous and elastic, rubber-like and plastomer-like characteristics. In plastomers, multiphase systems are due to crystallinity. The physical status of these polymers is characterized by their orientation, crystallinity, and crystal particle dimensions [10].

a. Crystallinity

Thermoplasts are generally crystalline materials at their application temperatures. Their excellent mechanical properties are due to crystallinity [24]. Tensile modulus and stiffness decrease as the crystallinity decreases [25]. Crystallinity gives the polymer a certain rigidity, and the molecular packaging produces opacity, due to the differences in the diffraction indices of the amorphous and crystalline parts. Polyethylene films are either blown or cast. Because of the different degrees of crystallinity of the two types of film, different tackifier loadings are used to achieve the same level of self-adhesivity (see also Chapter 8). For cast films, a level of 1–3% polyisobutylene (PIB) is required, and for blown films a level of 3–6.5% PIB [26].

As mentioned earlier, crystallinity also influences the optical properties. LLDPE forms large crystallites that results in surface irregularities and poor optical properties. This can be improved by making the crystallites smaller by using faster cooling. Crystallinity influences density and the modulus. Common HDPE has densities of $0.952–0.960$ g/cm^3. Materials of higher densitity are used when an optimum modulus (rigidity) is required. On the contrary, the impact resistance of C$_8$ LLDPEs is better. It increases with the MFI and decreases with the density. Elmendorf tear strength increases with the MFI and decreases with the density. Short chains along the molecules reduces the material's ability to crystallize, and so the more the short chains are, the lower the density.

The formation of nuclei of crystallization is a function of temperature. As the temperature decreases from the melting point, the kinetic effects become smaller and the chances for larger nuclei to form are greater. As the temperature is lowered, the average nucleus size increases and the critical size for crystal growth decreases. For most polymers there is an optimum temperature for crystallization. Polymers may exhibit one or more crystalline transitions at temperatures between T_g and the final melting temperature. The importance of such transformations is due to their possible effects on modulus and thermal, optical, and dimensional properties. Low MW chains possess higher mobility at a given temperature and therefore, are more easily crystallized. The crystal grows slowly with the formation of high MW chains with restricted mobility. Plasticizers aid crystallizability; fillers (solid-state) have the opposite effect. Low crystallinity may extend the rubbery plateau up to the melting point of the crystallites, giving a much broader application temperature (see Chapter 3, Figure 3.3). Medium crystallinity (e.g., LDPE) makes the material less flexible; high crystallinity (HDPE, polyethylene terepthalate (PET)) provides stiffness and temperature resistance. Bulky functional groups, like rings have a stiffening effect (see styrene-block copolymers (SBCs), PET) retaining the glassy plateau up to higher temperatures. Crystallinity due (partially) to incompatibility is displayed by organic antiblocking agents also. Such additives are organic substances incompatible with the polymer of carrier. During extrusion these materials tend to create spherulites on the surface of the films, acting as antiblocking agents [27].

Incompatibility between the carrier and an embedded additive related to molecular weight and structure can be used to buildup "working" constructions, where the incompatible additive is released from the carrier material. The best known examples are SAFs with embedded tackifying components. Another example is described by Brack [28]. The release agent is incorporated as an incompatible compound in the release binder. This film is cured to a flexible solid layer by irradiation. The abhesive substance migrates to its surface. In a similar manner, combining fluoropolymers with polyamic acid leads to antiadhesive coatings with improved performance, due to the spontaneous separation in two layers, the bottom layer being polyamic acid, the top layer being fluoropolymer [29].

Stretching, orienting the macromolecule (see postextrusion processing) produces crystallization and improves the mechanical performances of the film. Such behavior is exhibited by the manufacture of oriented PET, PP, and PVC [30]. The influence of the crystallization on the adhesion of a plastic film is more complex. The proportion of the crystals and their morphology influence the degree of contact and adhesion [31]. Using common industrial methods, it is not possible to manufacture plastic films without inducing order in the macromolecules. This is due to the big difference in the mobility of the macromolecules in molten and "frozen" material. Some raw materials for adhesives are known to exhibit partially plastomer-like structures that become ordered when cooled. These include styrene–diene block copolymers. When they are cooled, rigid polystyrene domains are formed. In this situation, at least theoretically, it would be possible to achieve oriented, ordered molecular structures by high shear, high speed cooling of such materials. This phenomenon was studied by de Jager and Borthwick [32]. To test whether shear induced orientation of a styrene–diene block copolymer-based hot-melt PSA (HMPSA) occurs, they measured the shear adhesion failure temperature (SAFT) and rolling ball tack of the films manufactured on samples cut parallel and perpendicular to the machine direction. They did not find shear-induced changes of the adhesive properties. Their results demonstrate the big differences between the rigidity of the structural order in partially crystalline plastomers and thermoplastic elastomers.

b. Other Structures

Multiphase structures can be built up in plastomers by crosslinking, fillers, or special synthesis. Heterophase polypropylene copolymers have been synthesized, that have a continuous polypropylene (PP) matrix and a discontinuous ethylene–propylene–rubber (EPR) phase. Such polymers exhibit tensile moduli of $300-1300 \, \text{N/mm}^2$ [33]. Traditional thermoplastic rubbers have heterophasic structure and undergo thermally reversible interactions among the polymeric chains. Thermoplastic olefinic rubbers generally consisting of PP–ethylene–propylene elastomer blends, show only weak interactions among the polymeric chains; they are not able to develop suitable elastic properties within a wide temperature range. By grafting PP onto EPR or introducing polar groups into the PP and EPR, these interactions have been strengthened. Heterophasic systems have been developed in which PP is the continuous phase and crosslinked ethylene–propylene rubber is homogeneously dispersed in the PP matrix [34].

In rubber-resin PSAs, two-phase systems were found by Wetzel [35] and Hock [36] in which the resin was distributed in the rubbery matrix. Later, such two-phase systems were synthesized in the same polymer by using block copolymerization. Block copolymers are multicomponent systems that exhibit phase separation on the microscale. Such systems can be manufactured using either common polymerization techniques or special methods. Polymerization of a gaseous monomer in an unswollen, noncrosslinked matrix, crystallization or precipitation of low or high molecular materials *in situ* of polymer systems, and polymerization of a viscous monomer or macromer in the presence of dispersed additives have been described as methods to obtain segregated polymers [37]. Well-known products include styrene block copolymers. For such polymers the effect of the polystyrene domain is similar to that of vulcanized rubber crosslinks in anchoring the network structure, but in addition it is effective in raising the rubbery modulus of the network by providing a perfectly adhered, hard particle reinforcement. The strength and stiffness of block copolymers depend on a variety of factors, including the chemical nature of the elastomer blocks and the interactions between the different block segments (see later in this chapter).

B. NONADHESIVE CARRIER MATERIAL

As discussed earlier, in a classic construction and for the major PSPs, the carrier material must ensure the solid web-like character of the product. Therefore, it is not adhesive. Nonadhesive carrier materials include natural polymers (paper, cellulose derivatives), synthetic polymers (polyolefins, olefin copolymers, polyesters, PVC, polyurethanes (PURs), etc.) metals, and composites with a film-like or textured structure. The best known of them is paper.

1. Paper-Based Carrier

Paper-based carrier materials have been used since the beginning of the production of PSPs. At first, labels and tapes used paper as the carrier material, because it was the best known and most widely available packaging and information carrying material. The disadvantages of paper are mostly due to the sensitivity of its base polymer, cellulose, humidity. Paper-based carrier materials differ in paper quality, geometrics, and surface properties. The development of laminating technology led to the use of homogeneous paper/paper and heterogeneous paper/plastic and paper/metal constructions as adhesive carrier materials also. Paper as carrier material is used as face stock and release liner for the main PSP classes. Polymer impregnated paper can be used for tapes [38], metallized paper and film/paper laminates are used for labels [39].

2. Polymeric Carrier Materials

The development of film forming polymers allowed the use of plastic films as packaging materials. Plastic films were first applied as a homogeneous monolayer, but eventually heterogeneous (plastic/plastic, plastic/paper, plastic/metal, etc., composites) and multilayer constructions were developed.

a. Homogeneous Plastic Carrier Materials

The use of common packaging films as carrier material in the early days of the PSP manufacture was possible due to the relatively low requirements concerning their mechanical properties and processibility and to the ability of PVC (the best known plastic film carrier material) to be plasticized and thus, to allow quite different mechanical properties as a homogeneous, relatively isotropic, polar (coatable) monoweb material.

As mentioned, some years ago PVC was the most used polymeric packaging and carrier material [40]. Labels, tapes, and protective films were manufactured using PVC as the carrier in the form of film or foam-like webs. Both, soft (plasticized) and hard PVC were used [41].

PVC displays excellent mechanical performances. Through plasticizing, it has been possible to regulate its mechanical properties, hardness and conformability so that it can be used as a homogeneous, isotropic, plastic or elastic monolayer material. It also has good thermal resistance. Due to its surface polarity, it displays adequate printability. Homo- and copolymers of vinyl chloride (with vinyl acetate, vinyl propionate, vinyl ether, or acrylic ester) have been synthesized [42]. Extruded PVC is the most used face stock material for screen printing. Therefore, PVC has been applied for the major PSPs as a universal carrier material. PVC has been proposed as face stock material for weather-resistant decals [41], for labels in pharmaceutical and electrical use [43], and for tamper-evident labels [44]. Various tapes with PVC carrier have also been manufactured [45–47]. PVC foam is used for double-coated tapes [48]. Soft PVC is applied as carrier material for decorative films [49]. Tapes for freezer use are manufactured with soft PVC carrier, which needs a primer coating for the rubber-based PSA [50]. PVC with silicon carbide as filler is used for electrically conductive tapes [51].

Environmental and economic considerations led to the introduction of polyolefins as carrier materials — first polyethylene and later PP and of olefin copolymers [52]. Polyolefins are the most important homogeneous (monolayer) materials used as carrier films. Polyethylene is the most common carrier material for protective films. Recycling is obviously much easier if the materials used for packaging are homogeneous. Therefore, in label manufacture there is a desire to use, to the extent possible, the same base materials for the labels as are used for the products to be labeled. In the early days of plastic carrier development, polyolefins were used as face stock material; later they were used for release liners also. It should be mentioned that the use of the nonpolar polyolefin films as adhesive carrier materials sometimes enables PSPs to be constructed without a coated release liner.

It should be mentioned that the introduction of polyolefins as film-like carrier materials led to diversification of the carrier materials having the same chemical composition. This was made possible by the synthesis of polyolefins that had almost the same monomer basis but were polymerized by different methods (see Chapter 8). Later, different film manufacturing methods (blowing/casting) and last but not least the orientation of the film (monoaxial or biaxial) allowed further diversification of the product range.

Different grades of polyethylene have been synthesized and applied. Various ethylene polymers are used in films. Low density polyethylene; linear low density polyethylene; high density polyethylene; high molecular weight, high density polyethylene (HMW–HDPE); medium density polyethylene (MDPE); ethylene copolymers; and functionalized ethylene co- and terpolymers in mono- and multilayer applications. Choice within polymer types include melt index, MWD, density, and comonomers (C_4, C_6, C_4, MeP-1, C_8, etc.).

Low density PE is a semi crystalline thermoplastic material with a complex chemical structure, formed by a repetition of $-CH_2-CH_2-$ groups with some ethylenic side-chains. The number and length of the side-chains determines the degree of linearity of the structure and therefore the crystallinity of the polymer. The addition of a second bulky or polar monomer, for example, VAc, destroys the regularity of the chains and changes the forces existing between them, preventing the free rotation of the segments of the chain around the carbon–carbon bonds, and increasing the distance between these chains.

In LLDPE the built-in voluminous side groups do not allow the growth of crystallites. Thus, a main amorphous structure is formed, having some crystallites also. The amorphous part of the polymer increases the impact resistance of the polymer. The optical properties of LLDPE films and their shrinkage in the cross direction are the main disadvantages. Linear low density polyethylene has a higher melt viscosity than LDPE and is difficult to process [53]. It has a melting point situated between 15 and $10°K$ higher than that of LDPE [54]. Very low density polyethylene (density under 0.918 g/cm^3) was used first together with LLDPE.

New ethylene–octene copolymers with a statistically well defined polymer structure at densities as low as 0.860 g/cm^3 (compared to 0.910 g/cm^3 using conventional technology) and having a fractional melt index (FMI) of more than 100 have been developed using special single-site metallocene catalysts. These are linear polymers that possess a narrow MWD and a large number of long branches. Long-chain branching provides shear sensitivity, that is, high processibility [55].

New, so-called "metallocene resins" have been synthesized [56] that display a combination of elastomeric and thermoplastic properties. These polymers, with a density of $0.895–0.915$ g/cm^3 and narrower MWD than conventional LLDPE, give films with better tensile strength, excellent puncture resistance, and improved optical quality, but they are difficult to process because of their higher viscosity, lower melting point, and lower melt strength. Due to the presence of long-chain branching, these polymers will process at slightly lower pressure than LLDPE. They are free of low molecular weight oligomers which can produce chill roll plate-out in the manufacture of cast film. They can be processed with conventional equipment for standard LLDPE; however, the die design and downstream equipment could make a big difference. A barrier screw with a grooved feed section, as for LLDPE, should be used. Metallocene polymers give a tacky film. Very low density polyethylene with a density of less than 0.915 g/cm^3 is not crystalline, is flexible, and exhibits auto-adhesion also.

Polyethylene is used as film-, foam- and fiber-based material or as laminating component of heterogeneous carrier materials. The following critical characteristics are used in evaluating polyethylene for various applications: melt index (which is inversely related to MW), density, MWD, and degree of side branching [57]. Bimodal (bimodal distributed molecular weight) HDPE grades allow HDPE levels to be increased in mixtures with LDPE up to 30%, and exhibit greater stiffness.

Labels, tapes and protective films are manufactured with polyethylene carrier. Polyolefin films can also be used for release liner [58]. Single-sided [59,60] and double-sided coated polyethylene tapes have been designed for polyethylene bag manufacturers [61]. Polyethylene foam is suggested for tapes also. Double-coated, crosslinked polyethylene foam tapes have been manufactured [62]. Polyethylene is proposed as carrier for textile labels [63] (see also Chapter 8).

PP has the lowest density of any commercial plastic material. PP materials can be nonoriented, mono-oriented and bi-oriented [64]. Oriented PP film ($30–50$ μm) is applied for packaging tapes, and oriented LDPE ($80–110$ μm) for insulation tapes [65]. Bi-oriented PP is based on cast coextruded film. PP is also used for common tapes [66] and extensible tapes [67]. As carrier material for tapes PP can have a silicone release coating [68]. PP is also used as carrier material for labels, plastic bottles and containers [69] (see also Chapter 11). Copolymerization of PP reduces its tacticity and the mean length of stereoregulated sequences, thus decreasing its melting point

[70]. Random and block copolymers have been synthesized. Copolymers with ethylene and butene are more flexibile [71]. Special polymers that can be calendered have been manufactured. The raw material has to be ductile, antiblocking, with low depolymerization, low melting point, and adequate MWD. Calendered PP has a higher stiffness and modulus, better cuttability, and no slip. Fabrics and nonwovens based on PP are extrusion-coated with EMAA or EVAc [72].

The comonomers for ethylene can be considered as plasticizing (e.g., vinyl acetate, methyl acrylate, isobutyl acrylate, ethyl acrylate, and n-butyl-acrylate), polar (acrylic acid, methacrylic acid, and maleic anhydride), and reactive comonomers (acrylic acid, methacrylic acid, maleic anhydride, and monoethyl maleate) [73]. Such copolymers have the following general formula

$$-[-CH_2-CH_2-]_x-[-CH_2-\underset{\underset{R}{|}}{CH}-]_y-[-\underset{\underset{\underset{OH}{|}}{C=O}}{CH}-\underset{\underset{\underset{OH}{|}}{C=O}}{CH}-]_z- \qquad (5.5)$$

where R is vinyl acetate, methyl acrylate, ethyl acrylate, etc., y has a value of $0-40\%$, z has values of $3-20\%$, and M is C_1-C_{20}.

Polar copolymers of ethylene were produced in high pressure reactors more than 30 yr ago. They can be extruded as conventional films and used as carriers or self-adhesive films. As discussed earlier, the crystallinity-related properties decrease with increasing polar monomer content. VAc copolymers have low modulus (see Table 3.10). They have been applied as a heat seal layer in extrusion coating and coextrusion. Fire-retardant compositions based on EVAc with special fillers, $Al(OH)_3$ or $Mg(OH)_2$, have been proposed [74]. Ethylene acrylic copolymers are applied as a tie layer for oriented polypropylene (OPP) and PET film or as heat seal films [75].

EVAc copolymers with a VAc content upto 42% have been grafted with styrene, vinylidene chloride and styrene-maleic acid. Polyethylene and EVAc have been grafted with VAc. The VAc units have quite different types of influence in the main chain or branches. Short polyvinyl acetate branches are compatibile with the amorphous phase of the partially crystalline system; longer PVAc branches buildup an other amorphous region [76]. EVAc copolymers having more than 70% VAc are polymer plasticizers for PVC [10]. Segmented EVAc possess segments with high and low vinyl acetate contents [77].

As raw materials on natural basis, used first in the packaging industry, cellulose derivatives have been applied as carrier material for PSPs also. Cellulose acetate and hydrate have been used for labels and tapes. Different types of cellulose hydrate (Zellglas) are known. These are generally hydrophilic films plasticized by polyhydroxy derivatives. Their humidity balance is very sensitive to drying conditions. Later top-coated qualities were manufactured to improve their water resistance, weldability, and printability [78]. Cellulose acetate has better chemical and water resistance but is sensitive to the plasticizers in printing inks, and to electrostatic charging. Clear, destructible acetate film can be used for tamper-evident labels [79]. Acetate films have been introduced as carrier material for self-adhesive tamper-evident products. A 50 μm cellulose acetate film modified to be a brittle film, with a low tear strength, may be used as overlaminate, seal, or label, as an alternative to PVC [80]. It is claimed to be useful for applications where labels are designed to cover an existing graphic element or product copy. Insulation tape with a cellulose acetate carrier is used for wire covering by telephone manufacturers. Tapes of cellulose hydrate or PVC have been designed for sealing boxes [46]. Polyvinyl isobutyl ether and polyvinyl alkyl ether can be used for medical tapes on cellophane [81].

As a material that has high thermal resistance, good mechanical properties and good aesthetic quality, polyester was introduced mainly for labels and as a release liner. Polyester is used as a carrier for labels, cosmetics, toiletries, pharmaceuticals, chemical products, and shrink sleeves [82]. Special tapes and protective films also use polyester as carrier material. Double-faced mounting tapes based on PET are used for electronics products [83]. Polyester is the face stock material for labels, pharmaceutical and electrical applications [43] and for splicing of hard-to-stick materials. Metallized polyester has been used for labels and tapes [79]. Polyester fabric is proposed as label face stock material [84].

Polyurethanes are recommended as carrier material for foam tapes and self sticking-clips [85]. A polyurethane self-sticking, double-backed foam tape, should also be fire- and mildew-resistant.

Because of their excellent mechanical properties and thermal resistance, polyamides have been used as carrier material, mainly for tapes. They are applied as film or as a reinforcing web, built into a carrier film. For instance, pipe wrapping tapes and repair tapes are flexible plastic composites reinforced with nylon cord and have to adhere to surfaces of metal, wood, rubber, or ceramic. They are designed to be resistant to punctures and tears [43]. TPEs based on polyamides have also been manufactured. Block amide–ether amide (Peba) copolymers based on combinations of soft, flexibile polymer chain segments of polyethers with a high melting point, and stiff segments of polyamides, possess a T_g of $-50°C$ which allows their use at low temperatures [86]. The polyamide/polyether ratio can be varied between 80/20 and 20/80. Therefore, flexible plastomers can also be obtained [87]. Polybutadiene–aromatic copolyamides give transparent flexible films [88]. The tensile strength and modulus of elasticity of such copolymers increases with the polybutadiene level.

Modified polyethyl methacrylate can be used as protective film for different substrates such as PVC, ABS, polystyrene, wood, paper, and metals [89]. Polar macromolecular compounds manufactured by polycondensation or polyaddition, for example, polyesters, polyamides, polyimides and polyurethanes are also used as carrier material [90]. Polyimides have been proposed for transparent pressure-sensitive sheets coated with silicone-based adhesives. The pressure-sensitive laminate resists 8 h at 200°C [91].

b. Heterogeneous Plastic Films

Composite materials can also be based on plastic films. These are combinations of chemically or physically different polymers that are built together via adhesion. This adhesion is achieved technologically by lamination or coextrusion. In some cases other thermoplasts or duroplasts (e.g., polyurethanes) can be coated on the back side of the adhesive carrier material [92]. Films composed of polyethylene and polystyrene have been manufactured and used as carrier for labels.

C. CARRIER MATERIALS WITH ADHESIVITY

Adhesive carrier materials are self-adhesive films. A self-adhesive film combines the characteristics of a mechanically stable monoweb with the adhesivity of a liquid PSA, that is, it is a noncoated face stock material having surface adhesivity. It should be noted that for some carrier materials from this product range, which at normal temperature show a thermoplast-like behavior, surface adhesivity exists only at elevated temperature or pressures or after a physical pretreatment of their surface. Such carriers are generally films based on nonpolar raw materials. Other carrier materials based on polar polymers, exhibit surface adhesivity (with or without surface treatment) at high temperatures. Some carrier materials incorporate viscoelastic or viscous nonpolar or polar components that display surface adhesivity at room temperature.

Self-adhesivity of the carrier material is known from the practice of nonself-adhesive, that is, adhesive-coated carrier materials too. In some cases in order to increase adhesive anchorage on the carrier material its adhesivity has to be improved. This obviates the need for primers; the adhesion of the carrier material is improved by the bulky inclusion of adhesive components. Biaxially oriented multilayer PP films that adhere well to the adhesive coating have been manufactured by mixing the PP with particular resins.

Generally, in this case in order to prepare an adhesive tape which can easily be drawn from a roll without requiring an additional coating on the reverse side, at least two different layers of different compositions are coextruded, and the thin back layer contains an antiadhesive component. As a tackifier, nonhydrogenated styrene polymer, α-methylstyrene copolymer, pentadiene polymers, α- or β-pinene polymers, rosin derivatives, terpene resins, and α-methylstyrene–vinyltoluene copolymers are preferred. In this case self-adhesivity of the plastomer has been achieved chemically, that is, by tackifying. Physical methods can also be used.

Self-adhesivity of plastic films is useful for other applications too. Biaxially oriented polypropylene (BOPP) films are produced with a thickness of 40–50 μm. Thick BOPP films cannot be manufactured economically with the common stretching processes. For these purposes, several thin films are combined (via heat lamination) to a multilayer film [93]. As shown in [94], a polyolefin laminate can be manufactured by laminating together two corona treated surfaces at a temperature that is lower than the softening temperature of the films.

1. Nonpolar Carrier Materials

Commercial polyolefins are used for the manufacture of common nonadhesive carrier materials. Special polyolefins can be applied as pressure-sensitive carrier materials. Their adhesivity is due to their chemical composition, to a special surface treatment and to special application conditions. It should be noted that such materials are not self-adhesive in the classic sense. They behave like a self-adhesive material only at elevated temperatures (near their melting point) and at elevated pressure.

Very low density polyethylene films manufactured by casting (i.e., with a greater amorphous content) and having a low film thickness (conformability) display self-adhesivity at room temperature. Such films are used as "cling" films (see also Chapter 8). It is well known, that VLDPE with a density of 0.885 g/cm^3 is self-adhesive. It is difficult to process for blown film because its self-adhesivity. The self-adhesivity of polyolefins can be improved by addition of viscous (nonpolar) and viscoelastic components such as polymers of isobutene. A cold stretchable self-adhesive film is based on a ethylene-α-olefin copolymer (88–97% w/w) and 3–12% w/w of polyisobutylene, atactic polypropylene, cis-polybutadiene and bromobutyl rubber [95]. The polymer has a density lower than 0.940 g/cm^3. The film exhibits an adhesive force of at least 65 g (ASTM 3354-74). Atactic polypropylene (2–9%) with a molecular weight of 16,000–20,000 has been suggested together with polyvinyl acetate (9–18%) and a tackifier resin for a removable adhesive composition [96]. The adhesion of polybutadiene as an SAF has been improved by modifying it with isopropylazo-dicarboxylate [97].

2. Polar Carrier Materials

Some polar monomers known as adhesion promoters from the manufacture of common viscoelastic raw materials for PSAs can also be used for the synthesis of thermoplasts, that display self-adhesion on certain adherend surfaces under well-defined conditions. These monomers are vinyl acetate and acrylics and are used as comonomers with ethylene.

a. Ethylene Copolymers

As discussed earlier, polar ethylene copolymers have been manufactured by copolymerization of ethylene with polar vinyl and acryl derivatives. The upper limit for the polar comonomers is about 40% and is limited by the mechanical properties and manufacturing features of the film [7]. Aprotic comonomers (with ester, ether, anhydride, and oxyrane functional groups), protic comonomers (with hydroxy, carboxy, and amide groups), and ionic comonomers (salts) have been polymerized. Polar ethylene copolymers can also be produced by polymer-analogous reactions or grafting. According to [98], the polyolefin surface of the carrier can be polarized by graft polymerization using electron beam (EB) radiation with a dosage level greater than about 0.05 Mrad. Ethylene polymers grafted with a carboxylic reactant or ethylene–vinyl/acrylic copolymers are suggested for adhering propylene polymers to polar substrates [99]. Hydrolysis of EVAc copolymers and neutralization are commonly used to obtain polyvinyl alcohol and ionomers. Acrylic and maleic copolymers are grafted (via processing also) on the PE backbone [100]. Acid-functionalized polymers may undergo crosslinking. Neutralized, ionic polar copolymers differ from acrylic ones by their resistance to solvents, toughness and heat sealing properties, owing to their ionic network. Both are used as adhesion promoters. Copolymerization of ethylene with acrylic acid leads to softening of the polymer. As shown by Kirchner [101], the high modulus of HDPE (of about $1600 \, kg/cm^2$, ASTM 638-58 T) decreases to $430 \, kg/cm^2$ for an EAA copolymer having 19% polar comonomer and the softening point decreases to 54°C.

Ethylene–vinyl acetate copolymers contain two domains [102]; there is a crystalline region of polyethylene and an amorphous region of EVAc. As the level of VAc increases, the amorphous region increases. The melting point and tensile strength decrease. A correlation has been developed that shows the dependence of the comonomer content (mol.%) and the melting point of the copolymer [103]. The number of comonomer molecules is more important than their size. The copolymers of ethylene and vinyl acetate range from typically thermoplastic materials (similar to low density PE) to rubber-like products. The copolymerization of ethylene with vinyl acetate increases the density, clarity, permeability, solubility, environmental stress cracking resistance, toughness (specially at low temperatures), compatibility with other polymers and resins, acceptance of fillers, and the coefficient of friction (COF). In contrast, rigidity, softening point, and surface hardness (modulus) are decreased. Such characteristics are related to film conformability and self-adhesion. As known, self-adhesion is due to chain mobility (see also Chapter 3, Section I). Taking as the molecular index for polymer flexibility the freedom of macromolecular chains to freely rotate, given as $(\lambda^2/\lambda_{r,i}^2)^{1/2}$, where λ is the chain length expressed as average quadratic length for a chain in theta solvents and $\lambda_{r,i}$ the chain length capable of rotation, the bonding ability of different plastomers or elastomers can be compared [104]. As stated, rubber has the lowest value (1.5) for the above ratio compared with 1.8 for polypropylene [105] and 2.3 for polyvinyl acetate [106] or 1.7 for polyisobutylene [107,108]. In self-adhesive films the versatility of these materials is in the following order: natural rubber > polyisobutylene > polyvinyl acetate.

Ethylene–ethyl acrylate (EEA) copolymers have been proposed for use as removable clear protective films on metals [109]. In comparison with EVAc, it is possible to reduce the formulating level (and the costs) of EEA by 3–35%.

EVAc copolymers have been used for a number of applications. In the early 1970s they were suggested for heat-activated adhesive bonding [110]. Copolymers of EVAc ensure sealing temperatures 40°C lower than sealing temperatures for PE. Such polymers were tested for sealable packaging films also. EVAc copolymers have also been used as sealants for PP. The polymer with more than 28% vinyl acetate can be sealed by high frequency. Copolymers having less than 5% w/w EVAc are used in thin films to regulate their mechanical properties. Copolymers with 6–12 %VAc are used to regulate mechanical properties, for tougher films at low temperatures, and for films with higher impact resistance and greater stretch properties. The sealing temperature of an EVAc copolymer film with 12% VAc is 130–150°C; it decreases as VAc content increases. The melting point of EVAc copolymers with 10–14% VAc content lies between 91 and 96°C [111]. Commercial EVAc types contain 10–40% VAc. A 32% content of VAc in EVAc gives partially crystalline polymer, and at 40% vinyl acetate, a complete amorphous polymer is formed [112]. It is to be noted that their properties are strongly influenced by the molecular weight of the polymer and the number of side chains. Copolymers with 15–18% VAc are proposed mainly to improve the sealing properties and stress cracking resistance of polyolefins at low temperatures. Copolymers with 18–30% VAc are suggested for adhesive- and wax-based coatings. EVAc copolymers are also used as plasticizers or flexibilizers for PVC [113]. Copolymers of ethylene with vinyl acetate and hydroxy functional monomers are applied as the sealing layer in coextruded films [114]. Polymers with a level of 15% VAc possess PE-like poperties. Polymers having 15–30% VAc display PVC-like properties, polymers with more than 30% VAc are elastomers [115]. Their elongation increases with VAc content, especially up to 15% VAc content. The optimum tensile strength is achieved for a content of 20–30% VAc. The clarity of EVAc films increases with the VAc content; however, films with less than 15% VAc are generally manufactured because of tendency to block. Blocking is reduced by slip agents, antiblocking agents, and cooling during manufacturing. EVAc foams are also manufactured. It can be seen that vinyl acetate copolymers are adequate raw materials for nonadhesive carrier films (where the low content of VAc allows the production of a tougher, softer film), for self-adhesive films having higher vinyl acetate content, and for semi pressure-sensitive and pressure-sensitive adhesives (HMPSAs) with a high polar monomer content. It is evident, that for an adhesive formulation the vinyl acetate content should be much higher, and generally plasticizers should be included. According to Litz [109], an HMPSA formulation on an EVAc basis contains about 35.5% elastomer, 30–50% tackifier, 0.2% plasticizer, 0.5% filler, and 0.1–0.5% stabilizer. The vinyl acetate content does not influence the melt viscosity. The peel value can be increased by increasing the VAc content. As plasticizers, phtalates, phosphates, chlorinated polyphenols, and liquid rosin derivatives have been used. As tackifiers, rosin derivatives, hydrocarbon resins, and low molecular weight styrene polymers have been suggested. Zinc oxide, calcium carbonate, titanium dioxide, barium sulphate, and organic resins have been recommended as fillers. Films based on linear polyurethanes can be hot laminated [90]. Peelable protective films for metals have also been manufactured on a PVC copolymer basis [116].

II. MACROMOLECULAR BASIS AND CHEMICAL COMPOSITION OF THE ADHESIVE

The adhesive used for PSPs has to exhibit pressure-sensitivity. As discussed earlier (see Chapter 3, Section I), pressure-sensitivity is given by a special rheology. Viscoelastic behavior allows pressure-sensitivity. Viscoelasticity can be the result of a built-in special chemical or macromolecular basis, but it can be achieved by formulation also. In this case the elastomeric and viscous components are mixed in order to give the desired balance of the viscoelastic properties. It is well known that the chemical composition of the adhesive includes nonadhesive components, which are required to ensure the fluid-state characteristics of the adhesive and its storage resistance. The adhesive must be fluid for coating. In some cases the adhesive is transferred (during manufacture and application of PSP) onto the final carrier or substrate as a solid-state component (e.g., transfer coating, calender coating, or transfer tapes). The use of transfer tapes allows the choice of materials that cannot be applied as fluids [115]. The chemical basis of the PSAs is described in detail in [1–3]. Therefore, in this chapter only the pressure-sensitive product-related aspects of the raw materials are discussed.

A. ELASTOMERIC COMPONENTS

The elastomeric components of the adhesive are rubber or rubbery products. Some of them display a low level of adhesivity also. In order to display viscoelasticity they need a viscous component (tackifier). As for plastomers used for carrier materials, the chemical composition, molecular weight and macromoleclar characteristics of the elastomers determine their applicability for PSPs. Polar functional groups, comonomers, side-chain length, crosslinking, molecular weight, and molecular weight distribution influence the rheology of the adhesive [117]. Unlike plastomers, which possess adequate mechanical properties due to their rigidity given by crystallinity, elastomers have a deformable crosslinked elastic construction. Thermoplastic elastomers are also elastic structures where the connection points of the network are given by molecular, segmental associations.

1. Molecular Weight

The molecular weight, the functionality of the monomers and the multiphase structure of the macromolecular compounds used for PSPs play a primary role in the control of the product characteristics.

a. The Role of Molecular Weight

It should be accentuated that the role of molecular weight (MW) as a macromolecular characteristic is more complex for adhesives than for plastomers used for PSPs. The molecular weight can be decisive for the use of the (chemically) same component as base elastomer or tackifier. For instance, for insulation tapes high molecular weight PIB (MW 100,000) is used as base elastomer and low molecular weight PIB (MW 1500) as tackifier [118]. As known from the practice of sealants, solvent-based low molecular acrylates are used as plastic sealants, whereas water-based acrylates (with higher MW) can be used as elastoplastic sealants. They have the ability to absorb vibrations up to 20% [119]. According to Köhler [120], the first PSPs were manufactured with relatively low MW polymers. For such PSAs, destruction products of rubber together with soft resins and low molecular weight polyvinyl ether have been used.

Increasing the molecular weight by crosslinking can lead to a self-adhesive polymer playing the role of the carrier material in a composite structure manufactured by radiation crosslinking of acrylates [121]. As stated in [122], the same chemical composition with different molecular weights allows a polymer to be used as carrier and adhesive also. An acrylic acid–ethyl acrylate copolymer with a molecular weight of 40,000 can be used as carrier material for an acrylic acid–ethyl acrylate adhesive (with a molecular weight lower than 12,000).

Polymers are not homogeneous on a molecular scale. Generally, the macromolecular compounds used for PSPs are linear or crosslinked. Linear molecules are typically threadlike structures. In crosslinked polymers separate molecules really do not exist. There are weak and strong regions in the polymers. The weak regions or imperfections may consist of chain ends, which are not entangled, and regions in which chain segments are oriented perpendicular to the direction of the stress. Strong regions include chain entanglements and regions where chain segments are oriented parallel to the stress. When a load is applied, the entangled chains will orient along the stress direction. (For instance, in rubber, stress-induced crystallization may appear.) The weak regions may form submicroscopic cracks.

The segments of the polymer molecule are in constant vibrational motion. This motion together with the imperfect packing of the molecules causes the existence of a free volume in the macromolecular structure (see also Chapter 4). Filling this free volume with low molecular substances such as plasticizer, changes the polymer's physical properties (e.g., coefficient of linear expansion) and mechanical properties. Thermal motion and free volume depends on the temperature. At a temperature great enough to produce chain mobility (T_g) free volume (and total volume) increases, the polymer becomes rubbery. The mechanical properties (tensile strength and elastic modulus) decrease above the T_g. Creep, impact resistance, and permeation increase above the glass transition temperature (see also Chapter 3, Section I). Such molecular mobility above T_g allows a pressure-sensitive adhesive to bond. Molecular flexibility can be controlled by crosslink density, glass transition temperature, fillers, plasticizers, and stabilizers. Molecular mobility causes debonding also. Disentanglement of linear molecules or melting of a crystalline order produces creep, that is, a slow deformation and slow bond failure. Chain entangling accounts for 75% of the equilibrum modulus, even at a high degree of crosslinking [123].

As discussed earlier the buildup of ordered, multiphase structures can modify the chain mobility. Such structures include fillers, crosslinks, crystallites, and molecular associations. Fillers modify the T_g. The T_g of PVC increases with the level of filler [124] (see also Chapter 3, Section I).

Crosslinking reduces creep, because the polymer segments are immobilized (see also Chapter 6). For systems with limited mobility, weak bonds between the chains or within the chains may rupture. For strongly immobilized structures rapid bond failure may be manifested as crack. Such systems do not display creep. As seen from the correlation giving the real stress (σ) in an extended rubbery sample as a function of the length of the sample before and after extension (expressed as the ratio λ of the length) the temperature (θ) and the end-to-end distance (r) of an uncrosslinked polymer chain [125],

$$\sigma = KT\left(\frac{\langle r^2 \rangle}{\langle r^2 \rangle_0}\right)(\lambda^2 - \lambda^{-1}) \qquad (5.6)$$

the stress depends on chain length, and chain length depends on the temperature (the term $\langle r^2 \rangle/\langle r^2 \rangle_0$ takes into account this dependence). This means that molecular weight strongly influences debonding. As shown by Hamed and Hsieh [126], with a short contact time the bond strength (tack) is primarily due to the diffusion of short macromolecular chains. At longer contact times higher molecular weight chains influence bonding. As discussed in Chapter 3 the plateau modulus is related to the entanglement molecular weight. The fraction of the polymer that has the molecular weight less than twice the entanglement molecular weight (M_e) will act as a plasticizer [127]. A physical model correlating polymer viscoelastic behavior to PSA performance has been developed to explain the relationship between T_g, M_e and PSA performance [128]. Acrylics are inherently tacky due to their high entanglement molecular weight without the need of adding tackifier [129]. An entanglement model can be used to estimate the critical molecular weight using the characteristic ratio of the polymer, the number of bonds per monomer and the monomer molecular weight. Recent investigations concerning cavitation takes into account the effect of different material models (e.g., Neo-Hookean and Money Rivlin) for the strain energy [130] and the role of the MWD [131]. Commercial acrylic adhesives, have a very broad spectrum of relaxation times, and in particular low MW fractions provide the fast relaxation times necessary to achieve a large real contact area within a short contact time. With monodisperse polymers one could not have a fibrillar (and therefore, dissipative) and adhesive (and therefore no residual polymer on the surface) fracture, they are useless for the PSAs [132]. However, the role of MWD in pressure-sensitive behavior is not sufficiently cleared. Studies on PIB showed that tack and adhesive fracture energy decrease monotonically for MW $>$ 45 kg/mole [133,134]. However, PIB does not fulfill the Dahlquist criterion (see Chapter 4) and does not show the fibrillation phenomenon typical for PSA. On the contrary, mixtures of PIB of high and low MW (where the low MW of 1.5 kg/mole is well below M_e (9 kg/mole fo PIB) show cavitation and fibrillation in tack experiments at mixing ratios around 80/20 (by weight) of short to long chains [135]. The more PSA-like blends of polyisobutylene (of low and high molecular weight) containing longer chains do not show a strong dependence of the adhesive fracture energy W_a on MW, temperature and mixing ratios especially as M_v exceeds a critical value (1000 kg/mole). This seems to be in contrast to the pronounced tack maximum at MW/M_e around 4 observed by Tobing et al. [136].

Compatibility is a key parameter for the mixing of formulation components. It depends on chemical nature and molecular weight. The mixing enthalpy of macromolecular compounds depends on their chain length. The enthalpy decreases with increasing chain length. On the other hand, with an increase in temperature the mutual solubility decreases because the entropy end enthalpy effects are working against each other [137]. The free mixing enthalpy ΔG_{mix}

is given as a difference of the enthalpy (ΔH_{mix}) and entropy ($\tau \Delta S_{\mathrm{mix}}$), where τ is the mixing temperature:

$$\Delta G_{\mathrm{mix}} = \Delta H_{\mathrm{mix}} - \tau \Delta S_{\mathrm{mix}} \qquad (5.7)$$

Elastomers with broad molecular weight distribution show better performance than materials with a narrow MWD. The concept of tackifying resins functioning as solid solvents for portions of the elastomer has been used to explain the need for a broad MWD. Resins will dissolve the lower molecular weight fractions of the elastomer and increase tack. The undissolved higher molecular weight fractions give the mechanical strength (see also Chapter 8). Natural and synthetic elastomers may be used as elastic component. Their mutual compatibility and their compatibility with the tackifiers depend on their molecular weight. Changes occur in the molecular weight distribution during aging, but coating does not influence it significantly [138]. Compatibility plays different roles for various pressure-sensitive formulations. For classic rubber-resin recipes where the resin is dissolved in the elastomer matrix of natural rubber (or vice versa at higher resin levels), the resin should be compatible with the rubber in order to achieve adequate adhesive properties. For block copolymers with segregated structures, resin mixtures are used, each of the components being compatible with one of the polymer blocks. For tackified plastomers non-compatible viscous components are suggested to allow their migration to the film surface. Tackifier compatibility plays an important role in understanding tackification mechanism [139,140].

b. Postregulating the Molecular Weight

Regulating the molecular weight of elastomers during the manufacture of PSPs is a common operation. In order to prepare high solid content solutions, the elastomer is masticated, that is, its molecular weight is reduced. As discussed in Chapter 3, the properties of the elastic network depends on the crosslinking density and molecular weight between crosslinked portions. Mechanochemical destruction may change both. Through mastication the Defo elasticity of rubber is decreased, and the flow properties of the elastomer may also change [141]. It is a way to regulate the peel force also. This procedure is usual for tapes and protective films [142]. For such products the rubbery network is rebuilt via crosslinking to achieve high cohesion or low adhesion. A too long mastication leads to excessive degradation of the polymer (dead calendering). Natural rubber for tapes is crosslinked [143]. According to Mueller and Türk [144], rubber-resin adhesives have been used almost exclusively for removable price labels. Natural and synthetic rubber and polyisobutylene have been used as the main raw materials. The rubber has been calendered (masticated).

It is well known that crosslinking is brought about by the synthesis of carboxylated rubber latices. The molecular weight of a crosslinked network cannot be compared with the molecular weight of a linear system. For crosslinked systems the molecular weight alone is not sufficient for polymer characterization. It is also necessary to evaluate the gel content.

Postsynthesis crosslinking is a common procedure for solvent- and water-based (WB) PSAs on different chemical basis using built-in or external (chemical) crosslinking agents. Hot-melts are also crosslinked by radiation curing. The crosslinking modifies the molecular weight of a sufficiently high molecular weight polymer in order to limit chain mobility and to achieve higher modulus according to correlation (Equation 3.22) (see Chapter 3). In other cases crosslinking is used to buildup a polymer or to achieve a long enough chain for cohesion. At low enough crosslink densities equilibrium moduli below the rubbery plateau modulus can be obtained, but such networks are weak due to the presence of long ineffective chain ends. Crosslinking of solvent-based PSAs plays a special role allowing fine regulation of the adhesion–cohesion balance and of the environmental resistance of PSAs; therefore, its science and technology will be described in a separate chapter (see Chapter 6). Regulation of the molecular weight via crosslinking is a necessity for low cohesion solid-state acrylics also [143]. Balanced pressure-sensitive properties are achieved by ultraviolet (UV)-initiated postcrosslinking (after coating) of the formulated adhesive. The curing of monomers as well as of polymers has been carried out to achieve or to improve pressure-sensitivity. For instance, in [145] the curing of hot-melts using 60–80 kGy radiation (EB) energy dosage is described. The properties of polyisoprene-based PSAs crosslinked by using electron beam radiation were studied by Yarusso et al. [146]; they stated that high gel content could compensate for low MW. Curing of acrylic polymers with UV light is discussed in [147–149]. Both procedures may lead to higher molecular weight associated with a nonlinear molecular structure. It can be concluded that postregulation of the molecular weight is really a postregulation of the molecular structure also. Pressure-sensitive products need composite structures, and such structures are formed by polymacromerization and polysegregation. Crosslinking as formulation possibility is described in detail in [150]. As stated crosslinking is the commonly used technology in off-line synthesis of PSAs.

2. Polymacromerization and Polysegregation

In principle there are two ways to achieve viscoelastic network structures for PSPs: polysegregation and polymacromerization. As discussed earlier for plastomers used as carrier materials for PSPs, the characteristic, ordered multiphase structure is mainly the crystallinity. Crosslinked and filled products can also be manufactured. On the contrary, it is generally assumed that a large number of amorphous polymers feature a definite degree of structural microheterogeneity: elementary domains formed by the macromolecules may associate or form first fibrils and then superdomains from which a three-dimensional network is built up. Such systems are not in equilibrium in character and remain capable of further ordering which takes place when the polymer has facilitated mobility.

For PSAs more elasticity and molecular mobility are needed. In this case elastomer structures including crosslinked networks, molecular associations, fillers, and in certain cases crystalline portions also have been developed. For such structures the molecular weight between entanglements plays a primary role. As in (crosslinked) plastomers, supplemental crosslinking (by covalent bonds) of the noncovalently segregated structures is possible. Like vulcanization of natural rubber, curing of synthetic elastomers and TPEs can buildup new segregated structures. For instance, such crosslinking can be achieved using radiation curing.

a. Buildup of Segregated Structures

It has been shown that polymer performance can be enhanced by the formation of heterogeneous systems in which one polymer exists above its glass transition temperature, while the other exists below its T_g at room temperature [151]. Thus a composite results at room temperature having one component that is glassy, whilst the other remains rubbery or rubbery elastic. Varying the relative amounts of polymers in this blend (which can be considered as an interpenetrating network) will alter the properties. The produced material can range from a reinforced rubber to a high impact plastic. Rubber-resin adhesive with the high T_g resin and low T_g elastomer or self-adhesive films based on a mixture of plastomer, elastomer and tackifier resin may be considered structures of such type. Chemical reactions can lead to interpenetrating networks also. Crosslinking reactions in the adhesive layer increase the T_g and buildup segregated macromolecular constructions. Crosslinking is responsible for the control of the size of the phase domains in the network and hence, the properties exhibited by the material.

For crystalline polymers the existing order is the main parameter that determines their performance; the chemical structure is secondary. Crystalline structures are nonadhesive. Unlike classic, nonpressure-sensitive adhesives where crystallinity builds up after the formation of bonds improves the joint strength (e.g., polychloroprenes), crystallinity has a negative effect on the usability of PSAs. Molecular weight can influence the tendency to crystallize also, leading to so-called semi pressure-sensitive products, which have a time-limited self-adhesivity (see also Chapter 8). After crystallization, the self-adhesivity of the elastomer disappears. For products where diffusion of viscous components affects the bonding properties (i.e., tackified self-adhesive films) the time-dependent self-adhesivity is a function of the molecular weight and crystallinity. In some cases crystallinity is a designed characteristic of the adhesive raw material. For instance, segregated polyurethanes have a crystalline structure.

Thermoplastic elastomers with diblocks and triblocks styrene–butadiene–styrene (SBS), styrene–isoprene–styrene (SIS), and styrene–ethylene–butylene–styrene (SEBS) have ordered amorphous domains and segregated incompatibility, but multiblock systems may also have crystalline segments. This morphology replaces the crosslinked network of common rubbers, but due to its low deformability, thermoplastic elastomers like SBCs have only a medium degree of elastic recovery, in comparison with classic rubbers. They can be considered as intermediate to elastoplastic and viscoelastic compounds.

Segregated structures are built up as block copolymers where molecular associations form amorphous incompatible domains (e.g., SBCs). In such cases the main problem is to find an adequate "blocky" monomer that allows high temperature resistance as well as low temperature processibility. According to Equation (3.25) (see Chapter 3), in a first approximation such "blocky" units work in an elastomer matrix, like a chemically inert filler (Figure 5.2). Their modification (if possible) with the use of other compatible viscous fillers (tackifiers) allows no further strengthening of the network. Such rubber-resin adhesives are considered similar to filled systems having plasticizer or tackifier particles embedded in a rubbery matrix or vice versa (Figure 5.3). Such systems can also be examined by using Equation (3.25), that is, given that the synthesized macromolecular composite is embedded like a filler in another formulated composite. The choice of compatible resins, (which are low molecular compounds) allows tackifying with a minimum loss of cohesion.

In certain cases segmentation alone does not provide self-strengthening. This is achieved by postcrosslinking. In other cases the thermal limits of strengthening by association have to be improved. For styrene block copolymers with unsaturated midsequences, if strengthening is necessary it can be made by using the reactivity of the rubbery (isoprene, butadiene, etc.), "bridges," that is, by crosslinking (Figure 5.4). By such reactions it is possible to create a covalently bonded network within the elastomeric midphase in order to reinforce the physically existing network (which disappears over the T_g of polystyrene) [152] (see Figure 5.4).

As seen in Figure 5.4 such crosslinking can be carried out with or without a supplemental multifunctional crosslinking monomer. It can be done by classic or radiation-induced polymerization. Radiation-induced crosslinking depends on the molecular weight and its distribution. The first

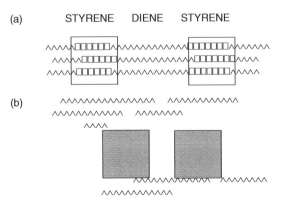

(a) STYRENE DIENE STYRENE

(b)

FIGURE 5.2 Schematic presentation of segmented copolymers. (a) Block copolymer chain; (b) polymer chain considered as blend of homopolymer blocks acting as filler.

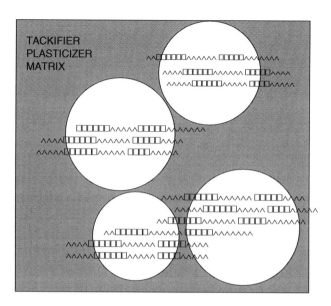

FIGURE 5.3 Systematic presentation of formulated adhesive based on block copolymers, considered as matrix or filler in a reinforced system.

radiation-curable styrene block copolymer (Kraton D-1320X) was processed like a common thermoplastic elastomer, but it was also able to crosslink. This was a star-shaped SIS block copolymer. Because of its high molecular weight low radiation doses are needed to produce crosslinking. Because of its star structure, partial crosslinking does not modify (essentially) the flexibility of the molecules.

Curing of acrylics is viable too (see Chapter 6). For instance, one patent [153] describes the curing of copolymers of diesters of unsaturated dicarboxylic acids with acrylics in a composition

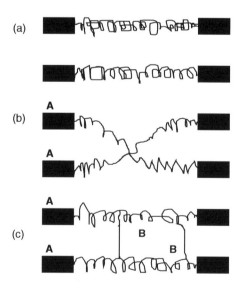

FIGURE 5.4 Schematic presentation of the uncrosslinked and crosslinked block copolymer. (a) Uncrosslinked chains of a block-copolymer; (b) crosslinking of a block copolymer by the midblock sequence, without supplemental multifunctional crosslinking agent; (c) crosslinking of a block copolymer using supplemental multifunctional crosslinking agent.

having a multifunctional crosslinking agent is described. Such polymers possess a T_g between 30 and 70°C below the temperature of use. In this case curing is used to improve adhesive properties, particularly shear. This composition can be cured chemically or by any convenient radiant energy. The molecular weight of the formed polymers differs according to their MWD. For a narrow MWD, a average molecular weight of 100,000, and for a broad MWD a molecular weight of 140,000 is required to enable the desired response to electron beam curing. In general, constituents having a molecular weight of less than 30,000, are nonresponsive to electron beam radiation. An M_w/M_n ratio between 4.2 and 14.3 is preferred. Cohesive strength is proportional to the concentration of the hard monomer (giving a polymer with T_g higher than -25°C). When a multifunctional monomer is used, its concentration should be preferably about 1–5% by weight. Its presence enables reduction of the dosage level, at least for EB curing.

Epoxidized star-shaped block copolymers can also be UV cured [154]. Polyisoprene and polybutadiene with a segmented structure have been synthesized as UV-curable adhesive raw materials, that are processible as warm-melts [155]. Although segmented, the new polymers do not present microdomain phase separation. They are viscous liquids near room temperature. Such behavior is obtained by keeping the polymer molecular weights low and MWD narrow. These polymers are functionalized oligomers that are bi- or multifunctional, and their star-shaped structure is used to increase the possible functionality, not to allow phase segregation. They possess two different kinds of functionalities: epoxy groups and hydroxyl groups. Epoxy groups may undergo ring-opening polymerization under acidic conditions. When multiple epoxides are located on a polymer, poly-ether crosslinking results. Epoxy polymerization can be initiated by UV light also when a sulfonium salt is used. The photolysis of the initiator generates a Brönsted acid that protonates the epoxide. The protonated epoxy group may react with another hydroxylated compound to form polyether linkages. Therefore, mixtures containing multifunctional oligomers (e.g., epoxidized star polymers, polymers having one or more hydroxy groups, or hydroxy and epoxy groups) allow multiple ways to regulate the reaction and form crosslinks. In such systems the star-shaped polymer is diluted by the monofunctional polymer.

It has been shown that the method of crosslinking influences the flexibility of the polymer network and the flow and mechanical performance of the macromolecular compound [156]. For instance, the modulus of polybutadiene networks made by hydrosilation-crosslinking differs from the modulus of radiation-crosslinked polymer. The effect of interaction of the junction points is weaker than for polybutadiene crosslinked by radiation [157]. It is known that in crosslinked systems the rheological and mechanical performances of the system are determined by chain mobility, which is influenced by the molecular weight of the polymer as a whole and the molecular weights of the polymer sequences. The molecular weight between crosslinks, M_c, depends on the chain length between functional groups. In experimental polymers, according to [155], molecular weights are larger between crosslinkable sites on the polybutadiene blocks, than the much smaller molecular weight of the polyisoprene blocks. Both together give the so-called mechanically effective molecular weight M_{cme}. The star-shaped polymer may carry a large number of epoxides per molecule to promote electron beam curing at low dosage level. In such polymers there are over 100 epoxy groups located on the polyisoprene endblocks. Such an abundance of functionality leads to overcure. Therefore, these polymers exhibit no tack. As stated in [155], in comparison with polymers that have the longest distance between crosslinks of a molecular weight of about 2000 Da, polymers with PSA properties should have an M_{cme} of at least 3000–4000 Da. A cured blend of mono-ol polymer, epoxidized mono-ol polymer, and hydrogenated tackifying resin exhibits aggressive tack and SAFT values which exceed 175°C without failure. This example illustrates the problem of functionality, that is, how the balance of linear and crosslinked segments influences the final properties. In this case chemical functionality, and in the case of styrene block copolymers the presence of associative styrene blocks as mono or diblock (on both ends of the sequence), affects the network buildup. As stated by Holden and Chin [158], a new SEBS with 30% triblock sequences has a tensile strength of 350 psi in comparison with the tensile strength of 142 psi for a

20% triblock copolymer. The chain length used for cross bridges between the main backbones must have a minimum length to ensure deformability. Such crosslinking sequences must have a MW greater than 10,000 [159,160]. According to Kerr [161], for UV-curable polyacrylate with styrene side groups, for example, (2-polystyryl methacrylate, long-chain C_{14} diol diacrylates should be used as crosslinker. The crosslinking "density" is also important. According to Havranek [162], in slightly crosslinked model networks, the T_g for long-bridge networks does not depend on the molar ratio of functional groups; in polymers with a high level of short chain crosslinkings it is a function of this ratio.

All hydrocarbon polymers undergo chain scission during degradation. This reaction competes with crosslinking as the termination step of the autoxidation process [163]. Since chain scission and crosslinking may occur simultaneously, the oxidative effect on viscosity and adhesivity will depend on the balance of the two competitive reactions. Formulations based on SIS degrade by chain scission; for SBS-based compounds degradation takes the form of crosslinking. Chain scission and crosslinking occur in the multiblock polymer formulation and their balance leads to increased bond strength, at relatively constant viscosity and service temperature. Electron beam radiation causes the crosslinking of the polybutadiene-α-methylstyrene block copolymer via the butadiene block and its degradation via the α-methylstyrene blocks [164].

b. Polymacromerization

Low molecular polymers are easily processed. They display the advantages of bulky polymers (no migration, no volatiles, etc.) and low viscosity liquids. Their buildup to a high molecular weight polymer via polymacromerization is a relatively simple process. Classic pressure-sensitive formulations have been based mostly on rubber and resin. Natural rubber has been used as an elastomer. Generally, natural rubber has been masticated, that is, depolymerized to achieve a relatively uniform molecular weight and better solubility. In many cases the molecular weight has been restored by using crosslinkers. The synthesis of ready-to-use, inherently tacky raw materials has allowed control of the molecular weight without the need of depolymerization. However, crosslinking has been used to achieve better cohesion. Developments in macromolecular science have led to the technology of macromer manufacture. Such relatively low molecular weight polymers can be postpolymerized (postcrosslinked) by using various polymerization or curing techniques. Some years ago, the coating of pressure-sensitive raw materials was synonymous with their physical and mechanical transformation. In the future, coating will be associated with the chemistry of macromer transformation (polymacromerization). As discussed later, macromers have been synthesized in almost each raw material class. Their starting molecular weight can be varied. Their final molecular weight can also vary depending on how they are "assembled."

As discussed in [1], formulations with 100% solids have been developed as radiation-curable compositions based on crosslinkable monomers and reactive diluents. Prepolymers with better coatability have been favored. It is possible to manufacture a prepolymer which is polymerized by radiation [165] (see also Chapter 8). A patent discloses the preparation of a low molecular weight spreadable composition to which may be added a small amount of catalyst or polyfunctional crosslinking monomer prior to the completion of the polymerization by heat curing. The prepolymer is more viscous and easily applied to a support. A crosslinkable acrylic formulation is a combination of nontertiary acrylic acid esters of alkyl alcohols and ethylenically unsaturated monomers that have at least one polar group; it may be substantially in monomer form or may be a low MW prepolymer or a mixture of prepolymer and additional monomers, and it may also contain other substances, such as a photoinitiator, fillers, and crosslinking agents (such as multifunctional monomers) [153]. The monomers may be partially polymerized (prepolymerized) to a coating viscosity of 1–40 Pa sec before the fillers are added. A prepolymer having a viscosity of (0.3–20 Pa sec) can be synthesized by UV-initiated polymerization. The crosslinking agent is added, and the mixture is coated and polymerized by UV to yield the final product [166,167].

Considering crosslinking as a partial stiffening of the macromolecular sequences, the same effect can be obtained by polymacromerization. It is possible to strengthen the macromolecular structure by partially stiffening it, using macromers to achieve the desired conformability. Polymeric monomers, macromonomers or macromers are useful as reinforcing monomers. Generally, the macromers should have a T_g above 20°C. Representative examples of such macromers are polystyrene, poly-α-methylstyrene, polyvinyltoluene, and polymethyl methacrylate. For instance, a methacrylate terminated styrene macromer is used for medical adhesives for application to the skin. Macromers can be included by the polymerization of vinyl monomers to give better flow, coating, and leveling properties. Ethylene–acrylic acid copolymers have been included in terpolymerization formulation of emulsion based vinylic monomers for tapes [168]. Unlike the synthesis of segregated polymers which are ready-to-use raw materials and need no chemical transformation, macromers require chemical reactions to be transformed into pressure-sensitive raw material or finished product. As presented in Figure 5.5, thermally initiated or radiation-initiated polymerization of a monomer can be carried out to obtain an oligomer or a high polymer. Either can be postprocessed later. For the oligomer this is a necessity; for the polymer this is a possibility. Generally, both are postprocessed as formulated product. The polymerization of the monomer (M) leads to an oligomer (O) or to a polymer (P). After its formulation the oligomer can be polymacromerized to a ready-to-use adhesive product (A_{R1}).

The polymerization can lead to a high polymer also, which (after formulating) can be postprocessed by crosslinking to a ready-to-use adhesive product (A_{R2}).

Generally, ease of processing is the main argument for preparing oligomers instead of polymers, but the possibility of achieving segregated structures is becoming more and more important. The main problems are related to the nature, molecular weight, and reactivity of the macromers. Their reactivity should be examined in the formulated mixture (see later, UV-cured acrylates). The methods used for the synthesis of the macromers are various, and radiation-curing is preferred

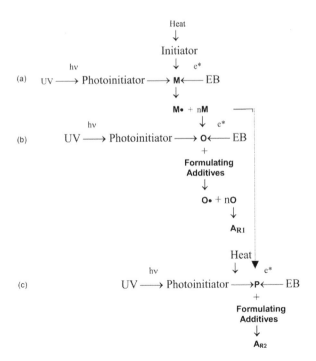

FIGURE 5.5 The manufacture of pressure-sensitive products by polymerization, macromerization, and postpolymerization. (a) Polymerization of a monomer to an oligomer (O) or to a polymer (P); (b) polymacromerization of an oligomer; postpolymerization.

for their postprocessing. Electron beam curing has the advantage that it does not require special initiators or initiating functionalities. Since, most of the monomers commonly employed do not produce initiating species with as sufficiently high yield upon UV exposure, for such photopolymerization it is necessary to introduce a photoinitiator to start the polymerization. Among the different possible types of reactions involved in the initiation of photopolymerization the most important are: (i) radical formation by photocleavage; and (ii) radical generation by hydrogen absorbtion.

Benzoine derivatives, benzylketals, acetophenone derivatives, hydroxy-alkylphenones, acyl-phosphine oxides and substituted α-amino-ketones belong to the group of initiators which undergo photocleavage. Because of their absortion in the long wavelength domain, acylphosphine oxides allow the use of white pigmented formulations also. Radical generation by hydrogen abstraction is characteristic for benzophenone or tioxanthone. Certain aromatic carbon compounds undergo a Norrish type I fragmentation when exposed to UV light:

$$AR-(CO)-\overset{|}{\underset{|}{C}}-X \xrightarrow{h\nu} AR-(CO)^{\cdot} + {}^{\cdot}\overset{|}{\underset{|}{C}}-X \qquad (5.8)$$

Efforts have been made to synthesize polymeric compounds that bear side moieties capable of α-cleavage after irradiation with UV light to generate free radicals. Such compounds are polymeric α-hydroxyalkylphenones, structure (5.9), with MW less than 2000 [169], for example, hydroxy-[4-(1-methyl-vinyl)]-isobutyrophenone, structure (4.10) [170]:

$$\begin{array}{c} CH_3 \\ | \\ -[-\overset{|}{\underset{|}{C}}-CH_2-]_n- \\ | \\ AR(CO)-C(CH_3)_2-OH \end{array} \qquad (5.9)$$

$$\begin{array}{c} CH_3 \\ | \\ -[-CH_2-\overset{|}{\underset{|}{C}}-]_n- \\ | \\ O{=}COCH_2CH_2O-AR(CO)-C(CH_3)_2-OH \end{array} \qquad (5.10)$$

Photoreduction of benzophenone via a charge transfer complex leads to a free radical which reacts with the monomer [171]:

$$(AR)_2C{=}O+:\overset{|}{\underset{|}{N}}-\overset{|}{\underset{|}{C}}H- \longrightarrow (AR)_2{}^{\cdot}C-OH+\overset{|}{\underset{|}{N}}-\overset{|}{\underset{|}{C}}{}^{\cdot} \qquad (5.11)$$

$$\overset{|}{\underset{|}{N}}-\overset{|}{\underset{|}{C}}{}^{\cdot}+CH_2{=}CH-COOR \qquad (5.12)$$

According to Scheiber and Braun [172], radiation cured PSAs are based on: (1) SIS multiarm block copolymers; (2) liquid rubbers with acrylate functionality; and (3) acrylate prepolymers.

The UV curing of "prepolymers" was began with the curing of PVC plastisols, polymer dispersions in plasticizer. If the plasticizer contains an unsaturated polyfuntional acrylic, it is possible to cure it with UV radiation [173]. A large variety of functionalized prepolymers are now available, making it possible to create networks with tailor-made properties. Acrylic monomers are among the most widely used curable systems because of their high reactivity, moderate cost, and low volatility. Acrylated polyester oligomers have been suggested as raw materials for PSA. Such products can be processed as HMPSAs [174,175]. For better radiation curing these products have acrylic side chains. For instance, a UV-curable polystyrene has been developed in which the end groups are methacrylate [176]. In another case a polyacrylate with styrene branching (2-poly-styryl)-methacrylate has been suggested [161]. For such oligomers a long chain C_{14} diol diacrylate has been proposed as crosslinking agent:

$$-C_4H_9-(CH_2-CH_2)_n-CH_2CH_2-O-C-\overset{CH_3}{\underset{O}{\overset{|}{C}}}=CH_2 \qquad (5.13)$$

PSA raw materials that polymerize without initiator were synthesized also. Benzocyclobutenone acrylamide monomer has been incorporated into polymers; its postinduced covalent crosslinking under UV radiation proceeds without the need of any initiator [177].

3. Natural Elastomers

Of the natural elastomers, natural rubber (NR) was the first to be used as raw material for PSA. The so-called rubber-resin adhesives (mixtures of natural rubber with tackifier resins) were the first commercial PSAs. According to Mueller and Türk [144], removable labels have been formulated on a basis of rubber-resin adhesive. The rubber has been tackified with soft resins and plasticizer. NR-based adhesive solutions and styrene block copolymer-based HMPSAs are the main formulations for common tapes. To achieve high shear resistance crosslinked formulations are used. Freezer tapes are coated with an adhesive based on natural rubber, styrene butadiene rubber, regenerated rubber, rosin ester and polyether [60]. This adhesive is also crosslinked. Crosslinked rubber-resin formulations are the most often used adhesives for protective films. Such formulations are soft, and their adhesivity can be regulated by mastication, tackification and crosslinking. Generally, using natural rubber as the main formulation component has the advantage that it is easy to regulate the adhesive properties by regulating the molecular weight. Crosslinked natural rubber formulations are softer than certain common noncrosslinked acrylic recipes. In some cases the softness of the masticated elastomer allows "solid-state" mixing and coating (for tapes) of the formulation components. Natural rubber can be formulated together with polar elastomers and reactive resins (phenolic resins). The crosslinked adhesive displays temperature- and solvent-resistance [178]. The reaction is carried out by mastication of the components. The high molecular weight portion of natural rubber is insoluble in solvents; therefore, natural rubber must be milled to a Mooney viscosity of 75 or below for the complete solubility (at 10% solids) [179]. The milled sample has a narrower molecular weight distribution. Many natural rubber solvent-based formulations contain a crosslinking agent. The process involves breaking down the natural rubber and then building it back up. It is evident that the level of the crosslinking agents and tackifiers used depends on the degree of mastication.

   NR latex is applied for low tack removable formulations, especially for tapes and envelopes [23]. A special domain of natural latex-based formulations is the manufacture of application tapes with paper carrier. Adhesives on NR latex basis are not compatible with other adhesives [180]. Readhering and removable adhesive in a solid stick form (glue stick, applicator crayon) can also be formulated on natural rubber basis [181]. Such products are removable unless the user applies a very heavy coating. A NR latex, a tackifying agent, and a gel-forming agent are the main components of the formulation. A friction reducing component and antioxidant may also be included. Rosin ester and 4,4-butylidene BIS are suggested as tackifier and antioxidant. Gelling agents are aliphatic carboxylic acid salts. NR latex used at a level of 10% or less improves shear resistance of CSBR [182], and also improves the aging characteristics. SBR latex crosslinks and hardens during oxidative aging, whereas NR latex undergoes chain scission and softens, thus a balance is obtained. Pattern or strip coating allows the use of nonremovable raw materials for removable applications. NR latex coated as a contoured surface can be used on paper to give removable protective web that is adequate for acrylic items [183].

4. Synthetic Elastomers

Synthetic elastomers are macromolecular products that display rubbery elasticity. This behavior is the result of their special chemical and macromolecular structure. NR-like stereoregulated polydienes, diene–styrene random and block copolymers, and thermoplastic elastomers on styrene–diene, styrene–olefin, ethylene–propylene–diene, acrylic, etc., basis have been synthesized end tested for use in PSPs. Unlike NR, certain of these products can also be used for noncoated PSPs.

a. Butene Polymers

Butylene and isobutylene polymers and copolymers are used as primary elastomer or as tackifier. Polyisobutylene tackified with rosins is suggested as an elastomer for tapes [184,185]. Low molecular weight PIB is used as a tackifier for butyl rubber or acrylic PSA formulations. According to [186], polyisobutylenes are liquids or solids with molecular weights ranging from 36,000 to 58,000 (liquids) or from 800,000 to 2,600,000 (solids). Nuclear magnetic resonance has disclosed that the isobutylene units are polymerized as head-to-tail sequences [187]. Polyisobutylene (MW 3000) is liquid, whereas high molecular weight PIB (MW 200,000) is a clear solid. Such polymers can be used as solutions or water-based dispersions [185]. There are several tape, label or protective film applications that use these compounds as the base elastomer or tackifier.

Aluminum film coated with a tackified polyisobutylene-based adhesive is used for street marking. The release layer for this adhesive is PVA (10–40%), silicone (20–50%), mineral oil (20–50%), and sodium acetate (10–12%) [184]. Masking tapes are used during the varnishing of cars [188]. They have to adhere to automotive paint, e-mail, steel, aluminum, chrome, rubber and glass and resist the high temperatures used for drying and curing of varnishes. For such tapes NR latex tackified with PIB is suggested. Polybutylene has been used as sealing tapes [189] and medical tapes [190]. Polyisobutylene with a molecular weight of 80,000–100,000 is proposed for medical tapes because it does not adhere to the skin [184]. Polyisobutylene is also employed as base elastomer and tackifier for insulation tapes [118]. Tackified natural and synthetic rubber and polyisobutylene have been used for removable labels [144]. Polybutylene polymers are suggested for hot-melts. They display good compatibility with aliphatic resins, atactic polypropylene and block copolymers. They are not polar, and therefore the adhesives formulated with them have good adhesion on unpolar surfaces. Polybutylenes are saturated; therefore, exhibit aging resistance. Such materials have a lower rate of crystallization than EVAc [191] (this behavior is very important for self-adhesive films). Low molecular weight butene polymers play a special role as tackifiers for polyolefin carrier materials used as self-adhesive films (see later).

Butyl rubber is a copolymer of isobutylene with isoprene (less than 3% isoprene) and has a molecular weight of 300,000–400,000 [180]. The different elastomer grades differ in unsaturation, molecular weight and type of stabilizing agent used. Butyl rubber is manufactured as latex also. Halogenated (chlorinated or brominated) grades of butyl rubber have been synthesized. Butyl rubber (MW 6000–20,000) does not need mastication to dissolve, as many natural and synthetic rubbers do. It can be crosslinked. Partially crosslinked butyl rubbers (terpolymers with divinyl benzene) are commercially available. The crosslinking improves the flow resistance and these polymers are used in sealing tapes. A precrosslinked butyl rubber can also be used instead of partially crosslinking in the mixer [192]. The degree of partial crosslinking prior to mixing with other components may vary between 35 and 75%. The proportion of partially crosslinked rubber in the composition varies inversely with the percentage of crosslinking. p-Quinone dioxime, *p*-dinitrosobenzene, phenolic resins, and similar compounds may be added as crosslinking agent. Butyl rubber is clear and tacky, it is not fragile after aging, and it may be used together with PIB as a tackifier for medical tapes.

b. Diene Copolymers

The first possibility to achieve natural rubber-like properties is to "copy" the chemical composition and structure of NR, that is, to synthesize macromolecular compounds having as base monomer isoprene, butadiene, etc. with a (more or less) stereoregulated structure. Such products have been manufactured as solids, that can be processed as solvent-based adhesives.

Polar (halogenated) diene monomers have also been polymerized, that have special characteristics such as nonflammability. As an example, there are nonflammable tapes with an adhesive based on polychloroprene rubber [193]. Copolymers of chloroprene with other polar monomers are synthesized as water-based dispersions too. Carboxylated neoprene dispersions are used for special temperature-resistant, nonflammable formulations.

c. Styrene–Butadiene Random Copolymers

Rubber-like properties have been achieved for styrene–butadiene random (SBR) copolymers also. These products have been used as raw materials for solvent-based PSA. Styrene–butadiene rubbers contain a residual double bond that enables further polymer chain growth (crosslinking), to form a network with improved mechanical properties without apparently effecting the T_g. Unfortunately, they are hard and their tackification is difficult. Styrene–butadiene rubbers possess better aging stability than natural rubber. They harden during aging [58]. Generally polymers with a low level of styrene are suggested for PSAs, mainly for tapes. Freezer tapes are coated with an adhesive based on styrene–butadiene rubber (with 12% styrene) [54].

Styrene–butadiene rubber is not recommended for medical tapes [189]. As mentioned earlier, styrene–butadiene latex is used as top coat for paper [4].

Carboxylated styrene–butadiene (CSBR) copolymers have been synthesized and used as aqueous raw materials for PSA. The acid comonomer works as a stabilizing agent for the aqueous dispersion [194]. The styrene content, molecular weight, and gel content are the main quality parameters for such dispersions. Gel and styrene content harden the polymer and makes its tackification difficult. With a higher styrene content the latex does not respond well to the addition of resin. Above 30% gel content, CSBR tack decreases rapidly [182]. Shear resistance increases with the gel content. Deep freeze properties of formulations are improved if the butadiene level of the polymer is greater than 80%. NR latex used at a level of 10% or less improves the shear resistance of CSBR and also improves the aging characteristics. Quick stick and high temperature shear are required for tape application [195]. Such properties may be achieved by using CSBR. A screening formulation for tapes includes CSBR, coumarone-indene resin, hydrocarbon resin, plasticizer, and thickener.

The T_g and modulus of acrylic latices can be adjusted more easily by polymerizing with different monomers, at various monomer concentration. Therefore, in comparison with CSBR, acrylates can be used with a low level of tackifiers or with none [196]. On the other hand, multiphase structure given by sequence distribution and crosslinking ensures special properties. Thermoelastic, multiphase styrene–butadiene copolymers have been prepared by radical emulsion copolymerization of styrene with styrene–butadiene copolymers having a molecular weight of 180,000–320,000 and polydispersity of 3–6 [197]. Such graft copolymers exhibit a gel content of 81%, an increase of 30% in tensile strength and 20% in elongation, and a decrease of 80% in the residual elongation in comparison with the ungrafted polymer.

d. Styrene Block Copolymers

The elastic component of the adhesive is a natural or synthetic elastomer. NR is a well-defined polymer of isoprene. Isoprene and other diene monomers have been polymerized to form new synthetic rubbers. In a quite different way, block copolymers have been obtained also, in which elastic and thermoplastic properties have been combined. Styrene-block copolymers have been produced. Styrene olefin block copolymers with an ABA sequence distribution and styrene endblocks (see Equation 5.14) have been synthesized since 1965.

$$-[-CH-CH_2-]_x-[-\overset{|}{C}-CH=CH-\overset{|}{C}-]_y-[-CH-CH_2-]_x- \qquad (5.14)$$
$$\underset{AR}{\overset{|}{}} \qquad\qquad\qquad\qquad\qquad\qquad \underset{AR}{\overset{|}{}}$$

$$A \qquad\qquad\qquad B \qquad\qquad\qquad A$$

where $x = 200-500$ and $y = 700-1500$.

Sequenced block copolymers of styrene with different dienes and olefin comonomers have also been produced. First polymers with unsaturated middle segments (butadiene and isoprene) were developed, then products with saturated middle sequences (having about 30% styrene) [198].

According to Gergen [199], styrene–diene block copolymers appeared on the market in 1965, styrene–olefin block copolymers in 1975, and hydrogenated block copolymers in 1983. Styrene block copolymers used for hot-melts are based mainly on SIS block copolymers with about 15% styrene [200]. Hydrogenated midblocks were first used in 1972, radial polymers in the mid-1970s, crosslinkable technology in 1979, and interpenetrating networks in 1984 [201]. In bisequenced copolymers, the elastomeric midblock can be polybutadiene (SBS) or polyisoprene (SIS); in trisequenced polymers, polyethylene–butylene (SEBS) and polyethylene–propylene (SEPS) sequences are built in [152]. Commercially available copolymers in this range have a styrene content of 14–25% w/w. These polymers are applied in an amount of 25–50% by weight, preferably 30–40% in the PSA formulation.

According to Rader [202], thermoplastic elastomers can be classified as: (i) block copolymers, (ii) thermoplastic polyolefins, and (iii) thermoplastic vulcanizates. Actually the styrene block copolymers are the most important TPEs as raw materials for pressure-sensitive products. The main thermoplastic elastomers [203] are styrene block copolymers, olefin copolymers, polyurethanes, copolyesters, and polyetheramides. Segmented EVAc copolymers have also been synthesized [204].

During the development of styrene block copolymers, various products were synthesized. The thermal resistance and cohesion of the copolymers have been improved; their reactivity and functionality have been changed order to allow better postprocessing. These modified styrene copolymers have been used as macromers. The degree of crosslinking and chain flexibility have been regulated by controlling the end groups and the molecular weight of the sequences. Functionalizing the middle and end sequences has also been tried (Figure 5.6).

Uncrosslinked and unentangled polymer chains do not contribute to the strength of the elastomer. Therefore, pure triblock SIS copolymers have also been manufactured. Pure (100% triblock) polymers have a high T_g endblock [205]. Diblock units work as low cohesion diluting agents in the composition and decrease the cohesion. Johnsen [206] discussed the development of pure triblock SBCs with styrene-α-methyl styrene endblocks, giving higher T_g and better cohesion. α-Methyl-styrene has a higher T_g, 170°C, and therefore, provides better thermal properties than styrene.

According to Johnsen [206], for SIS a 1/2 ratio of styrene/α-methyl-styrene and a molecular weight of 15,000 for the endblock are preferred. Generally the hard segment size influences the melt transition temperature of TPEs [207]. The T_g of polystyrene is function of its molecular weight up

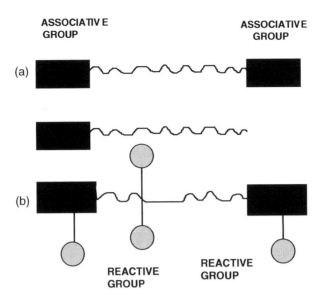

FIGURE 5.6 Schematic presentation of the macromolecular sites of a styrene–diene block copolymer able to be modified in postprocessing. (a) Unmodified triblock and diblock molecules with associative and elastic segments; (b) functionalized segments.

to an MW of 20,000. The higher MW of SIS allows its use where higher cohesive strength and elevated temperature performance are desired, or more tackifier resin should be used (where economy is a prime consideration).

Johnsen [208] examined the influence of the triblock/diblock ratio also. The principles of formulating such polymers have been described by Dehnke and Johnsen [201]. SIS triblocks and SI diblocks have been manufactured separately and mixed together in a desired ratio which allows the diblock level to be controlled precisely. Common SBCs are manufactured by using the coupling process. Commercial polymers contain 15–20% diblocks. Rubber theory states that uncrosslinked and unentangled polymer chain ends do not contribute to the strength of an elastomer. A pure block exhibits a sharper phase separation as multiblocks and consequently greater heat stability and creep resistance [209]. A range of copolymers with up to 42% diblocks have been synthesized. Such polymers are recommended for label PSA. Increasing the diblock content improves peel and tack, by a static shear of over 180 h, which corresponds to the label requirements. SAFT decreases with the diblock content. The number of molecules associated decreases at the end of the molecule. Viscosity decreases increasing the diblock content also (from about 40,000 to 21,000 mPa sec) for a standard formulation with an SIS/resin/oil/antioxidant ratio of 100/150/25/1 [210]. Roos and Creton [211], studied the fibrillation behavior of SIS with various (from 0 to 54%) SI content. They stated that differences due to the change of the SI content occur in the frequency range of 5×10^{-3} to 5×10^{-1} rad sec^{-1}, where an increase in the SI content leads to a decrease the elastic modulus G' and to an increase in the loss modulus G''. In this range the blends are more dissipative if there is more SI. A broadened glass transition is observed and a second relaxation process appears at low frequencies (high temperatures) while the SI content is increased; the lower the SI content, the pronounced the strain hardening.

The saturated midblock thermoplastic elastomers are commercial ABA-type block copolymers in which the polyisoprene or polybutadiene are hydrogenated. The block copolymers with a saturated midblock segment have an M_n of about 25,000–300,000. As shown in [212], the elastomer sequence of a rubbery block copolymer must be designed to adequately suppress crystallinity without raising the T_g at the same time. A saturated olefin rubber block has to be obtained with the lowest possible T_g, and the best rubbery characteristics. The rubbery elasticity is maximized when the elastomer segment is completely amorphous. In commercial SEBS block copolymers the best rubbery performances are obtained with rubber blocks having a T_g of about $-50°C$ and no detectable crystallinity. The strength and stiffness of the block copolymer depend on the chemical nature of the segments and their mutual interaction. Block copolymers having the same styrene content, about the same block molecular weight, but different types of midblocks, display different mechanical properties. The SEBS polymer has the highest strength and modulus. The stress–strain behavior of such rubbery polymers can be adequately modeled by the Mooney–Rivlin equation as an elastic network (considering the trapped entanglements as finite crosslink junctions) or using the Guth–Gold equation relating the stress level (modulus) to the hydrodynamic effects of the styrene blocks considered as filler (see Chapter 3, Section I).

In the classic segregated TPE construction, the butadiene (isoprene) units supply the elastic parts of the network and the associated aromatic rings of styrene (or styrene derivatives) provide the connection points of this network. The whole material works as a (spherical particle or short fiber) reinforced composite. It is evident that supplemental "bridges" between the chains can be formed if the elastic part of this construction is functionalized also. Such transformation is made via copolymerization with vinylbutadiene or 3,4-isoprene units.

ABA block copolymers and AB block copolymers of the same type of composition in another geometry, star-shaped block copolymers or radial block copolymers, have also been developed [210]. According to the structure and placement of the side chains, branched, radial teleblock and multiarm star block copolymers are known [213]. Various polymer structures are described by the term "branched," "radial," and "star." "Branched" is a general term indicating a nonlinear structure, which may contain various polymeric subunits appended to various places in the main polymer chain. "Radial" generally refers to the branched structures obtained by linking individual polymer segments, to yield a polymer mixture having four or fewer armes joined centrally. The term "star" describes the structure of a multiarm polymer with copolymer arms which are joined together at a nucleus formed of a linking group. Nonterminating coupling agents are preferred as linking agents for star structures. A desirable melt viscosity can be obtained selecting only star block copolymers with six or more arms [213]. Multiarm styrene–isoprene star diblock copolymers with 18 arms having various styrene contents and styrene block weights have been synthesized also [214]. The effect of the number and molecular weight of the arms influences the solid-state morphology of SIS block copolymers [215]. Polymers with 2–18 arms and molecular weights of 100,000, 23,000 and 33,000 have been synthesized. Structures with more than eight arms are bicontinuous and ordered. Such segregation of the side chains is common for long and flexible chains in linear polymers too (Figure 5.7).

Radial and teleblock copolymers with isoprene based branches are used for PSA tapes [216], but butadiene-based armed polymers have been proposed also [217]. The star-shaped (radial) SBS copolymers developed by Phillips (Solprene 416× and 417×) have styrene contents of 30 and 20%, respectively, and molecular weights between 140,000 and 180,000 [218]. The grade 416× with more styrene has a lower molecular weight but higher tensile strength (200 vs. 90 kp/cm^2) and lower elongation (720 vs. 920%) than 417×. The copolymer with lower styrene content exhibits better peel and tack values. According to Lau and Silver [213], the formulation with tackified star block copolymer gives a higher shear resistance (at least 1000 min according to ASTM-D3654, overlap shear to fiberboard at 49°C). The mechanical properties of a radial block copolymer are better. The tensile strength for a radial styrene block copolymer is higher (22 MPa) than for a linear styrene block copolymer (15 MPa), and the 300% modulus is higher also (5 vs. 2.5 MPa). Elongation at break is smaller (620 vs. 950%).

The elastomeric midblocks and the polystyrene end domains are specially designed to be thermodynamically incompatible and to form a two-phase morphology [219]. The polystyrene blocks at the end of molecules form an associative crosslinked structure. If the polystyrene level is lower than 50%, polystyrene domains are formed because of incompatibility [220]. A physical

(a)

(b)

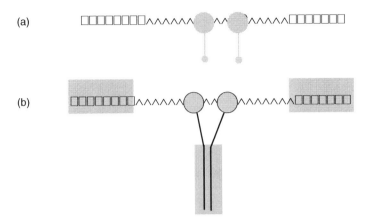

FIGURE 5.7 Side chain segregation depending on the flexibility and length of branches. (a) Short branches; (b) domain buildup by long branches.

network is built up that disappears only above the T_g of polystyrene. The compatibility of the polystyrene endblocks and of the elastomeric midblocks depends on their molecular weights and their volume fractions in the mixture. Phase separation in the radial block copolymer is higher owing to the increased organisation of the molecule. Therefore, physical crosslinking is also higher. The better connected pure polystyrene domains ensure higher elasticity of the product [221] (see also Chapter 3). Star-shaped SBCs have a (apparently) higher molecular weight. Therefore, they are more reactive to radiation-curing. The first radiation curable styrene block copolymer was processed like a common thermoplastic elastomer, but it was also able to crosslink. This was a star-shaped SIS block copolymer. Because of its high molecular weight, a low radiation dose was needed to produce crosslinking. Because of its star structure, partial crosslinking does not modify (essentially) the flexibility of the molecules.

Styrene–isoprene block copolymers are more expensive than natural rubber. Alternatives are SBS and high styrene content (20–25%) SIS polymers. The adhesive for medical tapes has to be resistant to light and heat, show skin tolerance and be permeable to water vapor. Such an adhesive may also include styrene copolymers. Common HMPSA systems for tapes are based on SIS, a resin which is partially compatible with the midblock of the SBC, and a plasticizer (oil). Tack properties of SBC are given by the midblock. Its mobility is increased adding oils and tackifiers to the thermoplastic elastomer. A compatible resin is more effective than oil. The other part of the polymer sequence, the styrene endblock is responsible for the cohesion and mechanical properties and influences the processing conditions. Its hardness is reduced by oils which are compatible with the styrene domains. Unfortunately, oils reduce cohesion [221]. According to Jacob [222], SIS block copolymers are used exclusively for tapes; as a less expensive alternative SBSs can be used, but they are less suitable because of their greater rigidity and lower compatibility with tackifier. The branched isoprene midblock is more flexible and more compatible than the butadiene (butene) block. In some cases the viscosity of SBS-based HMPSA formulations may change during storage, that is, these products are less aging-resistant. The type of oil used (mineral oil or process oil) also influences the stability of the HMPSA [223].

Carboxylated styrene–butadiene block copolymers can also be used [224]. Styrene–acrylate block copolymers have also been synthesized [225]. They can be crosslinked via methacrylic end groups. Styrene block copolymers are used as compatibilizers (2–8% w/w) for the manufacture of plastic carriers and the recycling of plastics [226]. To improve the radiation curability of SIS block copolymer, products with a low styrene content (less than 15%) have been developed. These products can be crosslinked in a mixture with acrylic monomers and tackifier resins, giving adhesive with improved temperature resistance and cohesion. Holding power values of more than 30 h have been obtained at 70°C (PSTC-7) [227].

The special segregated structure of SBCs displays a crystal-like network of styrene domains. Such ordered structure (actually crystalline order) ensures the excellent mechanical properties of some plastomers. Several attempts have been made to copy this structure, that is, to prepare block copolymers with domain-like structures (see Acrylic Block Copolymers later). The main problems in preparing such polymers are related to the synthesis of blocks and to the ability of the chain segments to buildup ordered structures. If it is possible for a polymer chain to remain active ("alive") and to grow for a long time, it should be possible to form long segments of two or more monomer molecules in the same polymer chain. The first example of living polymers was disclosed in 1956 by Szwarc [228] in the copolymerization of styrene with isoprene using sodium napthalene as anionic initiator. Rather sophisticated block copolymers have since been synthesized using lithium alkyls as initiator. Complex, bifunctionally anionic initiators like 1,4-*bis*-(4-(1-phenylvinyl)phenyl butane have also been tested [229].

Isotactic polypropylene has a T_g near 0°C, therefore it becomes brittle when refrigerated. This deficiency was overcome in part by blending isotactic PP with rubber. It has been supposed that incompatibility in blending would be decreased if the rubber molecule were bonded chemically to PP. In some cases such block copolymers have been synthesized with the help of Ziegler–Natta catalysts [230]. Our investigations in the late 1960s using soluble metallocene catalysts allowed the synthesis of polar monomer-based copolymers with short blocks [231]. Theoretically, if we wish to synthesize tailored block copolymers, we have to control certain characteristics of the polymer chain: the molecular weight and molecular weight distribution of each segment; the order of segments of specific monomers; the number of segments in the monomer chain; the composition and order of monomers in each segment; and the linearity, stereochemistry and microstructure of the olefins and diolefins in each segment. In practice, the usability of such a copolymer depends on the interactions between its segments which influence its processibility and mechanical properties.

Styrene and alkylstyrenes are unique monomers, because they can be polymerized by means of many mechanisms and the bulky aromatic side chain of the vinyl unit may act as a functional group because of its electrons. Associations are also possibile. Similar associations based on dipole–dipole interactions have been found in chemically crosslinked polymers also. In copolymers of oligomethylene dimetacrylic esters with oligo(ethylene glycol) derivatives, interactions between the soft, elastic "bridges" between the crosslinking portions allow the formation of an ordered structure at 10–20°C above the T_g [232].

Statistical (random) copolymerization of butadiene and styrene has been used as synthesis procedure of block copolymers also. In this case the segregation in the polymer is achieved by built-in ionic groups. Matsumoto and Furukawa [233] copolymerized styrene and butadiene with maleic acid (2%) and transformed the acid reaction product in salts. The macromolecular ions buildup a star-shaped structure that substantially improves the mechanical properties of the polymer. Because of the dissociation of the ionic bonds at higher temperatures only, such polymers display good temperature resistance (up to 150°C). In comparison, a thermally reversible network based on hydrogen bonds (in experimental butadiene block copolymers having urazol groups), although it produces a longer rubber-like plateau of the modulus, is more labile and displays poorer mechanical performances [234]. Star-shaped block copolymers of styrene and ethylene oxide with two arms of polystyrene blocks and two arms of polyoxyethylene have also been prepared [235]. Star polybutadienes with four arms have been included between linear polybutadiene segments [236].

The behavior of SBCs illustrates that the segregated and ordered structure is a condition for excellent mechanical properties. Bocaccio [237] synthesized a sequenced styrene–isoprene copolymer by depolymerizing natural rubber (MW 50,000) and grafting styrene onto the NR chains to produce a thermoplastic elastomer. Because such compounds are only slightly segregated, the mechanical characteristics of the polymers are not as so good as those of SIS or SBS. On the contrary, the bond strength of HMPSA formulated with these TPEs are comparable with those of compounds based on commercial SBCs.

The special characteristics of styrene-based segmented copolymers forced the development of similar structures based on other monomeric segments. Alternating block copolymers have been prepared from aromatic polyethersulfones and poly(dimethylsiloxane). Phase separations in rubbery polydimethysiloxane and vitrous polyethersulfone domains have been observed [238]. TPEs based on hard fluorelastomer segments and soft fluoro resin segments have also been developed [239]. Star-branched polyamides (nylon 6) retained part of their mechanical integrity even after melting, indicating the presence of enhanced entanglements [240]. In certain cases, properties other than mechanical performance have also been improved. For instance, dotted block copolymers of poly (3,4-diisopropylidene cyclo-butene) with polynorbornene exhibits electrical conductivity of 10^{-4} S/cm [241].

Polymers with segregated structure can also be prepared by grafting. Grafting of polytetra-hydrofuran chains onto polybutadiene leads to polymers having a two-phase structure of amorphous rubber and crystallites of polyether [242]. Thermoplastic elastomers with hydrogen bonding have been prepared from polybutadiene modified with 4-hydroxyphenyl-1,2,4-triazoline-3,5-dione [243]. The formed urazole groups act via hydrogen bonding, building thermally reversible crosslinkings. Polymerization of vinyl chloride in the presence of modified Ziegler–Natta catalysts comprising long-chain organoaluminum compounds (e.g., polystyrene- or poly-isoprenyl-aluminium dichloride) leads to block copolymers.

The styrene–diene block copolymers prepared in various ways may have quite different compositions and structures. Because of their hard segments such polymers are always harder than natural rubber (Table 5.2). Their modulus values are much higher than the common values (1×10^5 and 2×10^4 [253]) for PSAs. They behave like a filled natural rubber. Although the performance of such materials can be easy regulated by changing the molecular weight, styrene content, and sequence buildup, for pressure-sensitivity they need to be tackified.

Due to their multiphase structure, the tackification of SBCs supposes special interactions with each of the polymer phases, that is, unlike natural rubber, SBCs require tackifier mixtures. An additive may be compatible with both the elastomeric phase and the styrenic phase, compatible with only one of these two phases or compatible with neither phase [158]. In general, resins compatible with only the elastomeric phase or with only the styrenic phase provide the best balance of properties. Resins and oils that are compatible with the elastomeric phase soften the final product, while resins that are compatible with the styrenic phase harden it. Both reduce the viscosity and affect the phase with which they are compatible. It is evident, that in such multiphase compounds the concentration of the tackifier resin influences the morphology of the polymer also. According to Meyer [254], styrene block copolymers can have five different morphologies, depending on the styrene content. For styrene concentrations of <15%, 15–25%, 35–65%, 65–85 and >85%, spherical fibrillar and lamellar structures are formed.

For postcrosslinked systems based on block copolymers the behavior of tackifier resin (absorbancy) towards radiation has to be taken into account also. For such systems the use of an unsaturation index U_f of the formulation has been proposed [152]:

$$U_f = \sum_i^t w_i U_i \qquad (5.15)$$

where i is a particular oligomer in the formulation, w_i the weight fraction of oligomer i, U_i the unsaturation index of oligomer i, and t the total number of ingredients.

e. Polyacrylate Rubbers

High molecular weight elastomers based on acrylates have been produced by emulsion polymerization. Such polymers do not contain carbon–carbon double bonds as polydienes do. Their

TABLE 5.2
Mechanical Characteristics of Segregated Elastomers

Elastomer		Mechanical Characteristics			
Type	Characteristics	Tensile Strength (N/mm^2)	300% Modulus (N/mm^2)	Shore(A) Hardness	Reference
Natural rubber	Uncrosslinked	1–5	—	—	[244]
	Uncrosslinked; filled with 50% carbon black	24	10	—	[245]
Polybutadiene[a]	Uncrosslinked	0.3–0.5	—	—	[130]
Polybutadiene	Crosslinked	4–5	—	—	[130]
SBS copolymer	Linear; 31% styrene content	31	3.8	71	[246]
SBS copolymer	Radial; 30% styrene content	27	4.5	71	[247]
SBS copolymer	Triblock	30–40	—	—	[130]
SBS copolymer	Full triblock; 29% styrene content	31	—	65	[248]
SBS copolymer	Radial; styrene/butadiene ratio 40/60; MW 120,000	22	—	85	[220]
SBS copolymer	Linear; multiblock; styrene/butadiene ratio 43/57, MW 70,000	15	—	80	[220]
SBS copolymer	Melt index (g/10 min) <1	35	3	75	[249]
SBS copolymer	Melt index (g/10 min) 6	35	3	65	[249]
SBS copolymer	Melt index (g/10 min) 10	25	1	35	[249]
SBR copolymer	—	15.5–22	7–11	—	[250]
SEBS copolymer	20% Triblock	282	30[b]	—	[158]
SEBS copolymer	30% Triblock	490	—	72	[158]
SEBS copolymer	65% Triblock	34	27	—	[158]
SEBS copolymer	29% Styrene content	63	—	83	[219]
SEBS copolymer	28% Styrene content	49	—	—	[219]
SEBS copolymer	14% Styrene content	34	—	—	[219]
SIS copolymer	18% Styrene content	27	—	39	[251]
SIS copolymer	Fully coupled triblock	4500	—	—	[208]
SI copolymer	Fully diblock	20	—	—	[208]
MPR[c]	Halogenated polyolefin TPE	12.1	3.7	60	[252]
MPR[c]	Halogenated polyolefin TPE	13.1	4.5	70	[252]
MPR[c]	Halogenated polyolefin TPE	13.4	7.2	80	[252]

[a]Nonsegregated elastomer used as comparison.

[b]100% modulus.

[c]MPR is melt processed rubber.

backbone includes a saturated alkyl and alkoxy acrylate (95–99%) and (1–5%) curing monomers having a chlorine or hydroxy functionality in the side chain.

Products with carboxy, epoxy, and isocyanate functionalities have also been developed. The main polymer chain is based on ethylhexyl or butyl acrylate. These polymers can be used for HMPSAs (reactive hot-melts), solvent-based PSAs, pressure-sensitive sealant tapes, and sealant caulks. They are used in tackified (40% resin) or plasticized formulations. Polyisobutylene can also be added. Their quick stick values are low; therefore, a common tacky PSA can also be added to the formulation (50 parts PSA to 100 parts acrylic rubber). The peel value of the tackified formulation on aluminum (3–5 ml PSA coated on PET) is about 120–140 oz/in. They can be cured with isocyanates or UV-curing agents (e.g., *p*-chlorobenzophenone). Polyacrylate rubber-based formulations can include high loadings of fillers (200 parts hydrated alumina as filler per 100 parts rubber). The mill processability and rheology of acrylic rubbers depend on the crosslinking agent used for their cure [255]. Acrylic rubbers based on ethylacrylate-methacrylic acid–allylglycidyl ether and ethyl acrylate or glycidyl methacrylate can be crosslinked by using onium salt catalysts [256]. Acrylic rubber latices having glass transition temperature of $-20°C$ to $-60°C$ with balanced mechanical characteristics and excellent hysteresis properties contain N-methylol acrylamide as crosslinking monomer and preferably itaconic acid [257]. Peelable protective films (30 μm) comprising methacrylate-(ethylene–methyl methacrylate) copolymers, an inorganic filler, and slip have been blow-molded. Such films are useful for protecting rubber articles and show a peel strength of 165 g/25 mm, in comparison with 500 g/25 mm for ethylene–ethyl acrylate copolymers [258].

f. Acrylic Block Copolymers

According to Beaulieu et al. [259], the first acrylic hot-melts having a common linear, non-sequenced (block-less) structure were introduced in the late 1970s. They were characterized by a modest balance of PSA properties and low cohesion at elevated temperatures [146]. These polymers show no rubbery plateau and exhibit a linear relationship between modulus and temperature. In the early 1980s, thermally reversible acrylic HMPSAs were introduced [260]. Such formulations are harder (Williams plasticity number 2.7) than the early common acrylics and exhibit better cold flow resistance. The new, phase-separated acrylic block copolymers display (like SBCs) a thermally reversible crosslinking. As the material cools the viscosity doubles with a 28°C temperature drop. As the temperature increases, the crosslinking mechanism is deactivated and the modulus drops. Early acrylic HMPSAs exhibited a viscosity of 30,000 cP at 350°F; the new polymers have 8000–30,000 cP under the same conditions.

The new acrylic HMPSAs exhibit peel values of 62–63 oz/in. (PSTC-1, 25 μm coating thickness on PET), loop tack of 2.5–30 oz and rolling ball of tack of 5–7 in. The shear values (according to PSTC-7, 1.2 in. × 1 in. × 1 kg) are 15–76 h at room temperature. As seen, the rolling ball values are (in comparison with a common value of 1–2.5 cm) to high, that is, the material is not tacky enough and the shear values at normal temperature are also low. According to Sanderson [260], such values ensure adequate shear resistance. In reality, they are not enough to satisfy practical requirements.

The hard blocks of such acrylic block copolymers can contain styrene derivatives also to achieve segregation by incompatibility and/or association. Segmentation and chemical crosslinking are also used. Acrylic hot-melt adhesives based on acrylic block copolymers containing zinc carboxylates have been prepared by Enanoza [261]. Such HMPSAs with good melt flow and cohesive strength contain zinc carboxylates and copolymers of methacrylic ester (with C_{1-14} non tertiary pendant group) with polar comonomers and 2-(polystyryl)–ethyl methacrylate macromer. Although theoretically segregation may appear due to the styrenic component, the real elastical network is built up by crosslinking with zinc acetate. Drzal and Shull [262] used as model adhesive a triblock copolymer with poly(n-butyl-acrylate) midblock and poly(methyl methacrylate) endblocks. A filler volume fraction of 0.5 substantially stiffens the adhesive, while also reducing the overall deformation.

Polyurethane–acrylate hot-melt adhesives have been prepared from polyisocyanate–polyol copolymers and unsaturated monomers (alkyl acrylates and vinyl esters) [263]. AB block copolymers containing methacrylic acid and/or methyl methacrylate blocks (methyl methacrylate–styrene block copolymer, tert-butyl methacrylate–styrene block copolymer and tert-butyl methacrylate–methyl methacrylate block copolymers) have been prepared by selective cleavage of methacrylic esters [264]. Low molecular weight block copolymers have also been prepared from acrylonitrile, methacrylic acid and butadiene [265]. Blocky butyl acrylate–styrene copolymers prepared via step emulsion polymerization showed a broad transition region in DSC, in comparison with statistical copolymers, but two well-resolved relaxations associated with the glass transitions of styrenic and acrylic sequences [266,267]. The foregoing examples show that there have been many attempts to prepare segregated acrylic block copolymers; however, only a few experimental products promise industrial applicability. According to [268], in 1988 no commercially available acrylic hot-melts have been produced. Actually, such hot-melts are available as warm-melts which need postcuring.

g. Silicones

Elastomers based on silicones can also be used as raw materials for PSPs [269–273]. In comparison with organic polymers silicones are fluids at high molecular weight (100,000) and crystallize at very low temperatures ($-50°C$); they exhibit a T_g of $-120°C$ [272]. Silicone adhesives are temperature-resistant and can be coated on apolar substrates [274]. Double side-coated mounting tapes used in electronics have silicone adhesive on one side and acrylic adhesive on the other [275]. For temperature-resistant tapes tackified silicones (water-based dispersions of polydimethylsiloxane) have been suggested [276]. High temperature methylsilicone PSAs provide elevated thermal stability. Silicone adhesive with silicone release has been used for freezer tapes [277]. A fluid permeable adhesive useful for transdermal therapeutic device to human skin for periods up to 24 h is based on an acrylic, urethane, or elastomer PSA mixed with a crosslinked polysiloxane [278]. The therapeutic agent passes through the adhesive into the skin. PSAs made with silicone curable at low temperatures have been developed from rubber-like silicones containing an alkenyl group, a tackifying silicone resin, a curing agent and a platinum catalyst [279]. Such adhesives for heat-resistant aluminum tapes cure in 5 min at 80°C. The adhesion of silicone PSAs can be improved using high molecular weight dimethylsiloxane or dimethylsiloxane-diphenylsiloxane copolymer gum and resins having $R_3Si_{0.5}$ units ($R = C_{\leq6}$ hydrocarbyl) and SiO units, with resins having a polydispersity of ≤2.0 [280]. Emulsified silicone PSAs could be blended with rubber latex and acrylic emulsion PSAs to improve their high temperature shear performance and adhesion to low energy surfaces [281].

h. Polyurethanes

Polyurethanes are synthesized as plastomers, elastomers or viscoelastic compounds. PSPs use polyurethanes as rubber, viscoelastic PSA, top coat, and ink or film/foam-like carrier material [282]. Suitable adhesives are based on the chemistry of polyurethanes, that is, on the chemical reaction between isocyanates and polyesters or polyethers having hydroxylic terminals (see also Chapter 8). The reactivity of isocyanates is used for crosslinking reactions also. The crosslinking degree regulates the hardness, tension yield and elastic modulus of the polymer. Its thermal stability depends also on its chemical composition. Polyurethane adhesives are built up generally from hard and soft sequences. The soft sequences contain long polyols, polyethers and polyesterdiols; the hard sequences include polyisocyanates and short chain polyols or diamines. Thermodynamically these segments are incompatible, that is, there is a two-phase morphology. The molecular weights of soft sequences are about 600–3000. Primary alcohols and hexamethylene diisocyanate are the preferred base components. Polyetherurethanes are more sensitive to thermal oxidative degradation. The resistance to hydrolysis is more pronounced for polyesterurethanes [283].

Polyurethanes were first raw materials for classic solvent-based adhesives. Later, water-based and hot-melt adhesives were formulated also. For instance, for PSAs, a glycerine–propylene oxide copolymer (with a molecular weight of 8000) has been suggested for use as soft segment together with a hard segment of toluene diisocyanate–propylene oxide (with a molecular weight of 850). The reaction of the components occurs *in situ*, that is, on paper, after coating [284]. A two-package type of PSA, that is, an adhesive having two components based on PUR contains a polyol component (MW 500–100,000) and a polyurethane having a terminal, free isocyanate group obtained by the reaction of an isocyanate component having oxadiazine-2,4,6-trione ring and a polyol component having an average molecular weight of about 62–500 [285]. Photopolymerizable polyurethane precursors having latent catalysts as salts of organometallic complex cations form Groups IVB, VB, VIB, VIIB, and VIIIB have also been prepared [286].

Thermoplastic elastomers based on polyurethanes have also been manufactured. Such thermoplastic polyurethanes (TPUs) are synthesized via polyaddition of polyisocyanates and polyols [see Equation (5.16)].

$$
-[-\underset{\underset{O}{\parallel}}{C}-\underset{\underset{H}{\mid}}{N}-X-\underset{\underset{H}{\mid}}{N}-\underset{\underset{O}{\parallel}}{C}-O-(CH_2)_a-O-\underset{\underset{O}{\parallel}}{C}-\underset{\underset{H}{\mid}}{N}-X-\underset{\underset{H}{\mid}}{N}-\underset{\underset{O}{\parallel}}{C}-]_{x}-[-O-R-O]_{y}-[-\underset{\underset{O}{\parallel}}{C}-\underset{\underset{H}{\mid}}{N}-X-\underset{\underset{H}{\mid}}{N}-\underset{\underset{O}{\parallel}}{C}-O-(CH_2)_a-O-\underset{\underset{O}{\parallel}}{C}-\underset{\underset{H}{\mid}}{N}-X-\underset{\underset{H}{\mid}}{N}-\underset{\underset{O}{\parallel}}{C}-]_{x}-
$$

(5.16)

The hard segments of such "block copolymers" are built up from polyisocyanates and a chain extender, and the soft sequences are polyols. The soft blocks are generally polyether or polyester with low T_g value and weak intermolecular forces, but hydroxy-terminated polybutadienes or polydimethylsiloxanes can also be used. As the polyisocyanate aromatic and cycloaromatic diisocyanates such as 4,4′-diphenylmethanediisocyanate (MDI) or hexamethylenediisocyanate (HDI) can be used [287]. As the polyol, a polyester (e.g., butandiol-1,4-polyadipate or ethanediol-butanediol-polyadipate) has been proposed [288]. The composition of these TPEs is given by the general formula of multiblock copolymers

$$-(AB)_n-$$

(5.17)

where n should be higher than 10. The soft sequences have a relatively low molecular weight of 600–4000. Such polymers possess a modulus of elasticity of 10–700 MPa. As mentioned earlier, such multiblock TPUs include a spherulitic crystalline phase. The crystalline regions reinforce the polymer. If more voluminous globular parts are synthesized as hard domain, the mechanical performance of the elastomer can be improved. As reported by Kunii [289], the introduction of polyoxy(tetramethylene glycol) in the polyethylene oxide and polypropylene oxide sequences without modification of the global sequence length of a common propylene oxide/ethylene oxide polymer increases its tensile strength. Polyurethane ureas synthesized with ethylene and propylenoxide, MDI and ethylendiamine display improved thrombo-resistance [290]. Thermoplastic polyurethanes based on ring-opening polymerization of δ-valero-lactone and ε-caprolactone are biodegradable [291]. Such polymers possess moduli of elasticity of 0.12–3.83 and elongations of 60–2000%. α-Alkylolefin-urethane block copolymers have also been synthesized [292]. Such copolymers are based on α-alkylolefin polymers, bearing NCO reactive groups, polyisocyanates, or isocyanate prepolymers, and chain extenders. Generally, PUR derivatives are used for medical PSPs.

i. Propylene Copolymers

Amorphous polypropylene based HMPSAs have been known for many years [293]. Atactic polypropylene (with a molecular weight of 15,000–60,000) can be used as base elastomer for PSAs [293]. The selection of suitable comonomers and polymerization conditions led to the development of olefin polymers that are pressure-sensitive in the neat form [294]. Such polymers have a T_g of −25°C. In their formulations no oil is needed (no migration and less skin irritation are given by common tackifiers). They display good thermal stability (less than 10% viscosity break after 100 h at 177°C), low color and odor, lower density and the ability to be modified with most olefin-compatible ingredients. According to Clubb and Foster [295], amorphous polypropylene obtained as by-product has a glass transition temperature of −14°C. Amorphous polypropylene obtained by direct synthesis shows a T_g of −12°C. A mixture of 75–85% atactic polypropylene with a molecular weight of 15,000–60,000 and 15–25% terpenephenol tackifier resin with a molecular weight of 1200 and a melting point of 115°C has been suggested as an HMPSA [296]. A propylene–butene amorphous polyolefin (APO) copolymer exhibits a T_g of −16°C. A patent [297] discloses a PSA formulation based on amorphous polypropylene, styrene block copolymers, and tackifier. The formulation discussed by Wakabayashi and Sugii [297], is based on polypropylene, styrene block copolymer, tackifier, and wax and displays special electrical properties.

Atactic polypropylene has been manufactured as by-product in the synthesis of isotactic polypropylene. Other products are prepared by the copolymerization of propylene with low molecular weight alkenes, for example, ethylene or 1-butene. Allen et al. [298] gives a raw material formulation that includes ethylene (10–30%), propylene (65–90%) and optionally, a C_{4-8} a olefin (≤15%). The tackiest polymers have been synthesized by using hexene and octene as comonomers. The nature of the comonomer influences the crystallization rate of the product, that is, its shelf life. Copolymers having short-chain comonomers are used for "semi pressure-sensitive" adhesive. Such products lose their tack after a certain time because of crystallization. Their adhesive properties can be improved by using tackifier (hydrocarbon, terpene, and rosin resins) and plasticizers (polyisobutylene, polybutene or mineral oil) and thermoplastic elastomers of styrene (3–7% w/w SIS or SEBS). HMPSA compositions based on ethylene–propylene copolymers (with 20–50% propylene) aliphatic tackifiers, and napthetinc oil, applied on paper (with 20 μm thickness) showed similar tack but better weatherability and die-cutting performance than SBC-based formulations [299]. An HMPSA based on atactic polypropylene includes PP with a molecular weight of 15,000–60,000 and terpene resins with a molecular weight of 1200 [297].

Atactic polypropylene can be crosslinked by using gamma radiation [300]. In a quite different manner from crystallization, where the "condensed" crystalline order, that is, the crystalline network causes a plastomer-like behavior, that is, the loss of pressure-sensitivity, the crosslinked elastic network exhibits rubber-like properties. Such crosslinked atactic polypropylenes are elastomers. Oxidatively degraded ethylene–propylene copolymers are useful as softeners for rubber-resin adhesives for tapes [301].

j. Polyesters

Thermoplastic elastomers can also be prepared as block copolymers based on esters. A thermoplastic copolyester may have the formula [252]:

$$-[-O-(CH_2)_a-O-\underset{\underset{O}{\|}}{C}-AR-\underset{\underset{O}{\|}}{C}-]_x-[-O-(CH_2CH_2\ CH_2CH_2O)_b$$

$$-O-\underset{\underset{O}{\|}}{C}-AR-\underset{\underset{O}{\|}}{C}-]_x-[-O-(CH_2)_a-O-\underset{\underset{O}{\|}}{C}-AR-\underset{\underset{O}{\|}}{C}-]_x-$$

$$(5.18)$$

where $x = 1–1.1$; $y = 7–10$; $a = 4$ and $b = 12–16$. Such copolymer can have hard segments of butylene terepthalate and flexible segments of polyalkyleneoxide terepthalate [302]. If polytrimethylene terepthalate blocks are reacted with acrylonitrile–butadiene copolymer blocks, a multiblock copolymer is formed that has crystalline portions due to the hard terephtalate segments. Such polymers resist temperatures up to 125°C. Over 150°C the unsaturation of the polybutadiene segments allows supplemental crosslinking [303].

PSAs based on radiation-curable polyesters include macromers (liquid, saturated copolyesters with one terminal acrylic double bond per 3000–6000 molecular weight units) that have glass transition temperature of -10 to $-50°C$. These macromers can be coated at elevated temperatures (100°C), having a viscosity of 2000 to 10,000 mPa s [304].

B. Viscoelastic Components

The use of elastomers used as raw materials for adhesives led to the formulation of viscoelastic materials displaying elastic and viscous properties without the need for supplemental viscous components. A broad range of comonomers can be used to synthesize such polymers. Olefinic, vinyl, acrylic, and maleic comonomers have been used. Water-based PSAs based on polyacrylate latices, ethylene–vinyl acetate copolymer latices, and tackified rubber latices (natural rubber, functionalized or nonfunctionalized styrene–butadiene) have gained a significant share of the total PSA market for labels, tapes protective films and miscellaneous products. Benedek and Heymans [1] and Benedek [2,3] give a detailed description of the common viscoelastic raw materials used for PSAs. The present discussion concerns the product-related features of these raw materials.

1. Acrylic Copolymers

The most important viscoelastic polymers used as raw materials for PSAs are copolymers of acrylic and methacrylic esters. Copolymers of acrylic monomers (long side chain derivatives of acrylic and methacrylic acid) have been synthesized as raw materials for PSAs. They are used for almost all PSP classes, mainly for labels and tapes but also for protective films. Generally, common acrylics exhibit balanced adhesive properties (see also Chapter 7), which allow their use as is or tackified, primarily for labels. Formulations in this application field require relatively small changes in their adhesive properties. Their versatility to be used with low levels of tackifier and additives has been imposed by their coating on migration-sensitive PVC and paper carrier in the first period of their application history. Crosslinking is the main possibility of tailoring them for special applications (see Chapter 6). As can be seen from the thermomechanical plot of an amorphous polymer (Chapter 3, Figure 3.3) there is a plateau (of the modulus) corresponding to rubbery/elastical behavior of the material as a function of the temperature. The existence of this plateau and its position (T_g) depends on the molecular weight of the polymer. Acrylics possess a very broad range ($-85°C$ to $+185°C$) of T_g values [305]. If the molecular weight and molecular weight between entanglements of the polymer do not attain a critical value (M_c), no rubbery elastic plateau appears. This means, for low molecular weight polymers a "pressure-sensitive" bonding may appear but "pressure-sensitive" debonding properties does not exist. In this case the tack of the polymer is not sufficient and the cohesion of the material is low. This is the case of low molecular weight acrylics developed for hot-melts.

To understand how and why a polymer will perform, one must know the macrostructure as well as the microstructure. Characterization of the macrostructure requires knowledge of the weight average molecular weight (M_w), number average molecular weight (M_n), molecular weight distribution and branching. Polyacrylates are amorphous polymers. Their properties depend on their molecular weight and chain mobility. For high molecular weight acrylics, the melting point of the polymer is higher than the depolymerization temperature. Therefore, their molecular weight is limited. For hot-melts only relatively low molecular weight polymers can be synthesized and they need a postapplication increase of their molecular weight [146]. In this situation some years ago HMPSAs based on acrylics were considered only as hypothetical products [306], and in 1994, Wabro et al. [58] stated that the synthesis of acrylic block copolymers has not yet been carried out.

Acrylics have to be crosslinked to reduce their chain flexibility, that is, bonding performance. This requirement appears for removability (see also Chapter 7 and Chapter 8). Partial crosslinking of water-based acrylics can be carried out during their synthesis. For instance, acrylate–vinyl copolymers with about 3% tetraethylene glycol diacrylate have been polymerized using macromolecular stabilizers (protective colloids) to produce adhesives with better flow (near Newtonian), shear of 500 min, peel strength of 5 pli and tack of 750 g [307].

For special applications hydrophylic compositions are needed. Polar functional groups ensure hydrophilicity. Increased polarity and functionality improve bonding performance. To regulate them (i.e., to ensure removability), crosslinking may be necessary. In some cases porous or water-absorbent macromolecular structures are needed. Acrylic PSAs have been used for many years for medical and surgical applications [308–310]. Acrylic adhesives have been proposed for surgical tapes [311,312]. The acrylic adhesive for medical tapes has to be resistant to light and heat, show skin tolerance and be permeable to water vapor. Polyacrylate-based water-soluble PSAs are suggested for medical PSPs (operating tapes, labels, and bioelectrodes) [313].

The chemical basis for crosslinking is given by built-in unsaturation or functional groups. The free radical crosslinking of multifunctional monomers or macromers with acrylic groups can lead to network structures (Figure 5.8). Such functionality is used to enhance the crosslinking ability of various raw materials. Polyfunctional acrylates are suggested for the crosslinking of ethylene–propylene rubber [314]. As stated by Sasaaki [315], the EB curing of SBS using a dosage of

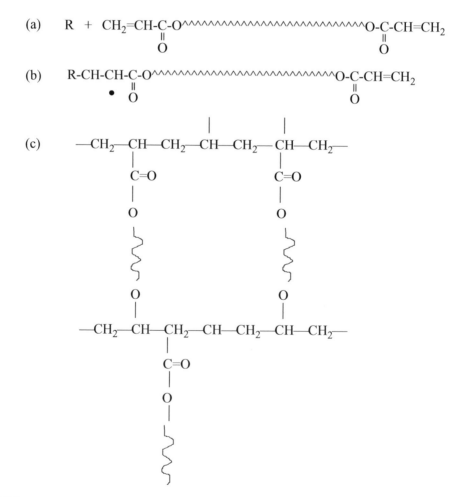

FIGURE 5.8 Buildup of a crosslinked network based on acrylic functionality. (a) Reaction of a free radical with a linear macromer; (b) linear macroradical; (c) molecular network.

89 kGy without crosslinker leads to a polymer with 47 h shear at 70°C; with a multifunctional acrylate crosslinker and a dosage of 30 kGy, more than 1440 h of shear is obtained.

It is well known that the main acrylic pressure-sensitive copolymers contain acid groups. They allow crosslinking, and can modify the work of adhesion and the interfacial energy [316], and the water resistance of the polymer [317]. The increase of the level of carboxy groups allows a cross-linking mechanism as in CSBR. Special acid-sensitive organic crosslinkers or multivalent metallic salts and oxides may be used for their curing. For instance, a special crosslinkable carboxylated acrylic with a high T_g ($-25°C$) has been developed [318] for high shear, low tack applications and for contact adhesives. Mounting tapes and double faced foam tapes are made using this adhesive.

The (neutralized) carboxylic groups may provide water solubility for solvent-based acrylics. Tack and heat stability together with water solubility or dispersibility are required for splicing tapes in papermaking and printing. For such products water solubility is given by a special composition containing vinyl carboxylic acid (10–80% w/w) neutralized with an alkanolamine [319]. The irritation caused by the removal of a medical tape applied to the skin was overcome by including in the adhesive certain amine salts [320], making it water-washable. As stated by Wistuba [11], carboxylic functionality in such compounds also influences their adhesion. The adhesion coefficient (K_{adh}) depends on the number of carboxy groups according to the correlation:

$$K_{adh} = k(COOH)^n \qquad (5.19)$$

where $n = 0.59$, $k = 0.87$ N/mm and (COOH) = Mole COOH per 1000 polymer molecules.

Generally, the crosslinking formulation also contains other multifunctional monomers. In order to regulate the elasticity of the crosslinked network, difunctional monomers having various chain lengths may be used. Copolymers of n-butylacrylate with acrylic acid and glycidyl methacrylate, N-vinylpyrrolidone, methacrylamide, acrylonitrile, and methacrylic acid have been tested as PSAs [320]. Such polymers have a molecular weight average of 200,000–1,500,000 and an M_w/M_n ratio of 2.0–4.0 [321], that is, a broad molecular weight distribution.

Mostly solvent-based or 100% solids acrylics are suggested for such special applications (see Chapter 6). Their manufacture includes the synthesis of the main polymer and its crosslinking. Thermally-induced or radiation-induced polymerization are used to manufacture the main PSA. Thermally-induced or radiation-induced curing may be used for its crosslinking also. The adhesive can be synthesized as a thermally noncrosslinkable prepolymer and its recipe can be modified before crosslinking (Figure 5.5). It has been suggested that a prepolymer be manufactured that is then polymerized by radiation [156]. A low molecular weight spreadable composition is prepared, to which may be added a small amount of catalyst or polyfunctional crosslinking monomer prior to completing the polymerization by heat curing.

Postapplication crosslinking is proposed to improve shear or to decrease peel resistance, manufacturing a storable crosslinkable adhesive. Postcrosslinking is achieved by thermally initiated or photoinitiated reactions [322]. A latent crosslinking agent has to be used. As crosslinking agent a lower alkoxylated amino formaldehyde condensate having C_{1-4} alkyl groups (e.g., hexamethoxymethylmelamine) has been proposed. A prepolymer can be synthesized by UV-initiated photopolymerization, then the crosslinking agent is added, the mixture is coated and polymerized by UV to the final product.

According to Harder [323], HMPSA on acrylic basis has been prepared via solution polymerization. Relatively low molecular weight, low viscous polymers have been synthesized and isolated by evaporating the solvent and degassing of the polymer as solid-state raw materials for HMPSA. In such low molecular weight polymerizations, gel effect may appear, allowing a high reation rate. The gel effect is an autocatalytic phenomenon occurring at high conversions rates and is marked by a reduction in the termination of polymer chain radicals. It is used currently to improve the yield of industrial acrylic polymerizations [324]. The termination rate is decreased because the increased viscosity at high conversion makes it more difficult for two polymer chain radicals to come together to produce a termination reaction. Such polymers can be coated at 90–140°C and postcrosslinked. The HMPSA can be crosslinked by using EB and UV curing. Chemical crosslinking (by NCO-containing systems) is also possible.

Both operations, the preparation of the prepolymer and its postpolymerization can be effected by radiation-induced polymerization also. Photopolymerization of the monomers neat, without any diluent that needs to be removed after the polymerization provides processing advantages [325]. The liquid monomers are applied to the carrier which is an endless belt, the polymerization is completed thermally in an oven or by UV light. A patent [326] describes the polymerization of acrylic monomers direct on the carrier to manufacture a tape. The acrylic ester PSA formulation is a combination of nontertiary acrylic acid esters of alkyl alcohols and ethylenically unsaturated monomers having at least one polar group which may be substantially in monomer form or may be a low MW prepolymer or a mixture of prepolymer and additional monomers; it may further contain other substances such as a photoinitiator, fillers, and crosslinking agents (such as multifunctional monomers). The acrylic esters generally should be selected primarily from those that as homopolymers possess some PSA properties. Diacrylates as crosslinking agents are used from about 0.005 to 0.5% by weight. Typically, the crosslinking agent is added after the formulation of the prepolymer (see Chapter 6). As photoinitiator 2,2-dimethoxy-2-phenyl-acetophenone has also been proposed.

If UV light-induced photopolymerization is used, there is a choice between external and built-in photoinitiators. Generally, benzophenone type initiators are suggested. The use of benzophenone is known from the light-induced crosslinking of ethylene–propylene copolymers [327]. Benzophenone or deoxybenzoin have been tested for UV-photocrosslinking of butadiene–styrene copolymers also [328]. Benzyldimethylketal can be used also [152,329]. For instance, pressure-sensitive adhesive tapes whose adhesion can be decreased by UV irradiation are prepared from an elastic polymer, a UV-crosslinkable acrylate, a polymerization initiator, and a methacrylate photosensitizer containing NH_2 group [330]. An ethylene glycol–sebacic acid–terepthalic acid copolymer (having a T_g of about 10°C), a tackifier resin, dipentaerythritol hexaacrylate as crosslinking monomer, benzyldimethylketal as photoinitiator, and dimethylaminoacrylate have been polymerized. The resulting adhesive has been coated on a 100 μm PVC carrier primed with a modified acrylate. Such an adhesive displays a peel resistance of 140 g/25 mm before irradiation and 52 g/25 mm after UV irradiation.

The initiating mechanism via benzophenone containing systems and the main photoinitiating systems have been described in a detailed manner by Röltgen [331]. The photoinitiator absorbs energy of light and $n \rightarrow \pi$ or $\pi \rightarrow \pi^{\nu}$ electron transfers occur. The excited electron is in the singlet state. It can return to the basical status or go into a more stable triplet status. The photopolymerization involves the participation of an excited (singlet or triplet) form of the monomers or oligomers. Light absorption is needed to achieve this excited status. Chromophoric molecular parts are required to ensure light absorption. The exciting energy can be transmitted from a photosensibilisator also. The photoiniziated polymerization occurs via free reactive species (radicals or ions) which initiate thermal polymerization (see Structure 5.11). Short-lived intermediates (biradicals, carbenes, nitrenes) can also be formed that cause crosslinking. The photolysis of the initiator occurs via intramolecular dissociation (benzoin derivatives, benzyle and acetophenone ketals, peroxides, perhalogenides, benzoiloxyme esters, etc.), interchain hydrogen abstraction (aromatic ketones such as benzophenone, thioxanthones, quinones), or photoinitiated transfer of electrons (ketones with amines, aromatic onium derivatives with electron donors, etc.). The wavelength of the light should be between 300 and 800 nm. The efficacy of the initiating expressed as liters per mole per centimetre is between 10 and 50,000 (benzophenone having values of 100–200) [332].

The aromatic ketones absorb UV radiation to form a triplet excited state. These molecules can abstract hydrogen radicals from the polymer and create free radical sites on it. Benzophenone may be used together with amines where initiating radicals are generated by photoreduction of benzophenone by the amine, via formation of an intermediate complex between a photoexcited (triplet-state) benzophenone molecule and a ground-state polyamine molecule [333]. If many polymer layers are manufactured by photopolymerization, their macromolecular characteristics are regulated by the concentration of the crosslinking agent and of the photoinitiator (see Chapter 8). The use of external photoinitiators has the disadvantage that they interact with atmospheric oxygen. By quenching the (excited) photoinitiator reacts with oxygen and will be deactivated; active polymeric radical sites are scavenged by oxygen molecules to give stable peroxides. As demonstrated by Decker [334], the exposure time for UV-curing in air is of 0.2 sec and in nitrogen, 0.03 sec. Therefore, crosslinking by photosensitive monomers, with built-in photoinitiator has also been tested.

Photopolymerization as a generally useful finishing technology has been developed for various applications. Copolymerizable photoinitiators have been studied for UV curing of epoxy acrylates and hexanediol diacrylates [335]. Photoinitiators for waterborne UV-curable coatings can contain oligomeric hydroxyalkylphenyl ketone as photoinitiator [336]. Mono-ethylenically unsaturated aromatic ketones can be used for crosslinking PSA compositions with UV light [337]. Such monomers are copolymerized in the adhesive [132]. Preferred monoethylenically unsaturated aromatic ketone monomers have the following general formula:

$$
\begin{array}{c}
R\!-\!C\!-\!Ar\!-\!Y\!-\!Z \\
\underset{O}{\overset{\|}{}} \quad \underset{(X)_n}{\overset{|}{}}
\end{array}
\tag{5.20}
$$

where R is lower alkyl or phenyl, provided that R may be susbstituted with one or more halogen atoms, alkoxy groups or hydroxy groups; X, the halogen, alkoxy, or hydroxyl; n an integer from 0 to 4; Y, a divalent linking group; and Z, an alkenyl or ethylenically unsaturated acyl. The aromatic ketone monomer is free of ortho aromatic hydroxyl groups. Particularly preferred monomers are the acryloxy benzophenones, for example, *para*-acryloxy-benzophenone. Unfortunately, by the polymerization of such monomers side reactions may appear (see Chapter 8).

Crosslinking via polymer-bonded benzophenone groups has been studied on model polyimides containing benzophenone as well as alkyl-substituted diphenylmethane groups in the main chain. As stated, the crosslinks are formed upon UV irradiation through hydrogen abstraction by triplet benzophenone from the alkyl groups acting as hydrogen donors and subsequent coupling of the formed radicals. The quantum yield of the process is low. This is attributed to a specific energy dissipation process that operates at the reactive site, namely reversible hydrogen exchange between benzophenone and the hydrogen donor [337]. There is a correlation between the activity and structure of polymeric photoinitiators containing side chain benzophenone, [e.g., poly(4-vinylbenzophenone) or 4-acryloyl-oxybenzo-phenone] chromophores. It has been confirmed the main activation mechanism in UV polymerization of acrylic monomers by benzophenone-containing polymeric photoinitiators is intramolecular hydrogen abstraction by excited benzophenone moieties [338]. Those photoinitiator systems show high activity, where both benzophenone and amine groups display a large conformational mobility.

Such built-in photosensitizers have been tried for styrene–butadiene copolymers also. A styrene–butadiene copolymer has been modified by benzoyl or phenylacetyl groups. The built-in functional groups served for the postcrosslinking of the polymer [339]. Mannich polymers can be used as phosensitizers for onium salts for the polymerization of vinyl monomers too [340].

Electron beam-crosslinkable acrylics contain a multifunctional monomer such as pentaerithritol triacrylate to allow curing with low energy use [341].

Such adhesives display good high temperature shear strength. As stated by Braun and Brügger [342], most EB curable PSAs are acrylic functionalized; they are based on: (i) postcrosslinking of SIS HMPSA; (ii) acrylate functionalized polyester prepolymers; (iii) acrylate functionalized liquid rubber; or (iv) acrylate prepolymers.

Acrylate-based PSAs containing acid or acid anhydride groups and glycidyl methacrylate coated on a flexible carrier can be crosslinked thermally between 60 and 100°C using a zinc chloride catalyst [325]. Instead of acrylate copolymers with monomers containing epoxide groups and compounds containing an acid anhydride group, mixtures of two separately prepared copolymers can also be also used, for example, an acrylate copolymer containing epoxide groups and an acrylate copolymer containing anhydride groups. Crosslinking catalysts (0.05–5% w/w) are added to the polymer. The polymer coated on a carrier is crosslinked (after solvent evaporation) 2–7 min at 60–100°C. Zinc or magnesium chloride, monosalts of maleic acid, organic phosphoric or sulphonic acid derivatives, oxalic or maleic acid can be used as the catalyst. Tertiary amines and Friedel-Crafts catalysts accelerate crosslinking. Compared to common uncrosslinked rubber-based PSA (with a holding power of some minutes at 100°C), the holding power of these adhesives attains 20 h at 150°C and they are more resistant to water and solvents.

Principally, the same comonomers can be used as crosslinking agents for water-based formulations. Adhesives for diaper tapes have to display high shear. Therefore, crosslinking monomers are added, (e.g., N-methylol-acrylamide) to increase their shear resistance [343]. Unsaturated dicarboxylic acids may be included as crosslinking agents [344] together with polyfunctional unsaturated allyl monomers [345], and di-, tri-, and tetrafunctional vinyl crosslinking agents [346]. The polyolefinically unsaturated crosslinking monomer enhancing the shear resistance should have a T_g of $-10°C$ or below. Shear resistance is improved by using a special stabilizer system, comprising a hydroxypropyl methyl cellulose and an ethoxylated acetylenic glycol.

Crosslinking leads to a blend of polymers having different molecular weight. Such pressure-sensitive macromolecular compounds of different molecular weights can be prepared by other methods also. According to numerous patents (see later), PSA synthesized via suspension polymerization are used together with solution- or emulsion-polymerized adhesives. The phase structure of PSA mixed (with tackifier) in monomer state and solution polymerized is different than that expected from the phase diagrams of solution blended systems. The lower molecular weight of acrylate and some reactions of acrylate with rosin caused the system more compatible [347]. As stated in [348], in acrylic copolymers, the EHA content increases vs. conversion differently as a function of the polymerization method. It is possible to determine a critical fraction of EHA for which there is a transition between adhesive and cohesive failure. The critical fraction depends on the polymerization process.

The PSA surface can have a special discontinuous shape ensured by the PSA itself, if it is a suspension. Thus spherical contact sites are formed [148,345]. The anchorage of the spherical PSA particles needs a continuous matrix. In this case, the PSA particles synthesized by suspension polymerization and the adhesive used as matrix are of different molecular weights. Suspension polymerization leads to lower degrees of polymerization than the emulsion procedure [349].

2. Vinyl Acetate Copolymers

Polyvinyl acetate homopolymer has been used as solution- and water-based dispersion for classic adhesives. Its copolymerization with comonomers that allows a lower T_g leads to raw materials for PSAs. Vinyl acetate–acrylate, vinyl acetate–maleinate and ethylene–vinyl acetate copolymers are the main representatives of this class of raw materials. Copolymers of vinyl acetate have been used as raw material basis for classic adhesives. Vinyl acetate–acrylate copolymers and vinyl acetate–ethylene copolymers have been manufactured as raw materials for solvent- and water-based pressure-sensitive adhesives. Ethylene–vinyl acetate copolymers have been developed for so called "semi pressure-sensitive" adhesives also and also as primers for extrusion coating [350].

As discussed by Benedek in [2,3,351], and by Benedek and Heymans in [1], because of their unbalanced adhesive properties and limited tackification ability, vinyl acetate copolymers are not competitive with acrylics for use in labels. However, copolymers of vinyl acetate with long-chain acrylic comonomers are raw materials for PSAs. Vinyl pyrrolidone–vinyl acetate copolymers are used for water-soluble PSAs [352]. Adhesives for medical tapes have also been formulated on vinyl acetate copolymer basis [353]. Ethylene–vinyl acetate–vinyl chloride copolymers can also be used for medical tapes [352]. The use of ethylene as an inexpensive low T_g comonomer allows the synthesis of new raw materials based on vinyl acetate, maleinates, and acrylics for PSAs. EVAc copolymers were first marketed in the early 1960s [354]. The first European EVAc-based pressure-sensitive aqueous dispersion was Vinnapas EAF 60, manufactured by Wacker [355]. The so-called combination adhesives are made as blends of PSA and nonpressure-sensitive EVAc dispersions (70/30) [356]. Such formulations used as flooring adhesives contain a PSA on EVAc basis and a nonpressure-sensitive EVAc dispersion, resins, fillers, and additive. They can be applied as one-side flooring adhesives after a drying time and display excellent thermal stability [357]. So-called legging adhesives [358] are also formulated. Here the addition of a resin allows the regulation of the legging (see Chapter 3, Section I). Removable flooring adhesives are also required. Such adhesives are removable with or without water (see Chapter 11). The PVA content of EVAc dispersions allows their easy removal with water. Systems that can be removed without water are also available. Here an EVAc-based primer is applied on the substrate to facilitate removability. High shear pressure-sensitive EVAc emulsions resistant to high temperatures are used for labels in automotive industry, and for tiles and packaging tapes [358]. Generally, EVAc dispersions are used for flooring adhesives and as tackifiers for contact adhesives; for tapes and labels; and adhesive films [359]. Slightly pressure-sensitive EVAc copolymers with a minimum film forming temperature (MFT) of about $0°C$ and T_g of $-20°C$ have been suggested as one-side flooring adhesives also [360]. EVAc copolymer emulsions useful as carpet adhesives have good compatibility with SBR emulsions [361]. Ethylene–vinyl acetate copolymers with improved plasticizer-, temperature- and shrinkage-resistance and good cohesion have been proposed for flooring adhesives [362].

Special segmented EVAc copolymers have been developed as raw materials for HMPSA. Such polymers have a higher VAc content (36%) and can be considered elastomers [363]. They exhibit improved aging stability and adhesion on soft PVC [364]. Their adhesive characteristics are not as good as those of SBCs. A peel adhesion of 14.5–24.0 N/25 mm (PSTC-1, 180°C), rolling ball tack of 11–50 cm, Polyken tack of 2.5–5.6 N (300–600 g), and shear adhesion of 1–7 h (23°C) have been achieved on paper with an adhesive coating thickness of 25.4 μm. The comparative examination of the adhesive characteristics of HMPSAs based on different segregated copolymers shows that SBCs exhibit the best performances (Table 5.3) although they are less tacky than PSAs based on natural rubber.

A hot-melt PSA on EVAc basis contains about 35.5% elastomer, 30–50% tackifier, 0.2% plasticizer, 0.5% filler, and 0.1–0.5% stabilizing agents [111]. The VAc content does not influence the melt viscosity. The peel value can be increased by increasing the VAc content. As plasticizer, pthalates, phosphates, chlorinated polyphenols, and liquid rosin derivatives have been used. As tackifier rosin derivatives, hydrocarbon resins, and low molecular weight styrene polymers have been suggested. Zinc oxide, calcium carbonate, titandioxide, barium sulphate and organic resins have been recommended as filler. High molecular weight PVAc dispersions are used for hot sealing at elevated temperatures (120°C), low molecular ones at temperatures lower than 60°C [363]. A polyvinyl acetate polymer can be used used as filler in UV-curable formulations [320]. Water-soluble hot-melt adhesives can be formulated with vinyl pyrrolidone–vinyl acetate copolymers [368,369].

Pressure-sensitive cold sealing adhesives have been manufactured from EVAc copolymers with 55–85% ethylene, tackifiers, and special micro-sphere fillers [370]. Such adhesives can include natural rubber latex (2–32 parts and vinyl acetate copolymers (8 parts) [371]. As filler, micronized pyrogenic SiO_2 has been proposed. Mixtures of EVAc (30–45% with 15–50% VAc content) with 55–70% chlorinated polypropylene (73–85% chlorine) have been suggested as adhesives for removable protective films [372].

TABLE 5.3
Adhesive Characteristics of Pressure-Sensitive Formulations Based on Thermoplastic Elastomers

| Chemical Composition | | | Adhesive Characteristics | | | |
Elastomer Type	Level (pts)	Tackifier Level (pts)	Tack (cm)	Peel Resistance (N/25 mm)	Shear Resistance (h)	Reference
SEBS	100	115	4	15	100	[219]
SEBS	100	120	7	12	>83	[365]
SBS	100	125	2.5	6	>150	[222]
SIS	100	125	1.5	5	>150	[222]
EVAc	100	217	21	8	5	[366]
EVAc	100	150	13	24	7	[359]
AC	100	—	12	62	15	[260]
NR	100	125	1.5	7.5	1	[367]

Note: 180° peel on stainless steel and room temperature statical shear were measured.

3. Polyvinyl Ethers

Polyvinyl methyl-, ethyl-, and isobutyl ethers are used as viscoelastic raw materials for PSAs. They are soluble in common solvents such as alcohol, acetone, butyl acetate, ethyl acetate, toluene and some of them are also soluble in water [373]. Polyvinyl ethers with acrylics are used for tapes [374]. Polyvinyl ethers are known as raw materials for wet tapes also [375]. An adhesive for medical tapes contains polyvinyl ether with a K-value of 50–130 [178]. Polyvinyl ether is recommended for medical tapes alone or with polybutylene and titanium dioxide as filler. Such compounds are coated warm without solvent. A porous adhesive is achieved by cooling [186]. Polyvinyl isobutyl ether is a component of skin-compatible acrylic PSAs also [376]. Polyvinyl ether can be used for masking tapes during the varnishing of cars [178]. The aging resistance of PVE can be improved using phenol and pinene derivatives [377]. Polyacrylates tackified with water soluble PVE have been proposed for water-soluble splicing tapes [378]. Because of the lack of sensitivity to atmospheric humidity, polyvinyl methyl ether has been added to tape formulation of to avoid curling [367].

C. VISCOUS COMPONENTS

At first rubber-resin formulations were used for PSAs. To achieve viscoelastic behavior, the elastic component (rubber) had to be mixed (formulated) with a viscous component. Viscoelastic raw materials can also be blended with such viscous components to regulate their rheology. Either macromolecular or micromolecular compounds can be used as viscous components. The best known are resins and plasticizers; both are used as tackifiers.

1. Tackifiers

Tackifiers are raw materials used in PSA formulations in order to improve the tack (see Chapter 7 and Chapter 8). However, they also change other adhesive properties. Generally, they improve the peel resistance and decrease the shear resistance (cohesion) of the adhesive. Other nonadhesive properties, (e.g., coatability, cuttability, aging resistance) are influenced by tackifiers also (see Chapter 11). Tackifiers have to be viscous. Therefore, their molecular weight is limited. Generally, low molecular weight resins and relatively high molecular weight solvents are used as tackifiers in PSA formulations. Tackification, tackifiers and plasticizers for PSAs are described in detail in

[1–3]. This chapter discusses some of their PSP-related features only. Tackification of PSPs is more than tackification of the adhesive. It can provide an adhesive character for the carrier too. As described in the previous chapters, certain PSPs contain built-in pressure-sensitive components. Such (extruded) products are tackified plastics. For their tackification, high polymers as well as common tackifiers are used (see also Chapter 8).

a. Resins

The range of resins used as tackifiers has been continuously expanded over recent decades. At the beginning of the manufacture of PSPs, mostly natural raw materials were used. Natural rubber was blended with natural (rosin) resins. Later, modified natural resins (rosin derivatives) with improved aging resistance were used as the viscous component. Some decades ago, hydrocarbon-based synthetic resins were produced. The addition technology of these resins has also been developed. Initially, resin solutions were developed, later water-based resin dispersions. Such products have to display tackifying ability, but other properties are also needed. For water-soluble formulations, the resin also has to be water-soluble [379].

Molecular weight and molecular structure have a special importance in processes where mutual compatibility of the components is required. Compatibility depends on the molecular weight and chemical structure of the tackifier. The compatibility of the resin plays a special role for block copolymers where the polymer sequences have a quite different chemical nature (and structure), and depending on the choice of tackifier, on its compatibility with one of the sequences or both, different adhesive (and processing) performances can be obtained. Resins are defined by generic type, softening point, color, and molecular weight. Tackifier resins are situated in the molecular weight range (300–3000) where T_g strongly depends on molecular weight [380].

Crystallization, oxidative destruction or polymerization can lead to changes in the tackifying or adhesive properties of the resin. Nonesterified rosin is recommended as tackifier for hot-melt adhesives, provided it is stabilized against crystallization and oxidation [381]. Freezer tapes are coated with an adhesive based on natural rubber, styrene–butadiene rubber, regenerated rubber, and rosin ester [143].

According to Donker [200], styrene block copolymers used for hot-melts are mainly SIS block copolymers with about 15% styrene. Alternatives are SBS and high styrene (20–25%) SIS polymers. Special solvent blends are suggested to identify resin compatibility for such elastomers. For SIS block copolymers modified C_5 hydrocarbon resins, hydrogenated rosin esters, and alkylaromatic hydrocarbon resins are suggested [381]. Common HMPSAs for tapes are based on SIS plus a resin that is compatible with the midblock of the SBC, and a plasticizer (oil). Tack properties of SBC are given by the midblock. Its mobility is increased adding oils and tackifiers to the thermoplastic elastomer. A compatible resin is more effective than oil. The other part of the polymer sequence (the styrene endblock) provides the cohesion and mechanical properties and influences the processing conditions. Its hardness is reduced by oils that are compatible with the styrene domains [382]. One way to reduce the processing viscosity of the HMPSA is to increase the processing temperature (upper limit 190°C) or to decrease the melting temperature of the polystyrene domains. According to Hollis [381], 80 Pa sec is the preferred maximum viscosity for high speed coating. Using a mid- and endblock-compatible resin for SIS, this viscosity value can be obtained with a lower resin level (at rubber/resin ratio of about 1:1).

It is known that common HMPSA formulations incorporate a high level of plasticizer, usually a naphthenic oil or a liquid resin. Most tackifiers used for HMPSA are aliphatic, midblock-compatible resins that do not influence the styrenic domain. Oils are present to act on the polystyrene domain (being compatible with these groups) and thus soften it. For instance, a partially fumarized or maleinized, disproportionated rosin ester is used as tackifier together with a plasticizer and a thermoplastic rubber for PSA [383]. The use of plasticizers results in a number of disadvantages including long-term degradation of the PSA. For tapes a mixture of two resins is suggested, with one being midblock-compatible, and the other compatible with the endblocks. If resins are used that are partially compatible with the styrene domains, (e.g., rosin esters, modified terpene resin, or a modified aliphatic resin) two effects — middle block mobility increase and end block softening — are achieved. Such a resin can have an aromatically modified aliphatic hydrocarbon as its basis with a higher molecular weight. Fully saturated petroleum resins with a well defined number average molecular of 400–800, a softening point of 40–70°C, and a T_g lower than 45°C have been developed for use in HMPSA formulations based on thermoplastic elastomers without plasticizer oils. According to Donker et al. [383], resins synthesized from the C_5 feedstock having a major portion of piperylene and 2-methyl butenes with a defined diolefin/monoolefin ratio and a weight average molecular weight (M_w) of 800–960, and M_n of 500–600 (M_w/M_n ratio at least 1.3) are clear, exhibit a softening point of about 57°C and a T_g of about 20°C, and need no oils in the formulation with saturated midblock thermoplastic elastomers.The saturated midblock thermoplastic elastomers are commercially ABA type block copolymers where the polyisoprene or polybutadiene is hydrogenated. The block copolymers with the saturated midblock segment have an M_n of about 25,000–300,000.

Rosin derivatives have to be disproportionated to decrease the conjugated double bond concentration. Such bonds are sensitive to oxygen attack [384]. For instance, a common acid resin used for water-based tackifier dispersions displays a narrow molecular weight distribution ($M_w/M_n = 1.19$) and T_g of 10°C ester resins have a broader MWD (1.19–2.48) and somewhat lower T_g (4–6°C). They display better compatibility and aging resistance. At low temperatures, loss of wetting and a jerky debonding are observed for formulations with the acid resin. The molecular weight of the acid resin is about 400; that of the resin esters is situated between 1000 and 1300. Such (small) differences in the molecular weight and the increased tendency of acid resin to crystallize can cause incompatibility at low temperatures. The crystallization of rosin acids can also result in the formation grit [385]. Such crystals are insoluble in water and alkalis. Crystallization depends on the rosin type. Tall oil rosin has the highest tendency to crystallize. Gum rosin (from living trees) crystallizes due to its higher acid content.

The nature of the resin affects radiation curing of PSA formulations. Curing by UV light is influenced by the unsaturation index of the tackifier resins [156]. The formulation of EB-crosslinked, SIS block copolymer-based PSAs has been discussed by Ewins and Erickson [386]. The effect of resins and plasticizer in a standard formulation having 100/100 phr of rubber (Kraton D1320X) resin or plasticizer has been evaluated. Electron beam-cured thermoplastic elastomer-based tackified HMPSA formulations have been studied by Nitzl and coworkers [387] also. They stated that when radiated at high dosage level of 20 kGy, unsaturated resins give adhesives with increased temperature resistance. Resins can be used as curing agents in classic formulations also. Butyl rubber-based formulations use brominated phenolic resins as the crosslinking agent.

b. Plasticizers

Plasticizers were originally used as viscous components for plasticizing PVC. Their use made it possible to regulate the hardness and mechanical properties of PVC. As an (undesired) side effect, its surface adhesivity was also changed. Such chemical compounds have been used for adhesives too, to give them a permanent adhesive character (pressure-sensitive adhesivity), or to regulate their adhesion–cohesion balance to achieve removability. Plasticizers can be incorporated in the peelable pressure-sensitive adhesives to soften the adhesive and thus to improve peelability. However, care may be needed, as some plasticizers can have a tackifying effect on adhesive polymers and this may limit the amount that can be used. The use of plasticizers in printing inks makes the dried film more flexible and pliable [388].

The chemical nature of plasticizers differs according to the type of elastomers or plastomers to be plasticized. For polar hydrophilic macromolecular compounds, polar, hydrophilic plasticizers are choosen. Macromolecular, polyester-based plasticizers have been suggested for soft

PVC to be used as carrier for tapes. For a tape carrier the adhesive has been formulated with natural rubber, butadiene–styrene rubber, and terpene or phenolic resin as tackifier [389]. Dioctylphthalate (a common micromolecular plasticizer for PVC) has been tested as plasticizer or tackifier for tape adhesives on natural rubber and chlorinated rubber [178], and had also been used for medical tapes [390]. In SBCs aromated plasticizers like dioctylphthalate associate with the polystyrene domains and work like aromatized oils, softening the polystyrene domains. Therefore, if plasticized PVC is used with such adhesives, it should be coated with a barrier layer.

Plasticizers can be used in water-soluble formulations also [178,391]. According to [392], water-soluble compositions used for splicing tapes contain polyvinylpyrrolydone with polyols or polyalkyglycol ethers as plasticizer. Ethoxylated alkylphenols, ethoxylated alkylamines, and ethoxylated alkylammonium derivatives have also been used as the water-soluble plasticizer.

The molecular weight of the plasticizer influences its water dispersibility. A maximum or a minimum molecular weight is desired depending on the use of the plasticizer. Water-soluble plasticizers whose molecular weight does not exceed 2000 (e.g., polyethylene glycols with a molecular weight of about 200–800) are preferred for water-soluble splicing tapes [390]. For resin dispersions with softening points above the boiling point of the water, plasticizers with a well-defined molecular weight should be added [385].

Low molecular weight polyisoprenes (35,000–80,000) having a low T_g (-65 to $-72°C$) have been used as viscosity regulators for EVAc-based HMPSAs [393]. They replace common plasticizers, improve the migration resistance and low temperature adhesion. Dibutylmethylene-*bis*-thioglycolate has been suggested as plasticizer for special rubbers [394]. Special plasticizers as lanoline, sesame oil, vaseline, and ricinus oil are used in adhesives for medical tapes [195]. Plasticizer oils for block copolymers have to fulfill the following requirements [395]; no solubility in the polystyrene, domains, compatibility with the elastomer segments, low volatility, low density, and good aging performance.

D. Plastomers

The synthesis and development of pressure-sensitive hydrogels based on crosslinked-plasticized plastomers led to a new class of pressure-sensitive products (see Chapter 4 and Chapter 9). Hydrogels based on blends of high molecular weight polyvinylpyrrolidone (PVP) with short-chain polyethyleneglycols (PEG) reveal a pressure-sensitive character of adhesion within a narrow range of PVP–PEG compositions; their adhesion is a result of hydrogen bonding of both PEG terminal hydroxy groups to the carbonyls in PVP repeating units. The synthesis of such PSA is carried out by mixing nontacky hydrophilic polymers with short chain plasticizers, which bear complementary functional groups at their ends. As long chain complementary hydrophilic polymers, poly(*N*-vinyl amides), particularly poly(*N*-vinyl caprolactam) (VCap), NVP- and VCap-copolymers, polyacrylic acid, etc., can be used [396]. Tacky blends can be obtained by mixing PVP with hydroxyl terminated or carboxyl terminates short chain plasticizers, ethylene glycol and its polymers, with MW 200–600 g/mole, low MW propylene glycol, alkane diols propane diol up to hexane diol. The highest peel strength has been found for PVP with glycerol which possesses the maximum of density of hydroxyl groups per molecule. Water is an appropriate plasticizer to impart adhesion to PVP blends with carbonic acids.

E. Other Components

The processing of the PSP components and of the finished product and its end-use or recycling require different additives. Such products can have various chemical basis. The right selection of fillers, antiblocking agents, antioxidants, antistatic agents and processing aids depends on the type of processing as well on the properties that must be achieved. Some of these additives are used as technological aids, whereas others are constituents of the finished product.

1. Final Constituents

Various materials (solubilizers, fillers, antioxidants, etc.) are needed as additives for PSP formulations. Some are adhesive, but others are not. Most are inherent components of polymer formulations; therefore, they are used for both the carrier material and the adhesive.

a. Solubilizers

The achievement of water solubility or dispersibility requires special additives. Generally, such substances have hydrophylic functional groups that increase water solubility and water absorbtion. Some are macromolecular derivatives (e.g., plastomers or resins), and others are micromolecular products. Solubilizers can be elastomers, plastomers, viscoelastic materials or viscous additives. Depending on their influence on the pressure-sensitivity of the product, they can be considered as main adhesive components or as special fillers.

Macromolecular solubilizers have polar (hetero) atoms in the main polymer backbone or as branches. They may possess the usual hydroxy, amine, carboxy, etc., polar goups that provide the hydrophylicity. Vinyl ether copolymers [397] and neutralized acrylic acid–alkoxyalkyl acrylate copolymers [398] have been suggested as macromolecular pressure-sensitive solubilizers. Polyacrylate-based water-soluble PSAs are suggested for medical PSPs (operating tapes, labels, and bioelectrodes) [313]. Suitable water-soluble plasticizers include polyoxyethylenes and rosin derivatives containing carboxyl groups. Compounds like C_4–C_{12} alkylacrylates give adhesion by having a low T_g; vinyl carboxylic acids and hydroxyacrylates display hydropylic properties. Their combination possibilities are numerous. About 30 patents are mentioned by Czech [313] that concern only water-soluble acrylics.

As mentioned earlier, some of these components exhibit adhesive properties, whereas others do not. Water-soluble waxes are not pressure-sensitive [399]. Other macromolecular compounds are only slight adhesive. Water-soluble hot-melt adhesives can be formulated with vinyl pyrrolidone and pyrrolidone–vinyl acetate copolymers [368]. Polyalkylene oxides and polyalkyleneimines can be used as solubilizer also. For instance, a hydroxy substituted organic compound (alcohols, hydroxy substituted waxes, polyalkylene oxide polymers, etc.) and a water-soluble N-acyl-substituted polyalkyleneimine obtained by polymerizing alkyl substituted 2-oxazolines can be also added. Such water-releasable HMPSAs contain 20–40% polyalkyleneimine, 15–40% tackifying agent, and 25–40% hydroxy substituted organic compound [369].

A repulpable splicing tape specially adapted for splicing of carbonless paper contains a water-dispersible PSA based on acrylate–acrylic acid copolymer, alkali, and ethoxylated plasticizers. A polar hydrophylic polyamide-epichlorohydrine crosslinker may also be included in its formulation [400]. Polyfunctional aziridine derivatives (1–3%) can be used as crosslinker also (see Chapter 6). Such compounds work at room temperature [401]. Polyacrylic acid blended with polypropylene glycol has been suggested too [402]. Amine-containing soluble polymers and soluble plasticizers are proposed in [403] (see also Chapter 8).

b. Fillers (Pigments and Extenders)

Common fillers are used mainly as inexpensive PSP components. They are used for carriers as well as for adhesives. In some cases fillers are included in the recipe to achieve opacity or special mechanical properties. Fillers in adhesive formulations increase viscosity and improve tensile strength but can cause problems due to their abrasiveness [404].

Generally fillers are inorganic, low molecular materials. Fillers are described in a detailed manner in [405–407]. Fillers in carrier materials influence the bulk and surface characteristics. Regulation of the mechanical properties is the most important function of fillers in carrier materials. Optical and electrical characteristics are also affected. Surface absorptivity, porosity, and polarity can be controlled by fillers. In the range of mechanical characteristics stiffness, tensile strength/extensibility, and tear resistance are the main properties regulated by fillers (see also Chapter 8 and Chapter 3, Section II).

The influence of fillers on the mechanical characteristics of polymers has been the subject of extensive study. Filled structures are used on the macromolecular scale also (see Chapter 3, Section I). Plastomer–plastomer, plastomer–elastomer, and elastomer–viscous component mixtures have been examined as filled systems. The effect of fillers on the Young's modulus of filled formulations can be calculated by using the Guth–Gold, Smallwood, Brinkmann, Zosel, or other equations [408]. The most important effect of excessive filler levels in carrier- or adhesive formulations is the loss of the main mechanical or adhesive characteristics. Such effects are used to regulate the tear resistance of carrier materials, or to reduce the adhesivity of pressure-sensitive products. For instance, easy tear breakable hank tapes are used in the same finished product as easy tear paper tapes. The tape is applied per hand. Such tapes are easily torn or split during handling. The carrier is highly filled and has a low elongation and easy break force. Highly filled pressure-sensitive layers ensure removability or repositionability.

The main natural white pigments for paper are kaolin, limestone, and talcum. Synthetic products such as calcium carbonate, calcium sulfo-aluminate, titanium dioxide, calcium silicate, sodium aluminum silicate, barium sulfate, and zinc oxide have been developed also. The most used filler for paper is kaolin. Calcium carbonate is less expensive and allows fast ink penetration. Calcium carbonate can be used to increase the porosity of the paper, which is very important for heat set rotary offset printing [409]. Precipitated silicates regulate the ink absorption and increase opacity [410].

Special fillers ensure that the plastics in which they are used will be nonflammable. Aluminum hydroxide, silica, antimony oxide, halogenated compounds, and phosphor derivatives are the main additives used to impart nonflammability. Aluminum hydroxide can be used up to 180°C, and magnesium hydroxide up to 340°C [411]. Barium methaborate can replace 50–75% of the antimony trioxide used [412].

PVC with silicon carbide as filler has been suggested for electrically conductive tapes [413]. Metallized ceramic bubbles can be used for electrically conductive plastics or adhesives. Such materials are proposed for protection against EMI (electromagnetic interference) or RFI (radiofrequency interference) [414]. Metallic fillers have been developed for medical tapes (embedded on one side of the carrier material) to avoid the appearance of static electricity and sparks when the tapes are slit or cut [415]. Highly conductive silver-clad hollow microspheres exhibit low resistivity at loading levels of slightly more than 0.1 g/in.^2 [416].

Carbon black fillers are used to impart electrical conductivity also (see Chapter 3). A high carbon black content may make processing difficult and requires previous drying. The conductivity achieved with carbon black depends on the additive concentration and on the morphology of the product [417]. Special composite systems consisting of electrically conductive carbon black and rubber are also known as pressure-sensitive rubbers, because of the dependence of their conductivity on compression [418]. Antistatic webs that prevent the accumulation of dust are manufactured by using carbon black filled polyurethanes [419]. In some cases the filler should ensure electrical conductivity and nonflammability without affecting the conformability and extensibility of the carrier material. Such formulations are required for automotive electrical tapes. A patent [420] describes the use of special fillers for such formulations. Photochromic and thermochromic masterbatches can also be applied. Such fillers cause the finished article to change color when exposed to sunlight or a change of temperature, respectively [421].

As in plastomer formulation (see earlier), in classic adhesive formulation fillers are used mainly as inexpensive recipe components. As stated in [422], extenders or fillers have been regarded as the least important ingredients in the adhesive or sealant formulation, and in the past, almost any type of calcium carbonate was used as filler. Active fillers served primarily to stiffen the material; inactive fillers were added to reduce the costs of the formulation, but they can also act as "builders," that is, they may have a decisive influence on the stiffness [121]. The importance of fillers as raw materials is quite different for labels, tapes, and protective films. As discussed in [1,3], PSA formulations for labels contain only a low level of fillers. Because of the negative effect of chemically inert fillers on the pressure-sensitivity, this statement is also valid for tapes and protective films. Because of their generally negative influence on adhesive properties, fillers can only be used for well-balanced formulations or for products having a high coating weight. Therefore, they are suggested mainly for the formulation of PSAs for tapes. On the other hand, there are several active fillers that positively modify some features of the adhesive or end-use behavior such as shear resistance, cuttability, or tear strength. These requirements are important mainly for tapes. Therefore, fillers are used as adhesive components mostly for tapes. Fillers can improve the processing characteristics of an adhesive raw material blend. As stated in [423], adding CaO to a polychloroprene-based raw material formulation reduces the mixing (kneading) time by 90%.

As mentioned earlier, fillers generally do not possess adhesive properties. They have the disadvantage of being inert. In some cases they can contain moisture, which is not inert, and may interact with other components of the formulation, (e.g., moisture-cure polyurethane formulations). Most good quality fillers have a moisture content of 0.2% less [421]. In order to achieve this level, special calcium carbonate fillers can be given two kinds of surface treatments — one to make them opaque and the other to impart hydrophobicity. According to [205], a level of 15% $CaCO_3$ is used for carpet tapes coated with HMPSA on SBC basis. For HMPSA on EVAc basis, zinc oxide, calcium carbonate, titanium dioxide, barium sulfate and organic resins have been recommended as filler. These adhesives require only a low level of filler. They contain about 35% elastomer, 30–50% tackifier, 0.2% plasticizer, 0.5% filler, and 0.1–0.5% stabilizing agents. As plasticizers, pthalates, phosphates, chlorinated polyphenols and liquid rosin derivatives have been used. As tackifiers, rosin derivatives, hydrocarbon resins, and low molecular weight styrene polymers have been suggested [109]. In a so-called combination, adhesives made as blends of PSA and nonpressure-sensitive EVAc dispersions (70/30), higher level of filler (up to 40%) and additives are used [356]. Filled ethylene–vinyl acetate copolymer emulsions useful as carpet adhesives can have a solid content as high as 85% [362].

Fillers have been proposed as viscosity regulating agents also. The cold flow of the adhesive in adhesive-coated PSPs depends on its viscosity and coating weight. Therefore, for tropically resistant tapes a low coating weight and high filler loading are recommended. For these tapes the use of silica (0.01–0.03 μm) is suggested as filler to increase the viscosity of the adhesive [424]. Polyvinyl ether is used for medical tapes with titanium dioxide as filler for coating a highly viscous warm mass, without solvent. Since, tapes used for automobile interiors have to resist high temperatures, it is recommended that, they may be manufactured from crosslinked acrylics with filler [425].

The adhesive properties of polychloroprene-based PSAs can be improved with chloroparaffins and fillers [426], for which 5% w/w zinc oxide and 12% w/w magnesium oxide, respectively, are used as filler. In this case the filler interacts as a reactive component with the base elastomer [427] and improves the cohesion of the adhesive. In a quite unusual manner, zinc oxide may increase both tack and cohesive strength in formulations with butyl rubber [428]. Zinc oxide has been used as a reactive filler in PSAs for medical tapes [165]. It is also suggested for use as an active filler for medical and insulation tapes because of its good resistivity. In such formulations zinc soaps are formed. For instance, 15–55 parts butyl rubber untackified or tackified with 7.5–40 parts PIB and crosslinked with 45 parts zinc oxide is a suggested recipe. Such formulations give 180° peel values of 0.73–1.81 kg/25.4 mm. The zinc oxide-induced crosslinking is also of interest in FDA-sensitive applications. Silica and ZnO increase the aging stability of adhesives on rubber (natural or synthetic) basis [429,430]. They also affect the coating viscosity. As absorbing pigments they can change the mechanical properties by the photooxidation of ethylene–propylene copolymers [431].

There are special formulations, in which the filler improves the application performance of the product and allows a carrier-less construction. Such fillers are needed for transfer tapes used as carrier-less high cohesion mounting tapes. Foam-like pressure-sensitive tape can be manufactured with glass microbubbles as filler [432] using. Dark glass microbubbles are embedded in a pigmented adhesive matrix. The glass microbubbles have an ultraviolet window that allows UV polymerization of the adhesive composition. The pressure-sensitive tape filled with glass microbubbles has a foam-like appearance and character and is useful for purposes that previously required a foam-backed PSA tape. The adhesive matrix may include 0.1–1.15% w/w carbon black also, without affecting the UV polymerization. Glass microbubbles as fillers have the advantage of higher distortion temperatures.

Transparent microbubbles are included in the composition of one foam-like tape [433]. In this case the glass microbubbles are embedded in a polymeric matrix 5–65% by volume. The thickness of the pressure-sensitive layer should exceed three times the average diameter of the microbubbles in order to enhance the moving of the bubbles in the matrix. Optimum performance is attained if the thickness of the PSA layer exceeds seven times the average diameter of the bubbles. The bubbles may be colored by adding metallic oxides. The average diameter of the stained glass microbubbles should be of 5–200 μm. Bubbles with a diameter greater than 200 μm would make UV polymerization difficult. The die imbedded in the PSA matrix should have a UV window also. Thick, triple-layered adhesive tape, with filler material in the center layer can also be manufactured. As fumed silica may be used as filler [433]. Microspheres made from borosilicate glass have been suggested as filler with a good stiffening effect [434]. Special, expandable microspheres based on a copolymer of vinylidene chloride and acrylonitrile, that encapsulate a liquid blowing agent have been used to improve paper stiffness [435]. Similar products can be used for the adhesive. Pressure-sensitive cold sealing adhesives have been manufactured from EVAc copolymers with 55–85% ethylene, tackifiers, and special microsphere fillers [436]. A closure tape with crosslinked acrylate terpolymer is based on a reinforced web and a reinforcing filler. This type of adhesive is useful for bonding the edges of heat-recoverable sheets to one another to secure closure. An amine–formaldehyde condensate and a substituted thrihalomethyltriazine are used as additional crosslinking agent [437].

The influence of fillers on the tack/peel and on the peel buildup has to be taken into account also. Special fillers are used in order to minimize the adhesive/adherend contact in order to achieve removability. Such filler can be rigid, nonadhesive, elastic or viscoelastic. As an example, microbubbles of glass have been been used as filler for a repositionable tape [438] and sheet labels [439]. Removable pressure-sensitive tapes containing resilient, hollow polymeric microspheres (20–66.7%) of a thermoplastic, expanded acrylnitril–vinylidene copolymer having a diameter of 10–125 μm, density of 0.01–0.04 and shell thickness of 0.02 μm in an isooctyl acrylate–acrylic acid copolymer have been prepared [440]. The particles are completely surrounded by the adhesive, having a thickness of at least 20 μm. When the adhesive is permanently bonded to the backing, and the exposed surface has an irregular contour, a removable and repositionable product is obtained and when the pressure-sensitive adhesive forms a continuous matrix, that is strippable bonded to the the backing (see transfer tape), having a thickness higher than 1 mm. This product is a foam-like transfer or foam tape. The (40 μm) cellulose acetate-based tape could be removed from paper without delaminating it.

As detackifying filler polymer (PSA) particles having a diameter of 0.5–300 μm can be used also [441]. Such pressure-sensitive filler particles added as detackifyers, are synthesized via suspension polymerization. The T_g of the particle should be lower than 10°C. They allow the use of special nonexpensive release coatings. Modified starch or starch and a water soluble fluorine compound (water soluble salts of perfluoroalkyl phosphates, perfluoroalkyl sulfonamide phosphates, perfluoroalkoxyalkyl carbamates, perfluroalkyl monocarboxylic acid derivatives, perfluoroalkylamines and polymers having as skeleton epoxy, acrylic, methacrylic, fumaric acid, vinyl alcohol, each of which contains a C_6–C_{12} fluorocarbon group on a repeating unit of the polymer) can be used as release layer on paper for an adhesive containing a minute and spherical elastomeric polymer [442]. The adhesion buildup can be reduced according to [443] using a powderlike surface coating of the PSA layer with particles having a diameter less than 10 μm, where the particles are imbedded in the PSA having two thirds from their diameter over the PSA level. Special fillers are used as detackifier agents mixed with the PSA. Their chemical composition may be various. For instance, a PSA for mounting tapes includes a detackifying resin comprising a caprolactone polymer (1–30% wt.) [444].

Common fillers for electrical conductivity are carbon black, graphite, carbon fibers, metallic powders, metalloxides, metalic fibers and metallized glass bubbles. Metallic powder is used as filler for electrically conductive adhesives. Electrically conductive adhesive tapes have been manufactured by coating of an acrylic emulsion containing 3–20% nickel particles onto a roughed flexible material, (e.g., Ni foil) [445]. Medical tapes with gelled PUR adhesive layer can include metallic powders [446]. As conductive filler fine powders of tin oxide, zinc oxide, titanium oxide and silicon based organic compounds have been applied [447]. It should be mentioned that the notion of coagulum or "grit" is used in correlation to fillers also. Such particles are included in carbon black used as filler [24].

Water can be used as conductive filler too [448]. Compositions based on acrylic acid polymerized in a water soluble polyhydric alcohol and crosslinked with a multifunctional unsaturated monomer contain water which affords electrical conductivity. Electrical performances can be improved adding electrolytes to the water. Such products can be used as biomedical electrodes applied to the skin. Starch is used also as water absorbant filler.

As shown in [449] electrically conductive (higher than 10^{-5}S) transparent PSA layer can be achieved using a special electrolyte. As known electrically conductive PSAs can be manufactured as gels containing water, NaCl or NaOH [450]. Such gels are useful in detecting voltages of living bodies. Such polymers based on maleic acid derivatives, have a specifical resistivity of 5 KΩ (1 Hz).

Flame retardant adhesives are prepared using halogen derivatives. A chlorine or bromine flame retardant with 65% or more chlorine or 80% or more bromine has been suggested [451]. A formulation with 1–30 parts by weight of the inorganic flame retardant and 0.1–20 parts by weight of silica powder has been recommended. As known decabrominated compounds are not allowed for food. Common, inorganic fillers can be used as flame retardant also. Sealant tapes can comprise 200 parts hydrated alumina as filler per 100 parts rubber, 50 parts carbon black and 50 parts plasticizer. The efficacity of inorganic fillers as flame retardants decreases as follows [452]: $Al(OH)_3$, $Mg(OH)_2$, caoline, limestone, zincoxide. Their activity is based on the increase of the enthalpy of gases built up [453].

Pigmented adhesives for labels offer better opacity [454]. For such pigments light absorbtion up to 385 nm is required in order to avoid adhesive dammage. Such protection is given generally for 1 yr Middle European climate. Because the release solution is transparent, converters may add an ultraviolet-detectable agent [455]. Special paints, inks and light sensitive additives can be incorporated in the adhesive. For instance, squarylium type sensitizing dyes, for visible light-sensitive (laser-activated) antitheft labels have been developed [456].

Thermally activatable PSAs contain a thermoplastic resin [457]. The formulation of hot seal PSA may include polyethylene (10–50% wt.) with a molecular weight of 1000–10,000 [458].

Fillers influence the crosslinking of adhesives and carrier materials also. The small difference in crosslink density of certain filled and unfilled polyurethanes is attributed to the presence of an active pigment, which may affect the distribution of crosslinking by localization of crosslinks in network systems [156]. Fillers influence the drying time also. Coatings containing increased level of fillers may be dried with a higher speed. A detailed descriprion of fillers is given in [459].

c. Antioxidants

Generally each of the PSP components contain antioxidants. They are used in order to improve the weathering, thermal and light resistance of the product components (elastomers, resins, plastomers).

Their chemical composition may differ according to the product component, its end-use and formulation technology. Antioxidants react with free radicals formed during the oxidation process. Their role is to delay the oxidative reactions. The most common antioxidants are napthol and phenol derivatives, oximes and aromatic amines [42]. For instance, for natural rubber-based adhesives 0.05–2% neozone may be used [113]. Generally, a HMPSA formulation contains 0.2–2.0% by weight of an antioxidant. High molecular weight hindered phenols, multifunctional phenols such as sulfur and phosphorous containing phenols are suggested. Such antioxidant like 2,2-thiodiethyl-*bis* [3-(3,5-di-*tert*-butyl-4-hydroxy-phenyl)] propionate (known under the commercial name of Irganox 1035) deactivates the singlet oxygen and reacts with alkoxy radicals. Aging of SBC based PSAs may be related to the incompatibility of the two phase system and to the unsaturation that exists in styrene midblocks [138]. For such compounds stabilizers based on salts of di (alkyl, phenyl, or benzyl)thiocarbamic acid have been recommended [460]. The performance of such stabilizers may be further enhanced by utilizing in conjunction known sinergists such as thiodipropionate esters and phosphites. Antioxidants for hotmelts should fulfill the following requirements: low volatility, to provide better maintenance, to reduce viscosity changes during processing, to delay skin formation and not to impart color. For SIS block copolymers in HMPSAs amines, phenols, phosphites, and thioesters are used as antioxidants. Oxybenzoquinones and oxybenzotriazoles are added as UV stabilizers [461]. For saturated midblock thermoplastic elastomers (commercial ABA type block copolymers) the formulations include a phenolic antioxidant 1–5% by weight.

The photoinitiated degradation of polymers is a radicalic reaction caused by light of 290–400 nm [462]. As protective agents UV absorbers, quenchers, hydroperoxide destroying additives and radical scavengers are used. According to [222] for tapes, one of the most important performances is open face aging. (This is not a crosslinking phenomenon according to the author [222].) Common tackifiers used for SBS show after a week of aging an important loss of tack (and later) of peel. For screening of resins for SBS a standard formulation with 200 parts SBS, 125–150 parts resin, and 25–50 parts oil, and 1 part antioxidant has been used. For SIS a 100/100/10/2 resin/SIS/oil/antioxidant composition has been suggested [463]. Special antioxidants should be used for EB-cured SBCs, because BHT and ZBTC inhibit the crosslinking. Phosphites work better [386].

As known, a storage stability of 6 months is given for tackifier dispersions [279]. A typical resin dispersion contains 40–60% resin, 2–20% plasticizer, 2–10% surfactant, 0–3% stabilizer/thickener, 0–0.5% antioxidant, 0–0.01% biocide. BHT, that is, (2,6 di-*tert*-butyl-*para*-cresol), tetrakis (methylene(3,5-di-*tert*-butyl-4-hydroxyhydrocinnamate)methane can be added to the resin dispersions to avoid color degradation and prevent oxydation [464]. As antioxidant for polyvinyl-pyrrolidone based formulations distearyl-pentaerythritoldiphosphite is suggested, for acrylics 4,4-butylidene-BIS has been recommended [369,464].

Antioxidants are used to prevent chemical denaturation of the carrier material in time under the environmental influences also. They prevent its oxidation. However, side effects can appear also. Some antioxidants can cause gel sensitivity, due to the stabilization of free radicals which are released. Thus they can produce large gel particles in the die [10]. Another problem is discoloration caused. No antioxidants should be postadded because of a possible yellowing of LDPE. If LLDPE is used as outer layer for coex, more antioxidant has to be added in order to avoid material destruction due to prolonged residence time [15]. The level of antioxidants depends on the film thickness also. For thinner films higher level should be used.

d. Antistatic Agents

In order to avoid accumulation of electrical charges on the carrier surface antistatic agents may be used as coated or built-in additives [465] (see also Chapter 8). Antistatic additives are classified as internal or external according to their application [466]. Such compounds possess an apolar and a polar part. This one may be a nonionic functional group (OH), for example, glycerinmonostearate, an anionic (alkylsulfonate) or a cationic one (quaternary ammonium salts). Fatty acid choline ester chloride is BGVV allowed. Antistatic agents reduce the common 10^{16} to 10^{17} Ω resistance of plastic films to 10^{10} Ω [467] (see Chapter 3). External antistatic agents are used at a level of $10-100$ mg/m^2 for different plastics. Internal antistatica may have different levels of application (0.05–8%) according to the plastic nature; conductive fillers are added at a level of 5–30% [466,468].

External antistatic agents possess the disadvantage that they need a supplemental processing (coating) step. On the other side they display the following advantages: they are working instantaneously; the influence of air humidity is less important; low level (0.01 g/m^2) is used; one side application is possible.

Internal antistatic agents must be soluble in the molten polymer. They (e.g., ethoxylated alkylamine) migrate on its surface and work together with the humidity of the air (buildup a thin conductive layer) [469]. Their disadvantages are the following: the crystallinity of the polymer influence their migration; pigments and migratory additives affect their migration; the temperature influences their migration; heat and chemical reactions influence their concentration and stability. They have to be used in higher concentrations. Such agents exhibit a decrease of the antistatical properties in contact with humidity and oxygen [470]. A too high level of antistatic agents influence the anchorage of printing inks. Generally, the resistivity which can be obtained when migrating additives are used is not lower than 10^9 Ω.

Ethoxylated amines are the best known compounds [27]. Unfortunately, they have a very low compatibility in masterbatches and a strong amine odor. For good antistatic behavior they need a relative humidity of 40%. Cationically active compounds with imidazole as quaternary group display antistatical effect also [471]. Derivatives of 1-alkyl-3-alkoxymethyl-imidazoliumchlorides have been tested as antistaticum for LDPE. Antistatic PP films contain N,N-bis-(β-hydroxyethyl)-stearylamine, or stearylamide or glycerinestearate [472].

Soluble, electrically conducting poly(alkoxythiophenes) may be used as antistatic agents also. Due to the $sp^2 pz$ hybridization of the orbitals in their aromatic rings, the free moving electrons assure (if dopped) sufficient conductivity. Polyethoxythiophene is used together with bonding agents (5 wt.% active substance). Antistatic agents like polyethoxythiophene can be coated on the film surface [470]. Polyethoxythiophene is coated as solution in organical solvents (4–8% solids content) using a gravure roller (with a coating weight of 1–4 g/m^2). Polypyrrole has been coated on plastic surfaces, displaying a surface resistance of 10^6 Ω [473]. Polyanilines have been suggested as built-in or coasted (100–400 nm) layer for improving of the electrical conductivity of polymers [474].

e. Antiblocking Agents

Blocking is a "not desired adhesive bonding of different surfaces" [475]. Antiblocking agents are special fillers (see also Chapter 8). Antiblocks prevent blocking, autoadhesion between layers. Such agents are used in carrier and adhesive formulation also. They are added to an adhesive (or carrier) formulation to prevent the adhesive coating (or carrier) made therefrom adhesing to its backing, when the adhesive coated/uncoated carrier material is rolled or stacked at ambient or elevated temperatures and relative humidities [476]. Blocking can be caused by heat or moisture, which may activate latent tack properties of the adhesive/carrier composition. Excessive surface treatment can cause blocking also. Excessive tackiness can require the use of additives or special processing techniques to facilitate ease in handling. For instance, by film manufacture good chill roll release is needed [477].

Antiblock additives are small (few microns) particles of hard materials such as silica and talc which protrude from the film and prevent the surface reaching close contact, which would cause blocking. Their nature and concentration depend on the nature of the base plastomer. In some cases the additives used as opacity agents can be used as antiblocking agents also [478]. Matteing agents have been suggested as antiblocks for PVC. For LDPE almost exclusively silica is added as antiblock [479]. Ultra high molecular weight PE may be used as antislip additive for LDPE also [480]. Linear low density polyethylene some times has more low molecular weight species and will require more antiblock additive. Processing aids concentration depends on the level of antiblock and other additives. Processing additives require a higher percentage of antiblocking agents and may influence the adhesion on the film surface [481].

f. Slip Agents

Such additives reduce the friction during transport and processing of plastic carrier materials. By blown films slip tends to bloom more to the outside of the bubble [482]. According to Grünewald [481], slip agents affect adhesion and surface treatment. Slip additives influence printability also. Their presence as undesirable boundary layer can disturb printability and coatability [42]. Recommended levels of slip agents are 1000–2000 ppm.

In LDPE and LLDPE the following slip agents are the most commonly used: unsaturated C_{18} primary amide (oleamide, MW 275), unsaturated C_{22} primary amide (erucamide), unsaturated C_{36} secondary amide, and unsaturated C_{40} secondary amide. The last one is having a molecular weight of about 600. As known migration is inversely proportional to the molecular weight [27]. The molecular weight of such additives is a compromise between the need to migrate to the surface within a resonable period of time and their blocking tendency. Amides with higher melting point have less effect on the wettability after corona treatment.

The level of slip additives needed depends on the application, film thickness and treatment intensity [483]. Thin glossy films generally have poorer slip and block more. Friction increases with the density of the base polymer. For cast film the surface quality of the chill roll influences the antislip concentration also. Secondary amides with better thermal resistance are particulaly suitable for cast film. Slip agent migration is influenced by the laminate components. Compatibility of amides with EVAc, EBA or EEA copolymers influence their efficacity. For such polymers metallic soaps are added as dispersing aid [27]. If the polarity of colaminated, extruded films is higher, slip agent migration increases [481]. Printing inks without slip agent may influence slip agent migration too.

The temperature resistance of slip agents is important also. Erucamide (ESA) is more efficient and has higher temperature resistance than oleamide (OESA). Its efficiency is less affected by the adhesive. For polymers processed at higher temperatures ESA is suggested [481]. Slip, antiblock and antistatic agents are interactive. Their combination can cause effects that are difficult to predict. Other additives such as UV absorbers and UV stabilizers, antioxidants, processing aids and some pigments can also strongly interferate. Fluorocarbon elastomers are the main processing aids used for polyethylene [484–486]. Such compounds work as slip agents between the molten polymer and the metallic surface of the processing equipment.

g. Printing Inks

Printing inks can have different chemical compositions. Their composition depends on the following parameters: carrier nature and geometry (web or sheet base material); type of printing process (flexography, litography, gravure, etc.); type of printing press (production speed and drying characteristics); finish required (matte or glossy); end-use application (color detail required, other processing steps, e.g., cutting, thermoforming, laminating, etc.); run length, sequence of ink application. Ink properties, such as COF, scuff resistance, gloss, and pigment light fastness are important characteristics for printing and film processing [487].

Special inks and coatings are used in combat theft and forgery, for example, thermochromic inks, photochromic inks, optically variable inks, reactive, metameric, infrared and luminescent inks [488]. Thermochromic inks change color at certain temperatures. This change is normally irreversible, but reversible inks have been developed also [485]. The so-called chiral nematics (liquid crystals) are applied by screen printing. These are water-transparent inks applied over a black background. They allow an invisible printed subject to become visible simple by touching. The most used temperature range is 27–33°C.

Photochromic inks change color with light. Optically variable inks display different colors when viewed from different angles. Reactive inks change color when reacted with special chemicals. Metameric inks change color using different light sources. Metallic inks change color when photocopied. Infrared inks show or do not show their color under infrared light. Luminescent inks continue to glow in darkness.

As stated by Nitschke [489] the trend to water-based compositions exists for printing inks also. New printing ink systems without solvents, with reduced solvent level and with "new" solvents have been developed. UV-cured printing ink systems, high solid ink systems and water-based inks have been manufactured [490]. It should be accentuated that because of the control of printing/drying properties of the inks via solvents, a general (full) replacement of solvents with water is not possibile. Water-based printing inks are really water containing or water dilutable systems. This is a big difference in comparison to water-based adhesives. Water based ink systems possess a lower running speed [489]. Water-based printing inks bond according to a physical or chemical process [490].

Physically, bonding systems are based mainly on styrene–acryl copolymers. Chemically, bonding binder systems are based on acrylated polyester prepolymers and acrylics. Vehicle systems embody the properties required by printing process, drying and end-use [42]. The polymer is the main part of the vehicle. Low molecular, narrow MWD polymers are used as vehicle. Solvent release and good block resistance are required. For rotative screen printing PVC-based inks (plastisols, foamed plastisols) and water-based dispersions are used [491]. Water-based screen printing inks have an excellent anchorage on PVC based pressure-sensitive products [492]. Binders for rotogravure printing are based mostly on acrylics. Special clays are designed for rotogravure with coarse pigments, calcinated clays, talcum, montmorrilonite are used with styrene butadiene rubber latex [493]. The water-based systems are stabilized via surfactants and protective colloids, but surfactant-free systems have been developed also [494].

Reactive printing inks as hot-melts and UV curable, water-dilutable printing inks and lacquers have been developed also [495]. They are curable after drying. Generally, radiation curing systems are based on binder (vehicle) which cure with UV light or electron beam [496]. Electron beam curing is too expensive for printing inks. Fewer than 50 of such (offset) printing units have been installed in the 1990s in the world [497], and actually there are no industrial developments in this field in Europe. Less expensive combined presses with UV flexo and rotary screen printing and cold UV systems have been introduced. UV-crosslinkable systems can cure via radicalic or ionic mechanisms. As known free radicalic UV systems contain acrylics as base macromolecular compound. Cationic systems include low viscous epoxides. Both are 100% solids in comparison with common solvent-based inks, having 30% solids. Radicalic systems react by light and during light action only. They can lead to an almost totally crosslinking [498]. Systems crosslinked by free radicalic mechanism use a liquid vehicle-based on multifunctional acrylics. Allyl and vinyl derivatives are also applied. These are di- and trifunctional monomers or oligomers. Tetrafunctional derivatives give films which are (due to the advanced crosslinking) too rigid. Polymers modified with acrylics are also used. They exhibit the advantages of the base compound (epoxides, polyesters or polyurethanes) and the reactivity of acrylates. Epoxyacrylates, amine modified polyetheracrylates and urethaneacrylate oligomers have been developed [486]. Modified natural raw materials (oils, fatty acids) were introduced too. The viscosity of the reactive prepolymers must be reduced using reactive diluting agents (di- and trifunctional acrylics). Tripropyleneglycoldiacrylate and dicyclo-pentenyl oxyethylacrylate are the most recommended difunctionally reactive monomers. Trimethylolpropane-triacrylate is used as monomer with more functionality. The adhesion between carrier and ink is improved by hydroxyl groups.

Seng [499] discussed the printing of polypropylene tapes using UV-cured systems. Ionic and free radical polymerization are compared. Cationic initiators are onium salts which give by photolysis an acid. Such onium salts include a diazonium, oxonium, sulfonium or iodonium cation and an anion of the following range [500]: ClO_4^-, BX_4^-, PX_6^-, AsX_6^- or SbX_6^-, where X is chlorine or fluorine. Good results are obtained using cycloaliphatical epoxides (chain opening polymerization via oxyrane rings). The ionic UV systems possess the following advantages: no inhibiting by oxygen, less shrinkage, good adhesion on metals and less toxicity. Unfortunately, such systems are less reactive, cure slowlier (curing is not finished after radiation, postcuring is necessary) and have lower penetration. Therefore, dual cure systems are proposed. First a peroxidic initiation, later UV-curing is carried out. As UV sensibilizers benzophenone derivatives are used. As discussed earlier, such compounds have been suggested for pressure-sensitive adhesives also.

2. Technological Additives

Materials used to allow mixing of formulation components, coating and drying of adhesive (or other PSP components) are technological additives. Some of them (like solubilizers) may be final constituents of the finished product also.

a. Solvents

Solvents are temporary components of the adhesives, primers, release coatings and printing inks. Evaporation rate and solvent retention influence the properties of the coating. Solubility parameter is a fundamental property of chemical compounds. Such numerical constants can be calculated or measured. Such parameters are of great help by the choice of the solvent.

Some years ago toluene, xylene, perchlorethylene and naptha blends have been the most used solvents for adhesives [500]. Special petrol fractions have been recommended for rubber-based adhesives [501]. Rubber-resin PSAs on solvent basis are manufactured with hydrocarbon solvents. Pressure-sensitive adhesives based on synthetic rubber uses special petrol fractions as solvent also [501]. The common manufacture procedure of tapes applies as solvent petrol fraction (boiling range 60–100°C) [502]. Although generally, aliphatic solvents are suggested, aromatic compounds can also be used. Such derivatives influence the viscosity. Adding of aromatic solvents to SBR-based PSA formulations results in viscosity increase on aging [503]. For printing inks the use of hydrocarbons is limited to rotogravure for solvent-based silicones hexane, toluene or xylene can be used. According to [144] for removable price labels rubber/resin adhesives are recommended. For such solvent-based adhesives petrol toluene, acetone, ethylacetate have been suggested as solvent. Turpentine has been applied for medical tapes as plasticizer-solvent [308]. Toluene replacement is a special problem [504]. High solvency hydrocarbon solvent systems with a high content of napthenic species provide a good combination of solvency and evaporation characteristics suitable for PSA formulation, using either styrene block copolymer or solvent-based AC. Because the solvent system has a low affinity for water, it may be an ideal candidate in currently used solvent capture and recycling equipment. Isoparaffinic solvents are suggested for natural rubber, butyl rubber, polyisobutylene, EPR and polyisoprene [505].

Ethylacetate, special boiling point petrol, n-hexane or n-heptane, toluene, acetone and isopropyl acetate are used as solvents for PSAs [506]. Typical solvent blends for acrylics contain ethyl acetate (54%), isopropanol (35%) and hexane (15%), or ethyl acetate (100%), or toluene (60%), hexane (39%) and methanol (1%).

For printing inks the rate of development of internal stresses depends on the type of solvent used and the rate of its evaporation. The slower the evaporation of the solvent, the lower the internal stress of the dried layer [42]. This influence plastic film shrinkage and curl also (see also Converting Properties). Alcohols are the primary solvents for flexographic inks, they are used in crosslinkable solvent-based acrylic PSAs also and may improve the solubility of sulfosuccinates in water-based

PSA dispersions. Esters are used because of their enhanced solubility as main solvents for solvent-based acrylic PSAs, for crosslinking agents and as component in solvent blends for printing inks (see also Chapter 6).

The addition of special solvents to water-based acrylic dispersions improve their adhesive characteristics [507]. Solvents with dielectric constant higher than 80 have been added to copolymers of C_{1-14} methacrylates with ethylenic unsaturated carboxylic acids. Such formulated double side coated PET tape showed an increase of about 10–20% of the peel.

Adequate choice of the solvent can increase the pot-life of two component crosslinked adhesives. Using t-butanol as solvent instead of ethyl acetate the pot-life of formulations based on copolymers of C_{4-12} of alkyl(meth)acrylates and unsaturated monomers having functional groups reactive to NCO crosslinked with hexamethylenediisocyanate, isophoronediisocyanate, hydrogenated MDI, hydrogenated xylenediisocyanate, lysine di- or triisocyanate and their derivatives increases the pot-life of the formulation from 0 to 8 h [508].

For solvent-based PSA the choice of the solvent used is determined by the polymerization. Blends of acrylic and vinyl esters can be copolymerized to achieve medical PSAs. Polymer solutions in isopropyl acetate (with a viscosity of 300–10,000 cP) are produced [509]. Cycloaliphatic, hydrocarbon-based solvents, ketones and alcoholic mixtures can also be used [308].

In certain cases the use of the solvents is imposed by coating problems. Wetting problems can be corrected through addition of cosolvents. Water-solvent mixtures have lower surface tension and may swell the dispersion particles giving a more continuous dried film [510].

In some cases the solvents are needed after coating and finishing of the PSP, if a posttreatment is carried out. According to [308] the tack free surface is achieved by superficial crosslinking of the adhesive by polyvalent cations (Lewis acid, polyvalent organometallic complex or salt). The crosslinking penetration depends on the polymer nature and solvent applied. Crosslinking can be achieved by dipping also. Water or water-miscible solvents should be used. Solvents are described in detail in [459], their recycling is discussed in [511].

b. Primers

The common manufacture procedure of tapes includes in-line primer coating, release and adhesive coating [502]. As classical primer solutions (in toluene) of butadiene–acrylnitryl–styrene copolymers have been preferred. For solvent-sensitive SPVC films primer dispersions based on Hycar Latex and Acronal 500 D (1/1) have been suggested by Reipp [502]. Tatsuno et al. [512] suggested as primer a mixture of maleated polypropylene and acrylic resin. The use of such primer increases the bonding strength from 0.5 to 2.4 kg. Chlorinated polyolefin–polyurethane adhesive have been proposed as primer for tapes also [513]. Epoxy containing silanes and vinyltriacetoxysilanes have been suggested as primer for silicone coating on paper [514].

As solvent-based primer a mixture of (90–50%) chlorinated polyethylene (with 40–75% chlor) and (1–50%) chlorinated polypropylene (with 25–35% chlor) has been proposed [515]. Tackified cyclized rubber has been suggested as primer for polypropylene [516]. For polyester films primers based on a reactive component, (e.g., chlorinated and fluorinated acetic acid or benzophenone) and a film forming polymer, (e.g., PVC, polyvinylformal, etc.) have been proposed [417]. For polypropylene a mixture of 20–60% chlorinated rubber (60% chlor), 80–40% EVAc copolymer (25% VAc), and 1–15% chlorinated polypropylene has been suggested [518]. For a polymeric film for silicone release applications the primer composition contains 25–75% glycidoxysilane and 75–25% polyester [519]. Solvent-based one component polyurethanes or polyimines can be used as primer ("tie-coat" adhesive) too [520,521]. Maleic anhydride and vinylacetate grafted butyl rubber have been suggested as primer for cellophane tape [522]. As illustrated by the earlier examples (Table 5.4) the chemical basis of primers is various. Because of the role of primers as anchoring intermediate between two chemically different layers their composition assumes reciprocal affinity and reactivity towards both components.

TABLE 5.4
Chemical Basis of Primers

Base Components	Primer Type	Carrier	Application	Reference
Butadiene–acrylonitrile–styrene copolymer	SB	Paper, film	PSA coating	[502]
Polypropylene graft copolymer–polyacrylate	SB	PP	PSA coating	[512]
Chlorinated polyolefin–polyurethane	SB	Film	Tapes	[513]
Chlorinated polyethylene–chlorinated polypropylene	SB	Film	Tapes	[515]
Chlorinated polypropylene–EVAc blends	SB	PP	Tapes	[518]
Tackified cyclized rubber	SB	PP	Tapes	[516]
Blends of PVC with chemical reagents	SB	PET	Tapes	[517]
Carboxylated rubber-polyacrylate	WB	SPVC	PSA coating	[502]
Epoxidized silanes–vinyltriacetoxysilane	SB	Paper	Siliconizing	[514]
Polyester–glicidoxysilane	SB	Film	Siliconizing	[519]

As discussed in detail in [459], technological additives include wetting agents (surfactants and cosolvents), neutralizing agents, viscosity regulators (thickeners and viscosity reducers), solubilizers, humidifiers, antimigration agents and cuttability agents also.

III. MACROMOLECULAR BASIS AND CHEMICAL COMPOSITION OF THE ABHESIVE

As discussed earlier, most PSPs require an abhesive layer between the pressure-sensitive layer and the (self-wound) back side of the product. In some special cases the low level of tack/peel and the unpolar character of the carrier material allow a release layer-free construction, but in most cases, a release layer is required. This layer may be a built into or coated adhesive layer.

A. Coated Adhesive Components

The coated adhesive layers are based on special chemical compounds having an unpolar or closed surface. Excepting some low molecular compounds from the range of fatty acid derivatives the most chemicals used as release coat are macromolecular compounds (e.g., silicones, polyvinyl carbamates, etc.). They can be classified according to their raw materials in nonsilicone and silicone-based compounds. Due to their excellent release characteristics silicones are the best known adhesive materials. Due to the need for a separate solid-state liner for labels and certain tapes the manufacture of siliconized carrier materials is actually an industry *per se*. It is not the aim of this book to discuss the complex problems of this domain. Some of them have been described in [1,2]. The number of products and applications where such very pronounced dehesive characteristics may disturb the manufacture and application of the product or they are not necessary is growing. On the other hand, as discussed earlier, there are plastomers used as solid-state laminate component which exhibit release performances. There are processing aids for plastomers, (e.g., slip, antiblock, matteing agents, etc.) which display adhesive properties too. There are special profilated carrier materials, (e.g., pleated, crepe paper, embossed PVC, etc.) which possess adhesive characteristics. Therefore, the importance of nonsilicone-based release materials is growing.

1. Nonsilicone Release Agents

As mentioned, for some products there is no need for such a high level of adhesive performance as given by silicones. On the other hand, for adhesive coatings with a very low tack and peel, the use of silicones would give laminates having an unsufficient bond strength. In such cases the abrasion sensitivity and low anchorage of silicones would influence the peel level also. For these applications other than silicone release layers are recommended. Various products have been developed for release layer using CMC, alginates and shellac [523]. As a release with good printability for tapes having siloxane, polyurethane, acrylics or modified rubber based PSA, coated on polyamide or glass fiber mates, polyesters, nitrile rubber, neoprene, and chlorosulfonated polyethylene have been suggested [524]. For medical tapes nitrocellulose, polyethylene, rubber, proteins and silicones have been used as release layer [142]. Copolymers of maleic acid with acrylic and vinyl monomers, styrene, vinylpyrrolidone and acrylamide have been proposed as release coating for tapes [525]. Aluminum film is used as street marking tape coated with a tackified polyisobutylene-based adhesive. The release layer for this adhesive is PVA [526].

The release mechanism involves low polarity surface or low surface tension, incompatibility of the release surface and adhesive surface [527]. Generally, the abhesive raw materials are high polymers bearing nonpolar or/and voluminous side chains or functional groups. The abhesive effect of low polarity is well demonstrated by silicones and fatty acid derivatives. Sterical and low molecular mobility effects are illustrated by fluorine, phosphorous or carbamide derivatives.

Werner complexes, complexes of basical chromchloride with a fatty acid rests, for example, stearine, palmitine or lauric acid, metalcomplexes of fluorinated compounds have been tested as release agent. Chromium complex salts in isopropylalcohol are suggested, film building substances can be added also, mixtures of silicones with chrome complex salts can be used also. Such composition contain chrome complex salt and silicone resin emulsion 1/1, or chrome complex salt and silicone rubber emulsion plus hardener, or silicone resin, silicone rubber and crosslinking agent [528].

Silicone release coating which stratify when backed, comprises hydroxy functional polyester containing fatty acid, crosslinking agent and silicone emulsion [529]. Polyester coated with stearic acid solutions in chloroform and treated with aqueous cuprichloride solution displays release properties (peel strength of about 10% for an acrylic adhesive tape).

According to [530], a release layer for paper carrier is based on modified starch or starch and a water-soluble organic fluorine compound, like salts of perfluoroalkylphosphates, fluoroalkysulphonamide phosphates, perfluoroalkoxycarbamates, perfluoroalkyl monocarboxylic acid derivatives, perfluoroalkylamines and polymers having as a skeleton acrylic acid, methacrylic acid, vinyl alcohol, etc., each of which containing C_6–C_{12} fluorocarbon groups on a repeating unit of the polymer (see Figure 5.9).

Vinyl esters having long pendant groups containing tetrafluoroethylene oligomer and hexafluoropropylene oligomers can be used as release coatings [531]. In a similar manner, combining fluoropolymers (hexafluoropropylene–tetrafluoroethylene copolymers) with polyamic acid leads to antiadhesive coatings with improved performance, due to the spontaneous separation in two layers, the bottom layer being polyamic acid, the top layer being fluoropolymer [29]. Favorable conditions for stratification have been created adjusting the concentration of fluoropolymers to 32–40 vol.% and increasing their particle size (water-thinned dispersion) to 1 μm. Nonporous, thin $(CF_x)_n$ coatings where x is having values of 0.3–2.0 with controlled wetability and good adhesion to plastics have been manufactured by plasma polymerization deposition of C_2F_6—H, C_2F_6—C_2F_4, CF_3—H and CF_4—C_2F_4 mixtures or C_2F_4. Such deposits possess a surface tension of 18 dynes/cm.

Polyvinylcarbamate is the main product used for nonsilicone-based release coatings (see also Chapter 8). It is applied for film coating. Kinning [532] has studied the release behavior of acrylic tapes from release coating based upon long alkyl side-chain polycarbamates (e.g., polyvinyl-*N*-octa-decyl carbamate (PVNODC) and polyvinyl-*N*-decylcarbamate (PVNDC). For special applications vinyl acetate polymers can be used as release also. For tapes an abhesive layer based on cellulose derivatives and PVAc has been suggested [46]. In certain cases acrylic polymers are adequate release agents too. Methyl methacrylate, styrene, butyl acrylate, and methacrylic acid

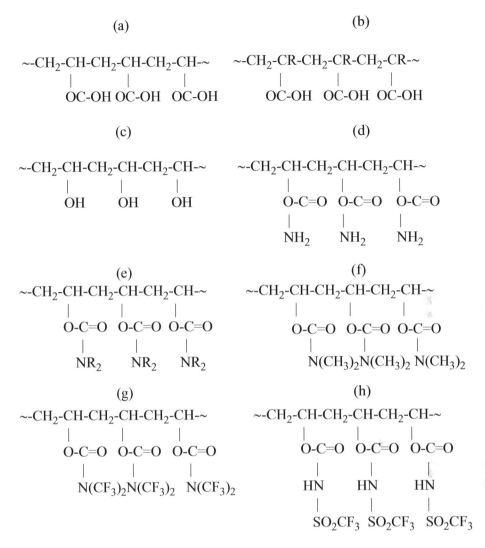

FIGURE 5.9 Buildup of some nonsilicone release agents. (a) Acrylic base polymer; (b) alkyl acrylic base polymer; (c) polyvinyl alcohol base polymer; (d) polycarbamate based on polyvinyl alcohol; (e) alkyl-substituted polycarbamate; (f) methyl-substituted polycarbamate; (g) perfluoromethyl polycarbamate; (h) perfluoroalkylsulfonamide polycarbamate.

have been grafted with the same monomers giving a polymer with T_g of $-80°C$. The neutralized product showed release properties [533,534].

2. Silicone Release Agents

The most used release coatings having a general applicability are those based on silicone polymers. Such products have been developed since many years for grease proof and release papers and-antiblocking coatings. Polymethylsiloxanes are used as soft block segments in high polymers or as oligomers in surface active agents. Silicone release is used mostly for labels but it has been tested for other products too (see also Chapter 8). Such formulations include silicones *per se*, together with other abhesive or non abhesive polymers or silicone-modified polymers. The main goal of development is to increase the ability to release efficiently "dial-a-release"– or to create differentials, which is the ratio of the base polymer release to a higher release due to high release additives (HRA) in polydimethylsiloxane (PDMS) polymer at a given application velocity (0.005 m to 5 m/sec) [535]. The industrial solution to release control is the inclusion of a methyl treated silicate in a manner analogous to tackifier. The role of the methylsilicates functioning as HRAs to modify the viscoelastic properties of PDMS was highlighted in [536]. It has been shown that the methylsilicate effectively reduces the segmental mobility of the flexible PDMS backbone. In PDMS-based silicone networks, adhesive or nonadhesive performances can be modulated either by PMDS chemical modification either by adding additive molecules [537]. The additive molecules are called MQ molecules. The network density increases (M_c decreases) as the MQ additive molecules amount increases in the network. The silicone network exhibit a vitrous plateau, which does not depend on MW and additive amount. For instance, dimethylpolysiloxane and methyl hydrogen polysiloxane based coating is used as release liner for stickies [48]. Diorganopolysiloxanes have been suggested as water-based release dispersion by Fau [538]. Such release coatings contain 10–70% wt. of polyvinylacetate and 2–20% of PVA. The T_g of the polymer is situated between -10 and $+35°C$. Mixtures of silicones with vinyl polymers [539] and ionomers [540] have been proposed as antiblocking surface treatment solutions for elastomers.

Ultraviolet light-cured silicones have been proposed for narrow webs, for broad webs EB curing has been recommended [541]. Because EB-curing occurs at low temperatures, these types of silicone release coatings could be applied and cured on heat-sensitive liners such as polyester, polypropylene and polyethylene films and polyethylene-coated papers without distorting such liners. Radiation (UV light)-curable release materials include reaction products of hydroxy group containing esters of pentaerythryitol and methacrylic acid and alkoxysilyl-terminated siloxanes [542]. Tapes with good adhesion and with release properties which decrease after irradiation have been prepared by treating substrates with silicone release agents (containing SiH and vinyl groups) on one side and butylacrylate acrylic acid copolymer on the other side, of a 60 μm corona-treated PP film. After UV irradiation of the release side (30 sec, 1000 W/m^2) the adhesive strength was 150 g/50 mm [543].

B. BUILT-IN ABHESIVE COMPONENTS

As discussed earlier, some nonpolar carrier materials exhibit abhesive properties. Such performances can be improved if special chemicals are built-in the bulky carrier material. These compounds may be macro- or micromolecular. It is well known from the practice of packaging materials that their processing and use on packaging machines require a low friction coeffiecient. Therefore, so-called slip agents are built in. Tapes have to allow low unwind resistance. Protective films, laminated on the surface to be protected must enhance the high speed processing. Therefore, in certain cases during film manufacture macromolecular or micromolecular compounds having abhesive properties are added to the base polymer (see earlier slip and antiblocking agents).

Micromolecular abhesive components are chemically similar to some low molecular compounds used as abhesive for web-coating. For instance, adhesive tapes with silicone PSA are coated with a release layer based on secondary amines, long chain fatty acid and vanadiumoxytrichloride [544]. Such formulations can be embedded in the carrier too. Macromolecular abhesive components are polymers having an unpolar, closed surface. It is well known, that highly crystalline polymers like HDPE are difficult to be coated. Macromolecular abhesive components can be embedded in common plastomers or elastomers also, to improve their abhesivity. The most known adhesive macromolecular compounds are silicone derivatives. They can be used as additives too. For instance, polydiorganosiloxanes are suggested as embodiments in coextruded polypropylene carrier materials with tackified top layer [91]. A dimethylpolysiloxane having a kinematic viscosity of at least 100 mm^2/sec at a temperature of 25°C is suggested (0.3–2.0% bw.) as such additive. The release layer of the coextrudate is having a thickness of 0.5–10 μm. The total thickness of the tape carrier film is about 15–50 μm. The removability of acrylic adhesives for tapes is improved by adding small amounts of organofunctional silanes (e.g., methacryloxypropyltreimethoxy-silane) [545].

REFERENCES

1. I. Benedek and L.J. Heymans, *Pressure-Sensitive Adhesives Technology*, Marcel Dekker Inc., New York, 1997, chap. 5.
2. I. Benedek, *Pressure-Sensitive Adhesives and Applications*, Marcel Dekker Inc., New York, 2004, chap. 5.
3. I. Benedek, *Pressure-Sensitive Formulation*, VSP, Utrecht, 2000.
4. R. Stockmeyer, *Deutscher Drucker*, (13), 42, 1988.
5. D.W. Aubrey, G.N. Welding, and T. Wong, *J. Appl. Chem.*, (19), 2193, 1969.
6. E. Seiler and B. Göller, *Kunststoffe*, (10), 1085, 1990.
7. H. Hub, *Composition, Characteristics and Application of Polar Ethylene Copolymers, Polyethylene and Copolymer Resin and Packaging Markets*, December 1–4, Maack Business Service, Zürich, Switzerland, 1990.
8. J.F. Kuik, *Papier und Kunststoff Verarbeiter*, (10), 26, 1990.
9. R.M. Ward and D.C. Kelley, *Tappi J.*, (6), 140, 1988.
10. E.B. Parker, *PE Copolymers and Their Application, Polyethylene and Copolymers Resin and Packaging Market*, in Proceedings of the Speciality Plastics Conference 87, December 1, Maack Business Service, Zürich, Switzerland, 1987.
11. E. Wistuba, *Kleben und Klebstoffe*, Sonderdruck, TI/ED 1665d, September 1993, BASF A.G., Ludwigshafen, Germany.
12. W.A. Fraser, *Novel Processing Aid Technology for Extrusion Grade Polyolefins*, in Proceedings of the Speciality Plastics Conference '87, on Polyethylene and Copolymer Resin and Packaging Markets, December 1, Zürich, Switzerland, 1987.
13. R. Hauber (Hans Neschen GmbH and Co. KG, Bückeburg, Germany), DE 42, 02, 070, 1993.
14. Proceedings of Polyethylene '93, October 4, Maack Business Service, Zürich, Switzerland, Session III, 1993, p. 6.
15. W.J. Busby, *Processing Problems with PE Films*, in Proceedings of PE '93, Session VI, 3–3 Maack Business Service, Zürich, Switzerland, 1993, p. 3.
16. R. Nurse, HDPE Applications, PE Developments, in Proceedings of Polyethylene '93, October 4, Session 3, Zürich, Switzerland, 1993, p. 1.
17. Packard Electric, Engineering Specification, ES-1881.
18. P.J. Flory, *Principles of Polymer Chemistry*, Cornell University Press, Ithaca, 1953, in T. Nishi and T.T. Wang, Eds., *Macromolecules*, 8, 908, 1975.
19. T. Nishi and T.T. Wang, *Macromolecules*, 8, 908, 1975.
20. M. Avella and E. Martuscelli, *Polymer*, 29(10), 1731, 1988.
21. *Eur. Plast. News*, (19), 19, 1993.
22. W.W. Bode, *Tappi J.*, (6), 133, 1988.
23. Kimberly Clark Co., U.S. Patent, 799,429, *Coating*, (1), 9, 1970.
24. H.G. Drössler, *Kaut. Gummi, Kunststoffe*, (1), 28, 1987.
25. G. Bodor, *A polimerek szerkezete*, Műszaki Könyvkiadó, Budapest, 1982, p. 85, 147.
26. *Deutsche Papierwirtschaft*, (3), 133, 1987.
27. R.C. Rombouts, *Proceedings of the Speciality Plastics Conference*, Maack Business Services, December 1, Zürich, Switzerland, 1987, p. 401.
28. K. Brack (Design Coat Co., U.S. Patent), 4,288,479, 1981, *Adhes. Age*, (12), 58, 1981.
29. V.D. Babayants, V.V. Kolesnitschenko, and S.G. Sannikov, *Lakokras. Mater. Ikh. Primen.*, (3), 45, 1988, *CAS, Coatings, Inks Relat. Prod.*, 17, 9, 1988.
30. E. Steffens, *Plaste u. Kaut.*, (3), 171, 1969.
31. J. Patschorke, *Adhäsion*, (2), 49, 1970.
32. D. de Jager and J.B. Borthwick, *Thermoplastic Rubbers for Hot-Melt Pressure-Sensitive Adhesives-The Processing Factors, Shell Elastomers, Thermoplastic Rubbers*, Technical Manual, 1994, TR 8.11, p. 5.

33. W. Neißl and H. Ledwinka, *Kunststoffe*, (8), 577, 1993.
34. S. Danesi and E. Garagnani, *Kaut. Gummi, Kunststoffe*, (3), 195, 1984.
35. J. Wetzel, *ASTM Bull.*, 221, 1957.
36. C.W. Hock, *J. Polym. Sci.*, C3, 139, 1963.
37. M. Kriszewski, *Acta Polym.*, (1/2), 37, 1988.
38. *Etiketten-Labels*, (3), 10, 1995.
39. Minnesota Mining and Manuf. Co., St. Paul, MN, USA, US Patent, 3,129,618, *Adhäsion*, (6), 14, 1985.
40. *Adhäsion*, (11), 482, 1967.
41. R. Davis, *Fassson Facts Int.*, (1), 2, 1969.
42. R.M. Podhajny, *Conv. Packaging*, (3), 21, 1986.
43. *Adhes. Age*, (11), 10, 1986.
44. *Labels Labell.*, (3/4), 82, 1994.
45. *Adhäsion*, (6), 14, 1985.
46. B. Hanka, *Adhäsion*, (10), 342, 1971.
47. *Paper, Film, and Foil Converter*, (5), 31, 1973.
48. Minnesota Mining and Manuf. Co., St. Paul, MN, USA, U.S. Patent, 3,129,618, *Adhäsion*, (2), 79, 1966.
49. G. Fuchs, *Adhäsion*, (3), 24, 1982.
50. H.K. Porter Co. Inc., Pittsburg, PA, U.S. Patent, 3,149,997, *Adhäsion*, (2), 79, 1966.
51. Allmänna Svenska Elektriska AB, DBP 1,276,771, *Adhäsion*, (2), 79, 1966.
52. *Etiketten-Labels*, (3), 9, 1995.
53. *Kaut. Gummi, Kunststoffe*, (6), 564, 1985.
54. H. Münstedt and H.J. Wolter, *Kunststoffe*, (10), 1076, 1990.
55. *Eur. Plast. News*, (19), 19, 1993.
56. *Eur. Plast. News*, (19), 22, 1996.
57. S. Füzesséry, *W. European PE Film Market and Applications, by Melt Index, Density and Comonomers Type*, in Proceedings of Polyethylene '93, Maack Business Service, October 4, Zürich, Switzerland, 1993.
58. K. Wabro, R. Milker, and G. Krüger, *Haftklebstoffe und Haftklebebänder*, Astorplast GmbH, Alfdorf, Germany, 1994, p. 48.
59. E. Djagarowa, W. Rainow, and W. Dimitrow, *Plaste u. Kautschuk*, (1), 28, 1970.
60. E. Djagarowa, *Plaste u. Kautschuk*, (9), 678, 1969.
61. *Adhes. Age*, (12), 8, 1986.
62. *Adhes. Age*, (1), 43, 1985.
63. *Etiketten-Labels*, (5), 136, 1995.
64. D. Djordjevic, *Tailoring Films by the Coextrusion Casting and Coating Process*, in Proceedings of the Speciality Plastics Conference '87, on Polyethylene and Copolymer Resin and Packaging Markets, December 1, Maack Business Service, Zürich, Switzerland, 1987.
65. B. Kunze, S. Sommer, and G. Düsdorf, *Kunststoffe*, 84(10), 1337, 1994.
66. *Handling*, (5/6), 12, 1994.
67. Minnesota Mining, Manuf. Co., St. Paul, MN, USA, British Patent, 1,102,296, *Coating*, (7), 210, 1969.
68. K. Nakamura and Y. Miki (Nitto Electric Industrial Co. Ltd.), Japanese Patent, 6,386,786, 1988; *CA Selects Adhesives*, 21, 5, 1988, 109: 111779 z.
69. *Etiketten-Labels*, (5), 6, 1995.
70. P. Galli, F. Millani, and T. Simonazzi, *Polym. J.*, 17, 37, 1985.
71. *Kunststoffe*, (7), 611, 1992.
72. B.H. Gregory, *Extrusion Coasting Advances — Resins, Processing, Applications, Markets*, in Proceedings of Polethylene '93, on The Global Challenge for Polyethylene in Film, Lamination, Extrusion, Coating Markets, October 4, Maack Business Services, Zürich, Switzerland, 1993.
73. E. Eastmann, *Ethylene Polymer Compositions for Hot Melt Adhesives, Coating*, (10), 368, 1988.
74. M. Gebauer and K. Bühler, *Kunststoffe*, (1), 21, 1992.
75. *Kunststoff J.*, (4), 60, 1985.
76. A. Barbero and A. Amico, *A New Performance ULDPE/VLDPE from High Pressure Technology — Potential Applications*, in Proceedings of Polyethylene '93, October 4, Zürich, Switzerland, 1993.

77. F.R. Baker, *A Unique Ethylene/Vinylacetate Copolymer for the Adhesives Industry, Tappi*, Hot Melt Adhesives and Coating Short Course, 2 May, 1982, Hilton Head, SC, USA.
78. K. Taubert, *Adhäsion*, (10), 379, 1970.
79. *Adhes. Age*, (8), 8, 1983.
80. *Converting Today*, (11), 9, 1991.
81. DBP 1,079,252, in *Coating*, (6), 185, 1969.
82. *Etiketten-Labels*, (35), 21, 1995.
83. *Adhäsion*, (11), 37, 1994.
84. *Etiketten-Labels*, (5), 136, 1995.
85. *Adhes. Age*, (12), 38, 1987.
86. P. Barot and J. Goletto, *Kaut. Gummi, Kunststoffe*, (10), 967, 1986.
87. G. Deelens, *Kaut. Gummi, Kunststoffe*, (10), 967, 1986.
88. D. Ogata and M. Kakimoto, *Macromolecules*, 18, 851, 1985.
89. *Kunststoff Information*, (1020), 2, 1991.
90. *Coating*, (6), 154, 1974.
91. T. Nakajima, K. Oda, K. Azuma, and K. Fujita (Nitto Electric Ind. Co.), Japan Patent, 6,327,579, 1988, *CAS, Siloxanes Silicones*, 17, 5, 1988.
92. Minnesota Mining and Manuf. Co., St. Paul, MN, USA, DBP 1,486,514.
93. *Papier und Kunststoff Verarbeiter*, (2), 18, 1996.
94. Milprint Overseas Corporation, Milwaukee, WI, USA, U.S. Patent, 1,504,556, *Coating*, (8), 240, 1972.
95. A. Haas (Société Chimique des Charbonnage-CdF Chimie, France), U.S. Patent, 46,24,991, 1986, in *Adhes. Age*, (5), 26, 1987.
96. Sun Oil Co., U.S. Patent, 3,342,902, *Coating*, (8), 244, 1969.
97. J.C. Chen and G.R. Hamed, *Rubber Chem. Technol.*, 60 (2), 319, 1987.
98. T.J. Bonk and J.T. Simpson (Minnesota Mining and Manuf. Co., St. Paul, MN, USA), EPA 0,120,708, November 19, 1987.
99. S. Schmukler, J. Machonis Jr., and M. Shida (Chemplex Co., Rolling Meadows, IL, USA), U.S. Patent, 4,472,555, 1985.
100. C.I. Simionescu and I. Benedek, *Angew. Makromol. Chem.*, 106, 1, 1982.
101. C. Kirchner, *Adhäsion*, (10), 398, 1969.
102. C.L. Gueris and E. McBride, *Ethylene Copolymers for Hot Melt Adhesives for Adhesion to Difficult Plastics*, in Proceedings of the 16th Munich Adhesive and Finishing Seminar, Munich, Germany, 1991, p. 72.
103. J. Brandup and E.H. Immergut, *Polymer Handbook*, 3rd ed., John Wiley and Sons, New York, 1990.
104. H. Wagner and P. Flory, *J. Am. Chem. Soc.*, 74, 195, 1952, in I.C. Petrea, *Structura Polimerilor*, Ed. Didactica si Pedagogica, Bucuresti, 1971, p. 75.
105. F. Danusso, G. Moraglio, and G. Gianotti, *Rend. ist. lombardo, Sci. P.I., Classe Sci., mat. e nat.*, A93, 666, 1959, in I.C. Petrea, *Structura Polimerilor*, Ed. Didactică şi Pedagogică, Bucuresti, 1971, p. 75.
106. A. Schultz, *J. Am. Chem. Soc.*, 76, 3422, 1954, in I.C. Petrea, *Structura Polimerilor*, Ed. Didactică şi Pedagogică, Bucuresti, 1971, p. 75.
107. M. Kuwahara, M. Kaneko, and J. Furuichi, *J. Phys. Soc., J.*, 17, 568, 1962, in I.C. Petrea, *Structura Polimerilor*, Ed. Didactică şi Pedagogică, Bucuresti, 1971, 75.
108. I.C. Petrea, *Structura Polimerilor*, Ed. Didactică şi Pedagogică, Bucuresti, 1971, p. 75.
109. R.J. Litz, *Adhes. Age*, (8), 38, 1973.
110. *Coating*, (8), 246, 1972.
111. J. Verseau, *Coating*, (11), 330, 1972.
112. *Adhäsion*, (3), 83, 1974.
113. J. Gerecke, F. Zachaeus, and R. Wintzer, *Plaste Kaut.*, 32, 332, 1985.
114. *Polyethylene*, Technical Information, SP0237,1991,11, Neste Chemicals International NV-SA, PE Marketing Department.
115. K.F. Schroeder, *Adhäsion*, (5), 161, 1971.
116. Takdust Products Co., *Rubber and Plastics Age*, (London), (6), 559 1968, *Coating* (1), 14, 1987.
117. M.A. Krecenski, J.F. Johnson, and S.C. Temin, *J. Macromol. Sci., Rev. Macromol. Chem. Phys.*, C26, 1986, 143.

118. A.B. Wechsung, *Coating*, (9), 268, 1972.
119. A. Wolf, *Kaut. Gummi, Kunststoffe*, 41(2), 173, 1988.
120. R. Köhler, *Adhäsion*, (6), 147, 1968.
121. US Patent, 4223067, in D.K. Fisher and B.J. Briddell EP 0426198 Adco Product Inc., Michigan Center, MI, USA, A2, 1991.
122. K.S. Lin (Avery Int. Co.) PCT WO 88 02014, 1988, *CAS, Adhesives*, 20, 4, 1988.
123. *Kaut. Gummi, Kunststoffe*, 40(2), 110, 1987.
124. *Kaut. Gummi, Kunststoffe*, 39(5), 450 1986.
125. W. Gleim, W. Oppermann, and G. Rehage, *Kaut. Gummi, Kunststoffe*, (6), 516, 1986.
126. G.R. Hamed and C.H. Hsieh, *J. Polym. Phys.*, 21, 1415, 1983.
127. M. Coleman, J. Graf, and P. Painter, *Specific Interaction and Miscibility of Polymer Blends*, Technomic Publishing Co., Lancaster, PA, 1991.
128. H.W.H. Yang, *J. Appl. Polym. Sci.*, 55, 645, 1995.
129. A. Zosel, *Int. J. Adhes. Adhes.*, 18, 265, 1998.
130. V. Muralidharan, C.Y. Hui, J. Dollhofer, C. Creton, and Y.Y. Lin, *Surface Tension and Machine Compliance Effects on Cavity Growth in Soft Materials*, in Proceedings of the 27th Annual Meeting of the Adhesion Society, February 15–18, Wilmington, NC, USA, 2004, p. 299.
131. S.P. Bunker, and R.OP. Wool, *Pressure-Sensitive Adhesives from Renewable Resources*, in Proceedings of the 24th Annual Meeting of Adhesion Society, February 25–28, Williamsburg, VA, USA, 2001, p. 174.
132. H. Lakrout, C. Creton, D. Ahn, and K. Shull, *Adhesion of Monodisperse Acrylic Polymer Melts to Solid Surfaces*, in Proceedings of the 23rd Annual Meeting of the Adhesion Society, Myrtle Beach, SC, February 20–23, 2000, p. 46.
133. M.A. Krecenski and J.F. Johnson, *Polym. Eng. Sci.*, 29(1), 36, 1989.
134. C. Creton, *Processing of Polymers*, H.E.H. Meijer (Ed.), VCH, Weinheim, 1997, Chap. 15, p. 707–741.
135. N. Willenbacher and A.E. O'Connor, *Effect of Molecular Weight and Temperature on the Tack of Model PSAs from Polyisobutene*, in Proceedings 24th Annual Meeting of Adhesion Society, February 25–28, Williamsburg, VA, USA, 2001, p. 378.
136. S.D. Tobing and A. Klein, *J. Appl. Polym. Sci.*, 79, 2230, 2001.
137. *Chemie in unserer Zeit*, 21(2), 52, 1987.
138. D.J.P. Harrison, J.F. Johnson, and W.R. Yates, *Polym. Eng. Sci.*, 22(14), 865, 1982.
139. W.H. Wetzel, *Rubber Age*, 82, 291, 1957.
140. M. Fujita, M. Kajiyama, A. Takemura, H. Ono, and H. Mizumachi, *J. Appl. Polym. Sci.*, 70, 771, 1998.
141. *Kaut. Gummi, Kunststoffe*, 40(8), 34, 1987.
142. *Coating*, (6), 184, 1969.
143. Minnesota Mining and Manuf. Co., St. Paul, MN, U.S. Patent, 1,594,178, *Coating*, (8), 240, 1972.
144. H. Mueller and J. Türk, EP 0,118,726, BASF AG, Ludwigshafen, Germany, 1984.
145. U.S. Patent, 3,321,451, in S.E. Krampe and C.L. Moore, EP 0202831A2, Minnesota Mining and Manuf. Co., St. Paul, MN, USA, 1986.
146. D.J. Yarusso, J. Ma, and R.J. Rivard, *Properties of Polyisoprene Based PSAs Crosslinked by Electron Beam Radiation*, in Proceedings of the 22nd Annual Meeting of the Adhesion Society, Panama City Beach, FL, USA, February 21–24, 1999, p. 72.
147. G. Auchter, J. Barwich, G. Rehmer, and H. Jäger, *Adhes. Age*, (7), 20, 1994.
148. S.D. Tobing and A. Klein, *Synthesis and Structure Property Studies in Acrylic Pressure-Sensitive Adhesives*, in Proceedings of the 24th Annual Meeting of Adhesion Society, February 25–28, Williamsburg, VA, USA, 2001, p. 131.
149. K.H. Schumacher and T. Sanborn, *UV-Curable Acrylic Hot-Melt for Pressure Sensitive Adhesives- Raising Hotmelts to a New Level of Performance*, in Proceedings of the 24th Annual Meeting of Adhesion Society, February 25–28, Williamsburg,VA, USA, 2001, p. 165.
150. I. Benedek, *Pressure-Sensitive Formulation*, VSP, Utrecht, 2000, Chap. 3.
151. US Patent, 3,725,115, in Y. Sasaaki, D.L. Holguin, and R. Van Ham, EP 0252 717A2, Avery International Co., Pasadena, CA, USA, 1988.
152. M. Dupont and N. De Keyzer, *Ultra Violet Light Radiation Curing of Styrenic Block Copolymer Based Pressure-Sensitive Adhesives*, in Proceedings of the 19th Munich Adhesive and Finishing Seminar, 1994, Munich, Germany, p. 120.

153. U.S. Patent, 4,069,123, in Y. Sasaaki, D.L. Holguin, and R. Van Ham, EP 0252 717A2, Avery International Co., Pasadena, CA, USA, 1988.
154. R. Hinterwaldner, *Coating*, (12), 477, 1995.
155. J.R. Erickson, E.M. Zimmermann, J.G. Southwick, and K.S. Kiibler, *Adhes. Age*, (11), 18, 1995.
156. M. Akay, S.N. Rollins, and E. Riordan, *Polymer*, 29(1), 37, 1988.
157. M.I. Aranguren and C.W. Macosko, *Macromolecules*, 21(8), 2484, 1988.
158. G. Holden and S. Chin, *Adhes. Age*, (5), 22, 1987.
159. J. Weidenmüller, *Deutsche Papierwirtschaft*, (3), T78, 1987.
160. R. Hinterwaldner, *Coating*, (7), 252, 1991.
161. S.R. Kerr, Heat seal macromolecular compounds, in F.Th. Birk, *Coating*, (9), 238, 1985.
162. A. Havranek, *Kaut. Gummi, Kunststoffe*, 39(3), 238, 1985.
163. J.L. Haldeman, *Adhes. Age*, (9), 35, 1984.
164. A.T. Govorkov, D.L. Muriskin, and Yu.N. Safonov, *Plast Massy*, (3), 28, 1987.
165. Y. Sasaaki, D.L. Holguin, and R. Van Ham, EP 0252717A2, Avery International Co., Pasadena, CA, USA, 1988.
166. Lehmann et al., U.S. Patent, 3,729,338, in D.K. Fisher and B.J. Briddell, EP 0426198A2, Adco Product Inc., Michigan Center, MI, USA, 1991.
167. US Patent, 4,181,752, in R.R. Charbonneau and G.L. Groff, EP 0106559 B1, Minnesota Mining and Manuf. Co., St. Paul, MN, USA, 1984.
168. G.R. Frazee (Johnson S.C. and Son. Inc.), EP 259,842, 1988, *CAS, Adhesives*, 20, 4, 1988.
169. *Coating*, (8), 268, 1987.
170. H. Bayer, *Adhäsion*, (5), 17, 1987.
171. K. Fuhr, *Defazet*, 31(6–7), 259, 1977.
172. M. Scheiber and H. Braun, *UV Vernetzbare Haftklebstoffe*, in Proceedings of the 19th Munich Adhesive and Finishing Seminar, 1994, p. 141.
173. C.R. Morgan, *Adhäsion*, (12), 25, 1983.
174. H. Huber and H. Müller, *Coating*, (9), 328, 1987.
175. J. Weidenmüller, *Deutsche Papierwirtschaft*, (3), T78, 1987.
176. S. Kerr, *Der Polygraph*, (5), 366, 1986.
177. K. Li, P. Mallya, P. Iyer, C. Kuang, and W. Wang, *Synthesis of Benzocyclo-Butenone Containing Polymers for Ultraviolet: Light Curable Pressure-Sensitive Adhesive Applications*, in Proceedings of the 24th Annual Meeting of Adhesion Society, February 25–28, Williamsburg, VA, USA, 2001, p. 368.
178. P. Beiersdorf & Co., AG, Hamburg, U.S. Patent, 1,569,882, *Coating*, (11), 336, 1972.
179. A. Zawilinski, *Adhes. Age*, (9), 29, 1984.
180. Adhesive L 1233, Product Data, Sealock, Andover, Hampshire, UK, 1997.
181. R.J. Shuman and B.D. Josephs, Dennison Manuf. Co., Framingham, MA, USA, PCT, WO 88/01636.
182. P. Green, *Labels Labell.*, (11/12), 38, 1985.
183. Rohm & Haas Co., Philadelphia, PA, USA, U.S. Patent, 3,152,921, *Adhäsion*, (3), 80, 1966.
184. *TNII Bumagi*, SSSR Patent, 300561, *Coating*, (1), 14, 1987.
185. *Coating*, (6), 185, 1969.
186. *Coating*, (1), 24, 1969.
187. Y.C. Chu and R. Vukov, *Macromolecules*, 18, 1423, 1985.
188. *Coating*, (12), 455, 1969.
189. *Eur. Adhes. Seal.*, (6), 23, 1995.
190. *Eur. Adhes. Seal.*, (6), 36, 1995.
191. W.H. Korcz, *Adhes. Age*, (11), 19, 1984.
192. E.G. Huddleston, U.S. Patent, 4,692,352, The Kendall Co., Boston, MA, 1987.
193. Johns Manville Corp., U.S. Patent, 3,356,635, *Coating*, (7), 210, 1969.
194. A.J. Liebermann, A.J. Kotova, V. Vedenov, V. Verchoyancev, and R.J. Mirkina, *Lakokras Mater., Ih. Primen.*, (3), 12, 1983.
195. R.G. Jahn, *Adhes. Age*, (12), 35, 1977.
196. *Adhes. Age*, (9), 24, 1986.
197. H.J. Neupert, M. Arnold, S. Klodt, U. Schoenrogge, B. Rothenhauser, and U. Rossow, VEB Chemische Werke, Buna, East German Patent, 247,903, 1987, *CAS, Emulsion Polym.*, 18, 2, 1988.
198. D.J. St. Clair and J.T. Harlan, *Adhes. Age*, (12), 39, 1975.
199. W.P. Gergen, *Kaut. Gummi, Kunststoffe*, 37(4), 284, 1984.

200. C. Donker, *Characterisation of Resins for Hotmelt Pressure-Sensitive Adhesives*, in Proceedings of the 16th Munich Adhesive and Finishing Seminar, Munich, Germany, 1991, p. 41.

201. M.K. Dehnke and K.E. Johnsen, Formulating flexibility of pure triblock styrene/diene thermoplastic elastomers, *Coating*, (5), 176, 1988.

202. C.P. Rader, *Kunststoffe*, 83(10), 27, 1993.

203. C.N. Smit, *Kunststof Ruber*, 41(2), 16, 1988.

204. R. Hinterwaldner, *Adhäsion*, (6), 11, 1984.

205. *Adhes. Age*, (11), 32, 1985.

206. K.E. Johnsen, *New Developments in Thermoplastic Elastomers*, in Proceedings of the TAPPI Hot Melt Symposium 85, June 16–19, Hilton Head, SC, USA, 1985.

207. S.V. Canewarolo and A.W. Birley, *Brit. Polym. J.*, 19, 43, 1987.

208. K.E. Johnsen, *Adhes. Age*, (11), 29, 1985.

209. E. Diani, A. Riva, A. Iacono, and E. Agostinis, *Styrenic Block Copolymers as Base Material for Hot-Melt Adhesives*, in Proceedings of the 16th Munich Adhesive and Finishing Seminar, 1991, p. 77.

210. J.A. Miller and E. von Jakusch, EP 0306232B, Minnesota Mining and Manu Co., St. Paul, MN, USA, 1993.

211. A. Roos and C. Creton, *Adhesion of PSA Based on Styrenic Block Copolymers*, in Proceedings of the 24th Annual Meeting of Adhesion Society, February 25–28, Williamsburg, VA, USA, 2001, p. 371.

212. *Kaut. Gummi, Kunststoffe*, 37(4), 285, 1984.

213. F.F. Lau and S.F. Silver, EP 0130087B1, Minnesota Mining and Manuf. Co., St. Paul, MN, USA, 1985.

214. L.J. Fetters, R.W. Richards, and E.L. Thomas, *Polymer*, 28(13), 2252, 1987.

215. D. Alward, D.J. Kinning, E.K.L. Thomas, and L.J. Fetters, *Macromolecules*, 19, 215, 1986.

216. U.S. Patent, 4,163,077, in F.F. Lau and S. F. Silver, EP 0130087B1, Minnesota Mining and Manuf. Co., St. Paul, MN, USA, 1985.

217. French Patent, 2,331,607, in F.F. Lau and S.F. Silver, EP 0130087B1, Minnesota Mining and Manuf. Co., St. Paul, MN, USA, 1985.

218. L.D. Jurrens and O.L. Mars, *Adhes. Age*, (8), 31, 1974.

219. E.H. Otto, *Allg. Papier Rundsch.*, 16, 438, 1986.

220. C. Parodi, S. Giordano, A. Riva, and L. Vitalini, *Styrene–Butadiene Block Copolymers in Hot Melt Adhesives for Sanitary Application*, in Proceedings of the 19th Munich Adhesive and Finishing Seminar, Munich, Germany, 1994, p. 119.

221. C. Donker, R. Lut, and K. van Rijn, *Hercules MBG 208 Hydrocarbon Resin: A New Resin for Hot-Melt Pressure-Sensitive (HMPSA) Tapes*, in Proceedings of the 19th Munich Adhesive and Finishing Seminar, Munich, Germany, 1994, p. 64.

222. L. Jacob, *New Development of Tackifiers for SBS Copolymers*, in Proceedings of the 19th Munich Adhesive and Finishing Seminar, Munich, Germany, 1994, p. 107.

223. *Coating*, (12), 480, 1995.

224. L.E. Jacob, A. Lepert, and M.L. Evans, U.S. Patent, 4,623,698, Exxon Research and Eng. Co., Forham Park, NJ, USA, 1986.

225. P.A. Mancinelli, J.A. Schlademann, S.C. Feinberg, and S.O. Norris, *Advancement in Acrylic HMPSAs via Macromer Monomer Technology*, in Proceedings of the PTSC XVII Technical Seminar, 4 May, Woodfield Shaumburg, IL, *Coating*, (1) 12, 1986.

226. G. Obieglo and K. Romer, *Kunststoffe*, 83(11), 926, 1983.

227. J.R. Erickson, *Experimental Thermoplastic Rubber with Improved Radiation Curing Performances for HMPSA*, in Proceedings of the TAPPI Hot Melt Symposium '85, June 16–19, 1985, Hilton Head, SC, USA, in *Coating*, (1), 6, 1986.

228. M. Szwarc, *Nature*, 178, 1168, 1956.

229. F. Bandermann, H.D. Speikamp, and L. Weigel, *Makromol. Chem.*, 186, 2017, 1985.

230. Z. Mo, L. Wang, H. Zhang, P. Han, and B. Huang, *J. Polym. Sci., Polym. Phys.*, 25(9), 1829, 1987.

231. C.I. Simionescu, I. Benedek, S. Ioan, M. Galin, and N. Asandei, *Rev. Roum. Chim.*, 17, 2003, 1972.

232. N.G. Matveeva, A.G. Kondratieva, E.S. Pankova, O.G. Selskaya, and E.S. Mamedova, *Kinetics and Mechanism of Polyreactions*, Akadémiai Kiadó, Budapest, 1969, Vol. 3, p. 161.

233. Y.S. Matsumoto, and J. Furukawa, *Kaut. Gummi, Kunststoffe*, (6), 541, 1986.

234. L.L. de Lucca Freitas and R. Stadler, *Macromolecules*, 20(10), 2478, 1987.

235. H. Xie and J. Xia, *Makromol. Chem.*, 188(11), 2543, 1987.
236. J. Roovers, *Macromolecules*, 20(1), 148, 1987.
237. G. Bocaccio, *Caoutch. Plast.*, 62(653), 83, 1985.
238. B.C. Auman, V. Percec, H.A. Schneider, and H.J. Cantow, *Polymer*, 28(8), 1407, 1987.
239. S. Kawachi, *Gummi Asbest, Kunststoffe*, (4), 162, 1986.
240. L.J. Mathias and M. Allison, *ACS Symp. Ser.* 367, 66, 1988, in *CAS, Crosslinking React.*, 21, 3, 1988, 109: 111261z.
241. T.M. Swager and R.H. Grubbs, *J. Am. Chem. Soc.*, 109(3), 894, 1987.
242. *Kaut. Gummi, Kunststoffe*, 41(1), 102, 1988.
243. L. de Lucca Freitas, J.B. Urgert, and R. Stadler, *Polym. Bull.*, 17, 431, 1987.
244. U. Eisele, *Kaut. Gummi, Kunststoffe*, 40(6), 539, 1987.
245. R. Jordan, *Adhäsion*, (5), 256, 1980.
246. Vector 2518D, Product description, Dexco Polymers, November 1992.
247. Vector 2411D, Product description, Dexco Polymers, November 1992.
248. Vector 8508D, Product description, Dexco Polymers, November 1992.
249. Cariflex TR-1000 Polymere für Klebstoffe, Beschichtungen und Dichtungsmassen, Technische Broschüre, RBX/73/8 (G), Shell Elastomers.
250. *Kaut. Gummi, Kunststoffe*, 40(1), 40, 1987.
251. Vector 2514D, Product description, Dexco Polymers, November 1992.
252. R. Koch, *Kaut. Gummi, Kunststoffe*, 39(9), 804, 1986.
253. D.W. Bamborough and P.M. Dunkley, *Adhes. Age*, (11), 20, 1990.
254. W.H. Meyer, *Chemie in unserer Zeit*, 21(2), 59, 1987.
255. N. Nakajima, R.A. Miller, and E.R. Harrel, *Internat. Polym. Process.*, 2(2), 88, 1987.
256. T.S. Yagishita, K. Hosoya, and N. Inagami, *Am. Chem. Soc. Div. Polym., Chem. Prep.*, 26(2), 32, 1985.
257. V. Stanislawczyk, EP 264903, Goodrich B.F. Co., 1988, in *CAS, Emulsion Polym.*, 18, 2, 1988.
258. K. Yamada, K. Miyazaki, Y. Oowatari, Y. Egami, and T. Honma, EP 257803, Sumitomo Chem. Co., Ltd., 1988, in *CAS, Coatings Inks Relat. Comp.*, 17, 6, 1988.
259. A. Hecht Beaulieu, D.R. Gehman, and W.J. Sparks, *Tappi J.*, (9), 102, 1984.
260. F.T. Sanderson, *Acrylic Hot Melt Pressure-Sensitive Adhesives*, in Proceedings of the Hot Melts—The Future is Now, Tappi Hot Melt Symposium, June 2–4, Toronto, *Coating*, (7), 175, 1980.
261. R.M. Enanoza, EP 259,968, Minnesota Mining and Manuf. Co., St. Paul, MN, USA, 1988.
262. P.R. Drzal and K.R. Shull, *Adhesive Properties of Model, Filled Elastomeric Adhesives*, in Proceedings of the 24th Annual Meeting of Adhesion Society, February 25–28, Williamsburg, VA, USA, 2001, p. 168.
263. National Starch and Chem. Co., Bridgwater, VA, USA, Japanese Patent, 6,306,076, 1988, in *CAS, Adhesives*, 19, 2, 1988.
264. D.E. Bugner (Eastmann Kodak Co., Rochester, NY, USA), *ACS Symp. Ser.*, 364, 1988 (*Chem. React. Polym.*), 276, in *CAS, Polyacrylates* (*Journals*) 9, 2, 1988.
265. T. Hamaide, A. Revillon, and A. Guyoz, *Eur. Polym. J.*, 23(10), 787, 1987.
266. E.V. Gruzinow, V.P. Panov, V.V. Gusev, V.N. Frosin, and V.S. Rytbchinskaya, *Vysokomol. Soed.*, B29, 373, 1987.
267. P. Cebeillac, D. Chatain, C. Megret, C. Lacabanne, A.S. Bernes, and P. Dupuis, *Makromol. Chem., Macromol. Symp.*, 20/21, 335, 1988.
268. *Druck Print*, (4), 18, 1988.
269. G.R. Homan and H.L. Vincent, EP 0,183,378, Dow Corning Corp., Midland, MI, USA, 1984.
270. J.D. Blizzard and T.J. Swihart, EP 0,183,379, Dow Corning Corp., Midland, MI, USA, 1984.
271. J.D. Blizzard and D. Narula, EP 0,183,377, Dow Corning Corp., Midland, MI, USA, 1984.
272. A. Tomanek, *Kunststoffe*, (11), 1277, 1990.
273. K.L. Ullmann and R.P. Sweet, *Silicone PSAs and Rheological Testing*, in Proceedings of the 22nd Annual Meeting of the Adhesion Society, Panama City Beach, FL, USA, February 21–24, 1999, p. 410.
274. F. Hufendiek, *Etiketten-Labels*, (1), 9, 1995.
275. *Adhäsion*, (11), 37, 1994.
276. *Adhes. Age*, (9), 8, 1986.
277. Mystik Tape Inc., U.S. Patent, 3,161,533, *Adhäsion*, (6), 277, 1966.
278. J.R. Pennace, C. Ciuchta, D. Constantin, and T. Loftus, WO8703477A, 1987.

279. Y. Hamada and O. Takuman, EP 253,601, Toray Silicone Co., Ltd., 1988.

280. B. Copley and K. Melancon, EP 255,226, Minnesota Mining and Manuf. Co., St. Paul, MN, USA, 1988.

281. D.F. Merril, *Int. SAMPE Symp. Exhib.*, 1988, 33, in *CAS, Adhesives*, 14, 2, 1988, 108: 222717b.

282. R. Milker, PUR PSAs, Pressure-Sensitive Adhesives and Adhesive Coating, Cowise, Management and Training Service, 1996, Amsterdam.

283. B.H. Edwards, *Polyurethane Structural Adhesives*, in Proceedings of the Adhesives '85, Conference Papers, September 10, Atlanta, GA, USA, 1985.

284. *Coating*, (8), 242, 1972.

285. M.K. Yamazaki and S. Kamatani, U.S. Patent, 4,471,103, Takeda Chemical Industries, Ltd., Osaka, Japan.

286. R.J. DeVoe and C.D. Lynch, U.S. Patent, 474577, Minnesota Mining and Manuf. Co., St. Paul, MN, USA, 1988.

287. C. Hepburn, *Progr. Rubber Plast. Technol.*, 3(3), 33, 1987.

288. B. Krüger, *Kaut. Gummi, Kunststoffe*, (10), 967, 1986.

289. N. Kunii, *Kaut. Gummi, Kunststoffe*, 39(6), 541, 1986.

290. N. Yamazaki, *Kaut. Gummi, Kunststoffe*, (9), 836, 1986.

291. C.G. Pitt, Z.W. Gu, P. Ingram, and R.W. Hendren, *J. Polym. Sci., Polym. Chem.*, 25(4), 995, 1987.

292. J. Franke, K.P. Meurer, P. Haas, and J. Witte, DE 3,622,825, Bayer A.G., 1988.

293. T. Wakabayashi and S. Sugii, EP 0285430B1, Minnesota Mining and Manuf. Co., St. Paul, MN, USA, 1988.

294. Sun Oil Co., U.S. Patent, 3,356,766, *Adhäsion*, (10), 303, 1972.

295. C.N. Clubb and B.W. Foster, *Adhes. Age*, (11), 18, 1988.

296. B.W. Foster, *A New Generation of Polyolefin Based Hot Melt Adhesives*, in Proceedings of the Tappi Hot Melt Symposium, June 7–10, Monterey, CA, USA, 1987.

297. U.S. Patent, 3686107, in T. Wakabayashi and S. Sugii, EP 0285430B1, Minnesota Mining and Manuf. Co., St. Paul, MN, USA, 1988.

298. G.C. Allen, J.B. Pellon, and M.P. Hughes, EP 251771, El Paso Products Co., 1988.

299. Y. Mizutani, T. Noguchi, H. Kuroki, and T. Imahaba, Japanese Patent, 62265379, Tosoh Co., 1987, in *CAS, Hot Melt Adhesives*, 11, 1, 1988, 108: 168847d.

300. L.C. DeBolt and E. Riande, *Makromol. Chem.*, 187, 2497, 1986.

301. H. Takao, H. Kuribayashi, and E. Usuda, EP 254002, 1988.

302. G.A. Luscheyki, F.M. Medvedeva, L.I. Voytesonok, M.K. Polevaya, and L.D. Pin, *Plast. Massy*, (10), 16, 1985.

303. B.M. Mahato, S.C. Shit, and S. Maiti, *Eur. Polym. J.*, 21, 925, 1985.

304. H.F. Huber and H. Müller, Proceedings of the 10th Radcure '86, Conference Proc., 1986.

305. P.P. Hoenisch and F.T. Sanderson, *Aqueous Acrylic Adhesives for Industrial Laminating*, in Proceedings of the Adhesives 85, Conference Papers, September 10–12, Atlanta, GA, USA, 1985.

306. Minnesota Mining and Manuf. Co., St. Paul, MN, USA, U.S. Patent, 2,926,105, *Coating*, (6), 185, 1969.

307. R.G. Frazee, EP 258,753, Johnson S.C. and Son, Inc., 1988, in *CAS*, 19, 3 1988.

308. S.E. Krampe and C.L. Moore, EP 0,202,831A2, Minnesota Mining and Manuf. Co., St. Paul, MN, USA, 1986.

309. US Patent, 288,416/RE 24906, in S.E. Krampe and C.L. Moore, EP 0202831A2, Minnesota Mining and Manuf. Co., St. Paul, MN, USA, 1986.

310. US Patent, 3121021, in S.E. Krampe and C.L. Moore, EP 0202831 A2, Minnesota Mining and Manuf. Co., St. Paul, MN, 1986.

311. British United Shoe Machinery Co., British Patent, 761,840, *Coating*, (6), 185, 1969.

312. *Druckprint*, (4), 19, 1988.

313. Z. Czech, *Eur. Adhes. Seal.*, (6), 4, 1995.

314. *Kaut. Gummi, Kunststoffe*, 40(4), 313, 1987.

315. Y. Sasaaki, European Tape and Label Conference, April 28, Brussels, Belgium, 1993, p. 133,

316. L. Li, C. Macosko, G.L. Corba, A. Pocius, and M. Tirrell, *Interfacial Energy and Adhesion Between Acrylic PSA and Release Coatings*, in Proceedings of the 24th Annual Meeting of Adhesion Society, February 25–28, Williamsburg, VA, USA, 2001, p. 270.

317. C.M. Miller and H.W. Barnes, *Factors Affecting Water Resistance of Latex-Based PSAs*, in Proceedings of the 24th Annual Meeting of Adhesion Society, February 28, Williamsburg, VA, USA, 2001, p. 153.

318. D.G. Pierson and J.J. Wilczynski, *Adhes. Age*, (8), 52, 1990.

319. U.S. Patent, 3865770, in F.C. Larimore and R.A. Sinclair, EP 0197662A1, Minnesota Mining and Manuf. Co., St. Paul, MN, USA, 1986.

320. P.K. Dhal, R. Murthy, and G.N. Babu, *Org. Coat. Appl. Polym. Sci., Proceed.*, 48, 131, 1983.

321. Sekisui Chemical Ind. Co. Ltd., Japanese Patent, 072,785,513, 1995, *Adhes. Age*, (5), 12, 1996.

322. U.S. Patent, 2925174, in R.R. Charbonneau and G.L. Groff, EP 0106559B1, Minnesota Mining and Manuf. Co., St. Paul, MN, USA, 1984.

323. C. Harder, *Acrylic Hotmelts Recent Chemical and Technological Developments for an Ecologically Beneficial Production of Adhesive Tapes State and Prospects*, in Proceedings of the European Tape and Label Conference, Brussels, April 28–30, 1993.

324. C.I. Simionescu, N. Asandei, and I. Benedek, *Rev. Roum. Chim.*, 16, 1081, 1971.

325. D.K. Fisher and B.J. Bridd, EP 0426,198 A2, Adco Product Inc., Michigan Center, MI, USA, 1991.

326. Belgian Patent, 675,420, in D.K. Fisher and B.J. Briddell, EP 0426198A2, Adco Product Inc., Michigan Center, MI, USA, 1991.

327. G. Geuskens and M.S. Kabamba, *Polym. Degrad. Stabil.*, 19(4), 315, 1987.

328. R. Schaller, M.G. Mertl, and K. Hummel, *European Polym. J.*, 23, 259, 1987.

329. J.N. Kellen and C.W. Taylor, EP 0246826A2, Minnesota Mining and Manuf. Co., St. Paul, MN, USA, 1987.

330. H. Kuroda and M. Taniguch, Japanese Patent, 6343987, Bando Chem. Ind. Ltd., 1988, in *CAS, Adhesives*, 20, 3, 1988.

331. H. Röltgen, *Coating*, (5), 124, 1985.

332. H.J. Timpe and H. Baumann, *Adhäsion*, (9), 9, 1984.

333. P. Gosh and A.R. Bandyopadhyay, *Eur. Polym. J.*, (11), 1117, 1984.

334. C. Decker, Radcure 85, in F.Th. Birk, *Coating*, (12), 278, 1985.

335. W. Baueumer, M. Koehler, and J. Ohngemach, in Proceedings of the 10th Radcure '86, Conference, 1986, in *CAS, Coatings, Inks Relat. Prod.*, 22, 3, 1988.

336. G.L. Bassi and F. Broggi (Fratelli Lambert S.p.A., Albizzate, Italy), *Polym. Paint Colour J.*, 178 (4210), 197, 1988, in *CAS, Coatings Inks Relat. Products*, 18, 2, 1988.

337. A.A. Lin, R.V. Sastri, G. Tesoro, E. Reiser, and R. Eachus, *Macromolecules*, 21(4), 1165, 1988.

338. C. Carlini, L. Toniolo, P.A. Rolla, F. Barigeletti, P. Bortolus, and L. Flamigni, *New Polym. Mater.*, 1(1), 63, 1987.

339. K. Hummel, R. Schaller, and M.G. Martl, *Polym. Bull.*, 17, 369, 1987.

340. R.J. DeVoe and S. Mitra, EP 260,877, Minnesota Mining and Manuf. Co., St. Paul, MN, 1988.

341. B.K. Bordoloi, Y. Ozari, S.S. Plamtottham, and R. Van Ham, EP 263866, Avery Int. Co., 1988, in *CAS*, 19, 4, 1988.

342. H. Braun and Th. Brugger, *Coating*, (9), 307, 1993.

343. G.W.H. Lehmann and H.A.J. Curts, US Patent, 40,38,454, Beiersdorf AG, Hamburg, Germany, 1977.

344. U.S. Patent, 3,257,478, W.E. Lenney, Canadian Patent, 1,225,176, Air Products and Chemical Inc., USA, 1987.

345. U.S. Patent, 3,697,618, W.E. Lenney, Canadian Patent, 1,225,176, Air Products and Chemical Inc., USA, 1987.

346. U.S. Patent, 3,998,997, W.E. Lenney, Canadian Patent, 1,225,176, Air Products and Chemical Inc., USA, 1987.

347. M. Kajiyama, *Phase Structure of PSA Prepared from Solution and Emulsion*, in Proceedings of the 24th Annual Meeting of Adhesion Society, February 25–28, Williamsburg, VA, USA, 2001, p. 283.

348. A. Aymonier, E. Papon, J.J. Villenave, and P. Tordjeman, *Direct Relation Between Copolymerization Process and Tack Properties of Model PSA's*, in Proceedings of the 24th Annual Meeting of Adhesion Society, February 28, Williamsburg, VA, USA, 2001, p. 280.

349. H. Gunesch and I.A. Schneider, *Makromol. Chem.*, 125, 213, 1969.

350. R.N. Henkel, *Paper Film Foil Convert.*, (12), 68, 1968.

351. I. Benedek, *Adhäsion*, (12), 17, 1987.

352. H. Monsey and A. Malersky, U.S. Patent, 4,331,576, 1982, in *Adhes. Age*, (12), 53, 1982.

353. Protective Treatments Inc., British Patent, 925,007, *Coating*, (6), 185, 1969.

354. F.M. Rosenbaum, *Adhes. Age*, (6), 32, 1972.

355. *Coating*, (2), 123, 1984.

356. H. Hintz, *Kunststoff Dispersionen für das Verkleben von PVC Bodenbelägen und Textiler Auslegeware*, in *Kunstharz Nachrichten* (Hoechst), (2), 18, 1983.

357. W.M. Stratton, *Adhes. Age*, (7), 21, 1985.

358. *Adhäsion*, (12), 283, 1983.

359. F.R. Baker, *A Unique Ethylene/Vinyl Acetate Copolymer for the Adhesives Industry*, in Proceedings of the Tappi Hot Melt Adhesives and Coating Short Course, May 2–5, 1982, Hilton Head, SC, USA.

360. A.A. Drescher, *Coating*, (5), 113, 1974.

361. *Mowilith VDM 1360 ca 55%*, Data Sheet, Hoechst A.G., F + E/polymerisate II, Frankfurt a.M., Germany.

362. J.G. Iacoviello, U.S. Patent, 4,735,986, Air Prod. and Chem. Co., in *CAS Adhes.*, 20, 3, 1988.

363. *Coating*, (2), 44, 1984.

364. W. Hoffmann, *Kunststoffe*, 78(2), 132, 1988.

365. *Coating*, (7), 186, 1984.

366. R. Koch and C.L. Gueris, *Allg. Papier Rundsch.*, 16, 460, 1986.

367. *Coating*, (1), 15, 1984.

368. Colon et al., U.S. Patent, 4,331,576, in W.L. Bunelle, K.C. Knutson, and R.M. Hume, EP 0199468A2, H.B. Fuller Co., St. Paul, MN, USA, 1986.

369. Morrison, U.S. Patent, 4,052,368, in W.L. Bunelle, K.C. Knutson, and R.M. Hume, EP 0199468A2, H.B. Fuller Co., St. Paul, MN, USA, 1986.

370. Y. Aizawa, Japanese Patent, 6,317,945, Cemedine Co. Ltd., 1988, in *CAS Adhes.*, 14, 4, 1988.

371. A. Lamping and W. Tellenbach, Swiss Patent, 664,971, Nyffeler Corti A.G., 1988, in *CAS, Colloids (Macromolecular Aspects)*, 21, 6, 1988.

372. Nitto Electrical Co., Japanese Patent, 24229/70, *Coating*, (2), 38, 1972.

373. *Coating*, (2), 45, 1969.

374. Beiersdorf AG, Hamburg, U.S. Patent, 15,698,888, *Coating*, (8), 240, 1972.

375. *Allgemeine Papier Rundsch.*, 18, 664, 1970.

376. R. Hauber (Hans Neschen GmbH & Co. KG), Bückeburg, Germany, DE 43 03616, 1994.

377. Minnesota Mining and Manuf. Co., St. Paul, MN, USA, Canadian Patent, 588869, *Coating*, (6), 185, 1969.

378. U.S. Patent, 3,441,430, in F.D. Blake, EP 0141504 A1, Minnesota Mining and Manuf., Co., St. Paul, MN, USA, 1985.

379. Z. Czech (Lohmann GmbH, Neuwied), DE 44 31053, 1994.

380. M.F. Tse and K.O. McElrath, *Adhes. Age*, (9), 32, 1988.

381. S.D. Hollis, *Non-Crystalizing Rosin. A Tackifier for Hot Melt Adhesives*, in Proceedings of the APPI Hot Melt Adhesives and Coatings, Short Course, Head, SC, June, 1982.

382. J. Rustige, *Adhäsion*, (10), 275, 1977.

383. C. Donker, R. Luth, and K. van Rijn, *Hercules MBG 208 Hydrocarbon Resin: A New Resin for Hot Melt Pressure-Sensitive (HMPSA) Tapes*, in Proceedings of the 19th Munich Adhesive and Finishing Seminar, Munich, Germany, 1994, p. 64.

384. G. Bonneau and M. Baumassy, *New Tackifying Dispersions for Water Based PSA for Labels*, in Proceedings of the 19th Munich Adhesive and Finishing Seminar, Munich, Germany, 1994, p. 82.

385. G.J. Kutsek, *Adhes. Age*, (6), 24, 1996.

386. E.E. Ewins, Jr. and J.R. Erickson, *Formulation to Inchance the Radiation Crosslinking of Thermoplastic Rubber for HMPSAs*, in Proceedings of the TAPPI Hot Melt Symposium '85, June 16–19, Hilton Head, SC, USA.

387. K. Nitzl, Th. Horna, and U. Hoffmann, *Strahlenhärtbare Hotmelts*, in Proceedings of the 16th Munich Adhesive and Finishing Seminar, 1991, Munich, Germany, p. 100.

388. S. Tsuchida, Y. Kodama, and H. Hara, U.S. Patent, 4622357, Arakawa Kagaku Kogyo Kabushiki Kaisha, Osaka, Japan, 1986, in *Adhes. Age*, (5), 15, 1987.

389. V.L. Hughes and R.W. Looney, EP 0,131,460, Exxon Research and Engineering Co., Florham Park, NJ, USA, 1985.

390. R.M. Podhajny, *Converting Packaging*, (3), 21, 1986.

391. Minnesota Mining and Manuf. Co., St. Paul, MN, U.S. Patent, 3,129,816, *Adhäsion*, (6), 271, 1965.
392. Protective Treatments Inc., British Patent, 814,001, *Coating*, (6), 185, 1969.
393. T.R. Mecker, *Low Molecular Weight Isoprene Based Polymers — Modifiers for Hot Melts*, in Proceedings of the TAPPI Hot Melt Symposium, 1984, in *Coating*, (11), 310, 1984.
394. *Kaut. Gummi, Kunststoffe*, 41(7), 662, 1988.
395. R. Jordan, *Coating*, (2), 37, 1986.
396. M.M. Feldstein, N.A. Platé, A.E. Chalykh, and G.W. Cleary, *General Approach to the Molecular Design of Hydrophylic Pressure-Sensitive Adhesives*, in Proceedings of the 25th Annual Meeting of Adhesion Society, and the Second World Congress on Adhesion and Related Phenomena, February 10–14, Orlando, FL, USA, 2002, p. 292.
397. British Patent 941,276, in P. Gleichenhagen and I. Wesselkamp, EP 0058382B1, Beiersdorf A.G., Hamburg, 1982.
398. U.S. Patent, 3,441,430, in P. Gleichenhagen and I. Wesselkamp, EP 0058382B1, Beiersdorf A.G., Hamburg, 1982.
399. U.S. Patent, 3,152,940, in P. Gleichenhagen and I. Wesselkamp, EP 0058382B1, Beiersdorf A.G., Hamburg, 1982.
400. F.D. Blake, EP 0141504A1, Minnesota Mining and Manuf. Co., St. Paul, MN, USA, 1985.
401. Crosslinker CX-100, Data Sheet, Zeneca Resins, Waalwijk, The Netherlands, 1997.
402. U.S. Patent, 2,838,421, in F.D. Blake, EP 0141504 A1, Minnesota Mining and Manuf. Co., St. Paul, MN, USA, 1985.
403. U.S. Patent, 3,661,874, in F.D. Blake, EP 0141504 A1, Minnesota Mining and Manuf. Co., St. Paul, MN, USA, 1985.
404. P. Dreier, *Adhes. Age*, (6), 32, 1996.
405. J. Hansmann, *Adhäsion*, (10), 360, 1970.
406. B. Mayer, *Coating*, (4), 109, 1969.
407. E. *Plastics News*, (4), 24, 1992.
408. K. Sato, *Rubber Chem. Technol.*, 56, 942, 1984.
409. A.T. Franklin, *Printing Trades J.*, (1044), 46, 1974.
410. O. Huber, *Wochenblatt für Papierfabrikation*, (17), 657, 1973.
411. *Kaut. Gummi, Kunststoffe*, (12), 1160, 1986.
412. *Kaut. Gummi, Kunststoffe*, (1), 6, 1986.
413. Allmänna Svenska Elektriska AB,DBP 1276771/1987.
414. *Coating*, (10), 373, 1986.
415. D.R.M. Harwey, U.S. Patent, 2,822,509, *Coating*, (6), 184, 1969.
416. *Adhes. Age*, (5), 6, 1985.
417. A.I. Medalia, *Rubber. Chem. Technol.*, 59, 432, 1986.
418. G. Langsley, *Int. Polym. Sci. Technol.*, 13 (1), 44, 1986
419. F. Shikata, Japanese Patent, 63,105,187, Teijin Ltd., 1988, in *CAS, Adhesives*, 22, 7, 1988, 109: 130717.
420. I. Benedek and E. Frank, DE 4421420A1, 1994.
421. *Eur. Adhes. Seal.*, (9), 22, 1987.
422. *Eur. Plastics News*, (11), 67, 1993.
423. T. Aryoshi, G. Fujiwara, T. Hayashi, and Y. Kaneshige, Japanese Patent, 6,341,648, Tosoh Co., 1988, in *CAS, Adhesives*, 22, 7, 1988, 109: 130634q.
424. B.B. Blackford, British Patent, 8,864,365, *Coating*, (6), 185, 1969.
425. U.S. Patent, 2,925,174, EP 0120708, p. 2.
426. V.G. Raevski et al., SSSR Patent, 304,264, *Coating*, (6), 185, 1969.
427. SSSR Patent, 349782,TNII Bumagi, *Coating*, (5), 122, 1974.
428. G. Grove, *Adhes. Age*, (5), 39, 1971.
429. Ankerwerk Rudolfstadt, DBP 1056787, in *Coating*, (6), 185, 1969.
430. Johnson & Johnson, US Patent, 2909278, in *Coating*, (6), 185, 1969.
431. R.P. Singh, J. Lacoste, R. Arnaud and J. Lemaire, *Polym. Degrad. Stabil.*, 20(1), 49, 1988.
432. G.F. Vesley, A.H. Paulson, and E.C. Barber, EP 0202938 A2, 1986.
433. U.S. Patent, 4,223,067, in G.F. Vesley, A.H. Paulson, and E.C. Barber, EP 0202 938 A2, 26, 1986.
434. *Adhäsion*, (12), 12, 1983.
435. Ö. Söderberg, *Paper Technol.*, (8), 17, 1989.

436. Y. Aizawa, Japanese Patent, 6,317,945, Cemedine Co. Ltd., 1988, in *CAS, Adhesives*, 14, 4, 1988.
437. T.J. Bonk, T.I. Cheng, P.M. Olsom, and D.E. Weiss, PCT, WO 87/00189, Jan. 15, 1987.
438. Vitta Corporation, U.S. Patent, 3,598,679.
439. *Druck Print*, (9), 65, 1986.
440. W.K. Darwell, P.R. Konsti, J. Klingen, and K.W. Kreckel, EP257984, Minnesota Mining and Manuf. Co., St. Paul, MN, USA, 1986.
441. Japanese Patent, 2736/1975, in H. Miyasaka, Y. Kitazaki, T. Matsuda, and J. Kobayashi, DE 3544868 A1, Nichiban Co., Ltd. Tokio, Japan, Offenlegungsschrift, 1985.
442. T. Shibano, I. Kimura, H. Nomoto, and C. Maruchi, U.S. Patent, 4624893, Sanyo Kokusaku Pulp Co., Ltd., Tokyo, Japan, 1986, *Adhes. Age*, (5), 24, 1987.
443. E. Pagendarm, Hamburg, DE 3632816 A1, Germany, Offenlegungsschrift, 1986.
444. J.W. Otter and G.R. Watts, U.S. Patent, 5,346,766, Avery International Corp., Pasadena, CA, USA, 1994.
445. R. Shibata, H. Miyagawa, Japanese Patent, 63 86785, Hitachi Condenser Co. Ltd., 1988, *CAS, Adhesives*, 20, 4, 1988.
446. M. von Bittera, D. Schäpel, U. von Gizycki and R. Rupp, EPA 0147588 B1, Bayer A.G., Leverkusen, Germany, 1985.
447. S. Maeda and T. Miyamoto, U.S. Patent, 5456765, Sumitomo Bakelite Co. Ltd., Tokyo, 1994, *Adhes. Age*, (5), 15, 1995.
448. Z. Czech and H.D. Sander, Lohmann GmbH, Neuwied, Germany, DE 4,219,368, 1994.
449. H. Eifert and G. Bernd, DE 4,228,608, Fraunhoffer Gesellschaft, München, Germany, 1992.
450. K. Takashimizu and A. Suzuki, Japanese Patent, 63 92683, Advance Co. Ltd., 1988, in *CAS, Adhesives*, 19, 5, 1988.
451. N. Hosoi and S. Azuma, U.S. Patent, 5,346,539, Sumitomo Electric Industries Ltd., Japan, 1994.
452. V.A. Uskov, V.M. Lalayan, A.K. Abysev, and N.Yu. Morozova, *Kauch. Resina*, (3), 8, 1986.
453. *Neue Verpackung*, (6), 72, 1991.
454. *Finat Labell. News*, (3), 29, 1994.
455. D. Lacave, *Labels and Labelling*, (3/4), 55, 1994.
456. Oji Paper Co., Japanese Patent, 07,325,391, 1995, in *Adhes. Age*, (5), 12, 1996.
457. Toyo Ink Manuf. Ltd., Japanese Patent, 07,292,344, 1995.
458. R. Hauber, DE 42 02 070, Hans Neschen GmbH and Co. KG, Bückeburg, Germany, 1993.
459. I. Benedek, *Pressure-Sensitive Formulation*, VSP, Utrecht, 2000, Chap. 4.1.2.
460. S. Toshiki, M. Nagano, H. Ito, and T. Miyaji, Japanese Patent, 6351443, Japan Synthetic Rubber Co., Ltd., 1988, in *CAS, Adhesives*, 20, 3, 1988.
461. S. Milton and C. Max, *Adhes. Age*, (1), 12, 1983.
462. K. Berger, *Gummi, Asbest, Kunststoffe*, (6), 318, 1985.
463. M.E. Ahner and M.L. Evans, *Light Color Aromatic Containing Resins*, in Proceedings of the TAPPI Hot Melt Symposium '85, June 16–19, 1985, Hilton Head, SC, USA, *Coating*, (1), 6, 1986.
464. R.J. Shuman and B.D. Josephs, PCT, WO 88/01636, Dennison Manuf. Co., Framingham, MA, USA.
465. U. Storb, *Allg. Papier Rundsch.*, (9), 244, 1986.
466. W. Kahle, *Pack Report*, (6), 87, 1991.
467. *Neue Verpackung*, (5), 63, 1991.
468. E. Fuchs, *Coating*, (3), 78 1969.
469. Lupolen B 581, Technical Booklet, BASF Kunststoffe, d/12.92, p. 33.
470. K.H. Kochem, H.U. ter Meer, and H. Millauer, *Kunststoffe*, 82(7), 575, 1992.
471. J. Pernak, A. Skrzypczak, E. Gorna, and A. Pasternak, *Kunststoffe*, 77(5), 517, 1987.
472. Japanese Patent, 55,151,046, *Hochmolekularbericht*, 57(1), 94, 1982.
473. K. Sirinyan and F. Jonas, DE 3625272, Bayer A.G., Ger. Offen., 1988, in *CAS, Coatings, Inks Relat. Prod.*, 17, 9, 1988.
474. *Incoblend*, Technical Information, 1997, Zippeling, Kessler, Ahrtensburg, Germany,
475. O. Schwab, *Adhäsion*, (5), 155, 1972.
476. M.W. Uffner and P. Weitz, U.S. Patent, 3,345,320, General Aniline and Film Co., New York, NY, USA, 1967.
477. R.M. Ward and D.C. Kelley, *Tappi J.*, (6), 140, 1988.
478. *Coating*, (11), 274, 1985.

479. E. Haberstroh, *Papier und Kunststoff Verarbeiter*, (8), 14, 1986.
480. C. Gondro, *Kunststoffe*, (19), 1084, 1990.
481. N. Grünewald, Proceedings of the Sp '89 on Additives for the Improvement of Polymers, Maack Business Service, Zürich, Switzerland, p. 148.
482. F.T. Kitchel, *Tappi J.*, (12), 156, 1988.
483. Dowlex. For Lamination Films. CH-254-052-E-288, Data Sheet, Dow, Horgen, Switzerland, 1997.
484. A. Rudin, A.T. Worm, and J.E. Blacklock, *J. Plast. Film Sheeting*, 1, 189, 1985.
485. C. De Smedt and S. Nam, *Plast. Rubber Process. Appl.*, 8(1), 11, 1987.
486. W.N.K. Revolta, *Kunststoff J.*, (9), 72, 1986.
487. *Labels and Labelling*, (3/4), 44, 1994.
488. G. Joannou, *FINAT Seminar*, March, 1987, p. 126.
489. H. Nitschke, *Papier und Kunststoff Verarbeiter*, (10), 8, 1987.
490. *Etiketten-Labels*, (35), 28, 1995.
491. K. Heger, *Coating*, (5), 180, 1987.
492. R. Hinterwaldner, *Coating*, (3), 98, 1993.
493. J. Dietz, *Coating*, (11), 296, 1982.
494. G. Bosch, *Der Siebdruck*, (6), 58, 1988.
495. J.F. LaFaye, J.P. Maume, J.M. Schwob, and R. Chiodi, *Tappi J.*, (12), 63, 1988.
496. *Coating*, (7), 246, 1993.
497. *Coating*, (3), 101, 1993.
498. A. Dettling and J. Burri, *Etiketten-Labels*, (3), 4, 1995.
499. H.P. Seng, *Coating*, (9), 324, 1993.
500. P. Penczek, *Adhäsion*, (1/2), 32, 1988.
501. *Coating*, (12), 362, 1970
502. H. Reipp, *Adhes. Age*, (3), 17, 1972, *Coating*, (6), 187, 1972.
503. G.L. Burroway and G.W. Feeney, *Adhes. Age*, (7), 17, 1974.
504. R.A. Tait, A. Salazar, W.C. Chung, R.J. Skiscim, and D.R. Hansen, *Toluene Replacement in Solvent Borne Pressure-Sensitive Adhesive Formulations*, in Proceedings of the PTSC XVII Technical Seminar, 4 May, Woodfield Shaumburg, IL, USA.
505. Esso Research and Eng. Co., U.S. Patent, 3,351,572, *Coating*, (7), 210, 1969.
506. R. Milker and Z. Czech, *Lösungsmittelhaltige Akrylhaftklebstoffe*, in Proceedings of the 19th Munich Adhesive and Finishing Seminar, Munich, Germany, 1994, p. 136.
507. Y. Moroishi, T. Sugii, and K. Noda, Japanese Patent, 6381183, Nitto Electric Ind. Co. Ltd., 1988, in *CAS, Adhesives*, 22, 6, 1988, 109: 130548q.
508. T. Sugiyama, N. Miyaji, Y. Ito, and T. Tange, Japanese Patent 62199672, Nippon Carbide Industries Co., Inc., 1987, in *CAS, Adhesives*, 14, 4 1988, 108: 222753k.
509. Pittsburgh Plate Glass Co., U.S. Patent, 3,355,412, *Coating*, (7), 274, 1969.
510. N.C. Smith, *Tappi J.*, (7), 106, 1986.
511. I. Benedek, *Pressure-Sensitive Adhesives and Applications*, Marcel Dekker Inc., New York, 2004, Chap. 9.5.
512. T. Tatsuno, K. Matsui, T. Masaharu, and W. Mitsui, Japanese Patent, 62285927, Kansai Paint Co., Ltd., 1987, in *CAS, Adhesives*, 14, 4, 1988, 108: 222753k.
513. T. Murachi, Japanese Patent, 63 61075, Toyoda Gosei Co., Ltd., 1988, in *CAS, Adhesives*, 20, 4, 1988.
514. H.J. Northrup, U.S. Patent, 3691206, *Coating*, (5), 123, 1974.
515. S. Denko, K.K. Tokyo, Japan, U.S. Patent, 1900521, *Coating*, (3), 60, 1974.
516. Mitsubishi Petrochemical Co. Ltd., Tokyo, Japan, DBP 1,277,217, in *Coating*, (9), 272, 1969.
517. Keuffel and Esser Co., Morristown, NJ, USA, U.S. Patent, 1,569,086, in *Coating*, (1), 31, 1978.
518. Kjin Co., Japanese Patent, 2,8520/70, *Coating*, (2), 38, 1972.
519. Hoechst Celanese, Sommerville, NJ, USA, EP 484,819, 1991, in *Coating*, (19), 354, 1992.
520. Pentacoll ET 691A, Data Sheet, Morton Adhesives Europe, London, UK, Adhesives and Coatings, 1995.
521. Montrek Polyethylenimine Products, Data Sheet, Dow, Horgen, Switzerland, 1994.
522. T. Kishi, Japanese Patent, 6,369, Sekisui Chem., Co. Ltd., 879, 1988, in *CAS, Adhesives*, 19, 6, 1988.
523. D.H. Teesdale, *Coating*, (6), 243, 1983.
524. A.K. Agarwal and C. Balian, CMR Ind. Inc., EP 258,974, 1988.
525. British Patent, 1,123,014, *Coating*, (7), 274, 1969.

526. TNII Bumagi, SSSR Patent, 300,561, *Coating*, (7), 368, 1969.

527. L. Bothorel, *Emballages*, (278), 372, 1972.

528. L. Dengler, *Coating*, (6), 143, 1974.

529. K. Kriz and T.H. Plaisance (De Soto Inc.), U.S. Patent 1878093, *CAS, Coating Inks Relat. Prod.*, 10, 2, 1988.

530. M. Omote, I. Sakai, and T. Matsumoto (Nitto Electric. Ind. Co.), Japanese Patent, 63,86,788, 1988, in *CAS*, 19, 4, 1988.

531. H. Shimitzu, F. Nemoto, I. Okamura, and K. Ito (Neos Co. Ltd., Toray Industries Inc.), Japanese Patent, 62,270,638, 1987, in *CAS, Coating Inks Relat. Prod.*, 11, 11, 1988.

532. D.J. Kinning, *J. Adhes.*, 60, 249, 1997.

533. T. Kikuta and H. Hori (Nippon Shokubai Kagaku Kogyo Ko. Ltd.), Japanese Patent, 6,333,482, 1986, in *CAS, Coatings, Inks Relat. Prod.*, 17, 10, 1988.

534. T. Kikuta and H. Hori (Nippon Shokubai Kagaku Kogyo Ko. Ltd.), Japanese Patent, 6,351,478, 1986, in *CAS, Coatings, Inks Relat. Prod.*, 17, 11, 1988.

535. G.V. Gordon, T.M. Leaym, M.J. Owen, M.S. Owen, S.V. Perz, J.L. Stasser, J.S. Tonge, M.K. Chaudhury, and K.A. Vorvolakos, *Resin–Polymer Interaction in Silicone Release Coatings*, in Proceedings of the 23rd Annual Meeting of the Adhesion Society, Myrtle Beach, SC, USA, February 20–23, 2000, p. 39.

536. T. Cosgrove, I. Weatherhead, M.J. Turner, R.G. Schmidt, G.V. Gordon, and J.P. Hannington, *Polym. Prep.*, 39, 545, 1998.

537. M. Brogly, L. Vrevin, G. Castelein, and J. Schultz, *Silicone Release Coating/PSA Adhesive Strength Modulation. How PMDS Based Network Composition and Structure Influence Nanoscale Properties of the Interface*, in Proceedings of the 24th Annual Meeting of Adhesion Society, February 25–28, 2001, Williamsburg, VA, USA, p. 134.

538. A. Fau, EP 0169098 B1/22, Rhone Poulenc Chimie, Courbevoie, France, 1986.

539. T. Horichi and M. Tanaka, Japanese Patent, 6,308,429, Uchiyama Kogyo Kaisha Ltd., 1988, in *CAS, Siloxanes Silicones*, 17, 7, 1988.

540. H. Yasui and M. Okamoto, Japanese Patent, 6,308,430, 1988, in *CAS, Siloxanes Silicones*, 17, 7, 1988.

541. *Coating*, (12), 344, 1984.

542. N. Hayashi (Toshiba Silicone Co. Ltd.), Japanese Patent, 62,292,867, 1987, *CAS, Coating Inks Relat. Prod.*, 11, 11, 1988.

543. K. Nakamura, Y. Miki, and Y. Nanzaki (Nitto Electric Ind. Co. Ltd.), Japanese Patent, 6,386,787, 1988, in *CAS, Adhesives*, 19, 5, 1988.

544. Norton Co., U.S. Patent, 3,508,919, *Coating*, (5), 130, 1974.

545. J.L. Walker and P.B. Foreman, National Starch and Chem. Co., Bridgewater, VA, USA, EP224,795 1987, in *CAS, Colloids (Macromolecular Aspects)*, 1, 5, 1988.

6 Developments in Crosslinking of Solvent-Based Acrylics

Zbigniew Czech

CONTENTS

Crosslinking as a possibility to buildup segregated structures was discussed in Chapter 5, Section II. The adhesive manufacture-related aspects of crosslinking are described in Chapter 8. Crosslinking allowed the synthesis of a new class of pressure-sensitive adhesives (PSAs) based on plastomers: the pressure-sensitive hydrogels (see Chapters 4 and 9). Curable raw materials for PSAs are discussed in detail in [1,2]. Adhesive formulation by the use of crosslinking is described by Benedek in Ref. [3]. The growing importance of pressure-sensitive products (PSPs) for special end-use (see Chapter 11), imposes the use of crosslinked solvent-based adhesives, mainly of acrylics (ACs). Owing to the broad range of raw materials and to various curing methods used, this domain became a technology *per se*. Unfortunately, because of the manufacturers-related, mostly patented technical solutions, there is no exhaustive scientific study of this field. In this chapter the developments in crosslinking of solvent-based acrylics will be described. The performance characteristics of PSAs, related to crosslinking, are described in Chapter 7. As discussed earlier, common crosslinking technology includes conventional, and radiation-induced crosslinking.

I. CONVENTIONAL CROSSLINKING OF SOLVENT-BASED ACRYLIC PSAs

Crosslinking is one of the most interesting processes, attractive for the chemist, informative for the physicist, and helpful for the user in the joint development of tailored PSAs. This applies mostly also for solvent-borne acrylic PSAs used increasingly for coating of labels, tapes, decorative films, and similar self-adhesive articles. For these products the PSAs must have well-defined properties. Besides an excellent surface adhesion, the PSAs should display improved stability against light, oxygen, moisture, and plasticizers, and their adhesion characteristics should be constant over a very large temperature range [4].

The properties of PSAs synthesized by copolymerization of acrylic monomers and formulated in organic solvent mixture are determined, to a great extent, by the kind and quantity of the crosslinking agent added to the polymers. As with molecular weight (MW), crosslinking influences the bulk properties of the film and builds shear, shrinkage, heat, and chemical resistance, while negatively impacting the tack and peel [5]. As discussed earlier (see Chapter 5, Section II) conventional crosslinking may occur physically or chemically.

Physical crosslinking plays a special role for thermoplastic elastomers (TPEs) and pressure-sensitive hot-melts based on such materials (see Chapter 5). In solvent-based systems physical crosslinking also appears. The PSA is also linked by hydrogen bonds when the polymerization carrier is removed. The hydrogen bonding existing between polymer chains in PSA systems is fully reversible. Additionally, van der Waals forces have to be mentioned, which are noticeable in all PSAs being free of crosslinking agents and are more or less connecting the molecule chains by secondary valence forces in a narrow spaced network [6]. The hydrogen bonds also belong to secondary valence crosslinking and are based on a dipole–dipole attraction of polar groups such as —COOH, —CONH$_2$, or —OH. If, therefore, such polar groups are present in certain distances in an overall nonpolar PSAs, the chains are more strongly connected at places where such groups are present in comparison to other places. As discussed in Chapter 4, plastomer-based hydrogels, a special class of PSAs was synthesized, where crosslinking between the main components of the polymeric system occurs via hydrogen bonds. The poly *N*-vinyl pyrolidone (PVP) solubility in liquid short-chain polyethylene gycol (PEG) is due to hydrogen bonding between terminal OH groups of PEG and carbonyl C=O groups in PVP repeat units as presented in Figure 4.2. As every PEG chain bears two reactive terminal OH-groups, the PEG acts as H-bonding crosslinker of longer PVP macromolecules. In common, acrylic copolymers crosslinking via hydrogen bonds only, does not allow significant increase of the cohesion. For such polymers chemical crosslinking is required.

This chapter discusses the main aspects of chemical crosslinking. As described in detail in [7] crosslinking is carried out off-line or in-line. Off-line crosslinking occurs mainly during the synthesis of raw materials for PSAs, in-line crosslinking is the result of formulation and is carried out generally during coating.

For the crosslinking of PSAs various compounds are used as internal or external crosslinking agents. Crosslinking includes the most important multifunctional chemical compounds and the suitable chemical methods, such as: monomers with crosslinking properties [6,8,9], crosslinking agents like metal acidesters [6,10], metal chelates [6,11–13], multifunctional isocyanates [6,14], polycarbodiimides [6,15], multifunctional propylene imines [6,16–21], and amino resins [6,22]. Crosslinking monomers are used in thermally or ultra violet (UV)-light induced [6,23–31] reactions (see Chapter 5, Section II.A).

According to the technology of their inserting in the polymer buildup, the chemicals used for crosslinking are classified as internal or external crosslinking agents.

The most commonly used internal crosslinking systems are based on multifunctional unsaturated monomers and monomers with crosslinking performance. By use of internal crosslinking agents, the crosslinker is incorporated into the polymer chain of acrylic PSAs through the polymerization process. This is a general method in the synthesis of crosslinkable raw materials for dispersed or 100% solids PSAs, and may allow crosslinking during synthesis of the polymer or postcuring without or with other crosslinking agents, thermally or light-induced. Natural rubber (NR) contains also crosslinkable, reactive sites. Significant progress has been made with the development of high performance pressure-sensitive acrylics which crosslink after drying.

In external crosslinking, curing occurs as a chemical reaction after formation of the adhesive film. The adhesive is formulated, that is, its main polymer contains built-in reactive sites and crosslinking agents. Formulation by crosslinking, the scope, theory, and basis of crosslinking were discussed in a detailed manner in Ref. [7]. The main external crosslinking agents used in solvent-based acrylic PSAs are metal acidesters, metal chelates, polyfunctional isocyanates, polycarbodiimides, polyfunctional propylene imines, and amino resins. The polymer incorporates carboxylic or hydroxyl groups, which react during the drying of the adhesive film. Such an adhesive has the advantage, among others, that no pot-life has to be considered during its use. After the addition of a stabilizer the PSA can be used over a long period of time. It should be noted that certain external crosslinking agents can be also used during polymer synthesis.

A. CROSSLINKING OF ACRYLIC PSAs BY CROSSLINKING MONOMERS

Crosslinked systems require crosslinking sites, which are statistically distributed in the polymer chain. Such systems are synthesized by conventional solvent-based polymerization of acrylate monomers, which are capable of thermal reaction with carboxyl groups or self-condensation [32] (see also Chapter 5, Section II.B). Such monomers include N-methylol acrylamide (NMA), N-(iso-butoxymethyl) acrylamide (IBMA), and methyl acrylamidoglycolate methyl ether (MAGME) [33–35].

$$CH_2 = CH - \overset{\displaystyle O}{\overset{\displaystyle \|}{C}} - NH - CH_2OH \qquad (6.1)$$

N-methylol acrylamide (N-MA)

$$CH_2 = CH - \overset{\displaystyle O}{\overset{\displaystyle \|}{C}} - NH - CH_2OCH_2CH \overset{\displaystyle CH_3}{\underset{\displaystyle CH_3}{<}} \qquad (6.2)$$

N-(iso-butoxymethylene)acrylamide (IBMA)

$$CH_2 = CH - \overset{\displaystyle O}{\overset{\displaystyle \|}{C}} - NH - \underset{\displaystyle OCH_3}{\overset{\displaystyle |}{CH}} - \overset{\displaystyle O}{\overset{\displaystyle \|}{C}} - OCH_3 \qquad (6.3)$$

methyl acrylamidoglycolate methyl ether (MAGME)

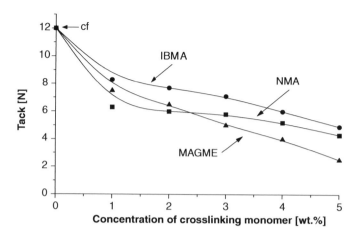

FIGURE 6.1 The effect of NMA, IBMA, and MAGME level on tack.

From these compounds NMA is a classical curing agent used for water-based (WB) adhesive and nonadhesive polymers also. In the 1960s it was discovered that the inclusion of a small amount of NMA into homopolymeric polyvinyl acetate formulations resulted in improvements in cohesion and environmental resistance [35].

The influence of the above crosslinking monomers on the main adhesive properties (tack, peel, and shear strength) of acrylic PSAs is illustrated in Figure 6.1 through Figure 6.3.

The data of Figure 6.1 through Figure 6.3 illustrate that the investigated crosslinking monomers have a negative influence on the tack of acrylic PSAs. The greatest decrease of tack was noted for NMA and MAGME. An increase of NMA, IBMA, and MAGME content decreases the peel adhesion measured at 20 and 70°C. This effect correlates the glass transition temperature (T_g) of monomers with crosslinking properties incorporated in the polymeric chain (see also Chapter 3). The higher the T_g of monomer (T_g of NMA +119°C, T_g of MAGME +78°C, T_g of IBMA +30°C) the lower the tack and the peel adhesion of the synthesized PSAs.

The substantial improvement of shear strength by application of crosslinking monomers has been shown earlier (see Chapter 5, Section II). The high cohesion of acrylic PSAs with crosslinking monomers is based on the reaction between the carboxylic groups from acrylic polymer and the functional groups of crosslinking monomers (Figure 6.4).

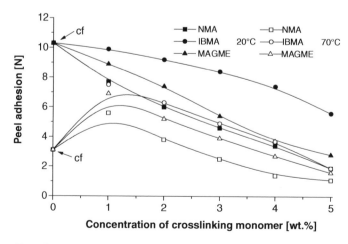

FIGURE 6.2 The effect of NMA, IBMA, and MAGME level on peel adhesion.

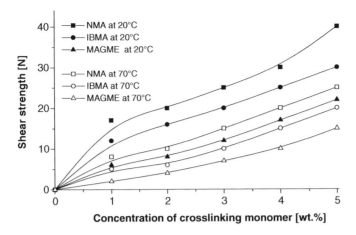

FIGURE 6.3 The effect of NMA, IBMA, and MAGME level on shear strength.

From our experimental data concerning the improvement of the shear with crosslinking mono-mers based on derivatives of *N*-substituted acrylamide, it can be concluded that the best thermal reactive crosslinking monomer is NMA: therefore, it has a practical importance as a second cross-linking agent for the manufacture of acrylic PSAs with postcrosslinking ability at high temperatures (see also Chapter 5, Section II.A).

The built-in reactivity of multifunctional monomers is used by the aid of crosslinking agents.

B. CROSSLINKING AGENTS

Crosslinking agents include chemically different compounds, for example, inorganic, organic, or mixed compounds having high reactivity. The main representatives of crosslinking agents used for solvent-based acrylics are the metal acidesters and the metal chelates.

1. Metal Acidesters

This group of crosslinkers includes organic compounds, such as titanates, zirconates, or hafnates. The main class of metal acidesters are alkyl esters of orthotitanic acid ($Ti(OH)_4$), commonly

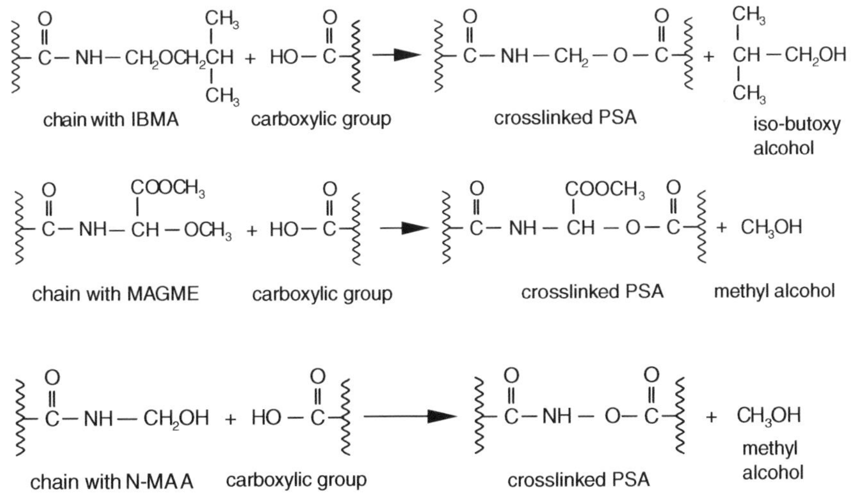

FIGURE 6.4 Crosslinking reaction between the carboxylic group and *N*-substituted acrylamides.

referred to as alkyl titanates or titanium alkoxides. They are suggested as crosslinking agents for PSA solutions [36]. Characteristic for these crosslinking systems is the addition of alcohol as a stabilizer after the polymerization process. They offer the potential of crosslinking at room temperature, merely by evaporation of the solvent.

Despite of the high reactivity of titanates, the shear strength (measured as slip on steel plate) of crosslinked acrylates with 0.5% w/w titanates, is not enough for practical use. After several hours under a load of 10 N acrylic PSAs crosslinked with titanates exhibit cohesive failure (Figure 6.5). The cohesion values obtained with titanates demonstrate that tetraisopropyl titanate and tetra-n-butyl titanate are the most efficient crosslinking agents. Tetraethyl titanate and tetraisobutyl titanate lead to moderate cohesion results.

2. Metal Chelates

The transition metals with a coordination number greater than 2, typically 4, 6, or 8 (such as Zn, Ni, Mn, Fe, Co, Cr, Al, Ti, or Zr) form with 2,4-pentanedione (acetylacetone) chelate complexes known as acetylacetones [37]. The hydrogen atoms of the methylene group —CH_2— accordingly the I-effect of the neighboring ketone group are very mobile. Therefore, the acetylacetones show the keto–enol tautomerism. The central hydrogen atom of the chelate rings is accessible to electrophilic substitution [38,39].

The metal chelates, which are reacted with the acrylic polymer chains containing carboxylic groups are particularly efficient as crosslinking agents. The general formula of a metal acetylacetonate is given in structure (6.4):

$$(RO)_a \diagdown M \Big[\begin{array}{c} O-C \diagup CH_3 \\ \diagdown CH \\ O=C \diagdown CH_3 \end{array} \Big]_c \qquad \begin{array}{l} \text{M – central metal atom} \\ a+b+c = \text{n-metal valence} \\ b = 0 \text{ or } 1 \\ R = \text{alkyl group} \end{array} \qquad (6.4)$$

Characteristic for these crosslinking systems with metal chelates, is the addition of alcohol as a stabilizer; after its evaporation (in the drying channel together with other solvents) crosslinking starts spontaneously. Therefore, acrylic PSAs including the above mentioned crosslinking agents are the so-called room temperature crosslinking PSAs.

As known, metal chelates based on titanium are also used as homogeneous catalysts for the polymerization of acrylics. Acrylic copolymers, vinyl acrylic copolymers, diene copolymers, and acrylic copolymers with heterocyclic compounds were synthesized using the cyclopentadienyl derivatives of titanium together with alkylaluminum cocatalysts [40–47].

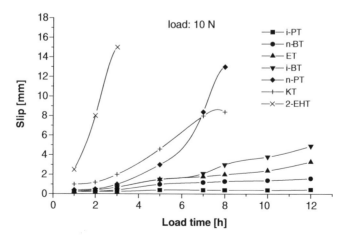

FIGURE 6.5 Shear strength of PSA measured as slip vs. time at a constant load.

a. Addition of Metal Chelates during the Polymerization

The preferred crosslinking agent family of metal 2,4-pentadiones for this crosslinking process is selected from aluminum acetylacetonate (AlAcAc), zirconium acetylacetonate (ZrAcAc), and iron acetylacetonate (FeAcAc). In practice, referring to the color, solubility in common organic solvents, and the best PSA performance values, only AlAcAc has any importance [48]. AlAcAc was diluted in the polymerization medium (ethyl acetate, at about 78°C) before the polymerization process started. After 3 h synthesis with a monomer content between 38.5 to 47.6 wt.% (gelling limit) the acrylic PSA solutions were cooled down to room temperature. The measured parameters were viscosity, tack, peel adhesion, and shear strength of the synthesized adhesives.

The change of the viscosity of PSA containing 1.0 wt.% AlAcAc with time (its pot-life) and the relationship between the viscosity and polymer content for two concentrations of AlAcAc (1.0 and 1.5 wt.%) are illustrated in Figure 6.6 and Figure 6.7.

The pot-life of solvent-based PSA acrylics crosslinked with AlAcAc depends on polymer content (Figure 6.8). With a 37 wt.% polymer content, the pot-life attains about 30 days, at 36 wt.% polymer content about 33 day, and at 35 wt.% polymer content about 46 days. The reduction of solids content to less than 33 wt.% avoids gelation of the investigated PSA acrylic systems. Figure 6.9 shows that the admissible polymer content for 1.5 wt.% AlAcAc is up to 36 wt.% and for 1.0 wt.% AlAcAc is 38 wt.%. Figure 6.10 illustrates the influence of AlAcAc concentration on tack, peel resistance, and shear strength values.

The increase of AlAcAc content increases the level of cohesion at 20 and 70°C. The other properties (e.g., tack and peel resistance at 20°C) are decreased. The best results with respect to overall performances are achieved with 1.0 to 1.5 wt.% of AlAcAc.

The significant advantage given by the dosage of AlAcAc during the polymerization process is the absence of the alcohol stabilizer. Thus, the organic solvent recovered in the drying oven (during coating and drying) can be reused as a polymerization medium.

b. Addition of Metal Chelates after the Polymerization

A preliminary crosslinking of acrylic PSAs has been performed with aluminum(III) acetylacetonate (AlAcAc), chromium(III) acetylacetonate (CrAcAc), iron(III) acetylacetonate (FeAcAc), cobalt(II) acetylacetonate (CoAcAc), nickel(II) acetylacetonate (NiAcAc), manganese(III) acetylacetonate (MnAcAc), titanium(IV) acetylacetonate (TiAcAc), zinc(II) acetylacetonate (ZnAcAc), and zirconium(IV) acetylacetonate (ZrAcAc).

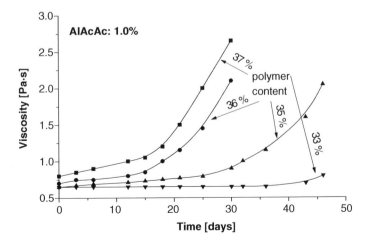

FIGURE 6.6 The change of PSA viscosity containing 1.0 wt.% AlAcAc in the time for two concentrations of AlAcAc.

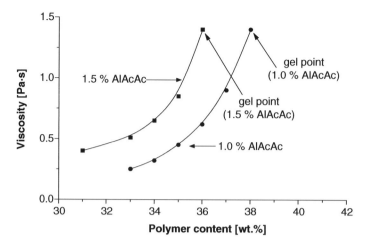

FIGURE 6.7 The relationship between the viscosity and acrylic polymer content for two concentrations of AlAcAc.

The influence of the selected metal chelates, using various amounts of 0.4, 0.8, and 1.2 wt.% as referred to the acrylic polymer solids, on the shear strength at 20 and 70°C of the acrylic PSAs is illustrated in the Figure 6.9 and Figure 6.10.

As shown in the Figure 6.9 and Figure 6.10 satisfactory properties of crosslinked acrylic PSAs have been obtained by use of AlAcAc, TiAcAc, ZrAcAc, FeAcAc, and MnAcAc. The best values of tack, peel resistance, and shear strength have been observed in case of AlAcAc and TiAcAc.

Figure 6.11 illustrates the crosslinking reaction between AlAcAc and the carboxylic group of the acrylic acid, built in the acrylic polymer chain.

In order to explain the crosslinking reaction of acrylic PSAs by AlAcAc, the following mechanism of crosslinking is to be assumed (Figure 6.12):

In the first step of the crosslinking reaction the carboxylic group provided by the acrylic acid reacts with AlAcAc (with its mesomeric form 2) (Figure 6.12) in the following way (Figure 6.13).

The acrylic acid acidity grows by conjugation of the carboxylic group with the neighboring double bond. The increased acidity of acrylic acid depends on the hybridization sp^2 of the

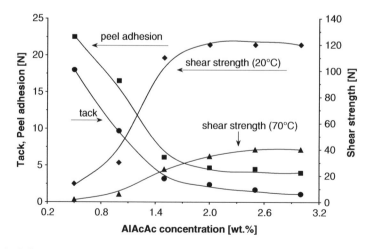

FIGURE 6.8 The influence of AlAcAc concentrations on tack, peel resistance, and shear strength.

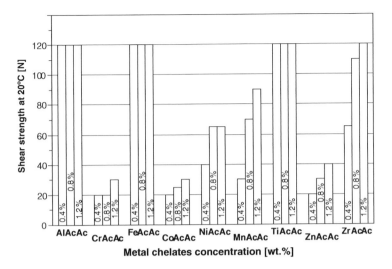

FIGURE 6.9 Shear strength at 20°C of PSA crosslinked with metal acetylacetonates.

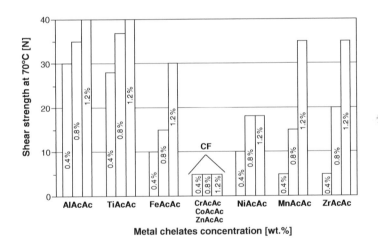

FIGURE 6.10 Shear strength at 70°C of PSA crosslinked with metal acetylacetonates.

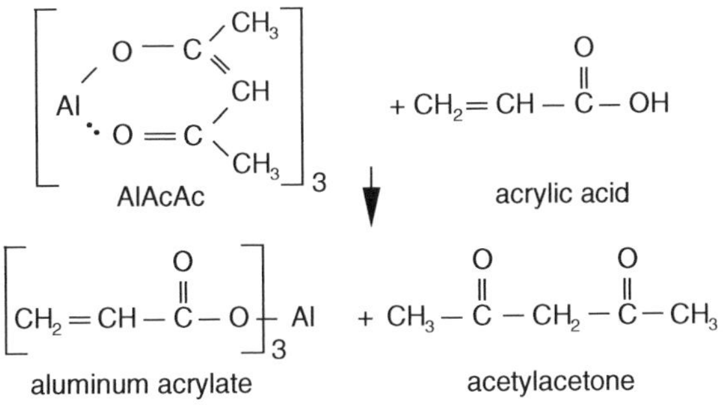

FIGURE 6.11 Reaction between AlAcAc and acrylic acid.

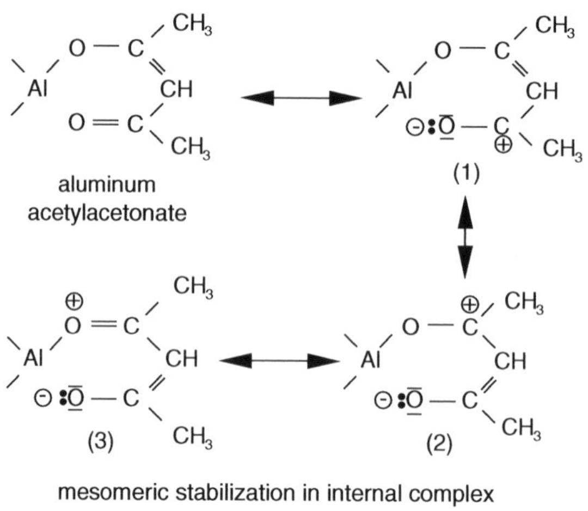

FIGURE 6.12 Intramolecular mesomeric stabilization of AlAcAc.

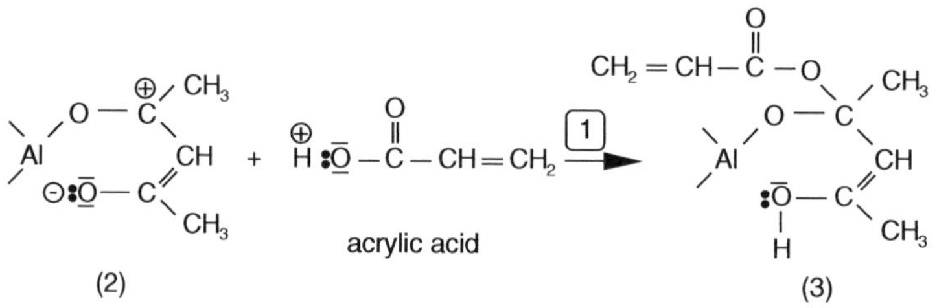

FIGURE 6.13 First step of the reaction between AlAcAc and acrylic acid.

unsaturated carbon atom, which strongly attracts the electrons like the unsaturated carbon atom with sp^3 hybridization. The carbon atom with hybridization sp^2 is a weak electron donor comparatively to a carbon atom with sp^3 hybridization.

In the second stage of the crosslinking of the acrylic PSA the following rearrangement proceeds, which is stabilized by the π electrons of the vinyl group provided by the acyl cation (Figure 6.14).

The carbanion (acyl cation) split-off is stabilized by the π electrons of the vinyl group or by the π electrons of the heteroatom (Figure 6.15).

The oxygen atom interacts with the neighboring electron attracting the cationic center of the carbon atom, due to its free electron pairs. The formed oxonium ion is unstable, and in the second step of crosslinking the following rearrangement proceeds (Figure 6.16).

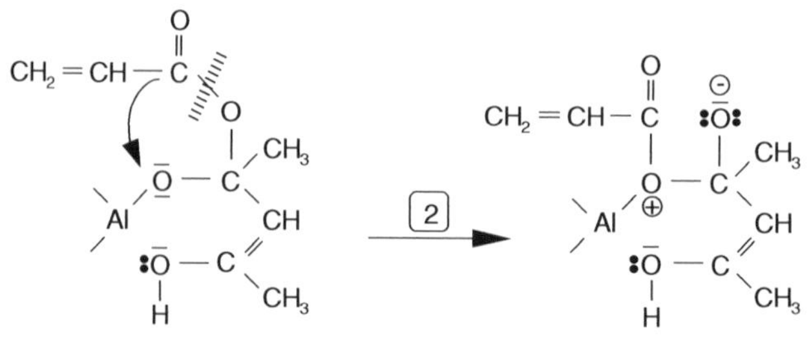

FIGURE 6.14 Second step of reaction between AlAcAc and acrylic acid (abstraction of the acyl cation).

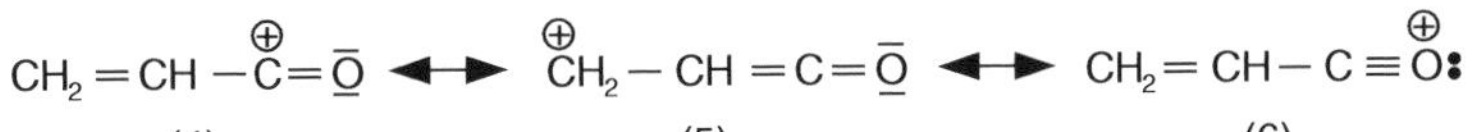

FIGURE 6.15 The acyl cation and its mesomeric structures.

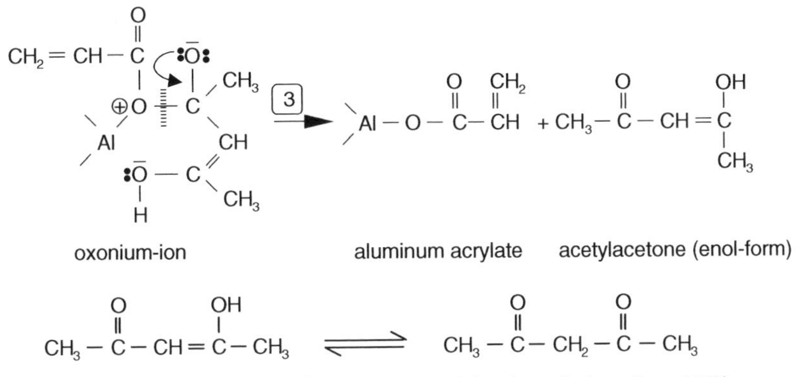

FIGURE 6.16 The end phase of the reaction between AlAcAc and the acrylic acid and the either forms of acetylacetone.

In the range of the studied metal acetylacetonates, TiAcAc, and AlAcAc present practical importance. Considering the chemical structure of TiAcAc, this is atypical compared with the other metal acetylacetonates (Figure 6.17).

In comparison with other metal acetylacetonates, TiAcAc has a higher reactivity to the carboxylic groups provided by the polymer. This is due to the reactive isopropoxy groups in the structure of TiAcAc.

The mechanism of crosslinking of PSAs is based on the initial step of the replacement of reactive isopropoxy groups with the carboxylic groups of acrylic polymer. An acidic hydrogen atom of the carboxylic group migrates to the isopropoxy group of TiAcAc. Isopropyl alcohol (Figure 6.18) is a by-product.

Such mechanism of crosslinking reaction is highly probable because of the higher reactivity of isopropoxy groups (esters of orthotitanic acid) in comparison with acetylacetonate structures (metal acetylacetonates). The intermediate product, which has two acetylacetonate groups (Figure 6.18) reacts with other free carboxylic groups of the polymeric chain (Figure 6.19) in the second step. Taking into account steric hindrance and the mesomeric stabilization of 2,4-pentandione groups, this reaction proceeds extremely slow like the reaction mentioned in the first step.

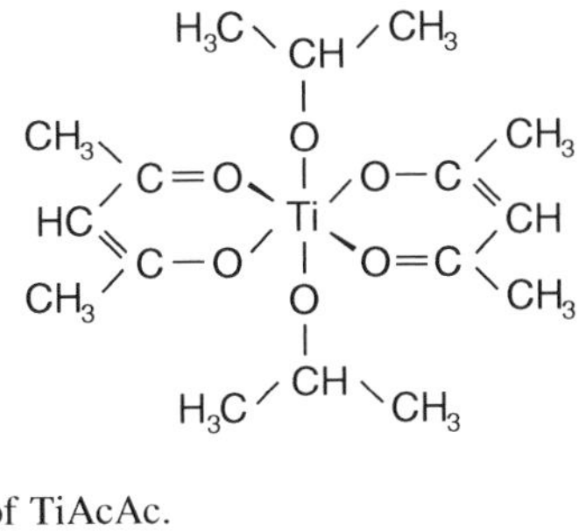

FIGURE 6.17 Chemical structure of TiAcAc.

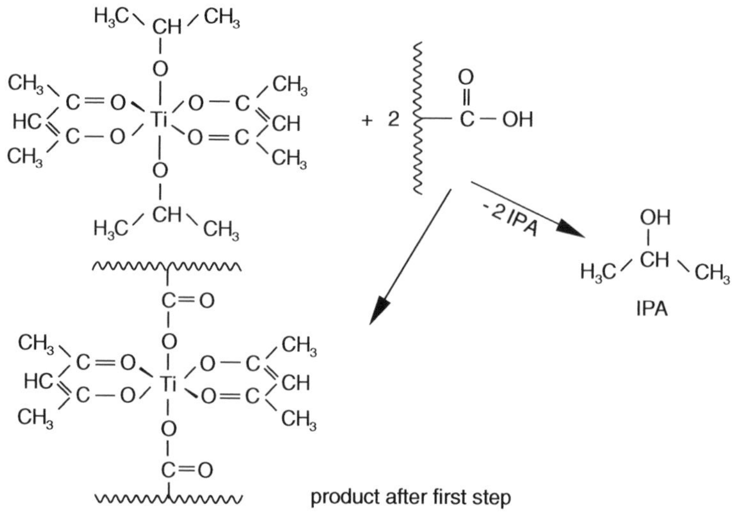

FIGURE 6.18 First step of the crosslinking reaction of acrylic adhesive polymer containing carboxylic groups by use of TiAcAc.

AlAcAc and TiAcAc have turned out to be the best metal chelates as crosslinking agents, useful for industrial crosslinking of solvent-borne acrylic PSAs. The acrylic PSA crosslinked with AlAcAc and TiAcAc displays the advantages of a high tack, high peel resistance, high shear strength, and excellent balance between adhesion and cohesion.

3. Multifunctional Isocyanates

The great reactivity of the multifunctional isocyanates is also used for crosslinking PSA acrylics containing hydroxyl or carboxyl in order to increase their inner strength [49]. Useful polyfunctional isocyanate crosslinking agents with the following formula (Figure 6.20) are described in Ref. [50].

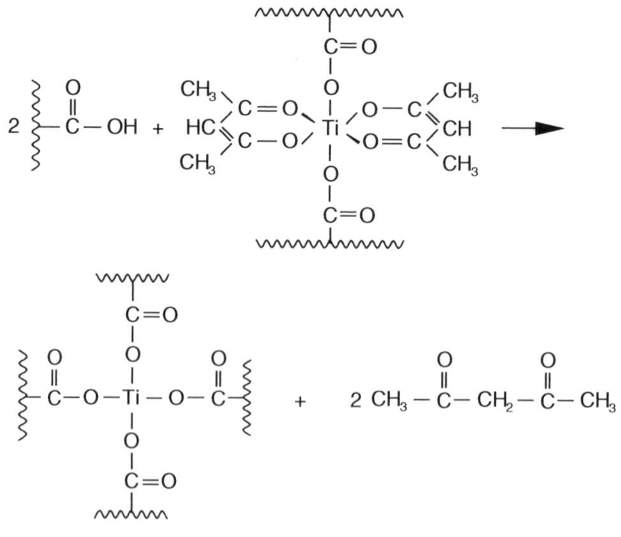

FIGURE 6.19 Second step of the crosslinking reaction of an acrylic adhesive polymer containing carboxylic groups by use of TiAcAc.

$$R-(N=C=O)_n$$

R = hydrocarbon group, n = 2 to 4, preferably 2 to3

FIGURE 6.20 General formula of multifunctional isocyanates.

Multifunctional crosslinkers include aromatic isocyanates such as toluene diisocyanate, aliphatic isocyanates such as hexamethylene diisocyanate or tetramethylene diisocyanate, cyclo-aliphatic isocyanates such as *bis*(4-isocyanato hexyl) methane, or heterocyclic isocyanates such as uretdione of 2,4-diisocyanato toluene [6,51].

During the crosslinking reaction there ensures an addition between isocyanate groups and active hydrogen atoms, and, hereby, no separation products are generated. The isocyanate groups can react with OH groups in the polymer chain to form a high molecular weight by a chemically crosslinked urethane structure (Figure 6.21).

In the absence of hydroxyl groups in dried coatings of acrylic PSA multifunctional isocyanates undergo the normal reactions with active hydrogen atoms of carboxylic groups of acrylic polymer chain, to form amides and further the crosslinking process [52] (Figure 6.22).

The use of polyisocyanate crosslinkers is recommended at a level of 0.1 to 5% based on the solid content of the solvent-borne acrylic adhesive. The application of these kinds of compounds is, however, limited in practice as the pot-life of the pressure-sensitive acrylic adhesives, being crosslinked with multifunctional isocyanates, is rather short, namely a few minutes, hours or days. As discussed in Chapter 8, isocyanates are preferred as crosslinking agents for PSAs for protective films. Table 8.14 lists the crosslinking agents used for various PSPs.

The typical performance characteristics of the crosslinked solvent-borne acrylics, such as peel resistance and shear strength using available multifunctional isocyanate are illustrated in Figure 6.23 through Figure 6.25.

It can be concluded from these experimental results, that all tested isocyanates have a negative influence on tack (Figure 6.23) and peel resistance at 20°C (Figure 6.24) of acrylic solvent-borne adhesives. Generally, low peel resistance and moderate shear strength have been achieved using multifunctional isocyanates as crosslinkers in solvent-borne acrylic PSAs. This justifies the use

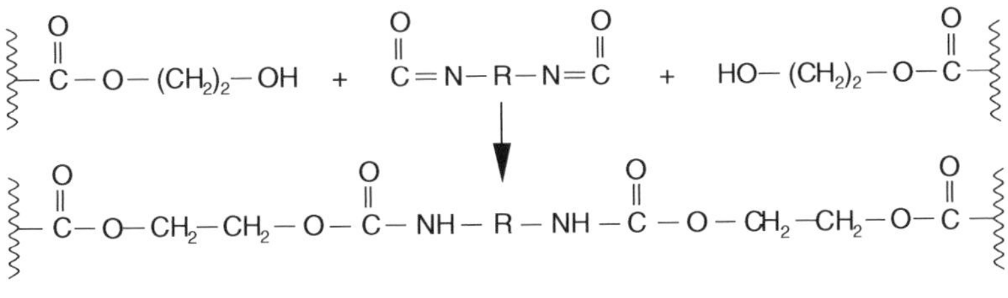

FIGURE 6.21 Crosslinking of acrylic polymers containing hydroxyl groups with isocyanates.

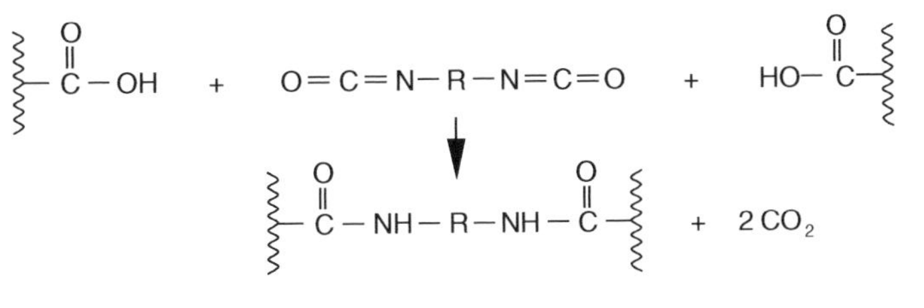

FIGURE 6.22 Crosslinking of acrylic polymers containing carboxyl groups with isocyanates.

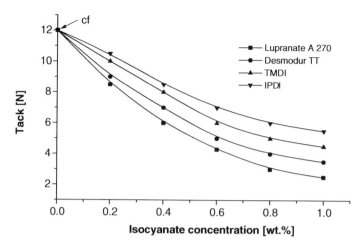

FIGURE 6.23 The effect of isocyanate amount on tack.

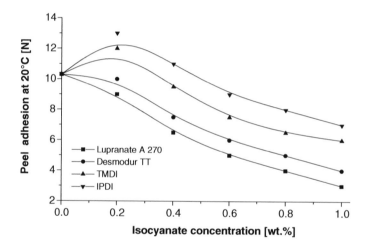

FIGURE 6.24 The effect of isocyanates amount on peel resistance at 20°C.

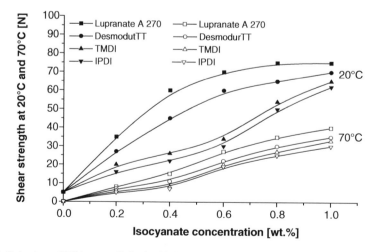

FIGURE 6.25 Cohesion of PSAs crosslinked with isocyanates at 20 and 70°C.

of multifunctional isocyanates as crosslinkers for acrylic adhesives in the case of protective films mainly. In fact, the most popular crosslinker for acrylic PSAs for protective films is IPDI.

Cycloaliphatic isocyanate (e.g., IPDI) and aliphatic isocyanate (e.g., TMDI) are inferior to the aromatic isocyanate Lupranate A 270 and the heterocyclic isocyanate Desmodur TT, although they ensure enough cohesion at 20 and 70°C (Figure 6.25) of acrylic PSA crosslinked with isocyanates. The superiority of aromatic and heterocyclic isocyanates compared with aliphatic and cycloaliphatic isocyanates used as crosslinking agents is obvious.

4. Polycarbodiimides

The aromatic polycarbodiimides having the following structure (Figure 6.26), render the reaction with nucleophilic carboxyl groups of polymer possible [53].

Several grades of multifunctional carbodiimides have been used in the coating industries, normally supplying the product as 50% active solids in ethyl acetate miscible solvents (e.g., glycol ether acetates). The crosslinking reaction involves the addition of the carboxylic acid group across the carbodiimide to form an N-acylurea (Figure 6.27).

The reaction mechanism between the carboxylic groups of the polymer chains and polycarbodiimide is shown in Figure 6.28 [6]. The short pot-life of maximal 24 h can be avoided by the addition of alcohol, preferably isopropanol achieving a pot-life of days or months.

The influence of the polycarbodiimide crosslinkers Permutex XR-5551 and Permutex XR-5580 on tack and peel resistance at 20 and 70°C is shown in Figure 6.29; their influence on the shear strength at 20 and 70°C is illustrated in Figure 6.30. After the addition of small concentrations of polycarbodiimides Permutex XR-5551 and Permutex XR-5580 an amelioration of the PSA properties, like tack and peel adhesion, has been observed for 3 wt.% of either polycarbodiimides. PSA layers crosslinked with polycarbodiimides are unusually tacky and aggressive. After exceeding a threshold of 3 wt.% polycarbodiimides, the tack and peel adhesion levels decrease.

5. Multifunctional Propylene Imines

Multifunctional propylene imines (aziridines) are functional methylaziridine derivates (Figure 6.31), which perform as very reactive, low energy crosslinking agents in carboxylated polymers [54].

The high reactivity of propylene imine crosslinkers can be ascribed to ring strain inherent in the terminal aziridine groups. Ring opening is acid catalyzed, proceeding initially via protonation of the tertiary nitrogen atom. Effective utilization of aziridine chemistry is very much dependent upon

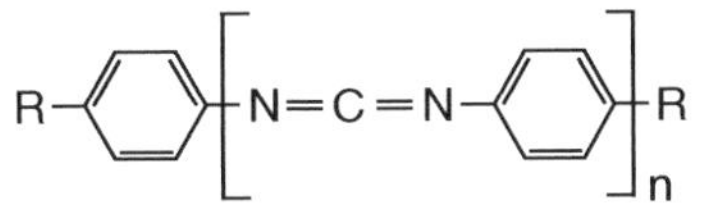

FIGURE 6.26 General formula of aromatic polycarbodiimides.

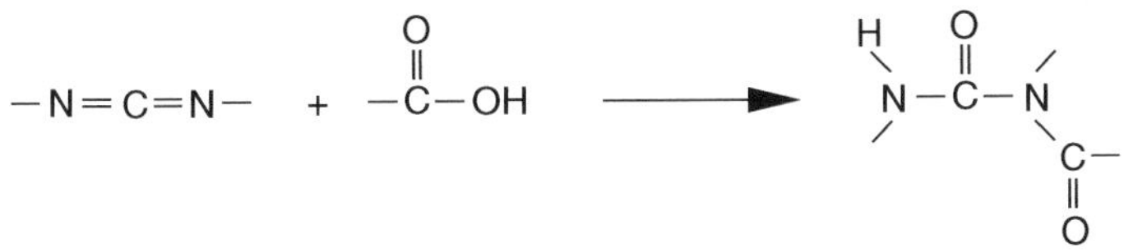

FIGURE 6.27 Reaction of carbodiimide with the carboxylic groups of acrylic polymers.

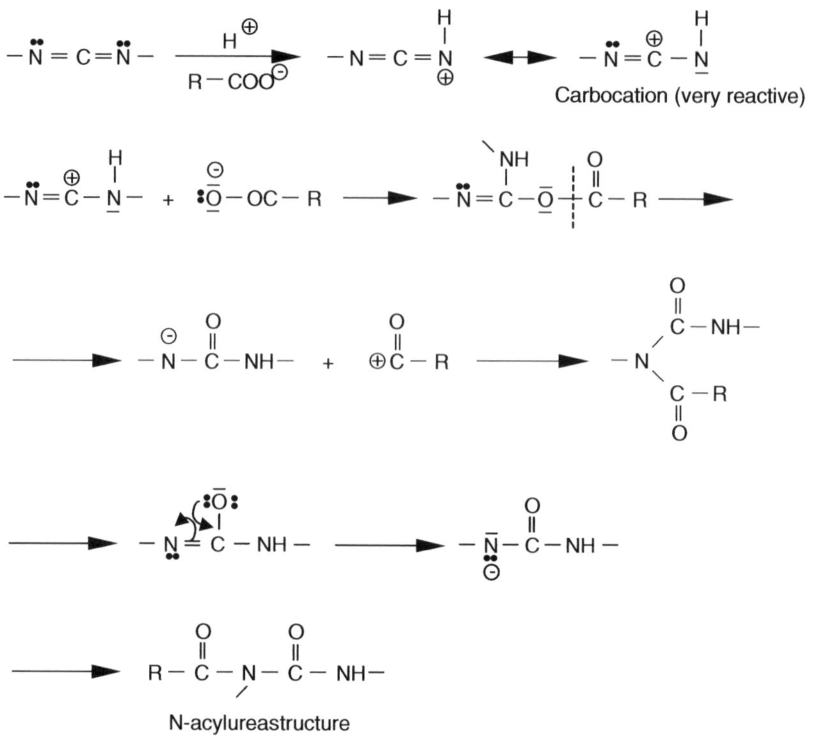

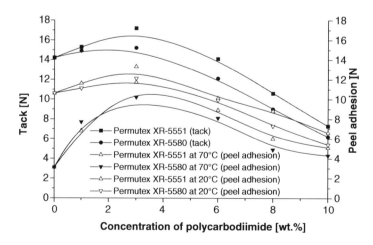

FIGURE 6.28 Mechanism of the reaction between carboxylic groups and polycarbodiimide during crosslinking of PSA acrylics.

the availability of an active H^+ to protonate the aziridine ring. The most common method used to protonate the aziridine ring is the reaction with carboxyl functionality in the backbone of the polymer system.

Crosslinking of acrylic PSAs can be carried out by the use of a propylene imine crosslinker, a molecule that has two or more *m*-ethylaziridine groups capable of reacting with the functional carboxylic groups on the polymer chain (Figure 6.32).

FIGURE 6.29 The influence of the polycarbodiimides on tack and peel resistance of an acrylic PSA.

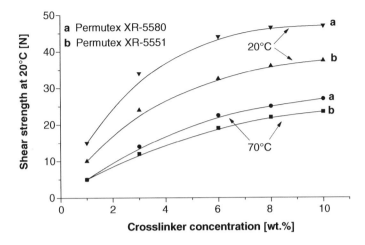

FIGURE 6.30 The influence of tested polycarbodiimides on the acrylic PSA cohesion.

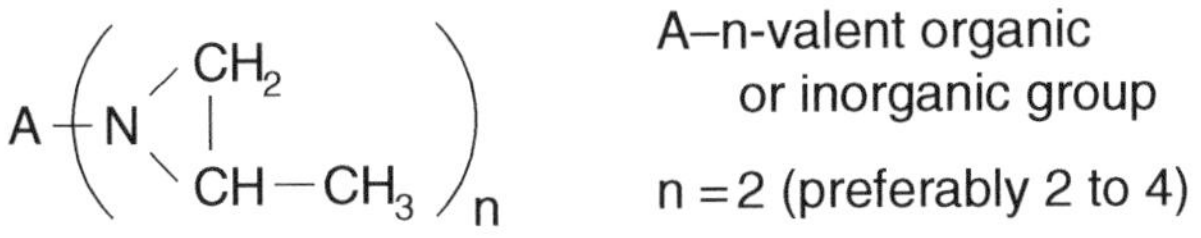

A $-$ n-valent organic
or inorganic group

n = 2 (preferably 2 to 4)

FIGURE 6.31 Chemical formula of more functional propylene imine crosslinkers.

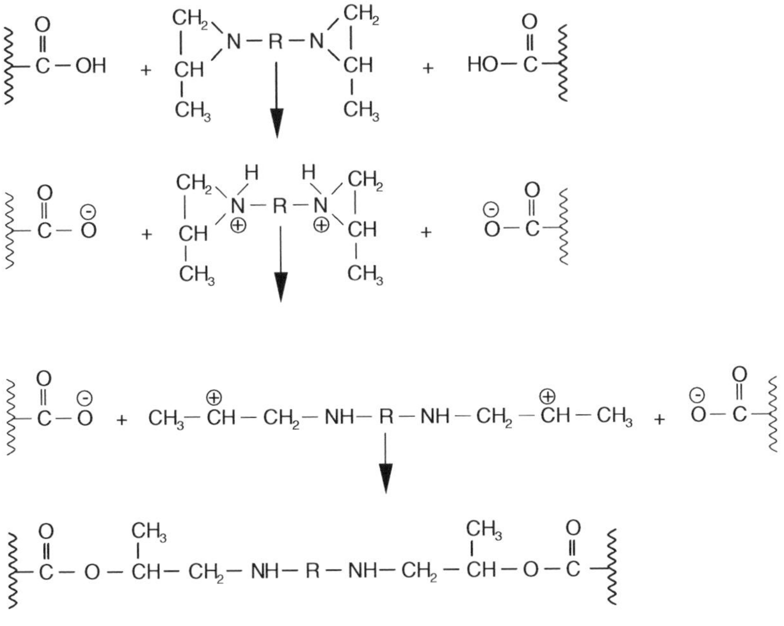

FIGURE 6.32 Crosslinking of a carboxylated PSA initiated by propylene imine.

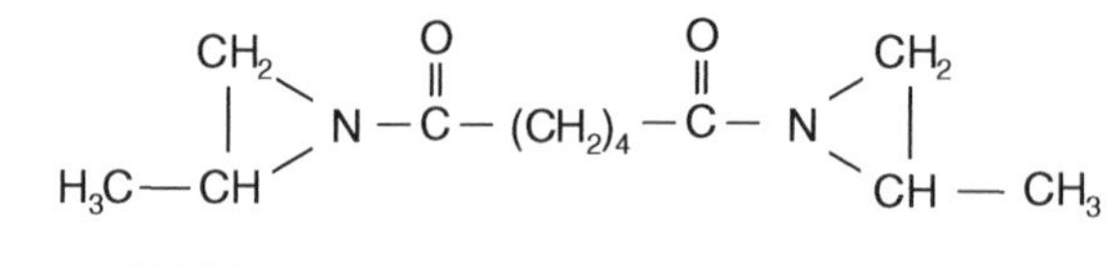

N,N'-bis-propyleneadipic acid amide (BPAA)

FIGURE 6.33 Aliphatic propylene imine.

The multifunctional propylene imines are the most important class of crosslinking agents for permanent self-adhesive coatings which crosslink rapidly at low temperatures [55–57].

For crosslinking, the polyfunctional propylene imines are added to the finished compounded coating, just prior to use. These pressure-sensitive acrylic adhesives have a pot-life of about 8 to 36 h. The greatest attention is attached to the multifunctional propylene imine derivatives shown in Figure 6.33 through Figure 6.42.

The influence of selected various multifunctional propylene imines as crosslinking agents at levels between 0.1 and 0.9 wt.% on tack, peel adhesion, and shear strength of solvent-based acrylic PSAs is presented in Figure 6.43 through Figure 6.46.

As it can be seen from Figure 6.43, tack is insignificantly reduced for a small increase of propylene imines concentration. The best crosslinking agent concerning tack was trismethylol-propan-*tris*-(*N*-methylaziridinyl) propionate (Neocryl CX-100). The aliphatic propylene imine

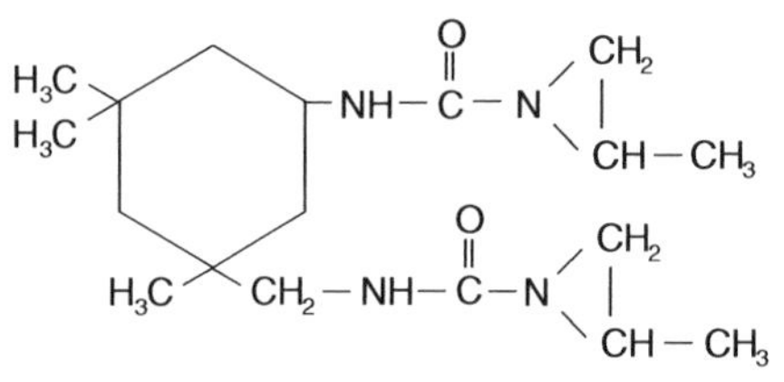

Isophoronedipropylene urea (IPPU)

FIGURE 6.34 Cycloaliphatic propylene imine.

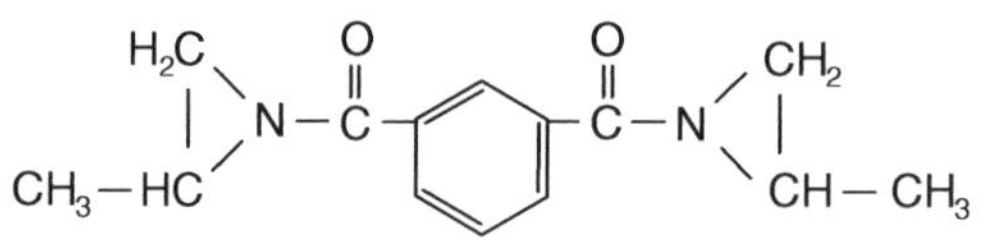

N,N'-Bis-propyleneisophtalic acid amide (BPIA)

FIGURE 6.35 Aromatic propylene imine.

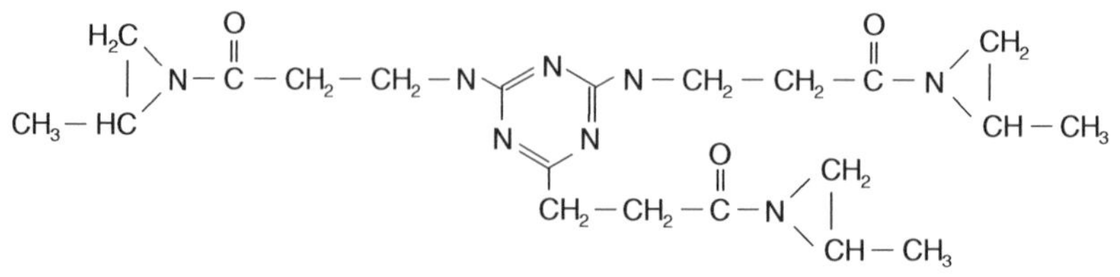

2,4,6-Tris(propylenepropionic acid amide)-1,3,5-triazine (TPAT)

FIGURE 6.36 Propylene imine based on *s*-triazine.

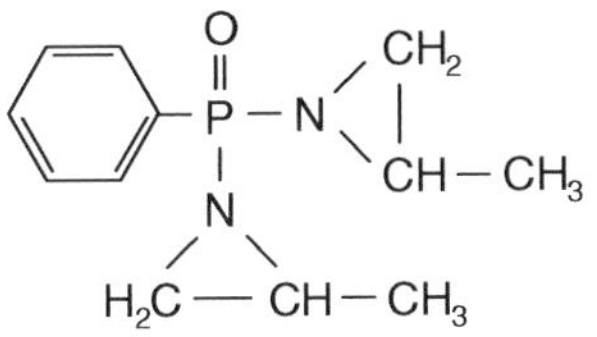

N,N'-Bis-propanephosphonic acid diamide (BPAD)

FIGURE 6.37 Propylene imine crosslinker containing a central heteroatom.

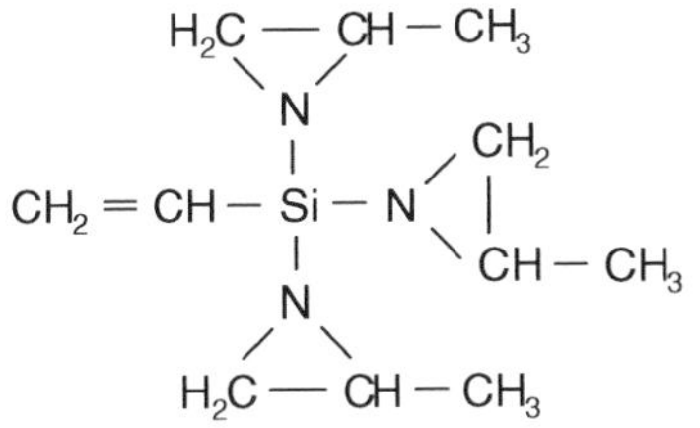

Vinyltrimethylaziridinylsilane(VTAS)

FIGURE 6.38 Propylene imine silane crosslinker.

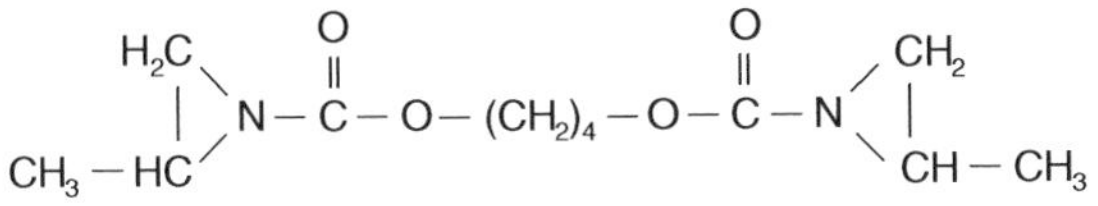

Butandiol-bis-propyleneimineformiate (BBPF)

FIGURE 6.39 Propylene imine formiate.

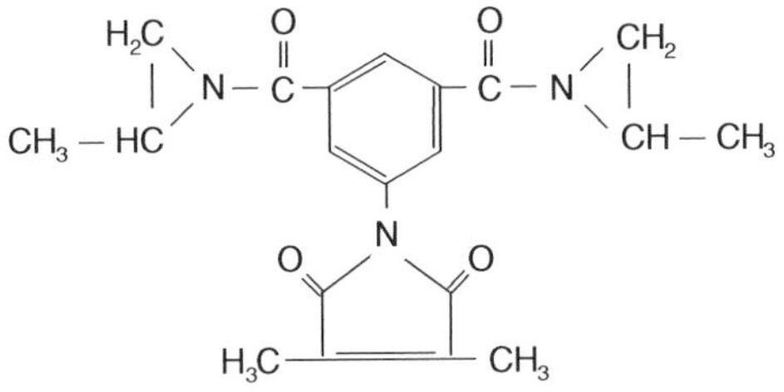

5-Dimethylmaleinimidyl-N,N'-bis-popyleneisoptalic acid amide (DMPA)

FIGURE 6.40 Heterocyclic propylene imine.

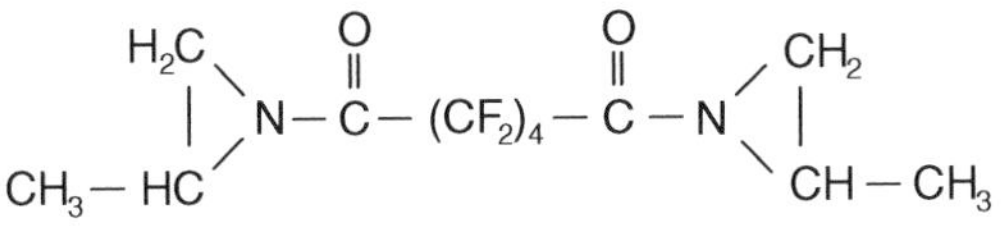

N,N'-Bis-propyleneoctafluoro adipic acid amide (BPFA)

FIGURE 6.41 Fluorinated propylene imine.

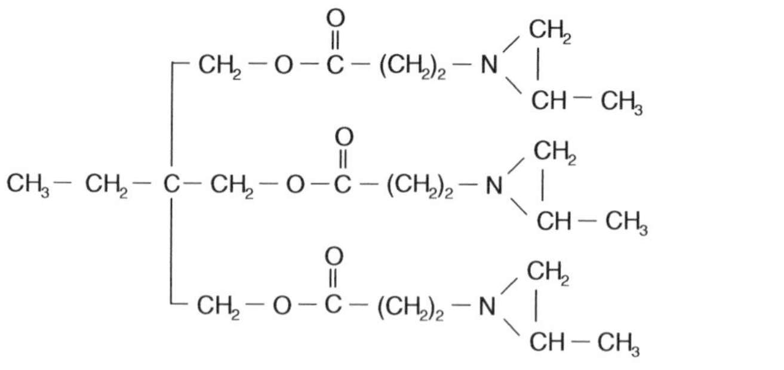

Trimethylolpropane-tris-(N-methylaziridinyl)-propionate (Neocryl CX-100)

FIGURE 6.42 Commercially available propylene imines with known chemical formula.

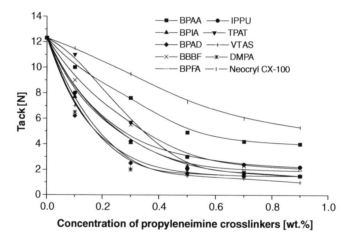

FIGURE 6.43 Tack dependence of propylene imine concentration.

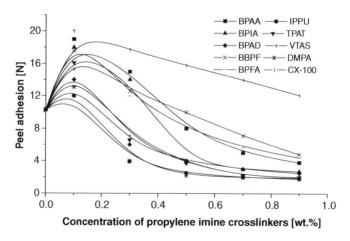

FIGURE 6.44 Peel adhesion dependence of propylene imine concentration.

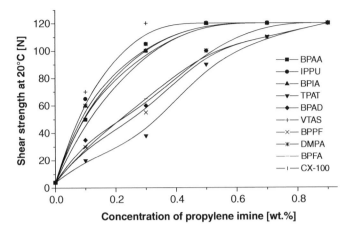

FIGURE 6.45 Shear strength at 20°C dependence of propylene imine concentration.

N,N'-bis-propyleneadipic acid amide (BPAA), and aromatic propylene imine crosslinker, *N,N'-bis*-propyleneisophtalic acid amide (BPIA), also led to good tack values.

In the case of all propylene imines for 0.1 wt.% a maximum of peel resistance was observed (Figure 6.44). The increase of propylene imines crosslinker concentration from 0.1 to 0.5 wt.% decreases the adhesion performance by a factor of about 4 to 5 in comparison with the initial peel resistance value. The acrylic PSAs crosslinked with propylene imines containing *s*-triazine structures have moderate peel resistance. As shown, propylene imine Neocryl CX-100 provides higher peel resistance. This class of crosslinkers is known as "soft crosslinking agents" producing adhesives with high tack and high adhesion.

The highest cohesion value at room temperature and 70°C was observed for aliphatic propylene imine Neocryl CX-100, cycloaliphatic propylene imine IPPU, and aromatic propylene imine BPIA for amounts of 0.3 wt.% or rather 0.5 wt.% (Figure 6.45 and Figure 6.46). As illustrated, Neocryl CX-100 supplies the best balance between tack, adhesion, and cohesion among the multifunctional propylene imine crosslinked solvent-based acrylic PSAs. Pot-life of PSAs containing multifunctional propylene imines is very short.

Because of the high reactivity of polyaziridines, a marginal disadvantage of the propylene imine crosslinking systems is their pot-life, which is limited to several hours, that is lower than

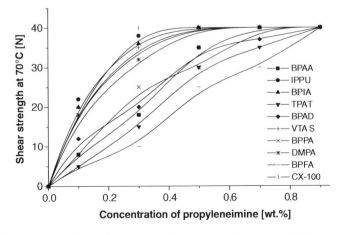

FIGURE 6.46 Shear strength at 70°C dependence of propylene imine concentration.

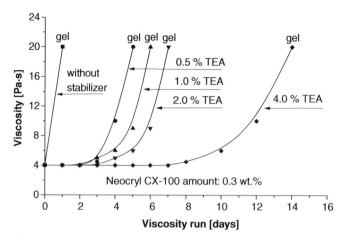

FIGURE 6.47 The influence of TEA content on viscosity of acrylic PSA containing Neocryl CX-100.

the pot-life of adhesives containing polyisocyanate. This problem can be solved by mixing the adhesive and the crosslinker directly prior to the coating step.

Focusing on the high reactivity of propylene imine crosslinkers, the addition of alcohol does not provide the same stabilizing effect as in the case of the conventional crosslinkers such as typical metal acetylacetonates. But there is a way out: the prolongation of pot-life up to several days, or even weeks, can be achieved by addition of volatile aminoorganic stabilizers to the propylene imine crosslinker containing polyacrylate. The concentration of such stabilizers ranges from 1 to 4 wt.%. The use of TEA enables to prolong the pot-life of a PSA containing Neocryl CX-100 (Figure 6.47).

The most efficient propylene imine crosslinker, Neocryl CX-100, was compared with the most-used metal acetylacetonate crosslinker for solvent-borne acrylic PSAs, TiAcAc. Peel adhesion at 20°C to different surfaces such as wood, polyamide (PA), polystyrene (PS), and low surface energy polymers, such as ethylene–propylene–diene multipolymer (EPDM), polyethylene (PE), and polypropylene (PP) is illustrated (Figure 6.48).

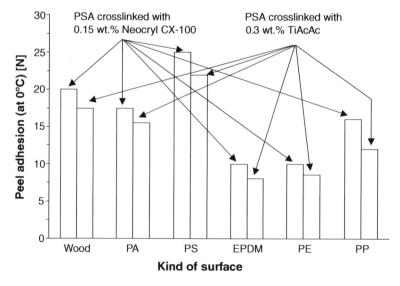

FIGURE 6.48 Adhesion to selected substrates of acrylic dispersions crosslinked with propylene imine Neocryl CX-100 and the metal chelate TiAcAc.

A comparison of Neocryl CX-100 and TiAcAc crosslinkers on the basis of the measurement of their peel adhesion to selected surfaces demonstrates that the first crosslinker is superior to the second. The use of Neocryl CX-100 yields much higher values of peel adhesion in such crosslinked solvent-borne acrylic PSAs to wood, PA, PS, EPDM, PE, and PP.

6. Amino Resins

As an interesting class of crosslinking agents, for the thermal crosslinking of solvent-based acrylic PSAs including carboxyl-, hydroxyl-, or amide groups, a wide range of numerous amino resins has been on offer during the last year which can be classified into four main groups, as melamine–formaldehyde resins [58,59], benzoguanamine resins, glycoluril resins, and urea resins [60–62].

Amino resins are characterized by their reactive end groups and enable a controlled cross-linking reaction and a precise adjustment of the required adhesive properties. It is their special feature that their crosslinking speed is practically zero at room temperature, while it increases exponentially at elevated temperatures between 105 and 150°C. That means, that the crosslinking ensues only during the drying of the PSA layer in the drying oven [63]. The chemistry of these multifunctional resins is extremely complex and they can be generally classified as fully alkylated, partially alkylated, and high imino resins. In general, the principle of the crosslinking reaction of acrylic solvent-borne PSA containing carboxylic groups with amino resins containing methoxy groups is presented in Figure 6.49 [64].

In order to obtain a crosslinking effect the amino resins need only weak catalyzing. In most cases, a low acid value of the basis resin is already sufficient, and the reactivity of the alkoxyalkyl groups is relevant for the crosslinking speed. The temperature time profile of the crosslinking reaction will also be improved by incorporating strong acid catalysts such as p-toluenesulphonic acid or dodecyl-benzenesulphonic acid into the pressure-sensitive formulation. The selected amino resins are listed in Table 6.1.

The use of amino resins allow to find the most suitable system for solvent-borne acrylic PSAs without losing important properties in comparison with other tested crosslinking systems. The performance of crosslinked acrylic PSAs using the mentioned amino resins are presented in Figure 6.50 through Figure 6.52.

The obtained results of the tested crosslinked acrylic PSAs show that, the benzoguanamine derivative, butoxylated glycouril, and butoxylated urea with methylol group decrease all of the adhesiveness performance (tack and peel adhesion) compared with the melamine resin crosslinkers

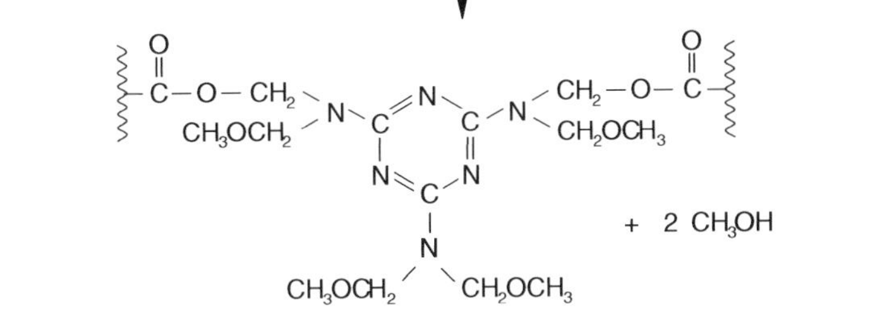

FIGURE 6.49 Crosslinking reaction of PSA acrylics containing carboxylic groups with a melamine-formaldehyde resin.

TABLE 6.1
Examined Amino Resin Crosslinkers

Amino Crosslinker	Nonvolatile (wt.%)	Type of Solvent	Viscosity at 23°C (Pa s)	Chemical Formula Basis
Cymel 303 (highly methylated melamine resin)	≥98	Free	3.6–6.0	Hexamethoxymethyl melamine
Cymel 370 (partially methylated melamine resin)	88	i-Butanol	5.1–10.2	Methoxymethyl methylol melamine
Cymel 1123 (highly alkylated benzoguan-amine resin)	≥98	Methanol, ethanol	3.8–10.2	Methoxymethyl ethoxymethyl benzoguanamine
Cymel 1170 (highly n-butylated glycouril resin)	98	n-Butanol	1.7–6.0	Tetrabutoxymethyl glycoluril
Dynomin UB-24-BX (n-butylated urea resin)	63	n-Butanol xylene	1.7–2.6	Butoxymethyl methylol urea

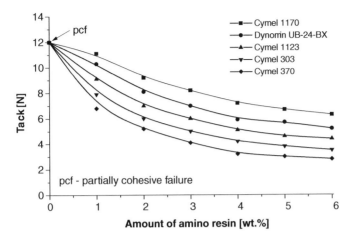

FIGURE 6.50 Tack of crosslinked PSA vs. amino resin concentration.

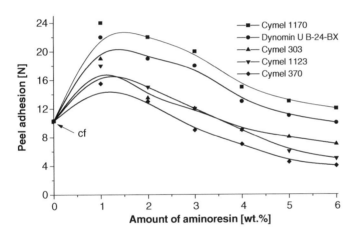

FIGURE 6.51 Peel adhesion of crosslinked PSA vs. amino resin concentration.

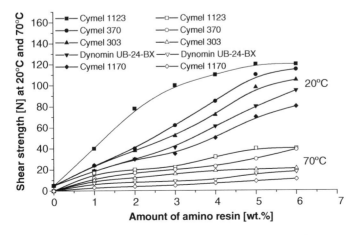

FIGURE 6.52 Shear strength of crosslinked PSA vs. amino resin concentration.

Cymel 303 and Cymel 370. This statement can be explained with the higher functionality of melamine resins (6) in comparison with functionality of benzoguanamine (4), glycouril (4), or urea resins (4). The relatively low reactivity of tetrabutoxymethyl glycouril Cymel 1170 and butoxymethyl methylol urea Dynomin UB-24-BX comparatively with methoxymethyl ethoxymethyl benzoguanamine Cymel 1123 demonstrates once more the higher reactivity of methoxy and methylol groups in crosslinking carboxylic groups of the acrylics. The product crosslinked with benzoguanamine resin crosslinkers show better thermal and shear performance than those with amino crosslinkers. Benzoguanamine resins are even better in these properties, but they are more expensive than melamine resins. Therefore, a mixture of melamine and benzoguanamine resins may be used to achieve a cost–benefit compromise.

The lowest possible crosslinking temperature necessary to finish the crosslinking process and to ameliorate the shear strength of crosslinked acrylic PSA, has been investigated. This parameter was tested and illustrated in Figure 6.53 for 3.0 wt.% Cymel 303 during 10 min in the drying oven.

Below 120°C the crosslinking reaction does not occur completely. A temperature increase lets the carboxylic groups of the polymer chain react with the methoxy groups of the amino crosslinker.

II. PHOTOCROSSLINKING OF SOLVENT-BASED ACRYLIC PSAs

The use of ultraviolet (UV) radiation to cure solvent-based acrylic PSAs is an alternative method to the conventional crosslinking (see Chapter 5, Sections II.A and II.B). Acrylic systems to be photocured require a photoinitiator to induce crosslinking.

The most important properties of the UV-crosslinked acrylic PSAs, such as tack, peel resistance, and shear strength, can be controlled by the photoinitiator amount, UV-crosslinking time, and UV dosage. The solvent-borne UV-crosslinkable acrylic PSA is coated directly and after removing the solvents the adhesive film is crosslinked by UV-irradiation. The high UV doses required at production speeds of about 200 m/min, are obtained by using six or more UV-lamps with an output power of about 200 W/cm for each one.

Films formed by UV-crosslinked acrylic adhesive become more cohesive as more radiant energy is applied, while their tack and peel adhesion decrease. The effect of the UV radiation dose (mJ/cm^2) on the adhesive properties and cohesion is shown schematically in Figure 6.54. As it can be seen, both curves pass through a plateau within the same range of UV doses, and this plateau thus represents optimum conditions for the curing that has been referred to as the "curing window." To achieve optimal pressure-sensitive performance with UV-crosslinkable adhesives, it is necessary to find process settings that lead to balanced values of tack, peel adhesion, and shear strength for the preferred application.

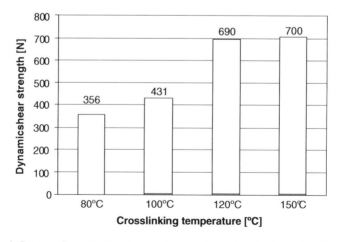

FIGURE 6.53 The influence of crosslinking temperature on dynamic cohesion using Cymel 303.

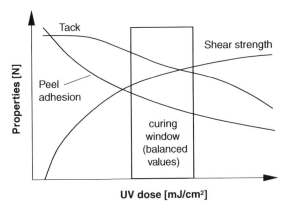

FIGURE 6.54 Tack, peel adhesion, and shear strength as a function of UV dose.

A. UV-Light

UV radiation is widely used in industrial processes (see Table 8.27). The UV equipment, like UV-lamps and UV-lasers, emits light in the UV-region from 100 to 400 nm of the electromagnetic spectrum. The UV-radiation spectrum used for the crosslinking of PSAs comprises the wavelengths between 200 and 400 nm and is subdivided into UV-A (315–400 nm), UV-B (280–315 nm), and UV-C (200–280 nm) (see also Chapter 5, Section II.B). For the manufacturing of radiation crosslinked PSA acrylics the UV-A band width between 315 and 400 nm is used.

1. UV-Lamps (Spectral)

The crosslinking with UV light can be carried out directly after the application of the PSA or after passing a cooling zone. UV-lamps are categorized as low, medium, and high pressure UV-lamps [6]. Mercury lamps (low, medium, high pressure) are used with powers between 80 and 120 W/cm, which includes UV-stations with six and more UV-lamps as state of the art with a power of 120 to 260 W/cm. This type of UV-lamp possesses the right wave-length bands in their emission spectrum as needed for an efficient UV-crosslinking of photoreactive PSAs.

2. UV-Laser (Monochromatic)

Regarding the PSA technology, the excimer laser has the highest chance to be used. The excimer laser was invented in 1975 and is considered, in comparison to Nd:YAG-Laser, the youngest laser with high radiation power. Still, only few practical related application areas for excimer lasers are found due to the relatively short development time. Excimer lasers are very intense pulsed light sources, which radiate UV wavelengths between 170 and 351 nm (depending on the noble gas–halogen mixture used), and provide a high energy, in the ns-range of radiation times.

The advantage of UV-radiation from excimer lasers lies primarily on the tailored UV-light spectrum. The excimer laser with monochromatic UV-light offers, in comparison with commercial UV lamps with spectral UV light, significant advantages in the application possibility of chemical synthesis of macromolecules, particularly in laser-induced polymerization [24] or laser-initiated crosslinking [65].

The application of laser equipment, in comparison with conventional UV lamps, provides the following advantages:

- A monochromatic wavelength instead of a spectral one guarantees high reaction speeds.
- The distance between the reflected laser beam and the PSA surface that is radiated is not relevant in comparison with the distance of UV lamp/PSA layer when conventional UV equipment is used.

- The laser beam is monochromatic and contains no IR parts (no additional heating of the PSA layer or the coated substrate).
- UV crosslinking of "thick" PSA layers (up to 3 mm).
- No problems with the UV crosslinking of PSA containing suitable tackifiers, plasticizers, fillers, and other additives.

B. Photoinitiators

The UV-crosslinking of diverse acrylic PSA films is based on the photoinitiation of radical cross-linking reaction. The ultraviolet crosslinking technique calls for the use of a photoinitiator to be added to the PSA system. The photoinitiator is, therefore, one of the key components in a UV-cross-linking, and the outcome of such a polymerization is critically dependent on the choice of the photoinitiator, including its chemical nature and the amount.

In recent years, there have been many new developments in the synthesis and photochemical studies of novel photoinitiator molecules with more desirable properties such as higher activity coupled with greater speed and low migration rate to the surface of the cured coating, in order to ameliorate shear strength and minimize toxicity where food contact is important [66–69]. There are two basic categories of photoinitiators used in actual practice for initiating radical cross-linking. The first group involves type I photoinitiators and the second group type II photoinitiators (see Chapter 5, Section II.A). A detailed description of photoinitiators is given in Ref. [70].

1. Photoinitiators of Type I (α-Cleavage Photoinitiators)

The first group includes type I photoinitiators. As discussed earlier, the mechanism characterizing the classes of these photoinitiators is photofragmentation that generates radical pairs through a highly efficient α-cleavage process. This class includes aromatic carbonyl compounds that are known to undergo a homolytic C—C bond scission upon UV-exposure. The benzoyl radical is the major initiating species, while the other fragment may, in some cases, also contribute to the initiation. Of all the investigated radical species, the benzoyl radical is the most effective radical in UV crosslinking, although the substituted benzyl radical also initiates some crosslinking. The photoinitiators of type I undergo a direct photofragmentation process in the excited state into free radicals. Depending on the structure of the molecule, scission may occur in the α or β position (see Structure 5.8–Structure 5.12).

2. Photoinitiators of Type II (Hydrogen Atom Abstracting Photoinitiators)

The H-type photoinitiators produce initiator radicals via intermolecular hydrogen abstraction. These photoinitiators require hydrogen-donors, such as amines, thiols, or alcohols, for generating radicals, which are effective in initiating polymerization. The intermolecular H-abstraction photo-initiators include benzophenone and its derivatives such as xanthone, thioxanthone, and 4,4'-*bis*-(*N*,*N*'-dimethylamino) benzophenone, as well as benzyl and quinones. The photoinitiators in the type II undergo a primary process of hydrogen atom abstraction from the environment (R–H), which may be the resin itself or a solvent, to produce a ketyl radical. The photoreductive ability of the environment is also an important factor and is related to the carbon–hydrogen bond strength of the species donating the hydrogen atom.

3. Multifunctional Conventional Photoinitiators

This group of substances consists of saturated photoinitiators, which contain at least two photoreactive structures in the molecule and form crosslinkage with the PSAs by UV radiation [6,71]. The most typical advance, however, is the development of multifunctional benzophenones (Table 6.2). Here,

TABLE 6.2
Saturated Photoinitiators Type I and Type II Used for UV-Crosslinking of Acrylic PSAs

Photoinitiator	Chemical Formula	Chemical Name
HBBF		Hexanediole-1,6-*bis*-benzo-phenoxyformiate
TBPO		*tris*-Benzo-phenyloxy phosphinoxide

the anchoring in the polymer takes place at protruding alkyl side-chains like in typical H-abstractors. One of the major investigations in this field is the development of nonmigratory photoinitiators especially where food contact is required.

During UV-exposure of the intermolecular benzophenone derivatives, H-abstractor structures are excited and react with the neighboring C—H positions of the polymer side-chains.

The UV-crosslinking mechanism of solvent-based acrylic PSAs containing photoreactive multifunctional benzophenone derivatives has been thoroughly investigated and is presented schematically in Figure 6.55 [6].

Multifunctional H-abstractors, based on benzophenone derivatives, brought a better performance (similar to their photocleavable photoinitiator equivalents) because they had the potential to produce two or more different radical initiating species. UV-crosslinkable acrylic PSAs containing one or multifunctional hydrogen atom abstractors can give way for middle performance self-adhesives, that are produced without environmental problems.

4. Photosensitive Crosslinking Agents

For some PSA applications crosslinking is also required, particularly where it is desired to increase the cohesive strength of the adhesive without unduly affecting its compliance. This can be achieved using trichloromethyl-*s*-triazines, which function simultaneously as photoinitiator and photo-crosslinker, that have absorption maxima in the UV-A area at the wavelength of about 330 to 380 nm [72]. An important class of photocrosslinking agents are chromophoric substituted *bis*chloromethyl-*s*-triazines [73], which function simultaneously as photoinitiator and photo-crosslinker. Examples of such examined *s*-triazine photoreactive crosslinking agents are presented in Table 6.3.

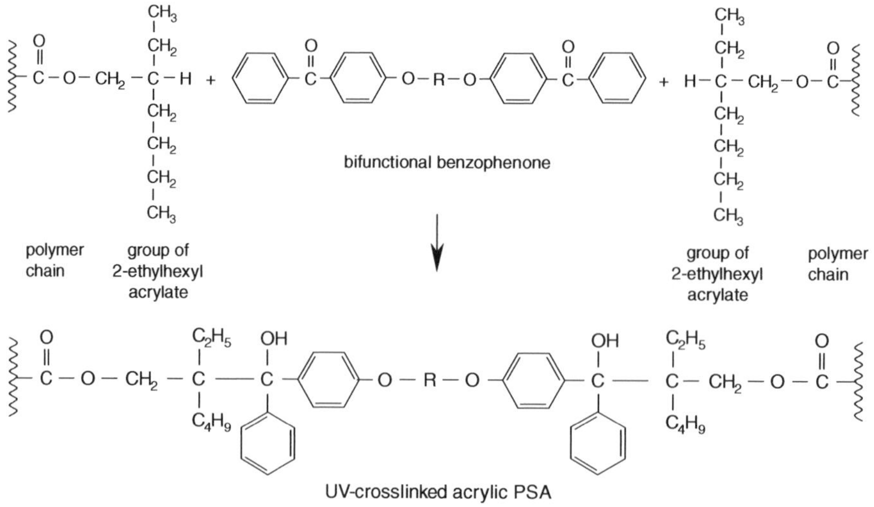

FIGURE 6.55 UV-initiated crosslinking reaction of PSA with multifunctional benzo-phenones (R = organic group).

TABLE 6.3
Investigated s-Triazine Photoreactive Crosslinkers

Photoinitiator	Chemical Formula	Chemical Name
XL-353	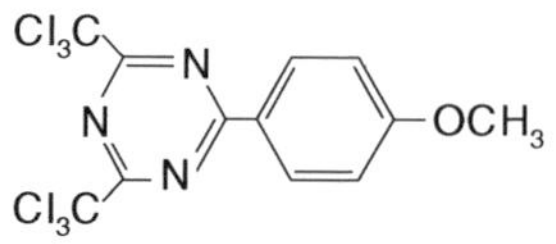	2,4-*bis*-Trichloromethyl-6(4-methoxyphenyl-*s*-triazine
BMP-*s*-T	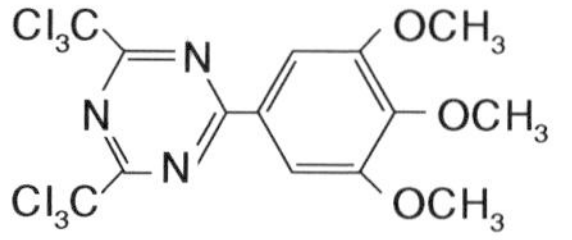	2,4-*bis*-Trichloromethyl-6(3,4,5-trimethoxyphenyl)-*s*-triazine
BN-*s*-T	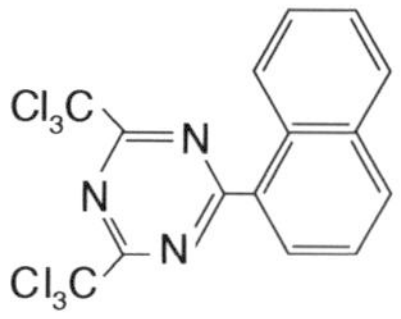	2,4-*bis*-Trichloromethyl-6(1-naphthyl)-*s*-triazine
BMN-*s*-T	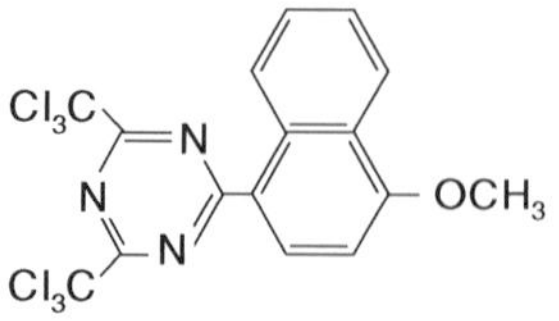	2,4-*bis*-Trichloromethyl-6[1-(4-methoxynaphthyl)-*s*-triazine
MOST	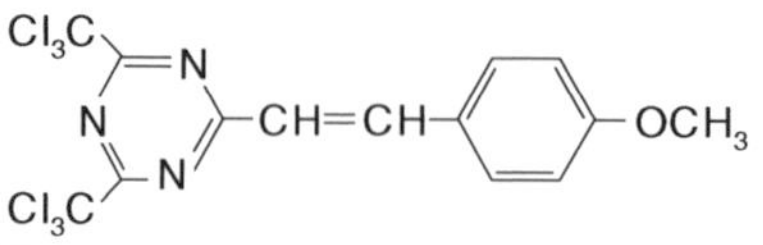	2,4-*bis*-(Trichloromethyl)-6-*p*-methoxystyryl-*s*-triazine

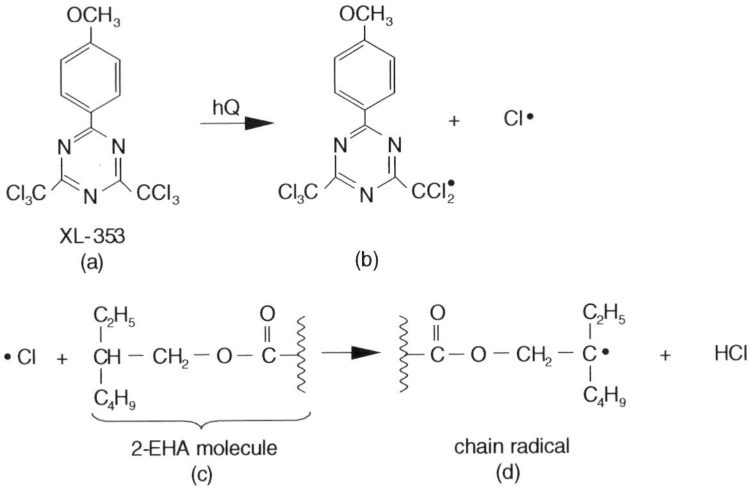

FIGURE 6.56 Mechanism of *s*-triazine photoreactive crosslinker photolysis.

The efficiency of *bis*-trichloromethyl-*s*-triazines as photoinitiators in radically initiated photo-crosslinking systems, as well as photoacid generators, may be explained by the mechanism outlined in Figure 6.56.

Electronically excited fragments (**a**) suffer homolytical transformation to the radical (**b**) and a chlorine atom, which abstracts hydrogen from a donor (**c**) resulting in the formation of hydrogen chloride and the radical (**d**). While hydrochloric acid serves as the catalyst in chemical amplification systems (**d**), it may be used to start a chain crosslinking reaction (Figure 6.57).

The data concerning the peel resistance show that photoreactive *s*-triazine crosslinkers increase the peel resistance of UV-crosslinked acrylic PSAs a little bit more effectively than other previously examined multifunctional unsaturated photoinitiators. The maxima of the peel resistance were noticed for about 0.8 wt.% of the *s*-triazine photoinitiators (Figure 6.58). The highest peel adhesion was observed for 2,4-*bis*-trichloromethyl-6(3,4,5-trimethoxy-phenyl)-*s*-triazine (BMP-*s*-T).

The shear strength after UV-crosslinking is directly proportional to the concentration of the photoreactive *s*-triazines (Figure 6.59). During the UV-induced curing reaction, the elastomeric acrylic PSA chains react with each other to form chemical crosslinks. At a certain stage, after application of the photoreactive *s*-triazine crosslinker 2,4-*bis*-trichloromethyl-[1-(4-methoxy-naphtyl)]-*s*-triazine (BMN-*s*-T), a very strong 3-dimensional chemical network is built up. Unfortunately,

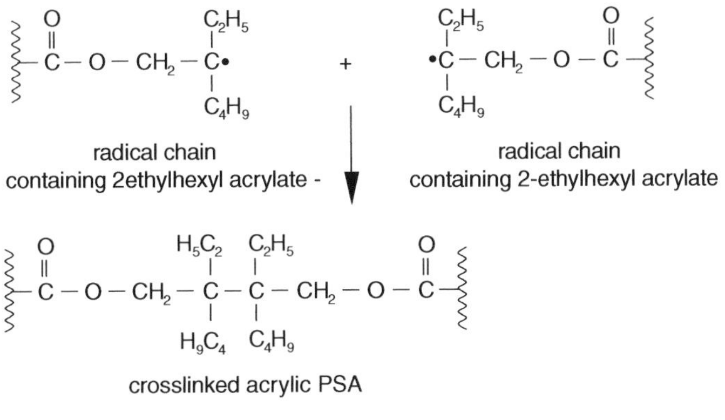

FIGURE 6.57 UV-crosslinking of acrylic PSAs by use of photoreactive *s*-triazines.

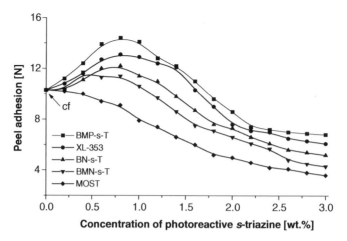

FIGURE 6.58 Peel adhesion of a UV-crosslinked acrylic PSA as a function of the *s*-triazine photoinitiator amount.

the measured temperature resistance (shear strength at 70°C) is very low for all tested *s*-triazine derivatives.

5. Unsaturated Copolymerizable Photoinitiators

Alternatively, the photoinitiator can be polymerized into the backbone. The idea to use a photo-initiator suitable for copolymerization in the process of UV radiation evolved about 30 years ago, however, it was not utilized at that time. Photoinitiators suitable for polymerization should have good solubility, should react completely in the polymerization process, should be high temperature resistant, and should not form photolytic fragments which tend to migrate with a strong specific odor after the UV radiation. The most typical directions include ethene unsaturated UV reactive functionalized acrylated, vinylated, allylated, acrylamidated, or vinyloxylated chromophores of type I and type II initiators (Figure 6.60) [74].

The spacers and substituents of the basic photoinitiator molecule influence the photoreactivity of the photoinitiator suitable for polymerization, as well as the effectiveness of the intermolecular

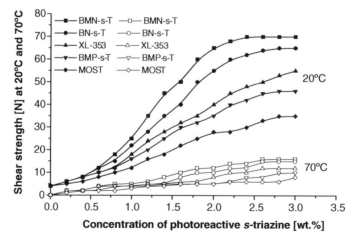

FIGURE 6.59 Shear strength of a UV-crosslinked acrylic PSA as a function of the *s*-triazine photoinitiator amount.

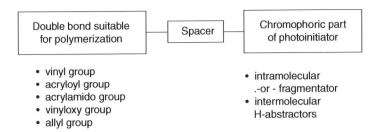

FIGURE 6.60 Schematic of photoinitiators suitable for polymerization.

crosslinking reaction by the mobility of the chromophoric part. The attached photoinitiator-double bond at the polymer chains with regard to the attached photoinitiator, gives a synergistic effect [12].

Copolymerizing the unsaturated photoinitiator into the backbone of the acrylic PSA copolymer allows crosslinking of the acrylic PSA with UV radiation, after formation of the copolymer. Preformed polymer structures that crosslink directly under the influence of UV energy require special photosensitive groups to effect network formation (Figure 6.61) (see also Chapter 5, Section II.B). The most important photoinitiators suitable for polymerization are shown in Table 6.4.

The performance qualities for example, tack, peel resistance, and shear strength of solvent-based acrylic PSAs, containing the unsaturated photoinitiators listed in Table 6.4, at a level of about 0.1 to 2.0 wt.%, are illustrated in Figure 6.62 through Figure 6.65. As illustrated by Figure 6.62, a small increase of the tack of the UV-crosslinked acrylic PSA can be achieved by the use of the unsaturated photoinitiator 2-hydroxy-1-[4-(2-acryloyloxy-ethoxy)phenyl]-2-methyl-1-propanone (ZLI 3331) copolymerizable into the acrylic polymeric chain. It can be seen from the Figure 6.63, that 4-acryloyloxy benzophenone (ABP) was the most efficient H-abstractor. An increase in concentration of the acryloyloxy-photoinitiator ABP to 0.1 wt.%, can lead to a significant effect on peel resistance. This is a consequence of the effect of "light" crosslinking

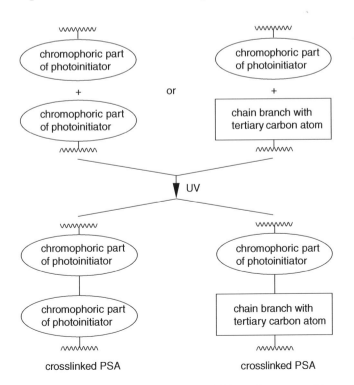

FIGURE 6.61 Photocrosslinking by use of unsaturated photoinitiators incorporated into polymer chain.

TABLE 6.4
Typical Photoinitiators Suitable for Polymerization of Acrylics

Photoinitiator	Chemical Formula	Chemical Name
PAC		Phenyl-(1-acryloyloxy)-cyclohexyl ketone
ZLI 3331		2-Hydroxy-1-[4-(2-acryloyloxyethoxy) phenyl]-2-methyl-1-propanone
ABP		4-Acryloyloxy benzophenone
AB		α-Allyl benzoine
ACDB		4-Acrylamidocarbonyl dioxy benzophenone
p-VB		p-Vinyl benzophenone
BVCN		4-Benzophenylvinyl carbonate

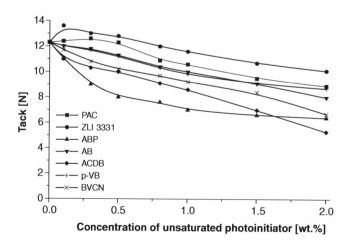

FIGURE 6.62 Tack of UV-crosslinked acrylic PSAs containing unsaturated photoinitiators.

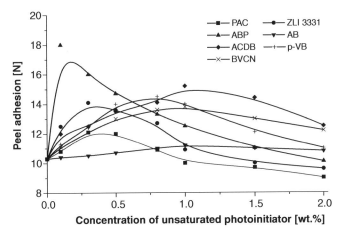

FIGURE 6.63 Peel adhesion of UV-crosslinked acrylic PSA containing unsaturated photoinitiators.

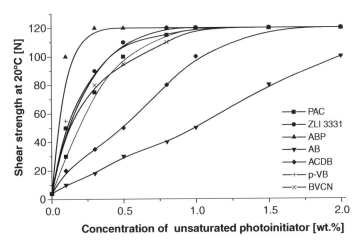

FIGURE 6.64 Shear strength at 20°C of UV-crosslinked acrylic PSA containing unsaturated photoinitiators.

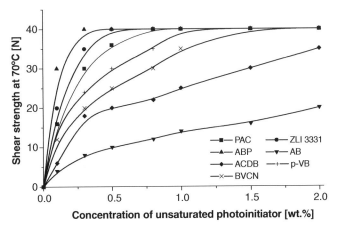

FIGURE 6.65 Shear strength at 70°C of UV-crosslinked acrylic PSA containing unsaturated photoinitiators.

using H-abstractors and small amounts of adhesive layer for peel adhesion. For higher concentrations of all unsaturated photoinitiators the peel adhesion decreases.

When acryloyloxy-photoinitiators ABP, ZLI 3331, and PAC were used in amounts between 0.1 and 0.8 wt.%, crosslinked acrylic PSA films with excellent shear strength (measured at 20 and 70°C) were obtained. The shear strength results in Figure 6.64 and Figure 6.65 show that the incorporation of unsaturated acryloyloxy-photoinitiators of type I and II into an acrylic PSA polymer chain strongly raises, unforeseen, their cohesion after UV exposure. The presence of an acryloyloxy-photoinitiators ABP, ZLI 3331, and PAC molecules in the acrylic PSA chain led to the development of good pressure-sensitive properties.

For the tested hydrogen atom abstractor acryloyloxy-photoinitiators (e.g., ABP) compared with other evaluated acryloyloxy-photoinitiators of type I, this phenomenon can be explained by a longer "living time" of formed radicals from ABP after UV exposure of the acrylic adhesive layer. In the case of H-abstractor ABP, more possibilities exist for the stability of a free biradical due to mesomeric forms (Figure 6.66).

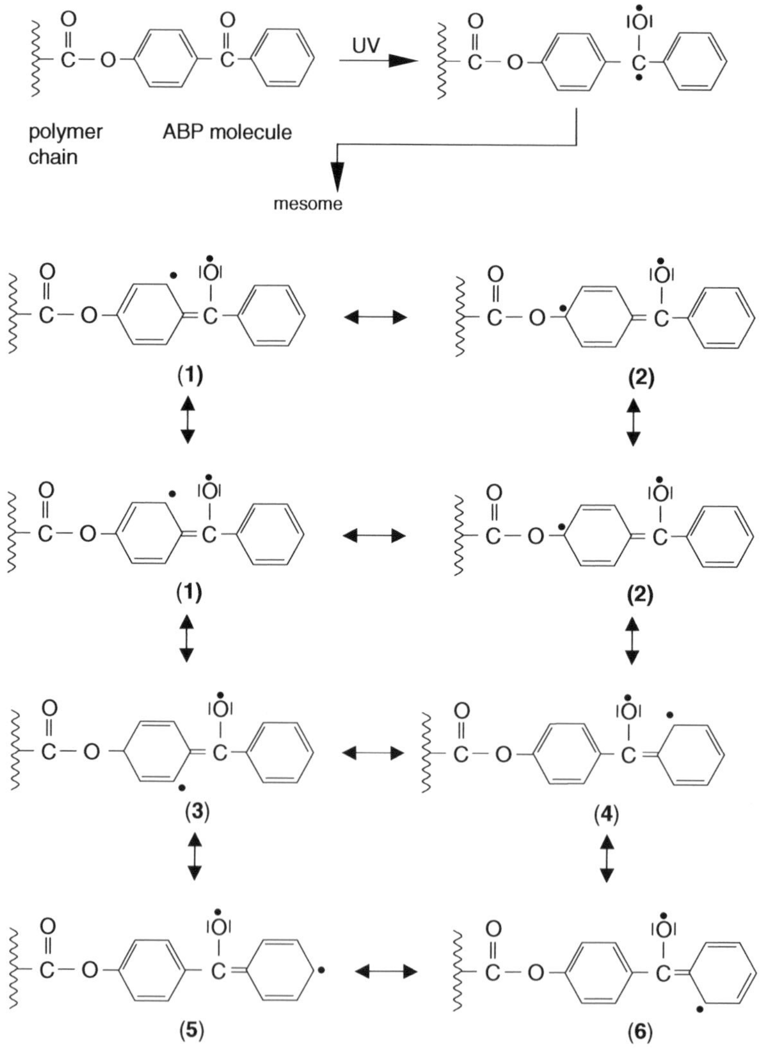

FIGURE 6.66 Mesomeric forms of radicals arisen from ABP incorporated into the polymer backbone after UV radiation.

The fewer mesomeric forms of free radicals there are, the shorter is their "living time" and with this shortening their reactivity declines. It must be noted that in the case of incorporating into the polymer backbone, H-abstractors form free effective biradicals during UV exposure, whereas by using α-cleavage photoinitiators they form only effective monoradicals for crosslinking reaction of acrylic PSAs.

Probably the decisive role, regarding the explanation of greater reactivity, is the greater efficiency of ABP under investigated hydrogen atom copolymerizable abstractors in many UV initiated crosslinking processes and is played positively in a great part by the inductive effect of the carbonyloxy ester group. In a lot of UV-crosslinked acrylic PSAs containing ABP, this effect stimulates the photocrosslinking reaction (Figure 6.67).

Another example (Figure 6.67), is valid for other photoinitiators with H-abstractor character, with a chemical structure containing alkylene or alkyleneoxy groups between ester groups and with a radical connected aromatic ring. These barrier groups are the reason for weakening the induction effect and with it the reduction of the H-abstractors crosslinking activity.

6. Crosslinking Mechanism of UV-Crosslinkable Acrylic PSAs

The crosslinking mechanism of photoreactive acrylic PSAs, containing photoreactive benzophenone derivatives incorporated into polymer backbone, has been thoroughly investigated. It is presented schematically in Ref. [75]. During UV exposure the intermolecular benzophenone derivative H-abstractor structures are excited and react by hydrogen abstraction with the neighboring tertiary carbon atom positions of the polymer side chains. Other carbon atoms in the polymer backbone do not contribute significantly to the crosslinking of UV-crosslinkable acrylic adhesives. The reason for this behavior lies on the chemical structure of the polymer chain and of the photoreactive group in the H-abstractor ABP. This phenomenon can be explained for the combination 2-EHA and ABP via possible UV initiated reaction between tertiary carbon atoms of the 2-ethylhexyl residue and the benzophenone groups, or between benzophenone groups from the H-abstractor ABP incorporated into the polymer backbone and tertiary carbon atoms in the polymer chain (Figure 6.68).

7. Photoinitiators Suitable for Addition

A new group of photoinitiators possesses a special buildup [66]. One side of the photoinitiator is a conventional chemical group, which tends to addition reaction or another kind of reaction with the carboxyl or hydroxyl groups of the polymer chain and of the other photoreactive side group (Figure 6.69).

The photoinitiator suitable for addition is generally used in amounts ranging from about 0.01 to 5.0% by weight of the total polymerizable composition. Photosensitive PSAs are known where the

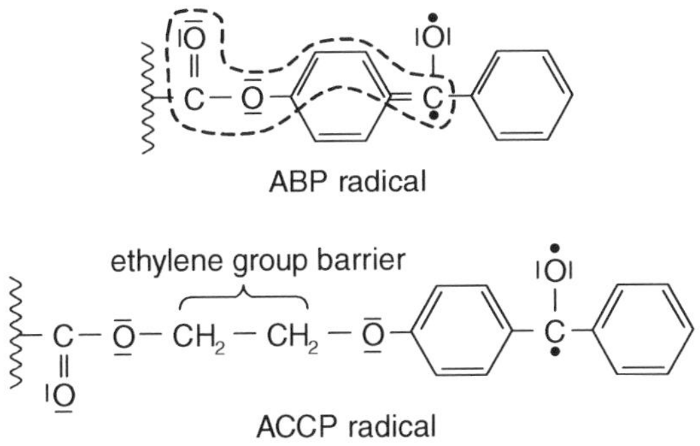

ABP radical

ethylene group barrier

ACCP radical

FIGURE 6.67 The induction effect of ester group on crosslinking efficiency by using ABP.

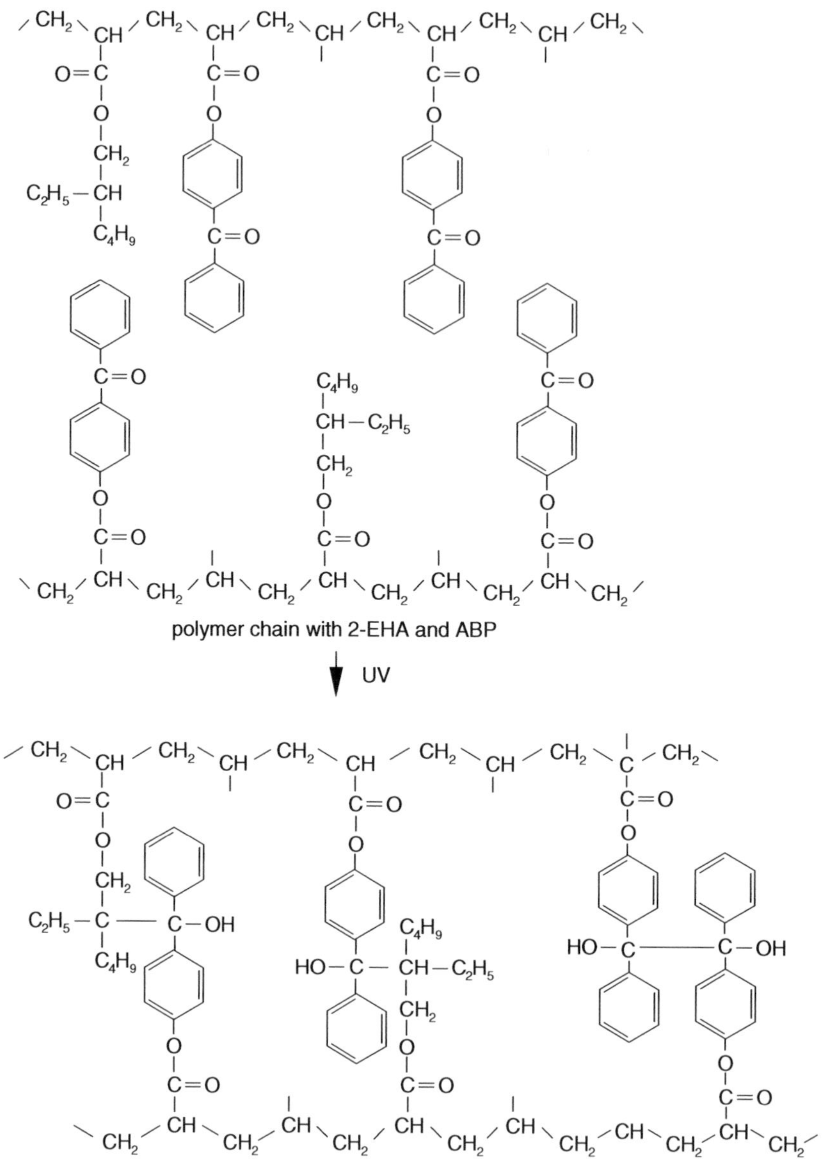

FIGURE 6.68 UV-induced crosslinking of acrylic PSAs using the copolymerizable H-abstractor ABP.

photoinitiator suitable for addition is incorporated into a UV-crosslinkable composition, which brings about the crosslinking of the adhesive with attendant enhancement of the cohesive strength of the adhesive composition. Examples of such photoreactive photoinitiators are presented in Table 6.5.

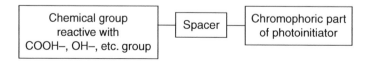

FIGURE 6.69 Structure of saturated photoinitiators suitable for addition.

TABLE 6.5
Additionable Photoinitiators

Photoinitiator	Chemical Formula	Chemical Name
PCB	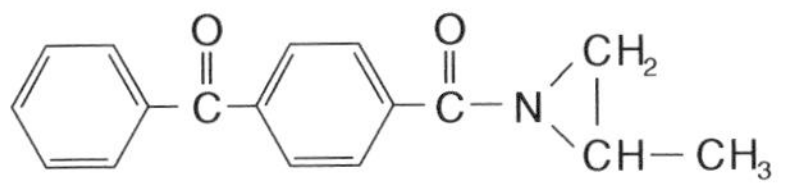	4-Propylene-iminecarbonyl benzophenone
PCF	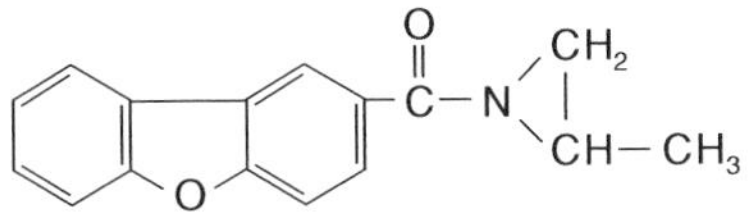	2-Propylene-iminecarbonyl-9-fluorenone
ClPCB	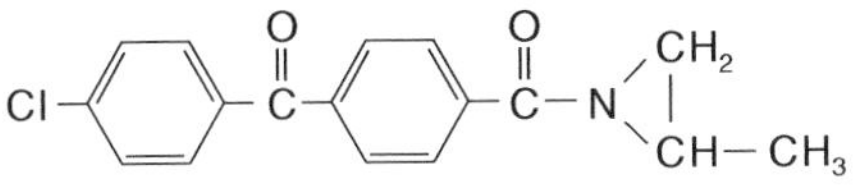	4-Chloro-4′-propyleneiminecarbonyl benzophenone
PCA	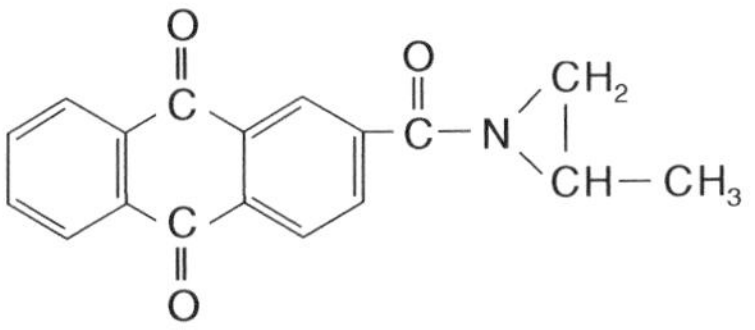	2-Propylene-iminecarbonyl antraquinone

Photoinitiators suitable for addition are taken from a new group of saturated compounds. One side of the photoinitiator's chemical structure is a conventional photoreactive group (e.g., benzophenone or antraquinone), on the other side there are groups that tend to addition reaction with carboxyl groups, such as compounds from the aziridine group [24].

The incorporation of photoinitiators carboxyl groups suitable for addition to the acrylic polymer chain runs after the following reactions (Figure 6.70).

For 4-propyleneimine-carbonyl benzophenone (PCB) and 2-propylene-iminecarbonyl antraquinone (PCA) were subjected to the UV-initiated crosslinking of acrylic PSAs containing chemical structures outlined in Figure 6.71.

Novel additionable photoinitiators for UV-curable solvent-based acrylic PSA formulations can overcome the challenge of fulfilling the demands of conventional UV-crosslinkable PSAs.

C. UV-INITIATED CROSSLINKING OF SOLVENT-BORNE ACRYLIC PSA USING UV EXCIMER LASER

The light emitted from a laser has many properties that distinguishes it from the light emitted from conventional light sources. Among the unusual properties a high degree of collimation, a narrow spectral line width, good coherence, and the ability to focus to an extremely small spot. They offer the capability for pulsed short wavelength UV lasers with very high peak power. Most of the excimer lasers involve molecules of the noble gases, which generally do not form stable chemical compounds [76].

The used xenon fluoride XeF pulse excimer laser LPX 210 which emits monochromatic UV light of 351 nm wavelengths has been characterized by the following parameters:

Maximal emitted pulse UV energy: 320 mJ
Pulse durations: 30 ns
Maximal pulses frequency: 100 Hz
Number of pulses: variable

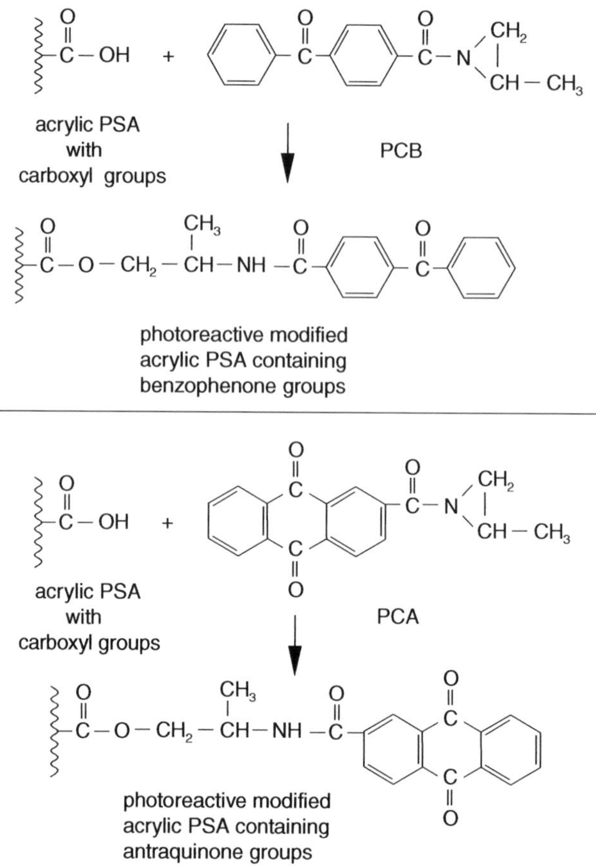

FIGURE 6.70 Transformation of carboxyl groups with photoinitiators tending to addition reaction.

The influence of most efficient unsaturated photoinitiators ABP, ZLI 3331 and phenyl-(1-acryloyloxy)-cyclohexyl ketone (PAC), on the peel resistance and shear strength of acrylic PSAs is illustrated in Figure 6.72 and Figure 6.73. The peel resistance results show for all tested unsaturated photoinitiators an influence of the type of photoinitiators and their concentration in the synthesized acrylic adhesives. UV-initiated crosslinking of acrylic PSA containing ABP proceeded quickly when an excimer laser was achieved.

Similar cohesion results (measured at 20°C, after UV excimer laser exposure), were obtained with all acrylic PSAs containing the mentioned copolymerizable photoinitiators (Figure 6.73). A faster and more complete crosslinking was achieved by setting the photoinitiators amounts above 0.1 wt.% ABP, above 0.5 wt.% ZLI 3331, and above 0.8 wt.% PAC. The observed shear strength results at 70°C have proved the maximal measured values for similar photoinitiator contents.

III. THE INFLUENCE OF SELECTED CROSSLINKING AGENTS AND PHOTOINITIATORS ON SOLVENT-BASED ACRYLIC PSAs SHRINKAGE AFTER CROSSLINKING

Shrinkage is one of the most important converting properties of pressure-sensitive products, and is described in detail in Chapter 10. Here we are discussing the crosslinking related aspects of

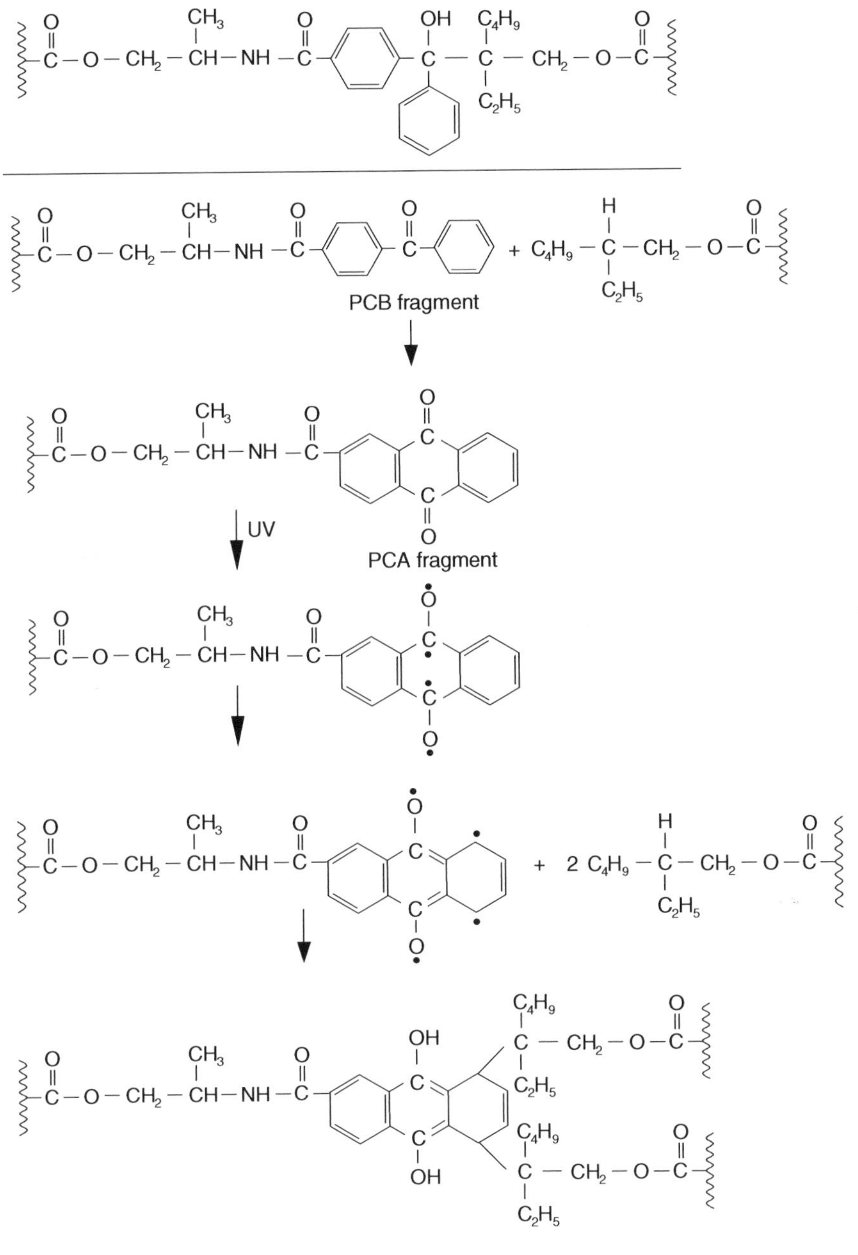

FIGURE 6.71 UV-initiated crosslinking reaction of acrylic PSAs containing additionable photoinitiators based on benzophenone or antraquinone derivatives.

shrinkage only. The influence of the external crosslinkers AlAcAc and TiAcAc concentration on shrinkage of the solvent-based acrylic PSAs containing AlAcAc and TiAcAc is illustrated in Figure 6.74. The increase of the metal chelate crosslinking agents AlAcAc and TiAcAc amounts corresponds with the decrease of shrinkage of acrylic PSAs. By using of AlAcAc in comparison to TiAcAc the best shrinkage resistance was observed. For this reason the TiAcAc in the PSA technology is becoming more and more replaced by AlAcAc as crosslinker for solvent-borne acrylic PSAs.

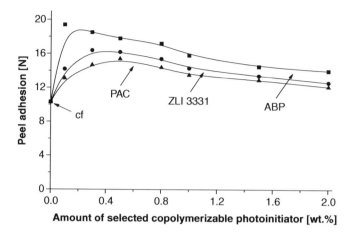

FIGURE 6.72 Peel adhesion of acrylic PSA containing selected unsaturated photoinitiators after crosslinking using UV excimer laser.

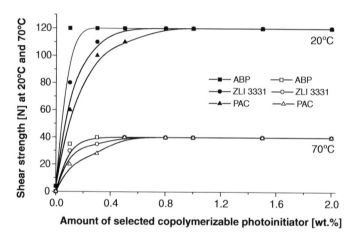

FIGURE 6.73 Shear strength of acrylic PSA containing selected unsaturated photoinitiators after crosslinking using UV excimer laser.

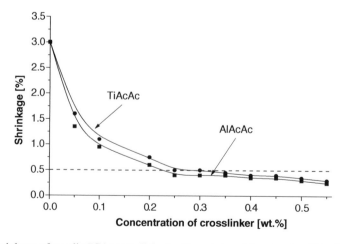

FIGURE 6.74 Shrinkage of acrylic PSAs crosslinked with metal acetylacetonates AlAcAc and TiAcAc.

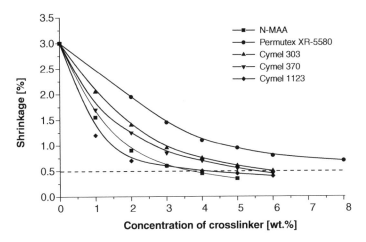

FIGURE 6.75 Shrinkage of acrylic PSAs crosslinked with *N*-MAA, Permutex XR-5580 and selected Cymel types.

The shrinkage profiles of solvent-borne acrylic PSAs crosslinked by using various crosslinking agents such as *N*-methylol acrylamide (NMA), polycarbodiimide (Permutex XR-5580), or amino resins (Cymel 303, Cymel 370, and Cymel 1123) are shown in Figure 6.75.

In Figure 6.76, the influence on shrinkage of acrylic PSAs is shown for IPDI and the propylene imine crosslinkers BPIA, and trimethylolpropane-*tris*-(*N*-methylaziridinyl)-propionate (Neocryl CX-100) at for this type of crosslinkers usual concentrations between 0.1 and 0.9 wt.%. The lowest shrinkage values are obtained using the tested propylene imines as crosslinkers in amounts above 0.2 wt.%. By using 0.3 wt.% of Neocryl CX-100 an excellent shrinkage level of 0.2% was observed. In comparison to the selected propylene imine the shrinkage, measured for acrylic PSAs, crosslinked with IPDI is entirely unsatisfactory.

Figure 6.77 illustrates the effect on shrinkage of acrylic PSAs containing varying amounts of unsaturated photoinitiators: ZLI 3331, ABP, ACDB, BVCN, and saturated additionable PCB.

The shrinkage of solvent-borne acrylic PSAs after UV exposure tremendously decrease, suggesting continued crosslinking activity under UV exposure. Above 0.3 wt.% ZLI 3331 and 0.3 wt.% ABP, the shrinkage values correspond to less than 0.45% and 0.3%. This has shown,

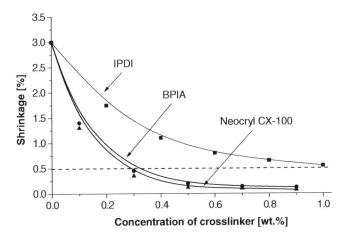

FIGURE 6.76 Shrinkage of acrylic PSAs crosslinked with selected diisocyanate and selected propyleneimines.

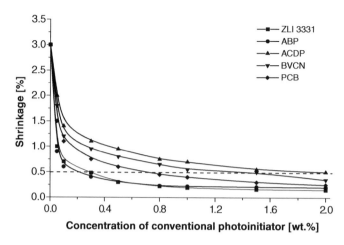

FIGURE 6.77 Shrinkage of acrylic PSAs containing selected copolymerizable and additionable photoinitiators after UV-induced crosslinking with a UV lamp.

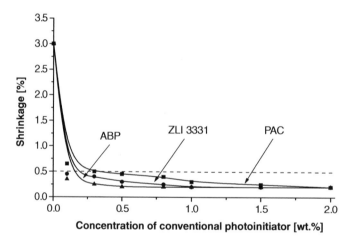

FIGURE 6.78 Shrinkage of acrylic PSAs containing copolymerizable photoinitiators after UV crosslinking by use of an excimer laser.

that ZLI 3331 and ABP are a good alternative to other unsaturated photoinitiators. The best copolymerizable photoinitiator referring to very low shrinkage of UV-crosslinked solvent-borne acrylic PSAs was ABP. At already 0.1 wt.% ABP content, an acceptable shrinkage of 0.45% was achieved.

Figure 6.78 demonstrates that the shrinkage resistance of acrylic PSAs crosslinked with a UV excimer laser has been improved in comparison to acrylic PSAs crosslinked with a UV lamp.

Above all, the lowest shrinkage values are measured for ABP. For 0.1 wt.% ABP an excellent low shrinkage of 0.35% was achieved. The including of UV crosslinking by application of an excimer laser shows "dramatically" reduced shrinkage of acrylic self-adhesives.

CONCLUSIONS

The crosslinking process of solvent-borne acrylic PSAs has recently attracted considerable attention, due to their practical and potential application in various industrial branches to comprehensive self-adhesive technical and medical products (see Chapter 11). A wide variety of different cross-linking agents, different photoinitiators, and crosslinking methods influences the most important properties of acrylic PSAs such as tack, peel resistance, shear strength, and shrinkage.

Solvent-borne acrylic PSAs with high performance: good tack, good peel resistance, high shear strength, and high shrinkage resistance can be formed by crosslinking with selected metal chelates: TiAcAc and AlAcAc. Novel products, which can be prepared by utilizing propylene imine, have been proposed as a new generation of crosslinking agents. These multifunctional propylene imine crosslinkers are used, especially in the US, for the manufacturing of self-adhesive labels, pro-tective films and tapes with very high cohesion, high tack, high peel resistance, and excellent shrinkage resistance. They are applied in small amounts as indispensable components for solvent-borne acrylic PSAs, used actually in Europe for production of double-sided mounting tapes with an excellent balance between adhesive and cohesive properties, excellent shear strength at higher temperatures, and excellent low shrinkage performances.

The technical and economical advantages of UV initiated crosslinking, as well as the increasing limitations of volatile organic compounds, forced the development of unsaturated α-cleavage photoinitiators; on the other hand, new applications required unsaturated H-abstractor photoinitia-tors with improved properties and without by-products after UV crosslinking. The UV-initiated crosslinking of acrylic PSAs containing unsaturated photoinitiators incorporated into the polymer backbone can be carried out in air. In some cases, the adhesive and cohesive performances for UV-crosslinkable acrylic PSAs are as good as for acrylic PSAs crosslinked with conventional crosslinking agents.

Especially, ABP, the first copolymerizable photoinitiator, manufactured according to own method in tonnage quantity, needs very specific crosslinking conditions to satisfactorily perform the important properties of photoreactive acrylic PSAs after UV induced crosslinking. The new solvent-borne UV-crosslinkable acrylic PSAs containing small amounts of ABP also exhibit excellent low shrinkage after UV-exposure, unlimited pot-life and are stable at high temp-erature. Production rates can, therefore, be increased, and the manufacture costs of UV-cured PSPs can be minimized. The UV-crosslinkable acrylic PSAs containing ABP are used for production of a wide range of self-adhesive medical products (see Chapter 11).

A new generation of patented additionable photoinitiators can be used for the synthesis of UV-crosslinkable acrylic self-adhesives. These acrylic PSAs show good tack, good peel resistance, very high shear strength at 20°C, moderate shear strength at 70°C, and low shrinkage after UV-exposure.

The advantage of using an excimer laser in the UV-A area as a monochromatic UV source, instead of a spectral UV lamp, for UV-initiated crosslinking of solvent-borne photoreactive acrylic PSAs, is that it is a very efficient crosslinking method in PSP technology for manufacturing of self-adhesive products with high performance. Acrylic-based PSA systems, which are cross-linked using UV excimer laser, offer the formulator a wide freedom from limitations.

REFERENCES

1. I. Benedek and L.J. Heymans, *Pressure-Sensitive Adhesives Technology*, Marcel Dekker Inc., New York, 1997.
2. I. Benedek, *Developments in Pressure-Sensitive Products*, Marcel Dekker Inc., New York, 1999.
3. I. Benedek, *Pressure-Sensitive Formulation*, VSP, Utrecht, 2000.
4. R. Milker and Z. Czech, *Vernici Eur. Coat.*, 2, 49, 2002.
5. M. Ooka and H. Ozawa, *Progr. Org. Coat.*, 23, 325, 1994.
6. Z. Czech, Ed., *Vernetzung von Haftklebstoffen auf Polyacrylatbasis*, Politechnika Szczecinska, Szczecin, 1999.
7. I. Benedek, *Pressure-Sensitive Formulation*, VSP, Utrecht, 2000, chap. 3.
8. R. Milker and Z. Czech, 9. Münchener Klebstoff- und Veredelungsseminar, München, Germany, 1984.

9. S.J. Chen, DE Patent 38 10 307, 1990.

10. G. Feld and C. Cowe, *The Organic Chemistry of Titanium*, Butterworth, London, 1965.

11. C.C. Anderson and R. Maska, U.S. Patent 3,769,254, 1973.

12. L.W. McKenna and W. Mass, U.S. Patent 3.886.126, 1975.

13. G.C. Michael, DE 2,649,544, 1976.

14. R. Boyack and R. Rao, WO 86/04591, 1986.

15. W. Brown and T. Waterborne, Higher-Solids and Powder Coatings Symposium, New Orleans, USA, 1994.

16. G. Canty and R. Jones, EP 0 206 669, 1986.

17. A.J. Hoffman, U.S. Patent 3,393,184, 1968.

18. Technical Data from Arsynco, USA, 1988.

19. W. Traynor and M. Martin, EP 0 286 420, 1988.

20. M. Iqbal, U.S. Patent 5,208,092, 1990.

21. D.W. Kenneth and L.A. Meixner, U.S. Patent 5.296.277, 1992.

22. J.W. Rehfuss, EP 0 544 183, 1992.

23. R. Milker and Z. Czech, *J. Appl. Polym. Sci.*, 86, 1354, 2002.

24. Z. Czech and F. Herrmann, WO 93/20112, 1993.

25. U. Hofmockel, 17. Münchener, Klebstoff- und Veredelungsseminar, München, Germany, 1992.

26. M. Köhler, *Merck Kontakte*, 3, 115, 1979.

27. S.P. Pappas, *J. Am. Chem. Soc.*, 95, 484, 1973.

28. E. Smit, 11. Cowise-Symposium, PSA and Adhesive Coating, Amsterdam, Holland, 1996.

29. M. Köhler, Radure '86, Baltimore, USA, 1986.

30. R.W. Oemke, AFERA-Jahrestagung, Amsterdam, Holland, 1991.

31. H. Brackemann, *Chem. Ing. Tech.*, 7, 596, 1987.

32. B. Wen Chiao, EP 0 213 317, 1986.

33. E. Levina and H.L. Wolfson, U.S. Patent 5,214,094, 1991.

34. S.D. Tobing and A. Klein, *Synthesis and Structure Property Studies in Acrylic Pressure-Sensitive Adhesives*, Proceedings of the 24th Annual Meeting of Adh. Soc., Feb. 25–28, 2001, Williamsburg, VA, USA, p. 131.

35. W.B. Armopur, *N*-Methylolacrylamide-Vinylacetate Copolymer Emulsions, U.S. Patent 3,041,301, 1962.

36. L.W. McKenna, U.S. Patent 3,886,126, 1975.

37. K. Flatau and H. Musso, *Angew. Chem.*, 82, 390, 1970.

38. J.P. Colmann, *Angew. Chem.*, 77, 154, 1965.

39. B. Bock, *Angew. Chem.*, 84, 239, 1971.

40. C. Simionescu, N. Asandei, I. Benedek, and C. Ungurenaşu, *Eur. Polym. J.*, 5, 449, 1969.

41. C. Simionescu, N. Asandei, I. Benedek, and C. Ungurenaşu, *Eur. Polym. J.*, 6, 925, 1969.

42. I. Benedek, N. Asandei, and C. Simionescu, *Eur. Polym. J.*, 7, 995, 1971.

43. C. Simionescu, I. Benedek, and N. Asandei, *Eur. Polym. J.*, 7, 1549, 1971.

44. I. Benedek et al., Romanian Patent, 56852/31.03.1970.

45. I. Benedek et al., Romanian Patent, 54,213/1969.

46. I. Benedek et al., Romanian Patent, 59,125/14.10.1971.

47. I. Benedek et al., Romanian Patent, 59,674/07.01.1975.

48. Z. Czech, *Polimery*, 7/8, 561, 2003.

49. *Technische Informationsblätter*, Nr. 5 der Fa. Bayer, Germany.

50. H. Perrey and M. Dollhausen, EP 0 071 142, 1984.

51. A. Everaerts and J. Malmer, WO 93/13148, 1993.

52. J. Dormish, *Adhes. Age*, 2, 17, 1996.

53. B. Suthar, *Polym. Internat.*, 34, 267, 1994.

54. G.R. Canty and E. Jones, EP 0 206 669, 1986.

55. J.P. Kealy and R. Zenk, EP 0 100 146, 1983.

56. G. Canty and R.E. Jones, EP 0 206 669, 1986.

57. D.H. Klein and R.H. Cramm, U.S. Patent 3,993,716, 1970.

58. Technical Data Polyfunctional Aziridines, Flevo Chemie, Holland, 1999.

59. W. Kleeberg and H. Ahne, DE 23 08 576, 1973.

60. Z. Czech, *Polym. Int. Sci.*, 52, 347, 2003.
61. Technical Booklet, *Resimene High Solids Melamine Crosslinkers*, Monsanto, USA, 1995.
62. Technikal Booklet, *Amino Resins*, BASF, Ludwigshafen, Germany, 1999.
63. R. Milker and Z. Czech, 23. Münchener Klebstoff- und Veredelungsseminar, München, Germany, 1998.
64. Technical Booklet, *Carboset Resins*, BF Goodrich, USA, 1981.
65. Z. Czech, W. Blum, and F. Herrmann, WO 94/14853, 1993.
66. Z. Czech, DE 44 33 290, 1994.
67. Z. Czech, DE 44 47 615, 1994.
68. Z. Czech, DE 195 01 025, 1995.
69. Z. Czech, DE 195 01 024, 1995.
70. I. Benedek, *Pressure-Sensitive Formulation*, VSP, Utrecht, 2000, chap. 4.
71. Z. Czech, *J. Appl. Polym. Sci.*, 87, 182, 2003.
72. L.M. Clemens and J.A. Martens, U.S. Patent, 4,181,752, 1980.
73. G. Vesley, WO 93/13149, 1992.
74. A. Boettcher and G. Rehmer, DE 38 44 444, 1988.
75. G.F. Vesley, U.S. Patent 4,391,687, 1983.
76. *Handbook of Laser Science and Technology*, Vol. 2, CRC Press, Boca Raton, 1985.

7 Adhesive Properties of Pressure-Sensitive Products

István Benedek

CONTENTS

In principle, pressure-sensitive products (PSPs) are characterized by their adhesive, converting and application performance characteristics. Table 7.1 summarizes the main performance characteristic of the PSPs and their components. It is evident, that for different pressure-sensitive products the converting and end-use behavior have a specific character. It is astonishing that the general adhesive properties of various product classes must be examined as specific properties also. For different PSPs the level of the instantaneous and dwell time dependent peel force, the permanent or removable character of the bond and/or its deposit free destruction are of quite different degrees of importance and are evaluated in a different way. For labels and tapes, instantaneous adhesion, tack and peel are the most important adhesive properties. For the major part of these products, removability is a special characteristic. For protective films it is essential.

With the development of the applications of PSPs the requirements for different product classes became more complex. Asthetic and chemical resistance considerations imposed the use of plastics as solid-state carrier materials. Environmental considerations forced changes in film carrier composition. Polyvinyl chloride (PVC) has been replaced, mainly by polyolefins. Recycling required improving of paper quality too. Short fiber papers were introduced. Due to the development of plastics, especially of thin polyolefin films, for the packaging industry, a need developed for adhesion and anchorage of PSPs on nonpolar, nongeometric, and nonstatic items. The broad range of new, unknown substrates requires the pressure-sensitive adhesion to be substrate-independent. For different applications, peel regulation should ensure instantaneous permanent adhesion (e.g., permanent labels and tapes) as well as a controllable permanent adhesion (e.g., repositionable labels) or nonpermanent adhesion (e.g., removable labels, tapes, and protective films) (Figure 7.1).

I. DEFINITION AND CHARACTERIZATION OF ADHESIVE PROPERTIES

Adhesion onto the substrate surface under slight or high pressure is a general characteristic of PSPs. It is achieved with products having quite different buildup, under different application conditions for various end-use times, and different deapplication conditions and methods. Generally, adhesion is evaluated in terms of adhesive properties. The adhesive properties are characterized by general adhesive performance characteristics and special product dependent properties.

A. GENERAL ADHESIVE PERFORMANCE

The characterization of the adhesion of pressure-sensitive adhesives (PSAs) using common, standard adhesive properties is described in detail by Benedek and Heymans [1] and Benedek [2]. This

---

TABLE 7.1
Major Performance Characteristics of PSPs

Adhesive properties
   Adhesion: Tack, peel resistance
   Cohesion: Shear resistance

Converting properties
   Coating properties: Printability, laminability
   Confectioning properties: Cuttability, die-cutability

End-use properties
   Aging resistance
   Chemical and water resistance
   Bonding/debonding properties
   Processibility

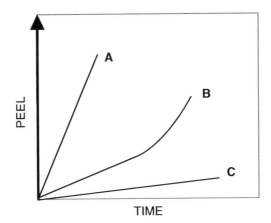

FIGURE 7.1 Schematic presentation of the adhesion buildup over time for the main types of PSPs. (A) Permanent PSP; (B) repositionable PSP; (C) removable PSP.

chapter reveals the special features of the adhesion, related to the pressure-sensitive products. As is known, adhesive properties are defined by tack, peel, and shear resistance of the joint. Tack is the ability to bond instantaneously. Peel is the strength of the joint, tested by debonding, using a stress direction other than that of the laminate. Shear is the internal cohesion of the adhesive tested by shear stresses parallel to the joint direction. Tack, peel, and shear are interdependent. They are controlled by the adhesion–cohesion balance. This balance depends on the material and application or test conditions. Theoretically, it is a function of only the material. In practice, because of the non-Newtonian flow of the macromolecular compounds used as chemical bases for PSAs, this balance depends on time and temperature, that is, on actual test and application conditions (see also Chapter 3, Section I). In the end-use of the product, the adhesion–cohesion balance is a function of the rate of product application and deapplication. Moreover, because of the influence of the solid-state aminate components (substrate and carrier) on the rheology of PSA and vice versa, the adhesion–cohesion balance depends on the buildup, on the manufacturing parameters of the PSP and also on the substrate. According to an old definition [3]: "pressure-sensitive adhesives give instantaneous, mechanically removable, nonspecific adhesion, generally without damaging the substrate." According to Chang [4] PSAs are adhesives "which require no activation by water, solvent or heat … they have a sufficient holding and elastic nature." Except for labels, there are numerous pressure-sensitive products which do not give sufficient instantaneous adhesion. Such products do not have balanced adhesive properties and may require activation. Nonspecific adhesion can be given by PSAs applied to one side (e.g., labels), double side (carrier and substrate) coated PSA, (e.g., envelopes) or double side (carrier and carrier) coated cold seal adhesives (Figure 7.2). Pressure-sensitive cold sealing adhesives have been manufactured from tackified ethylene–vinyl acetate (EVAc) copolymers with 55–85% ethylene, tackifiers and special microsphere fillers [5]. It is evident that auto-adhesion as a special case of adhesion presents a simplified model for the study of pressure sensitivity. For such a model contact building is controlled by diffusion of the same material on both sides of the interface. Physical adhesion can develop from adequate wetting at interface or by diffusion interpenetration of segments across the interface when this is thermodynamically possible. For noninteracting polymer–polymer pairs the only driving force for diffusion is entropic in origin and very small. According to Dormann [6], cold seal is "the use of a soft adhesive sealing medium with or without temperature and pressure." Auto-adhesive tack is "the ability of two similar materials to resist separation after their surfaces have been brought into contact for a short time under a light pressure" [7]. Auto-contact (self-adhesion) is characteristic for contact adhesives. Common contact adhesives are applied under pressure. Lap shear adhesion of acrylic (AC) contact adhesives increases with the pressure [8].

Pressure

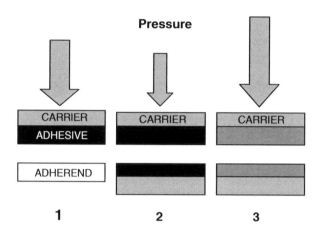

FIGURE 7.2 Pressure-dependent contact buildup for different products. (1) Bonding of PSA-coated PSP on different adherend; (2) bonding of PSA-coated PSP on itself; (3) bonding of cold sealing adhesive-coated carrier on itself.

A common PSA does not have to bond on itself. It has to be xenophilic. Taking into account the above definition current (one-side coated) PSA applications must have another bonding mechanism. A more complex adhesion mechanism is given by macromolecular compounds that exhibit diffusion but are chemically different. Studying the fracture energy of polystyrene (PS) and poly-monomethacrylate (PMMA) adhesive joints attained by contact molding of polystyrene and PMMA sheets, it has been stated that in such cases the van der Waals interaction was sufficient to create strong adhesive joints [9]. Such cases do not take into account the influence of the surface and bulk characteristics of the solid-state components. According to another definition pressure-sensitive adhesives are coated "onto rigid (relative to the adhesive, N.A.) backings" [10]. This statement excludes auto-adhesive tack and self-adhesive soft carriers.

The definitions given for pressure-sensitive adhesives illustrate the fact that "classic" pressure-sensitive products (e.g., labels) having excellent instantaneous tack or peel on other substrates can be considered as a special case only. Certain PSAs have to be applied via auto-adhesion and under pressure. On the other hand, pressure-sensitive products are not the same as pressure-sensitive adhesives. Pressure-sensitivity of PSPs is the result of a complex structure with or without a PSA. Therefore, its characterization is more difficult. The standard test methods for characterization of the adhesive properties have been developed for PSAs. They are discussed in detail in [11]. In certain cases such methods do not fulfill the practical requirements for PSPs. Therefore, in this chapter we examine whether it is possible to evaluate the adhesive performance characteristics of PSPs with unbalanced adhesive–cohesive properties and with or without PSA, and if so, how.

As mentioned earlier, tack, peel and shear resistance are the main parameters of the adhesion (Figure 7.3). Normally, tacky pressure-sensitive adhesives are used in the manufacture of a variety of articles such as adhesive labels, tapes and other materials that are intended to be easily attached to another substrate by the application of a slight pressure alone. Such PSAs are adhesives which do not require specific activation to render them adhesive.

In practice, depending on the end-use requirements, that is, high application speed (e.g., label) or durable high strength joint (e.g., mounting tape, transfer tape) and removable adhesion on large surface items (e.g., protective films) the balance of the tack, peel and shear may be different (see Figure 7.3). Problems may also arise from the evaluation of adhesives having balanced tack-peel-shear. The definition of an "adequate" peel, tack or shear resistance is relative and the test methods used are more or less versatile for a given PSP class.

According to Pasquali [12], common PSAs exhibit a peel level of 50–800 N/m (g/cm). As is known, reducing the peel value makes PSP application difficult. Adhesives with peel values lower than 150 N/m gives almost no adhesion (except for some special adherends such as marble, glass or

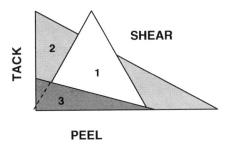

FIGURE 7.3 The main parameters of the adhesion and their balance for different PSPs. (1) Label; (2) tape; (3) protective film.

metal). On the other hand, it is also difficult to delaminate large, adhesive-coated surfaces. Products having a common, medium peel value of 350 N/m would require 350 N force for delamination of a sample 1 m wide. Highly aggressive PSAs are those having peel adhesion values in the range of 60–100 N/100 mm (15–25 N/25 mm). Tackified acrylic block copolymers for hot-melts show excellent peel values of 17–30 N/25 mm [13]. The peel of thermoplastic elastomer (TPE)-based removable formulations should be of 2.5–40 oz, and the initial peel 1.0–16 oz (using 1 in. × 6 in. strips on stainless steel). Values above these limits can result in paper tear. Tackified formulations based on amorphous polypropylene (PP) exhibit peel resistance of 31 N/in. (Pressure-Sensitive Tape Council (PSTC-1), steel), loop tack of 25 N/in.[2] and shear resistance of 2.5 h (PSTC-7) [14]. Adequate selection of the comonomers and polymerization conditions led to the development of olefin polymers which are pressure-sensitive in the neat form. Such polymers possess a peel value of 2.5 N/in. (PTSC-1) quick stick of 1.2 g/in. (PTSC-5) and shear resistance of 20 h [15]. Preferred skin adhesives exhibit initial peel values of 50–100 g and final peel values (after 48 h) of 150–300 g/ 25 mm [16]. A cold stretchable self-adhesive film (SAF) based on a tackified ethylene-α-olefin copolymer displays an adhesive force of at least 65 g (ASTM 3354-74) [17]. Tack values of 16–24 N/ 25 mm are acceptable for use on diapers [18].

For carboxylated butadiene rubber (CSBR), holding time values of 116 h at 75°F and 120°F (PSTC-7) are, according to [19], excellent. Mueller and Tuerk [20] proposed a water-based (WB) acrylic formulation having physically or chemically crosslinkable comonomers like acrylo- or methacrylonitryl, acrylic or methacrylic acid, N-methylolacrylamide and plasticizer (10–30% wt./wt.). Such formulations give peel resistance of 1–6 N/20 mm on different substrates (at debonding speed of 75 mm/min). An acrylate-based PSA, crosslinked between 60 and 100°C using a zinc chloride catalyst, shows holding power of 20 h at 150°C compared to common uncrosslinked, rubber-based PSA (with holding power values of some minutes at 100°C) [21]. According to Midgley [22], an arbitrary shear failure time of 20 h (1.25 cm overlap, 2.5 cm width, 1 kg at room temperature) was adopted as adequate value in practice. Shear values of 12–680 min (on steel, according to PSTC) are acceptable only for labels. Packaging tapes need higher shear [23]. A special crosslinkable, carboxylated, water-based acrylic displays room temperature shear values of 5000–60,000 min at 8.8 psi (0.5 × 0.5 in., 1000 g). A shear adhesion failure temperature (SAFT) value of 240°C has been obtained which compares with the shear performance of the highest quality commercial solvent-based acrylics [24]. Unfortunately, these values were obtained without tackifier and the corresponding peel and tack values are very low (180° peel, PSTC-1, on polyethylene terephtalate (PET), 1.1 lb/in., quick stick 0.7 lb/in.). Formulations with acrylic PSAs for permanent label application show (min) shear values of 8 min (tackified) to 380 min (nontackified); for tackified styrene–butadiene–rubber (SBR) values of 60–100 min are obtained [25]. As illustrated by the above test values, although standard methods exist, such values have to be evaluated always with respect to the specific formulation.

Different tack methods are adequate for different PSP [26]. With the probe tack test the load is regulated, as in the lamination of protective films. Polyken tack is less strongly affected by the resin softening point (it is applied under load!) during tackifying. Loop tack simulates the real application conditions of labels, which are blown onto a surface by using air pressure. Rolling ball (RB) tack is more complex. It is friction-related, but the correlation between polymer friction and its viscoelastic properties by bonding/debonding are not clear.

A well-found study of the dependence of the adhesive performances on the experimental conditions has been carried out by Bonneau and Baumassy [27], who used different water-based tackifier dispersions (Dermulsene, DRT). As stated previously, the optimum resin concentration for the best peel value depends on the peel angle. For $90°$ peel on polyethylene (PE) the highest value is obtained at the highest resin loading (ca. 35%), for $180°$ peel, best values are obtained with less resin (30 phr). This means, the common peel test methods used for tapes ($90°$ peel) and labels ($180°$ peel) gives quite a different evaluation of the same formulation.

Tack and peel are characteristics used for evaluation of instantaneous or short time behavior. They can be estimated or exactly measured and if necessary rapidly improved. Shear measurement is a possibility to test the internal cohesion of the adhesive. Quick stick and high temperature shear are required for PSAs for tapes. For labels creep is not so important, because high load bearing capability is not required. Except for some special applications for PSPs where a high level of shear resistance is necessary, in most cases the shear resistance is taken into account as a cohesion-dependent component of the adhesion–cohesion balance only, which can be evaluated more easily, than other cohesion-dependent converting or application properties, such as cuttability and die-cuttability.

Cuttability and die-cuttability depends on the adhesive also. A mathematical correlation between the various parameters influencing cuttability is given in [28]. Cohesive, tack-free elastomers or duromers can be more easily cut than tacky, low viscosity materials. Therefore, one can suppose that the cohesion of the adhesive and its shear resistance allows characterization of cuttability. Practically, shear tests and cutting or die-cutting operations are carried out at quite different stress rates and temperatures. Supplemental problems arise from the lack of correlation between the shear values measured at room temperature and cuttability. Room temperature tests are too slow and the dispersion of the test result is too high. Hot shear measurements give better results, but there is no linear correlation between the measured values and cuttability. The situation is more complex for water-based dispersion where moisture content of the laminate (degree of drying) influences the shear resistance also, or for thermally crosslinking compositions, where hot shear results are denatured by the crosslinking of the product. Generally, high surfactant concentrations in the formulation lowers the peel and in many cases the shear also. Excess moisture in the carrier paper can be troublesome and it is preferred that the papers have a moisture content per weight of about 4% or less [26]. The interdependence of cuttability parameters with the adhesive properties was discussed in detail in [28].

As discussed in Chapter 3, the T_g and the modulus are base characteristics that make it possible to predict the adhesive performance. Equations can be derived which express the effect of molecular weight (MW), plasticization, degree of crosslinking and copolymerization on the second-order transition temperature. Such equations may be reduced to form equations derivable from free volume theory [29] (see also Chapter 4). Although the possibility of computer-based tape and label design has been demonstrated by Kaelble [30] two decades ago, the evaluation of the adhesive performance needs a thorough experimental study. Dynamic mechanical analysis (DMA) measurements are useful for the base elastomer, but can be denatured by the composite structure of the adhesive and adhesive–carrier interactions. Such investigations were proposed in the 1960s by Gramberg [31]. Later the value of the storage modulus at tan δ peak as a function of the tan δ peak value was successfully used by Bamborough and Dunckley [32] to predict the application window for bookbinding hot-melts. Bamborough [33] proved the applicability of the application window values stated by Chu [34] for solvent-based, hot-melt and water-based PSAs. It has been shown that for natural rubber (NR) and styrene–butadiene copolymers the addition of a compatible resin allows us to bring the system's values into the application window. For styrene block copolymers (SBCs) at room temperature the loss tangent peak temperature is lower than that of natural rubber or styrene–butane–rubber (SBR) and the storage modulus is greater by two decades than that of natural rubber (see also Chapter 3). Therefore, the concentration required to bring the system in the application window is much higher, but possible if an oil is added as well. Unfortunately, for water-based PSAs, such simple prediction of the adhesive performances is impossible. As stated by Bamborough [33], the loss tangent peak temperature (i.e., the interdependence of storage modulus and loss tangent peak temperature) for water-based adhesive systems is not a reliable predictor of PSA performances (see also Chapter 3). As discussed in [35], advances in contact physics allowed a better correlation of the macroscopic bonding–debonding with the rheological parameters and macromolecular characteristics of PSAs; unfortunately, these investigations are focused on the simplest adhesives, without taking into account the mutual interactions in pressure-sensitive products as systems. On the other hand, as illustrated by the development of pressure-sensitive hydrogels (see Chapter 4 and Chapter 9) a new base theory of pressure-sensitivity is required.

B. SPECIAL ADHESIVE PERFORMANCES

Table 7.2 lists the main adhesive characteristics and (standard or special) test methods for evaluating them. Examination of these methods on the basis of the industrial experience shows, that the actual definition of test methods for the adhesive properties (developed originally for labels and tapes produced according to the classical PSA coating technology) are not adequate for the whole PSP domain. The standard adhesive characteristics describes the adhesive performances of PSPs having a balance between tack or peel and cohesion. Labels, tapes, and some protective films belong to this group of products. On the other hand there are some special application fields, where bonding and debonding are influenced more by the application/deapplication conditions and the standard test methods cannot characterize the real behavior of the PSP. In such cases it would be better to describe the end-use tack and the end-use peel, which are complex functions of the application conditions.

1. Application Peel and Tack

As discussed earlier, an ideal PSA, by definition gives instantaneous tack and peel on a rigid substrate. Certain pressure-sensitive products must adhere to soft substrates and provide enough adhesion after a given time under special application conditions. Under such conditions "dwell time" means time of forced contact and strongly depends on the coating weight. The importance

TABLE 7.2
Main Adhesive Properties and Test Methods for Their Evaluation

Characteristic	Principle of Measurement	Method
Adhesion tack	Rolling adherend	Rolling substrate (rolling ball, rolling cylinder) or rotating substrate (toothed wheel)
	Peel measurement	Flexible carrier (loop tack) or rigid carrier (Polyken tack)
Bond strength	Peel resistance	Standard carrier and substrate quality (adhesion peel, self-adhesion peel); standard peel angle ($90°$ and $180°$) and rate
Cohesion	Shear resistance	Static, standard debonding angle, standard debonding force, and standard contact surface; room temperature and high temperature; dynamic

of the peel buildup for the strength of the adhesive joint is well known. Generally, such an increase in the adhesion occurs under well-defined static (storage) conditions having time as sole parameter. For certain products (at least in the first period of this time) supplemental parameters can also enchance contact and peel buildup.

a. The Influence of the Coating Weight

The role of the coating weight for adhesive coated PSP is discussed in detail in [1,2,36]. It has been stated that adhesion generally increases with coating weight and a critical coating weight is necessary to achieve adequate adhesive properties. On the other hand, an excessive coating weight can negatively influence removability, cuttability and shear resistance. As discussed in [2], the adhesive properties of a PSA depends on the coating weight. Peel resistance increases with the coating weight (Figure 7.4) up to a certain value. The form of this dependence is a function of the adhesive, substrate and application conditions.

As presented in Figure 7.5 in a simplified form, a plot of peel vs. coating weight can be considered as a diagram having a point of inflexion. The dependence of peel resistance on the coating weight before the point of inflexion can be considered as a linear one (Figure 7.5).

In a first approximation, in the linear domain of coating weight values (OL) the peel resistance (P) increases with coating weight (C_w). This increase can be characterized by the angle α of the plot and/or the critical coating weight (C_{wcr}), according to the correlations:

$$P = f(C_w) \tag{7.1}$$

$$P = \alpha C_w \tag{7.2}$$

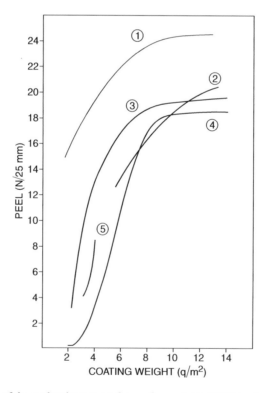

FIGURE 7.4 Dependence of the peel resistance on the coating weight. (1) Water-based, tackified acrylic PSA, 180° peel on glass; (2) water-based, tackified acrylic PSA, 180° peel on glass; (3) water-based, tackified acrylic PSA, 180° peel on polyethylene; (4) water-based, tackified acrylic PSA, 180° peel on polyethylene; (5) tackified CSBR dispersion, 180° peel on glass.

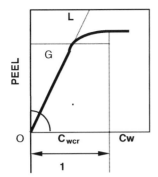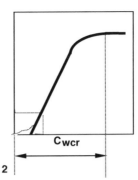

FIGURE 7.5 Schematic presentation of peel dependence on coating weight. (1) Dependence function composed of a coating weight dependent and coating weight independent part at high coating weight values; (2) dependence function composed of a coating weight "independent" part at low coating weight values and a coating weight dependent part at high coating weight values.

Generally α depends on the adhesive nature and substrate and application conditions:

$$\alpha = f(\text{adhesive nature, substrate surface, application conditions}). \tag{7.3}$$

The application conditions are characterized by time and temperature, therefore:

$$\alpha = f(\text{adhesive, carrier, substrate, time/temperature}) \tag{7.4}$$

For an ideal adhesive, α should be zero, that is, the adhesion should not depend on the coating weight. On the other hand, an examination of Figure 7.4 and Figure 7.5 shows that the peel resistance attains a measurable value only over a certain coating weight. Contact buildup needs adhesive flow. For given application conditions surface roughness, carrier conformability and adhesive cold flow are the main parameters of contact buildup. Therefore, the dependence of the peel force on the coating weight can be described by a simplified plot, with the linear part of the dependence shifted from the origin to higher coating weight values (Figure 7.5), according to:

$$P = \alpha C_w - \beta \tag{7.5}$$

where β is a conformation factor. The value of this factor is a function of the parameters influencing the contact buildup between the adhesive and substrate surface and the so-called "free flow" region of the PSA. This means, for too low a coating weight or too rough a surface (see Application of Protective Films, p. 574) no adhesion builds up. It is evident that rough, nonpolar surfaces and harder adhesives on rigid nonconformable carrier materials lead to higher values of β; in such cases the critical coating weight is higher (see Figure 7.4 also). It is well known from industrial practice that the coating weight depends on the mechanical characteristics of the carrier (Table 7.3). Such results can be explained with the recent theory of Creton et al. [37]; that is, supposing that full contact buildup is referred to as self-healing and in general the self-healing time depends on the viscoelastic properties of the material, the aspect ratio, the spacing of the asperities, and the surface properties.

As can be seen from Table 7.3, higher carrier thickness requires the use of higher coating weights to achieve better contact buildup. Therefore, it can be supposed that:

$$\beta = f(\text{adhesive, carrier, substrate, time/temperature}) \tag{7.6}$$

TABLE 7.3
Dependence of the Coating Weight Values (g/m²) on Carrier Thickness

Code	Carrier Thickness (μm)							
	50		70		80		100	
	TH	PR	TH	PR	TH	PR	TH	PR
1	2.80	3.60	2.40	2.70	—	—	—	—
2	2.80	2.86	2.80	2.83	—	—	—	—
3	4.50	3.97	—	—	4.50	4.13	—	—
4	4.50	4.80	—	—	—	—	4.50	3.86
5	4.50	3.95	—	—	—	—	4.50	3.84
6	4.50	4.10	—	—	—	—	4.50	4.02

Note: TH represents theoretical values; PR represents measured values.

Figure 7.4 illustrates that a permanent acrylic PSA is coated for labels with a coating weight situated above the critical value. Table 7.4 presents the coating weight domains for the main PSP classes. As can be seen from this table, tapes may have very high coating weight values. On the other hand, separation and protective films may have coating weight values that are lower than common coating weight for lacquers or printing inks. It is evident that the coating weight values strongly depends on the type of the adhesive. As stated in [38], solvent-based acrylics are coated for different product classes with the following coating weight values: 5–10 g/m² for protective films; 25–40 g/m² for film-based tapes; 50–100 g/m² for foam tapes; and 20–30 g/m² for labels and tags. For labels, peel is regulated by formulation, not by coating weight. Labels can be considered as a special case of PSPs where coating weights are relatively high (above the critical domain, see Figure 7.6).

Higher coating weights are used for tapes. These weights are needed for high tack and instantaneous peel for adhesives having high cohesion. The high coating weight of tapes can be apparently reduced for applications where conformability and cohesion are required. Such apparent reduction is made by using special fillers (see later). For an adequate peel value a high coating weight (above the critical value) is needed. The critical coating weight value depends on adhesive and surface qualities and application conditions. For rough surfaces, high coating weight values are generally needed. Although these values are necessary for rapid peel buildup, later they may disturb the application of the adherend-PSP assembly. Therefore, a postreduction of the initial coating weight may be necessary. Such an apparent reduction of the coating weight, that is, of the mobility of the adhesive layer, can be achieved by modifying the rheology of the adhesive. As is well known, one way to achieve this modification is by crosslinking (see also Chapter 3, Section I and Chapter 6). Another possibility is to use fillers.

TABLE 7.4
Coating Weights of Selected PSPs

Pressure-Sensitive Product	Coating Weight Range (g/m²)
Label	5–30
Tape	1.5–500
Protective film	0.5–20
Separation film	0.5–2.5

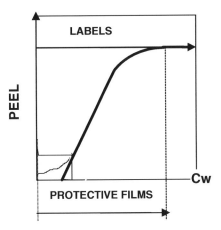

FIGURE 7.6 Typical coating weight domains for labels and protective films.

Generally, crosslinking of the adhesive is carried out before and during its coating (see Chapter 6), but it can also be achieved after the application of the adhesive coat. Peel can be postregulated by crosslinking after use of the laminate [39]. In this case the storable, crosslinkable adhesive is formulated for postapplication crosslinking, which is proposed to improve shear resistance and/ or to allow easier delamination of the applied product. Postcrosslinking is achieved using free radical-initiated [40] or photoinitiated reactions [41]. Postcrosslinking leads to adhesionless surfaces, that is, to easier delamination. This is a development of the adhesive properties as in the case of the so-called semipressure-sensitive adhesives, where crystallization causes the time-dependent loss of adhesivity (see later). According to Charbonneau and Groff [39,42], the procedure allows delicate electronic components to be removed from the tape, even though they were nonremovable before crosslinking. A latent crosslinking agent has to be used. The tape has a high coating weight and is conformable to rough, uneven surfaces (coating thickness of 0.5–1.5 mm). To facilitate immediate adhesion to rough and uneven surfaces a resilient foam backing can be used (see also End-Use of Tapes, p. 553).

Such foam carriers coated either one or both sides can have an overall thickness of 0.1–2.0 mm [43]. Glass microbubbles can be incorporated also to enhance immediate adhesion to rough and uneven surfaces [44]. As is known from formulating practice, microbubbles have been suggested as filler for the adhesive layer (see also Chapter 5). They can be applied to reduce the contact surface, that is, for removability (see later) or to improve the cohesion of the adhesive (see Chapter 3, Section I). Such improvements are necessary for tapes of rough surfaces where adhesive conformability needs a very high coating weight (one that shows pronounced cold flow) and/or a soft foam-like carrier. A foam-like pressure-sensitive tape is manufactured according to Vesley et al. [45] by using glass microbubbles as filler. The microbubbles are imbedded in an adhesive matrix. Pressure-sensitive tapes filled with glass microbubbles have a foam-like appearance and character and are useful for purposes previously requiring a foam backed PSA tape. As illustrated by the above product application related examples, although standard coating weight values exist for various PSPs, such values have to be evaluated always with respect to the specific formulation, that is, the conformability of the adhesive layer. The latter can be modified by formulation, coating technology or adequate carrier.

Tapes with thick adhesive layer (0.2–1.0 mm) have been prepared by photopolymerization (UV) of acrylics. The PSA matrix may contain glass microbubbles [44]. Microbubbles are included in the composition of a foam-like tape according to [43]. They are embedded in a polymeric adhesive matrix, containing at least 5% (5–65% v/v) filler. The thickness of the pressure-sensitive layer should exceed three times the average diameter of the microbubbles, to enhance the migration (flow) of the bubbles in the matrix under applied pressure instead of breaking. Such a formulation enables flow and buildup of an intimate contact with rough and uneven surfaces, while retaining the foam-like character. Optimum performance is attained if the thickness of the PSA layer exceeds seven times the average diameter of the bubbles. This construction ensures an apparent reduction of the coating weight and an apparent improvement of the cohesion of the adhesive layer. To increase cohesion of the adhesive layer, polar monoethylenically unsaturated monomers (less than 20%) are included in the acrylic polymerization recipe and photoactive or heat-activatable crosslinking agents are added [46,47]. It is evident that the reduced flow properties influence the peel values negatively. This can be seen from the peel measurement method (T-peel is tested) and the peel values. The cohesive (tensile) strength of the uncrosslinked adhesive is 2550 kPa (elongation 840%); that the crosslinked formulation may attain 6000 kPa (depending on the crosslinking conditions).

In some cases the conformability of the PSP is achieved by the manufacture of a carrier-less construction [48]. The adhesive can be crosslinked with a multifunctional, unsaturated monomer. A prepolymer (precursor) is synthesized, that is polymerized to the final product using ultraviolet (UV) radiation [49]. According to Larimore and Sinclair [48], a carrier-less PSP can be obtained (the adhesive layer has adequate strength to permit it to be used without a carrier material) that is more conformable. The product exhibits an adhesive surface and an adhesion-free surface. The tack-free surface is achieved by superficial crosslinking of the adhesive by polyvalent cations (Lewis acid, polyvalent organometallic complex, or salt). The crosslinked portion of the adhesive has greater tensile strength and less extensibility. The crosslinking penetration depends on the type of polymer, and the solvent, that is, a crosslinking gradient can be achieved as for the UV-photopolymerized products (see later). Thick PSA tapes (0.2–1.0 mm) have been prepared by UV-photopolymerization of acrylics [44]. The formulation can be polymerized as a thick layer (up to about 60 ml), which may be composed of a plurality of layers, each separately photopolymerized. The length of polymerization zones and the density of lamps in these zones affect the manufacture. When the thick layer is sandwiched between two thinner layers, the thick layer is often referred to as a carrier, but it has pressure-sensitive properties also. The support or carrier layer is 25–45 ml thick, it conforms well to substrates that are not flat.

As stated in [50], even a high molecular, highly viscous 100% polyacrylate processible as hot-melt is not of high molecular weight to display balanced pressure-sensitive properties. For such a product the PSA definition given by Chang [4] is not valid. Therefore, the PSA should be crosslinked. A special acrylic raw material for acrylic hot-melt PSA (HMPSAs) has been developed in order to be cured photochemically [50–52]. The photoinitiator is built in the polymer. The uncrosslinked product does not have balanced adhesive properties. Because of its low molecular weight it is very tacky but does not have sufficient shear. The pressure-sensitive properties are achieved only by crosslinking. Another aspect of its crosslinking is its anisotropy. The radiation is partially absorbed, partially transmitted, and partially reflected. The maximum radiation level is given at the top of the adhesive layer. The crosslinking degree is the maximum on this side. Therefore, direct and transfer coating give different adhesive characteristics. A uniform, isotrope adhesive layer can be achieved using an UV-transparent face stock (backing) and irradiation on both sides. A transfer coated PSA (with 15% tackifier resin) gives 25 N/25 mm peel resistance (on steel) after irradiation, and 27.5 N/25 mm quick stick, respectively 18.7 N/25 mm peel on polyethylene, and 44 min shear resistance (1.2×1 in.^{2}, 500 g). The adhesive layer irradiated on the face stock material gives 20 N/25 mm peel on steel, 16.8 quick stick, and 6.2 N/25 mm peel on polyethylene and (surprisingly) a quite different shear value also (77 min). Unfortunately, the shear values are not comparable for the given experiments. These examples illustrate that balancing of the adhesive properties is achieved in many cases by product buildup.

Protective films use low coating weight (Table 7.5). Such products are always situated in the critical domain (see Figure 7.6). Peel is regulated for protective films by controlling coating weight. For low coating weight values applied generally for removable products (e.g., special

TABLE 7.5
Coating Weight Values for Common Adhesive-Coated Protective Films

Adhesive		Coating Weight (g/m^2)	
Code	Type	Theoretical Value	Practical Average Value
1	Acrylic	1.70	1.80
2	Acrylic	2.40	2.57
3	Acrylic	2.80	3.00
4	Acrylic	4.00	4.60
5	Acrylic	5.70	6.30
6	Acrylic	7.00	7.10
7	Acrylic	12.00	10.10
8	Rubber-resin	1.90	2.26
9	Rubber-resin	2.90	3.34

Note: Crosslinked, solvent-based acrylic and masticated, crosslinked, solvent-based rubber-resin PSAs were used.

labels or protective films), situated in the critical coating weight domain where peel strongly depends on the coating weight, tolerances are more important. As can be seen from Table 7.6 the average coating weight tolerance values for protective films range from -16% to $+14\%$, which means, such products also have a broader peel value distribution.

The main adhesive formulations used for protective films are crosslinked compositions. Therefore, their cold flow is limited also. Thus it can be supposed that, in contrast to labels and tapes, the contact building of protective films has to be achieved at very low tack and peel values. Admitting that instantaneous peel and tack are interdependent, it is evident that low tack protective films need quite different application conditions in order to improve the initial peel value. Later, the buildup of adhesion increases the peel resistance up to the value required in practice.

As seen from the relations 7.4 and 7.6 both α and β depends on the adhesive, carrier, substrate, time and temperature. For a given application of the same PSP (same adhesive, coating weight and carrier) on a given substrate, α and β can be considered as parameters that depends only on time and temperature. Therefore, in this case the peel value depends on the application conditions only, that is, laminating pressure, time and temperature:

$$P = f(p_l, t, \theta) \tag{7.7}$$

TABLE 7.6
Coating Weight Tolerances for Protective Films

Coating Weight (g/m^2)		Tolerances (g/m^2)		
Theoretical	Average	Extreme Values		Average Values
		Minus	Plus	
1.6	1.86	0.2	0.8	0.26
1.7	1.77	—	0.2	0.07
1.7	1.85	—	0.2	0.15
2.2	1.90	0.4	—	0.30

TABLE 7.7
Peel Buildup for Protective Films

Code	Peel Resistance (N/25 mm)	
	Instantaneous	After 24 h
1	2.50	2.61
2	3.50	3.57
3	4.30	4.33
4	7.50	7.69
5	14.00	14.31

where p_1 is the laminating pressure, t is the laminating time, and θ is the laminating temperature. The comparison of correlations 7.1 and 7.7 shows that, for the application of classic PSPs (tapes and labels) the intrinsic adhesive properties of the product (regulated by adhesive nature and coating weight) are more important for peel buildup. For protective films the application conditions are determinant. In some cases, the self-adhesive film (SAF) has to display good adhesion properties at very high temperatures and good blocking properties at elevated temperatures. For instance, a heat-sensitive EVAc film gives good adhesion when heated for 20 sec at 120°C and good blocking resistance during 30 min at 50°C and 1 kg/cm^2 load [53].

It should be mentioned that coating weight regulation is a problem for adhesiveless PSP constructions also. For such coextruded products the thickness of the self-adhesive layer is the regulating parameter having the role of the coating weight.

b. Peel Buildup/Peel Gradient

For classic, PSA-coated pressure-sensitive products the buildup of adhesion is dependent on dwell time. For instance, an aluminum carrier and a tackified CSBR are used as insulating tape in clima-technics [54]. The peel value of this tape builds up to 37 N/25 mm after 24 h. As discussed in Chapter 5, acrylic HMPSAs with thermally reversible crosslinking (segregated structure) have been developed [55]. For such formulations the buildup of peel values can also be observed. The initial peel value of 38 oz/in. increases after storage for 1 week at 50°C to 68 oz/in. For extruded self-adhesive products also, time-dependent adhesion buildup can be observed. The adhesive properties of such self-adhesive films with a built-in adhesive component (e.g., polybutylene tack-ified polyolefin) depend on the diffusion of the viscous component (through the carrier) to the product surface. Such a diffusion is a function of time. As is known from plastic film processing, that blocking of PVAc films depends on pressure, temperature and time. Table 7.7 illustrates the peel buildup over time for protective films. As can be seen from this table, peel buildup is a general phenomenon for products of different coating weights.

As shown in the data of Table 7.8, such peel buildup also occurs for highly crosslinked products with very low initial peel (e.g., separation films). Like instantaneous peel, peel buildup is also a

TABLE 7.8
Peel Buildup for Separation Films

Product Code	Laminate Peel (N/10 mm)		
	Storage Time		
	24 h	1 Week	2 Week
1	0.08	0.21	0.47
2	0.03	0.10	0.21

TABLE 7.9
Dependence of the Peel Buildup on the Adherend Surface Quality

Product Code	Protected Surface	Peel (N/25 mm) Application Time					
		Nominal	24 h	1 Week	2 Weeks	3 Weeks	4 Weeks
1	Polished aluminum	0.50	0.40–0.65	—	—	—	0.45–0.65
2	Polished aluminum	0.40	0.20–0.40	—	—	—	0.30–0.45
3	Aluminum	1.20	0.50–1.50	0.70–1.80	0.70–1.80	—	0.75–1.80
4	Stainless steel	4.00	3.20–4.00	4.00–4.20	4.10–4.40	4.30–4.50	4.30–4.50

function of the substrate quality (Table 7.9) and is greater on glossy surfaces. Peel buildup also characterizes self-adhesive films (Table 7.10). Such films are warm laminated; therefore, peel buildup is tested at elevated temperatures.

As illustrated by data of Table 7.11, peel buildup is a general phenomenon for pressure-sensitive products having different adhesives, carriers, geometries and buildup. Thus, the control of the removability is really the regulation of the peel buildup.

Acrylics display the disadvantage of compliance failure, that is, adhesion buildup. As discussed earlier, in some cases adhesion buildup can be balanced by a decrease in adhesion due to the reduction in the chain mobility caused by crosslinking. For instance, adhesion buildup has been reduced by crosslinking of the acrylic with dimethylaminoethyl–methacrylate [56] or with poly-isocyanate [57]. Gobran [58] used superimposed adhesive layers having different gradients of shear creep compliance to avoid adhesive buildup.

c. Static and Dynamic Dwell Time

According to [59], for protective films a difference exists between adhesion on the surface and bonding on it. Similar behavior has been observed for certain tapes where the dwell time from the moment of application of the tape to the buildup of the maximum peel adhesion has been called "time to conformation" [60]. Sometime ago, for better wetting-out and to increase the contact surface between adhesive and adherend, pressure and solvents were used for application of tapes. In such cases the "pure" (dwell) time dependence of the bond formation changed with the simultaneous change of the stress rate, temperature and chemical affinity. For special print transfer elements (like Letraset), the low level of application tack is essential for storage and applicability. Such products need very high application pressure [61]. Adhesive laminating under pressure is a common operation in other domains too. For instance, lineary pressure in laminating by gravure printing attains 200 N/cm (for a pressing cylinder of ϕ160 mm); wet and dry laminating use pressure of 300 N/cm. Solventless laminating applies a pressure of 270 N/cm [62]. It is known from PSA characterization that for the

TABLE 7.10
Peel Buildup for SAF

Code	Peel (N/10 mm) Instantaneous	2 d at RT	4 h at 80°C
1	0.15	0.10	0.22
2	0.70	0.75	0.40
3	0.80	0.85	0.30

Note: EVAc-based protective film on PC substrate was tested.

TABLE 7.11
Peel Buildup for Selected PSPs

	Peel Resistance									
	Dwell Time									
Product	**0**	**1 h**	**3 h**	**1 d**	**1 Week**	**2 Weeks**	**4 Weeks**	**8 Weeks**	**Unit**	**Reference**
Tape, repositionable	26.7	—	36.2	—	—	—	—	—	oz/in.	[50]
Tape, removable	5.65	7.15	—	6.15	6.50	—	5.25	—	N/25 mm	[51]
Label, water-based	13.00	—	—	—	—	—	—	12.50	N/20 mm	[52]
Label, removable	2.00	—	—	—	—	3.00	—	—	N/25 mm	[53]
Protective film	—	1.4	—	1.0	2.3	—	—	—	N/25 mm	[54]
Tape, paper	0.057	—	—	—	0.065	—	—	—	kg/cm	[55]

FINAT Test Method (FTM)-4 high speed release test the samples are placed between two flat metal plats under a pressure of 6.87 kPa to ensure good contact. In such cases the static (forceless) and dynamic dwell times differ. After studying the effect of contact pressure on the adhesive properties, Johnston [63] stated that the PE film, unlike other films does not attain a plateau (as a function of the contact pressure) but continues to climb in tack value due to its high extensibility.

The built-in tack of a PSP and the application conditions (application pressure and temperature) allows the instantaneous adhesion of the product on a substrate. For high tack products (e.g., labels, tapes and some protective films) a light application pressure if any at room or low application temperatures are used. For PSPs with very low tack and instantaneous peel (e.g., protective films) the bonding depends on the applied pressure and temperature also. As known, foam-like transfer tapes also used under pressure to increase initial bond strength [64]. According to Kuik [65], the peel resistance of heat seal polar ethylene copolymers depends on the corona treatment, coating weight and aging. The peel strength increases from 50 to 150 g/25 mm if the surface tension varies between 0 and 32 dynes/cm. The peel value (on PET) increases with time. Here, contact buildup is influenced by other than rheological parameters (e.g., functionalization degree of the surface) also. All these products can be considered as working below the critical domain of coating weight where the conformation factor is determinant for laminating. The importance of the laminating conditions is illustrated by the data of Table 7.12 for a protective film used

TABLE 7.12
Peel Resistance Values as a Function of the Test Conditions

	Peel Values (N)			
Sample	**Test Method 1**		**Test Method 2**	
Width (mm)	**Carrier 1**	**Carrier 2**	**Carrier 1**	**Carrier 2**
25	0.90	1.34	0.75	1.35
25	1.05	1.43	0.71	1.28
100	0.78	1.35	0.65	1.16
100	0.99	1.45	0.71	1.26
100	1.13	1.39	0.58	0.87

Note: Water-based acrylic coated clear (carrier 1) and pigmented (carrier 2) PE film was warm laminated on PMMA plate. Method 2 uses double sided heating (plate and film).

TABLE 7.13
Rolling Ball Tack Values for Protective Films

Code	Coating Weight (g/m²)	Carrier Thickness (μm)	Peel (N/10 mm)	Rolling Ball Tack		
				Ball Diameter (mm)	Mean Value (mm)	Tolerances (mm)
1	12.6	154–156	0.46	4.5	212	14
				11.1	>400	—
				14.3	>400	—
2	7.6	100–108	0.29	4.5	267	33
				11.1	>500	—
				14.3	>500	—
3	3.4	50–55	0.20	4.5	290	45
				11.1	>500	—
				14.3	>500	—

for PMMA plates. As can be seen, the importance of the laminating conditions for the bonding characteristics is greater than that of the sample geometry.

Tack is one of the standard adhesive properties. The use of tack as criterion for evaluation of the instantaneous adhesion requires a measurable tack value. Although for protective films in some cases such measurable (low) tack value is given (Table 7.13), it cannot be used for characterization of the application performance of the product. This tack (T) does not work like a normal tack. It does not ensure sufficient instantaneous adhesion on the product surface. Therefore, the product is laminated under pressure. In such cases it would be better to use an application tack (T_a), with a known dependence on the lamination pressure p (and/or temperature θ):

$$T_a = f(T, p, \theta) \tag{7.8}$$

Johnston [68] measured tack as tear energy (E_t) according to the following correlation:

$$E_t = AC^{mx} \tag{7.9}$$

where A and m are constants including the effect of load and dwell time and x is the separation rate of the probe.

If tack is considered (according to Bates [66]) as the ratio of the separation energies at optimum (E_{SO}) and infinite ($E_{S\infty}$) dwell time:

$$T = E_{SO}/E_{S\infty} \tag{7.10}$$

and the dependence of such energies on the application conditions is taken into account also, we actually have an application tack that is generally valid for different PSPs. It should be mentioned that instantaneous tack of certain self-adhesive PSPs depends on the surface treatment degree and age also. For such products the definition of the separation energy at optimum time is more complex (see also Chapter 8).

d. Stress Rate-Dependent Peel

As discussed earlier, crosslinking is used to modify the adhesion–cohesion balance. The use of crosslinked formulations may change the debonding mechanism also. According to Charbonneau and Groff [39], for UV-crosslinked formulations the characterization of the adhesive performances needs a special peel measurement method. A so called "cleavage peel" value is determined (Fisher Body Test Method TM 45-88). A breakaway cleavage peel value and then a continuing cleavage peel value are measured. The continuing cleavage peel value is about 50–60% of the initial value. The existence of two different peeling behaviors, that is, a first high level starting peel value which supposes a high rate of the peeling force, and a second, normally, steady-state peeling, denotes that the adhesive behaves like a multiphase mixture having a high strength (but low contact) component and a relatively soft adhesive matrix with common adhesive flow. Verdier and Piau [67], investigates the mechanism of cohesive failure for various adhesives and test conditions. They state that, for crosslinked adhesives more complex three-dimensional flow patterns are obtained; at very high velocities stick release is observed.

The dependence of the peel value on the peel rate has been discussed in [2]. The increase in peel resistance with stress rate generally can be considered as a consequence of the time/temperature superposition principle (see also Chapter 3, Section I). As is known from practice, for medical tapes the rate of removal is very low (15 cm/min) [68]. Such low speed delamination provides better removability, whereas high speed delamination may change the peel resistance. For instance, crosslinked silicone pressure-sensitive adhesives can be contacted with a release based on highly crosslinked silicones [26]. High speed (600 in./min) stripping of this release gives 25–30 g release force in comparison with low speed stripping (12 in./min) which gives 9–16 g. High speed stripping values do not depend on storage time. Which means, at high speed both components (crosslinked and uncrosslinked) behave like elastic materials (no energy dissipation) but energy is stored. Soft (e.g., freezer tape) adhesives increase the release force with increasing separation speed more than permanent ones [69].

As stated by Eckberg [70] there is no instrument yet devised capable of mimicking hand removal of a label from a liner. It is generally not agreed that such products are best tested at low peel rate, defined as <100 in./min. The peel release profile is the variation in release force as a function of peel speed.

It should be mentioned that the dependence of the peel value and of the break nature on the peel rate is a common phenomenon observed by delamination of PSPs during their converting and application also. Such dependence causes the so called "inversion" of the adhesive break when tapes are unwound at too low temperatures (lower than 10°C) or with too high a running speed [71] (see also Chapter 10).

Common PSA labels or tapes are usually designed to give a well-defined peel resistance under test conditions where delamination occurs at a standard debonding speed. For high quality special products (e.g., removable label) uniform delamination over time, for example, a uniform peel resistance value, is required. However, there are other applications, that is, debonding conditions for several tapes and protective films where peel force uniformity during debonding is not a quality criterion and variable peel resistance values are needed during debonding. It should be accentuated that in such applications peel should vary as a function of the debonding speed. Moreover, a nonlinear dependence is required.

Diaper closure tapes are used as refastenable closure systems for disposable diapers, incontinence garments, etc. Such systems allow reliable closure and re-fastening. The tapes have to exhibit a maximum peel force at a peel rate between 10 and 400 cm/min, a log peel rate between 1.0 and 2.6 cm/min. Tapes with the maximum value of these parameters have been found to be strongly preferred by consumers. It is supposed that the higher peel forces at low rates discourages the wearer from peeling open the closure, whereas the lower peel force needed at greater peel rates helps the user remove the closure without tearing the back sheet of the product. According to Miller and von Jakusch [72] the presence of a maximum of the peel force as a function of the peeling rate strongly depends on the formulation. When no high T_g endblock re-inforcing resin is used, no maximum appears in the peel force vs. rate curve. The peel force increases monotonically with rate, until the peel force exceeds the strength of the tape carrier material. The rate for maximum peel force depends on the concentration of the high T_g resin, varying between 120 and 225 cm/min. The re-openability increases with the level of high T_g resin also. Although the nature of the substrate (different polyethylene grades or polypropylene) influences the value of the peel force (10–20%) it does not influence the presence of a maximum as a function of the peel rate, demonstrating that such effect is due to the viscoelastic behavior of the adhesive or tape carrier material. As stated in [73], for the same adhesive composition 180° peel test leads to much higher (300%) peel resistance. Figure 7.7 illustrates the form of the peel curve for a common label delamination, for a diaper closure system and for a special zip-peel protection film.

As seen from Figure 7.7 unlike label delamination, for special tapes and protective films peel resistance should present a barrier value (cleavage peel), followed by a rapid decrease of the peel resistance. The time and temperature dependent nature of the pressure-sensitive macromolecular formulations usually allows an increase of the peel value as a function of the delamination speed and leads to an increase in debonding resistance at higher speeds. A decrease of the debonding force at high speeds supposes a quite different debonding mechanisms than common relaxation. According to [73] at increased rates of testing a slip stick region appears, which is characterized by a regular oscillation of the peel force. It may involve a regular oscillation between rubbery and glassy response of the adhesive. It may alternate storage and dissipation of the elastic energy in the stretched backing film.

In end-use of special protective films nonuniform delamination behavior, that is, variation of the peel force during debonding of the protective film is a major requirement. Manual delamination of the film from large substrate areas is facilitated by stepwise debonding; after a first high force delamination peeling occurs at a much lower debonding resistance. According to our experience such a delamination is the result of a special elastic carrier material, or a special PSA formulation or both. The PSA is formulated as a blend of low T_g and high T_g adhesives. Recently, Sharma et al. [74] studied the debonding of elastic films. A rigid surface initially in contact with a soft elastic film, upon withdrawal debonds by formation of a pattern of well defined spacing consisting of areas of intimate contact and interfacial cavities. Sharma et al. [74] suppose that the formation of cavities engenders extremely high stresses near the column edges leading to the peeling of contact zones at much smaller average stresses than the adhesive strength. Although the authors state that the wavelength of the surface patterns does not depend on the elasticity of the film, our industrial experience does not confirm this hypothesis.

The existence of a high starting peel value has been observed by other crosslinked adhesive formulations also. A peel test (90° peel) of such a (crosslinked) adhesive tape measures first the

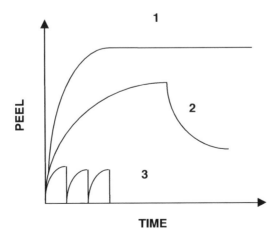

FIGURE 7.7 Typical peel diagram for three types of PSPs. (1) Label; (2) diaper closure tape; (3) special protective film (arbitrary scale).

force required to start the breaking of the bond ("initial breakaway peel") and then the force needed to continue the bond breaking, the "initial continuing peel" [75]. A pluck test (90° peel, with a slow peel rate of 2.5 cm/min) is carried out also. Initial continuing peel is lower (about 30%) than breakaway peel. It is evident that, such peel behavior is the result of a special adhesive/product buildup. Bonneau and Baumassy [28] found under the same peeling conditions different peeling behavior for water-based formulations with acid resin and with resin esters (Dermulsene, DRT). The acid resin used displays narrower molecular weight distribution (MWD) ($M_w/M_n = 1.19$) and higher T_g (10°C), the ester resins have broader MWD (1.19–2.48) and somewhat lower T_g (4–6°C). For the acid resin the debonding is jerky (stick-slip phenomenon) for the other resins the debonding is quite steady. According to the authors this compartment is due to the "better wetting of the substrate" by the ester resins. Due to the greater modulus of the adhesive at low temperatures, a loss of wetting and a jerky debonding is observed for the acid resin. The resins were of similar softening point and compatibility and the molecular weight of the acid resin used was about 400. The resin esters have a molecular weight of 1000–1300. For a given softening point and concentration of the resin, the higher the increase of the tan δ peak temperature, the better the compatibility (see also Chapter 3, Section I). Bonneau and Baumassy [28] found no difference by DMA study, between the compatibility of the tested resins although in practice results have been different. It can be supposed, that because the tendency of acid resin to crystallize, an apparently crosslinked, incompatible system is built up displaying different peel behavior.

The use of classic peel measurement does not allow perfect characterization of practical adhesion behavior for the whole product range of PSPs. As discussed in [76], for test of zip-peel products the use of a modified tensile strength equipment is suggested with a spring balance (Prohaska-device) having the resistance of the maximum peel force of the product. Debonding resistance characterized by peel is a force measured under well-defined conditions (speed, temperature, strain). For the majority of PSPs the application parameters (although different from those used for quality tests) are constant. This means, for these products peel resistance can be used as normed method for laminate bond strength. Good product quality means desired and uniform peel force level. A uniform peel force level means constant peel value during test independently from the specimen length, that is, testing time. On the other hand, it is obvious that there are some special products, where the peeling force may change during debonding and therefore, the standard test method is not enough to characterize the real behavior of the product.

The adhesion force tested as peel is a so-called bonding/debonding peel, that is, it is assumed that the debonding resistance is an index of the bonding force. For certain products it is more important to know not only the final value of the peel-stress function but also the form of this dependence (Figure 7.8).

As can be seen from Figure 7.8, for the so-called "zip peel" the peel resistance should not be uniform during delamination. The peel should show an increase up to a maximum value followed by an almost zero level, in order to allow slight and rapid delamination of large coated surfaces. As mentioned earlier, such behavior can be achieved designing a special adhesive or carrier film rheology and composition. It can be supposed that the two-phase peel is due to energy storage in a highly elastic carrier (which cumulated with the external force causes high speed debonding) or to the destruction of a crosslinked adhesive network where hard crosslinked and soft viscous parts of the network exhibit different debonding resistance. The first hypothesis accentuates the special role of plastic-based carrier materials.

e. Carrier Deformation

As schematically presented in Figure 7.9, the use of a very stiff adhesive (e.g., a foamed, crosslinked adhesive) or of a rigid carrier material can lead to changes in the peeling nature of

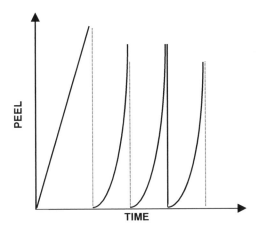

FIGURE 7.8 Typical zip-peel (saw-tooth) plot, illustrating variation of the peel force during debonding (arbitrary scale).

delamination (peel is transformed into butt peel). On the other hand, as discussed earlier (see Chapter 3, Section I) a deformable carrier can grade the value of the peel.

It is supposed that in the first step of delamination the debonding force causes a deformation of the carrier material (ε_c), followed by the deformation of the adhesive layer (ε_a) and bond failure (B_f) (Figure 7.10).

It can be supposed that for a PSP having a nondeformable carrier material the peel force acts on the adhesive bond instantaneously; for a product with deformable carrier first the film deforms. In reality, deformations of both carrier and adhesive occur simultaneously and lead to bond failure. The debonding time (t_{db}) is given by the time required for the deformation of the carrier (t_{Dc}), the time necessary for adhesive deformation (t_{Da}) and the time for bond failure (t_{Bf}):

$$t_{db} = t_{Dc} + t_{Da} + t_{Bf} \tag{7.11}$$

The debonding force is also the sum of the forces required for extension of the carrier material and of the adhesive layer, and for bond failure. These processes occur over time, being partially superposed (see Figure 7.10). The peel force (P) required to destroy the adhesive bond is the

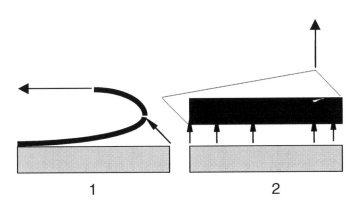

FIGURE 7.9 Peel degradation by nondeformability or excessive deformability of the carrier material. (1) Soft deformable carrier material; (2) rigid nondeformable carrier material.

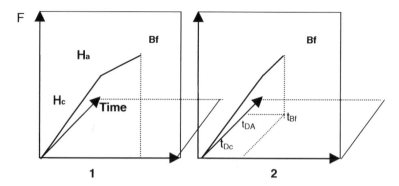

FIGURE 7.10 Schematical presentation of force transfer by debonding. (1) Force transfer via carrier to adhesive; (2) simultaneous carrier and adhesive deformation.

sum of the forces necessary to deform the carrier material (F_{Dc}) and the adhesive (F_{Da}), and to destroy the bond (F_{Bf}):

$$P = F_{Dc} + F_{Da} + F_{Bf} \qquad (7.12)$$

Correlation (7.12) suggests that the peel force required for pressure-sensitive laminates based on deformable carrier material may be higher than that required for a nondeformable carrier. According to Dunckley [23], for common papers paper strain appears at 1500 g/in. debonding force. The real forces in the paper are higher because they are partially absorbed by the deformation of the paper. Yamauchi et al. [77], reported peel forces and failure modes as function of peel rates and paper types. Pelton et al. [78], state that roughness, porosity and the extent of adhesive penetration dictate whether paper failure occurs with peel. According to Zhao et al. [79], paper failure is common for peeling from paper, and results in an unstable region in the peel curve. The governing paper property in the paper failure region is the paper internal bond strength.

As discussed in [2], soft carrier materials allow a low level of peel force. This is possible due to their energy absorbtion and time and temperature dependence of the stress transfer. As is known from work with removable labels, high speed debonding leads to paper tear. Good removability requires that the adhesive have energy absorbtion properties (i.e., flow, relaxation of its macromolecular components). This process needs time. The deformation time of the carrier material can be (at least partially) used for adhesive deformation too. Therefore, it would be more correct to admit that in correlation (7.12) the time necessary for deformation is the sum of the time required for processes occurring in part simultaneously. Thus the time for debonding is given by the time for deformation (τ_D) and bond failure,

$$\tau_{db} = \tau_D + \tau_{Bf} \qquad (7.13)$$

where the deformation time is a function of the mechanical characteristics and geometry of the plastic film carrier and rheology of the adhesive layer. As is known from the testing of the mechanical properties of carrier materials, the mechanical resistance of the plastic carrier material is higher than the force necessary for adhesive bond failure. This force is higher than the forces required for adhesive flow. As discussed in Chapter 3, Section II, in evaluating the carrier quality for labels its dimensional stability plays an important role. This is evaluated generally by measuring the forces necessary to achieve a given (low percentage) elongation, supposing that machining forces will not exceed this value. As can be seen from Table 3.6 the value of this force is generally much higher than the bond strength for removable presssure-sensitive laminates.

This means, the deformable carrier material works like a nondeformable one, transmitting the debonding force almost instantaneously and at high level on the adhesive–substrate interface.

According to Kim et al. [80], the measured peel value is an "engineering strength per unit width." For thin films it does not represents the true interface adhesion strength. The peel resistance may represent the product of the interface adhesion and other work expended in the plastic deformation of the film. The contribution of the plastic deformation of the film to the peel strength is found in some cases to be many times higher than the true adhesion. The major controlling factors in the peel strength are the thickness, modulus, yield strength, and strain hardening coefficient of the film, the compliance of the substrate, and the interface adhesion strength.

In reality both, the plastic carrier and the PSA are (at least partially) viscoelastic materials, and their mechanical characteristics depend on the material, geometry, time and temperature. The peel force required to delaminate the pressure-sensitive construction includes a component required for elongation work. Variations in peeling energy due to the deformation of the stripping member, which in turns changes with speed and other parameter should be taken into account.

The change of the stress rate acting on the bond by delaying the forces due to carrier deformation enhances debonding at lower peel forces. Thus the resultant peel is a complex function of carrier deformation (ε_c):

$$P = f\left(\frac{\varepsilon_c^{\gamma}}{\varepsilon_c^{\delta}}\right) \tag{7.14}$$

where γ and δ are exponents taking into account the influence of the deformation work on the peel force. For thick, mechanically resistant or elastical materials the value of δ is low, lower than γ. For plastically deformable thin films, δ increases and the carrier deformation can lead to the decrease of the peel force (Table 7.14). For a given material and given delaminating conditions the geometry of the film is the main parameter influencing the carrier deformation. One can suppose that for each material used there is a critical thickness. Under this limit the deformation of the plastic film decreases the delaminating force. Therefore, the decrease of the carrier thickness for pressure-sensitive products, especially for protective films presents a complex problem.

The above considerations have a general character, independent of the product buildup (construction of the adhesive layer). It has been shown by Kuik [65] for heat-sealable polymer films, that the 90° peel is the result of adhesion ad energy loss. Energy loss is due to the deformation of the plastic film during peeling and this deformation is proportional to the resistance against dimensional change and material thickness. Therefore, it is obvious that carrier deformation (depending on

TABLE 7.14
The Influence of Plastic Carrier Deformability on the Peel of PSPs

| Code | Carrier Characteristics | | | 180° Peel Resistance (N/25 mm) | |
	Thickness (μm)	Tensile Strength (N/mm²)	Elongation (%)	Nominal	Measured
1	23–31	27	419	1.2–1.5	0.15
2	32–37	23	415	0.4–0.6	0.02
3	40–48	31	532	2.5–3.7	2.0
4	42–46	33	558	1.2–2.2	1.1
5	42–47	28	494	2.5–3.0	2.2
6	58–67	18	259	2.5–3.0	2.4
7	65–75	18	348	2.5–3.0	2.7

geometry and composition) will have a general influence on bonding and debonding of PSPs. According to Boutilllier [81], for an extensible polymer film the higher the work of adhesion the higher is the dissipation energy (due to the polymer deformation) and the higher is the peel energy. For a self-adhesive film (EVAc), the dissipated energy, deformation energy, and the dissipated energy per volume increase with the thickness of the polymer. In the classical field of labels this is less evident, but elastic/plastic carrier deformation has to be taken into account always by winding of tapes as a factor influencing bond strength. It must be taken into account by delamination of protective films, where carrier deformation modifies the peel force, that is, removability. It is evident that carrier deformation and adhesive deformation are a function of the stress rate. As discussed earlier, removability is a relaxation phenomenon and relaxation needs time.

In the above examples (and according to the data of Table 7.14) the viscous and viscoelastic tensile deformation of the carrier (parallel to the direction of the peeling force) is a main factor influencing the transfer of the debonding force. Carrier deformation by flexural stress influences it conformability by bonding and deformability (peeling angle) by debonding. The deformation of the carrier characterized by its tensile strength is more pronounced for self-adhesive films whose formulation is soft in comparison with adhesive-coated PSPs. The reduced conformability of a carrier by bonding can be balanced by the conformability of the adhesive for adhesive-coated PSPs only.

The flexural deformability brought about by debonding is common for adhesive-coated as for adhesiveless PSPs. Removability of SAFs is strongly influenced by their stiffness. As stated by Kuik [82], for polar terpolymers of ethylene with maleic anhydride in blends of low density polyethylene (LDPE), two different influences coexist. A terpolymer with a good hottack (i.e., high polar component content) gives excellent sealability at low temperatures. The addition of LDPE in the terpolymer increases its peeling strength. This is due to modification of the stiffness of the film. The higher the density of PE the higher the increase of adhesion.

For special PSPs the carrier conformability and deformability imposes the use of special test methods for evaluating adhesion. For applications where combined stresses act on the adhesive joint, the test methods have to be adapted to the actual conditions of product application. Such case is that of the self-adhesive (tackified polyethylene) stretch films where so-called peel cling and "lap cling" are used for product characterization (Figure 7.11). Peel cling is really a T-peel

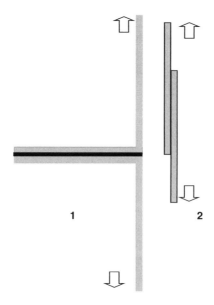

FIGURE 7.11 Schematic presentation of peel cling (1) and lap cling (2) measurement for self-adhesive films.

test where self-adhesion is tested. T-peel is used for flexible adherends [83,84]. According to [85], peel cling needs to be high enough to stop the loose end of the film from unwinding on the pallet. Its optimum value is about 2.4 N/m.

Lap cling needs to be sufficiently high to maintain the integrity of the wrapped load. Its optimum value is about 5 N/2.5 cm. Using adhesiveless, soft PSPs for this test, the measured data will strongly depend on the stiffness, conformability and elasticity of the film. Lap cling is measured like shear resistance, parallely to the joint. In the case of lap peel slip friction appears and tensile stress combined with shear stress act on the adhesive joint. Therefore, the value of lap cling is higher than the value of peel cling. Adhesion considered as friction was investigated for adhesive-coated PSPs also. Mizumachi [86] studied the adhesion of PSAs by rolling ball tack, and used for its characterization. The coefficient of friction (COF), considered as the sum of rolling friction caused by compressive deformation of the substrate material and of friction caused by extensional deformation of the substrate.

Due to the slow buildup over time of an adhesive–substrate contact the relative value of aged peel cling (in comparison with its initial value) is much higher than that of the aged lap cling [87]. Generally, peel cling increases with the level and molecular weight of the tackifier (polybutene). As is known, the molecular weight of the tackifier resin and have a similar influence on the peel of adhesive-coated PSPs. On the other hand lap cling increases with the level of polybutene, but decreases as the molecular weight of polybutene increases due to decrease in its ability to migrate and its tackiness. Unlike the classic tackifying of elastomers, where the components must be compatible, the tackifying effect of polybutenes for polyethylene is due to the partial incompatibility of the two materials and the inherent tackiness of polybutenes. It is evident that tackifier diffusion also depends on the film thickness. This dependence also exists for slip migration [87].

A special method for evaluating the adhesive strength of adhesive coatings (e.g., SAF) exhibits similarities to the lap cling test [88]. In this case the adhesive strength (W) of a polymer coating on a metal substrate is determined by stretching the substrate and recording its elongation (ε) at the onset of debonding, followed by measuring the modulus of elasticity (E) of unsupported polymer film. The adhesive strength is calculated as εE. The results were comparable to the data obtained by peel test [89].

2. Removability

For certain applications, PSPs are required that display a low peel force and give clean, deposit-free separation from the substrate. Generally, removable adhesive joints are those that can be destroyed without damaging the solid-state components of the joint. Conversely, as is known from the work with pressure-sensitive labels and packaging tapes, permanent labels or tapes are those, that can be peeled off only by destroying the face stock material (labels) or substrate (tapes). That is, for permanent laminates the mechanical resistance of the adhesive joint (R_{aj}) is higher than the resistance of the solid components of the joint, of the carrier (R_c) and that of the substrate (R_s):

$$R_c < R_{aj} > R_s \tag{7.15}$$

The resistance of the adhesive joint is characterized by the internal strength of the adhesive (A_c) and adhesive bond strengths towards the carrier (A_{ac}) and substrate (A_{as}):

$$R_{aj} = A_c + A_{ac} + A_{as} \tag{7.16}$$

This means, both the internal cohesion of the adhesive layer and its adhesion (to the carrier, and to the substrate) should be higher than the mechanical resistance of the solid-state components:

$$R_c < A_c + A_{ac} + A_{as} > R_s \tag{7.17}$$

As mentioned earlier, for a removable joint during bond break the adhesive layer will be destroyed, leaving the solid-state components of the construction intact. In this case,

$$R_c \geq R_s > R_{aj} \tag{7.18}$$

Theoretically, this relationship ensures a damage-free separation of the adhesive bonded solid-state components.

For practical use, the failure place is important also. Generally, removability requires bond breaking at the adhesive–substrate interface (e.g., removable labels and protective films) or in some special cases at the interface adhesive–carrier (e.g., double-faced tapes). Thus,

$$R_c \geq R_s \geq A_{ac} > A_{as} \tag{7.19}$$

Correlation 7.19 is valid for protective films also, where bond failure should occur at the interface adhesive–substrate in order to ensure a deposit-free peel-off (delamination) of the protective film after use.

Peel resistance and removability can be regulated according to Pasquali [12] modifying the ratio between the adhesion surface (S_{ad}) and application surface (S_{appl}). For a removable product this ratio R_{su} should be smaller than 1:

$$R_{su} = \frac{S_{ad}}{S_{appl}} < 1 \tag{7.20}$$

This ratio can be modified by using partially coating of the PSA. Actually it is difficult to obtain pressure-sensitive products which whose peel value does not change in direct proportion with the adhesive surface area. Therefore, as suggested in [90] for adhesive balance regulation the adhesion surface area should be diminished (see also later, Formulation). It should be emphasized that reducing the peel resistance to a substrate causes a reduction in peel toward the carrier also. Such behavior can be illustrated by rolling ball test using adherent and nonadherent carrier material. Theoretically, the rolling ball tack (RB_{pa}) of a permanent adhesive (with good anchorage on paper) measured with a PSA sample coated on paper, must be better than its tack on siliconized paper (RB_{si}):

$$RB_{pa} \gg RB_{si} \tag{7.21}$$

As shown in Figure 7.12 for a PSA sample coated on release paper, the lack of adhesion between the adhesive layer and carrier material causes separation of the adhesive layer from the release paper during rolling of the ball. In this situation, the adhesion of the PSA on the ball will be influenced by the flexural modulus of the adhesive layer and its increased contact surface and time, that is, rolling ball tack on release paper is better than on paper carrier (see Table 7.15). On the other hand, for a transfer coated removable adhesive where the anchorage of the adhesive on the paper is lower, the rolling ball tack on paper or on release paper would have less different values:

$$RB_{pa} \cong RB_{si} \tag{7.22}$$

As illustrated by the data of Table 7.15 for a tackified permanent acrylic PSA the RB values on paper and release liner are very different. For a removable adhesive the RB values are similar. This emphasizes the importance of primer coating for removability also.

Debonding (peel-off) requires a separation force. As shown earlier, for removable joints this separation force should be higher than the force of adhesion at the adhesive–substrate interface.

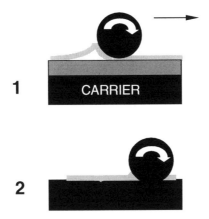

FIGURE 7.12 Delamination of the adhesive layer from the carrier during rolling ball tack test as a function of its anchorage. (1) PSA on release liner; (2) PSA on film carrier.

The adhesion at that interface is a function of the chemical nature of the components, but it also depends on the nature of the applied force. That means, generally for removability, both the force level and its time/temperature dependence are important.

The buildup of adhesion is another time-dependent parameter of removability. As illustrated in Figure 7.1, removable, repositionable and permanent adhesives vary in the rate of the adhesion buildup.

TABLE 7.15
Rolling Ball Tack on Adherent and Abhesive Carrier Material

PSA	Coated Thickness (μm, wet)	Rolling Ball Tack (cm)	
		Abhesive Carrier	**Paper Carrier**
Removable	60	4.0	3.5
		4.5	4.0
		4.0	3.5
	100	2.0	1.5
		2.0	1.5
		2.0	1.5
	200	1.5	1.5
		1.5	1.5
		1.5	1.5
Permanent, tackified	60	3.5	11.5
	100	3.5	13.0
	200	3.5	13.5
	60	2.5	3.5
	100	2.5	3.5
	200	2.5	3.5
	60	1.5	4.5
	100	1.5	5.5
	200	1.5	5.5
Permanent, nontackified	60	2.5	7.5
	70	3.0	9.0
	100	1.5	6.5

The bond strength of a viscoelastic material varies as strain rates and temperature vary. If this is true, removability is not an absolute performance characteristic but a relative one, that is, regulation of the debonding force and its rate allows the control of removability also. Kuik [65] studied the dependence of the adhesion energy as a function of the coating weight and stress rate, for polyethylene and polar ethylene copolymers and stated that the adhesion of self-adhesive thermally sealed films increases with the coating weight and decreases with the speed of debonding.

In the industrial practice, debonding force acts via the carrier material, that is, the carrier material will be tensed and peeled-off and debonding force is transmitted through this material. Therefore, as discussed earlier, the distribution of the debonding force as absolute and time dependent value will strongly depend on the plasticity and elasticity of the carrier material, that is, on its stress damping properties (see Chapter 3, Section I). This means, mutual interaction exists between the characteristics of the solid-state carrier material and adhesive properties. Both bond forming and debonding depends on the viscous flow of the solid-state carrier web.

Ideally, permanent PSAs failure is 100% cohesive, indicating that the maximum strength of the bond has been reached. In this case when failure occurs, a layer of adhesive remains on each adherend. However, specimens may exhibit adhesive failure initially and cohesive failure after aging or vice versa. The type of the carrier surface may also influence the test results. As an example, clay coated papers give paper tear if the superficial strength of the layer is not enough [91].

Repositionability and readherability are performance characteristics dependent on peel buildup. Both allow the debonding of the PSP after its application on an adherend and its re-bonding. Repositionability is an aging related characteristic; it needs a slow buildup of the adhesion on the substrate. In this case, peel buildup is delayed in time, but is not limited as an absolute value. Repositionable labels may also be permanent. Readhering labels are removable after their application independently of the aging time, which means, their peel buildup is of limited value. There are different possibilities to achieve removability (e.g., softening or crosslinking of the adhesive, regulating its geometry, etc.). The formulation related industrial possibilities for removability are described in detail in [92]. Readhering requires balanced adhesive properties and excellent bonding characteristics. Therefore, the only manufacturing technique that can be used for readherable products are those, that do not influence the instantaneous tack of the adhesive. Repositionable and readherable products are required in those PSP classes where instantaneous tack provides product applicability. For instance, abrasion resistant automotive decoration films with resistance to the impact of stones must be repositionable [93].

In some cases crosslinked compositions having a reduced adhesive contact surface are used for readherable products [94]. In such formulations pressure-sensitive polymer filler particles having a diameter of $0.5-300$ μm can also be used [94]. To improve the detackifying effect of the particles, the formulations may include an ionic low tack monomer [96]. The composition of the PSA polymer according to a Nichiban patent [95] includes a crosslinking agent also used for both the matrix polymer and particles. The matrix polymer covers the surface of the particles also. The particles are synthesized via suspension polymerization. The T_g of the particles should be lower than 10°C; these are elastic. Polyisocyanate derivatives, polyepoxides or aziridines are suggested as crosslinking agents. A low coating weight (7 g/m^2) is used. The readherability of the product on paper has been tested by measuring the peel after repeated lamination and delamination ($50-100$ steps). Readhering and removable adhesive in a solid stick form (glue stick, applicator crayon) is also used [97]. Such products are removable, unless the user applies a very heavy coat. A natural rubber latex, a tackifying agent, and a gel-forming component are blended. The adhesive-coated paper substrate may be readily lifted and removed from the contact surface and can be reapplied at least eight additional times to the paper adherend surface. The product must be coated along one edge with a strip of adhesive.

As mentioned earlier, removability is related to adhesion buildup, that is, to the change of the adhesive performances over time. Therefore, removability must always be examined with respect to aging also. A removable adhesive that ages poorly, looses its cohesive strength and will not peel cleanly. In some cases the adhesive properties change after the application of the adhesive. As is known, aging stability characterizes the resistance of the adhesive to such undesired changes. However, in certain cases the changes of the adhesive performances over time are inherent properties of the formulation used or they are designed by the formulator of the adhesive. Changes of the adhesive properties that are inherent for a given formulation are more pronounced for systems with an apparent character of the pressure-sensitivity, that is, with a limited shelf life of tack or peel. Such systems may lose their tack or peel due to the loss of flow properties. Generally, the flow properties disappear due to the buildup of a structural order (network) in the polymer. Therefore, crosslinking, crystallization, plasticizer migration, plasticizer volatilization can change the flow properties of a formulation. Postplasticizing due to absorption, migration of liquid laminate- or atmospheric components also affects the flow properties adversely.

For instance, propylene copolymers with short-chain comonomers are used for so-called semi-pressure-sensitive adhesives [17]. Such products lose their tack after a well-defined time because of crystallization. The nature of the comonomer influences the crystallization rate of the polymers. Their pressure-sensitivity can be improved by using tackifier or plasticizer and SBCs. Such loss of pressure-sensitivity may occur by crystallization of acid rosins also. Generally, a pronounced loss of pressure-sensitivity is caused by crosslinking too (see later), although for some self-curing acrylics (according to the supplier [98]) the balance of adhesive characteristics is not altered by the degree of curing. As discussed in Chapter 3, the influence of the crosslinking depends on the molecular weight between the network points also. The molecular weight between crosslinks may be larger than M_n of the base polymer. For a low M_c the increase of the crosslinking degree is not accompanied by an increase of the elastic modulus [99]. The crosslinking method also affects the M_c and crosslinking density (see Chapter 6). As stated in [100], the cross-linking of polyurethane (PUR) acrylates by electron beam (EB) restricted the mobility of the polymer chain less strongly, than chemical crosslinking.

As discussed earlier, plasticizer migration from self-adhesive films based on plasticized PVC may cause those films to lose adhesion also. In a similar manner, volatilization of the solvent (plasticizer) from pressure-sensitive formulations with high T_g base polymers may produce a loss of their adhesion. For instance, such formulations on acrylic basis have been used for self-adhesive letters (transfers). These are coated by means of printing on a relatively soft polyolefin carrier material and are applied on the substrate by pressing (embossing) the backside of the carrier. High T_g acrylates (known from printing ink technology) have been used for such formulations. Their pressure-sensitive adhesivity (flow) has been given by a very high boiling point (over 250°C) plasticizing solvent (e.g., bornyl acetate). For such products the volatilization of the solvent caused a semipressure-sensitive character of the product.

According to Czech [101], an inverse phenomenon appears in certain formulations used for splicing tapes. For water-dispersable, soft adjusted splicing tapes, the adhesion depends on humidity. The dependence of the adhesion (P) on the relative humidity (H_R) is given by the (empirical) formula:

$$P = A(1 \times 12/10 \ H_R) \times H_R^2/200 + 7 \qquad (7.23)$$

where A is a parameter depending on the composition of the water-soluble PSA [101]. Removability of PSAs is discussed in detail in [2]; formulation for removability is described in [102].

3. Adhesive Properties of Labels

Labels are a special class of pressure-sensitive products generally manufactured and supplied as laminate including at least two solid-state components. These solid-state components are bonded with a pressure-sensitive adhesive, and the surface quality of the release liner is regulated in a such manner as to allow a light separation of the components. The adhesion between the solid-state laminate components has to ensure the stability of the label. Therefore, the adhesive properties of label components are determined by the end-use adherence of the product on the substrate and its temporary adherence on the release liner. Unlike tapes and protective films, for labels the adhesion between the release liner and the PSA has to be regulated exactly.

The adhesion between the pressure-sensitive layer and the release liner depends on both components. It influences the application speed of the labels (roll labels) and therefore, it should be exactly controlled. Because of the low level of such adhesion it is sensitive towards the test conditions. It is well known, that peeling the label from the backing and peeling the backing from the label gives different peel values. Generally, peeling the backing from the label gives lower adhesion values. Depending on the mechanical characteristics of the solid-state components of the laminate and the type of silicone, a high speed release force test may or may not give a maximum of the peel force as a function of the debonding speed.

Labels need balanced adhesive properties. This balance refers mainly to sufficient tack and peel on surfaces that are difficult to bond. Cohesion is less important for labels. As stated by Chu [34], the application window for (permanent) labels is characterized by G' value of 20 kPa at room temperature. In comparison, tapes with high cohesive strength have G' values between 50 and 200 kPa.

According to Bonneau and Baumassy [27], the main requirements for adhesives used for labels are (1) good adhesion to untreated LDPE and retention of the adhesive properties over a wide temperature range. Labels exhibit only medium shear resistance. Shear values of 12–680 min (on steel, according to PSTC) are acceptable for labels only. Packaging tapes need higher shear [23]. Table 7.16 shows some typical values of adhesive properties of labels.

4. Adhesive Properties of Tapes

Unlike labels, tapes are self-wound laminates. Their delamination is influenced by the release properties of the back side of the carrier material, by the adhesive, and by the delaminating conditions. Compared to labels the adhesive properties of a tape on a given substrate are less influenced by the quality of the carrier material back, by the surface of the adhesive, and by application conditions (temperature, pressure, and speed). This is due to the different application conditions of labels and tapes. It should be taken into account that the adhesive on a label has to allow the label to

TABLE 7.16
Adhesive Properties of Labels

Value Grade	Adhesive	Carrier	Tack Value	Tack Method	Peel Resistance Value	Shear Resistance Value	Reference
Permanent	AC,WB	PET	40–45 oz/in.	LT	55 oz/in.	4–8 h	[103]
Permanent	AC, crosslinked	—	11 cm	RB	11 N/25 mm	—	[104]
Permanent	AC	PET	36–37 oz/in.^{2}	LT	37 oz/in.	3 h	[105]
	SBR	PET	54 oz/in.^{2}	LT	46 oz/in.	11 h	[105]
	EVAc	PET	54 oz/in.^{2}	LT	69 oz/in.	—	[105]
Permanent	AC, SB	PET	—	—	11 kg/25 mm	—	[106]
	AC, crosslinked	PET	—	—	10 N/20 mm	>24 h	[107]
Permanent	AC	PET	1100	Polytack	13 N/25 mm	8 min	[25]
Permanent	AC	PET	1350 g/in.^{2}	Polytack	25 N/25 mm	1 h	[25]
Permanent	AC, crosslinked	PET	—	—	12 N/25 mm	—	[108]
Removable	AC, crosslinked	PE	—	—	2 N/25 mm	—	[109]

adhere and bond (sometimes) on a given surface instantaneously and (sometimes) instantaneously. However, due to the special application conditions of a label (i.e., well-defined and relatively small dimensions, static adherend surface and/or very low shear force due to its own weight only), its adhesive has sufficient time to buildup the final bond force. Packaging tapes are applied on dynamic surfaces or on items with a trend to move, where the adhesive has to instantaneously balance forces of motion during application of the tape-work shear and peel forces in the case of a PSA layer. Under such conditions the bonding force has to instantaneously attain almost its full final value. Such behavior is enhanced by using high coating weights. As shown by Gent and Kaang [110], the pull-off force acting on a packaging tape is a resultant of peel, shear, flexural and tensile stress. It can be considered as a vector resulting from cling and lap peel debonding. Therefore, the detachment energy is lower than the energy obtained from peeling. According to Schroeder [60], the adhesion of a tape depends on its tack, instantaneous peel, peel buildup and wetting-out of the substrate surface.

As mentioned earlier, in the first period of development of the pressure-sensitive tape technology, dwell time was referred to as a conformation time [60]. Different tapes require different lengths of time to conform. For certain products that time has been as long as three months. Generally, tapes need high shear values, but the requirements for shear resistance differ according to the tape specialties. Because of the variety of carrier materials and substrates used and the different requirements fulfilled by tapes, as illustrated by the data of Table 7.17 and Table 7.18, their adhesive characteristics are also different.

Medical tapes require tack, peel and good shear adhesion [131]. It is well known that for many adhesive formulations for tapes in order to prepare high solid content solutions the elastomer is masticated. Therefore, postcrosslinking is necessary. Advanced crosslinking causes low tack. Such high shear, low tack formulations are used for contact adhesives, mounting tapes and double faced-foam tapes [24]. To provide high tack and conformability, high coating weights are applied. For instance, for mounting tapes a 88 μm thick adhesive layer is coated, whereas for splicing tapes 50 μm adhesive thickness is used. For mounting tapes peel values on steel and on polycarbonate have been tested. For high temperature splicing tapes shear at 250°C has been measured. For mounting tapes the target values of 180° instantaneous peel/loop tack and 72°F

TABLE 7.17
Adhesive and Mechanical Characteristics of Some Special Tapes

| Application | Adhesive Performances Peel (N/10 mm) | | | | Mechanical Performances | | | |
	Backing	PVC	PE	SST	Break Strength (N/10 mm)	Elongation (%)	Tear Strength (N)	Reference
Marker tape	1.2	1.2	1.5	—	22.0	50–100	—	[111]
Thermally resistant tape	—	10.0	14.0	—	80.0	250	100.0	[112]
Easy tear tape	1.8	1.8	—	—	18.0	100	8.0–10.0	[113]
Flame-resistant tape	1.0	1.0	1.0	—	15.0	160–300	7.6–10.6	[114]
Harness wrap tape	1.0	1.2	1.2	—	15.0	1400	700.0	[115]
Building masking tape	1.0	—	—	5.7	11.0	400	—	—
Insulating tape	—	—	—	1.7	29.0	26	—	—
Insulating tape	—	—	2.0	6.0	—	—	—	[108]
Application tape	0.1	—	—	0.5	29.0	340	—	—
Closure tape	1.0	—	—	2.8	—	—	—	—
Carpet tape	—	4.1	—	—	—	—	—	—

TABLE 7.18
Adhesive Performance of Tapes

Product	Application	Adhesive[a]	Carrier[b]	Adhesive Performance			Reference
				Peel	Tack	Shear	
General	Packaging	—	PVC	230 g/10 mm	—	—	[116]
General	Packaging	AC, SB	—	27 lbs/333 yd^2	—	—	[117]
General	Packaging	AC, WB	Paper	1450 g/in.	—	142 h	[118]
General	Packaging	SIS, HM	—	0.7 kN/m	0.65 kN/m	74 °C	[119]
General	Packaging	SEBS, HM	—	800 g/25 mm	—	—	[120]
Special	Packaging heavy duty	—	PVCR	350 g/10 mm	—	—	[116]
Special	Packaging	Silicone	PET	4.6 N/10 mm	2.25 N/cm	>20 h	[121]
Special	Packaging	AC, CR HMPSA	—	5.7 N/10 mm	—	—	[122]
Special	Packaging	AC, CR	—	925 g/25 mm	—	—	[123]
Special	Bag closure	—	PE	50 oz/in.	—	—	[124]
Special	Closure	SBR, NR SB	PE	2.8 N/10 mm	—	—	—
Special	Insulating	—	Alu	800 g/10 mm	—	—	[125]
Special	—	SBC, HM	—	6.0 kg/30 min	—	—	[126]
Special	Heat-resistant	Fluoro-derivatives	—	22 g/10 mm	14 cm	—	[127]
Special	Mounting	AC, WB	PET	4.5 lb/in.	6.5 lb/in.	400 × 8.8 psi	[24]
Special	Splicing	AC, WB	PET	2.2 lb/in.	1.5 lb/in.	250 × 8.8 psi	[24]
Special	Removable	Rubber-resin	PET	40 g/10 mm	17 N/cm	60°C/5°C/min	[128]
Special	Outdoor	—	Plastic woven	7.0 lb/in.	—	—	[129]
Special	Masking	Butyl rubber	—	0.73 kg/in.	—	—	[85]
Special	Electric	AC	—	5.0 lb/in.	—	—	[130]

[a] AC, CR-crosslinked acrylic.
[b] PVCR-glass fiber reinforced PVC.

shear are 34.5/6.5/400. For splicing tapes, the corresponding values are 2.2/1.5/360. Here the shear resistance has been measured at 250°F. Aluminum carrier and a tackified CSBR coated with 40 g/m^2 are used as insulating tape in climatechnics [54]. Shear resistance after water immersion (storage) is tested for such tapes also. Shear values of about 600–1000 min (25 × 25 mm, 1000 g, steel) and peel values of 25–35 N/25 mm (20 min, steel) are required. Adhesives with excellent water resistance (humidity and condensed water resistance) are required for technical tapes and pressure-sensitive assembly parts in the automotive industry [54].

As illustrated by the applications listed above, various adhesive properties are required for tapes. They differ from common and special products. Therefore, the raw materials used for PSAs for tapes are also different.

As discussed earlier, a special acrylic raw material for acrylic HMPSA has been developed that can be cured photochemically. The uncrosslinked product does not have balanced adhesive properties. Because of its low molecular weight it is tacky but does not have sufficient shear resistance. The pressure-sensitive properties are only achieved by crosslinking [52]. Both the peel and shear resistance of this polymer are inadequate. The product needs tackifying and crosslinking. A low level of tackifier resin (8.5%) significantly improves the peel and tack but reduces shear resistance of the adhesive. For the unmodified polymer with a high coating weight of 50 g/m^2 an instantaneous peel adhesion of only 7.10 N/25 mm has been achieved. The shear value is about 7.3 h (23°C, 2 kg, 5 cm^2). The tackified formulation gives 10.5 N/25 mm as peel value. A transfer coated PSA (with 15% tackifier resin) provides after irradiation 25.6 N/25 mm peel resistance (on steel) and 27.5 N/25 mm quick stick, respectively 18.75 N/25 mm peel on polyethylene and 44 min shear resistance (1.2 × 1 in.^{2}, 500 g). The adhesive layer irradiated on the face stock material gives 20 N/25 mm peel on steel, 16.87 N/25 mm quick stick, 6.25 N/25 mm peel on polyethylene, and a shear value of 77.5 min. Unfortunately, the shear values given by the authors are not comparable. According to [43] such formulation may be an alternative to solvent-based adhesives for tapes and labels.

As stated by Jacob [132], styrene–isoprene–styrene (SIS) block copolymers are used exclusively, as raw materials for HMPSAs for tapes. As a less expensive alternative styrene–butadiene–styrenes (SBSs) may be formulated, but they are not so good. As discussed by Jacob [130], the styrene-butadiene-styrene (SBS) formulations are significantly softer and their peel adhesion values are lower. A proposed formulation of SBS and tackifier for a PSA used for double-faced carpet tape having a high coating weight of 40 g/m^2 exhibits relatively low adhesive characteristics (peel strength of 16.76–21.25 N/25 mm, loop tack of 17–18 N/25 mm, rolling ball tack of 3–5 cm, and shear resistance on steel of more than 100 h). According to Donker et al. [133], high cohesion is required for tapes. For packaging tapes a peel value of 18 N/25 mm (on steel), loop tack values of 20–30 N/25 mm and rolling ball value below 5 cm are preferred. For packaging tapes a flap test is carried out also. This gives a combination of shear and peel resistance and tack with the preferred value being 100 min. Shear adhesion to carton at elevated temperature (40°C) is tested for packaging tapes too. The preferred value for this test is 100 min.

As shown by Gerace [134], tapes may be debonded by cleavage and peel. Gerace [134] presents an empirical graphical method that depicts adhesive performance over a wide range of conditions (pull rate and temperature) that are typical for tapes. Lines of equivalent adhesion were drawn to describe the degree of adhesion of pressure-sensitive tapes used to bond exterior trim moldings to automobiles over all combinations of pull rate and temperature. Each adhesive tape was used to bond a 0.5 in. vinyl bar to a substrate painted with 27% dispersion lacquer. Since, the vinyl bar has been semirigid, the separation mode of the bar was a hybrid of cleavage (rigid member) and peel (flexible member). The peak load values for cleavage peel are shown as constant value curved lines called "isocleaves."

Urahama and Yamamoto [135] used a rolling adhesive moment tester for the evaluation of the adhesive properties of tapes. They correlated the rolling adhesive moment to the peel and proved the applicability of the time–temperature superposition principle. The effect of temperature on the 180° peel force can be predicted from the profile of the velocity spectrum of the rolling adhesive moment (in the temperature range of $-20°C$ to $+40°C$). The correlation coefficients to mechanical properties of the adhesive illustrated that the mechanical properties of the acrylic adhesives and those of rubber-based adhesives each had different effect on adhesion.

As discussed earlier, labels need a fine regulation of their adhesion to the release liner. In the case of tapes the control of the unwinding resistance is required. As known, the unwinding resistance depends on the adhesive and carrier characteristics (see Chapter 10). In the case of labels the surface characteristics of the top side of the face stock material have no effect on label delamination. For tapes this (printed or lacquered) surface must exhibit release properties. Its abhesive (release) and adhesive (printing ink anchorage) characteristics must be balanced [136]. As for labels the dependence of peel resistance on the release (carrier back side) as a function of the debonding speed plays an important role for tapes.

5. Adhesive Properties of Protective Films

In the application of protective films, adhesive properties should ensure a fast, low pressure, full surface adhesion, independent of the type of surface, laminating pressure, web thickness and application temperature. While the laminate is functioning, its adhesive properties should ensure the monoblock character of the laminate (no debonding). At the time of separation of the protective film, its adhesive properties should allow easy, high speed debonding of the film.

As discussed earlier, protective films must be removable. Removability can be achieved by formulation of the adhesive and design of the adhesive joint. The choice of an adequate raw material with a built-in removable character, the regulation of this character by way of formulation with viscous components (improvement of the energy-absorbency), and the improvement of this character by crosslinking (hindering of uncontrolled flow, and reduction of tack) are common modalities for regulating the removability of the adhesive. On the other hand, improving the anchorage and stress distribution (by surface treatment, primer coating, and direct coating) are current modalities for controlling the removability via adhesive joint design. Other possibilities concern the regulation of the coating weight and the coating geometry (see later).

As can be seen from the above evaluation of possible ways to achieve removability, both crosslinking and the reduction of the coating weight involves strong reduction of tack. As mentioned earlier, the use of untackyfied formulations (acrylics), or tackifier formulations with low tackifier level (in comparison with that suggested for labels or tapes) provides low tack (see also Chapter 8). The use of high melting point resins (rubber-resin formulations) and of hard comonomers (acrylics) diminishes tack also.

Protective films exhibit very low tack values (see Table 7.13). Therefore, the adhesion–cohesion balance, and the corresponding tack-peel-shear diagram can by simplified for protective films to a two parameter plot (as shown in Figure 7.11), were the adhesive properties are regulated via peel (i.e., the final strength of the joint) and shear; but shear itself is not a primary requirement, it is only a side effect of the high degree of crosslinking. Therefore, because of the built-in character of the shear, the peel remains the sole adhesion parameter. Thus (at least theoretically), the adhesive properties of the protective films are controlled via peel, and PSAs for protective films are formulated for peel values only. As can be seen from the product literature [137,138] protective films are classified according to their adhesive strength, that is, peel. Table 7.19 presents the peel resistance values of common protective films.

As known from the industrial practice, the peeling of removable PSAs depends on the peel force value and removability. Both a low force separation and an adhesive (residuum)-free debonding are required. The first is characterized by the peel value; the second depends on the subjective examination of the failure mode and place. The peel force is measured in a laboratory on standard test surface and under standard conditions. The values measured are adequate to characterize a product, but are only indicative with respect to the use of this product for a given surface and the real peel value from this surface.

TABLE 7.19
Peel Values of Some Common Protective Films

Product Characteristics		Peel Resistance (N/10 mm)		
Adhesive	Carrier Thickness (μm)	Peel on SS	Tolerance	Peel on Carrier Back Side
Rubber-resin	50	0.12	0.03	—
Acrylic	50	0.20	0.07	0.01
Acrylic	50	0.50	0.20	—
Acrylic	50	0.60	0.20	0.10
Rubber-resin	50	0.60	0.20	—
Rubber-resin	80	0.90	0.30	—
Acrylic	80	1.10	0.30	0.30
Acrylic	50	1.10	0.30	—
Rubber-resin	80	1.10	0.30	0.10
Rubber-resin	90	1.10	0.30	—
Rubber-resin	90	1.10	0.30	0.30
Rubber-resin	90	1.20	0.30	—
Acrylic	50	1.40	0.30	0.40
Rubber-resin	110	1.40	0.40	0.50
Rubber-resin	80	1.40	0.40	0.50
Acrylic	70	2.50	0.50	1.70

Note: The protective films listed above are coated with solvent-based PSA excepting the last column ("Acrylic") position; 180° peel on stainless steel (SS) was measured.

It is well known from the work with PSAs, that the value of the peel force differs according to the chemical nature of the substrate and its polarity. In the case of protective films this phenomenon is more complex. Substrate surfaces that are chemically same can have quite different degrees of roughness. Moreover, although application (laminating) is carried out under pressure to compensate the lack of the tack, it influences the peel value too (see Table 7.12). Table 7.20 lists the main parameters affecting the peel value of protective films.

TABLE 7.20
Parameters Influencing the Peel Resistance of Protective Films

Bonding Parameters			Debonding Parameters		
Product Parameter	Substrate Parameter	Application Conditions	Product Parameter	Substrate Parameter	Delaminating Conditions
Adhesive characteristics	Chemical nature	Temperature	Adhesive characteristics	Substrate stiffness	Temperature
	Roughness	Pressure			Rate of delamination
Coating weight	Physical treatment	Time	Carrier stiffness		Laminate age
Adhesives age	Cleaness		Carrier deformability		
Carrier conformability	Surface age		Carrier elasticity		

As can be seen from Table 7.20, the substrate has a primary influence on the bond strength of protective films, the adhesive age, and adhesive too. Therefore, along with the chemical nature of the surface to be protected, the surface roughness and the application conditions, the age of the surface and that of the adhesive layer (and protective film) can act as supplemental parameters. Surface aging has especially high influence with physically treated or lacquered items (Table 7.21).

It is known from the tests of PSAs that adhesives of the same chemical nature exhibit different levels of peel on surfaces that are chemically different. Generally, there is an univocal correspondence between the surface quality (nature) and peel value, that is, the different adhesives will always give higher peel values on steel or glass than on polyethylene. Because of the strong influence of the surface roughness, carrier deformability, and laminating conditions, for protective films this unequal correlation is not always valid.

In the application of labels and tapes, the peel on standard polar and unpolar surfaces serves as a sure application guide, that is, peel values determine the choice of product for a given application. As can be seen from Table 7.22 and Table 7.23, protective films having the same peel value are recommended for quite different applications, according to the roughness and chemical nature of the product surface to be protected (see also Chapter 11).

It is known that rubber-resin PSAs (having a much lower T_g), are softer than acrylics. Therefore, such products are recommended for rough surfaces, where intimate contact between adhesive and substrate is difficult. In this case products having the same standard peel are suggested for different applications. On the other hand, certain metals catalyze the aging of natural rubber, that is, in such cases rubber-resin PSAs cannot be used. The above examples illustrate that parameters other than peel also enter into the choice of a protective film for a given application. Peel resistance, lamination, and working conditions influence the performance and thus the choice of protective films. In such a situation the manufacturer of protective films should have a broad range of PSAs on different chemical bases, that have the same standard peel value in order to fulfill the different application requirements.

The peel value can be regulated via formulation and coating technology. "Formulation" means the choice of appropriate raw materials and the working out of a recipe. Regulating peel through coating technology means making the choice of laminate components, coating weight, coating technology, and drying conditions.

Working out a recipe involves testing it in both uncrosslinked and crosslinked formulations. The latter is affected more by adhesive manufacturing technology, primer coating and drying conditions. The choice of the laminate components has to do with the manufacture of a coating with or without a top coat (primer). The use of a primer improves the anchorage of the adhesive on the carrier film and reduces the free flow (uncrosslinked) zone of the adhesive, that is, improves its

TABLE 7.21
Influence of Aging on the Peel Resistance of Protective Films

Code	Coating Weight (g/m^2)	Product Age (months)	Peel Resistance (N/10 mm)		
			Stainless Steel	Lacquer 1	Lacquer 2
1	4.0	1	0.90	1.05	1.10
	4.0	13	1.20	1.38	1.47
	4.3	16	1.19	1.50	1.65
	4.5	17	1.33	1.47	1.53
2	7.0	1	1.50	1.60	1.76
	6.7	15	1.63	1.71	1.80
	7.2	17	2.16	1.99	2.00
	5.1	22	1.83	1.79	1.89

TABLE 7.22
Application Domains of Protective Films as a Function of Their Adhesivity

Product Characteristics		Adhesivity				
Type of Adhesive	Carrier Thickness (μm)	Very Low	Low	Medium	High	Very High
Acrylic	50	—	3	4	10	—
Acrylic	70	1	—	—	13	—
Acrylic	80	—	—	—	11	—
Acrylic	100	—	2	—	9	—
Rubber-resin	50	—	—	8	16	—
Rubber-resin	60	—	—	6	—	—
Rubber-resin	80	—	—	7	—	—
Rubber-resin	100	—	—	5	12	15
Rubber-resin	130	—	—	—	—	14

Note: The application domains are described in Table 7.23.

removability. As illustrated by Table 7.24 the use of a primer allows the coating weight to be increased (giving more conformability), avoiding adhesive transfer.

Using the same raw materials (but different balances of the components) the same peel level can be achieved with a recipe with or without primed. However, generally for higher coating weights primer-containing coatings are manufactured. As common practice the adhesive and the primer may have the same formulation components, the primer being more crosslinked. The use of a crosslinked contact cement as a primer and in the adhesive enhances adhesion which in turn improves peelability [139]. For instance, the primer increases the bonding on the carrier from 0.5 to 2.4 kg [140].

As discussed earlier (see Figure 7.4) the peel force increases as coating weight increases, which means that, (at least theoretically) for the same adhesive formulation the same coating weight gives the same peel force value. This statement is generally valid for adhesives coated with different coating techniques in the domain above the critical coating weight (see Figure 7.5) also. However, in the industrial practice the situation is more complex because of the use of different coating devices (see Chapter 8) for very low coating weight of crosslinked adhesives. As can be seen from Table 7.25, different coating techniques can, for the same nominal coating weight but different practical coating weight values, lead to the same peel value as a result of differences in the geometry of the adhesive layer. In conclusion, unlike "normal" coating practice (average coating weight values of $15-20 \text{ g/m}^2$ of an uncrosslinked adhesive), where for a given adhesive there is a clear relationship between coating weight and peel value, the influence of the coating device should also be taken into account in the manufacture of protective films.

II. REGULATING THE ADHESIVE PROPERTIES

Because of the various end-use requirements for PSPs there is a need to regulate their adhesive characteristics. The main factors influencing the adhesion are:

1. The properties of the adhesive
2. The properties of the carrier (face stock, liner) material
3. The construction and geometry of the laminate
4. The coating weight
5. The coating technology
6. The age of the laminate

TABLE 7.23
Application Domains of Protective Films

Code	Application Domain	Comments
1	Glossy plastics; moderate processing conditions	Glossy plastic (PMMA, PC, polystyrene) plates; easy peel
2	High glossy metal surfaces	Stainless steel, aluminum, copper and copper alloys; high gloss
	Glass; moderate processing conditions	Glass surfaces
3	Polished metal surfaces	Stainless steel, aluminum, copper and copper alloys polished
	Glass; moderate processing conditions	Glass surfaces
4	Lacquered surfaces with high gloss	Lacquered (polyester, PVDF, AC) surfaces; >60% gloss
	Glossy metal surfaces	Stainless steel, aluminum, copper and copper alloys; glossy; processing, transport, storage
5	Glossy metal surfaces	Stainless steel, anodized aluminum
	Glossy PVC	Glossy PVC; processing, transport, storage
6	Glossy metal surfaces	Stainless steel, anodized aluminum, glossy
	PVC	PVC
7	Glossy and semimatte metal surfaces	Stainless steel glossy and semimatte, anodized aluminum semimatte
	Matte plastics	Plastics (PMMA, PVC, polystyrene); processing
8	Structured, rough surfaces; processing of profiles	Structured rough metal surfaces; processing of profiles and thick coils
9	Lacquered surfaces	Lacquered (polyester, PVDF, AC) surfaces
	PVC profiles	PVC profiles and plates
	Matte structured laminates	Matte structured plastic laminates; processing
10	Bursted, glossy metal surfaces	Stainless steel, aluminum, copper and copper alloys bursted, with a gloss of 20–60%
	Lacquered surfaces	Lacquered (polyester, PVDF, AC) surfaces
	Matte laminates	Matte composite laminates; slight processing conditions
11	Bursted and matte metal surfaces	Stainless steel and aluminum, bursted; copper and copper alloys matte
	Lacquered surfaces with medium gloss	Lacquered (polyester, PVDF, AC) surfaces with 20–60% gloss
	Matte laminates	Matte composite laminates; processing, transport, storage
12	Bursted and matte metal surfaces	Stainless steel bursted, aluminum anodized matte
	Matte, structured PVC	Matte, structured PVC plates and items; processing, transport, storage
13	Lacquered, matte surfaces	Lacquered, matte (polyester, PVDF, AC) surfaces
	PVC profiles	PVC profiles and plates
	Matte, structured laminates	Matte, structured composite laminates
14	Thick coils	Thick metal coils
	Heavy building elements	Heavy building elements of stainless steel and aluminum
	Powder coated surfaces	Epoxy and polyester powder coated surfaces difficult, complex processing
15	Structured, rough surfaces	Structured, rough stainless steel and aluminum surfaces
	Powder coated surfaces	Epoxy and polyester powder coated surfaces; processing
16	Structured, rough surfaces	Structured, rough stainless steel and aluminum surfaces; difficult, complex processing

TABLE 7.24
Coating Weight for Primed and Unprimed Protective Films

| | Adhesive Coating Weight (g/m^2) | | | |
| | Primed | | Unprimed | |
Product Code	Theoretical	Practical	Theoretical	Practical
1	4.00	4.60	3.70	3.75
2	—	—	4.50	4.28
3	4.50	6.60	4.50	3.97
4	4.50	4.30	4.50	4.20
5	4.50	4.80	4.50	4.15
6	5.50	4.96	4.50	4.10
7	5.50	7.30	4.50	3.95

It is evident that the relative importance of these parameters depends on the manufacture technology used for the PSP, whether it is produced with or without adhesive and the complexity of its buildup. The adhesive properties of a classic PSA-coated pressure-sensitive product are determined mainly by the adhesive properties of the PSA. Unfortunately, such characteristics cannot be defined independently from the composite structure of the product. As described in [1], there are numerous possible ways to regulate the adhesive properties by means of the chemical composition of the adhesive, the choice of carrier, and the manufacturing method. This chapter discusses the special features of the control of the adhesive performance characteristics related to the main PSPs.

A. REGULATION OF THE ADHESIVE PROPERTIES WITH THE ADHESIVE

The adhesive properties of a PSP depends on its components, its buildup and the manufacturing method. Of the product components the pressure-sensitive adhesive has the major influence on the adhesive properties of the product. Its composition and geometry are the most important parameters.

1. Regulating the Adhesive Properties of a PSP with the Chemical Composition of the Adhesive

The common way to regulate the adhesive performances of a PSP is to change its chemical composition. Although changes in the chemical formulation made to improve other than adhesive

TABLE 7.25
Influence of the Coating Device on Adhesive Properties

| | Coating Weight (g/m^2) | | Peel Resistance (180°, N/10 mm) |
Coating Device	Nominal	Measured	
Gravure cylinder, line nr. 30	2.0	2.2–2.4	1.5
Gravure cylinder, line nr. 40	2.0	2.2–2.4	1.5
Meyerbar	2.0	1.6–1.8	1.5

Note: Peel on SST was measured.

properties can influence the adhesive properties too (see later in this chapter), formulation for purpose of regulating adhesive properties is the most important modality used to tailor them. It should be mentioned that the manufacturer of the PSP has a limited influence on the chemical composition of the adhesive. As discussed in Chapter 5, except for the new (mainly radiation-based) technology the adhesive is synthesized by special chemical firms. As discussed in detail in [102] formulation offers more freedom to tailor the adhesive properties although in some cases the number of available formulating additives is limited. For a classic, elastomer-based PSA, formulation with a viscous component was at first the only way to achieve viscoelastic properties and the basic pressure-sensitive characteristics. Later, with the synthesis of viscoelastic raw materials the need for such a formulation was eliminated. A PSA based on viscoelastic raw material is formulated to modify the adhesion–cohesion balance. For PSPs displaying permanent adhesivity regulation of the adhesion–cohesion balance means attaining maximum peel and tack without the loss of (to much) cohesion. For removable PSPs, the adhesive properties are characterized by a (time-dependent or virtually time-independent) low level of adhesion. For such products formulation has to reduce the peel force without changing the bond break character, that is, the joint should fail as adhesions break, at a well-defined place (adherend surface). Taking into account the dependence of tack on peel and that of the peel on cohesion, it is difficult to formulate removable PSAs with a high tack or medium tack and high cohesion. Reduction of the peel force reduces the tack and the cohesion as well. As discussed earlier (see correlation 7.11), the peel force is used for bond deformation and bond break, that is, adhesive (and carrier) flow and failure of the adhesive–adherend contact. Adhesive flow is facilitated by using a high level of viscous component. Bond failure is enhanced by low cohesion and contact hindrance, that is, soft or hard cross-linked adhesive formulations. This means, certain removable formulations are not tacky enough or their low cohesion can lead to cohesion failure and deposition of the adhesive on the adherend surface. Therefore, adhesion and cohesion are generally well-balanced only for permanent labels. The most important special requirements, for example, removability and adhesion on nonpolar surfaces, require modifying this balance to reduce the peel level (removability) or to improve the anchorage (PSA for nonpolar surfaces). Although having quite different formulations, label PSA recipes provide adhesion–cohesion balance for both classes of products, which ensures common converting and application properties.

For tapes, the adhesion–cohesion balance is shifted to higher cohesion values. However, because of their higher coating weight, such products can be characterized by using the common evaluation criteria, that is, tack, peel and shear resistance. In quite a different manner protective films do not possess an usual adhesion–cohesion balance. The situation is more complex for adhesiveless products (see also Chapter 8).

a. Tackifying

As discussed in detail in [141] formulation by mixing includes tackification, cohesion regulation, and detackification. Tackification uses high polymers, resins and plasticizers. The regulation of tack and peel by the use of viscous components (resins and plasticizers) is the main formulation method. It is used mostly for labels and tapes. In the case of protective films rubber-based formulations are tackified only. Tackifying, that is, modifying a raw material to achieve higher tack and peel or (rarely) better shear resistance, is a controversial technology where technical development and commercial considerations (base elastomer and tackifier price fluctuations) may impose global changes. As discussed in detail in [2], although basic assumptions have been accepted concerning the mechanism of tackification of elastomers and the characterization of the formulation variants according to adhesive properties and rheological parameters (DMA) has made important advances, only a few experimental data have been published, and in many cases results serve only as commercial arguments for the raw material suppliers. The data given by different authors for an optimum tackifier level are astonishingly different. For instance, according to Pierson and Wilczynski [24] a special, crosslinkable, water-based carboxylated acrylic has been developed. A room temperature shear test at 8.8 psi (0.5 × 0.5 in., 1000 g), and elevated temperature shear test at 4.4 psi (0.5 × 0.5 in., 500 g; 200 and 300°F) have been used for its characterization. Shear times of 83–1000 h at room temperature and a SAFT value of 240°C have been obtained which, according to the authors, compares with the shear performance of high quality commercial solvent-based acrylics. Unfortunately, these values were measured without tackifier, and the corresponding peel and tack values (of 1.1 lb/in., 180° peel, PSTC-1, on PET and quick stick of 0.7 lb/in.) were very low. To improve these values, tackifiers were used at a loading of 12–35% dry tackifier per total solids. The goal of tackifying was to increase peel and tack by 25%. At a 35% tackifier level, 72°F and 4.4 psi, only a shear value of 48 h was obtained. According to [22], higher softening point resins are less miscible with the base polymer. The highest resin level which contributes to the peel is about 50%, but for a high softening point resin, this level is no more than 39%. The best overall balance is obtained with high softening point resins at a level of about 30%. This statement was confirmed by Green [25] for CSBR also. A higher softening point resin gives optimum tack at a lower loading. Bonneau and Baumassy [28] state that a 25–35% resin loading is optimum for a standard acrylic latex. Dunckley [23] suggests that a cohesive acrylic latex (Acronal 80 D) needs 60–70 phr tackifier for paper tear; Bamborough [33] says 80%. Such values are unrealistic. Dunckley [23] asserts (as discussed earlier by Benedek [142]) that carboxylated latex needs higher tackifier level. He is speaking about soft loop tack of such tackified CSBR dispersions with very high tackifier loading and states that the shear resistance increases with the resin loading. Surprisingly, the shear resistance of the studied formulations attains at 100 phr resin level only the shear value of HMPSAs (as is known HMPSAs display low cohesion). Such behavior can be explained only by the use of special resins and experimental conditions. Generally, for CSBR more than 50% resin loading is recommended [22]. It should be noted that the increase of the shear resistance with the resin level was also demonstrated by Mancinelli [13], but in this case special block copolymers were used. Acrylic block copolymers having both soft an hard acrylic phases have been tackified. They possess a star-shaped radial structure. Peel values of 90–105 oz/in. have been found for such tackified acrylic block copolymers for hot-melts. According to [13], optimal adhesive properties are characterized by a tack of more than 500 g (adhesive thickness of 25 μm), peel of more than 90 oz/in. (face stock polyester), and shear resistance of more than 500 min (1/2 × 1/2 in., 1 kg). Although shear resistance increases with tackifier loading (formulations with 75–80% resin were used), Mancinelli states that the crossoverpoint from adhesive failure to adhesive delamination occurs in the 30–45 phr tackifier range. It should be mentioned that the higher tackifier level for such block copolymers that are harder (see Chapter 3, Section I) is usual. Acrylic hot-melt PSAs with thermally reversible crosslinking (like SBCs) are harder (Williams plasticity number 2.7) than the early common acrylics and exhibit better cold flow resistance [51]. The storage modulus of the acrylic block copolymer discussed by Mancinelli [13] has a plateau value of $10^{9.5}$ dyne/cm^2. The soft phase of the polymer has a T_g of $-45°C$; the hard phase displays a T_g of 105°C. It should be stressed that the tackifying of such multiphase polymers differs from the tackifying of formulations based on natural rubber-based or one-component viscoelastomers. Although in both cases the matrix-filler theory can be applied as a simplified explanation of the practical behavior of the tackified systems (see also Chapter 3, Section I), the tackified block copolymer-based system has multiple "filler components," the phase inversion of such systems produces different morphologies, and compatibility with the matrix and filler phase leads to quite different results, and except for some experimental resins, the resin sole does not impart processibility. Natural rubber possesses a network structure, and in such a network the "hard phase" is not tackified. In segregated SBC it is. Therefore, the tailoring of the adhesive properties of SBCs by tackifying is always associated with side effects which leads to pronounced decrease in other parameter values. However, the tackifying of multiphase systems can lead to self-adhesive materials also.

Regulation of the adhesive properties of HMPSAs can be considered in some cases a by product of the viscosity control. According to Donker et al. [133] 80 Pa sec is the preferred maximum viscosity value for high speed coating. One possible way to reduce the processing viscosity of the HMPSAs based on SBC is to increase their processing temperature (upper limit 190°C) or decrease the melting temperature of the polystyrene domains. As discussed earlier (see Chapter 3, Section I) the mobility of the midblock can be evaluated by means of dynamic mechanical analysis. It is related to the softness, viscous flow of the adhesive (i.e., to energy loss modulus), the position of the rubbery plateau modulus, and the ratio of the two moduli (storage and loss modulus), that is, tan δ. The position of tan δ related to temperature, the tan δ minimum temperature ($T_{\delta min}$), and the tan δ peak temperature ($T_{\delta max}$) characterize the rheological behavior. The melting of the polymer given by the disappearing of the end block domains is indicated by the $T_{\delta min}$ and the cross-over temperature of the loss and storage moduli (T_{cross}). Between $T_{\delta min}$ and T_{cross} the end domains soften and dissappear. The difference between $T_{\delta min}$ and T_{cross} gives information about the melt viscosity of the formulation. The contour lines run parallel with the melt viscosity and SAFT. There is a direct correlation between them. Most tackifiers used for HMPSAs are aliphatic, midblock-compatible and do not influence the styrenic domain. Oils act on the polystyrene domain (being compatible with these groups) and thus soften it. Unfortunately, they reduce cohesion. In practice a mixture of resins is used for HMPSAs formulated for tapes. One of the resin is midblock-compatible; the other is compatible with the styrene end blocks. Using a midblock- and endblock-compatible resin for SIS, the desired viscosity value can be obtained with a lower resin level (rubber-resin ratio of around 1/1). If resins are used that are partially compatible with the styrene domains (e.g., rosin esters, modified terpene resin or a modified aliphatic resin), both an increase of the midblock-mobility and softening of the endblock are achieved. Such resins may have a higher molecular weight aromatically modified aliphatic hydrocarbon basis. In a standard HMPSA formulation with 100 parts SIS, 120 parts resin, and 20 parts oil, such resins give similar $T_{\delta max}$ and loss modulus values but a lower T_{cross}, that is, a lower softening temperature for the endblock domains. As discussed earlier for packaging tapes, a peel value of 18 N/25 mm (on steel), loop tack values of 20–30 N/25 mm and rolling ball value below 5 cm are preferred. Such characteristics are achieved if the resin fraction does not exceed 0.53. An optimum value (100 min) for flap test is measured if the oil fraction does not exceed 15 phr. Shear adhesion to cardboard at an elevated temperature (40°C) is tested for packaging tapes also. It is to be noted that although Donker et al. [133] stress the decrease in shear by the formulating oil in common formulations, they state that shear adhesion to cardboard for a HMPSA formulated with a midblock- and endblock-compatible resin is not influenced significantly by the oil level. Midblock- and endblock-compatible resins allow a formulation with less resin and oil. A formulation with 105 phr resin and 12.5 phr of oil is suggested as an optimum. For an optimal PSA packaging tape formulation, $T_{\delta max}$ values of −6 to 0°C and loss modulus values of 72 and 92 kPa at the tan δ valley are measured. As a less expensive alternative to the use of SIS block copolymers for tapes, SBSs may be formulated. They are less suitable because of their greater rigidity (higher modulus) and lower compatibility with tackifier. Aging problems make them inadequate also. The branched isoprene midblock is more flexible and more compatible than the butadiene (butene) one. In some cases the viscosity of SBS-based HMPSA formulations may change during storage. Jacob [132] states that the SBS formulations are "at comparable resin loading" significantly softer and the peel adhesion values are lower. The theories about tackifier practice and mechanisms discussed in [132, 133] are quite different. Donker et al. [133] states the need for special tackifier resins acting on both (midblock and endblock) segments. Jacob [132] shows that most SIS tackifiers are compatible with both the aliphatic middle block and the styrene domains and thus decrease the cohesion. According to Jacob [132], to achieve the softness of polyisoprene segments without influencing the end block polystyrene segments in SBS, it is necessary to use special tackifiers that are compatibile with only the midblock and do not affect the styrene endblocks. A special hydrocarbon resin with a molecular weight of 700 (M_n) and a T_g of 42°C (ring and ball softening point of 85°C) has been developed. According to [132], increasing the oil level in an SBS formulation increases the peel and tack values but does not greatly decrease the SAFT value. According to [133], for

SIS-based adhesives shear resistance is given by the block copolymer characteristics and the oil–resin ratio does not influence it. According to Jacob [132], for a tackifier resin which acts on the midblock of the SBS block copolymer, oil also influences the rigidity (modulus) of this midblock. Therefore, there is a need for oil. Oil-free or plasticizer-free SBSs have not been used commercially. Donker et al. [133] state that SIS of the oil does not influence the midblock. Standard formulation with 200 parts SBS, 125–150 parts resin, and 25–50 parts oil has been used. As is known, the plateau of the storage modulus gives an indication about the rigidity of the formulation. A big difference in the value of the storage modulus (2×10 Pa) was found by Jacob [132] for a formulation with 100 parts resin without oil and one with 175 parts resin and 50 parts oil. The formulation with oil shows a higher modulus value (4×10 Pa) although both formulations have the same T_g. Both formulations give adequate room temperature tack. Tests have been carried out with different stress frequencies at different temperatures (-50 to $160°C$). A PSA formulation for double-sided carpet tape having 220 phr resin and 20 phr oil to 100 phr block (SBS) copolymer (although with a high coating weight of 40 g/m^2) displays relatively low adhesive characteristics (a peel strength of 16.7–21.2 N/25 mm), loop tack of 17–18 N/25 mm and rolling ball tack of 3–5 cm. New midblock-reactive hydrocarbon resins improve the shear values.

An index of the energy absorption is the value of the loss modulus. High loss modulus (peak) at the practical frequency of bonding helps adequate bonding. On the other hand, at the most higher debonding frequency, the adhesive (storage modulus) must exhibit elasticity. A direct correspondence exists between the loop tack, peel, and the loss modulus. The slope of the curve allows predictions of the peel and tack values depending on the formulation. The value of the modulus may influence special technological characteristics of tapes also. By faster unwinding of tapes (increased frequency) the noise of the winding increases because of the increasing peel adhesion at higher frequencies. The tape tends to break more often.

According to [18], phase separation in the radial block copolymer is higher due to the increased organization of the molecule. Therefore, physical crosslinking is also higher. The better connected pure polystyrene domains ensure higher elasticity of the product. The radial copolymer shows a longer linear portion of the rubbery plateau and lower values of tan δ. As is known, the loss tangent vs. temperature plot characterizes the cohesive strength. The lower the tan δ value, the greater the cohesive strength. Tack needs a low modulus for bonding but high modulus at the strain rates and elongations during bond breaking. Admitting the storage modulus value at low frequencies (less than 10^{-2} rad/sec) is related to wetting and creep (cold flow), while the value at high frequencies may be related to peel or quick stick properties (i.e., high speed debonding by tack/peel evaluation), both values are necessary for adhesive characterization. A parallelism (proportionality) has been found between the G' ratios at different frequencies and quick stick values (e.g., a $G'_{100/0.1}$ value of 60.9 corresponds to a quick stick value of 24.1 and a $G'_{100/0.1}$ of 5.5 gives a quick stick value of 15.3 N/15 mm). Increasing storage modulus ratios leads to higher quick stick values. In a similar manner decreasing tan δ ratios (at different frequencies) increases the quick stick values. It is evident that the high elasticity of a radial polymer displayed in mechanical properties affects the adhesive properties, reducing cold flow (wettability) and tack. Cohesion is better for this polymer. On the other hand, on nonpolar surfaces where physical contact between PSA and surface is more important, peel is strongly affected (30% difference).

b. Softening of the Adhesive

As discussed earlier, removability can be achieved by softening the polymer with plasticizer and/or other viscous components. This possibility is used in [143] for acrylic, with paraffinic oil and polyisobutylene added in to ensure repositionability. Internal crosslinking by special monomers (like N-methylolacrylamide) and the simultaneous use of plasticizers have also been suggested [144]. According to [20], removable price labels used almost exclusively incorporate paper carrier coated with rubber-resin adhesive. Natural and synthetic rubber and polyisobutylene, tackified with soft resins and plasticizer have been suggested. The rubber has been calendered and masticated. For these formulations the adhesion buildup on soft PVC and varnished substrates has been excessive after long storage times. No clear adhesion failure has been possible. Legging and migration is characteristic for such formulations. For certain removable (not pure PSA) adhesives legging is required. As shown by Hintz [145], the so-called combination adhesives made as blends of PSAs and nonpressure-sensitive EVAc dispersions (70/30) are used as one-side adhesives. They can be applied as one-side flooring adhesives after a drying time of about 20 min. So-called legging adhesives are also prepared. Here, the addition of a resin allows the regulation of the legging. Flooring adhesives of this type are removable with or without water. For systems removable without water an EVAc-based primer is applied on the substrate in order to allow removability [146]. Removable HMPSAs can contain up to 25% by weight of a plasticizing or extending oil in order to enable wetting and viscosity control. Butyl rubber (15–55 parts) untackified or tackified with PIB (7.5–40 parts) and crosslinked with zinc oxide (45 parts) is suggested for removable formulations [85]. Such formulations give (180°) peel values of 0.73–1.81 kg/25.4 mm. Their peel increases as the tackifier level increases, but at a tackifier elastomer ratio of 40/15 legging appears. As shown in Table 7.26, the value of peel resistance required, differs for different classes of removable PSPs. Therefore, components for both softening and hardening the adhesive are used.

c. Crosslinking of the Adhesive

The chemistry and technology of crosslinking have developed very rapidly in the domain of PSAs also. Some years ago a monograph [147] stated that "the introduction of crosslinking increases cohesive strength ... but reduces tack and solubility ... and thus finds only limited use in pressure-sensitive type adhesives." Actually, it would not be possibile to manufacture special tapes or protective films without crosslinking. A decrease in tack and increase in cohesion are the main consequences of crosslinking.

Some raw materials possess a built-in crosslinkable structure that works physically or chemically (see Chapter 5 and Chapter 6). The best known is natural rubber, but synthetic elastomers possess crosslinked structure also (e.g., SBR or CSBR). It should be noted that crosslinking resulted from the synthesis of several elastomers influencing their adhesive and processing performances. As discussed by Burroway and Feeney [148], the gel content of the SBR determines the processibility of the adhesive formulation. On the other hand, for the same tape formulation (100 parts rubber and 154.5 parts resin) it can change the value of aged tack between 1 and 3 cm, that of the peel between 48 and 53 g/cm, and that of the shear resistance between 10 and 93 min.

TABLE 7.26
Adhesive Characteristics of the Major Removable PSPs

| Product Class | Adhesive Characteristics | | |
	Tack (RB, cm)	Peel Resistance (180°, N/25 mm)	Shear Resistance (HS, h)
Label	2.5–10	0.5–2.5	1–50
Tape	1.5–20	0.5–5.0	1–100
Protective film	10–70	0.05–1.5	10–200
Form	5–10	0.05–1.5	1–20
Separation film	>100	0.02–1.0	10–200

Note: RB represents rolling ball; HS represents hot shear.

Physical crosslinking occurs in specially sequenced copolymers (mainly SBCs or other block copolymers). As discussed in Chapter 5, acrylic hot-melt PSAs with thermally reversible crosslinking (such as SBC) have been developed [55]. Such formulations are harder than the early common acrylics and show better cold flow resistance. This type of crosslinking disappears at elevated temperatures allowing the use of the formulation as a molten material. As discussed in Chapter 5, the cohesion given by physical crosslinking and its pronounced temperature dependence do not fulfill special end-use requirements. In such cases a supplemental network is built up by using chemically or physically initiated (radiation-induced) crosslinking.

Special built-in polar monomers (e.g., nitriles, hydroxy derivatives) can strengthen the polymer also. Chemical crosslinking supposes the presence of unsaturation or physically and/or chemically activatable multifunctional, reactive groups in the polymer. Chemical crosslinking can be initiated using external or built in initiators or crosslinking agents. The use of external crosslinking agents is a common practice for solvent-based acrylics, the synthesis of which presents opportunities for building in reactive groups and the presence of a homogeneous reaction medium in the finished product (dissolved polymer-solvent) without the interference of technological additives (e.g., surfactants, water, etc.) facilitates the easy regulation of the postmodification reactions. Generally, polymeric isocyanates (0.05–1.5%) are suggested as crosslinker in order to achieve reproducible peel values [36].

As shown in Figure 7.13, the use of crosslinking agents drastically (70–90%) reduces peel resistance. Such pronounced peel reduction is acceptable for protective or separation films only, where the unbalanced adhesive characteristics are improved by film conformability and lamination under pressure. Therefore, the theoretical possibilities to regulate the peel resistance are limited and vary in practical importance for each product class (Table 7.27 and Table 7.28). Because of the predominance of paper as the face stock material for labels, adhesive anchorage is better; therefore, in this case the use of a primer has a secondary role. Modification of the adhesive geometry is also limited, because of the high requirements for smoothness of the PSA layer (e.g., "no label look"

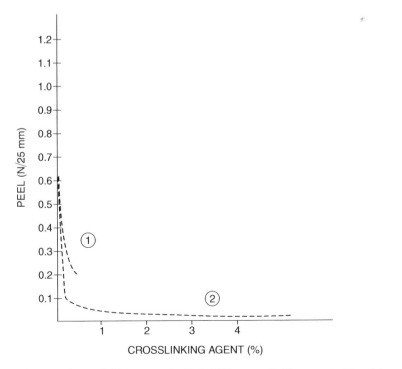

FIGURE 7.13 The influence of crosslinking on peel. (1) Aziridine crosslinking agent; (2) polyisocyanate crosslinking agent.

TABLE 7.27
The Use of Crosslinking for Regulation of the Adhesive Characteristics

PSP	Type of Adhesive	Nature of Crosslinking	Characteristic Improved
Protective film	Rubber-resin	Chemical, using supplemental crosslinking agents	Removability, anchorage on the carrier
	AC	Chemical, using supplemental crosslinking agents	Removability, anchorage on the carrier
	AC	Chemical, self-curing	Removability
Tape	Rubber-resin	Chemical, using supplemental crosslinking agents	Shear resistance, removability, peel resistance
	AC	Chemical, using supplemental crosslinking agents	Shear resistance, dimensional stability, removability, peel resistance
	AC	Chemical, self curing	Peel resistance
	TPE−resin	Radiation	Shear resistance, temperature resistance
		Radiation, using supplemental crosslinking agents	Shear resistance, dimensional stability, temperature resistance
	AC (oligomer) resin	Chemical, using supplemental crosslinking agents	Cohesion
		Radiation	Cohesion
Label	AC (oligomer) resin	Radiation	Cohesion
	Rubber-resin	Chemical, using supplemental crosslinking agents	Removability
	AC	Chemical, using supplemental crosslinking agents	Removability
	AC	Chemical, self-curing	Converting properties

products). As mentioned above, high degree of crosslinking is not possible for labels because of the required adhesion balance. Unlike labels, common tapes have a limited raw material basis; therefore, in this case the PSA geometry, coating weight, and crosslinking (to achieve high shear and peel resistance) are more important. The adhesive properties of common protective films are controlled by coating weight and degree of crosslinking.

The composition of the PSA polymer according to a Nichiban patent [93], although applied as patterned contact surface, includes a crosslinking agent that is used for both the matrix polymer and the particles, and the matrix covers the surface of the particles also. Mueller and Tuerk [20] proposed a water-based acrylic formulation having physically or chemically crosslinkable comonomers like acrylo- or methacrylonitrile, acrylic or methacrylic acid, N-methylolacrylamide and plasticizer (10–30% w/w) also. Special problems appear by the manufacture of acrylic hot-melts with respect to achieving enough cohesion and a low processing viscosity. Two approaches are known (see also Chapter 5). The first is to imitate the buildup of SBCs, that is, to prepare physically crosslinkable segmented copolymers. The second approach is to synthesize low molecular weight linear

TABLE 7.28
The Importance of Specified Regulation Parameters of Peel Resistance for the Main PSPs

Product Class	Control Parameter of the Peel Resistance				
	Adhesive	Coating Weight	Crosslinking	Primer	PSA Geometry
Label	1	2	3	3	3
Tape	3	1	2	2	2
Protective film	3	1	1	2	2

Note: 1, 2, and 3 denote the importance of the peel regulating parameter for a given PSP class, with 1 denoting the lowest importance.

random copolymers. In this case, postpolymerization is needed to increase the molecular weight of the coated product. As stated by Auchter et al. [50], even a high molecular, highly viscous 100% polyacrylate processible as hot-melt is not of high enough molecular weight to display balanced pressure-sensitive properties. Therefore, it should be crosslinked. The crosslinking possibilities of hot-melts are limited, because thermal crosslinking cannot be used. Radiation curing is one of the new modalities used to crosslink such compositions. It is possible to synthesize a prepolymer which is polymerized by radiation. A patent [149] discloses the preparation of a low molecular weight spreadable composition to which may be added a small amount of catalyst or polyfunctional crosslinking monomer prior to the completion of the polymerization. Unsaturated or saturated polymer chains are synthesized exhibiting polymerizability by radiation [150]. A monoethylenically unsaturated aromatic ketone can be used for crosslinking PSA compositions by UV radiation [151,152]. A special acrylic raw material for HMPSA has been developed that can be cured photochemically [52]. The photoinitiator is built into the polymer. The copolymerized photoinitiator is an acrylic ester with benzophenone terminal groups. The polymer itself before crosslinking is a highly viscous fluid (at room temperature its viscosity is about 10–20 Pa sec),which can be processed at 120–140°C. For such polymers the storage modulus can be used as a measure of the crosslinking density. The nonirradiated polymer has a storage modulus of about 10^3 Pa, which can be increased to 10^4 Pa by irradiation. In comparison, acrylic block copolymers with a soft and hard acrylic phase and a star-shaped, radial structure exhibit storage modulus with a plateau value of $10^{9.5}$ dynes/cm^2 [13]. Their viscosity depends on the shear rate but is much higher. At 10 sec^{-1} (177°C) a value of 248,000 cP, and at 100 sec^{-1} a value of 57,000 cP have been obtained. Melt viscosity of tackified compositions is less than 50,000 cP at 177°C.

As discussed in Chapter 5 and Chapter 6, for such systems the main crosslinking reaction is an addition of a hydrogen atom of the side alkyl group of an acrylic segment to the carbonyl group of the side benzophenone group of the vicinally polymer chain, that is, forming of a hydroxydiphenylpolyacryl-methane derivative. Although no unsaturation is present in the macro-molecule, that is, the system is inert to classic aging reactions, other intramolecular photoinitiated side reactions are possible, due to the built-in photoinitiator. To avoid such side reactions, low radiation levels should be used. Low radiation levels supposes the use of a low concentration of formulating components that absorb UV light, that is, the formulating freedom of such systems is limited by the classic tack-peel-shear interdependence and by the manufacturing procedure. For modifying the UV-crosslinkable polyacrylates, esterified highly stabilized rosin is suggested. A low level of such resin (8.5%) significantly improves the peel and tack, but reduces shear resistance. For the untackified adhesive with a high coating weight of 50 g/m^2, an instantaneous peel resistance of only 7.10 N/25 mm is achieved, that is, such a polymer must be tackified. The shear value is about 7.3 h (23°C, 2 kg, 5 cm^2). The tackified formulation gives 10.5 N/25 mm as peel value. In comparison, tackified acrylic block copolymers (with much higher tackifier loading) show peel values of 25–29 N/in. [13]. Optimal properties require a tack of more than 500 g (adhesive thickness of 25 μm), peel of more than 90 oz/in. (face stock polyester), and shear values of more than 500 min (1/2 × 1/2 in., 1 kg). The SAFT value attains 80–86°C. Formulations with 75–80% resin were used. It should be noted that for radial SBCs, G' values at different strain frequencies give quick stick values in the following range: A $G'_{100/0.1}$ of 60.9 corresponds to a quick stick value of 24.1, and a $G'_{100/0.1}$ of 5.5 corresponds to a quick stick value of 15.3 [17]. For UV-curable formulations according to [52], there is an upper limit to the amount of the resin that can be added. It is supposed that the resin acts as a diluting agent in the polymer matrix and reduces the number or contact (crosslinking) points. Therefore, for tackified formulations, the radiation dosage must be increased. For instance, a 8.5 % resin level requires doubling of the radiation dosage! The radiation dosage is the major parameter to influence the degree of crosslinking. For instance, for a soft deep freeze adhesive a dose of 3–5 mJ/cm^2 is suggested; with a dose of 20–25 mJ/cm^2 more cohesive, less tacky, removable PSA is synthesized [52].

Polymers with photoinitiator properties have been developed for different coatings [153]. The photoiniating polymer could be used as a base-coat that could then catalyze the curing of an UV-hardenable top coat. Thus coatings with controlled thickness and very low molecular weight can be applied. Compounds containing aromatic–aliphatic ketone groups and silicone are bonded to alkylene side chains of oligomers to prepare polymeric photoinitiators [154]. UV-curable acrylic esters and polyvinylbenzophenone as initiator have been used as adhesives on plasticized (35%) PVC [155]. These acrylic oligomers can be cured by electron beam radiation also [156].

In such crosslinked systems the flexibility of the crosslinking "bridges," that is, the nature of the oligomers used influence the adhesive properties. For instance, as stated in [157], with the use of polyisobornyl methacrylate instead of isooctyl acrylate the peel is increased (from 53 to 88 N/100 mm). Peel buildup also depends on the oligomer used.

Styrene–butadiene rubbers (random or alternative copolymers) contain a residual double bond that enables further polymer chain growth as well as crosslinking, to form three-dimensional network structures with improved mechanical properties without apparently affecting the T_g [24]. For such polymers internal networks (gel) can be formed during polymerization. Gel content influences the adhesive properties and the ways of regulating them. Higher gel content is an indication of higher molecular weight. Above 30% gel content, CSBR tack decreases rapidly. Shear resistance increases with the gel content. Pierson and Wilczynski [24] demonstrate that although the butadiene content decreased from 69.0 to 62.5% the increase in gel content from 37.0 to 45.0% reduces the loop tack from 1000 to 800 gr (FTM). Deep freeze properties of such formulations are improved if the butadiene level of the polymer is increased above 80%. For such raw materials the built-in crosslinking density (gel) has to be taken into account by regulation of the adhesive properties. It is well known that supplemental crosslinking of rubber-based adhesives builds up new "gel structure." In certain cases the primer used for such adhesives has the same chemical composition as the adhesive; only its curing agent level and degree of cross-linking are different.

It can be concluded that except for permanent labels with balanced adhesive properties for a medium coating weight, advanced crosslinking is a current technology for manufacturing PSPs (Table 7.29).

2. Regulating the Adhesive Properties of PSPs with the Adhesive Structure or Geometry

As discussed earlier, adhesive properties depend on the area of the contact surface between adhesive and substrate. This is a function of the adhesive geometry too. Adhesive geometry is characterized by the continuous or discontinuous character of the coated layer (due to the coating technique or its failure) and the coating weight. An increase in coating weight improves tack and peel and decreases removability. Therefore, regulating the coating weight can help to ensure a balanced adhesion–cohesion level for compositions with chemically enhanced cohesion (tapes with higher coating weight) or removable formulations. As is known from the formulation

TABLE 7.29
Crosslinked PSPs[a]

Product	Adhesive	Curing Agent[b]	Uncrosslinked Peel	Adhesive Characteristics Crosslinked			Reference
				Peel	Tack	Shear	
Tape	AC	UV	440	52	—	—	[155]
Tape	AC	UV	1500	15	—	—	[106]
Tape	AC	BA	—	1000	—	—	[158]
Label	AC	PI	—	120	—	—	[159]
Label	AC	ME	—	1200	—	—	[108]
Label	AC	UV	—	1050	17.8 cm	>24 h	[107]
Tape	AC	PI	—	630	18 cm	<1 mm	[120]
Tape	AC	PI	—	900	—	—	[123]
Tape	Silicone	BPO	—	1000	5.6–25 mm	—	[121]
Tape	Silicone	BPO	—	3500	24 cm	2 mm	[160]

[a]180° peel was measured on stainless steel in g/25 mm.

[b]BA-bifunctional acrylate; PI-polyisocyanate; ME-methylol functionalized monomer.

of protective films, the coating weight applied depends on the degree of crosslinking used for the adhesive. It also depends on whether a primer is used (see Table 7.21). Regulating the adhesive properties by controlling the coating weight can be achieved by knowing the interdependence of coating weight and adhesive performance for a given formulation.

For regulating adhesive performance by controlling the contact surface there are various possibilities. In such cases, the quantitative and shape-related characterization of the effective contact surface between adhesive and substrate is absolutely necessary to evaluate the adhesive properties. The contact surface can be reduced by using pattern-like contact points or pattern-like nonadhesive points (see Figure 2.13). The patent of Pasquali [12] suggests the use of an impregnated textile network, in which the adhesive-coated fibers act as the contact surface and so the adhesive contact surface can be regulated by modifying the dimensions of the network. The PSA surface can have a special discontinuous shape ensured by the PSA itself, as a suspension polymer [95]. Thus, spherical contact sites are formed. Pressure-sensitive polymer particles with a diameter of 0.5–300 μm can also be used as the filler [74]. Such pressure-sensitive filler particles are called for in [68]. Partial coating of the PSA leads to the same results [151,152]. Striped or drop-like coating of PSAs has been carried out for removability [161]. The adhesion buildup can be decreased according to [162] by using a powder-like surface coating of the PSA layer with particles having a diameter of less than 10 μm embedded in the PSA, with two-thirds of their diameter above the PSA level. The particles are pressed into the mass [162]. Contact surface can be reduced by pattern-like crosslinking also [163]. The same effect is obtained if the carrier is coated with alternate adhesive and release stripes, according to [164].

These procedures have the following disadvantages:

1. The formulation contains plasticizer (migration), thickener and surface active agents (migration, hydrophylicity).
2. The adhesion increases over time. This increase depends on the surface nature and the quality of the adherend. Therefore, a full surface coating does not ensure removability for all types of surfaces.
3. If the adhesive is coated in a discontinuous manner the adhesion increases over time on the contact sites, and so there may be high peel resistance in the longest contact direction.

Coating the PSA with spherical PSA particles (50–150 μm) does not display this behavior.

4. Unfortunately, the anchorage of the spherical PSA particles needs a continuous matrix and a high coating weight. On porous surfaces this coating weight is difficult to control. A supplemental primer is necessary. The spherical particles have to be coated as monolayer.

5. Removable coatings having alternate adhesive and abhesive stripes have to be achieved in two different coating steps, but with the same coating thickness.

Another procedure without such disadvantages uses pyramidal PSA-coated sites (30–600 μm base dimensions) coated by means of screen printing [150]. Rotary screen printing can be used, with a coating weight of $1-20 \, g/m^2$, using a screen geometry with 15–40 holes/cm, blade of 1.5–30 mm; blade thickness of 150–300 μm, and pressure of 2–6 mmHg. Regulation of the contact surface can be achieved inversely also, that is, by crosslinking the whole adhesive surface and making it sufficiently fragile to allow crack migration of the uncrosslinked adhesive [151].

The contact surface can be reduced by using expandable fillers also. According to [106], a photocrosslinked acrylic-urethane acrylate formulation used for holding semiconductor wafers during cutting and releasing them after cutting, exhibits a peel strength of 1700 g/25 mm without crosslinking and 35 g/25 mm with crosslinking. Adding an expandable filler to the formulation (6 parts to 100 parts adhesive) reduces the peel resistance. Such composition exhibits 1500 g/25 mm peel resistance before crosslinking and 15–29 g/25 mm after crosslinking. Such a releasable foaming adhesive sheet can be prepared by coating a blowing agent (e.g., a microcapsular type) containing PSA (30%) as a 30 μm dry layer [106]. This type of adhesive is removable and improves cuttability. Regulation of the adhesive properties with adhesive geometry is discussed in detail in Chapter 8.

B. REGULATING THE ADHESIVE PROPERTIES WITH THE CARRIER

In the classic manufacture technology for a PSP using a nonadhesive carrier material, that is coated with a PSA, the adhesive characteristics of the product are influenced by the surface and bulk properties of the carrier material. The influence of carrier material on the adhesive properties of PSAs is described in detail in [2]. As discussed in Chapter 2, unlike labels that have balanced adhesion–cohesion given by the PSA coating and where product design and manufacture minimizes the influence of the solid-state laminate components on the adhesive properties, in the manufacture of certain tapes and protective films, the conformability of the carrier and its mechanical resistance are taken into account as parameters of the adhesive properties.

The problems are more complex for extruded pressure-sensitive tapes or protective films where the diffusion characteristics of the carrier material can influence the migration of the pressure-sensitive components, that is, the adhesive properties of the final product. Mechanical stresses acting on the finished product (and depending on the product's own mechanical characteristics) can also influence adhesive diffusion. In some cases the manufacturing method used for the polymer film (blown or cast), and its crystallinity influence the migration of the pressure-sensitive component and the adhesive properties (see also Chapter 8). Similar problems may appear for special products having a (radiation) polymerized carrier material. According to [68], using UV photopolymerization of acrylics, the PSA can be synthesized as an adhesive layer supported by a nonadhesive layer (carrier or rigid body), or as double-faced systems. Such a carrier may be nonadhesive, although for certain applications the carrier itself has some pressure-sensitivity. Thick, triple layered adhesive tape, with filler material in the center layer can also be manufactured. Such a product is really a carrier-less tape and thus has better conformability and lower flexural resistance, both of which influence the adhesive properties. For instance, a foamed pressure-sensitive adhesive sheet prepared from an agitated acrylic emulsion according to [165], exhibits a higher peel strength (4.3 kg/cm) than the same adhesive coated on a PUR foam carrier material (1.5 kg/cm).

1. Influence of Carrier Material Surface Properties on Adhesive Properties

Coatability is characterized by wetting out and anchorage of the coated adhesive on the carrier material. Both are influenced by surface polarity and porosity. Special problems are encountered with nonpolar plastics where no chemical affinity exists between the polar PSA and the nonpolar, nonporous carrier surface. For textured, fiber-like porous surfaces, adhesive migration and breakthrough may improve anchorage but denatures the top surface of the face stock and at the same time change the coating weight. Fillers or surface-active agents built into the carrier material of protective films or tapes can act as release-enhancing agents, but they disturb the adhesive contact. Fillers can modify the relaxation spectra of the polymers and increase the relaxation time as a result of the retardation of relaxation in the vicinity of the filler surface [166]. Filler effects on rheology are examined in detail in Chapter 3.

2. Influence of Carrier Material Bulk Properties on Adhesive Properties

Paper-based carrier materials are humidity-sensitive. Their mechanical characteristics and isotropy may change as a function of the climate. Plastics are more temperature-sensitive. Both types of carrier materials can change dimensionally and in mechanical resistance as a function of environmental conditions. As is known, the mechanical properties of plastics depend on the test/force rate also (see also Chapter 3). Therefore, deformation of the carrier material during application (or test) may influence the adhesive properties also.

It can be seen that for PSPs coated with PSA the decrease in carrier film thickness increases their deformability and reduces their peel resistance. In practical use such films display sufficient adhesion to be processed together practice, the actual values of delamination force have to be taken into account. These result from the deformation of both the adhesive and the carrier. Laboratory tests using a standard carrier material or a re-inforced carrier material does not allow the correct evaluation of practical adhesion values. For adhesive-coated PSPs the bulk properties of the carrier have to be taken into account only for thin or soft plastic films. For pressure-sensitive webs manufactured by coextrusion (e.g., EVAc-based or hot laminating films) or for carrier-less transfer tapes manufactured by stiffening of the adhesive (using filling, crosslinking, foaming etc.) the exact separation of adhesive and carrier components is more difficult, sometimes impossible and therefore, it is not possible to use a standard carrier for laboratory measurements, which means, tests must be carried out with the finished composite. The product has to be evaluated as a whole. The regulation of the adhesive properties by means of the carrier is discussed in detail in Chapter 8.

C. REGULATING ADHESIVE PROPERTIES WITH MANUFACTURING TECHNOLOGY

The manufacturing technology used for PSPs also influence their adhesive properties. Adhesive manufacture and coating technology affects the adhesive performance, carrier performance, and product buildup. As discussed above, the carrier manufacture influences its mechanical, chemical, and surface properties. In special cases adhesive manufacture may be the manufacture of the finished product (carrier-less tapes, adhesive crayons, etc.). In certain cases, the carrier manufacture may be the production of the finished PSP (e.g., SAF). Generally, the manufacture of PSPs include both adhesive and carrier manufacture and their assembly into a finished product.

1. Regulating Adhesive Properties with Adhesive Manufacturing Technology

For PSPs having a carrier material with self-adhesive properties the method of mixing the components influences the adhesive properties. This is evident in the use of tackifying raw materials like polybutenes, where the mixing technology (solid-state or molten, etc.) influences the adhesive properties of the film (see also Chapter 8). Water-based dispersions can be formulated with different levels of supplemental (formulation-related) surfactants. Molten or solvent-based additives can be used. Special techniques allows the use of water-based formulating components without supplemental surfactant loading [54]. Hot-melts can be manufactured via batch processing or continuously. The thermal-oxidative influences on the base polymers are different for such procedures. Solvent-based rubber-resin adhesives can be manufactured with masticated or unmasticated rubber. The degree of mastication can vary according to the equipment used. The tackifying components can be mixed in solution, molten or in solid-state form. As illustrated by the above examples, the various steps of adhesive manufacture influences the adhesive characteristics of the product (see also Chapter 8).

2. Regulating Adhesive Properties with Adhesive Coating Technology

The coating technology of the liquid adhesive may influence the adhesive properties in various ways. The classic example is the regulation of adhesive properties via direct or transfer coating, or by different coating devices. Different anchorage, depth of penetration and geometry of the coated adhesive are achieved by the use of such technologies. As discussed earlier, the classic technology used for low coating weights and special adhesives leads to different peel resistance values for the "same" coating weight depending on the adhesive layer geometry (see Figure 7.11). Such a phenomenon would be unacceptable for labels where in certain cases the "roughness" of the release layer itself has to be improved. In the earlier example, the changes in the coating technology do not affect the adhesive. In other cases changes in the adhesive properties are due to its chemical modification. For instance, using the same coating device and other working parameters may alter the characteristics of HMPSAs. De Jager and Borthwick [167] studied the influence of the coating speed on SAFT and rolling ball tack for SBC-based HMPSAs. No significant effect was found, but the variation of the coating temperature within the interval of 120–180°C caused changes in the rolling ball tack, especially in the MD/CD ratio. A more pronounced temperature influence has been observed in working with an open melt mill or closed die. Changes in temperature between 140 and 180°C produced a decrease of the rolling ball tack from 17 to 3 cm for the open melt mill. For the closed die, the rolling ball tack (5 cm) did not change in this interval. In a similar manner the holding power changed drastically from 30 to 3 h for the open melt mill. For the closed system, the change was less important (from 22 to 17 h).

Nonclassic, special coating methods used for other products have a more pronounced effect on adhesive characteristics. As is known, for the manufacture of certain tapes the solventless, masticated bulk adhesive is warm coated on a textured carrier material in order to achieve high coating weight or to avoid volatiles. In certain cases, the carrier is impregnated with the adhesive. Some products are spray coated. Such coatings have a rough contact surface.

When radiant energy is used to improve the performance of PSAs or to manufacture them, the nature of the radiation energy, its level and the technical details of the equipment and procedure influence the adhesive and other properties of the PSPs. For instance, according to Martens et al. [168], a very precise exposure of $0.1-7$ mW/cm^2 optimizes the molecular weight of the resulting polymer when UV polymerization is used for tapes. As stated in [94], in the photopolymerization of acrylics with 280–350 nm UV light, the light intensity at the surface should be 4.0 mW/cm^2. According to Refs. [168,169], low radiation levels should be used for acrylic HMPSA prepolymers in order to avoid side reactions.

a. Regulating Adhesive Properties by Direct or Transfer Coating

Because of the quite different rheology of the liquid adhesive coated directly on the porous surface of a paper carrier material in comparison with a dried, solid-like, transfer-coated PSA, the anchorage of the adhesive layer is different for each procedure. Therefore, the adhesive properties of transfer- and direct-coated PSA are different too. Special examples of the change in adhesive properties as a function of the direct or transfer coating are given (as discussed above) for the UV-cured formulations in [42,50]. Because the radiation is partially absorbed, and partially transmitted, and partially reflected, the maximum radiation level at the top and bottom of the adhesive layer may differ. The crosslinking degree is the maximum on one side. Therefore, direct and transfer coating gives different adhesive characteristics (see also Chapter 8).

b. Regulating the Adhesive Properties with the Coating Device

Theoretically, the use of a coating device with a smooth coating cylinder gives a smoothly coated adhesive layer. The use of a gravure cylinder can give a discontinuous, point-like coating; the use of wire rod may lead to a coating having linear defects. A formulation required for Meyer bar differs from one coated with a reverse roll coater [24]. As discussed earlier, removability and/or readherability may require patterned coating. In this case the choice of coating device is determinant (see Chapter 8 also).

D. REGULATING THE ADHESIVE PROPERTIES WITH PRODUCT APPLICATION TECHNOLOGY

As discussed earlier, for PSPs that have balanced adhesive properties (i.e., enough tack and peel) the application conditions have less importance. It is mainly temperature that influences the applicability of the product (e.g., deep freezer labels). For low tack products the laminating conditions are determinant for the product application (see Chapter 11).

III. INTERDEPENDENCE ADHESIVE PROPERTIES AND OTHER PERFORMANCE CHARACTERISTICS

Pressure-sensitivity given by adhesive properties is the most important technical property of PSPs. As discussed in a detailed manner in [1,2] adhesive properties also influence conversion and end-use properties. For some pressure-sensitive products such as labels and tapes, where post-transformation of the adhesive-coated web is gaining more and more importance, the converting properties are also important. For other PSPs used as web-like, continuous products with or without or less postconverting, the end-use properties are more important. As discussed earlier, in some special product classes (tapes, protective films) pressure-sensitivity is the result of adhesive and end-use properties. Therefore, it can be stated that generally the adhesive properties influence in a determinant manner the converting or end-use performance characteristics of PSPs. This influence is discussed in Chapter 10 and Chapter 11.

REFERENCES

1. I. Benedek and L.J. Heymans, *Pressure-Sensitive Adhesives Technology*, Marcel Dekker Inc., New York, Basel, Hong Kong, 1997, chap. 6.
2. I. Benedek, *Pressure-Sensitive Adhesives and Applications*, Marcel Dekker Inc., New York, Basel, 2004, chap. 6.
3. R. Köhler, *Adhäsion*, (6), 247, 1968.
4. J.H.S. Chang, EP 0,179,628 A2, Merck & Co. Inc., Rahway, NJ, USA, 1984.
5. Y. Aizawa, Japanese Patent 6,317,945, Cemedine Co. Ltd., 1988, in *CAS, Adhesives*, 14, 4, 1988.
6. K. Dormann, *Coating*, (6), 150, 1984.
7. G.R. Hamed and C.H. Shieh, *J. Polym. Sci.*, 21, 1415, 1983.
8. D.R. Gehman, F.T. Sanderson, S.A. Ellis, and J.J. Miller, *Adhäsion*, (1), 19, 1978.
9. K. Chu and A.N. Gent, *J. Adhes.*, 25 (2), 109, 1988.
10. M. Toyama and T. Ito, *Pressure Sensitive Adhesives, Polymer Plastics Technology and Engineering*, Vol. 2, Marcel-Dekker, New York, 1974, pp. 161–230.

11. I. Benedek, *Pressure-Sensitive Adhesives and Applications*, Marcel Dekker Inc., New York, Basel, 2004, chap. 10.

12. J.C. Pasquali, EP 0,122,847, 1984.

13. P.A. Mancinelli, *New Developments in Acrylic Hot Melt Pressure Sensitive Adhesive Technology*, in Proceedings of the Tech 12, Advances in Pressure Sensitive Tape Technology, Technical Seminar, Itasca, IL, USA, May, 1989, p. 165.

14. M. Schleinzer and G. Hoppe, *Amorphe Polyalphaolefine als Basis Material zur Formulierung von HMPSA und Semi Pressure Sensitive Adhesives*, in Proceedings of the 19th Adhesive and Finishing Seminar, Munich, Germany, 1994, p. 120.

15. C.N. Clubb and B.W. Foster, *Adhes. Age*, (11), 18, 1988.

16. U.S. Patent 3,321,451, in S.E. Krampe and C.L. Moore, EP 0202831A2, Minnesota Mining and Manuf. Co., St. Paul, MN, USA, 1986.

17. A. Haas (Société Chimique des Charbonnage-CdF Chimie, France), U.S. Patent 624,991, 1986, *Adhes. Age*, (5), 26, 1987.

18. C. Parodi, S. Giordano, A. Riva, and L. Vitalini, *Styrene Butadiene Block Copolymers in Hot Melt Adhesives for Sanitary Application*, in Proceedings of the 19th Munich Adhesive and Finishing Seminar, Munich, Germany, 1994, p. 119.

19. R.G. Jahn, *Adhes. Age*, (12), 35, 1977.

20. H. Mueller and J. Türk, EP 0,118,726, BASF A.G., Ludwigshafen, Germany, 1984.

21. G.W.H. Lehmann and H.A.J. Curts, U.S. Patent 4,038,454, Beiersdorf A.G., Hamburg, Germany, 1977.

22. A. Midgley, *Adhes. Age*, (9), 17, 1986.

23. P. Dunckley, *Adhäsion*, (11), 19, 1989.

24. D.G. Pierson and J.J. Wilczynski, *Adhes. Age*, (8), 52, 1990.

25. P. Green, *Labels Labell.*, (11/12), 38, 1985.

26. J. Pennace and G.E. Kersey, PCT, WO 87/035537, Flexcon Co. Inc., Spencer, MA, USA, 1987.

27. G. Bonneau and M. Baumassy (DRT), *New Tackifying Dispersions for Water-Based PSA for Labels*, in Proceedings of the 19th Munich Adhesive and Finishing Seminar, Munich, Germany, 1994, p. 82.

28. I. Benedek, *Pressure-Sensitive Adhesives and Applications*, Marcel Dekker Inc., New York, Basel, 2004, chap. 7.

29. A.T. DiBenedetto, *J. Polym. Sci., Part B., Polym. Phys.*, 25 (9), 1949, 1987.

30. D.H. Kaelble, Proceedings of the Int. SAMPE Symp. Exhib., 33, 153 (1988), in *CAS, Adhesives*, 15, 2, 1988.

31. H. Gramberg, *Adhäsion*, 10(3), 97, 1966.

32. D.W. Bamborough and P.M. Dunckley, *Adhes. Age*, (11), 20, 1990.

33. D.W. Bamborough, *Water-Based Adhesives Dispersions for Labels: Prediction of Adhesive Performance Using Dynamical Mechanical Analysis*, in Proceedings of the 19th Munich Adhesive and Finishing Seminar, Munich, Germany, 1994, p. 96.

34. S.G. Chu, Visco-elastic properties of pressure sensitive adhesives, in *Handbook of Pressure Sensitive Adhesive Technology*, 2nd ed. D. Satas, Ed., Van Nostrand Reinhold Co., 1989.

35. I. Benedek, *Pressure-Sensitive Adhesives and Applications*, Marcel Dekker Inc., New York, Basel, 2004, chap. 3, Section 3.

36. I. Benedek, *Adhäsion*, (5), 16, 1986.

37. C. Creton, C.Y. Hui, and Y.Y. Lin, *Viscoelastic Contact Mechanism and Adhesion of Periodically Roughened Surfaces*, in Proceedings of the 24th Annual Meeting of Adhesion Society, February 25–28, Williamsburg, VA, USA, 2001, p. 21.

38. Duro-Tak, Druckempfindliche Klebstoffe, Specialty Adhesives, Product Data, National Starch and Chemical B.V., Adhesives Division, Zutphen, The Netherlands, 10/87, p. 9.

39. R.R. Charbonneau, and G.L. Groff, EP 0,106,559 B1, Minnesota Mining and Manuf. Co., St. Paul, MN, USA, 1984.

40. U.S. Patent 2,925,174, in R.R. Charbonneau and G.L. Groff, EP 0,106,559 B1, Minnesota Mining and Manuf. Co., St. Paul, MN, USA, 1984.

41. U.S. Patent 4,286,047, in R.R. Charbonneau and G.L. Groff, EP 0,106,559 B1, Minnesota Mining and Manuf. Co., St. Paul, MN, USA, 1984.

42. U.S. Patent 4,181,752, in R.R. Charbonneau and G.L. Groff, EP 0,106,559 B1, Minnesota Mining and Manuf. Co., St. Paul, MN, USA, 1984.
43. Canadian Patent 747,341, in R.R. Charbonneau and Gaylord L. Groff, EP 0,106,559 B1, Minnesota Mining and Manuf. Co., St. Paul, MN, USA, 1984.
44. U.S. Patent 4,223,067, in D.K. Fisher and B.J. Briddell, EP 0,426,198 A2, Adco Product Inc., Michigan Center, MI, USA, 1991.
45. G.F. Vesley, A.H. Paulson, and E.C. Barber, EP 0,202,938 A2, 1986.
46. U.S. Patent 4,330,590, in G.F. Vesley, A.H. Paulson, and E.C. Barber, EP 0,202,938 A2, 1986.
47. U.S. Patent 4,329,384, in G.F. Vesley, A.H. Paulson, and E.C. Barber, EP 0,202,938 A2, 1986.
48. F.C. Larimore and R.A. Sinclair, EP 0,197,662 A1, Minnesota Mining and Manuf. Co., St. Paul, MN, USA, 1986.
49. U.S. Patent 514,950, 1983, in F.C. Larimore and R.A. Sinclair, EP 0,197,662 A1, Minnesota Mining and Manuf. Co., St. Paul, MN, USA, 1986.
50. G. Auchter, J. Barwich, G. Rehmer, and H. Jäger, *Adhes, Age*, (7), 20, 1994.
51. S.D. Tobing and A. Klein, *Synthesis and Structure Property Studies in Acrylic Pressure-Sensitive Adhesives*, in Proceedings of the 24th Annual Meeting of Adhesion Society, February 25–28, Williamsburg, VA, USA, 2001, p. 131.
52. K.H. Schumacher and T. Sanborn, *UV-Curable Acrylic Hot-Melt for Pressure Sensitive Adhesives— Raising Hotmelts to a New Level of Performance*, in Proceedings of the 24th Annual Meeting of Adhesion Society, February 25–28, Williamsburg, VA, USA, 2001, p. 165.
53. K. Nobuhiro and T. Hirokazu, Japanese Patent, 63,152,686, Mitsubishi Paper Mills Ltd., 1988, in *CAS, Adhesives*, 25, 7, 1988.
54. A. Dobmann and A.G. Viehofer, Proceedings of the 19th Munich Adhesive and Finishing Seminar, Munich, Germany, 1994, p. 168.
55. A.H. Beaulieu, D.R. Gehman, and W.J. Sparks, Recent advances in acrylic hot melt pressure sensitive technology, in Proceedings of the Tappi Hot Melt Symposium, 1984, in *Coating*, (11), 310, 1984.
56. Gander, U.S. Patent 3,475,363, in J.N. Kellen and C.W. Taylor, EP 0,246,352 A2, Minnesota Mining and Manuf. Co., St. Paul, MN, USA, 1987.
57. Zang, U.S. Patent 3,532,652, in J.N. Kellen and C.W. Taylor, EP 0,246,352 A, Minnesota Mining and Manuf. Co., St. Paul, MN, USA, 1987.
58. Gobran, U.S. Patent 4,260,659, in J.N. Kellen and Charles W. Taylor, EP 0,246,352 A2, Minnesota Mining and Manuf. Co., St. Paul, MN, USA, 1987.
59. Technical Booklet, Tl-2.2-21d, November 1979, BASF, Ludwigshafen, Germany.
60. K.F. Schroeder, *Adhäsion*, (5), 161, 1971.
61. Letraset Ltd., London, U.S. Patent 1,545,568, in *Coating*, (6), 71, 1970.
62. *Coating*, (9), 301, 1993.
63. J. Johnston, *Adhes. Age*, (12), 24, 1983.
64. F. Altenfeld and D. Breker, *Semi Structural Bonding with High Performance Pressure Sensitive Tapes*, 3M Deutschland, Neuss, Germany, 2nd ed., February 1993, p. 278.
65. J.F. Kuik, *Papier und Kunststoff Verarbeiter*, (10), 26, 1990.
66. R. Bates, *J. Appl. Polym. Sci.*, 20, 2941, 1976.
67. C. Verdier, and J.M. Piau, *Understanding Peeling of PSAs by Use of Visualization*, in Proceedings of the 23rd Annual Meeting of the Adhesion Society, Myrtle Beach, SC, February 20–23, 2000, p. 116.
68. U.S. Patent 3,321,451, in S.E. Krampe and C.L. Moore, EP 0,202,831 A2, Minnesota Mining and Manuf. Co., St. Paul, MN, USA, 1986.
69. *Coating*, (4), 88, 1984.
70. R. Eckberg, *Silicone Release Coatings: Testing Pressure-Sensitive Adhesion*, in Proceedings of the 23rd Annual Meeting of the Adhesion Society, Wilmington, NC, USA, February 11–18, 2004, p. 307.
71. G. Meinel, *Papier und Kunststoff Verarbeiter*, (19), 26, 1985.
72. J.A. Miller and E. von Jakusch, EP 0,306,232 B1, Minnesota Mining and Manuf. Co., St. Paul, MN, USA, 1993.
73. M.A. Krecenski, J.F. Johnson, and S.C. Temin, *JMS-Rev. Macromol. Chem. Phys.*, C26(1), 143, 1986.
74. A. Sharma, V. Shennoy, and J. Sarkar, *Patterns, Forces and Metastable Pathways in Debonding of Elastic Films*, in Proceedings of the 23rd Annual Meeting of the Adhesion Society, Wilmington, NC, USA, February 11–18, 2004, p. 335.

75. D.K. Fisher and B.J. Briddell, EP 0,426,198 A2, Adco Product Inc., Michigan Center, MI, USA, 1991.
76. I. Benedek, *Pressure-Sensitive Adhesives and Applications*, Marcel Dekker Inc., New York, Basel, 2004, chap. 10.
77. T. Yamauchi, T. Cho, R. Imamura, and K. Murakami, *Nordic Pulp Paper J.*, (4), 128, 1988.
78. R. Pelton, W. Chen, H. Li, and M. Engel, *The Link Between Paper Properties and PSA Adhesion*, in Proceedings of the 23rd Annual Meeting of the Adhesion Society, Myrtle Beach, SC, USA, February 20–23, 2000, p. 120.
79. B. Zhao, L. Anderson, A. Banks, and R. Pelton, *Paper Properties Affecting Tape Adhesion*, in Proceedings of the 23rd Annual Meeting of the Adhesion Society, Wilmington, NC, USA, February 11–18, 2004, p. 391.
80. J. Kim, K.S. Kim, and Y.H. Kim, *J. Adhes. Sci., Technol.*, 3 (3), 175, 1989.
81. J. Boutillier, Proceedings of the 11th Munich Adhesive and Finishing Seminar, Munich, Germany, 1986, p. 4.
82. J.F. Kuik, Proceedings of the Polyethylenes, Copolymers and Blends for Extrusion and Coextrusion Coating Markets, Maacks Business Service, SP '89, Zürich, Switzerland, 1989, p. 107.
83. ASTM, D 1876-61T.
84. European Standard NMP 458, Nr.16-94D.
85. Hyvis/Napvis Polybutenes, BP Chemicals, Technical Booklet, PB 701.
86. H. Mizumachi, *J. Appl. Polym Sci.*, 30, 2675, 1985.
87. Hyvis, Napvis, Ultravis, Cling Properties, Technical Booklet, BP Chemicals, London, UK, 1995.
88. J.H. Glover, *Tappi J.*, 71(3), 188, 1988.
89. O.V. Stoyanov, V.F. Mironov, V.P. Privalko, R. Ya. Deberdeev, R.M. Kuzakhanov, and S.M. Minisalikhova, *Lakokras. Mater. Ikh. Primen.*, (1), 51, 1988, *CAS, Coatings, Inks Related Products*, 10, 2, 1988.
90. U.S. Patent 3.364063, in J.C. Pasquali, EP 0,122,847, 1984.
91. *Coating*, (1), 35, 1988.
92. I. Benedek, *Pressure-Sensitive Formulation*, VSP, Utrecht, 2000, chap. 5.
93. N.W. Malek, EP 0,095,093, Beiersdorf A.G., Hamburg, Germany, 1983.
94. H. Miyasaka, Y. Kitazaki, T. Matsuda, and J. Kobayashi, DE 3,544,868 A1, Nichiban Co. Ltd., Tokyo, Japan, Offenlegungsschrift, 1985.
95. U.S. Patent 3,691,140, 1972, in H. Miyasaka, Y. Kitazaki, T. Matsuda, and J. Kobayashi, DE 3,544,868 A1, Nichiban Co., Ltd. Tokyo, Japan, Offenlegungsschrift, 1985.
96. R.J. Shuman, and B.D. Josephs, PCT, WO 88/01636, Dennison Manuf. Co., Framingham, MA, USA.
97. K.W.M. Davy and M. Braden, *Biomaterials*, 8(5), 393, 1987.
98. Resins for Adhesives, UCB, Drogenbos, Belgium, 463-3.
99. U. Stirna, P. Tukums, N.P. Zhmud, and V.A. Yakushin, *Latv. PSR Zinat. Akad. Vestis Kim. Ser.*, (1), 69, 1988, *CAS, Crosslinking Reactions*, 13, 4, 1988.
100. M. Ando, and T. Uryu, *Kobunshi Ronbunshu*, 44(10), 787, 1988, *CAS, Polyacrylates (Journals)*, 5, 3, 1988.
101. Z. Czech, *Eur. Adhes. Seala.*, (6), 4, 1995.
102. I. Benedek, *Pressure-Sensitive Formulation*, VSP, Utrecht, 2000, chap. 4.
103. F.T. Sanderson, *Adhes. Age*, (11), 26, 1988.
104. H. Nakagawa, A. Baba, S. Furukawa, and F. Shuzo, Japanese Patent 92877, Nippon Shokubai Kagaku Kogyo Co. Ltd., 1987, *CAS, Adhesives*, 11, 6, 1988.
105. P. Mudge, *Ethylene-Vinylacetate-Based, Water-Based PSA*, in Proceedings of the Tech 12, in Advances in Pressure Sensitive Tape Technology, Technical Seminar, Itasca, IL, USA, May 1989.
106. E. Kazuyoshi, N. Hiroaki, T. Katsuhisa, K. Yoshita, and T. Saito (FSK Inc.), Japanese Patent 6,317,981, 1988, in *CAS, Adhesives*, 12, 5, 1988.
107. M.A. Johnson, *J. Plast. Film Sheet.*, 4(1), 50, 1988.
108. T. Tsubakimoto, K. Minami, A. Baba, and M. Yoshida, Japanese Patent, 6,335,676, Nippon Shokubai Kagaku Kogyo Co. Ltd., 1987, in *CAS, Adhesives*, 16, 1, 1988.
109. K. Akasawa, S. Sanuka, and T. Matsuyama, Japanese Patent 6,224,3669, Nippon Synth. Chem. Ind. Co. Ltd., 1987, *CAS, Colloids*, 4, 5, 1987.
110. A.N. Gent and S. Kaang, *J. Appl. Polym. Sci.*, 32, 4689, 1986.
111. Engineering Specification, ES M-2379, Packard Electric.

112. Engineering Specification, ES M-1881, Packard Electric.

113. Engineering Specification, ES M-2359, Packard Electric.

114. Engineering Specification, ES M-2147, Packard Electric.

115. Engineering Specification, ES M-4037, Packard Electric.

116. Industria Nastri Adesivi, Data Sheet, Modena, Italy, 1996.

117. Aroset, EPO 32216-APS-102, Data Sheet, Ashland Chemicals.

118. R.G. Czerepinski, and R. Gundermann, U.S. Patent 4,713,412, Dow Chem. Co., 1987.

119. *Coating*, (1), 8, 1986.

120. H. Satoh and M. Makino, DE 3740222, Nippon Oil Co. Ltd., Ger. Offen., 1988, *CAS, Adhesives*, 22, 4, 1988.

121. B. Copley and K. Melancon U.S. Patent, 878,816, Minnesota Mining and Manuf. Co., St. Paul, MN, USA, 1986, in *CAS, Siloxanes Silicones*, 14, 2, 1988.

122. R.M. Enanoza, EP 259968, Minnesota Mining and Manuf. Co., St. Paul, MN, USA, 1988, in *CAS, Adhesives*, 22, 3, 1988.

123. T. Sugiyama, N. Miyaji, Y. Ito, and T. Tange, Japanese Patent, 62,199,672, Nippon Carbide Industries Co. Inc., 1987, in *CAS, Adhesives*, 14, 4, 1988.

124. *Adhes. Age*, (12), 8, 1988.

125. J.P. Kealy and R.E. Zenk, Canadian Patent 1,224,678, Minnesota Mining and Manuf. Co., St. Paul, MN, USA, 1987.

126. M. Arakawa, S. Sakashita, H. Nagami, and T. Ono, Japanese Patent 63,142,086, Nitto Electric Ind. Co. Ltd., 1988, *CAS, Adhesives*, 25, 6, 1988.

127. A. Saburo, Japanese Patent 63,117,085, Central Glass Co. Ltd., 1988, *CAS, Adhesives*, 25, 7, 1988.

128. C.H. Hill, N.C. Memmo, and W.L. Phallen Jr., EP 259697, Hercules Inc., 1988, in *CAS, Adhesives*, 14, 5, 1988.

129. *Adhes. Age*, (3), 8, 1987.

130. *Adhes. Age*, (9), 8, 1986.

131. *Coating*, (6), 184, 1969.

132. L. Jacob, *New Development of Tackifiers for SBS Copolymers*, in Proceedings of the 19th Munich Adhesive and Finishing Seminar, Munich, Germany, 1994, p. 107.

133. C. Donker, R. Ruth, and K. van Rijn, *Hercules MBG 208 Hydrocarbon Resin: A New Resin for Hot Melt Pressure Sensitive (HMPSA) Tapes*, in Proceedings of the 19th Munich Adhesive and Finishing Seminar, Munich, Germany, 1994, p. 64.

134. M. Gerace, *Adhes. Age*, (8), 84, 1983.

135. Y. Urahama and K. Yamamoto, *J. Adhes.*, 25(1), 45, 1988.

136. O. Kazuto, H. Kenjiro, N. Akio, and N. Kiyohiro, Japanese Patent 63,483,82, Nitto Electric. Ind. Co. Ltd., 1988, in *CAS, Colloids (Macromolecular Aspects)*, 14, 5, 1988.

137. Poli-Film, America Inc., Cary, IL, Product Data Sheet, 1992.

138. Poli-Film für metallene Oberflächen, Poli-Film Verwaltungsgesellschaft mBH, Wermelskirchen, Germany.

139. D.J. Bebbington, EP 251,672, Wiggins Tape Group Ltd., 1988, in *CAS, Siloxanes*, 14, 4, 1988.

140. T. Tatsuno, K. Matsui, M. Takahashi, and M. Wakimoto, Japanese Patent 62285927, Kansai Paint Co. Ltd., 1987, in *CAS, Adhesives*, 14, 3, 1988.

141. I. Benedek, *Pressure-Sensitive Formulation*, VSP, Utrecht, 2000, chap. 3.

142. I. Benedek, *Adhäsion*, (12), 17, 1987.

143. DE-A-2407494, in P. Gleichenhagen, E. Behrend, and P. Jauchen, EP 0149135B1, Beiersdorf A.G., Hamburg, Germany, 1987.

144. Japanese Patent A 8231792, in P. Gleichenhagen, E. Behrend, and P. Jauchen, EP 0149135B1, Beiersdorf A.G., Hamburg, Germany, 1987.

145. H. Hintz, Kunststoff Dispersionen für das Verkleben von PVC Boden-belägen und Textiler Auslegeware, *Kunstharz Nachrichten (Hoechst)*, (19), 18, 1983.

146. I. Benedek and M. Mende, G9113755.1, Ebert Folien AG, Wiesbaden, Deutsches Gebrauchsmuster, 1991.

147. J.J. Higgins, F.C. Jagisch, and N.E. Stucker, Butyl rubber and poly-isobutylene, in *Handbook of Pressure-Sensitive Adhesive Technology*, 2nd Ed., D. Satas, Ed., Van Nostrand Reinhold, New York, 1989, chap. 14.

148. G.L. Burroway and G.W. Feeney, *Adhesives Age*, (7), 17, 1974.

149. Lehmann et al., U.S. Patent 3,729,338, in D.K. Fisher and B.J. Briddell, EP 0426198A2, Adco Product Inc., Michigan Center, MN, USA, 1991.

150. J.N. Kellen and C.W. Taylor, EP 0246352 A, Minnesota Mining and Manuf. Co., St. Paul, MN, USA, 1987.

151. U.S. Patent A 2,510,120, in P. Gleichenhagen, E. Behrend, and P. Jauchen, EP 0149135 B1, Beiers-dorf A.G., Hamburg, Germany, 1987.

152. German Patent A 2,535,897, in P. Gleichenhagen, E. Behrend, and P. Jauchen, EP 0149135B1, Beiersdorf A.G., Hamburg, Germany, 1987.

153. M. Koehler and J. Ohngemach, *Polym. Paint. Colour J.*, 178, 203, 1988.

154. Loctite Co., Japanese Patent 62,179,506, 1986, *CAS, Siloxanes Silicones*, 13, 1, 1988.

155. H. Kuroda, and M. Taniguchi, Japanese Patent 6343988, Bando Chem. Ind. Ltd., 1988, *CAS, Adhesives*, 21, 4, 1988.

156. A. Dobashi, T. Ota, and T. Uehara, Japanese Patent 6,368,683, Hitachi Chem. Co. Ltd., 1988, in *CAS, Crosslinking Reactions*, 14, 7, 1988.

157. W.J. Traynor, C.R. Moore, M.K. Martin, and J.D. Moon U.S. Patent 4,726,982, Minnesota Mining and Manuf. Co., 1988, in *CAS, Adhesives*, 17, 4, 1988.

158. Y. Ikeda, Y. Watanabe, and H. Tadenuma, Japanese Patent 6386777, Japan Synthetic Rubber Co. Ltd., 1988, *CAS, Adhesives*, 24, 3, 1988.

159. K. Hayashi, and K. Shimobayashi, Japanese Patent 63,118,383, Nitto Electric Ind. Co. Ltd., 1988, in *CAS, Adhesives*, 24, 5, 1988.

160. I. Murakami, Y. Hamada, and S. Sasaki, EP 269454, Toray Silicone Co. Ltd., 1988, in *CAS, Adhesives*, 19, 5, 1988.

161. T. Shibano, I. Kimura, H. Nomoto, and C. Maruchi, U.S. Patent 4,624,893, Sanyo Kokusaku Pulp Co., Ltd., Tokyo, Japan, 1986, *Adhes. Age*, (5), 24, 1987.

162. E. Pagendarm, DE 3632816 A1, Hamburg, Germany, Offenlegungsschrift, 1986.

163. U.S. Patent 3.865,770, in F.C. Larimore and R.A. Sinclair, EP, 0,197,662 A1, Minnesota Mining and Manuf. Co., St. Paul, MN, USA, 1986.

164. Japanese Patent A5695972, in P. Gleichenhagen, E. Behrend, and P. Jauchen, EP 0149135B1, Beiersdorf A.G., Hamburg, Germany, 1987.

165. A. Nakasuga and M. Kobari, Japanese Patent 6389585, Sekisui Chemical Co. Ltd., 1988, *CAS, Colloids (Macromolecular Aspects)*, 22, 5, 1988.

166. V.I. Pavlov, T.P. Muravskaya, R.A. Veselovskii, and M.T. Stadnikov, *Kompoz. Polym. Mater.*, 37, 14, 1988.

167. D. de Jager and J.B. Borthwick, *Thermoplastic Rubbers for Hot Melt Pressure-Sensitive Adhesives-The Processing Factors, Shell Elastomers, Thermoplastic Rubbers*, Technical Manual, TR 8.11, p. 5.

168. Martens et al., U.S. Patent 4,181,752, in J.N. Kellen and C.W. Taylor, EP 0246352 A2, Minnesota Mining and Manuf. Co., St. Paul, MN, USA, 1987.

169. K. Tatsuo, O. Nozomi, and T. Naomitsu, Japanese Patent 6333487, Nitto Electric. Ind. Co. Ltd., 1988, in *CAS, Adhesives*, 12, 5, 1988.

8 Manufacture of Pressure-Sensitive Products

István Benedek

CONTENTS

Considerable attention has been given to the general area of pressure-sensitive adhesive (PSA) manufacture [1,2], but a need now exists to review specific topics concerning the manufacture of pressure-sensitive products (PSPs) with and without PSAs. With the classic manufacturing procedure, PSPs are produced by coating a carrier web with a PSA. In this process, depending on the product class and application, multiple carrier and coating components can be used. Although paper and various special nonpaper, nonpolymeric materials have been developed, plastics are the main carrier materials for PSPs. In the early stage in the development of PSPs, carrier films used in packaging materials were applied. Later, other special plastic-based, web-like materials were produced and recommended as carrier matrials for PSPs.

In the manufacture of packaging films coating of web-like carrier materials were coated with adhesives to produce laminates, i.e., adhesive-bonded sheet-like composites. Later, similar coating and laminating technology was used in the production of PSPs. Actually, depending on the laminate components, type of adhesive, and lamination place, such adhesive-coated finished products are supplied as permanent or temporary laminates (e.g., laminated and overlaminating films, labels, tapes) or as a monoweb (tapes, protective films, etc.). For the latter products, laminating occurs during application (Table 8.1).

The manufacture of permanent laminates requires classical adhesives, which bond chemically. The coating of adhesives with permanent viscoelastic flow is the application domain of PSAs. Laminating is carried out because of the need to protect the tacky adhesive layer. The backside of the carrier (e.g., tapes or protective films) or a supplemental release liner (e.g., in a label or double-faced tape) protect the adhesive layer (see also Chapter 2).

The development of macromolecular chemistry allowed the synthesis of rubber-like or plastomer-like products which can be processed as film and (under well-defined conditions) exhibit self adhesivity and pressure-sensitivity (see Chapter 5). The performance of PSPs is characterized by their laminating, bonding, and delaminating characteristics. Laminating and delaminating refer to the application and deapplication of the product. According to the classic manufacturing technology, such characteristics are coating dependent properties. As illustrated in Figure 8.1 laminating to produce web-like materials is used for the manufacture of PSP components (e.g., carrier), for the manufacture of the finished product (e.g., label) or for the end-use application of PSPs (label, tape or protective film). In the case of labels, delamination of the liner precedes end-use lamination. In the case of protective films the lamination of the carrier components may represent the final step of the manufacture for a self-adhesive film (SAF). Special materials having intrinsic adhesive (and adhesive) properties allow the same performance without coating. Such raw materials lead to the

TABLE 8.1
Main Features of Laminated Products

Product	Buildup	Components	Type of Bond between Laminate Components
Packaging film	Multiweb	$Carrier_1$, adhesive, $carrier_2$ ($carrier_{n+1}$, $adhesive_n$)	Permanent
		$Carrier_1$, $carrier_2$ ($carrier_{n+1}$)	Permanent
Label	Multiweb	$Carrier_1$, adhesive, $carrier_2$, release liner	Temporary
		$Carrier_{n+1}$, $adhesive_n$, $liner_n$	
Tape	Monoweb	$Carrier_1$, adhesive	Temporary
	Multiweb	$Adhesive_1$, $carrier_1$, $adhesive_2$, release $liner_1$	Temporary
	Multiweb	$Adhesive_1$, release $liner_1$	Temporary
Protective film	Monoweb	$Carrier_1$, $adhesive_1$	Temporary
	Monoweb	$Carrier_1$	Temporary
	Multiweb	$Carrier_{n+1}$	Permanent

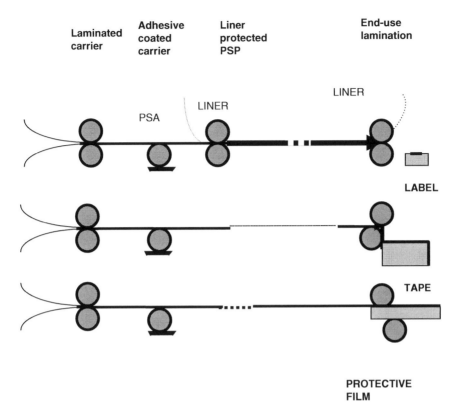

FIGURE 8.1 Lamination as a technological step in the manufacture of PSP components, of PSPs, and of their application.

manufacture of PSPs without PSAs and without the use of other than coating technology to buildup the adhesive-carrier assembly. Like common plastic films such products are made by extrusion (Table 8.2).

As seen from Table 8.2, the manufacture of PSPs by coating with a liquid PSA can be considered as a special case of the production of adhesive-coated PSPs. The manufacture of the PSPs by adhesive coating itself is a special case of the manufacture of PSPs. Generally the adhesive and

---

TABLE 8.2
Manufacturing Technology for PSPs

Pressure-Sensitive Product		**Manufacturing Technology**	
Label	Coating	Coating of liquid adhesive	Molten PSA
			Dispersed PSA
Tape	Coating	Coating of liquid adhesive	Molten PSA
		Coating of solid adhesive	Dispersed PSA
	Extrusion		
	Foaming-molding		
Protective film	Coating	Coating of liquid adhesive	Molten PSA
			Dispersed PSA
	Extrusion-physical treatment		

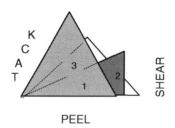

FIGURE 8.2 Schematic presentation of the tackification steps. (1) Untackified composition; (2) tackified soft composition; (3) tackified composition with balanced properties.

end-use characteristics of PSPs can be given by the carrier material also. It should be mentioned that the manufacture and application of such self-adhesive products are made possible by the special application conditions which may include elevated temperature and pressure (Figure 8.2). For PSPs with balanced adhesive properties peel buildup is controlled mainly by the dwell time (see also Chapter 7). For low tack self-adhesive films peel buildup is also regulated by the application temperature and pressure.

The manufacture of PSAs and the general problems of PSA coating are discussed in Ref. [2]. This technology can be considered as a special part of PSP manufacture. This chapter discusses the manufacture of PSPs as a whole, including the components which are necessary to buildup a PSP by means of classic coating method or by "plastics" processing.

In the classic manufacture of PSPs a PSA is coated on a carrier web. The production of the PSPs includes the manufacture and assembly of the individual components. Production of the components includes the manufacture of both the solid-state carrier material and the coating (adhesive, abhesive, primer, etc.).

I. MANUFACTURE OF COATING COMPONENTS

The main coating component is the PSA. Depending on its adhesive characteristics, on the type of carrier and on product geometry it should be protected by using different laminate constructions. Its delaminating is allowed by the abhesivity of the protective laminating component. Such abhesivity may be an intrinsic property of the solid-state material or is given by a release layer. The nature of this release layer (coated on the same carrier or laminated as a separate solid-state component) depends on the product construction (class) and application conditions. The release effect is given in both cases by a special, coated (or built-in) abhesive component. Other coating components are used to allow the anchorage of the adhesive, release agent, or printing inks. Additives improving some special performances (e.g., electrical conductivity, flammability, recycling) can be coated on or built into the carrier material too.

A. MANUFACTURE OF THE ADHESIVE COMPONENTS

With the rapid growth in PSPs the need to develop new adhesive formulations is ongoing. In-house coating is also gaining increasing attention. The adhesives used as coating component for PSPs are mainly PSAs. Some of them are viscoelastic compounds with built-in PSA properties; others have to be formulated to achieve the required rheology and adhesive characteristics (see Chapter 7). Their processing (depending on the coating technology, equipment, carrier material, end-use, and economic considerations), and their storage, application/deapplication and recycling require the modification of the properties of the raw materials used, that is, formulation. Therefore, the manufacture of the adhesive components include their synthesis and formulation.

1. Synthesis of Adhesive

In their first stage of development PSAs were formulated on the basis of natural macromolecular products. Natural rubber (NR) and natural resins were used. Some decades ago, synthetic elastic components (rubbers) and viscous raw materials (tackifiers, plasticizers, etc.) were developed and blended to produce (formulate) the adhesive. Advances in macromolecular chemistry allowed the synthesis of raw materials having built-in viscoelastic properties, that is, pressure sensitivity (e.g., acrylics (ACs), ethylene–vinyl acetate (EVAc) copolymers, carboxylated rubber, polyvinyl ethers, polyurethanes (PURs), polyesters) (see also Chapter 5). Some of the polymers used for PSAs are raw materials for other adhesives or plastics also. Their synthesis is a special, chemical/macromolecular technology. It is not the aim of this book to describe it.

It is to be noted, that in certain cases it would be desirable for the adhesive converter to carry out its own polymerization. Such necessity appears for low volume special products, or for adhesives where polymerization and converting technology have the same importance for the manufacture of the final PSP. For instance, to achieve a patterned contact surface (necessary for removability), the adhesive surface should have a special discontinuous shape given by the PSA itself, if it is a suspension [3]. Thus, spherical contact sites are formed. These are anchored on the carrier surface by the aid of a primer. The classic procedures concerning formulation for removability have some disadvantages (see later). The recipes contain plasticizer (migration), thickener and surface active agents (migration, bleeding, and softening). The adhesion increases over time. This increase depends on the nature of the surface and quality of the adherend. Therefore, full surface coating does not ensure complete removability. If the adhesive is coated in a discontinuous manner, adhesion increases over time at the contact sites and thus there may be high peel resistance in the longest contact direction. The use of an adhesive with spherical PSA particles (50–150 μm) avoids such behavior [4]. Unfortunately anchorage of the spherical PSA particles needs a continuous matrix. The matrix polymer has to cover the surface of the particles also. The particles have to be coated as monolayer. On porous surfaces such a coating weight is difficult to achieve. The special PSA particles (with a T_g lower than 10°C) are synthesized via suspension polymerization. The composition of the pressure-sensitive polymer according to this patent includes a crosslinking agent also used for both the matrix polymer and the particles. Polyisocyanate derivatives, polyepoxides or aziridines have been recommended as crosslinking agents (see also Chapter 6). A low coating weight (7 g/m^2) has been applied. Such product displays readherability. As seen from the above example, in this case the properties of the final product strongly depend on its synthesis.

Generally for the PSA converter the manufacture of the adhesive consists of its formulation. However, there are some special cases where the formulation is made to allow the adhesive synthesis of the adhesive by postpolymerization. Polymerization or postpolymerization use special multifunctional monomers or macromers and classic (chemical) or physicochemical polymerization techniques. Polyaddition, polycondensation, or crosslinking are carried out (see Chapter 5, Section II.A).

The manufacture of the adhesive can be carried out off-line or in-line (Table 8.3). Off-line manufacturing includes the mixing of the adhesive components followed by coating of the ready-to-use PSA. In-line manufacture of the adhesive consists of simultaneous coating and curing or postpolymerization. In this case first a special "ready-to-coat" mixture of polymerisable and crosslinkable monomers, oligomers, or polymers (e.g., radiation-cured hot-melts) is applied on a carrier. Such a reaction mixture is transformed after coating in "ready-to-use" adhesive. The postcoating synthesis of the PSA must be carried out by the converter.

a. In-Line Synthesis of the PSA

The simultaneous manufacture of the PSA and PSA laminate using radiation curing has been discussed in Ref. [2]. On the basis of the experience of radiation curing of thin coatings from the printing technology, in-line coating, and polymerization of the base monomers (actually of a blend of monomers, oligomers, and reactive diluting agents) have been carried out to manufacture

TABLE 8.3
Manufacturing Possibilities for the Adhesive

Synthesis		Formulation	
In-Line	Off-Line	In-Line	Off-Line
Synthesis from monomers	Synthesis from monomers (polymerization)	Formulation for adhesion	Formulation for adhesion
		Formulation for other properties	Formulation for other properties
Synthesis from oligomers (polymacromerization)	Synthesis from oligomers (polymacromerization)		
Synthesis from polymers (crosslinking)	Synthesis from polymers (grafting, crosslinking)		

PSA–coated laminate. The main problems of this technology had to do with chemistry, physiology, and rheology. The precise regulation of the polymerization (curing) as a function of the adhesive raw materials, layer thickness and carrier features is difficult. In the first stage of development side reaction products and effects (residues of unreacted monomers and initiators, carrier damage, etc.) created technical, environmental, and physiological problems. Additional technological aspects appeared also. The coating of such low-viscosity fluids requires special devices. Therefore, the use of prepolymerized or polymerized raw materials (reactive oligomers or polymers) and their postpolymerization have been preferred.

The full or partial postmanufacturing of the PSA is the result of the chemical development induced by the trend of solvent-free fabrication. Such formulations with 100% solids included hot-melts or radiation-cured reaction mixtures. Technological reason and (partially) insufficient progress in acrylic hot-melts [5] forced the development of oligomer and macromer-based curable formulations. It can be supposed that in the future polymacromerization will be used on a large scale (see also Chapter 5). The use of macromers is not new. It has been developed as graft copolymerization and crosslinking of plastomers. In water-based (WB) systems adhesive modification via macromers is also known. For instance, an acrylic PSA has been prepared by polymerization of an acrylic–vinyl monomer mixture in the presence of an alkali soluble or dispersible support resin having a number average molecular weight (MW) of 21,000–150,000. Such support resin is used as an emulsifier [6].

The raw materials for postpolymerization are common PSAs, PSAs with unbalanced adhesive properties or nonadhesive oligomers (see Table 8.4). Such compounds are manufactured via off-line synthesis. They are macromolecular compounds having a low to medium molecular weight and residual functionality, that is, they can be postpolymerized (or crosslinked). For this process the converter can choose to use either a classic polymerization technology (i.e., thermal, free radical-initiated polymerization, polyaddition, or polycondensation) or a radiation-induced reaction. It is evident, that in some cases this choice is limited by the chemical nature of the (pre)polymer. It is to be emphasized that in certain cases (i.e., acrylic hot-melt development) postpolymerization of an oligomer is a necessity because of the lack of suitable macromolecular compounds.

Photopolymerization of the monomers neat, without any diluent that needs to be removed after the polymerization provides processing advantages. Unfortunately if ultraviolet (UV) radiation is used, postcuring is needed for high coating weights. In a special case of off-line adhesive synthesis a hybride procedure combines the thermal postcrosslinking with radiation polymerization of the PSA. A patent to manufacture a tape describes the UV-induced polymerization of acrylic monomers direct on the carrier [6]. The liquid monomers are applied to the carrier using a doctor blade, roller coating, or spraying. The polymerization occurs in a tunnel from which oxygen has

TABLE 8.4
Macromer-Based PSPs

Product	Adhesive	Macromer	Ref.
Tape	Water-based acrylic	Acrylic–ethylene copolymer	[7]
Foam tape	Photocured acrylates	Isooctylacrylate-N-vinyl-pyrrolidone copolymer, poly(t-butylstyrene)	[8]
Tape	UV-curable acrylic esters, polymeric photoinitiator	Polyvinylbenzophenone	[9]
	UV-curable acrylic esters, polymeric photoinitiator	Poly(2-(p-(2-hydroxy-2-methylpropanoyl)-phenoxy(ethyl methacrylate)	[10]
	UV-cured butadiene rubber, polymeric photoinitiator	Aromatic–aliphatic ketones and silicone side chain substituted rubber	[11]
Tape	Ethylhexyl acrylate	Styrene–butadiene copolymer	[12]
Tape	UV-curable acrylic oligomer	Acrylic oligomers	[13]
Tape	Styrene–butadiene–acrylic	Tackified poly(styrene–butadiene) grafted with EHA	[14]
Tape	Acrylate	Acrylic oligomer, EB-cured with trimethylolpropane tris(thioglycolate) and 1,6-hexanediamine	[15]
Label	Acrylate	Acrylic oligomer, built-in UV initiator	[16]

been excluded. This patent uses a temporary carrier, an endless belt, that does not become incorporated into the final tape. The off-line step of the polymerization is completed in an oven. Such tapes have PSA on both sides of the carrier.

It is possible to prepare a prepolymer which is polymerized by radiation to a coatable product that is postpolymerized in-line. A patent discloses the preparation of a low molecular weight spreadable composition to which composition a small amount of catalyst or polyfunctional crosslinking monomer can be added prior to the completion of the polymerization by heat curing [17]. The viscosity of the prepolymer makes it easy to apply to the support. This procedure was development from an earlier, monomer-based in-line procedure using the UV photopolymerization (with light intensity of 4.0 mW/cm^2 at the surface) of acrylics [18]. The PSA can be produced as an adhesive layer supported by a nonadhesive layer (carrier or rigid body), or as double-faced systems. As stated in Ref. [19], PSAs cured with UV or electron beam (EB) radiation have not been a total success in performance or economics. In 1973, the costs for UV-crosslinkable inks and adhesives have were twice as higher [20]. In the last decades technical and economical advances have been registered. For instance, a 1995 patent eliminates the requirement for photoinitiators for UV-induced curing [21]. However, this method for synthesizing or modifying an adhesive should not be considered important for common PSPs. Very thin or very thick adhesive layers (e.g., protective layers, transfer tapes, etc.) are the suggested domain for this technology.

According to Ref. [22], postcuring is carried out to improve shear resistance. Copolymers of diesters of unsaturated dicarboxylic acids with acrylics (with a T_g of about 30–70°C below the use temperature), with a multifunctional (monomeric) crosslinking agent have been synthesized. This composition can be cured chemically or with any convenient radiation. The molecular weight of the first synthesized (pre)polymers should differ according to their molecular weight distribution (MWD). For a narrow MWD a weight average molecular weight of 100,000, for a broad MWD a molecular weight of 140,000 is required to enable the desired response to EB curing. M_w/M_n ratios between 4.2 and 14.3 are achieved. Generally, constituents having a molecular weight of less than 30,000 are nonresponsive to electron beam radiation [23]. The cohesive strength of the adhesive is proportional to the concentration of the hard monomer in the formulation. When a multifunctional monomer is used as crosslinking agent, its concentration should be preferably from about 1 to 5% by weight. Its presence enables the reduction of the dosage level for EB curing. A radiation dosage of 200 kGy is applied [22]. Thick PSA tapes (0.2 to 1.0 mm) have been manufactured by photopolymerization (UV) of acrylics [24]. Such an acrylic ester-based PSA formulation is a combination of nontertiary acrylic acid esters of alkyl alcohols and ethylenically unsaturated monomers having at least one polar group which may be substantially in monomer form or may be a low molecular weight prepolymer, or a mixture of prepolymer and additional monomers, and may further contain a photoinitiator, fillers and crosslinking agents (such as multifunctional monomers). The acrylic esters generally should be selected from those which possess PSA properties as homopolymers. Diacrylates are generally added as crosslinking agents (0.005–0.5 wt.%) after the synthesis of the prepolymer. As photoinitiator 2,2-di-methoxy-2-phenylacetophenone can be used. Unsaturated photoinitiators, built-in polymer are discussed in detail in Chapter 6, Section II.

A special case of adhesive synthesis is the manufacture of PUR-based PSAs used for special tapes, for example, medical tapes. Unlike rubber-based formulations which leave deposit, special PUR adhesives can be cleanly removed from the skin [25]. Here a one-shot or prepolymer procedure can be chosen in order to synthesize the adhesive.

Postapplication crosslinking can be considered a special case of off-line synthesis (finishing) of an adhesive. It is proposed to improve shear, or to ensure delamination (detachment) after use [26]. Such postcrosslinking is achieved using thermal, free radical [27,28], or photoinitiated reactions [29]. A latent crosslinking agent has to be formulated in the recipe. A low alkylated amino formaldehyde condensate having C_{1-4} alkyl groups (e.g., hexamethoxymethylmelamine) has been proposed as crosslinking agent. First an intermediate product (having a viscosity of 0.3–20 Pa sec) is synthesized by UV photopolymerization, then a crosslinking agent is added and the mixture is coated and polymerized by UV light to the final product. The product can also be used as a transfer tape [29]. By polymerization in mass a thick coating can be manufactured in one step; the use of a solvent-based or aqueous adhesive requires more than one coat if a thickness of more than 0.2 mm is desired. As discussed in detail in Chapter 5, in-line synthesis by photopolymerization of macro-mers and macromeric photoinitiators is possible. This procedure is of increasing importance in the manufacture of carrier-free tapes. As illustrated by the data of Table 8.4, most systems are based on acrylates but compounds containing aromatic or aliphatic ketone groups and silicone can be bonded to alkylene side chains of polymers; for example, butadiene rubber has also been hydrosilylated [11]. The in-line synthesis of the adhesive can be carried out in two steps to increase the pot-life of curable formulations. For instance, adhesives with a long pot-life are synthesized by coating the elastomers and the hardening agents on two different substrates, putting them together and crosslinking them. Corona treated 60 μm polyethylene (PE) has been coated with a mixture of 100 parts acrylic oligomer and 10 parts trimethylolpropane *tris*(thioglycolate), combined with a 20 μm film of PE coated with 0.5 g/m^2 1,6-hexanediamine, and irradiated with an electron beam of 5 M rad, to give an adhesive tape with a pot-life greater than 24 h and an adhesion of 410 g/25 mm [15]. In-line synthesis is discussed in detail in Ref. [30].

b. Off-Line Synthesis of the PSA

Off-line synthesis of PSAs is the common way to produce coatable adhesives. Such products can be macromolecular compounds having ready-to-use or ready-to-formulate adhesive properties or ready-to-postpolymerize reaction mixtures supplied for in-line adhesive synthesis.

2. Formulation of Adhesive

Formulation of PSAs has been discussed in detail in Refs. [31,32]. For most PSAs formulation is the final manufacture step, giving a product with the required processing and end-use properties. However, the same end-use properties may differ strongly according to product class. Therefore, the formulation of PSAs for PSPs displays numerous special features.

   Formulation depends on the chemical basis of the adhesive. As discussed in Chapter 5, one-component viscoelastic raw materials (AC, EVAc, PUR, etc.) and multicomponent compositions having separate elastic and viscous components can be used to design a PSA. Their choice also depends on the coating technology (physical state, coating method, coating device, etc.). Formulating is also influenced by the PSP product class. The main raw materials used for PSPs have been discussed in Chapter 5.

   Different classes of raw materials are used for various PSPs. For different raw material classes different formulating components and technology have been suggested. For natural rubber-based PSPs, formulation with resins has been the only way to achieve viscoelastic behavior and adequate pressure-sensitive properties. Hydrocarbon-based synthetic elastomers and polar synthetic elastomers (silicones, polyurethanes, polyesters, etc.) must also be formulated.

   In principle the scope of formulating is to develop a recipe with properties tailored to the product's end-use. One of the most important end-use criteria is the adhesive quality evaluated as the sum of the adhesive performance characteristics. On the other hand, formulation has to allow the use of the adhesive for the manufacture of PSPs (coating and converting properties) and to enhance the application of the final product (converting and other end-use properties). Ideally, the scope of the formulation is to fulfill the requirements of the entire spectrum of performance properties. Actually formulation gives priority to well-defined requirements.

a. Formulation for the Adhesive Properties of PSPs

As discussed earlier (see Chapter 7) the simplest way to regulate adhesive properties is by formulation. The main purpose of formulation is to regulate the adhesion–cohesion balance. Practically, it is a trial to achieve the required tack, peel, and bond break characteristics assuming that the formulated adhesive preserves its internal cohesion required for its converting and end-use. Therefore, in principle, the formulation of a common (label grade) PSA for the adhesive properties consists of its tackification or detackification. Although the raw adhesive possesses sufficient cohesion before tackifying (plot 1 in Figure 8.2), improvement of its tack and peel by compounding with tackifier resin or plasticizer may cause a pronounced decrease of its shear resistance (plot 2). The shear level required for a balanced adhesivity can be restored (plot 3) adding other formulation components to the recipe (e.g., high melting point resins, hard PSA, or crosslinkable components).

   Tackification is required mainly for permanent adhesives; detackification is needed for removable compositions. Tackification is a general formulating method. Detackification is applied only in special cases (see Table 8.5). As discussed in Ref. [31], the adhesive properties can be improved by tackification by (1) formulation with other viscoelastic components, (2) formulation with resins and plasticizers, or (3) formulation with special additives. Formulation with the aid of viscous components allows tackifying (regulation of tack and peel), removability, and easy processing.

TABLE 8.5
Adhesive Formulation with Detackifier

PSP	Adhesive	Detackifier	Ref.
Readherable PSA	Tackified polymer emulsion	Wax	[34]
Tape	Crosslinked acrylic	Tristearyl phosphite, 0.03–10%	[35]
Tape, sheet	EHA–VAc copolymer, crosslinked	Sorbitanetrioleate	[36]
Tape, mounting	Acrylic	Caprolactone polymer	[37]
Sheet	Acrylic	Ionic, low tack PSA particles	[38]
Tape	HMPSA	Metallic salts of fatty acids	

Tailoring of the chemical affinity of the adhesive towards the substrate surface depends on its chemical composition. Detackifying agents have to work on the bonding surface. As is known from the production and use of packaging materials, detackifiers (debonding agents) are used for "easy peel" closures. These work in the bulk sealing layer. Their efficacy is based on their incompatibility with the main sealing component. Therefore in this case debonding occurs in the bulk adhesive layer. Such detackifiers cannot be applied for removable PSPs where failure must occur at the adhesive/substrate interface. As a special case, the formulation of sealants should be mentioned. For sealants, cohesive failure is required [33]. According to Holden and Chin [33], in this case saturated hydrocarbon resins and saturated parafinic oil (up to 400%) are suggested to achieve the desired behavior. Detackification uses special additives, resins, crosslinking agents, fillers and abhesives (Table 8.5). For instance, a repositionable PSA for mounting tapes includes a detackifying resin comprising a caprolactone polymer (1–30 wt.%) [33]. To improve the detackifying effect of the elastic contact particles in a readherable product, such adhesive contains an ionic low tack monomer [38]. Detackification of hot-melt PSA (HMPSA) using metallic salt of a C_{14}—C_{19} fatty acid is described in [39]. Sorbitan fatty acid esters (e.g., trioleate) have been used for ethylhexyl acrylate–vinyl acetate copolymers crosslinked with isocyanates to improve their removability [36]. Such formulations exhibit 250 g/25 mm peel on stainless steel after 3 days storage at 60°C in comparison with 950 g/25 mm for a common crosslinked PSA. The removability of acrylic PSA is improved by small amounts of organofunctional silanes [39].

Tackification has been described in Refs. [31,40]. This chapter discusses the special aspects of tackification related to the product class (e.g., label, tape, or protective film). Tackification as discussed in Ref. [31], has been used to improve the adhesive properties of PSAs, that is, the performance of adhesive-coated PSPs. For the whole range of PSPs tackification consists of more than the modification of the adhesive; it also involves the carrier and other coating components (e.g., ink) also (Figure 8.3).

Tackification of the carrier is carried out for self-adhesive films. The mechanism can be quite different from that of the tackification of PSAs. As is known, PSA tackification assumes compatibility. Noncompatibility is taken into account only by the tackification of adhesives having composite, segregated structure (i.e., styrene block copolymers (SBCs)). In the range of SAFs some products are tackified with compatible tackifiers, but there are also products where tackifier incompatibility is used to achieve the final performance. Vinyl acetate and ethylene–vinyl acetate copolymers are compatible with polyethylene, low molecular weight polyolefins, and some

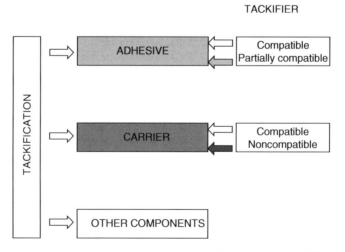

FIGURE 8.3 Schematic presentation of the general character of tackification for PSPs.

tackifier resins. Such compounds are used to prepare adhesive-free protective films. On the other hand, polybutylene is not compatible with certain polyolefines. Its incompatibility with and migration to the film surface are used to manufacture cling films. It is known that addition of compatible polymers may depress the melting point of crystallizable polymers blended with an amorphous polymer, according to the Flory–Huggins theory [41]:

$$\frac{1}{T_{\mathrm{m}}} - \frac{1}{T_{\mathrm{m}}^{\mathrm{o}}} = -\frac{RV_2}{\Delta H^{\mathrm{o}} V_1}\left[(\ln \phi_2/m_2) + (1/m_1 - 1/m_2)\phi_1 + \chi_{12}\phi_1^2\right] \tag{8.1}$$

where T_{m} and $T_{\mathrm{m}}^{\mathrm{o}}$ are the equilibrium melting points of the blend and homopolymer respectively, ΔH^{o} is the heat of fusion for the main polymer, V_1 and V_2 are the molar volumes of the repeat units m_1, ϕ_1 and m_2, ϕ_2 are the degrees of polymerization.

It should be taken into account that compatibility is temperature dependent. Thermodynamic miscibility may exist in the whole composition range at ambient temperature and phase separated regions may appear at high temperatures for compatible mixtures. The time–temperature superposition principle can be applied. In such systems at high frequencies, neither components has relaxed and both contribute to G'. At midfrequencies one component relaxes. At low frequencies, this component does not contribute to the modulus, and the second component which has still not relaxed, is responsible for the occurrence of a second plateau. This relaxation is similar to the relaxation observed in melts of mixtures of homopolymers having widely different molecular weights. At this frequency, the low molecular weight polymer may act as a low molecular weight solvent. Such binary blend exhibits a second plateau modulus GN_2^{o} similar to polymer–solvent mixtures:

$$GN_2^{\mathrm{o}} = Gn^{\mathrm{o}}(\phi^1) \times \phi^{2,5} \tag{8.2}$$

where Gn^{o} is the plateau modulus of the pure polymer, ϕ is its volume fraction. In the phase separated region a $\omega^{0.5}$ dependence of the modulus can be observed for some polymers. A similar $\omega^{0.5}$ power law has been reported for chemically crosslinked systems at the gel point. According to Refs. [42,43], the rheology of such systems should be discussed in terms of fractal dimensionality, that is, in terms of similarity of supermolecular order over a large distance scale.

It should be taken into account that tack and peel do not improve (always) in parallel; the required tackifier level depends on the nature (chemical and physical status) of the tackifier, and the peel-shear interdependence influences the level of the debonding force and the type of failure. It should be emphasized that the ideal adhesion–cohesion balance is different for labels, tapes, and protective films; therefore the scope of the formulation for the adhesive properties and the formulation possibilities for such products are also different. As illustrated by the data of Table 8.6, high tack and improved polyethylene peel are required (and achieved by tackification) for labels, whereas high shear and medium tack and peel are needed for packaging tapes. High tack and peel are obtained by using a low level (16–32%) of tackifier resin with soft acrylate (AC_1). A much higher resin level (45%) is necessary for styrene–butadiene–rubber (SBR) and NR–SBR mixtures for medium tack and peel. Therefore, such formulations are more suitable for tapes where high shear resistance is required. On the other hand, even mixtures containing cohesive acrylate (AC_2) and a low level (32%) of tackifier do not provide shear resistance competitive with that of formulations with SBR and SBR–NR mixtures with higher tackifier concentration.

Generally, the goal of tackification is to improve tack and peel and/or removability. This may be quantified by knowing the base performance of the adhesive and the chemical or macromolecular characteristics of the formulating components. For a high T_{g} ($-25°C$), crosslinkable, carboxylated acrylic, Pierson and Wilczynski [44] state the need to improve peel and tack by 25%. For such improvement the tackifiers have been tested at a loading of 12–35% dry tackifier per total solids. For a standard acrylic latex 25–35% resin loading is used [45]. As stated

TABLE 8.6
Tackification of Selected PSPs

PSP	Formulation Component	Concentration (pts, dry)	Peel (PE) 180° (N/25 mm)	Tack LT (N/25 mm)	Tack RB (cm)	Shear (RT, h)
Label, permanent	AC_1	100	9.8	20.0	—	10
	Tackifier 1	20				
	AC_1	100	10.5	22.0	—	4
	Tackifier 1	40				
Label, permanent	SBR	100	8.5	25.0	—	—
	Tackifier 1	110				
Label, permanent	NRL, SBR	100	5.0	11.0	—	—
	Tackifier 1	110				
Tape, packaging	AC_2, AC_1	100	4.4	6.5	12	200
	Tackifier 1	20				
	AC_2, AC_1	100	6.0	11.0	12	144
	Tackifier 1	40				
	AC_2, AC_1	100	8.0	12.5	20	120
	Tackifier 1	60				
Tape, packaging	SBR	100	8.5	9.0	8	>200
	Tackifier 1	40				
Tape, packaging	SBR	100	10.0	12.0	20	>100
	Tackifier 1	60				
Tape, packaging	NRL, SBR	100	6.5	10.0	20	>200
	Tackifier 2 + Tackifier 3	120				
Tape, masking	NRL, SBR	100	9.6	6.4	2	>200
	Tackifier 2 + Tackifier 3	100				

Note: Water-based formulations were used. AC_1 has a T_g of $-42°C$; AC_2 (self crosslinkable) has a T_g of $-58°C$. Tackifier 1 is a dismuted rosin (R&B 50°C); tackifier 2 is a terpenephenol resin (R&B).

in Ref. [46], the highest resin level that contributes to the peel is about 50%; however, for a high softening point resin a loading of less than 39% has been suggested. With high softening point resins the best overall balance is obtained at a concentration of around 30%. The T_g allows a forecast of the possible tackifier loading [47] (see also Chapter 3). A carboxylated acrylic latex needs higher tackifier level. Water-releasable HMPSAs contain 15–40% tackifying agent [48,49]. Polyacrylate rubber-based HMPSAs are tackified (40% resin) or plasticized formulations [50]. A common tacky PSA can be added to the formulation also (50 parts to 100 parts acrylic rubber).

Typical resin dispersions contain 40–60% resin and 2–20% plasticizer [51]. It is evident that when such tackifiers are used, plasticizers are also added in the formulation. Internal crosslinking of the adhesive using special monomers such as N-methylol acrylamide, together with softening via plasticizers has also been suggested [52]. As discussed in Chapter 5, in HMPSAs the oil used in the formulation is plasticizer and a viscosity-regulating agent. Formulating for shear resistance is a special domain, using crosslinking as main technical modality (see also Chapter 6). This technology is related mostly to the manufacture of tapes and protective film, so it will be discussed later.

As described in Chapter 7, formulating for removability achieves softening of the polymer with plasticizer and/or other viscous components. This is achieved for acrylic PSAs with paraffinic oil and polyisobutylene [38]. A repositionable transfer tape is manufactured by casting of a thin carrier film that is coated with polybutyl acrylate plasticized with dibutylphtalate [53]. According to Ref. [54], removable price labels used almost exclusively paper carrier, coated with rubber-resin adhesive. Natural and synthetic rubber and polyisobutylene, tackified with soft resins and plasticizer have been suggested for such compositions. A water-based acrylic formulation proposed by Mueller and Tuerk [54] includes physically or chemically crosslinkable comonomers like acrylo- or methacrylonitrile, acrylic or methacrylic acid, N-methylol acrylamide and plasticizer (10–30 wt.%) also. According to Jahn [55], a starting formulation for PSA for tape is based on a 70/17/10/10/0.45 part (dry) mixture of CSBR, coumarone indene resin, hydrocarbon resin, plasticizer, and thickener. A natural rubber latex-based readhering, removable composition contains a tackifying agent also [56]. HMPSAs generally contain two different grades of resin, endblock-compatible and midblock-compatible [57].

Removability is related to energetic (rheological) and chemical adhesion. To ensure a low peel resistance, the external force has to be balanced by the viscous and elastic (flow) deformation of the adhesive layer. Due to the self-deformation of the carrier material, reduced debonding energy is transferred at the adhesive/substrate surface (and adhesive face stock surface) where low adhesion should allow an adhesive (contact) break. In this situation formulation for removability has to allow energy absorbance (self-deformation) of the adhesive and low level adhesion on the adherend surface.

Adhesive deformability is achieved by increasing its viscous flow (e.g., using viscous raw materials or additives). Adhesion break at the adherend surface is achieved by reducing (physically or chemically) the adhesive-substrate contact. This results from the changes in the adhesive and carrier rheology (e.g., by crosslinking) or in the properties of the carrier/adherend surface. Table 8.7 lists the main technical methods used to achieve removability.

As seen from Table 8.7, some of the technical ways to achieve removability depend on formulation, whereas other modalities depend on the geometry of the adhesive layer, that is, they are (at least partially) a function of the coating technology.

According to Pasquali [58], contact surface reduction (the decrease in the ratio of contact surface to application surface of the adhesive) can be achieved by formulation that regulates the rheology of PSA and the chemical affinity between adhesive and substrate surface (see also Chapter 7). The control of the adhesive flow to reduce contact surface concerns the buildup of an internal structure in the bulk adhesive. Such a structure does not allow rapid adhesive flow and penetration of PSA in the "rough" substrate surface, that is, it hinders of the physical contact between adhesive and adherend. The same result is obtained by dividing the contact surface in small discrete areas.

TABLE 8.7
Technical Possibilities for the Manufacture of Removable PSPs

| Modification of Laminate Components | | | | Laminate Manufacture | |
| Modification of the Adhesive | | Modification of the Carrier | | | |
Adhesive	Nonadhesive	Bulk Modification	Surface Modification	Coating	Postlamination
Adhesive synthesis	Filling	Modification of carrier geometry	Tackification	Primer	Crosslinking
			Physical surface treatment	Coating method	
Crosslinking Plasticizing		Modification of mechanical properties	Mechanical surface treatment	Modification of the PSA geometry	

The rheology of a highly viscous and partially elastic structure is regulated by means of cross-linking. The contact surface of the adhesive is divided by formulation an inhomogeneous adhesive mass, with particles in the adhesive layer that do not buildup contact, because they are not adhesive, or not adhesive enough, or do not diffuse into the pores of the adherend. Such an adhesive layer can be produced using inert filler particles that are placed on the contact surface. These particles are partially adhesive or elastic or "fully" adhesive but too voluminous.

The principles of crosslinking of PSAs have been discussed in Refs. [31,59], which state that crosslinking can be chemically or physically (radiation). Crosslinking of acrylates is described in Chapter 6. It is well known that chemical crosslinking supposes the presence of reactive groups (polar functional groups or unsaturation) in the polymer. Ultraviolet light-induced crosslinking requires a chemical composition of this type; EB-induced crosslinking does not. Cross-linking due to built-in macromolecular reactive sites is generally the domain of solvent-based PSA. Here the long experience with the crosslinking of natural rubber can be called upon. The curing of NR uses external crosslinking agents (and heat). A slight mechanochemical destruction of the macromolecules of natural rubber (mastication) followed by postcrosslinking leads to products having balanced adhesive properties (labels). A more advanced mastication and crosslinking together with a higher coating weight allow the better shear and high tack and peel necessary for tapes. (Crosslinking may serve as a removability regulator also). A much higher degree of cross-linking and low coating weight lead to low tack, removable products, that is, protective films (see Figure 8.4).

As is known, the wet adhesive coating is dried in a drying channel. For uncrosslinked formulations, the drying channel serves for the volatilization of the liquid dispersing medium (solvents or water) used for the adhesive. For crosslinked formulations it serves for the crosslinking (via heat transfer) of the coated adhesive also.

As discussed in Ref. [3], acrylics display the disadvantage of compliance failure, that is, adhesion buildup. This disadvantage has been reduced by internal crosslinking agents (e.g., multifunctional comonomers) like dimethylamino-ethyl methacrylate, or with polyisocyanates [4] (see Section I.A.2).

Crosslinking as a means of regulating rheology has its limits. Such a high degree of crosslinking as that used for protective films is not practicable for labels. The application of external crosslinking agents for common functional water based polymers (e.g., carboxylated acrylics or styrene–butadiene rubber) causes a pronounced loss of tack and peel. This can be understood by considering the higher molecular weight, high content of hard comonomers, and advanced crosslinking degree of the particle surface. As stated in Ref. [60], the crosslinking possibilities of hot-melts are limited also; thermal crosslinking cannot be used.

For peel regulation via nonadhesive contact sites, fillers have been suggested (see also Chapter 5). For instance, according to Ref. [53], the PSA for repositionable transfer tape is softened and

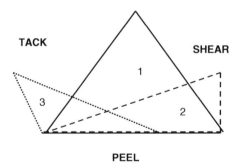

FIGURE 8.4 Schematic presentation of the adhesion–cohesion balance for masticated, tackified, and crosslinked adhesive formulations used for PSPs. (1) Label; (2) tape; (3) protective film.

filled with glass powder. The adhesive surface can possess a special discontinuous shape ensured by the PSA itself, if it is a suspension polymer. Spherical contact sites are formed [3], that are anchored on the carrier surface with the aid of a primer. PSA polymer particles with a diameter of 0.5–300 μm can be used as filler also [4]. Such pressure-sensitive filler particles are crosslinked. The composition of the PSA according to Ref. [4], includes a crosslinking agent for both the matrix polymer and the particles. Polyisocyanate derivatives, polyepoxides, or aziridines are suggested. Removable pressure-sensitive tapes containing resilient polymeric microspheres (20–66.7%) hollow, thermoplastic, expanded (acrylonitrile–vinylidene copolymer) spheres having a diameter of 10 to 125 μm, a density of 0.01–0.04 g/cm^3, and shell thickness of 0.02 μm, in an isooctyl acrylate–acrylic acid copolymer have been prepared [61]. The particles are completely surrounded by the adhesive, to a thickness of at least 20 μm. When the adhesive is permanently bonded to the backing and the exposed surface has an irregular contour, a removable and repositionable product is obtained, and when the PSA forms a continuous matrix, that is strippably bonded to the backing (see Transfer Tape), the product has a thickness of more than 1 mm. This product is a foam-like transfer tape or foam tape. The (40 μm) cellulose acetate-based tape could be removed from paper without delaminating the substrate. The contact surface can be reduced by using expandable fillers also. According to Ref. [62], a photocrosslinked acrylic–urethane acrylate formulation suggested for holding semiconductor wafers during cutting and releasing them after cutting, exhibits a peel strength of 1700 g/25 mm without crosslinking and 35 g/25 mm with crosslinking. Adding an expandable filler to the formulation (6 parts to 100 parts adhesive) reduces the peel resistance. This composition exhibits 1500 g/25 mm peel resistance before crosslinking and 15–29 g/25 mm after crosslinking. The releasable foaming adhesive sheet can be prepared by coating a blowing agent (e.g., of microcapsules) containing PSA (30%), as a 30 μm dry layer [63]. This adhesive is removable and improves the cuttability. Table 8.8 lists some typical examples of the use of microspheres in the formulation of PSPs.

b. Raw Material Dependent Formulating

The aim of formulation and its modalities vary widely depending on the raw materials. For the classic way to obtain PSA (i.e., using an elastic and a viscous component), formulation is the

TABLE 8.8
Microspheres Used in Formulation of PSPs

PSP	Carrier	Adhesive	Filler	Ref.
Tape	—	Acrylic	Adhesive copolymer microspheres, polymerized in suspension, φ 10–100 μm	[64]
Tape, releasable from paper	—	—	PSA particles, φ 50–100 μm	[65]
Tape, releasable from paper	—	—	Crosslinked, milled PSA, φ 50–100 μm	[66]
Transfer tape	Cellulose acetate	Isooctyl acrylate–acrylic acid	Acrylonitril–vinylidene copolymer, expanded spheres, φ 40 μm	[67]
Sealing tape	—	Acrylic, UV-cured	Silica microbubbles	[68]
Insulation tape	PET	Acrylic, UV cured	Glass microbubbles	[68]
Sheet releasable	—	Acrylic	Microcapsule with blowing agent	[62]
Sheet releasable	—	Acrylic, UV-cured	Microcapsule with blowing agent	[63]

manufacture of the adhesive by mixing of components. In the formulation of viscoelastic pressure-sensitive raw materials formulating consists only of tailoring of the PSA, that is, modifying a given "finished" adhesive to achieve other adhesive, conversion, or end-use properties. Formulation should be designed according to the chemical and macromolecular characteristics of the base elastomers or viscoelastomers. There are certain raw materials that can be used for different formulations. However, the chemical and macromolecular characteristics of these materials limit their preferred use to well-defined application domains. For instance, tackifiers work best with base polymers whose glass temperature is below $0°C$. Because vinyl acetate copolymers have a higher T_g, they do not work with tackifiers [69].

Formulations with natural rubber as base elastomer have to take into account NR's balanced adhesive properties (for labels), relatively low shear resistance (for tapes), poor solubility (need for milling) and softness in a highly crosslinked status (protective films). Current removable adhesives are rubber-resin-solvent-based and acrylic water-based recipes. Both of these formulations include high molecular weight polymers that reduce flow on a surface and prevent adhesion buildup [70].

Hot-melt PSAs contain lower molecular weight raw materials and a high level of micromolecular components. The adhesives may further contain up to 25% by weight of a plasticizing or extending oil to control wetting and viscosity. Hardness of formulations based on thermoplastic elastomers (TPE) is given by physical crosslinking only (see Chapter 5 also). This is temperature-dependent. Therefore, in formulations using such raw materials, their low cohesivity at higher temperatures should be taken into account. In comparison with tackification of natural rubber, where adequate compatibility and chemical reactivity of the elastomer allow the use of numerous types of resins (including reactive ones), for SBCs the choice of tackifier and that of tackifier level are more difficult. Because of the higher coating weight needed in their application field, economical considerations are also important. As mentioned, usual HMPSA formulations usually incorporate a high level of plasticizer, usually a naphtenic oil or a liquid resin. The use of plasticizers results in a number of disadvantages including long-term debonding. According to Ref. [71], resins synthesized from the C_5 feedstock having a major portion of piperylene and 2-methyl butenes, with a defined diolefin/monoolefin ratio, M_w of 800–960, M_n of 500–600 (M_w/M_n ratio of at least 1.3), a softening point of about $57°C$ and a T_g of about $20°C$, need no oils in formulations with saturated midblock thermoplastic elastomers. Saturated midblock thermoplastic elastomers include commercial ABA-type block copolymers where the polyisoprene or polybutadiene is hydrogenated. The block copolymers with a saturated midblock segment have an M_n of about 25,000 to 300,000. These thermoplastic elastomers are used in the formulation of adhesives for diaper closure tapes also [72]. It is desirable that the endblock resin have a softening point and glass transition temperature above those of the endblock and midblock of the block copolymer. But a liquid tackifier component is used also.

Polymers with built-in viscoelasticity possess a broader chemical basis for the manufacture of all adhesive classes (i.e., hot-melts, solvent-based, and water-based formulations). Some of them can be used as raw materials for noncoated PSP components also. As known, acrylics were developed as raw material for PSA in form of elastomers (rubbers) and viscoelastomers also. Random and block copolymers have been synthesized as acrylic rubber. Both can be applied for solvent-based or hot-melt adhesives. Viscoelastic acrylics are not suitable in their neat form as HMPSAs; they need postcrosslinking. Because of the similarities and differences in the properties of acrylic and styrene–diene block copolymers, their formulation differ. Acrylic block copolymers have both soft and a hard acrylic phase [73], and a star-shaped, radial structure. Their storage modulus has a plateau value of $10^{9.5}$ dyn/cm^2. Their soft phase has a T_g of $-45°C$, whereas that of the hard phase is $105°C$, which means that their soft phase is harder and their hard phase is softer than those for corresponding SBCs. Their viscosity depends on the shear rate. Tackified acrylic block copolymers for hot-melts exhibit peel values of 90–105 oz/in. The crossover point from adhesive failure to adhesive delamination is situated in the tackifier range of the

30–45 phr. Their shear resistance increases with tackifier level. This is unusual behavior. The shear adhesion failure temperature (SAFT) value attains 80–86°C. Their melt viscosity is less than 50,000 cP at 177°C. For such polymers formulations with 75–80% resin loading have been suggested. In comparison, the highest resin level which contributes to the peel of water based acrylics is about 50% and the best overall balance is obtained with high softening point resins at around 30% level [46].

Formulation as a function of the raw material basis (chemical composition of the raw materials) should take in to account the formulating additives. It is evident, that solvents used as dispersing media differ, according to the type of raw materials. As an example, isoparaffinic solvents are used for natural rubber, butyl rubber, poly-isobutylene, ethylene propylene rubber (EPR), polyiso-prene (see also Chapter 5).

A special aspect of formulating concerns the adhesives polymerized or cured by radiation (EB, UV, or mixed techniques) where formulations based on monomers or oligomers (reactive monomers, oligomers) or polymers (curable HMPSAs) can be used. In this case the use of adequate radiation intensity is determinative. Unfortunately, the required radiation dosage also depends on its absorption by the components of the recipe. As stated in Ref. [60], a special raw material for acrylic HMPSAs, developed for photochemical curing, requires tackifying. For this polymer the photoinitiator is built in the polymer. Although the classical way is to use external photoinitiators (e.g., benzophenone), built-in, monoethylenically unsaturated aromatic ketones can be used for radiation crosslinking of polyacrylates also [74] (see also Chapter 5 and Chapter 6). The built-in crosslinker makes it unnecessary the viscoelastomer to contain unsaturated polymerizable bonds (and therefore it displays improved aging properties), it does not produce skin irritation and its UV-induced polymerization is not affected by atmospheric oxygen [60]. In the classic technology, using an external photoinitiator, the liquid monomers are applied to the carrier with a doctor bladeor by roller coating or spraying and cured in a polymerization tunnel from which oxygen has been excluded. As shown in Ref. [75], photopolymerization of acrylics can be carried out with UV light having a wavelength of 280–350 nm (see also Chapter 6). According to Martens et al. [76], a very precise exposure of $0.1–7 \text{ mW/cm}^2$ optimizes the molecular weight of the resulting polymer. In the procedure given by Auchter et al. [60], the uncrosslinked product does not have balanced adhesive properties. Because of its low molecular weight it does not display enough cohesion and its tack must be improved also. Its PSPs are achieved only by tackifying and crosslinking.

The copolymerized photoinitiator is an acrylic ester with benzophenone terminal groups. The main crosslinking reaction is the addition of a hydrogen atom of the side alkyl group of an acrylic segment on the carbonyl group of the side benzophenone group of the vicinal polymer chain, that is, the formation of a hydroxy-diphenyl-polyacrylmethane derivative. Unfortunately, the build-in of this reactive group promotes side reactions. To avoid such reactions, low radiation levels should be used. To modify of UV-crosslinkable polacrylates, the use of an esterified highly stabilized rosin is proposed. Because of the absorbtion of UV radiation by the resin (cyclical aroma-tized structure) only a low level of the resin is suggested [60]. For the tackified formulation a resin loading of 8.5% improves the peel and tack but reduces shear resistance. Because the resin is a diluting agent in the polymer matrix and reduces the number or crosslink points there is an upper limit of the amount of resin that can be added, delimited by shear reduction and radiation absor-bance. Therefore, for tackified formulations the radiation dosage must be increased. Table 8.9 lists the main criteria for the choice of tackifier resins as a function of the raw materials used for the PSPs.

A similar technical method, the synthesis of a low molecular acrylic polymer (processible as a hot-melt) and its postcrosslinking have been proposed by Harder [77]. High concentration solution polymerization of acrylics has been carried out and relatively low molecular weight, low viscous polymers have been synthesized and isolated, by evaporating and degassing, as solid-state raw materials for HMPSA. According to Ref. [78], such polymers also include rest solvents (ca. 3%). The manufacture of highly concentrated acrylic polymer solutions is not new. High molecular polymers (used as fiber-like or film-like materials) were patented some decades ago [79,80].

TABLE 8.9
Technical Criteria for Usability of Tackifier Resins for Selected PSP Raw Materials

Raw Material	Usability Criterion
Natural rubber	Compatibility
	Reactivity
	Solubility
Styrene block copolymer	Compatibility
	Melting point
Radiation curable raw materials	Compatibility
	Melting point
	Absorbtivity
	Reactivity
Acrylics	Compatibility
	Melting point
	Solubility/dispersibility

The polymers described in Ref. [77] are not enough high molecular. Such low molecular weight is a need to be coatable between 90 and 140°C as a viscous liquid. Because of their low molecular weight such polymers have to be crosslinked after coating. They can be crosslinked by using EB- and UV-curing. Chemical crosslinking by NCO-containing systems can be carried out also. Their tackified formulation depends on the choice of the crosslinking method.

Generally, the choice of a raw material class determines the range of possible performances also. As discussed by Gerace [81], butyl tapes tested in comparison with acrylics for bonding exterior trim moldings to automobiles perform better at low temperature and pass cold shock tests. This behavior is due to the low T_g of the butyl system.

c. Coating Technology Dependent Formulation

Formulations may differ according to coating technology also. The pressure-sensitive layer can be coated on the carrier material as a dispersant-free, 100% solids component or as a liquid having a dispersed or solved adhesive in organic solvents or water. Hot-melts or extrusion-coated thermoplastic adhesive layers are applied in molten state. Thermoplastic elastomers combine the excellent processing characteristics of thermoplastic materials and a wide range of physical properties of elastomers. For tapes cold pressing of masticated natural rubber-based formulations is known also. Special compositions can be dry and cold sprayed.

Although with a 100% solid content, radiation-curable PSAs can also be coated in (at room or elevated temperature) liquid state. Because of the different abilities of the raw materials to be coated in molten or dispersed status their formulation also vary. Although the development of macromolecular chemistry allowed the synthesis of new elastomers and viscous components (resins), their number is limited, especially in the range of meltable compounds (i.e., raw materials for hot-melts) and water-based materials.

As discussed in Chapter 5, coating of a prepolymer is preferred for radiation-curing PSAs also. For instance, the viscosity of a radiation-crosslinkable low molecular acrylic is about 10–20 Pa sec. It can be processed at 120–140°C [60]. Another patent [82] states that the monomers should be partially polymerized (prepolymerized) to a coating viscosity of 1–40 Pa sec before use. The data of Table 8.10 illustrate the viscosity domain of low-viscosity oligomer-based, UV-curable

TABLE 8.10
Viscosity of Some Oligomer-Based HMPSA Formulations

Product	Chemical Composition	Viscosity	Temperature (°C)	Ref.
Tape, label	Acrylated polyester, UV-, EB-curable HMPSA	2000–10,000 mPa sec	100	[83]
Tape, label	Polyurethane–acrylate HMPSA	11,625 cP	120	[84]
Tape	Rubber-resin, in heptane solvent-based PSA	500 mPa sec	RT	[85]
Label	Styrene block copolymer, HMPSA	5000–80,000 mPa sec	175	[86]
Label	Acrylic, UV-curable oligomer	10,000–20,000 mPa sec	120–140	[60]
Tape	Acrylic, UV-curable oligomer	40,000 mPa sec	—	[82]

HMPSAs. As seen from these data, viscosity of these compounds is much higher than the usual viscosity of solvent-based adhesives, but significantly lower than that of common HMPSAs.

Generally hot-melts are polymer blends (elastomeric and viscous components). Styrene–butadiene block copolymers were the first rubbery raw materials to be coated in the molten status. Although later acrylics- and EVAc-based hot-melts were developed, styrene–butadiene block copolymers remain the most important raw materials for HMPSA. Like adhesive manufacture for natural rubber-resin based PSAs, in the actual stage of development TPE-based hot-melt formulation is really the manufacture of the adhesive. This manufacturing process is absolutely necessary to transform an elastic compound into a viscoelastic material of coatable viscosity. Therefore it may be concluded that, depending on the physical status of the coatable adhesive, formulating refer to the manufacture or modification of the adhesive. According to Hunt [87], because of the complex equipment and know how required for HMPSA formulation (in a different manner from water-based PSA formulation), special suppliers of ready-to-use HMPSA should be preferred.

Hot-melts have to be manufactured mixing the elastomeric and viscous components. The adhesive produced must also be tailored with respect to processibility. In reality, the choice of the tackifier resins influences the viscosity of the molten polymer also, but plasticizer and oils are necessary to regulate it. Low molecular weight polyisoprenes (35,000–80,000) having a low T_g (−65 to −72°C) have been used as viscosity regulators for EVAc-based HMPSAs [88]. They replace common plasticizers and improve migration resistance and low temperature adhesion.

Hot-melt formulation should take into account the temperature sensitivity of the carrier material. As is known, hot-melts may be coated on plastic carrier materials. Fabric, polyvinyl chloride (PVC), cellulose acetate, and polyester films have been coated with hot-melt for labels [89]. Because of the differences in monomer basis and manufacture procedure (polymerization, polycondensation, or polyaddition) of these polymers and films (blown, cast, nonoriented, or oriented, etc.), their temperature resistance is quite different (see Section II.B).

Formulation for hot-melts should also take the coating equipment, that is, the coating device into account. According to Kim et al. [90], the coating process has more of an effect on the peel resistance and tack of HMPSAs than the melting process. Different coating devices require different viscosity. The importance of processing viscosity is illustrated by the following list of quality criteria for HMPSAs given by Shields [91]: adequate adhesion on kraft paper and undulating surfaces, better adhesion on PE at −18°C, better tack, no migration, and no processing problems (no odor and low viscosity). The addition of low melting point resins to a hot-melt formulation with block copolymers (Kraton GX 1657) decreases the melt viscosity [92]. It should be mentioned that low viscosity resins may produce migration in hot-melt formulations based on block copolymers. The formulation of styrene block copolymers uses resins that are compatible with

one of the segments of the block copolymer. Such compatibility can be limited by special manufacture of the resin also. The insolubility of a rosin-based tackifier resin in the styrene blocks of
a hydrogenated SBR and SIR copolymer can be improved by its transformation into a coordination
complex having the formula: $M(ORes)_2 \cdot n (HORes)$ where $0 < n < 6$, M is the metal, and Res is
the resin [93].

Solvent-based PSAs have a broader raw material basis. They can be formulated with rubber-
resins, acrylics, polyurethanes, polyvinyl ethers and other materials. In this case formulating can
be either the manufacture of an adhesive (e.g., rubber-resin PSA) or the modification of adhesive
(e.g. AC, PUR, PVE). In both cases it is focused mostly on the design of the adhesive properties.
According to Hunt [87], solvent-based acrylics are used for special products such as no-label look
PVC labels and where optimum water resistance is required.

For water-based PSAs, formulation is mostly modification of the base adhesive (AC, EVAc,
PVE, etc.) but it is also focused on the adhesive and converting properties. In this case the
modification of the base adhesive is more complex, because of its built-in (bulk) chemical composition and its chemical composition as a dispersed system (having particles with a composite
structure). In the first stage of their development water-based PSAs were considered less aggressive
than solvent-based compositions. As seen from an old test standard [94], the coating weight
suggested for water-based PSAs was 25 g/m^2. In comparison, the value given for solvent-based
products was only 20 g/m^2. The chemical nature of the main polymer chain, its water-
activatable functional groups, and the chemical (or macromolecular) nature of the technological
(dispersion) additives also influence the formulation. As discussed in Ref. [95], the molecular
weight of the polymer and its T_g are the major variables that determine the PSA properties of
latices based on SBR. These are characteristics of the main polymer backbone. Other factors
that play a role in adhesive properties are the chemical structure of the additional monomers
used for synthesis, the choice of alkali for neutralization of carboxylic end groups, and the chemical
structure of the surfactant in the latex. For water-based dispersions the chemical basis of the main
polymer influences the technology used for its dispersion and the chemical and dispersion properties of the product. As a consequence the formulation is affected too. As shown in Ref. [96], vinyl
acetate polymers and copolymers with an acid pH give acid dispersions, that must be formulated in
acid domain, therefore only nonionic tackifiers should be used. Chloroprene latices have to be neutralized. The choice of neutralizing agent needs special know-how [97]. Formulation also depends
on the storage time. When water-based viscoelastic and viscous components are mixed (e.g.,
acrylics with tackifiers) the peel resistance increases with the storage time of the mixture (up to
a limit).

Coating technology and the particular coating device can require a certain viscosity and solids
content and influence the formulation. As an example, coating with an air knife needs a viscosity
of $100-300 \text{ mPa sec}$ [98] and solids content of about 30% (wt.) and the practical coating
weight (generally between 2 and 25 g/m^2) is $2-3 \text{ g/m}^2$ (dry). The maximum running speed
(250 m/min) decreases for films to 90 m/min. Formulation required for a Meyer bar differs
from a recipe to be coated with reverse roll coater [99]. According to Sanderson [100], acrylic
tape adhesives can be designed (i.e., produced) that differ in viscosity and foam generation as a
function of the coater used. An adhesive for rotogravure has a viscosity of $75-100 \text{ cP}$ and
foaming of $2-5\%$ in comparison with adhesives for a Meyer rod (viscosity of $300-800 \text{ cP}$,
foam generation of $10-20\%$) and an adhesive for knife over roll (viscosity of $4000-8000 \text{ cP}$,
foam of $10-30\%$). Spray-dried emulsion polymer powders are also known; for these raw materials
a special dispersing technology is required.

In Europe the large labelstock suppliers formulate their own water-based adhesives, but
ready-to-use formulations are also commercially available [87]. The types and grades of the
PSAs that are used vary according to geographical regions and countries. In Europe the proportion
of solvent-based, water-based, and hot-melt adhesives vary according to geographical area also. In
1989, the most water-based adhesives have been marketed in the northern area (UK, Iceland, and

Scandinavia). The southern region (France, Italy, and Spain) is leader in the use of HMPSA [99]. In the last years hot-melts are preferred in middle Europe too.

A special aspect of formulation is related to crosslinking. For common crosslinked formulations (e.g., PSA for tapes and protective films) the formulation should ensure thermal crosslinking in a short period of time of drying at elevated temperatures in drying channel (see Section III.B.4). The majority of crosslinked formulations give a crosslinked adhesive after coating and drying/ crosslinking/cooling of the pressure-sensitive web. However, there are certain formulations where crosslinkability is "stored," and crosslinking occurs later, during bonding or debonding of the PSP. For these products special coating technology and a crosslinking technology-dependent formulation are necessary. A superficial crosslinking or two-component formulation is suggested. Such technologies are described in Refs. [100,101]. According to Ref. [101], tapes with adhesion-free surface can also be manufactured too. The tack-free surface is achieved by superficial crosslinking of the adhesive using polyvalent cations (Lewis acid, polyvalent organometallic complex, or salt). This crosslinked portion of the adhesive has a greater tensile strength and less extensibility. In some cases the polymer should be crosslinked only along the edges of the adhesive layer. The tack-free edge avoids oozing and dirtiness. The crosslinking penetration depends on the polymer nature and on the solvent. Crosslinking can also be achieved by dipping. In this process water or water-miscible solvents should be used.

Acrylate-based PSAs containing acid or acid anhydride groups and glycidyl methacrylate coated on a flexible carrier can be crosslinked between 60 and 100°C by using a zinc chloride catalyst [102]. A special formulation is carried out. Instead of acrylate copolymers with monomers containing epoxide groups and compounds containing acid anhydride group, mixtures of two separately prepared copolymers may be used, for example an acrylic copolymer containing epoxide groups and an acrylic copolymer containing anhydride groups. Crosslinking catalysts (0.05–5 wt.%) are added to the polymer. The polymer coated on a carrier is crosslinked after evaporation of the solvent (2–7 min at 60–100°C). As catalyst zinc chloride or magnesium chloride, monosalts of maleic acid, organic phosphoric or sulphonic acid derivatives, or oxalic or maleic acid can be used. Tertiary amines and Friedel–Crafts catalysts accelerate crosslinking. The crosslinked adhesive is storable, that is, its adhesive properties are not changed by postcrosslinking during storage. Even heat-sensitive materials like plasticized PVC may be used as the carrier material for such adhesives. It is possible to make a prepolymer that is polymerized by radiation. A small amount of catalyst or polyfunctional crosslinking monomer may be added to the composition prior to the completing the polymerization by heat curing. Such prepolymers are more viscous and are easily applied to the support [17]. Formulation for adhesives coated as a radiation-curable mixture of oligomers or polymers should take into account the inhibiting or accelerating effect of the tackifier resin during curing. It has been shown that saturated hydrocarbon resins are more effective than their unsaturated counterparts in their ability to enhance electron beam crosslinking of styrene–isoprene block copolymers [103].

d. Formulation Depending on Product Class

A variety of PSPs are manufactured using various coating techniques, carrier materials, and different construction. Therefore, formulation differs according to product class also. Economic considerations also limit the use of certain raw materials. Except for special requirements and products, PSA for labels are formulated to achieve better peel and tack; those for tapes for to obtain better shear; and those for protective films to realize better removability.

For general application labels, formulation criteria imposed by their end-use depend on whether the labels will be produced on reels or sheets and on whether the product is to be permanent or removable. Other end-use requirements have to be taken into account for special labels, for example, increased or decreased water sensitivity or mechanical resistance.

Generally, cohesive formulations are used for tapes. Such formulations are crosslinked compositions. In adhesive recipes for tapes, natural rubber should be crosslinked [104]. For PE tapes, adhesives on natural rubber basis have been suggested [105]. In their formulation polyiso-butylene and rosin are added as tackifier. To afford better dissolution of the rubber and to reduce its molecular dispersity the elastomer is masticated, that is, depolymerized by calendering. After coating of the formulated adhesive, the molecular weight is "rebuilt" by crosslinking to achieve higher cohesion. According to Ref. [50], although acrylic rubbers can be solved without milling, mastication lowers their viscosity. Such adhesives can be cured using isocyanates, UV curing agents (e.g., p-chlorobenzophenone). Starting from the same raw materials formulation for labels, tapes, or protective films can be quite different. The mastication degree of the rubber, the tackifier level, and the degree of crosslinking may strongly differ.

In some cases mastication helps to realize a fully solventless mixing and coating technology. The adhesive is calendered on a continuous carrier material or in a porous one. Such, "press rolling" technology ensures solventless formulation and manufacture (no need for drying, no migration, no skin irritation), preferred for medical tapes. Rubber for medical tapes is masticated also [106]. According to Schroeder [107], the adhesives used for tapes have to display color stability, transparency, and chemical, thermal, and electrical resistance.

As stated in Ref. [108], adhesives for protective films must display adhesion that permits removal of the film. The adhesive properties of protective films are characterized by removability as a major requirement. This performance is achieved by an extreme reduction of the coating weight, but other formulation-related technical means are also to reduce peel resistance. The first formulating modality is the change of the adhesion-cohesion balance. Because of the require-ment for deposit-free removability, reduction of peel values cannot be achieved via softening of the adhesive (see Chapter 7). Another possibility of detackifying is crosslinking.

Softening of the "adhesive" was attempted earlier with the use of PVC plastisols (plasticizer gelled PVC compositions) to manufacture oil- and solvent-resistant, paper-backed adhesives for covering sensitive substrate surfaces. Softening of the plastic carrier materials leads to adhesive-free self-adhesive films (see Section III.A.2).

Like recipes for tapes, most natural rubber based formulations for protective films use mastica-tion. In this case calendering is used as one way to regulate viscosity and peel resistance. Because of the requirement for a low peel, a low coating weight is suggested for these products. Low coating weight can be exactly controlled by using low viscosity formulations. Because of the low peel requirement, a relatively low level of tackifier resins is recommended. A low level of tackifier resin means higher viscosity of the formulated adhesive. Therefore, mastication plays a special role in the manufacture of adhesives for protective films. As a consequence, postcrosslinking is very important also.

Removability criteria influence the adhesive choice and formulation for tapes too. Because adhesion builds up due to use of highly polar comonomers such as acrylic acid in acrylic PSAs, such components are not recommended for the synthesis of adhesives for masking tapes. Decorative films need special formulation in order to avoid an interaction (migration) between the components of the carrier material and the adhesive [109].

A special case of product-related formulation is given by the manufacture of pressure-sensitive hydrogels used for medical products (see Chapter 9).

e. Product Buildup-Dependent Formulation

The nature of the carrier material influences formulation. Heat-sensitive carrier materials need "low" temperature warm melts. These are hot-melts that can be used for temperature-sensitive applications. According to Ford [110], such products do not have a lower application temperature, but the plastic-to-liquid transition temperature has been dramatically reduced while the softening point has been held constant.

The carrier influence is given by its surface and bulk characteristics. Because HMPSAs are based on relatively low MW polymers, plasticizer migration affects them more. On the other hand, many healthcare applications exist, where it is desirable to coat a PSA on a porous web (see Chapter 9). A porous carrier is required and the adhesive should also be porous [111]. In this case the use of HMPSA may reduce adhesive penetration into the web.

The surface quality of the adherend can require a special adhesive geometry or/and coating weight. This can be achieved by tailored formulations. Crosslinked, filled, and multilayer compositions are used. According to Czech [112], UV curing of monomer–prepolymer mixtures is suggested for medical tapes with high coating weight. As stated in Ref. [113], EB-cured formulations can be used only for coating weights of less than 100 g/m^2.

The formulation of the adhesive depends on the printing method and also on the technology used for the face stock material. For instance, rating plates can be printed in sandwich printing, where on transparent materials are printed the reverse side by using mirror lettering [114]. Where the adhesive is to be used to form laminates wherein at least one surface is a printed surface, the presence of any residual surfactant can lead to discoloration or bleeding of the ink. This is a problem in applications such as overlaminating of books or printed labels, where the purpose of the outer surfacing film is to preserve the integrity of the printed surface.

f. End-Use-Dependent Formulation

End-use requirements for PSPs are various, therefore requirements towards PSAs also vary. According to a study conducted by the Skeist Laboratories [115], low temperature resistance for deep freeze price labels, balanced adhesion–cohesion for structural PSAs for holding and mounting, high shear and/or peel for packaging tapes adhesives, resistance to perspiration and body liquids for personal care products, and heat resistance for electrical and industrial uses are the main requirements towards the adhesives needed in future label production. Temperature-resistant adhesives are required for thermally printed labels and copy labels [116]. Solvent-based acrylics are recommended for durable mounting tapes, outdoor PSPs, medical tapes, protective tapes, and high performance decals [117]. Liquid resins are recommended for plasters, and multifunctional anhydrides and multifunctional epoxies are suggested as crosslinking agents [118].

Quite unlike labels where the adhesive properties are determinant for the end-use, for other product classes, the application and deapplication conditions have greater importance. There are special end-use requirements for a given application or for an application field. The actual application conditions affect formulation within the same product class also. For instance, for book-binding nonyellowing carrier materials and adhesives are preferred [119]. According to Fuchs [120], for decoration films, soft PVC carrier (coated with $18-28 \text{ g/m}^2$ HMPSA) is used. Some special pharmaceutical labels may have take-off (detachable) parts to allow multiple information transfer [121]. Parts of the label remain on the drug or on the patient. The detached part of label should have a medical adhesive so it can be applied to skin. Nonremovable and nontransferable constructions have been manufactured as antitheft labels (see also Chapter 11). In an application of this type a destructible, computer-imprintable vinyl film and a very aggressive PSA have been used. For such a product, the adhesive must bond in less than 24 h [122].

Water-resistant products are required for many applications. The water resistance of polymers depends on their polarity, crystallinity, and structure. The water resistance of dispersion-based coatings is primarily a function of their water-soluble additives [123] (see also Chapter 5). Theoretically, for water-resistant applications, water-unsoluble, solvent-based adhesives should be used. Water-based acrylics that can survive immersion in water have been synthesized also [124]. For several applications PSAs have to resist atmospheric moisture, but be removable with warm water [125]. Wine and champagne bottle labels have to be removable in water without supplementary addition of chemicals at 60°C [126,127].

Water-soluble formulations may have quite different raw material basis. They can be formulated by using common, commercially available polymers, or by synthesis of special macro-molecular compounds. Their formulation depends on the degree of solubility required and on the adhesive coating technology. Water-releasable labels and water-dispersible splicing tapes or medical tapes need quite different degree of solubility. Building solubilizing functional monomers into a polymer or adding of water-soluble agents to a formulation depends on the physical state of the adhesive also. Emulsion copolymerization of certain water-soluble monomers is difficult because of their tendency to homopolymerize in water. Therefore solution polymerization is used. On the other hand, water-soluble additives (surfactants) can be fed into water-based dispersions without a problem. The use of nontacky hydrophilic solubilizers in hydrophobic molten HMPSA requires special knowledge. Ideally both the main elastomer or viscoelastomer and the tackifier or plasticizer should be water soluble. In practice, such formulations are used only for difficult applications like splicing tapes or medical labels. For water-releasable labels the use of solubilizing additives is suggested [126,127].

A water-soluble PSA hot-melt is formulated on the basis of vinyl pyrrolidone–vinyl acetate copolymers. The composition also includes a free monobasic saturated fatty acid having an acid number higher than 137 [128]. According to Ref. [129], water-soluble HMPSAs for labels are formulated with 35–60% vinyl pyrrolidone or vinyl acetate–vinyl pyrrolidone copolymer together with free fatty acids. Such labels can be peeled off in 5 min by subjecting the edge of the label to a stream of water at 35°C. Fatty acids with 8 to 24 carbon atoms act as solubilizer in a concentration of 3% or less [130]. As discussed in Refs. [48,131], water-soluble hot-melt adhesives can be formulated with vinyl pyrrolidone-vinyl acetate copolymers and water-soluble polyesters. Poly-vinylpyrrolidone-based formulations exhibit poor thermal stability. A hydroxy substituted organic compound (alcohols, hydroxy substituted waxes, polyalkylenoxide polymers, etc.) and a water-soluble *N*-acyl substituted polyalkylenimine obtained by polymerizing alkyl substituted 2-oxazolines can be used with better results. Such water-releasable HMPSAs contain 20–40% poly-alkylenimine, 15–40% tackifying agent and 25–40% hydroxy substituted organic compound [48]. Water-soluble plasticizers whose molecular weight does not exceed 2000 are preferred (e.g., poly-ethylene glycols (PEGs) of molecular weight of about 200–800). Fillers with increased organophi-licity (e.g., fatty acid ester coated mineral extenders like stearate/calcium carbonate compound) are suggested. As a antioxidant distearyl pentaerythritol diphosphite is used. (As discussed in Chapter 4 and Chapter 9, polyvinylpyrrolidone plasticized with glycols can also be formulated to give pressure-sensitive hydrogels.) As a water-soluble base component polyethyloxazoline has been proposed for water-soluble hot-melts also [131]. Polyalkyloxazolines and water-insoluble polymers have been introduced for formulation of PSA displaying (alkaline) water-dispersibility, solubility, water activation and solubility with alkalies [49]. The recipe comprises a rubbery polymer, a polyalkylenimine, a functional diluent based on acid functional polymeric compounds, hydroxyl-substituted organic compounds, tackifiers (rosin acids) and plasticizers, waxes, and an acrylic acid copolymer to provide additional strength. Such a formulation for a repulpable general purpose HMPSA contains a thermoplastic polymer, (20–50%), polyalkylenimine (10–30%), diluent (2–90%), wax (5–40%), tackifier (85–40%), plasticizer (0–40%) and filler. As thermoplastic com-ponent vinyl polymers, polyesters, polyamides, polyethylene, polypropylene (PP), and rubbery poly-mers prepared from monomers including ethylene, propylene, styrene, acrylonitrile, butadiene, isoprene, acrylates, vinyl acetate, functionalized acrylates, and polyurethanes can be used.

According to Ref. [132], water-dispersible hot-melts are based on vinylester graft copolymers with water-soluble polyalkyleneoxide polymers. Vinyl carboxy acid based polymers, tackified with water-based resin dispersions are used too [133]. Water-dispersible HMPSAs can contain 25–40% styrene–isoprene–styrene (SIS), 25–40% tackifier, 25–10% fatty acid and 25–10% polyvinyl pyrrolidone (PVP) [134]. A proposed formulation for water-soluble HMPSA comprises 22, 39, 11, 14 and 14 parts of ethylene-acrylic acid copolymer, PVP, plasticizer, tackifier, and polyethy-lene–glycol, respectively. A simplified formulation contains 60, 30, and 10 parts of PVP, polyethy-lene glycol, and phtalate plasticizer [134].

According to Czech [135], a water-soluble adhesive for splicing tapes can be manufactured using (1) a polymer analogous reaction; (2) polymerization; or (3) polymerization and formulation.

By blending the PSA raw materials that have carboxyl groups with reactive hydroxylamine and neutralizing the mixture, it is possible to regulate the solubility of the polymer. The synthesis of a special PSA on the basis of aminoalkyl methacrylates or methacrylamide derivatives leads to water-soluble polymers also. Formulating the water-soluble viscoelastic components with water-soluble resins, gives better adhesivity and better water-solubility. Polyacrylates tackified with water-soluble PVE have also been proposed [136]. Amine-containing soluble polymers and soluble plasticizers have been proposed [137]. A water-activatable PSA containing a water-soluble tackifier (5–150 wt.%) is described in Ref. [138].

Paper-based tape recyclable as paper is manufactured using a water-soluble adhesive based on alkali-soluble acrylic acid esters and polyethylene or polypropylene waxes and alkali-dispersible plasticizer [139]. For splicing tapes which have to resist temperatures up to 220–240°C, carrier materials have to be water-dispersible, and the adhesive should be water-soluble, but resistant to organic solvents. No migration is allowed. Their adhesive has to bond at a running speed of 160 m/min. Under such conditions high instantaneous peel and shear resistance are required [140]. Water-soluble compositions used for splicing tapes include polyvinyl pyrrolidone with polyols or polyalkylglycolethers as plasticizer [140], vinyl ether copolymers [141], neutralized acrylic acid–alkoxyalkyl acrylate copolymers [142], and water-soluble waxes [143]. Such formulations must adhere to the paper substrate running at 1200 m/min and must be soluble in water at pH 3–9. For a water-soluble plasticizer, 20–70 wt.% ethoxylated alkylphenols, ethoxylated alkylamines, and ethoxylated alkylammonium derivatives have been proposed. For higher shear short cellulose fibers (1–5%) are added as filler.

Water-based PSA formulations that are removable with warm water contain 7–30% hydrophylic monomers [144]. Polymeric oxazoline derivatives can also be used in water-soluble adhesive compositions. A reaction product of an acrylic interpolymer and 2-oxazoline has been suggested [145].

Polyacrylate-based, water-soluble PSAs are recommended for medical PSPs (operating tapes, labels, and bioelectrodes) [137]. Operating tapes are used for fastening of cover material in surgical operations. Cotton cloth and hydrophobic special textile materials coated with low energy polyfluorocarbon resins are used as carrier materials, together with polyacrylate based water-soluble PSAs. Such adhesive has to be insensitive to moisture and insoluble in cold water. It has to display adequate skin adhesion and skin tolerance. The water-solubility of the product is reached above 65°C at a pH value higher than 9. Water-solubility is given by a special composition containing vinyl carboxylic acid (10–80 wt.%) neutralized with an alkanolamine [146]. Such tapes are used for biolelectrodes also because of their low electrical impedance and conformability to skin. Other compositions contain acrylic acid polymerized in a water soluble polyhydric alcohol and crosslinked with a multifunctional, unstarurated monomer. A prepolymer (precursor) can be synthesized also, which is polymerized to the final product using UV radiation, according to Ref. [106]. This polymer contains water which affords electrical conductivity. Electrical performances can be improved adding electrolytes to the water. Such products can be used as biomedical electrodes applied to the skin.

The adhesive for medical tapes has to be resistant to light and heat, show skin tolerance and transmit water vapour. The adhesive can contain acrylic, styrene copolymers, vinyl acetate copolymers and polyvinylether [147]. For medical tapes applied on skin, adhesives based on acrylics and natural rubber have been suggested [148]. To prepare high solids content solutions, the elastomer should be masticated [149]. According to Ref. [106], polyisobutene (MW 80,000–100,000) is used for medical tapes because it does not adhere to the skin. Physiologically compatible adhesives are based on polyurethanes, silicones, acrylics, polyvinyl ether and styrene copolymers also. These should be breathable, cohesive, conformable, self-supporting and nondegradable by body fluid [150]. Special polyurethanes have been suggested for medical tapes too. As shown in

Ref. [25], in a different manner from rubber formulations which leave deposit on the skin, and from acrylic formulations which may irritate the skin, PUR gels are inoffensive. As discussed in Ref. [78] generally the type of polyethylene glycole, its functionality (2–4), its structure (linear or branched), and its OH number influence its properties. For the polymer proposed in Ref. [25], the average functionality (F_1) of the polyisocyanates used being between 2 and 4 and the average functionality of the polyhydroxy compounds (F_P) being 3 to 6, the isocyanate number (K) is given by the following formula:

$$K = \frac{300 \pm X}{F_1 F_P - 1} + 7$$

where $X \leq 120$ (preferably $X \leq 100$) and K is between 15 and 70.

As discussed in Ref. [151], a fluid permeable adhesive useful for transdermal therapeutic device, to human skin for periods up to 24 h, is based on acrylic, urethane, or elastomer PSA mixed with a crosslinked polysiloxane. The therapeutic agent passes through the adhesive into the skin. Uninterrupted liquid flow through the adhesive has to occur over a prolonged period (6–36 h) and at constant rate. Acrylic PSAs in medical tapes may display adhesion buildup in the time, or weakening of the cohesive strength due to migration of oils. These disadvantages are avoided by crosslinking with built-in dimethylamino methacrylate or with isocyanate [152,153]. Table 8.11 lists some of the raw materials used for medical PSPs. As illustrated in the table, a very broad range of raw materials is proposed for the manufacture of medical tapes.

TABLE 8.11
Raw Materials and Manufacture of Medical Tapes

Product	Manufacture	Adhesive	Characteristics	Ref.
Surgical tape	Coating	Acrylic, HMPSA	Application temperature maximum 350°F	[146]
Tape	Coating	Acrylic, prepolymer UV-cured	Electrically conductive	[101]
Tape	Coating	Acrylic, solvent-based	Solution in isopropyl acetate, 3000–110,000 Cp	[154]
Tape	Coating	Acrylic solvent-based	Cold water insoluble, hot (65°C) water soluble	[137]
Tape	Coating	Acrylic crosslinked	—	[152,153]
Tape	Coating	Acrylic carboxylated solvent-based	—	[146]
Tape	Coating	Acrylic, PVE, PVAc	K 80–130	[146,147,149]
Tape	Coating	Acrylic, rubber-resin solvent-based	—	[147]
Surgical tape	Coating	i-Butyl acrylate–butylcarbamo methacrylate	—	[155]
Tape	Coating	Acrylic, PUR, PVE styrene copolymer	—	[156]
Tape	Coating	Rubber-resin, solvent-based	Solution in hexane (80%), toluene (3%), cyclohexane (17%)	[157]
Tape	Roll pressing	Rubber-resin	Zinc resinates, zinc oxide filler	[158]
	Roll pressing, extrusion	PIB	—	[105]
Tape	Coating	PUR	—	[25]

Such formulations are coated mainly solventless (roll-pressing, extrusion, 100% solids, etc.), but compositions soluble in organic solvents or water are suggested also.

Wet surface adhesives have to adhere on cold, condensed water coated adherend surfaces [159]. Water-resistant formulations are applied for different products. Tackified formulations made with a special technology (using the tackifiers own surfactants only), give better performances. In this case the solids content is very high (20% increase) in comparison with common formulations. Such adhesives with excellent water-resistance (humidity and condensed water-resistance) are required for technical tapes and pressure–sensitive assembly parts in the automotive industry [160].

Low temperature-resistant adhesives are required for deep freeze labels. As discussed in Chapter 3 and Chapter 5, sequence distribution strongly influences the rheology of block copolymers. Therefore for such applications diblock polymers are preferred [161]. For instance, for a low temperature HMPSA copolymers with A_1B_1 sequence distribution have been proposed, where A_1 is an aromated vinyl polymer block (molecular weight 12,000) and B_1 is a diene block with a molecular weight of greater than 150,000. Another patent [136] describes the use of natural rubber (50–75 parts), SBR (with 12% styrene), regenerated rubber (3–8 parts) and PVE (1–3%). As tackifier polyterpene resin 1–4% (MP 115°C), rosin ester (1–3%) and carboxylated alkylphenol resin (2–15%) are suggested. Such a crosslinked (with isocyanate) formulation can be used between −29 and +113°C.

Formulations used for contact adhesion (PSA on PSA) contain generally natural rubber. For instance, Ref. [162] proposes a recipe based on NR and resin (150/50). Filled NR-synthetic rubber mixtures can be used also [163].

3. Manufacturing Technology of the Adhesive

Manufacture technology includes both the procedure and the equipment used to manufacture the coating components of the PSPs. As discussed earlier, manufacture of the coating components generally consists of their formulation; therefore the manufacturing equipment is formulating equipment.

a. Formulating Equipment

The raw materials determine both the coating technology and the formulating equipment. Formulating is the manufacture of coating material by mixing the product components in solid, molten, or dispersed state. Certain raw materials are solid and others are liquid or have to be transformed into a liquid system. Some raw materials have to be mixed; other compounds are supplied in ready-to-use form. Certain raw materials are supplied in a micronized state, whereas others are predispersed. Therefore formulating equipment varies concerning the chemical basis of the adhesive, the coating technology, and the product classes.

Two-component formulations that include rubber-like products and tackifiers require that the components be micronized (cut, pelletized), mechanochemically destroyed (masticated, plasticized), and solubilized. To facilitate solvation, the surface area of the solid raw materials (e.g., elastomers) should be increased by physical size reduction. This can be accomplished by using a bale chopper or granulator; a sheet mill and slab chopper or a Banbury mixer. For instance, for the manufacture of an adhesive for pipe isolation tapes, based on butyl rubber a crosslinked elastomer is required. A cutting extruder is applied to mix the pellets of rubber (100 parts) with naphtenic oil (120 parts) and resin (4 parts, Super Bekacite 2000) [164]. The construction and hydrodynamic characteristics of the equipment may influence the processing performances of HMPSA. When polyacrylate rubber-based HMPSA are prepared high shear mixing is required [50]. Mastication depends on the temperature, rotating speed and clearance of the masticating cylinders [165]. According to Ref. [166], in the continuous processing of HMPSA (based on EVAc) there is a relationship between plug diameter and melt viscosity. Melt viscosity decreases with plug diameter.

b. Formulating Problems

General aspects of raw materials, such as storage conditions and warranty have to be taken into account. Special features concerning the compatibility of the components, and their order of addition are important also. For instance, for natural rubber latices and adhesives based on natural rubber latex, a shelf life of 3 months is given, when stored in sealed containers at normal temperature [167]. Water-based acrylics have a minimum storage time of 6 months. Natural rubber latices are not compatible with other adhesives and care should be taken to wash out the equipment [167]. Special formulating equipment is needed for crosslinked formulations. Curing of two component adhesives (or primers) starts immediately after the components have been mixed together. Therefore, only a limited time remains in which the adhesive can be applied. This period is referred as pot-life.

4. Manufacture of Adhesives for Labels

According to Ref. [168], the adhesive for labels must have the following qualities: good adhesion on critical substrate surfaces; chemical, temperature, and UV resistance; adhesion on different surface structures; easy labeling; removability; repositionability; noncorrosivity; plasticizer resistance; mechanical resistance; adhesion on wet surfaces; dimensional stability; transparency; instantaneous/final adhesion; uniform adhesion toward the release liner; no emission of gaseous components and FDA approval.

Evaluating these properties according to their priority, one can state that well-defined adhesive characteristics and the stability of these characteristics independently from the application and environmental conditions are needed. For labels, solvent-based, water-based, hot-melt, and radiation-cured adhesives can be used. As illustrated by statistical data, some years ago solvent-based adhesives have been preferred in label manufacture. In 1984, mainly (60%) solvent-based adhesives have been used for labels and tapes in the U.S.A. [169]. Now water-based formulations are the most common.

The choice of the adhesive formulation depends on its end-use. A preparation plant is necessary for the manufacture of the adhesive [170]. Its complexity is a function of the nature of the adhesives to be manufactured. The manufacture of adhesives for labels includes their formulation, production, and quality assurance. The adhesive for labels can be synthesized in-line also. Radiation-induced polymerization and curing can be carried out to achieve a PSA-coated label material.

Because of the balanced adhesivity required for labels, label manufacture is based on coating with PSA. The formulation of PSA has to ensure its coatability. The range of possible raw materials for PSAs for labels is very broad. Two-component rubber-resin or one-component viscoelastic polymer-based formulations can be coated in molten, dissolved, or dispersed state. Radiation curing has been recommended for liquid (dissolved) prepolymers or molten, high polymer-based formulations. Because of the special role of the face stock material as information carrier and the high surface quality required for labels only classical coating technologies using very fluid adhesive and giving smooth adhesive surface are suggested for label manufacture. Started with rubber-resin based PSA solutions, label coating now uses mostly acrylic water-based PSAs. Therefore, the formulation of PSAs for labels is mainly a formulation of water-based compounds. In practice it consists of blending liquid-state components. Water-based adhesives for labels have a 4-yr shelf life and can be used in the temperature range of -30 to $+70°C$ [171]. Generally, they conform with FDA regulation 21 CRF 175 105 and BGA XIV.

The adhesives for labels are tackified formulations having a resin level above 30 wt.% (paper labels). This is required by their high speed, low pressure application on difficult adherend surfaces. For instance, products wrapped in plastic film (PE, PP, PVC, PA, ionomer/PA, coated cellophane) such as refrigerated meat, need a PSA that can be applied at low temperature [172]. The service temperature of the film is -50 to $-150°F$. Tackified adhesives for labels are softer (more viscous) and therefore display a better wetting of nonpolar surfaces by low temperatures [168].

The dwell time, that is, the time necessary to build a full surface contact is shorter for such adhesives. According to Ref. [54], removable price labels use almost exclusively a paper carrier, coated with rubber-resin adhesive. Natural and synthetic rubber and polyisobutylene, tackified with soft resins and plasticizer have been suggested for such formulations. Gasoline, toluene, acetone, and ethyl acetate have been used as solvent for rubber-based adhesives. The elastomer is calendered and masticated. For such application Mueller and Tuerk [54] proposed a water-based acrylic formulation with crosslinkable comonomers like acrylo- or methacrylonitrile, acrylic or methacrylic acid, N-methylol acrylamide, and plasticizer (10–30 wt.%). For the plasticizer common commercial products (e.g., alkylesters of phtalic, adipic, sebacic, acelainic, or citric acid and propylene glycol methyl vinyl ether) have been suggested. If a two-component cross-linking water-based PSA is used, the mixing of components must be simple, and the proportions tolerance must be high for operator convenience. Pot-life at the coating station must exceed one working day. Foaming should be avoided.

Generally, the development of adhesive raw materials and formulated adhesives for labels is affected by requirements for more adhesivity or/and more cohesion. Requirements for more adhesivity include versatility for roll application, increased labeling speed, adhesion on nonpolar carrier material, formulation ability with resin ester based tackifiers, reduced coating weight and the use of solventless release. Requirements for more cohesion are related to cuttability and die-cuttability and bleeding. Figure 8.5 illustrates the influence of these parameters on the performance characteristics of label PSA, on an arbitrary scale.

5. Manufacture of Adhesives for Tapes

The volume of PSAs used for coating of tapes is higher than for labels (due to the higher coating weight). At the end of 1980's in U.S.A. the ratio of PSA manufacture for tapes compared to labels was 75/25 [173]; now it is about 65/35. As seen from Table 8.12 there is a very broad raw material basis for various tapes. Rubber-resin and acrylate-based formulations compete with special raw materials (silicones, PVEs, polyesters, radiation curable recipes, etc.). Monomer-, oligomer-, and

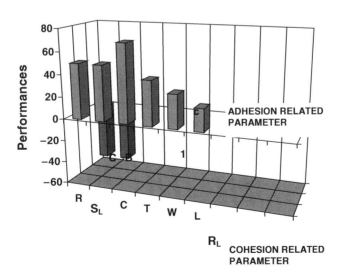

FIGURE 8.5 The influence of the adhesion- and cohesion-related functional parameters on the whole performance of PSA for labels. Adhesion-related parameters include reel application versatility (R), labeling speed (S_L), the use of nonpolar carrier materials (C) and ester based tackifiers (T), low coating weight (W) and solvent-free release liner (R_L).

TABLE 8.12
Chemical Basis of Tapes

Application	Chemical Basis		Ref.
	Base Elastomer	Other Components	
Packaging	NR, SBS	Terpenephenol resin	[174]
Packaging	NR, PIB	Rosin	[105]
Packaging	NR, polybutadiene	Rosin, crosslinker	[104]
Packaging	NR, SBS	Resin	[175]
Packaging	cis-Polybutadiene	Tackifier	[104]
Packaging	SIS, SBS, cis-polybutadiene	Resin, plasticizer	[176]
Packaging	AC	—	[177]
Packaging	EVAc, PE, HMPSA	Rosin, waxes	[178]
Packaging	SEBS, HMPSA	Hydrogenated hydrocarbon resin, polymethylpentene	[179]
Packaging	AC, WB	Epoxy-containing polyglycidyl compounds	[180]
Packaging	AC, HMPSA	Crosslinked with Zn-carboxylate	[181]
Packaging	Acrylated polyester, HMPSA, UV- or EB-cured	—	[83]
Packaging	AC, SB	Polyisocyanate cured	[182]
Packaging	NR	Aliphatic petroleum resin, oxidatively degraded EPR	[183]
Packaging, deep freeze	NR, SBR, PVE, regenerated rubber	Rosin ester, alkylphenol resin, crosslinker	[135]
Masking	AC	Terpenephenol	[184]
Masking	AC, oligomer, EB-cured	Trimethylolpropane tris(tioglycolate), hexanediamine	[15]
Insulating	Polybutadiene, SBC	Tackifier	[185]
Pipe wrapping	Polysiloxane	Tackifier	[186]
Pipe wrapping	Polysiloxane	Tackifier	[187]

polymer-based compositions are used. Tackifiers, plasticizers and crosslinking agents are the usual components of such formulations. Because of the wide range of materials, tape manufacturing methods vary widely also.

As discussed in Chapter 2 and Chapter 11, the main volume of tapes is used in packaging. Such tapes are relatively simple products which have to fulfill mechanical assembly requirements like a packaging material. These are low cost products where raw material price and productivity play special roles.

The most common tape formulations are rubber-based and are related to end-use. Special tapes can be manufactured on a rubber-resin basis also. Rubber has to be masticated or plasticized for many tape applications. Hot-melts have been used for labels, tapes and decals since the early 1970's [188]. Natural and thermoplastic rubbers provide a relatively nonexpensive raw material basis and hot-melt coating assures the highest coating speed. According to Fries [189], hot-melt technology consumes only 25% as much energy as a solution-based coating process. Therefore, several tapes are formulated with HMPSAs. Radial and teleblock copolymers with isoprene-based branches have been suggested for PSA tapes [190]. A formulation with a tackified star block copolymer gives a higher shear resistance (at least 1000 min, according to ASTM-D 3654, overlap shear to fiberboard at 49°C).

Water-based chloroprene dispersions exhibit improved shear performances and can be used for tapes. Silicone polymers have excellent electrical properties such as arc resistance, high electrical strength, low loss factors, and resistance to current leakage. These characteristics are very important for electrical insulating tapes. They polymers have low surface energy and excellent low and high temperature performance [191].

The formulation of the adhesive for tapes depends on the choice of carrier material also. For tapes with a PE carrier, PIB and NR have been proposed as elastomer [192]. Crosslinked butyl rubber based tapes for pipe isolation were manufactured since 1963. A mixture of PIBs with molecular weights of 100,000 and 1500 is applied. Rosin is added as tackifier. Tapes based on PE have been coated with adhesives based on natural rubber, polyisobutylene, tackifier resins (phenolformaldehyde, terpene, and coumarone resins), fillers (SiO_2, ZnO, talcum, etc). Air seal tape is made with butyl rubber (on ethylene propylene foam) and solvents [192]. Such formulations contain natural rubber (100 parts) with polyisobutylene, hydrocarbon resin (50–70 parts), and low molecular polybutylene (4–50 parts) and mineral oil. For insulation tapes, a mixture of polyisobutylene (100 parts) with a molecular weight of 1000, 120 parts of low molecular polyisobutylene, (molecular weight 1500), and 150 parts of rosin have been proposed.

For PVC carrier, natural rubber, chlorinated rubber, chloroprene, and perchlorvinyl have been suggested. Chlorinated rubber gives excellent adhesion on PVC but primers can also be used. Acrylonitrile–butadiene–styrene and acrylonitrile–butadiene rubber blends are recommended as primer [192]. Adhesive formulations based on chlorinated rubber (90 parts), include a plasticizer (29 parts), and a tackifier resin (35 parts). Such compositions can contain chlorinated rubber (60 parts) together with polyvinyl butyral (10 parts), phenolic resin (45 parts), and plasticizer (90 parts). Pipe insulation tapes with PVC carrier are coated with an adhesive based on natural rubber (crepe, 27 parts), ZnO (20 parts), rosin (22 parts), or hydrated abietic acid (8 parts), polymerized trimethyldihydrochinoline (2 parts) and lanoline (4 parts).

Double-faced foam tapes are designed for industrial gasketing application. Such tapes are based on closed cell PE foams, coated with a high tack, medium shear adhesive [193]. They can be applied as pipe insulation tape for gas pipe lines also [194]. Air seal tape is made with butyl rubber on ethylene–propylene foam. Tapes applied as corrosion protection for steel pipes contain butyl rubber, crosslinked butyl rubber, regenerated rubber, tackifier (polybutene or resins), filler, and antioxidants. Low molecular weight polyisobutylenes ($M_n < 600,000$) give better tack and peel [195].

Generally blends of different molecular weights are used. Faktis improves the creep resistance for such tapes. Insulating tapes for gas and oil pipe lines are manufactured by coextrusion of butyl rubber and a PE carrier.

According to Risser [196], butyl rubber, plasticizer, carbon black, and terpene phenol resins have been tested for sealing tapes. Polyacrylate rubbers can be used for HMPSAs, solvent-based PSAs and reactive hot-melts for pressure-sensitive sealant tapes and caulks [50]. Polyacrylate rubbers for these HMPSAs are tackified (40% resin) or plasticized. A common tacky PSA can be added to the formulation also (50 parts to 100 parts acrylic rubber). The composition can be cured using isocyanates or UV curing agents (e.g., p-chlorobenzophenone). The recipe can include a high loading of fillers. Such formulations can contain 200 parts of hydrated alumina as filler for each 100 parts rubber, 50 parts carbon black and 50 parts plasticizer. Polyisobutylene can be added also.

Deep freeze tapes are coated with an adhesive based on natural rubber, styrene butadiene rubber, regenerated rubber, rosin ester, and polyether [197]. Generally, SBR polymers used in the adhesive tape industry contain a gel, either a microgel or a macrogel [198]. A microgel is built into the polymer during synthesis by crosslinking between polymer chains within the latex particle. A macrogel results from crosslinking between particles, which builds up large masses of polymer, generally due to instability in the polymer. A macrogel can be broken mechanically to give microgel. When a macrogel is broken, the active radicals can react internally to produce a more tightly crosslinked microgel, or they may react externally to give viscosity instability; therefore the dissolution of SBR is a difficult process. A microgel yields an easily dispersed, low viscosity, smooth, shiny mastic; a macrogel gives a pooly dispersed, high viscosity and grainy mastics. Macrogel formation can be prevented by improving the rubber stability. Most of the viscosity increase occurs during the first few days of aging and depends on the solids level. The addition of aromatic solvents results in greater viscosity on aging. Although a stable viscosity could be achieved through proper selection of solvent/resin/rubber level and total solids, the characteristics of the elastomer are the major determining factor in viscosity stability.

Masking tapes are used in the varnishing of cars [199,200] (see also Chapter 11). They have to adhere on automotive paint, e-mail, steel, aluminum, chrome, rubber, glass, and to resist the high temperatures used for drying and curing of varnishes. Natural rubber latex, tackified with PIB is recommended for such tapes. Polyvinyl ether can be also used. A mixture of poly(vinyl-*n*-butyl-ether) having different molecular weights (1000–200,000) and (200,000–2,000,000) is suggested for such tapes [201]. Butyl rubber (15–55 parts) untackified or tackified with PIB (7.5 to 40 parts) and crosslinked with zinc oxide (45 parts) has been proposed also [202]. Such formulations give (180°) peel values of 0.73 kg/25.4 mm to 1.81 kg/25.4 mm. Peel increases with the increase of the tackifier level, but at a tackifier/elastomer ratio of 40/15 legging appears.

Water-based acrylics are recommended as adhesives for tapes also. Such PSAs are used for packaging tapes with OPP carrier [203]. CSBR has been proposed for high shear tapes [55]. The direct feed in of the resin in such formulations is suggested in Ref. [22]. Therefore a tackifier resin with MP < 95°C is recommended.

A special chapter of adhesive tape formulation concerns medical tapes. In this application field rubber- or polyvinyl ether-based formulations compete with special acrylics, TPEs, or poly-urethanes (see Table 8.10).

According to Refs. [204,205], PSAs for medical tapes are based on natural rubber or polyvinyl ether. The formulation contains natural rubber (100 parts), a terpenephenol resin (80 parts), and solvent (600 parts). Polyisobutylene is used for medical tapes [206]. This type of tackified polyisobutylene has been suggested for other tapes also [207]. It has been plasticized at 80°C. Butyl rubber (MW of 6000 to 20,000) does not need mastication to be dissolved, unlike many natural and synthetic elastomers. It is clear and tacky, it is not fragile after aging, and it can be used together with PIB as tackifier for medical tapes [208]. Air- and moisture-permeable nonwovens (polyethylene terepthalate (PET) nonwoven, embossed nonwoven) or air-permeable tissue (breathable face stock) are used for medical tapes coated with a porous, nonsensitizing acrylic adhesive for prolonged application of a medical device on the human skin. They minimize skin irritation due to their ability to transmit air and moisture through the adhesive system. The air porosity rate has to attain a given value (50 sec/100 cc/in.^{2}) [209].

The first adhesives for medical tapes were based on the reaction product of olive oil and lead oxide (lead oleate), plus tackifier resins and waxes [210]. Lead oleate was used as adhesive component for medical tapes later also [211,212]. Natural rubber, tackified with rosin, is recommended for adhesives for medical tapes [72]. The rubber for such tapes is masticated [213]. Polyvinyl ether is suggested for medical tapes alone or with polybutylene and titane dioxide as filler. These compounds are warm coated without solvent (via coating by calendering). By cooling of the roll-pressed mass, a porous adhesive is achieved [214]. Polyvinyl acetate, polyvinyl chloride, poly-ethylene, and ethylene copolymers can be used as adhesive for medical tapes too. Such compounds are hot-pressed in the fibrous carrier material.

According to Ref. [78], the PUR-based PSA used for medical tapes include a polyethylene polyol, polyisocyanate, resin, catalyst, and wetting agent. The type of polyethyene polyol, its functionality, its structure, and hydroxy number are the main parameters. As the isocyanate component isophorone diisocyanate, trimethylhexamethylene diisocyanate, hexamethylene diiso-cyanate, naphtylene-1,5-diisocyanate, etc., are recommended. Hydrocarbon resins, modified rosin, polyterpene, terpenephenol, coumarone-indene, aldehyde- and keton-based resins, and β-pinen resins are suggested as tackifier. As catalyst Sn(II) octoate is used. According to Ref. [78], the components of the formulation are mixed as follows: 100 parts polyethylene polyol, 50 parts resin, 1 part catalyst, 7 parts diisocyanate, and wetting agent 1 part. The resin is dissolved in the polyethylene polyol. This solution is blended with the isocyanate in an extruder, coated (closed blade) and crosslinked (120°C, 1.5–2.5 min). Two-component polyurethane PSAs are suggested for products with higher coating weights and thermally nonsensitive carrier materials.

Chemical-resistant tapes are based on silicones [189]. Linear polyamides are used for special tapes only [78]. Polymers based on isophtalic or adipic acid and 1,6-hexanediol with a molecular weight of less than 30,000 are preferred for hygienic hot-melts.

For adhesive-coated PSPs the cold flow of the adhesive depends on its viscosity and coating weight. Therefore, a low coating weight is recommended for tropically resistant tapes. For such tapes silica (particle diameter 0.01–0.03 μm) is proposed as filler in order to increase the viscosity of the adhesive. Starch is suggested also as water absorbent (see also Chapter 5).

A special, crosslinkable, carboxylated, high T_g (-25°C) acrylate has been developed for mounting tapes and double-coated foam tapes. This compound contains tackifiers at a loading of 12–35% dry weight [44]. For transfer tapes used at elevated temperatures, a solvent-based crosslinked acrylic is suggested. Formulations of transfer tapes that are conformable on uneven surfaces use fillers. Glass microbubbles can be incorporated also as filler to enhance immediate adhesion to rough and uneven surfaces [24]. Such tapes are produced by polymerizing *in situ* with UV radiation. Thick PSA tapes (0.2 to 1.0 mm) have been prepared by the photopolymerization (UV) of acrylics (see Section III.B.3). The filled layer can be laminated together with an unfilled one. The thinner layers are from about 1 to 5 mils thick.

Foam-like pressure-sensitive tape is manufactured according to Vesley et al. [215], by using glass microbubbles as the filler. Dark glass microbubbles are embedded in a pigmented adhesive matrix. The glass microbubbles have an ultraviolet window that allows UV polymerization of the adhesive composition. Foam-backed adhesive tapes are commonly used to adhere an article to a substrate. Pressure-sensitive tapes filled with glass microbubbles have a foam-like appearance and character and are useful for purposes that previously required a foam backed PSA tape. The adhesive matrix may include 0.1–1.15% by weight of carbon black also, without affecting the UV polymerization. Glass microbubbles as fillers display the advantage of higher distortion temperatures. The glass microbubbles (at least 5% by volume) are embedded in a polymeric matrix with a coating thickness of 0.2 mm. The thickness of the pressure-sensitive layer should exceed three times the average diameter of the microbubbles. This enables the buildup of an intimate contact with rough and uneven surfaces, while retaining the foam-like character. Optimum performance is attained if the thickness of the PSA layer exceeds seven times the average diameter of the bubbles. The bubbles may be colored with metallic oxides. The average diameter of the stained glass microbubbles should be between 5 and 200 μm. Bubbles having a diameter above 200 μm would make UV-induced polymerization difficult. The PSA matrix may also have a cellular structure. The die embedded in the PSA matrix should have a UV window also. In order to enhance the cohesion of the adhesive layer, polar monoethylenically unsaturated monomers are included in the acrylic polymerization recipe (less than 20%) and photoactive or heat-activatable crosslinking agents are also added. The monomers may be partially polymerized (prepolymerized) to a coating viscosity of 1–40 Pa sec before the microbubbles are added.

Nonflammable tape formulations must also include the carrier material and adhesive. Nonflammable tapes are coated with an adhesive with 0–40% polychloroprene rubber [216]. An adhesive for PVC-based tapes includes cold masticated nitrile rubber (100 parts), calcium silicate (40 parts), titanium dioxide (10 parts), oil (25 parts), coumarone indene resin (MP 45–55°C; 25 parts), chlorinated rubber (100 parts), and solvents (methyl ethyl ketone and toluene) [217].

Special end-use requires special adhesive formulation. For tear tapes crosslinked acrylic formulations are suggested with N-methylol methacryl-amide as comonomer. Tapes used for automobile interiors have to resist high temperatures [218]. Common rubber-resin-based adhesives do not meet these requirements, so crosslinked acrylics are used. In some cases the composition contains glass microbubbles [24,219].

Adhesives for diaper tapes have to display high shear [220]. Crosslinking monomers are added (e.g., *N*-methylolacrylamide) to increase shear resistance. Unsaturated dicarboxylic acids can be included also as crosslinking agents, polyfunctional unsaturated allyl monomers, di-, tri-, and tetrafunctional vinyl crosslinking agents have been suggested [221,222]. The polyolefinically unsaturated crosslinking monomer enhancing the shear resistance should have a T_g of $-10°C$ or below. Shear resistance is improved in water-based polymerization using a special stabilizer system, comprising a hydroxypropyl methyl cellulose and an ethoxylated acetylenic glycol.

Adhesives with excellent water resistance (humidity- and condensed water-resistance) are required for technical tapes and pressure-sensitive assembly parts in the automotive industry [155]. This can be achieved with water-based raw materials using the tackifier's own surfactants.

According to Ref. [223], PSA can be manufactured as powders also. A PSA based on a ABA thermoplastic elastomer with a softening temperature higher than about 23°C is cooled below $-20°C$ and pulverized; then a tackifier resin with a softening point of about 85°C is pulverized also. A dry blend of elastomer and tackifier is prepared. This powder can be dry coated and sinterized by heating of the adherend surface to 177°C.

PSA tapes whose adhesion can be decreased by UV irradiation are prepared from an elastic polymer, a UV-crosslinkable acrylate, a polymerization initiator and a methacrylate photosensitizer containing an NH_2 group [24]. An ethylene glycol-sebacic acid–terephtalic acid copolymer (having a T_g of about 10°C), a tackifier resin, dipentaerythritol hexaacrylate as crosslinking monomer, benzyldimethyketal as photoinitiator and dimethylamino acrylate as photosensitizer. This adhesive has been coated on a 100 μm PVC carrier, primed with a modified acrylate. It displays a peel resistance of 140 g/25 mm before irradiation and 52 g/25 mm after UV irradiation.

Contact hindrance for tape adhesives can be achieved using special fillers. For instance, a removable tape for paper contains PSA microparticles obtained from a crosslinked, milled PSA. The 50–100 μm particles are coated with 5 g/m^2 [224].

6. Manufacture of Adhesives for Protective Films

The adhesive for protective films is a composition without tack, therefore there is no need to achieve an adhesion–cohesion balance [225]. For PSA labels the nature of the adhesive is more important than its coating weight and crosslinking degree. As mentioned earlier, the adhesive properties of labels are tested and are stated for a standard coating weight. For tapes the coating weight is an important parameter to regulate parameter adhesive properties. For protective films both the coating weight and the crosslinking degree of the adhesive play important roles in the control of adhesive properties.

The choice of the adhesive for labels (excepting the economical end environmental arguments) is determined mainly by adhesive properties (by the peel value required for a given substrate). For tapes the same criterion is valid. Adhesives for protective films (rubber-based or acrylics) are suggested for certain applications according to their adhesivity and the softness of the composition. This means, that quite different adhesives may exhibit the same peel value as a function of the substrate nature and application conditions, and generally the specifications classify the products using such empirical, unexactly formulated performance indices as "light, medium, and strong" adhesion. Therefore unlike the label suppliers, manufacturers of protective films offer the "same" products (with respect to standard peel value) with adhesives that are chemically or physically different.

Generally the adhesives for protective films are crosslinked formulations. Therefore, it is evident that the coating machines for labels, tapes, and protective films exhibit different characteristics (see Section III.B.1.c). Rubber-resin solvent-based adhesives were first manufactured for protective films, both crosslinked and uncrosslinked Later solvent-based crosslinked and water-based crosslinked and uncrosslinked acrylics were tested. The recipes include masticated natural rubber and cyclized rubber, tackified with a low level of resins and crosslinked with isocyanates. Isocyanates

TABLE 8.13
The Main Raw Materials Used for Coated Adhesives for Protective Films

Base Polymer	Additive	Characteristics
Natural rubber	Resin Plasticizer Curing agent Solvent	Crosslinked
Styrene–butadiene rubber	Resin Plasticizer Curing agent Solvent	Crosslinked
Cyclized rubber	Resin Solvent	—
Polyurethane	—	Crosslinked
Polyacrylate	Crosslinker Solvent/Dispersant	Crosslinked/uncrosslinked

are recommended as curing agent for solvent-based acrylic adhesives, and polyaziridine or isocyanates (in a concentration of 0.3–2.0%) for water-based acrylics. Table 8.13 lists the main raw materials and manufacture technology for adhesives for protective films. Table 8.14 illustrates the use of crosslinking agents for different PSPs.

As seen from the Table 8.14 the most commonly used external crosslinking agents are isocyanate derivatives, applied in a concentration comparable to the level of internal multifunctional crosslinking monomers.

The most used solvent-based acrylates for protective films have a glass transition temperature of about −20°C. Such adhesives need crosslinking in order to improve internal cohesion. In the last decade or so, new products have been developed that do not require external crosslinking. Such formulations generally contain at least two components. The base polymer is harder

TABLE 8.14
Crosslinking Agents for PSPs

Product	Adhesive	Crosslinking Agent		Ref.
		Type	Concentration (%)	
Protective sheet	Acrylate	Polyisocyanate	5	[224]
Tape	Acrylate, with OH groups	Diisocyanate	0.5	[182]
Tape	Acrylate, tackified	Isocyanate	1.0	[226]
Tape, low temperature	NR, SBR, PVE, carboxylated alkylphenol resin	Isocyanate	0.1–5.0	[170]
Tape	Acrylate, Water-based	Polyethylene-glycol-diglycidylether	0.1–0.5	[227]
Tape, protective film	EHA-VAc, copolymer	Isocyanate	2.0	[228]
Tape, label	EHA-oligomer-VAc	Isocyanate	0.5	[229]
Label	Acrylate, carboxylated	N-methylol-acrylamide	0.1–10	[230]
Tape	Acrylate, carboxylated	Aziridine	0.1–10	[231]

(with a T_g of about $+7°C$). Such an adhesive is self-curing and exhibits better UV-resistance. The common base polymer for solvent-based acrylates is so soft that it is not possible to use it for Williams plasticity test. The new self-crosslinking products exhibit a Williams plasticity of 2–2.8 mm (PN).

7. Manufacture of Adhesives for Forms

Business forms are very complex, multilayer constructions, manufactured like multilayered labels. The formulation of the adhesive includes a range of permanent or removable adhesives having different, exactly controlled adhesivities. The special adhesives have very low adhesion (allowing the detachment of die-cut label parts, e.g., cards). They are generally formulated from very hard, high-T_g acrylates, high-T_g vinyl acetate–acrylate copolymers or low-T_g crosslinked acrylates.

The formulation of PSAs for various end-uses is discussed in detail in Ref. [240].

8. Formulation of the Primer

As discussed in Chapter 2 and Chapter 7, for certain applications top coats are required to improve the adhesion of the postcoated layers (e.g., adhesive, supplemental carrier, printing ink, release, etc.) to the carrier material. Primers do not have a universal chemical composition (see also Chapter 5). They differ with respect to coating technology, the nature of the solid-state surfaces to be coated, and the postcoated layer (Table 8.15).

a. Primers for Coated Adhesives

Adhesion promoters are used to improve the adhesion, the anchorage of a liquid postcoat (mainly adhesive, release layer, or printing ink) on a solid-state carrier material. Chemical primers provide a bond between substrate and adhesive. They are supplied as solvent-borne, water-borne, or

TABLE 8.15
Primers for PSPs

PSP	Carrier	Primer	Adhesive	Coating Method	Ref.
Tape	Cellulose, hydrate	Butyl rubber, grafted	SBC, poly-butadiene	Coating extrusion	[232]
Tape	—	NR, SBC, nitrile rubber, tackifier, epoxy resin	NR, crosslinked	Coating	[233]
Tape	PP	Acrylate, PP, maleated	—	—	[234]
Tape	PP	Chlorinated PP	—	Coating[a]	—
Tape	PVC	Acrylate, water-based	Acrylate	Coating	235
Tape	PVC	ABS	Rubber-resin	Coating	235
Tape	PVC	SBR	Chloroprene	Coating[b]	[187,236]
Tape	PVC	Rosin	Rubber-resin	Coating	[165]
Tape	PVC	PUR	Rubber-resin	Coating	[237]
Tape	PVC	Butyl rubber, isocyanate	Butyl rubber	Coating	[170]
Tape	PVC	CSBR, water-based	Rubber-resin	Coating	—
Tape	PP	Chlorinated rubber, chlorinated PP, EVAc	Rubber-resin	Coating	[238]
Tape	PP, PET	Silicone, NR, polyterpene resin	Rubber-resin	Coating	[239]
Protective film	PE, PP	NR, resin, isocyanate	Rubber-resin	Coating[c]	—

[a]Coating weight 0.2–0.5 g/m^2. [b]Coating weight 2.0–6.0 g/m^2. [c]Coating weight 2.0–1.5 g/m^2.

solid-state products. Available equipment can influence the choice of a primer. It determines whether the primer is to be applied in-line or off-line. Common application systems include rotogravure, transfer gravure, and smooth roll. Water-based primers are advantageous from the standpoint of compliance with environmental requirements.

Primers are recommended mainly for plastic film carriers [240]. On untreated PE surfaces common acrylics PSAs exhibit a peel force of only 4 N/cm^2 [241]. To improve the anchorage primers have to be applied [107]. They are used for labels, tapes, and protective films also. The common manufacture procedure of tapes includes in-line primer, release layer and adhesive coating [235]. As a classic primer, solutions (in toluene) of butadiene acrylonitrile–styrene copolymers have been preferred. For solvent-sensitive SPVC films primer dispersions based on Hycar Latex and Acronal 500 D (1/1) have been suggested by Reipp [235]. According to Wechsung [192], for PVC tapes coated with a chlorinated rubber adhesive, primers based on butadiene–acrylonitrile rubber and butadiene-styrene rubber (1:1), with a butadiene–styrene ratio of 1:8 have been proposed. For such products the primer layer has a thickness of 2–6 g/m^2 [236]. Tapes for deep freeze applications use soft PVC carrier, which needs a primer coating for the rubber-based PSA. According to Malek [237], for such tapes the carrier is primed with rosin. Abrasion-resistant automotive decoration films are made on PUR or EVAc/PVC basis. Their carrier is coated with a PUR primer [237].

The chemical nature of the primers varies. ABS, isocyanates, and silane derivatives can be coated as primers [236]. In some cases (e.g., protective films, tapes) primers are crosslinked versions of the base adhesive used to coat a carrier material. Thus, a primer for a polyethylene tape and with adhesive based on butyl rubber, consists essentially of a butyl rubber, a polyisocyanate, and an organic solvent [175]. On tapes with impregnated paper carrier the butadiene–styrene rubber may function as an abhesion promoter also. A primer based on SBR latex is coated on the carrier for removable PSA. As primer for polypropylene coating, a mixture of chlorinated rubber, EVAc copolymer and chlorinated polypropylene can be used [238]. Polar ethylene copolymers are suggested as adhesion promoters also [242]. The effect of primers based on chlorinated polyolefins depends on the coating weight and coating conditions [243]. For silicone tapes with PE or PET carrier a primer containing organosilicones, nitrile rubber, and polyterpene resin is recommended [239].

Generally, water-based primers are used for paper, whereas solvent-based primers are used for films. Water-penetrable carrier materials are primed with water-based solutions of resins or latex. Hydrophobic carriers are primed with solutions of nitrile rubber, or SBR [236].

A discontinuous adhesive layer can be preprimed in to provide a continuous matrix. As discussed earlier, the PSA surface can have a special discontinuous shape ensured by the PSA itself, if this is a suspension polymer [3]. Thus spherical contact sites are formed. These are anchored on the carrier surface with the aid of a primer. Theoretically primer thicknesses approaching a single molecular layer are required, but in reality the primer coating is 0.5–1.0 μm is thick. A release coating may also need a primer. According to Ref. [101], on paper, a barrier layer based on vinyl chloride, vinyl acetate, and vinyl alcohol is applied to paper before siliconizing.

b. Primers for Extruded Layers

Extrusion primers are used to improve adhesion between carrier layers manufactured via extrusion (see Section II). Various raw materials and recipes have been proposed as primer. Some of them act as primer for postcoating of the liquid adhesive also. The know-how for their manufacture and use comes from the manufacture of multilayer packaging films. As tie-coat or primer for the extrusion coating of PE and EVAc, carboxylated ethylene polymers can be used. Such top coats exhibit excellent anchorage on many substrates and are not sensitive to water, grease, oil, and many chemicals.

Coextrudable primers are known from the manufacture of laminated plastic plates or metal/plastic laminates [170]. Generally, they are ethylene copolymers with a melt flow index of 6–25. Higher MFIs allows better deep drawing properties. These compounds improve anchorage on PET, PVC, PP, low density PE (LDPE) or high density PE (HDPE), EVAc, PC, and styrene copolymers (HIPS, SAN, and ABS). Extrusion primers can be used as plasticizer barriers for soft PVC also. For such applications polymethyl methacrylate or PUR (3–10 g/m^3) can also be used. Ethylene–acrylic acid copolymers (EAA) with randomly distributed carboxy groups provide excellent adhesion characteristics when bonding with polar surfaces [244]. In EAA copolymers, the carboxyl groups disrupt crystallinity and also provide a site for H-bonding between molecules [245]. They can be used on paper too. EAA polymers are applied mainly to prime foil (0.5–1.0 lb/3000 ft^2). Such products are supplied as water borne dispersions also [246]. Ethylene–vinyl acetate copolymers have been introduced as primers for extrusion coating [247]. They can be used as solvent-based or water-based primers. According to Ref. [248], such formulation allows the anchorage of postcoated layers on cast polypropylene. EVAc copolymers ensure sealing temperatures situated about 40°C lower than those needed for PE [247].

The use of primers can be avoided if the adhesion of the carrier material is improved by the inclusion of adhesive components. Biaxially oriented multilayer PP films for adhesive coating have been manufactured to improve adhesion to the adhesive coating by mixing the PP with particular resins. In this case in order to prepare an adhesive tape which can easily be drawn from a roll without requiring an additional coating on the reverse side, at least two different layers having different compositions are coextruded and the thin back layer (1/3) contains an antiadhesive component.

B. MANUFACTURE OF THE ABHESIVE COMPONENTS

As discussed earlier (see Chapter 2), most PSPs need a release layer or a separate release liner. Different grades of release liner for labels, tapes, hygienical items (sanitary towels, sanitary pads, diapers) envelopes, etc. are manufactured [249]. In order to design their formulation and coating method, first their function has to be clarified. The release liner for label serves multiple functions [168]: protection of the adhesive layer, handling of the label, ensurance of lay flat during converting, and other functions during converting (e.g., optical guide).

According to Ref. [78], for tapes the following properties are important for a release liner: unwinding performance, exactly regulated release force toward the PSA, dimensional stability, tear resistance, environmental stability, and cuttability and die-cuttability.

Generally, the liner has to have trouble-free release and dimensional stability and perform well during die-cutting, perforating, prepunching and fan folding. Some PSPs (e.g., transfer tapes or label sandwich constructions) require release layers that exhibit different release forces.

Generally, the manufacture of the release liner includes the manufacture of the abhesive (formulation and technology), the manufacture of the release carrier material, and the coating technology. In some cases the release agent is embedded in the carrier material or the carrier material itself possesses release properties given by its chemical or physical characteristics or its geometry. In such cases the manufacture of the release liner consists of the manufacture of the carrier or its modification. As discussed in Chapter 5, special chemicals are used as abhesive components. Some are macromolecular compounds having a very complex production technology. Other products are common commercial polymers or micromolecular substances. It is not the aim of this book to discuss the synthesis of these components but their characteristics and their coating technology are briefly described.

Abhesive agents are used in different various domains. The processing of plastics needs such materials for the mold. In the packaging industry, release lacquers are used for cold seal adhesives. For instance, calcium behenate exhibits a strong lubricating action due to its long fatty acid chain. It is used for bottle blowing, film blown extrusion, and food packaging. It is applied as a main stabilizer, and lubricant (1–2 phr) in PVC processing. Silicone mold release agents are used in injection molding of plastics. Polyamide-based release layers have been developed for cold sealing.

Various types of release liners using silicones, CMC, alginates, and shellac have been produced [250]. A chemical release agent is applied together with a mechanical release effect. For instance, paper is coated with profilated natural latex as release layer [251]. Release coatings are also formulated with melamine resins. These formulations have limited pot-life. Self-crosslinking release coatings are used together with precrosslinked adhesives. The release layer is formulated and manufactured according to the end-use properties of the PSP, mainly according to the degree of abhesivity required during application of the PSP. Generally, abhesive agents having different chemical basis display different degrees of abhesion. Silicones are the sole polymers with a broad range of abhesion performances, that is, with general usability for all classes of PSPs.

1. Formulation of Abhesive According to Its Chemical Basis

Silicones are the most used abhesive compounds. Their main application field as solid-state release liner is in the manufacture of labels, but separate silicone-based abhesive components have been used for medical tapes also [252,253]. Double-side-coated mounting tapes and transfer tapes must have solid-state liner with multiple release layers (see also Chapter 2). For such products the release force generally depends on the following parameters [168,254]: quality of paper carrier, liner stiffness, silicone nature, silicone formulation, silicone coating weight, silicone affinity toward the adhesive, temperature, adhesive coating weight, and age of the PSA laminate.

a. Formulation of Silicones

The most used release coatings having general applicability are those based on silicone polymers. They were first manufactured for greaseproof and release papers. As release coatings, silicones, and silicone-modified elastomers have been developed. Solvent-based, water-based, and solventless conventional silicones and radiation-curable silicones are manufactured [255]. Such compounds are coated on paper and plastic carrier materials. Silicone release papers have been coated with $0.3-0.8 \text{ g/m}^2$ coating weight [256].

The technical problems concerning the use of silicones have been related to their formulation for controlled release, their dosage during coating as a very thin abrasion–sensitive layer, and their use for different adhesives and PSPs having different delaminating speeds. As is known, the release force depends on the adhesive. For instance, the same release gives 350 mN peel force with SBR-based PSA and about 150 mN peel force with acrylics [257]. Classic silicone coatings were not adequate for silicone PSA. Silicone PSAs can be laminated together with highly crosslinked silicone release layer [258,259]. Fluorosilicone coatings improve readherability [260]. Silicone–acrylates, have been developed as release coatings for radiation-curable film and paper [261]. The photoinitiated curing of silicones was discussed more than a decade ago by Timpe et al. [262].

b. Formulation of Other Compounds

For some PSPs there is no need for the high level of abhesive performance given by silicones. On the other hand, for adhesive coatings with very low tack and peel the use of silicones would give laminates with insufficient bond strength. In such cases the abrasion sensitivity and low anchorage of silicones would influence the peel level also. Another problem is silicone transfer which can occur from any release liner to any adhesive, and render the adhesive coating completely tackless. Trace amounts of migratable silicone can cause detackification. A rough silicone surface or silicone transfer can cause a loss in subsequent adhesion, which is the adhesion measured after an adhesive has been in contact with a silicone release liner. A stated in a recent survey by Eckberg [263], industrial siliconizing technology does not ensure the opportunity for corrective action until considerable quantity of bad product has been generated.

Therefore, for certain applications other than silicone release layers are recommended (see also Chapter 5). According to Ref. [264], the release performances of a carrier can be improved by using 10–15% polyacrylonitrile latex and 2.5–5% melamine–formaldehyde resin. Copolymers of octadecyl acrylate–acrylic acid or vinyl stearate–maleic anhydride have been proposed as release layers for tapes. Tristearyl-tetraethylenepenteneamine and polyvinyl octadecyl carbamate have also been suggested [78]. Modified starch or starch and a water-soluble fluorine compound (water-soluble salts of perfluoroalkyl phosphates, perfluoroalkyl sulfonamide phosphates, perfluoroalkoxyalkyl carbamates, perfluoroalkyl monocarboxylic acid derivatives, perfluoroalkylamines) and polymers having as skeleton epoxy, acrylic, methacrylic, fumaric acid, vinyl alcohol, (each of which contains a C_6–C_{12} fluorocarbon group on a repeating unit of the polymer) can be used as release layer on paper for an adhesive containing a minute spherical elastomeric polymer [265].

Tapes with an olefinic carrier material are noisy. Modification of the adhesive by addition of a mineral oil substantially reduces the noise level during unwinding of the tape, according to Galli [266]. Further improvement can be achieved by using a release coat and/or special treatment of the carrier back. Polyvinyl carbamate, polyvinyl behenate can be used as the release layer [267]. The common manufacture procedure of tapes includes the in-line primer coating, release coating and adhesive coating [235]. As an example, the backside of the tape carrier can be coated via rotogravure with a 0.15% solution of Hoechst Wachs V. Gasoline (boiling range 60–100°C) has been used as solvent. Blade coating and drying at 30–80°C have been proposed.

For special applications vinyl acetate polymers can also be coated as the release layer. Such abhesive layer based on PVAc has been proposed for tapes [268]. PVA is formulated with silicones as release agent for certain tapes. Maleic copolymers with polar vinyl and acrylic monomers such as styrene, vinyl pyrrolidone oder acrylamide can be used also [269]. Other water-based an solvent-based silicone-free release materials have been developed. These silicone-free release coatings need no postdrying curing [270]. Solvent coatings are free of migratable species, hydrophylic components which can cause humidity-induced variations in release performances and they can be dried faster. Water-based formulations have the tendency to foam and contain migratable species.

The polyvinyl carbamates are the reaction products of $C_{12–18}$ aliphatic isocyanates and polyvinyl alcohol. They adhere well to plastic films, as well as cellulosic substrates and are normally used for tackified rubber formulations. The carbamates exhibit the lowest initial release values but show the greatest peel buildup. These compounds are recommended for tapes [271], less than 20% alcohol can be added to their solution in toluene to improve their solubility. Although the readhesion (subsequent adhesion) values are unchanged after aging on carbamate, and polyvinyl ether release coatings, the rolling ball (RB) values are worse. This suggests the presence of low molecular weight migratable species in the release coatings. The DSC plot presented in the Figure 8.6 shows the presence of low molecular weight fraction with a lower melting point in

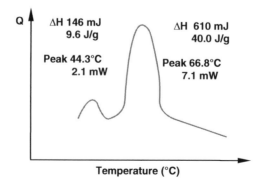

FIGURE 8.6 Differential scanning calorimetric plot of a carbamate-based release agent.

the release agent. Experimental results demonstrate that the higher is its concentration the lower the release degree. The presence of low molecular weight products in the carbamate-based release agent is the result of the multiple polymer-analogous reactions used for the synthesis of such compounds. The polyvinyl alcohol itself is the result of hydrolysis reaction of polyvinyl acetate that has a broad molecular weight dispersion. The reaction of the macromolecular alcohol with isocyanates leads to side products also.

Behenates and carbamates are applied as release layer for protective films. Polyvinyl carbamate is the main product used for nonsilicone-based release coatings, mostly for film. Other carbamate derivatives have been proposed also. Polyethylene imine octadecyl carbamate has been suggested for tapes with good rewindability [272]. Such craft paper based tapes are coated with PE. For instance, a 10 μm layer of PE with a density of 0.918 g/cm^3, and a 10 μm PE layer with a density of 0.944 g/cm^3 are laminated on PE and coated with the carbamate.

Acrylic copolymers based on fatty alcohol acrylate such as stearyl methacrylate are solved in hydrocarbon solvent and diluted to 1–4% solids for coating. Such release agents are used with rubber-resin adhesives as well as with certain acrylic adhesives, where a higher release value is acceptable, such as in self-wound tapes. Such compounds are not suitable for labels.

Polyamide resins are used for paper tape applications (and cold seal coatings). Such compounds are less expensive as carbamate but their aged release values are high. The polyamide release values are not significantly different form those of the uncoated polyester film. Octadecyl vinylether copolymers can be used with rubber-resin or acrylic adhesives, but are expensive. Polytetrafluorethylene (40–80%) can be embedded as an abhesive filler in a binder based on cellulose derivatives (e.g., nitrocellulose) [273].

The water-based release agents are crosslinkable or self-crosslinking compounds. Vinyl acetate and acrylic-based emulsions are used. These compounds have been suggested for tapes, but their adhesion inadequate for use in plastic films. As mentioned earlier with respect to silicones, the performance of any given release coating is dependent on the adhesive composition against which is measured. This statement is valid for nonsilicone release coatings also. Table 8.16 presents some nonsilicone release materials used for PSPs and their application.

2. Formulation According to Coating Technology

A special domain of release coating is the manufacture of radiation-cured abhesive materials. In this case the polymer displaying release performances is cured (polymerized) with radiation. Such polymers are silicone-based compounds. The first experiences with EB-cured silicones have been carried out with solvent-based silicones (e.g., for a coating weight of 1 g/m^2, with a dosage of 1 M rad). For such abhesive layers the release values increased in the time. As stated in Ref. [278], no damage of carrier has been produced by EB-curing. However, actually economical considerations preclude its general use.

The development of UV-curing in printing brought about its application as a common method for siliconizing also. In principle, cold and hot UV-curing systems are known [279]. Hot UV systems use IR and UV radiation together. Cold UV systems use only UV radiation alone for curing of the silicone coating. UV–IR systems were first commercialized in Europe for solventless silicones in 1984. UV-cured silicones have been proposed for narrow webs, and EB-curing for broad webs [280]. A special manufacture technology of a radiation-cured release layer is described by Brack [281]. In this case the release agent is incorporated as an incompatibile compound in the release binder. The film is cured to a flexible solid layer by irradiation. The abhesive substance migrates to its surface.

Diorganopolysiloxanes have been suggested as water-based release dispersions [280]. Such release layers have to adhere to polyethylene-coated paper and must display abrasion resistance. The formulation contains a level of 10–70 wt.% polyvinyl acetate and 2–20% polyvinyl alcohol [280].

TABLE 8.16
Nonsilicone Release Coatings

| PSP | Carrier | Release Agent | | | |
		Chemical Basis	Coating Weight (g/m²)	Release Force	Ref.
Tape	PVC 120 mm	VAc-VC copolymer, PVA-octadecyl isocyanate adduct	0.2	150 g/50 mm	[274]
Tape	—	Octadecyl acrylate–acrylic acid copolymer	—	—	[78]
Tape	—	Polyvinyl acetate copolymer	—	17.0 oz/in.	[268]
Flexible PSP	—	Methyl methacrylate–styrene copolymer	—	—	[275–277]
Tape	—	Hexafluorpropylene–tetrafluorethylene copolymer-polyamic acid	—	—	[4]
Tape	Paper	Natural rubber	1.5–2.0	70–100 g/25	[251]
Tape	—	Modified starch	—	—	[78]
Tape	—	Vinylstearate–maleic anhydride copolymer	—	—	[78]
Tape	—	Tristearyltetraethylene pentenamine	—	—	[78]
Tape	Paper/PE laminate	Polyethylenimine octadecylcarbamate	—	—	[272]
Tape	—	Polyvinyl octadecylcarbamate	—	—	[78]

Abhesive substances can be embedded in the carrier material also. Such carrier is manufactured for insulating tapes. Special insulating tapes have to adhere to refrigerated surfaces. A self-adhesive tape for thermal insulation used to secure a tight connection between heat exchange members (metal foils) and thermal insulating materials based on polyurethane foam is prepared by applying an adhesive (which is able to adhere to refrigerated surfaces) on an olefin copolymer carrier material that contains 0.2–2% fatty acid amides or silicone oils as the release agent and is bonded to the polyurethane [282]. Silicone rubber can also be included in the carrier material. According to Babayants et al. [283], 1 phr silicone rubber embedded in a 100 μm LDPE film reduces the release force from 350 to 70 g/25 mm.

3. Formulation According to Product Classes

As discussed earlier, different application technologies require different degrees of abhesion. Therefore abhesive choice and formulation depend on the product class. Labels and certain tapes need the use of low release force abhesives based on silicones. These abhesives can be modified to suit the product end-use. Generally siliconized paper has been suggested as release liner for labels, but special compounds can also be used [284]. According to Omote et al. [285], a release layer for paper carrier is based on modified starch or starch and a water-soluble organic fluorine compound, like salts of perfluoroalkyl phosphates, fluoroalkyl sulfonamide phosphates, perfluoro-alkoxy carbamates, perfluoroalkyl monocarboxylic acid derivatives, perfluoro-alkylamines, and polymers having as skeleton acrylic acid, methacrylic acid, vinyl alcohol, etc., each of which containing C_6-C_{12} fluorocarbon groups on a repeating unit of the polymer.

Copolymers of maleic acid with acrylic and vinyl monomers, styrene, vinyl pyrrolidone, and acrylamide are used as release coating for tapes [286]. Nitrocellulose, polyethylene, rubber, proteines, silicones have been proposed as release layer for medical tapes [287]. Aluminum film coated with a tackified polyisobutylene-based adhesive is applied as street marking tape. The release layer for this adhesive is PVA [288]. Water-based release coatings suitable for paper tapes have been developed for rubber-based adhesives. Such products are precrosslinked vinyl acetate copolymer emulsions containing variable levels of a functionalized additive for adjusting the release degree (e.g., ethyleneglycol diacrylate, or pentaerithritol triacrylate). Release values of 13.9–17.0 oz/in. are obtained. Silicone-modified release coatings are also used to give an initial release value of 8.7 oz/in. [289].

As mentioned earlier, polyvinyl carbamate is used as main release agent for protective films. It can be applied alone or in mixtures with chlorinated polyolefins. The melting point and MWD of the polymer have to be well-defined in order to ensure the required solubility and anchorage of the product. As known, polyvinyl carbamate exhibits low solubility; therefore generally solutions having less than 1% solids are coated. Dissolving, transport, and storage of the release solution have to be carried out above room temperature. Toluene is the best industrial solvent for this release agent.

Polyamide-based release agents are used for cold seal coatings. Blends of such agents with Teflon derivatives have been proposed as release agent for protective films also. Unfortunately their solubility, anchorage, and abhesion are too low. Table 8.17 presents data on some of the release agents used for tapes and protective films.

4. Technology of Abhesive Formulation

Quite unlike adhesive formulation technology, the formulation of abhesive materials is mainly a simple dissolution or dilution of special almost ready-to-use formulated chemicals. Except for some micromolecular compounds and the vinyl carbamate homo- and copolymers, which have to be used in solutions having low viscosity, raw materials applied as abhesives are supplied as

TABLE 8.17
Release Agents for Tapes and Protective Films

Product	Liner	Adhesive	Release Liner Buildup	Release Agent	Ref.
Tape	LDPE	Polyester acrylate	Embedded[a]	Silicone	[289]
Tape	LDPE		Embedded[b]	Methyl hydrogen siloxane + α-diolefin	[290]
Tape	—	Acrylate	Coated	PVA + synthetic microballoons	[291]
Tape, for porous substrate	PE laminated kraft paper	—	Coated	Polyethylenimine-octadecyl-carbamate	[272]
Tape, Double-side-coated foam	Paper, 60 lb	—	Coated	Silicone	[292]
Protective film	PVC/PE laminate	—	Coated	Silicone	[224]
Masking tape	Crepe paper	—	Coated	Silicone	[183]
Tape	Plastic film	Silicone	Coated	Fluoroalkylsilicone	[260]

Release force: [a]70 g/25 mm; [b]95 g/25 mm.

liquids. The equipment used for their formulation depends on the nature of the raw materials. Polyvinyl carbamate-based release materials are used as solutions with low solids content (0.6–1.5%) in toluene. Such solutions are coated warm, with a ceramic cylinder. Because of their low solubility they are stored warm and under continuous stirring.

II. MANUFACTURE OF THE CARRIER MATERIAL FOR PSPs

The carrier materials for PSPs can be classified in paper-based and nonpaper-based ones. Of the nonpaper-based carrier materials synthetic films are the most important. Generally, the carrier is the cost-expensive component of the laminate. According to Hunt [87], a decade ago laminate costs typically accounted for about 40% of selling price of labels. They now account for 70%. Table 8.18 lists the main carrier materials for selected PSPs.

A. CARRIER ON PAPER BASIS

Paper was the first carrier material used for PSPs. Various products, such as labels, tapes, and protective films were manufactured on paper basis. Later synthetic films replaced paper partially (e.g., in the manufacture of labels) or almost totally (e.g., in the manufacture of tapes and protective films) as the adhesive carrier material. In 1993/1994 films were used for 9% of the European self-adhesive products and paper for 38% [293]. After a decade films cover about 20% of carrier materials for PSPs.

1. Manufacture of Paper

Paper is produced by a specialized industry. Unlike plastics manufacturing, where some of the film producers are also converters (PSP producers), paper is supplied for converters as a separate raw material. It is not the aim of this book to discuss paper manufacturing. For paper converters, its processing characteristics and surface and bulk properties are more important. The surface characteristics of paper influence its processing and end-use also.

TABLE 8.18
Major Carrier Materials for PSPs

Label		Tape		Protective Film	
Paper	Uncoated Coated Laminated	Plastic film	Oriented Nonoriented Laminated Embossed Foamed	Plastic film	Nonoriented Oriented Laminated
Plastic film	Oriented Nonoriented Laminated	Paper	Coated Laminated Embossed Pleated Impregnated	Paper	Coated
Textile	Nonwoven Woven Laminate	Textile	Woven Nonwoven Impregnated Laminated	Textile	—
		Foam	—		
Special materials	Metal ceramics	Special materials	—	Special materials	—

2. General Requirements for Paper as Carrier Material

There are some general requirements with respect to paper quality as carrier material. These include its surface and mechanical characteristics. Surface characteristics influence mainly its coatability. Its bulk properties (lay flat, dimensional stability, and flexibility) influence its coating and converting properies (printability, cuttability, and labeling). Printability is also a function of paper surface quality. Nonimpact printing procedures require a relatively porous surface [294]. Different paper qualities such as uncoated, coated, cast-coated, and colored are used. Synthetic papers have also been introduced. These are polymer (mainly HDPE)-based materials [295]. Synthetic papers provide the printing properties of real paper and the increased durability and weatherability of a plastic film [296].

The surface quality of paper varies. In Europe coated papers possess higher gloss and density than in the United States [297]. The quality of coated papers depends on the coating procedure used, the coating weight, the drying process, and compatibility between coated layer and paper [298].

The nature of pigments used for surface coating of papers is very important for its coating performance. Various compounds, such as kaolin, clay, talcum, calcium sulfoaluminate, titanium dioxide, calcium silicate, sodium aluminum silicate, barium sulfate are used [298]. These pigments possess different surface energies and pH stabilities. In top-coated papers the surface layer is based on about 90% white pigment and 10% binder (vehicle). The binder influences the adhesion and anchorage of the top coat, its stiffness, porosity, and ink absorbance. This layer has a porous surface, with the diameter of the pores between 0 and 40 μm [299]. The average diameter of the pores for fine quality papers is 2–4 μm. Kaolin is the most used filler for paper. There are gravure printing and offset grade kaolins (having different particle diameter). Calcium carbonate is less expensive and allows fast ink penetration. Calcium carbonate increases the porosity of paper. This is very important for heat-set-rotary printing [300]. The mean coating weight of the top coat layer is about 15 g/m^2, that is, the layer has an average thickness of 12 μm and a pore volume of 3 ml/m^2. The printing and adhesion properties of the top coat depend on its chemical composition. Caseine, starch, PVA, acrylic, or styrene–butadiene latex are used as binder. Polymer dispersions have been used since 1949 [299]. It should be mentioned that the uniformity of the surface distribution of binders and pigment, that is, the coating structure influences the printing defects ("print mottle") [301]. As stated by Topliss [302], the main face stock materials have been the following: white-uncoated paper, white-coated paper, nonwhite paper, paper-backed foils, vinyls, and polyacetate. Above average growth has been foreseen especially for vinyls, polyester, and cellulose acetate. Growth in demand for coated paper has been higher than for uncoated material.

Papers for offset printing with 70 g/m^2 can be coated on both sides [300]. Calcium carbonate, clay, and satine white are used as the top layer. Low weight coated papers for gravure an roll offset printing, with at least 45 g/m^2 can be double-side-coated too. Gravure printing requires high gloss and no "missing dots."

The bulk properties of paper influence its dimensional stability. These depend on the environmental conditions also. Saturated and coated foudrinier paper can exhibit up to 10% weight gain and greater than 1% width gain over the range of 0–97% relative humidity [303]. The humidity content of the paper (the ideal dry status of paper) depends on its quality. Newspaper materials have a humidity content of 9%, a quality paper contains 6% water for the same external relative humidity of 9% [304]. Dimensional stability is a function of the geometrics and dimension of the carrier material. For labels, thin films (15–20 μm) and cardboard of 300 g/m^2 can both be used [305]. Narrow web (420–620 mm) flexo-printing machines can use webs with a thickness of 20–300 μm [306]. Printing methods influence the choice of carrier material requiring tailored surface quality and thermal resistance for the carrier. As shown by Strahler [307], most label printers prefer a working temperature of at least 35°C for UV-offset printing.

3. Special Requirements for Paper Used as Carrier Material for Different PSPs

Various PSPs have different end-use functions. Market requirements lead to well-defined characteristics of the label material. For instance, labels for lubricants have to display durability, visual
impact, and flexibility [308]. Toiletries and cosmetics require no-label look, high consumer
image, and stain resistance. The hardware and electrical markets need visibility and heat resistance.
For toys, nontoxic labels with a long life and visual impact are necessary. The chemical industry
requires carrier materials that are resistant to chemical attack, visibility and durability. Mechanical
stresses during use can be quite different. Therefore mechanical requirements for carrier material
are different also. A paper carrier material has to have certain special performances for use in
each PSP class. For instance, fragile pressure-sensitive materials either are very thin or have low
internal strength. Such materials based on paper or film are used for labels and tapes also. Different
face stock materials are suggested for envelopes, warning labels, sterile needle cartridges, etc.
[309]. A porous carrier can be required for medical or mounting labels and tapes. Paper for
release liner has to fulfill various mechanical- and surface-related requirements and should not
contain inhibitors for silicone curing agents (e.g., heavy metals, sulfo, and carboxy groups)
[310]. Fan folders are used for table labels; the machines used need a low paper weight of
67–68 g/m^2 [311]. Copy labels must allow go-through of the image to the release.

a. Paper as Carrier for Labels

Paper is the most used carrier material for labels. For most labels it is used as both the face stock and
the liner material. A 70–80 g/m^2 paper has been suggested as generally versatile carrier material.
Common label papers have a weight of 70–80 g/m^2 and pressure-sensitive labels 83–90 g/m^2
[312]. In 1965, paper was considered the sole carrier material for labels [313]. Now 80% of
labels have paper as carrier material [314]. The number of paper qualities used for the same
application is very large. For instance, for wine bottle labelling a single supplier of labels uses
25 different paper qualities coated with various water-based PSAs [315]. For labels different
types of materials, thermopapers, vellum and clay-coated papers are suggested [316]. Uncoated,
coated, cast-coated, and coloured papers are applied. First, cast-coated papers have been introduced. Now, machine-coated paper qualities are preferred. The proportion of satinated kraft
paper increases [317]. Low weight, high rigidity papers are required [318]. For table labels
printed via nonimpact printing, open porous papers are preferred [319]. Metallized paper and
film/paper laminates are used also. Metallized paper displays low curl in comparison with metallized film. According to Ref. [320], paper, fabric, vinyl, acetate, and polyester films are suggested as
carrier material for HMPSA-coated labels. Fluorescent paper labels have been introduced in 1958
[321]. Latex-impregnated paper is proposed for weather-resistant labels, calendered (90 g/m^2)
paper for opaque labels, paper having high absorbtivity for labels printed with so-called cartridge
inks, oil- and fat-resistant paper for thermal printing [322]. According to Senn [323], 55–100 g/m^2
papers should be used for paper laminates and 12–180 g/m^2 papers for film laminates. Because of
the roughness of the paper it is difficult to achieve release values less than 450 g, therefore in this
case polyethylene-coated papers are recommended [324]. Such papers have release forces which
can be regulated between 20–400 g.

   According to Ref. [319] the main requirements for a paper carrier material for labels can be
listed as high gloss, high surface strength, good wettability, and good printing ink anchorage.
Paper as face stock material for labels should display special mechanical and surface characteristics. Adequate mechanical characteristics are required to ensure its dimensional stability and
special behavior during cutting and labelling. Surface characteristics should allow its coatability
(printing and varnishing). Pressure-sensitive labels displaying security peel-off must have either
exceptional strength to prevent removal, or be so fragile, that they cannot be removed in one
piece [325]. On the other hand, high speed converting machines solicitate the label material
mechanically. Electronic confectioning machines for labels cut, die-cut, punch, and perforate the

paper with a running speed of 200 m/min [304]. Electronic data systems use primarily a paper engineered for strength, and with high absorbency to capture ink [326]. Special papers like radiation luminescent papers have also been manufactured [327].

The mechanical and surface characteristics of paper can be improved by impregnating, coating and laminating. Latex-impregnated paper is a lower cost alternative to PVC [328]. HDPE usually blended with LDPE is used on photographic paper to avoid curl [329]. Paper-backed aluminum foils and metallized face material have the highest opacity (100%). Latex-impregnated and plastic-coated papers possess an opacity degree of 84–88% [330]. Plastic-coated paper attains a gloss degree of 96%.

Special paper is required for direct thermal coating. Common organic direct thermally printable papers include a colorless leuco dye and an acidic color developer [331]. These are coated and held onto the surface of the paper with a water-soluble binder. During printing the two components melt together and react chemically to form the color. In order to limit this image to the heated area, inorganic fillers (e.g., calcium carbonate, clay, etc.) are used. The problem of image stability (protection against chemicals) is solved by applying a topcoat as a transparent film forming layer. It is possible to also apply a barrier coat to the underside of the paper to prevent adhesive or plasticizer migration from the opposite side. Therefore, these papers are classified as nontop-coated (nonsmudge-proof) and top-coated (smudge-proof) papers. Nontop-coated papers are applied for price weight labels of perishables and quick sale items, for tags, and tickets. Top-coated papers are suggested for frozen food and long-life labels. Latex-impregnated paper is used for data system applications requiring moisture resistance and improved flexibility. They provide smudge resistant ink absorbtion [332].

b. Paper as Carrier for Tapes

At the beginning of the tape manufacture, paper was the only material used as carrier. According to Stott [333], paper-masking tapes were used in the 1920s for car varnishing. Crepe paper coated with crosslinkable silicone adhesive has been applied up to 180–200°C. Different types of paper, dimensionally stable or deformable, tear-resistant and fragile materials have been used. Some papers are water-resistant and others are water-soluble. Paper for splicing tapes has to resist temperatures up to 220–240°C and be water-dispersible, but resistant to organic solvents. It must also be migration resistant. The tear resistance of the tape has to be as high as that of the paper to be bonded [18,135]. Extensible, deformable paper is required for medical tapes [334]. Glossy and crepe papers are used for tapes. For instance, a PSA tape for protection of printed circuits, based on acrylics and phenolic terpene resin as adhesive (with a 100–200 g/m^2 siliconized crepe paper as temporary carrier) is manufactured by coating the adhesive mass first on a siliconized poly-urethane release liner and then transferring it under pressure (10–30 N/cm^2) on the final carrier material. Decal premasking tapes are made from a latex saturated paper which is coated with a PSA. In most cases as low adhesion adhesives are utilized, the decal premask/application tape does not require a release coating. Such papers are recommended where delamination resistance but clean removal is required; temperature resistance, high abrasion resistance, markability and translucency, dimension stability, and wet rub resistance are needed. For special tapes, kraft paper is reinforced with a mid polypropylene film [335].

c. Paper as Carrier for Protective Films

Actually there are only a few special protective papers. Corrosion protective papers [336] and weathering resistant papers are used [337]. Self-adhesive wall cover consists of a layer of fabric having a visible surface, a barrier of paper that has one surface fixed to another fabric layer, a PSA coated on the barrier paper and a release paper [338]. Protective masking papers have a base paper of 49–147 g/m^2, Elmendorf (CD) tear of 43–327 g, and elongation of 2.7–5.0%.

d. Paper as Carrier for Release Liners

According to Hufendiek [168], the paper used for release liners can be normal, densified, glassine, plastic-coated, or clay-coated paper.

The weight of the release liner depends upon the method of label conversion, and the type of die-cutting and tooling as well as the strength of the sheet. According to Reinhardt et al. [339], silicone penetration in paper decreases as the density of the paper increases. Low weight (65 g/m^2) open papers allow silicone penetration, the level of penetrated silicone and its distribution in paper influence the release force. Base materials for release liner in 1987 were calendered paper (50.5%), clay-coated paper (21.3%), polymer-coated paper (6.7%), other crafts (12.9%), and plastic films (2.0%) [340]. Different release degree is achieved for a double-side-coated release liner by using the same release component but one-side with polyethylene-coated release paper [341].

Polycoated-craft, clay-coated craft, and densified craft papers have been proposed for use in the manufacture of release liners [342]. For siliconizing (release liner) various paper qualities with weights of $40-220 \text{ g/m}^2$ have been suggested. Satinated (glassine), clay-coated, single- and double-side polyethylene-coated papers, and normal kraft papers are used [343]. Conventional glassine paper liners account for about 60% of the market [87]. Clay-coated papers are widely used in the computer sector, because of their good lay flat and handling. Polyethylene extrusion-coated papers combine the strength of long fiber papers with the smooth nonabsorbent surface of film layers. Release liners in weights from 68 to 150 g/m^2, silicone coated on either or both sides, with carrier of super calendered, clay-coated and poly-coated paper are supplied [344].

The liner plays an important role in the functionality and cost of most PSPs. Because the release liner is discarded after the label is applied, the development of more environmentally friendly materials is important. As stated by Allen [345], there is a trend to use for face stock material and release liner the same paper qualities. According to DeFife [346], the most important features of the release liner are tear strength, dimensional stability, lay flat, and surface characteristics.

Tear strength of the carrier material is important for label converting and dispensing. Good calliper control and liner hardness (lack of compressibility) are the main parameters. Dimensional stability, the ability to maintain the original dimensions when exposed to high temperature and stresses is required for print-to-print and print-to-die registration (see also Chapter 10). If the release liner stretches under heat and tension, graphics will be distorted. Liner stretch can affect the dispensing of the label also. Special attention should be given to the influence of super-calendered base papers with different density and closeness.

The smoothness of the release liner affects several performance features [346]. The adhesive surface is a replica of the release surface. If it is rough the level of initial adhesion will be reduced. If rough label is laminated with a clear face stock material label clarity will be negatively influenced. If the liner is too smooth air entrapment during dispensing can become a problem. The roughness of the label backside can affect web tracking on the printing press. Smooth liners can weave causing cross-web-print registration. Also, in laser printer applications, the roughness of the liner plays a critical role in sheet feed ability and tracking through the printer. Single-side, double-coated label papers have been developed for special surface quality requirements [347].

High-speed table labels require lay flat; the weight of the release paper should be between 67 and 85 g/m^3 [348]. A 80 lb kraft liner is used for repositionable labels in order to avoid curl; a 78 lb kraft release liner is applied for thick (18 mil) embossed vinyl [349]. Lay flat is very important for sheet labels or pin perforated and folded labels. Labels that are butt cut, laser printed, and fan folded must lie flat to ensure proper feeding and stacking. Paper liner lay flat is affected by the liner material's resistance to humidity, dimensional stability at different temperatures, and stiffness (thickness). The use of radiation-cured or low temperature heat-cured silicones produces less changes in the humidity balance of paper. Glassine liners are harder (densified) and are used for high-speed dispensing [350]. To reduce silicone consumption, silicone coaters require high and uniform gloss and low porosity of the paper surface [351].

Satinating reduces paper porosity (the specific gravity of paper increases from 0.90 to 1.20 kg/dm^3) and its permeability to air decreases to one tenth of the initial value. Fillers increase the specific gravity of paper to 1.25 kg/dm^3. The components of the pigments and binder for topcoated papers must be free of substances that can poison the silicone catalysts [351]. According to Moser [350], the most important characteristics of glassine silicone papers are (i) their ability to be siliconized, (ii) machining properties, (iii) dimensional stability and lay flat, (iv) die cuttability, and (v) storage performance. Theoretically these characteristics are influenced by the porosity, surface structure, and surface energy of the material. Measurable characteristics are the length of penetration, porosity, smoothness, gloss, solvent hold out, and surface tension. As stated by Moser [350], porosity is the main parameter for solvent-based siliconizing and surface structure is the principal factor for solventless siliconizing of such papers. Surface tension does not influence the versatility of the paper to be siliconized.

Generally, paper thickness and stiffness affect the processing of the laminate. As discussed in Ref. [352], these are the main parameters for cuttability. According to Wabro et al. [78], low weight (70 g/m^2), low thickness (60 μm) siliconized papers are recommended for laminates that are not confectionated (e.g., decals and foam-based products). The standard paper grade for die-cut products has a weight of 95 g/m^2 and a thickness of 80 μm. Polyethylene-coated papers have a weight of 120 g/m^2 and a thickness of 100 μm. Tear-resistant release liners with improved cuttability are manufactured by using paper having a weight of 110–145 g/m^2 and thickness of 120–155 μm [78].

For one-side splicing tapes the siliconized paper liner remains on the adhesive and has to be recyclable also. Carrier-free splicing tapes must also be water soluble. The liner for transfer tapes has to allow detachment of the adhesive, that is, it has to display different degrees of release either side [135]. Siliconized paper is also used as a pleated liner for tapes.

B. SYNTHETIC FILMS AS CARRIER

According to Schroeder [107], fiber-like materials and textiles (water-resistant and water-soluble) based on cellulose, plastic films (soft and hard PVC, cellulose derivatives, PET, etc.), foams (open pore PUR, closed pore PVC, closed pore polychloroprene, etc.) are used as carrier materials for tapes. The paper and nonwovens can be impregnated with water-soluble and water-insoluble resins in to achieve a reinforced or film-like character; the foam carriers are coated with adhesive one or both sides.

A decade ago, labels and tapes have been the main market areas for film carrier materials. Oriented PVC (60%) with a thickness of 25–70 μm, OPP (40%) with a thickness of 25–40 μm, and PET (2%) with a thickness of 23 and 36 μm have been used [353]. According to Ref. [354], the growth areas for film carrier materials for labels are beverages (45%), personal care (20%), pharmaceuticals (15%) and on-demand digital applications. Among the nonpaper materials, synthetic films are the most important carrier materials. PVC, polystyrene, PET, metallized materials, and synthetic papers are used. PVC, PP, PE, PET, PUR, Zellglas, and polymonomethacrylate (PMMA) can be coated with HMPSA. Soft PVC is used as carrier material for decorative films [120]. PTFE is used as carrier for splicing transfer tapes [107]. Special carrier materials as polyester, anodized aluminum, stainless steel, and ceramics resist up to 1400°C [355].

Synthetic films were introduced as carrier material for PSPs because plastic films have certain technical and commercial advantages over paper. There are certain application fields like toiletries, and pharmaceuticals, where film labels are used almost exclusively. Some special pharmaceutical labels may have take-off (detachable) parts also in order to allow multiple information transfer. Parts of the label remain on the drug, or on the patient [356]. Such labels use a film carrier. Labels for textiles are manufactured on PE basis [357].

1. General Requirements for Synthetic Films as Carrier Material

Independently of their use for a given product class (label, tape, protective film, etc.) or their raw material basis, synthetic film carrier materials have to satisfy certain requirements. These films must have oil and grease resistance, moisture resistance (dimensional stability, UV stability, high edge-tearing resistance, chemical resistance, brilliance and transparency, flame resistance (self extinguishing), and recyclability. Printing films must have color and surface versatility, opacity, UV-stability, fire-protection provision, shrink reduction, deep drawing qualities, and impact strength at extreme temperatures [358]. The general requirements for synthetic films as carrier material concern their web-like character and surface quality. Related to surface quality, there are requirements to ensure coatability (adhesive coating and printing) and machinability (during converting, application, and in applied state). Blocking, smoothness, hardness, and electrical conductivity must be regulated. Good sliding properties, optical properties, static resistance, scratch resistance, and shock resistance are also required, as are lay flat, minimal thickness tolerances, no wrinkles, no gel, and surface tension higher than 40 dyn/cm.

Depending on their use for certain product classes and the real application case of a product, synthetic films used as carrier material have to meet certain special requirements. When selecting a suitable raw material for carrier manufacture, consideration must be given to the desired film properties, the processing method to be applied, and not least, to economic factors. As an example, linear low density PE (LLDPE) is not foldable or stiff. It is difficult to cut or die-cut, and to punch, because its high elasticity. Its extensibility is a disadvantage for the converter [359]. In choosing the material for release liner it should be taken into account that the stiffness of the laminate is the sum of the stiffness of the components (taking into account the reinforcing effects; see also Chapter 3). Therefore for thin face stock materials a heavy release liner should be used, and for thin release materials a heavy face stock. There are some applications where the thickness of the liner is given by technical functions, for example, copy labels with a self-copyable release liner. White, thin ($50 \ g/m^2$) release paper is recommended for self-copyable products [324]. For instance, for name plates with a 100 μm stiff face stock film, a thin release paper is suggested. Table 8.19 summarizes the main requirements for synthetic films used as carriers for different PSPs. As seen from this table, labels need aesthetically high-quality face stock material, because their primary use is as carriers of information. Tapes for fastening are web-like products that require excellent mechanical properties. Protective films have to be able to conform to and deform with the product to be protected.

The development of carrier materials has been influenced by requirements for more film or more paper carrier material. Factors forcing the film carrier development take into account its better surface quality (S), excellent aesthetics (E), and chemical resistance (Ch), its versatility as release liner (R) and the general development of film-based packaging materials (Pa) which needs the same label material as used for the packaging film. The improved cuttability (Cu),

---

TABLE 8.19
Main Technical Requirements for Plastic Film Carrier Materials

Label	Tape	Protective Film
Coatability	Coatability	Coatability
Printability	Mechanical resistance	Conformability
Cuttability and die-cuttablity	Tearability	No elasticity
	Foldability	Processibility
Chemical and environmental resistance	Chemical and environmental	Chemical and environmental resistance

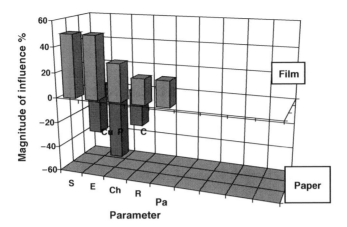

FIGURE 8.7 Technical and economical parameters influencing the face stock development.

better processibility (P), and lower costs (C) enhance paper carrier development. Figure 8.7 illustrates (on an arbitrary scale) the estimated magnitude of influence of such parameters on the development of label face stock.

2. Special Requirements for Synthetic Films as Carrier Materials

As discussed in Chapters 2, 10, and 11 carrier materials used for different product classes need different special properties. Their performance characteristics are determined primarily by the manufacture and application of the PSP.

a. Requirements for Plastic Films for Labels

Plastic films used as face stock material for labels are coated as webs and laminated and converted as webs. Reel labels are dispensed (labeled) in web form. Therefore they have to satisfy the general quality requirements for machining of web-like materials. On the other hand, labels are materials used for information transfer, that is, having an aestethic character. Their message is transfered onto their surface by using various printing and lacquering techniques. It is to be supposed that the requirements concerning the mechanical properties for dispensing are higher than those for coating and converting, and dimensional stability-related requirements for printing are more severe than those arising during conversion. Therefore, the main requirements for carrier materials used for labels concern their dimensional stability and surface quality. The film has to be stiff enough for dispensing and flexible enough to conform to various container shapes, and to resist repeated deformations [360]. For label converters, the stock must be able to meet printing and die-cutting specifications for computer imprintables. For end-users the label has to withstand harsh environments and rough handling. This means that plastic films having a pronounced rigidity, and surface quality are recommended as carrier for labels. Such films must allow good register control, easy die-cutting, and reliable high-speed dipensing. The specialist for label design and the manufacturer of the labels should know which polymers and plastic film manufacturing methods produce plastic film suitable for label application. Caliper, gloss degree, opacity, and corona treatment are the main performance characteristics of synthetic face stock materials [330].

Polyvinyl chloride was the first film-like synthetic material used as a carrier for labels. Thin PVC is used as carrier for tamper-evident labels [361]. Polyvinyl chloride exhibits the desired tear and temperature resistance for computer applications. It has excellent UV-stability [362]. The main disadvantages of PVC are its caliper variation, shrinkage, and plasticizer migration. In the last decade PE and PP replaced PVC.

Filled HDPE can be used as face stock material for water-resistant dimensionally stable labels that can be printed by different methods [363]. They have to posses a good chemical resistance also. For battery labels BOPP has been suggested [364]. PE carriers used for labels require 1-yr treatment stability [365]. Common HDPE has densities of $0.952-0.960$ g/cm^3. Higher densities are recommended when an optimum modulus (rigidity) is required [366]. Faster cooling gives less stiffness due to the formation of smaller crystallites. Like polyethylene-coated paper, polyethylene laminates can be used for improved humidity-resistance [367]. Mixtures of polystyrene and polyethylene are used as face stock material for labels also. As a face stock material polystyrene displays adequate printing properties, die-cuttability, and labeling ability [368].

A typical example for specials requirements for a label carrier film is given by name plates. Such film with printability via flexo-, letterpress-, and computer printing methods must have a matte, metallic appearance, like that of anodized aluminum. It has to display the handling characteristics of hard aluminum. No curl should occur after removal. A heat resistance of 150°C is needed. The matter surface has to absorb ink, maintaining sharp graphic contrast, good color intensity, and excellent legibility. It has to be printable with a range of common or special computer printers. Abrasion, chemical, and environmental resistance are needed. Opportunities for development of new films are given in the fields of security, computer, battery, speciality, and promotional labels [308].

Bioriented polypropylene (BOPP) is the fastest growing face stock material with sheet and reel fed options, printable in almost any format. It provides the opportunity for significant down gauging from PE. Unfortunately its dispensability is not the best [369]. Label-Lyte label films (white and transparent two-sided materials for roll stock lamination) from based on PP may be uncoated or coated; the uncoated film has a thickness of $19-24$ μm (mono); or 60 μm (three layer coex). The coated film as mono, possesses a thickness of 50 μm; the coex film is 60 μm thick.

Comparison of various polymers concerning their flatness, transparency, rigidity, printability, density, cuttability, temperature stability, and price, gives different ratings. Thus PVC offers the best cuttability and printability compared with PP and PET; Polypropylene has good density, temperature stability, and price; and PET is the most rigid and transparent film material [358].

b. Requirements for Plastic Films for Tapes

The carrier for tapes has to support and absorb the stresses during application in order to allow bonding of the adhesive onto the surface. The carrier material is chosen according to the nature, magnitude, and direction of the forces. It is important to know the yield and tear resistance of the material and elongation in the machine direction, and in cross-direction. It is imperative to know if the carrier splits, parallel to the machine direction. Dart drop resistance and foldability are required also. Elmendorf tear resistance also plays an important role (see Chapter 3, Section II). The ability of a film to retain a crease is an important property in packaging and is also required for some special tapes [370].

In the mid-1960s, cellulose hydrate, cellulose acetate, PVC, crepe paper, and fabric were the main carrier materials for tapes coated with solvent-based adhesives [313]. In the 1980s, PVC, PET, and OPP were the most used carrier materials for tapes [107,371]. Now, nonoriented, oriented, mono- or multilayer films, and film/film or film/fabric reinforced laminates are recommended for tapes. Fiber-reinforced films (applied for wet adhesive tapes [372] are used for PSA tapes also. Electrically conductive polyolefins having a high loading of carbon black (30 wt.%) have been proposed for insulation tapes. Chemically pretreated polyethylene (octane-based LLDPE) is recommended for tapes [192,373]. Biaxially oriented polypropylene with a thickness of $30-35$ μm is suggested also [192]. Its costs $20-30\%$ less than PET [364]. Shrinkable adhesive tapes based on polyolefins are manufactured by extrusion. As central section of polypropylene and ethylene–propylene copolymers are coextruded [374].

c. Requirements for Plastic Protective Films

Protective films are conformable, laminated covers that have to pass through the same processing and storage cycle as the protected item. Such films are only temporary and do not contribute to the value of the finished product. Therefore, protective film carriers are generally common materials although they have to be soft, mechanically resistant, thin and deformable, with a balanced plasticity elasticity, medium temperature resistance, and a low manufacturing cost.

3. Film Manufacture

The most raw materials for films are produced from synthetic monomers, via polymerization (polyaddition or polycondensation). The polymers are processed as thermoplast to produce a film. The physical (bulk or surface) transformation of the film is the next step in product manufacture. Table 8.20 summarizes the main manufacturing procedures for carrier films.

Most plastic films are produced by extrusion. An other method is casting from solution or dispersion. In this procedure the dissolved and dispersed film raw material is cast on a conveyor belt and dried. A high-quality film, free of tensions is produced by casting. This method is used to manufacture PVC (4–100 μm) film with defined, low tensile elongation [375]. Cellulose acetate, butyrate, propionate, and ethyl cellulose can be processed by casting [376]. Such cast or support-based coating lines manufacture a continuous strip from a liquefied polymer mass. In comparison with extrusion coating, the liquefied material to be coated generally contains liquid components, and the coating is only temporary. Upon solidification the film layer becomes a self-supporting material and it is stripped off the support base before being wound to form a roll of cast film material. There are two types of support bases, reusable materials and endless belts. Roll coaters, and curtain or flow box coaters are used, depending on the dispersing medium, the

TABLE 8.20
The Main Manufacturing and Postprocessing Procedures for Plastic Carrier Films for PSPs

Process	Main Procedure		Variants
Manufacture procedure	Extrusion	Blowing	Mono
			Coex
		Flat die	Mono
			Coex
	Calendering		
	Casting	Solvent-based	
		Radiation cured	
	Sinterizing		
	Impregnating		
	Polymerization in bulk		
	Lamination	Homogeneous	Adhesiveless with adhesive
		Heterogeneous	Adhesiveless with adhesive
Postprocessing	Orienting	Monoaxial	
		Biaxial	
	Physical treatment	Corona	Athmospheric
			Special
		Plasma	Inert
			Reactive
		UV	
	Chemical treatment		
	Lacquering		

viscosity of the material to be coated, and the thickness of the end product. A special polymer film having the characteristics of paper can be manufactured using the technology of plastisols, by precipitation of polypropylene dissolved in dichloromethane [377].

Calendering has been developed also. It has been applied specially for manufacture of PVC films. Generally thermoplasts having a thermoplastic domain with a viscosity of $10^2–10^3$ Pa sec and a broad softening range of $10–30°C$ can be processed by calendering [377,378]. Copolymers of vinyl chloride, vinyl acetate, cellulose acetate, polyethylene and its copolymers, polypropylene, and polybutene can be transformed into film via calendering. Films with a minimum thickness of $70–100$ μm can be manufactured using this procedure. Polypropylene films with a thickness of $100–800$ μm are manufactured by calendering. The raw materials for calendered PP film have to meet special requirements; they have to be processed at about $200 \pm 10°C$. ZN-polypropylene has a MWD of $6–13$, but it can be degraded to achieve a narrower MWD ($M_w/M_n < 6$) MWD which enhances calendering. Slowly crystallizing polypropylene is suggested for calendering. Such films exhibit higher stiffness (crystallinity of the calendered film is about 62%, in comparison with the crystallinity of 82% for the extruded films) and lower tensile strength and are recommended for furniture [379].

The principle of extrusion is well known. An endless screw rotates in a heated cylinder, which leads the molding materials forward, to be compressed plastified and homogenized. In front of the cylinder there is a tool to form the molding material. The molten plastic is molded through a die.

In the main manufacture procedures a flat die is used, to mold a flat film ("cast" film) or a circular die is used to mold a tubular film. The flat film is calandered (via cooled chill rolls) to achieve the final dimensions at normal temperature. The tubular film is blown up with air to stretch it to the final dimensions and to cool it.

In choosing a polymeric raw material for plastic film carrier the most important criteria are the following [380]: mechanical properties (tensile resistance, elongation, and shrinkage), chemical resistance, permeability (oil, gas, and water), sealability, surface quality, stiffness, thermal resistance, adhesion, coefficient of friction (COF), deep drawability, and melting temperature. From the above listed characteristics the resultant mechanical properties (resistance to different stresses, elongation, and stiffness), the properties related to dimensional stability (shrinkage and deep drawability), the properties related to surface quality (coefficient of friction and sealability), and the performances related to thermal resistance (melting temperature, sealability, and warm deep drawability) have a special importance for carrier materials used for PSPs also.

Roll down properties of tapes are very important (see also Chapter 10). Biaxially oriented, multilayer PP films for adhesive-coated tapes have been modified to improve their adhesion to the coating, by mixing the PP with particular resins. In this case in order to prepare an adhesive tape which can easily be drawn from a roll without requiring an additional coating on the reverse side at least two different layers having different compositions are coextruded, and the thin, back layer (1/3) contains a release component.

a. Blow Procedure

In this procedure the molten polymer mass is brought to the blowing head and is constricted to a thickness of about 2 mm between the exterior and interior wall. If necessary, several layers may be put together in the lowing head to buildup a multilayer coextruded film. When the molten material leaves the blowing head, the film is blown with air impinging at a certain angle under pressure. The diameter of the bubble is mechanically controlled with an adjustable aperture [381]. The turbulence of the air produces a tube of special shape. At a certain height (distance from the head) there is a freezing zone, where the molten material is cooled. Bubble cooling may use either conventional external air rings with blowers, or internal cooling. In processing of polyethylene (PE) molten ($120°C$) is cooled to $60–80°C$. The wall tolerances are balanced within $\pm 5\%$ by (1) rotating the entire blowing area or (2) turning the tube around with a rotating lay flat.

Haul-off rates are dependent upon extruder output, gauge, and bubble diameter. These factors together with melt temperature, cooling rate, die gap, and blowup ratio (BUR), control the degree of molecular orientation in the film. This in turn determine the shrinkage characteristics of the film in both the machine and cross-directions. Of the machine parameters, the neck-in influences the orientation. The material is oriented in the machine direction. The neck-in is the ratio of the constriction of the bubble to the diameter (BUR). The greater the constriction, the higher the BUR and the better the isotropy of the film [382].

Narrowing the die leads to a closer balance of transverse and machine direction tear. The advantages of narrower die gaps — that they provide a thinner melt cross section and allow for improvement of various film properties, gauge control, cooling and output — are well known. On the other hand, large die gap means thicker polymer layer, poor heat transfer, longer hot temperature time, better relaxation, and lower anisotropy [381]. Processing with a long neck (for HDPE) is associated with a high BUR with more freezing height. A calibration unit for better cooling, a large die gap, and a lower operating temperature for better bubble stability are required. A high melt index film with low temperature processing conditions blown with very high degree of orientation may give excellent stiffness [383].

Low-density polyethylene is considered as the easiest material to process into blown film [362]. LDPE is highly branched, gives much entanglement, possesses a broad molecular weight distribution, and exhibits reduced mechanical properties. The high degree of branching of LDPE promotes easy extrusion and bubble stability. The narrow MWD of LDPE reduces the ease of extrusion and bubble stability.

Special internal bubble cooling (IBC) systems are required to ensure a trouble-free start-up, in processing HDPE [384]. High-density polyethylene requires high blowup ratio and high frost line heights, that is, bimodal grades. The frost line height at which the melt is normally considered to be frozen is important. Lowering the frost line by faster cooling gives smaller crystallites (like chill roll cooling) and normally better optical properties, but does not allow relaxation before freezing [366]. Faster cooling gives less stiffness due to the smaller crystallites. HDPE has a long melt memory, so the die must be designed to allow time for any melt strain to disappear. The stiff film has a tendency to crease and wrinkle so that nip height, collapsing geometry, and tension are critical.

In the early 1970s linear low-density PE, with no long branches, little entanglement, a narrow MWD and improved mechanical properties was introduced [385]. There are fundamental rheological differences between LDPE and LLDPE. Linear low-density polyethylene is less viscous at low shear rates and more viscous at higher shear rates than LDPE. Extrusion equipment is quite different for different PE types. Straight grooved barrel extruders have been developed for HDPE. Helically grooved barrels developed originally for high molecular weight HDPE have been used for LLDPE also. Barrier screws may be used for different PE types [386]. Grooved barrels were developed in the 1970s. Such devices are not affected by the high back pressures produced by the die. With conventional screws, HDPE and LLDPE produce high pressures and high abrasion. To reduce abrasion by the material, the barrier screw was introduced, which has a homogenization section where the molten and solid-state materials are separated. Any solid is left in the plasticizing unit, because the channel has a decreasing volume. With LDPE and LLDPE a BUR of 1.5–2.5 is used, for HDPE higher BUR is needed. Films with thicknesses of 20–300 μm can be processed. For low friction or tacky film applications roller framers with aluminum or nonmetallic roll are needed [384].

The most important polymer variables to be taken into account by PE processing are the melt index, density, MWD, and long-chain branching [387]. Optimized operating conditions include polymer temperature, output rate, BUR/bubble stability, thickness, and frost line height.

b. Flat-Die (Cast) Procedure

This procedure, also called chill roll method includes one or more chill roll (cooling drums or rollers) units as well as chill roll water bath process. Polishing and smoothing rollers do not remove enough heat from the coating; therefore additional reverse side chilling of the web with water bath is required. Generally for the cast procedure the extrusion temperature is higher. For low-density polyethylene conventional blowing techniques operate at 160–220°C; cast techniques use melt temperatures of 200–260°C [388]. A temperature profile for cast PET film should be situated about 10°F higher than the blown profile [370]. The cast procedure allows better cooling of the molten material by using a cooling cylinder with controlled temperature (± 10°C) [389]. The exact control of the cylinder speed, the regulation of the rotation speed of the casting and cooling cylinder, give a better film quality. The finish of the film depends on the smoothness of the roll and its relative velocity. Films with a thickness of 0.01–0.3 mm can be manufactured (3000 mm width, 1000 kg/h) using the flat die procedure. The cooling determines the morphology of the film and the mechanical and other properties (e.g., transparency, gloss, tensile strength, internal tensions, and dimensional tolerances) [390]. Productivity also depends on cooling and product thickness. For instance, for a film with a thickness of 0.1 mm a cooling temperature of about 36°C is achieved with a running speed of 120 m/min. For a 0.5 mm film the speed decreases to 40 m/min; and for a 1.0 mm film the running speed is only about 20 m/min. For high speed lines one or more rolls are often used to increase heat removal capacity [391].

When casting film the melt is brought in to close contact with the chill roll by the use of an air knife, vacuum box or electrostatic pinning [366]. Film polished on both sides and less than 200 μm thick is difficult to manufacture by cast procedure; large roll gap forces are required [392]. The running (rotation tolerances) of the glossy cylinder are at least 5 μm. That is, a change of 10 μm in the clearance should be taken into account, which means a tolerance of at least 10 μm for a 100 μm film. Cast film can be oriented monoaxially; a 1/7 stretching ratio is suggested [393]. Oriented PP film (30–50 μm) and oriented LDPE film (80–110 μm) are recommended for insulation and packaging tapes [391].

c. Coextrusion

Generally the properties of plastic films can be improved by means of (1) coextrusion, (2) extrusion of polymer blends, and (3) orientation of the film. Downgauging and improving of the film tolerances are supplemental technical possibilities. Coextrusion consists of directly combining several plastics in the die, during extrusion. For semicompatible or almost compatible polymers (e.g., LDPE/LLDPE) the mechanical properties of coextruded films are close to those of films made from blends [395].

Coextrusion has economic and technical advantages. As shown by Hensen [380], the increase of the number of layers (e.g., splitting a layer) provides economic benefits. A reduction of up to 40% raw material cost is achieved by using this procedure. Coextrusion also allows some mechanical properties to be improved independently from the nature of the layer. Generally coextrusion increases the material stiffness. For instance, for HMHDPE a 20% improvement in stiffness has resulted from coextrusion [380]. Thus, the stiffeness/raw material costs ratio increases by about 25%.

In coextrusion at least two different plastics are plasticized separately and the melts are extruded through multiple-layered dies (slot or annular). The plastics are brought together in the die orifice itself or shortly after. Coextrusion casting and coextrusion coating are possible [394].

The main possible coextruded film constructions based on polyolefins and used for packaging films have been listed by Verse [395] as LDPE/LDPE, LDPE/EVAc, LDPE/HDPE, LDPE/HDPE/LDPE, LLDPE/HDPE/LLDPE, and LDPE/EVAc/PP. Combination of two LDPE layers offers the advantage of different colors and foamed/nonfoamed layers. The use of EVAc can impart special mechanical or adhesive characteristics. A symmetrical buildup of the layers ensures lay flat (no curl) and the buildup of a medium layer containing recyclate. In PE/PP coextruded films EVAc is the adhesion primer. Generally a coextruded film containing three different layers is curl-free. Four-component films may display curl if the components have different rheological/mechanical properties. Films produced with three-layer coextrusion uses HDPE as the outside layer, HDPE or PP as the middle layer and EMAc as the internal or heat seal layer [396].

The release layer can be coextruded also. For instance, polydiorgano-siloxanes are suggested as fillers in coextruded polypropylene carrier materials with tackified top layer. A dimethylpolysiloxane is preferred (0.3–2.0 wt.%) as additive. The release layer of such coextrudate is has a thickness of 0.5–10 μm. The total thickness of the tape carrier film is about 15–50 μm. Generally polyethylene embodied in a polypropylene carrier acts as release agent.

d. Blown Coextrusion

The high branching of LDPE promotes easy extrusion and bubble stability. The narrow molecular weight LLDPE is less stable than LDPE and tends to form gels if the material stagnates in the die. In coextrusion, residence time increases in the die especially in the outer layer. When tubular coextrusion is used, problems arise with lay flat and a small rolling up effect of the composite films due to the different behavior of the melts during drawing, the solidification ranges, and the different heat expansion coefficients. The typical multilayer film is built up with three layers referred to as the print skin, core layer, and inner skin; the core can contain recyclate, and may have a soft texture top surface with a paper-like back [397].

e. Cast Coextrusion

Melt fracture and interfacial stability are the main phenomena delimiting the possibilities of cast extrusion. Increasing layer thickness and decreasing the extrusion rate or increasing the die gap opening and decreasing side layer melt viscosity may improve the performance characteristics of the film [391]. Coextruded outer skin layers of HDPE are used to afford excellent chill roll release for LDPE at high speeds [320].

f. Comparison Blown Film/Cast Film

It can be concluded that the choice of a polymer and film grade for the manufacture of carrier material depends on many factors. It is evident that most extruders are specialized equipments with a barrel designed for a given PE type (LDPE, HDPE, or LLDPE). The use of another grade is in some cases possible, but problems concerning the productivity and quality of the product may arise. Assuming that the extruder allows the use of all raw materials, the next problem is the choice of production parameters. The first problem is the choice of a die in order to determine the final BUR ratio. As discussed earlier, large BUR allows balanced MD/CD properties, but lower productivity. The choice of the die gap is also complex. It influences residence time of the material, and thus, gels building up and haze may occur. The choice of the cooling place and rate are also important. In order to allow relaxation of the internal tensions for HDPE in the (HDPE, LLDPE) melt, for such materials long neck extrusion (with a delayed cooling of the film) is preferred. The higher sensibility of LLDPE related to shear and residence time is taken into account by the choice of composition of the outer layer. Narrow gaps generally result in better machine direction tear resistance, dawn gauging potential, and orientation balance in the film. However, these benefits are achieved at the cost of adding processing aids to the polymer in order to prevent melt fracture [385]. According to Feistkorn [398], the advantages of cast films are (1) high transparency, (2) high stiffness, (3) different (textured) surface qualities, and (4) adhesivity on one or both sides.

Thickness control is more difficult for blown film, than cast film, because of the lack of control during cooling. Control is better when the melt strength is higher, so using lower temperatures, lower melt index, and LDPE rather than LLDPE improve thickness control. As mentioned earlier, lower temperatures are recommended to avoid gel buildup also [366]. According to Djordjevic [391], the physical properties of a blown coextruded film are better than those of a cast coex film, but the optical properties of a cast coex film are better than those of a blown film. Blown films never look as brilliant as the corresponding flat film produced by slot-die process. This is due to the always present streaks caused by the annular die and the uneven surface appearance. To achieve optimum optical properties for a PE film, the frost line should be kept low [399]. Wrinkles may be caused by bubble collapse, due to high friction on the slats [385].

The minimum thickness of blown films is about 6 μm, that of cast films about 10 μm, an that of biaxially oriented films (depending on the raw material) about 2–5 μm [380]. The maximum winding speed is situated at 20–140 m/min for blown films, 120–400 m/min for cast films, and 280–350 m/min for biaxially oriented film.

Blown film equipment designed optimally for processing of one grade of polyethylene is generally not suitable for processing other types of polymers. The higher viscosity of LLDPE at typical working shear rates makes its processing more difficult [366]. The ratio of investations costs for blown vs. cast film equipment is 1:5–1:10 [400].

According to Ref. [401], water quenched blown film is competitive with cast film, with high quality, balanced orientation and deep draw thermo-formability. The high clarity achieved by the system is due to the higher heat transfer rates which limit the time the polymer is in the crystal-growth temperature range; the downward blowing water-quenched line is 2.5 times faster as upward blowing film.

4. Main Synthetic Film Materials

Various synthetic (polymer) materials are used as carrier for PSPs. Films are the most known carrier materials, but foams, fabrics, and plastic profiles are used also. According to Fust [402], in 1984 about 14% of label carrier materials were manufactured on plastic basis and in 1990 the proportion of plastics attained 40%. In 1996 the main film label carrier materials [403] were polyethylene (41%), polyvinyl chloride (30%), polyethylene terephtalate (13%), polypropylene (8%), and polyimides (4%). Reactivity, ecology, handling, and printability were the main evaluation criteria for their use [395]. According to Ref. [369], BOPP is the fastest growing face stock material with sheet and reel fed options, printable in almost any format. Most of synthetic carrier films are thermoplastic materials. Table 8.21 lists the main synthetic films used as carrier material for PSPs and their manufacturing technology.

a. Polyethylene

Polyethylene is produced by specialized firms using different process technologies (high pressure tubular process, high pressure autoclave process, gas phase process, slurry process, solution process, etc.) and different catalysts (oxygen, peroxides, heterogeneous, or homogeneous organometallic compounds). The properties of the polymers are different. Different grades of polyethylenes are used for films and as carrier material for coated or laminated webs. For pressure-sensitive products LDPE, medium density PE (MDPE), HDPE, and LLDPE have been suggested.

LDPE gives a tough, conformable film, with good thermal characteristics. It is the most common film used as carrier material for protective films, and meets about 75% of the market requirements. Medium density polyethylene is slightly stiffer than LDPE. It exhibits medium elongation characteristics.

HDPE is the stiffest polyethylene. It has more rigidity and better barrier properties than the other grades. It is manufactured by running the process at high BUR (more than 3:1) using a high stalk bubble profile. This is necessary to cope with the intrinsic tendency of HDPE films to develop excessive machine direction orientation leading to anisotropy in film (mechanical properties) [404]. HDPE blown films can exhibit very low machine direction tear strength. Such films

TABLE 8.21
The Main Plastic Films Used as Carrier Material for PSPs

Plastic Film	Manufacturing Procedure	PSP Application
LDPE	Extrusion, blowing	Protective film, label, tape
	Extrusion, chill roll	Label
	Oriented	Label, tape
	Coextruded with HDPE, PP, LLDPE	Label, tape, protective film
HDPE	Extrusion, blowing	Label, tape
VLDPE	Extrusion, blowing	Tape
	Extrusion, chill roll	Protective film
PP	Extrusion, blowing	—
	Extrusion, chill roll	Label
	Oriented, monoaxially	Label, tape
	Oriented, biaxially	Label, tape
	Coextruded with PE	Label, tape, protective film
LDPE–HDPE blends	Extrusion, blowing	Label, tape, protective film
	Extrusion, chill roll	Label, protective film
EVAc–PE blends	Extrusion, blowing	Label, tape, protective film
	Extrusion, chill roll	Label, tape, protective film
Polystyrene–PE blends	Extrusion, chill roll	Label
Polystyrene	Extrusion, nonoriented	Label
	Extrusion, oriented	Label
Polyester	Extrusion, oriented	Label, tape, protective film
PVC	Extrusion	Label, tape, protective film
	Extrusion, oriented	Tape, label, protective film
	Casting	Label
	Calendering	Label, tape
Polyamide	Extrusion, blowing	Label, tape
	Extrusion, chill roll	Label, tape
PUR	Extusion, blowing	Label, tape
	Extrusion, chill roll	Label, tape
Cellulose acetate	Extrusion	Label, tape

can be split very easily. The high stalk blowing of the bubble and large BUR ratios are intended to impart some biaxial orientation to the film and balance the mechanical properties. Transverse direction orientation is achieved by delaying the blowup until the melt has cooled somewhat. On the other hand, the delayed blowup of the bubble makes it unstable. To solve this problem special air rings, and mechanical bubble stabilizers (iris technology) are used. The best results are obtained with blown HDPE using high BURs, and high frost kine heights; to achieve these very broad, preferably bimodal grades are needed [366]. This can be realized by using faster cooling or nucleation with LDPE blends. Because film blocking is not a problem with HDPE, the bubble collapsing frame sits closer to the die than in LDPE or LLDPE extrusion. Another problem which arises from the high BUR is that the diameters are kept on the low side. A small diameter means low specific die rate (output per unit time per die circumference). A special PE having high melt flow index, low processing temperature, and high degree of orientation gives very stiff films [405]. A modulus of 400 N/mm can be achieved for this polymer at a density of 0.930 g/cm^3.

LLDPE exhibits better elasticity than other grades and a medium stiffness. Therefore for carrier materials used for protective films mixtures of LLDPE with LDPE are recommended. Their ratio affects the properties of the final product. LLDPE melt fracture must be avoided to obtain best optical properties. The higher melt temperatures needed for LLDPE are dangerous because gel building. Higher temperatures promote oxidation, and oxidation is a source for gel particles. The physical properties of LLDPE compared with LDPE may be listed as follows: higher tensile strength and elongation, better puncture resistance, superior toughness, better machine direction–cross-direction balance, and deep drawability. LLDPE has been introduced to the U.S. market in 1977 [406]. Copolymers of ethylene with butene (gas-phase polymerization), hexene or octene (liquid-phase polymerization) have been manufactured. Copolymers with hexene give LLDPE with a low MFI, that displays better tear resistance. Such films may be used as self-adhesive materials also [407].

Very low-density polyethylene (VLDPE) is defined as a linear LDPE with a density lower than $0.915 \, g/cm^3$ [383]. Common VLDPE has a density of $0.905–0.915 \, g/cm^3$. It may be oriented also. The orienting ratio used decreases with increases in the density (from $1/8$ to $1/6$). VLDPE displays excellent mechanical and optical characteristics (less than 2% haze). It is proposed for films (carrier for tapes) and tissues. VLDPE and ULDPE with densities below $0.900 \, g/cm^3$ are used where super high flexibility and auto-adhesion are required [366].

The main market segment in the European polyethylene film market are [408] food packaging, nonfood packaging, consumer, industrial, agricultural, hygienic, and medical, and special film markets. In the industrial films segment, which includes construction and protection films, antistatic films, surface protection films and adhesive films, labels, and sleeves are the most important applications. In the range of special films, lamination films, oriented films and tapes, foiled films and perforated films are the main applications. For tapes, mainly oriented films are recommended.

b. Polypropylene

Homopolymers with high crystallinity and block copolymers exhibiting melting range of 160–165°C have been prepared; the melting range of random copolymers is situated at 135–160°C [409]. The temperature resistance and elasticity of PP are higher than those of PE; therefore PP is used for higher temperature applications. It can be sterilized, but it becomes brittle at +5°C [376]. Polypropylene offers better clarity, better barrier properties, higher impact strength, and reduced memory effect. Polypropylene has a very low solvent retention value (from printing inks). It is more expensive than PE and because its built-in elasticity it is more difficult to handle after manufacture. Compared with PE, polypropylene is more grease- and chemical-resistance. Is difficult to compound polypropylene with PE. There are some patents that recommend the use of a compatibilizer, such as ethylene–propylene elastomer, to improve blend homogeneity.

Polyolefins can be used as nonoriented or mono- or biaxially oriented films. Oriented films are applied mainly for tapes. Polyvinyl chloride and BOPP have been the main plastic carrier materials for tapes [410]. Rigid calendered PVC has been used as mono- or biaxially oriented material, with a plain or embossed surface, and a thickness of 28–70 μm. BOPP has been applied as a blown or cast film, biaxially oriented, with a gauge of 25–38 μm. BOPP films are also now produced with a thickness of 40–50 μm. With the common stretching process thick BOPP films cannot be produced economically. For these purposes it has been suggested that several thin films be combined to form a multilayer film (via heat laminating). According to Ref. [411], the idea of putting a series of BOPP films one upon the other in order to achieve an increased film gauge is not a new one. In 1985 a patent was granted that covers the production of plates made of heat sealable BOPP films. The performance characteristics of thick BOPP films can be compared with those of PVC, amorphous PET, or cellulose acetate film. Because of its low density, polypropylene offers the advantage of increased yields per unit area. The yield per unit area with PVC is about 53% lower than for PP; for PET it is 47% lower, and for cellulose acetate 43% lower than for polypropylene [411]. Normally the extruded film is stretched to six times its initial length, and its thickness decreases. Such warm laminating is influenced by temperature, pressure, and time. With an increase in speed much air bubbles are entrapped; therefore more pressure (maximum 250 N/cm) is needed. A heated, rubber-coated press, and heated laminating steel cylinder are used.

As discussed in Ref. [412], oriented polypropylene film is thermofixed, that is, shrinkage (recovery of original dimensions) is avoided up to about 110°C. However, certain solvents (e.g., toluene) can cause "deblocking" of the shrinkage mechanism at lower temperatures, that is, the loss of dimensional stability. Polypropylene can be combined with nonolefinic films too. For instance, for a pressure-sensitive label having a wrinkle resistant lustrous, opaque facing layer for application to collapsible wall container (squeeze bottles), the carrier has thermoplastic core layer, having an upper and lower surface and voids, and a void-free thermoplastic skin layer fixed to the upper surface and optionally to the lower surface of the core layer, and discrete areas of PSA. As core the layer, a blend of isotactic polypropylene and polybutylene terephtalate is used, as the skin layer, polypropylene, and as the adhesive, circular dots of HMPSA [413]. With appropriate formulations the range of olefin-based raw materials can be used for the manufacture of carrier films for different PSPs (Table 8.22).

c. Polyvinyl Chloride

PVC is polymerized in emulsion or suspension. Emulsion PVC (EPVC) contains stabilizers (surfactants, protective colloids); therefore its electrical properties are less good, but its resistance against static electricity is better [414]. Hard PVC plates having a thickness of less than 1 mm are called films [415]. According to DIN 4102, PVC is nonflammable. Films of PVC have been manufactured via extrusion, calendering and casting.

During calendering the particles of EPVC (having a diameter of about 0.1 μm) are pressed together in a calender and later syntherized at higher temperatures [415]. Casting is based on precipitation of the polymer from a solvent. Because of the lower tensions induced in the film during manufacturing, cast and calendered PVC have better performance characteristics as carrier material for PSPs than extruded PVC.

Cast vinyl marking films are used for producing labels, emblems, stripes and decorative markings for trucks, automobiles, and other equipment. They are durable, conformable, and dimensionally stable, and withstand weather and handling conditions. They can be processed by screen printing, roll coating, steel rule die-cutting, thermal die-cutting, and premasking.

TABLE 8.22
DSC Phase Transformation Characteristics of Some Olefinic Raw Materials Used as Carrier for PSPs

Product	Material	ΔH (mJ)	ΔH (J/g)	Peak (°C)	Peak (mW)
Warm laminated protective film for plastic plates	LDPE, 50 μm	934	114.5	111.5	7.7
Warm laminated protective film for 0.3 mm coils, BA	C_8-LLDPE, 30 μm	679	165.6	127.7	7.9
Label carrier film	MDPE, LLDPE, 80 μm	751, 348	97.2, 45.1	117.9, 127.0	9.7
Deep drawable protective film	LDPE, PP, 70 μm	104, 443	17.3, 73.4	109.9, 154.7	0.9, 2.9
Hot laminated protective film	LDPE, 120 μm	930	111.8	112.3	10.9

Nonoriented and biaxilly-oriented, embossed, printable PVC (25 μm) has been used as carrier material for tapes [371,410]. Rigid calendered PVC has been applied as mono- or biaxially oriented material, with a plain or embossed surface, and a thickness of 28–70 μm [416]. Calendered PVC is used for plotter film also.

Formulation of PVC includes plasticizers, fillers, antiblocking agents, and lubricants. Micromolecular and polymeric plasticizers are used (e.g., saturated polyesters) [417]. Fillers for opacity and antiblocking agents are added [418]. The level of opacity additives (0.5–5.0%) depends on whether the PVC is hard or soft. For antiblocking purposes, only 0.5–1.0% of such an agent is added. Calcium behenate exhibits a strong lubricating action due to its long fatty acid chain. It is used as a main stabilizer lubricant in PVC field. For rigid PVC film, the raw materials may be either emulsion or suspension PVC. Suspension PVC is primarily used for small rolls, for office and household use, and for long rolls for automated packaging [410]. At the end of the 1980s suspension PVC made up less than 5% of the total market. Some years ago rigid PVC exhibited the advantage (in comparison with PP) that it could undergo direct dyeing during film production.

d. Polyethylene Terephtalate

Biaxially oriented PET for the packaging sector has good dimensional stability and thermal and mechanical resistance, as well as good transparency and luster. This film possesses an excellent tear resistance and good resistance against oil and chemicals [414]. It can be heat sterilised. A range of films are offered: transparent film with thicknesses of 15–50 μm, in standard, chemically treated, corona-treated (52 dyn), and opaque grades, white film having a thickness of 12–36 μm, and other colors [419]. PET has to be oriented or crystallized to obtain tough usable articles. Amorphous nonoriented PET is too brittle. Chemically toughened PET (copolymer) films have been developed also [370]. Elongation for the chemically toughened PET is higher than for oriented PET. Polyethylene can be coextruded with PET using a coextrudable adhesive. Thin PET film (2–20 μm) can be manufactured at a productivity rate of less than 720 kg/h [377]. Cast PET film having a thickness of 20–300 μm is manufactured with a productivity of less than 5500 kg/h [380]. Polyester is used as carrier for labels for cosmetics, toiletries, pharmaceutical, and chemical products, and for shrink sleeves [420,421]. It is applied as release liner for tapes also [78]. PET is available in both thermosetting and thermoplastic forms. It is unsuitable for heat sealing and thermal die-cutting. Oriented polyester can be thermoformed. Raw, top coated, and pretreated grades are available.

e. Polystyrene

Films of polystyrene can be used as carrier materials for PSPs. They can be either nonoriented or oriented. Nonoriented polystyrene films display clarity and excellent printability. Such films are notch-sensitive and susceptible to web breaks [422]. Unoriented polystyrene films have relatively little elongation, that is, little deformation during printing. Such films can be used between −40 and +150°F. Unfortunately they can buildup a static charge. Polystyrene possesses excellent printability, but low chemical and weathering resistance [416].

In the last decade, biaxially-oriented polystyrene (OPS) films have been developed. These films posses a thickness of 0.2–0.6 mm [422]. They are made from cast polystyrene film oriented in a two-step procedure. The orientation degree is 2.5–6.5. The advantages of OPS are the following:

1. They can be high gloss or matte film with excellent transparency.
2. They have good antistatic properties.
3. They have good rigidity even with low thickness.
4. They have low specific gravity (1.05 g/cm^3), i.e., good yield per kg.
5. They are nontoxic and nonhygroscopic.
6. They are stable at extreme temperatures (−60 to +70°C).

OPS is suitable as antistatic support film for transfer letters and symbols and adequate for printing by litho, letterpress, screen, and flexo printing. Polystyrene has been introduced as carrier for labels [423]. It complies with food industry regulations, including FDA and BGA requirements. No-label look labels have been manufactured from OPS for lubricant containers. Labeling of toiletries, cosmetics, as well as motoring, gardening, cleaning, and other consumer products use this material [424]. These labels can be recycled together with the labeled product. The availability of transparent and opaque white, matte or glossy films gives greater flexibility for graphics. New laser markable film grades of PS were developed for single or multilayered, blown or cast film. Butadiene–styrene copolymers for sheet extrusion replace PET [425].

f. Polyamide Films

Carrier materials based on polyamide can be used also. They display abrasion resistance, and resistance against oil [414]. Such films posses good printability. Polyamide 6 has been considered as the best of all polyamide grades. 6,6-Polyamide is also used, because its higher melting point and greater stiffness [426]. Blown and cast PA films are produced. For the manufacture of PA blown film, medium to high viscous polyamides are used. For cast film a low viscosity PA is processed. Amorphous polyamides are glass clear, can be mixed with copolyamides, PET and ethylene vinyl hydroxide (EVOH) and possess a good oxygen barrier [427].

g. Cellulose Acetate Films

Extruded transparent cellulose acetate film has been developed by Eastman in 1967 [428]. Cellulose acetate films exhibit reduced shrinkage (0.6% at 71°C, 24 h), and low triboelectricity and have good printability. They can be used as carrier material for self-adhesive tapes and tamper-evident products. For instance, a 50 μm cellulose acetate film has been modified to be a brittle film, with a low tear strength. The film has to be die-cut and skeleton stripped. For cellulose acetate films no pretreatment is required for normal printing. They can be used as overlaminate, seal or label, as clear white and computer imprintable white films. Such films are an alternative to PVC also [429]. Cellulose acetate has been introduced as carrier for labels [424].

h. Cellulose Hydrate

Cellulose hydrate (Zellglas, Cellophane) can be used as carrier material for PSPs also. Cellophane is a natural material with excellent dead-fold and antistatic properties and a wide range of colors but is not moisture proof [430]. There are different grades characterized by sealability (S), flexibility (F), lacquering (P, D, and X), moisture proof (M), etc. [431].

i. Ethylene–Vinyl Acetate Copolymers

EVAc copolymers with a density of 0.930 g/cm^3 have been suggested as a high gloss, clear non-adhesive carrier films [432]. The ethylene copolymers such as EVAc, ethylene-butyl acrylate (EBA), and EMA are miscible with LDPE [433]. The use of VAc copolymers for carrier films is enhanced by their flame-retardant properties also. Halogen-free flame-retardant noncorrosive materials (FRNC) are preferred for insulating of wires. Such materials have to resist 20–180 min. Flame-resistance increases with VAc content [433].

j. Polyurethanes

Linear polyurethanes can be processed as cast or blown films. Such films having a melting range of 75–170°C can be hot laminated without adhesive [434]. For instance, a PSA tape for protection of printed circuits, based on acrylics and phenolic terpene resin, with a 100–200 g/m^2 siliconized crepe paper, is made by coating the adhesive mass first on a siliconized polyurethane release liner and then transferring it under pressure on the final carrier material [184]. Breathable components for wound care include polyurethane films and foams [435].

k. Polycarbonate

Polycarbonate is supplied as films and sheets and is used for membrane switch graphics, decals, product identification, name plates, etc. It can transmit more light than PET and retains its transparency in thicker films and sheets.

l. Polyacrylate

Cast acrylate carrier is solvent- and temperature-resistant between -40 and $+180°C$, and resist to heat up to $+300°C$ for 60 sec; it is used for labels in electric and electronic domains [436].

m. Laminating

A quite different technology is the manufacture of film laminates (homogeneous or heterogeneous concerning their raw materials, component geometry or structure). The range of such products is very broad. Thirty years ago flexible packaging used over 600 different types of laminations [437]. In the recent decades their diversity and manufacturing methods have been strongly developed.

5. Manufacture of Foams

Generally, web-like foams or foamed films are manufactured by firms specialized in foam manufacturing or production of special films. Foams with good tensile strength and elongation are manufactured by mixing LDPE, blowing agents, additives, peroxides, other polyolefins, and extruding grinding, and foaming [438]. A technology for the production of expanded polypropylene sheets (EPP) with a density down to 0.6 g/cm^3 was also developed [425]. Iononomer foams having high tensile strength have been developed also [439]. Fire-resistant composites comprise plastic films (e.g., polyimide) and *in situ* foamed silicones [440], or polyamide-polyimide foams [441]. The most used foams are PUR based. A special industry has developed for their manufacture.

6. Formulation and Manufacture of Synthetic Carrier Films for Labels

The requirements for chemically and aesthetically improved labels have been forced the introduction of plastic films as face stock materials for labels. As discussed earlier plastic films used for labels have to display stiffness and dimensional stability. Such properties can be achieved through the choice of an appropriate raw material, the choice of an appropriate film processing technology or both. The choice of an appropriate raw material means, formulation on the basis of a high molecular weight polymer, that is processible as thermoplast, to give a balance of stiffness and flexibility and plasticity and elasticity, in order to allow cuttability (plasticity), labeling (flexibility and plasticity), and printing (dimensional stability).

   The choice of an adequate raw material is limited by the range of available polymers and the compatibility of the mixture components. The choice of the processing technology refers to the extrusion technology and postextrusion processing technology (orienting). A combination of such variants, that is, the use of a raw material compound (e.g., polystyrene/polyethylene), processing via multilayer coextrusion and orientation of the film allows the desired properties to be achieved. According to Hufendiek [168], the choice of the carrier for labels is influenced by several factors: the nature and surface of the substrate, the end-use environment (weathering and applications climate), mechanical requirements, printing methods, processing conditions, and special requirements (e.g., FDA, BGA approval).

The main carrier materials used for labels are [168,442]: PET, OPP, LDPE, HDPE, polystyrene, PC, PA, cellulose acetate, PVC, polyacrylate, and laminates. According to Hufendiek [168], the choice of a plastic carrier material for labels depends on the following parameters: internal or external use; postprintability and writabilty, temperature resistance, chemical resistance, weatherability, abrasions resistance, labeling ability, and stiffness.

It is known that stiffness influences labeling ability also. The choice of label materials is a function of the application field requirements and economical considerations too. Multilayer materials are being used with filled medium layers for economical reasons [443,444]. According to Ref. [445], the carrier materials for labels can be classified as monolayer (paper, films), coated monolayer (coated, printed, metallized paper and foils), and composite. The broad range of quite different carrier material grades required for labels having the same end-use is illustrated by wine bottles labels. Such labels use glossy, matte, antique, and aluminum laminated papers with permanent and washable adhesives [446,447].

a. Polyvinyl Chloride for Labels

Some years ago PVC was the most commonly used carrier material for labels. For special products (e.g., sheet labels for outdoor application) cast PVC has been used [448]. According to a forecast concerning the label production and usage to 2000, the annual growth of film carrier material consumption is estimated as 12–14% for PET, 13–17% for PE, 18–22% for OPP and minus 3–5% for PVC [449]. In reality the forecast for the decrease of PVC consumption was not confirmed in the last years because of the special application fields of this material and due to developments of special, phtalate-free films [450]. Soft PVC 80–120 μm is used for self-adhesive labels. Films with thickness lower than 60 μm are made from HPVC; and for large surface products calendered PVC is used [451].

b. Polyolefin Films for Labels

Polyolefin films are the most used plastic carrier materials for labels. They are used as the carrier for the face stock and for the release liner also. Some years ago nonoriented blown polyethylene films have been used only as label carrier material only. Later biaxially-oriented, cast, and coextruded PE films were developed. Cast LDPE, oriented LDPE, LDPE-polystyrene blends, LDPE–LDPE laminates, and coextrudates can also be used. The main disadvantage of LDPE as carrier material for labels is its limited stiffness. Different processing methods, conditions, and material combinations have been suggested to eliminate this disadvantage. For instance, cast PE used in the cosmetics and toiletries [452]. Polyethylene/polystyrene film (clay-coated for ink receptivity) is developed as carrier for labels too. From the range of polyethylenes, HDPE is the most adequate material for labels. Where a combination of high stiffness and puncture resistance is required, that is, for as a substitute for cardboard and paper (synthetic paper) bimodal HDPE is used [453]. Bimodal HDPEs are tailor made products for coextrusion and for the production of labels [454]. Crosslinking of PE improves its dimensional stability. A crosslinked PE exhibits an elongation of 175% at a tension of 20 N/cm^2 [455]. Cross-direction mono-oriented films display low cross direction shrinkage and are used for labels [456]. Polyethylene lamination films are usually produced by the blown film process [457].

The high stiffness required for labels can be achieved by orienting the PE film. Films made using a machine direction orientation exhibit outstanding stiffness in the machine direction and flexibility in the cross-direction Lay flat by printing is completed by on-pack wrinkle-free squeezability [87]. For most applications an isotropic material is necessary, therefore biaxially-oriented films should be used. A special product is manufactured that has different lamination angles of the oriented film. This is produced as 75–215 μm film [458,459]. Cross-laminating of two films in which the orientation of the molecules runs at an angle of 45° with the machine direction leads to outstanding physical and mechanical properties surpassing those of traditional materials

[459]. For instance, the ultimate tensile load of such a 50 g/m^2 film can attain 30 N/10 mm (DIN 53455), and its Elmendorf tear resistance is about 17 N (ASTM D 1922). Biaxially-oriented and cross-laminated films have also been manufactured.

According to Waeyenbergh [442], PP films for labels have to fulfill the following requirements: usability for transparent and opaque face stock, high mechanical strength (yield and stiffness), moisture and chemical resistance, good weathering characteristics, good printability (matrix printability also), good embossing and die-cutting characteristics, ability to be combined with other films, ability to be metallized and recyclability. Chemical resistance is related to the end-use of the labels and to the environment [445]. For application on containers, resistance to organic solvents, lubricants and oils, and to humidity, steam, and water are required.

Nonoriented, and mono- and bioriented thermo-fixed PP films are available. Oriented films give higher yield at lower gauges. For instance, a PP film having a thickness of 20 μm can be manufactured giving 56.2 m^2/kg film; a biaxially-oriented product having a thickness of only 12.5 μm can be used for a 89 m^2/kg film [460]. Coated pearlized, white opaque, and metallized PP films have been manufactured as carriers for labels [461]. Pearlized, white film may give a plastic-coated paper effect. Such (coextruded) films can incorporate white pigment in the core or voids in the common "cavitated" films. Transparent films having a thickness of 40 or 50 μm and opaque/white films with a gauge of 28, 35, 40, 50, and 60 μm have been suggested for labels. Coextruded white, metallized, nonlacquered, and acrylic lacquered-oriented polypropylene films as carrier for labels are manufactured with a thickness of 21–60 μm [462]. Polypropylene films provide good machine performances and when coated a glossy surface for high quality print. Polypropylene is used as carrier for labels for plastic bottles, containers, as matte, white, pretreated material [463].

Very stiff polypropylene (with a modulus of 2400 N/mm^2) and low modulus polypropylene (with a modulus of 100 N/mm^2) can be used as carrier materials for labels and tapes [464]. The stiffness of common materials used for tapes (e.g., OPS, PET, and PVC) is situated at 2500–3500 N/mm (E-modulus). Higher stiffness required for labels can be achieved with special PP homopolymers. Filled polypropylene or heterophasic copolymers exhibit a modulus of 3000 N/mm^2. Biaxially-oriented polystyrene replaced PVC, but has since been replaced by PET and PP [429]. BOPP and nonoriented transparent PP with a thickness of 60 μm are recommended for no-label look labels [463,464]. Biaxially-oriented PP films are used in cosmetics and other bottle labels [465]. Generally films are made with a thickness of 30–80 μm. Films having a thickness of 50 μm are suggested for high-speed labeling [465]. Such films possess a surface tension of about 52 dyn/cm. Because of their higher stiffness, lacquered films may have a lower thickness. For instance, a nonlacquered film with a thickness of 100 μm corresponds to a lacquered material with a thickness of 90 μm only. Top-coated superwhite opaque PP films with a thicknesses of 50 and 60 μm have been developed for labels. The top coating of such films can be printed by letterpress, UV silkscreen, litho, UV-flexo, and thermal transfer processes. Top-coated PE and PP films display the advantage of better printability and environmental resistance. About 15–25% of polyolefin films used in Europe as label carrier are top-coated. Such films have to be lacquered via rotogravure in order to achieve a high quality of the coating.

In order to improve the mechanical characteristics (stiffness and tensile strength) of the label carrier film, coextrudates can also be used. Such films are built up with a middle layer of LLDPE (e.g., with a thickness of 36 μm) and two external LDPE layers (e.g., with a thickness of 22 μm).

Label films have to display excellent cuttability. That is, the cutting forces in both direction MD/CD are the same (see also Chapter 10).

c. Polyester Films for Labels

Computer printable top-coated PET films are used for stiff labels such as name plates. For the production of high-quality labels metallized and print lacquered polyester films are suggested [466].

Due to its thermal resistance and dimension stability PET is the most important carrier material used in hot stamping and for holograms. As known, the hologram on a special photoresist is coated with a nickel layer, this is the master embossing tool. As a transfer film for hot stamping, 19–23 μm PET is used. This is a metallized film coated with a special lacquer that can be embossed. During hot stamping the transfer temperature is about 120–160°C. Diffraction films are made using the same procedure (they contain two-dimensional images only). Pressure-sensitive holograms can be laminated, cut, and overprinted. For pressure-sensitive holograms a 50 μm PET film has been proposed.

Common, commercially available polyester films are manufactured with a thickness of 12–100 μm [467]. Such film may be top-coated. According to Ref. [393], the bulky colored white film allows more application flexibility because white pigments may change the adhesive performances. Generally PET is chemically pretreated. Common PET exhibits a contact angle of 54° with water (in comparison with 66° for PP) [468]. For writability special coatings are used [469]. For transparent films a solution of precondensed melamine resins, cellulose ether or esther, containing a dispersed inorganic powder, has been patented as a writable coating [470]. Polyester laminated with polypropylene can be used in medical applications where hot steam (134°C) sterilization is applied [471].

d. Polystyrene Films for Labels

PE/PS polymer blends have been developed as recyclable carrier for labels, which complies with European food contact regulations. Such films display squeezability, printability, roll conformity, cuttability, and dispensability [453,472]. Polystyrene films were the first "no-label look" materials.

e. Laminates for Labels

Film/film and film/paper laminates are suggested as carrier material for labels. Humidity-resistant papers are made by PE coating (extrusion coating or laminating). Polyethylene (LDPE or HDPE) is coated on paper by extrusion coating (with a coating weight of 20–30 g/m^2). By laminating LDPE is applied on paper with a coating weight of 10–20 g/m^2 [465]. For extrusion coating a PE grade with a density of 0.915 to 0.925 g/cm^3 and MFI of 190/2.16 is recommended [473]. For special end uses, where transparency of a partially delaminated multilayer label is required, paper/film laminates with the PE film as medium layer are suggested [474]. A white opaque polypropylene film overlaminated with PET is used for labeling of soft drinks. It has a very low coefficient of friction to metal, which allows trouble-free high-speed application [475].

f. Other Face Stock Materials for Labels

Metallized paper and laminated metallic films are used for labels [476]. Since 1981 there has been a trend to replace aluminium/paper laminates with metallized papers. Such papers are less humidity-sensitive and their demetallizing offers new decoration possibilities [477]. Aluminum base layers are used for special laminates for security film technology [478].

Conformable and porous carrier materials are suggested for medical labels and tapes. Such materials include nonwoven fabric, woven fabric, and medium-to-low tensile modulus plastic films (PE, PVC, PUR, low modulus PET, and ethyl cellulose). For conformability the films should have a tensile modulus of less than about 400,000 psi (in accordance to ASTM D-638 and D-882) [479]. Preferred carrier materials are those that permit transpiration and perspiration or tissue or wound exudate to pass through them. They should have a moisture vapour transmission of at least 500 g/m^2 over 24 h at 38°C (according to ASTM E 96-80), with a humidity differential of at least about 1000 g/m^2. A conventional polyethylene terephtalate film has an approximate value of 50 g/m^2 of moisture vapour transmission. Cellulose acetate can be used as porous carrier material [480]. The coating technology of textiles and manufacture of impregnated carrier materials is described by Witke [481–483]. The manufacture process includes coating with dispersed or dissolved plastics and resins or with solid-state coating components.

Carrier manufacture for labels may include coating also. For certain label manufacturers the surface quality (gloss, adhesive anchorage) should be improved chemically. Therefore the film has to be coated with a lacquer. The top-coated film is converted into a pressure-sensitive laminate and later processed, printed on the lacquered side as a narrow web. For such a carrier a high coating quality is required. For instance, such quality is given by gravure printing using a line number of 44.

7. Formulation and Manufacture of Synthetic Carrier Films for Tapes

As discussed earlier, the first plastic films for tapes were introduced from the range of common packaging films. Their development occurred in parallel to the general development of raw materials, film manufacturing, and transformation methods. The development of other web-like polymer-based products (fabric, nonwoven, foam, etc.) influenced the development of the packaging industry and of the carrier materials also. Hard and soft PVC, cellulose hydrate, polyethylene, cellulose acetate, and polypropylene have been the most used carrier materials for tapes [484]. Generally oriented films have been applied.

The procedure used for postforming the carrier is more important for tapes than for labels. For instance, a carrier for tapes for low temperature application may have transverse cuts, or holes [485]. Easy tear, breakable tapes allow easy tear or split of the tape during handling. Easy tear breakable hank tapes are used in the same finished product as easy tear paper tapes. These tapes are applied by hand. They are used for hanking and spot tapes when fast, easy removal is required. They are highly filled, having a low elongation and easy break force. PVC/PVAc has been proposed also [486].

A large variety of homogeneous and composite materials are suggested as carriers for tapes depending on the end-use requirements. HPVC, SPVC, PP, PE, PET, PETP, polyimide, PET/nonwoven, PET/glass fiber, paper/PET/glass cloth, PE/EVAc foam, PE foam, PU foam, Al, paper, cloth, and nonwovens are used as carrier for tapes. For instance, electrically insulating tapes are based on PET, polyimide, PET/nonwoven, glass cloth, paper, PET/glass fiber, and other carriers. High temperature (300°C) resistant special labels and tapes for printed circuit board industry use polyimide films as carriers also [487]. Such polyimide materials have been recommended for transparent pressure-sensitive sheets coated with silicone-based adhesives. The pressure-sensitive laminate resists 8 h at 200°C [488].

High temperature nonsilicone-type adhesive systems based on high temperature resistant rubber on PTFE or polyester backing are used as aircraft PSAs. Such products are suggested for holding composite layup pieces in place during bonding [489]. Polyvinyl isobutyl ether and polyvinyl alkyl ether on cellophane are used for medical tapes [490]. Biaxially-oriented cast films are recommended as carriers for packaging tapes [491].

a. Polyolefins for Tapes

Narrow MWD HDPE is raw material for flat film for monofilaments and tapes [452]. Such products require excellent mechanical properties during and after stretching. Therefore a medium MWD polymer is used for blown film manufacture and also for tapes. Mechanical strength and elongation are required for these products. Therefore, the blown film extrusion of HDPE is carried out at high BURs to reduce the imbalance of the mechanical properties of the film. Transverse direction orientation is achieved by delaying the expansion of the bubble (special air rings and bubble stabilizers are necessary, collapsing frame is situated closer) until the melt has cooled somewhat. Polyethylene used for tapes is chemically pretreated [189]. Polyolefins are suggested for medical tapes also [492–494]. Using a special method a monoaxially-oriented PE film may be sealed (laminated under pressure, without adhesive) on another PE or PP film in order to obtain improved mechanical characteristics [387]. Oriented LDPE (80–110 μm) is proposed for insulation tapes, oriented PP (30–50 μm) is recommended for packaging tapes [393].

LLDPE and mixtures of LDPE/LLDPE are recommended for tapes also. Such formulations make use of the deformability of LLDPE and processibility of LDPE. The differences between the molecular structure of LLDPE and LDPE and the narrow MWD are responsible for the differences in rheological behavior during processing. LLDPE has a higher viscosity under shear. This viscosity is due to the friction of the polymer with the metal, and to shear on the surface of metal [396]. Extensibility and deformability are general requirements for certain mounting tapes also (see Chapter 11). The elasticity of LLDPE is used to avoid blocking during unrolling. Such tapes have a coextruded backing layer of LLDPE [495]. Ethylene acrylic copolymers are amorphous, low modulus compounds. Such products are recommended as blown or cast film carrier for special, elastic tapes [392].

Polypropylene can be synthesized that have quite different mechanical properties. As mentioned earlier, very stiff polypropylene (modulus 2400 N/mm^2) and low modulus polypropylene (100 N/mm^2) are suggested as carrier materials for labels and tapes. Flame-proof grades are manufactured also. Diaper closure tapes are used as refastenable closure systems for disposable diapers, incontinence garments, etc. [496]. These products include two- or three-tape systems. The two-tape systems include a release tape and a fastening tape. The fastening tape comprises a carrier material such as paper, polyester, or polypropylene. The preferred material is polypropylene (50–150 μm) with a finely embossed pattern on each side. Such tapes allow reliable closure and refastenability from an embossed, corona-treated polethylene surface used for the diaper cover sheet.

Coextrusion of PP is carried out in order to allow different colors, different surface properties (adhesion and sealability) and to improve the mechanical properties. Biaxially-oriented multilayer PP films for adhesive coating have been manufactured in order to improve their adhesion to the adhesive coating by mixing the PP with particular resins. The resins are added up to 25% by weight. Coextruded ethylene/propylene can be stretched for use as tape carrier also [497].

For tapes and protective films the backside of the self-adhesive (coated or uncoated) products can work with a supplemental release coating [498], or with an appropriate carrier material with low level of adhesivity (e.g., polyolefins) as an uncoated liner [499]. The abhesive agent can be coated on the backside of the carrier material as a separate layer or embedded in the carrier material. If coextruded structures are used, the total thickness of the support film is about 15–50 μm, whereas the thickness of the abhesive layer is about 1.0–5.0 μm. For instance, a polyorganodisiloxane, particularly a dimethylsiloxane with a well-defined viscosity, can be used as abhesive component. Such an abhesive substance is added to the layer in an amount from 0.2 to 3.0 wt.%. The second layer of a biaxially-oriented multilayer polypropylene film that faces the adhesive has a thickness of less than one third of the total thickness of the adhesive tape and contains the antiadhesive substance [500].

Masking tapes can have special carrier constructions to allow conformability. According to Lipson [501], a masking tape of polypropylene has a stiffened, wedge shaped, adhesiveless longitudinal section of PET extending from one edge, with an accordion-pleated structure, to conform to small radii. Some masking tapes have to be flexible, stretchable and contracting, capable of maintaining the contour and the curvature in the position in which they are applied [502]. They are suitable for protecting a surface on a substrate that has irregular contours. Such tape carrier materials can contain VAc copolymers. For instance, smooth, white crosslinked polyolefin foams having an average cell diameter of 0.50 mm, and a yellowness index of 20.4 have been prepared [503]. To produce a smooth carrier material, polymer compositions with high water absorbtion can be extruded also. Such a film contains a mixture of EVAc (80 parts), absorbent resin (10 parts) and calcium oxide (10 parts), and displays a water absorbtion of 50–100 g/g [504].

b. Polyvinyl Chloride for Tapes

Hard (rigid) and soft (plasticized) PVC have been suggested as carrier for tapes. Such films are the most used carrier materials for packaging tapes. About 80% of the tapes are packaging tapes. PVC does not need release and is not noisy [442]. Monoaxially- or biaxially-oriented films are applied with a thickness of 3–90 μm, transparent or clear, glossy or embossed [505]. Embossed films have a thickness of 28 μm in comparison with 40 μm for glossy films. In the last years the thickness of the PVC film has been reduced from 40 to 25 μm for packaging tapes; for thicknesses of less than 30 μm embossed film is suggested, which is used as mechanical release agent also.

Soft PVC may contain up to 50% plasticizer [189]. It is applied for electrical tapes and decor films [506]. Copolymers of vinyl chloride serve as clear tape for packaging and household. Polyvinyl chloride films with a thickness of 30 and 50 μm are carrier materials for packaging tapes [170]. Freeze tapes are based on primed soft PVC [507]. Such tapes for deep freeze application are manufactured with soft PVC carrier, which needs a primer coating for the rubber-based PSA that is used. Generally PVC used as carrier for tapes has been primed [415]. Seal tapes based on PVC carrier are printable. Their sandwich printing is possible also (flexo printing for the top side, and gravure printing for the backside between film and adhesive). Such tapes are used for closing [508]. Polyvinyl chloride with silicium carbide as filler is recommended for electrically conductive tapes [509]. PVC has been introduced as carrier for medical tapes too [510–514]. Cover tapes for bathroom are made on special PVC basis, with a bacteriostate (which inhibits bacterials growth) incorporated in the vinyl formulation. Such tapes are embossed to produce a secure nonslip surface. They are coated with a repositionable adhesive which builds up adhesion from 26.7 to 36.2 oz/in. [515]. For such tapes a 78 lb kraft release liner is used.

The development of PVC substitution by OPP as carrier material for tapes is illustrated by the following data. In 1979 about 80% of the tapes have been manufactured with PVC carrier material. In 1995 about 37% were PVC based only. Now PVC covers about 25% of the carrier materials for tapes. The polyvinyl chloride carrier material represented 60% from the manufacturing costs of a packaging tape [516]. According to Ref. [517], in 1989 the costs for the OPP carrier represented 43% of the global costs. The coated components (primer, release, adhesive) accounted for 27%.

c. Other Films for Tapes

Double-sided mounting tapes of PET having silicone adhesives on one side and a silicone release on the other side are used in electronics [518]. Cellulose acetate and hydrate have been introduced as carrier for clear tapes in household and office [519]. Electrically conductive adhesive tapes have been manufactured by coating an acrylic emulsion containing 3–20% nickel particles onto a rough flexible material (e.g., Ni foil) [520]. High temperature-resistant polyimid films have been proposed for insulating tapes. Such films are flame- and temperature-resistant between −269 and +400°C.

d. Fiber-Like Carrier Materials for Tapes

Fiber-like, fabric (textile) materials were the first carrier webs used for PSPs [519]. Actually such products are taken for applications with high mechanical requirements. Fiber-like materials (water resistant and water soluble) based on cellulose, combined with plastic films (soft and hard polyvinyl chloride, cellulose derivatives, polyester), foams (open pore polyurethane, closed pore polyvinyl chloride, closed pore polychloroprene, etc.) are suggested as carrier for tapes [107]. The paper and the nonwovens can be impregnated with water-soluble and water-insoluble resins; foam-like carriers are coated with adhesive on one or both sides. As separating solid-state components glossy or crepe papers and nonwovens coated or impregnated on one or both sides, are taken. Fiber-based plastic carrier materials, cellulose-based nonwoven, polyethylene-coated cloth, polyamide cloth, siliconized cloth are used for various tapes. A special two-layer nonwoven laminate is applied for tapes [521,522]. Combinations of materials such as chemically-bonded cellulose with cotton and synthetic fibers or thermally-bonded polypropylene (50%) with cellulose (50%) have been proposed for special tapes. Impregnation is a general method for strengthening the carrier for tapes [523].

Woven and nonwoven materials have been suggested as carrier for medical tapes [524,525]. Air- and moisture-permeable nonwovens (PET nonwoven, embossed nonwoven), or air-permeable tissue (breathable face stock) are used for medical tapes. The carrier has to meet air transmission criteria. For such tapes cotton cloth has also been proposed as carrier material [147]. A special conformable carrier material contains a layer of nonwoven web of randomly interfaced fibers bonded to each other by a rewettable binder dispersed throughout, and at least one additionally layer of the same composition. The fiber of the additional layer is laid directly on the first layer prior to adding the second layer [521]. Cotton cloth and hydrophobic special textile materials coated with low energy polyfluorocarbon resins are used as carrier materials for surgical tapes.

Reusable textile materials as Goretex, polyamide, polyethersulfone/cotton, 100% cotton, or other hydrophobized carrier materials can also be applied. Porous PE can also be used as carrier for medical tapes [526]. PVAc, PVC, PE, and copolymers are also recommended as adhesive for medical tapes. The adhesive is hot-pressed in the fibrous carrier. Such materials possess the advantage of preventing contact of small fibers and wound.

Electrical tapes have to possess high dielectric strength and good thermal dissipation properties. They are used for taping generator motor and coils and transformator applications, where the tape serves as overwrap, layer insulation or connection and lead-in tape [527]. The carrier is woven glass cloth impregnated with a high temperature-resistant polyester resin or a combination of impregnated woven polyester/glass cloth. This last variant is recommended where conformability is required. For insulating tapes resistance to delamination and tear resistance of impregnated fabric or nonwoven-based tapes are required. Heat-crosslinkable, pressure-sensitive insulating tapes are manufactured using an adhesive based on aromatic polyester coated on a textile carrier [528].

Mounting tapes have been manufactured with woven plastic carrier materials also. Such tapes can have a woven plastic backing [529]. As nonwoven material Tyvek can be used also. This is an HDPE fiber based carrier material [530,531]. It possesses the opacity of paper and the tear resistance of fabric. Generally there are two different types of nonwovens. Certain materials are manufactured with a special resin as adhesive, others (e.g., Du Pont's Reemay polyester or Monsanto's Cerex nylon) do not have an adhesive matrix. Thermoplastic acrylic films can be laminated with nonwovens also [532]. Binders and binding techniques for nonwovens are discussed in Ref. [533]. Filament-reinforced tape is used for fiber-optic cables, and is useful for binding optic cable components [534]. Such tapes have a carrier of PET and a PUR low adhesive component on the back [535]. Cloth-based tapes with good dimensional stability and easy tear property are manufactured laminating synthetic resin layer on one side of cloth with warp of 10–25 yarns and werft of 10–20 yarns, and coating adhesive on the other side [536].

e. Plastic Foams as Carrier for Tapes

Foamed polypropylene films can be used to replace expensive satin and acetate films for tapes [410]. These products are films having a relatively low foaming degree and the geometrics of common film carrier material. Common foam carrier materials are thicker, of lower density, softer, and more elastic. Rubber, neoprene, PUR, PP, PE, and soft PVC foam are applied for double-sided tapes [537]. Double-side-coated foam tapes are designed for industrial gaskets. They are based on closed cell PE foams, coated with a high tack medium shear adhesive [538]. Foam as carrier (PUR) alone or with film is proposed for medical tapes for fractures; the adhesive is sprayed on the web [539]. Laminated structures of PET foam are also used. Masking tape based on foam is discussed in Ref. [540]. According to Ref. [78], PUR and PE foams are the most used foam-like carrier materials for tapes. Polyethylene foams are aging resistant and can be applied up to 90°C. PUR foams resist temperatures up to 150°C, but are destroyed by UV light. Tensile strength, stiffness, density, flexural modulus, flammability, and impact resistance are the main characteristics of the foams used as carrier for tapes [541]. Generally, MDI-based PUR foams from the United States are softer and have a low specific gravity than European products [542]. Foams with high mechanical strength are manufactured from water-based polybutadiene dispersions [543].

f. Other Carrier Materials for Tapes

Film/film laminates, film/plastic laminates, film/paper laminates, and metallized film are applied as carrier materials for tapes also. Such materials are manufactured by specialized firms. Metallic foils are used for mounting tapes [544]. For instance, a special mounting tape for instrument housing assembly is designed like a tear type with transfer adhesive. The tape consists of a metallic foil strip of predetermined thickness and an acrylic transfer adhesive. Metallic foils are suggested for tamper evident products also. Aluminum film is proposed as street marking tape. Aluminum carrier and a tackified CSBR coated with 40 g/m^2 is used as an insulating tape in air conditioning [160]. The converting characteristics of aluminum films are described in detail in Ref. [545]. Paper containing asbestos fibers and impregnated with an elastomer has been suggested for double-sided tapes [546]. Silicone-impregnated paper can also be used as carrier material. For instance, a pressure-sensitive adhesive tape for protection of printed circuits, based on acrylics and phenolic terpene resin with a 100–200 g/m^2 siliconized crepe paper is made by coating the adhesive mass first on a siliconized polyurethane release liner and then transfering it under pressure on the final carrier material [176].

8. Formulation and Manufacture of Plastic Films for Protective Films

The formulation and manufacture of protective films differ according to their buildup, adhesive-coated or self-adhesive film. General quality requirements as lay flat, narrow thickness tolerances, no wrinkles and no gel particles are valid for both. As seen from the Table 8.23 the carrier materials for protective films have lower mechanical performance characteristics than the films used for labels or tapes.

a. Polyolefins as Nonadhesive Carrier for Protective Films

Polyethylene is the most used carrier material for protective films [107]. The first PE-based polyolefin films with UV protection (black color) for outdoor use were introduced into the

---

TABLE 8.23
Typical Mechanical Characteristics of the Main PSP Carrier Materials[a]

Property	PSP Carrier		
	Label	Tape	Protective Film
Tensile strength (N/mm^2)	22/19	220	18/11
Elongation by break (%)	350/420	25	450/650
Force at break (N/cm)	20/17	120	17/10
Elongation by maximum force (%)	280/300	25	300/550
Force by 5% elongation (N/cm)	19/17	80	11/11

[a]MD/CD values were measured for an 80 μm polyolefin film.

TABLE 8.24
Dependence of the Adhesion on the Mechanical Strength of the Carrier Film[a]

Tensile Strength F_{max} (N)	Peel Resistance on SST (N/10 mm)
8.0	No adhesion
10.0	0.20
12.0	0.43
14.0	0.65
16.0	0.80
18.0	0.95
20.0	1.10

[a]Polyolefin film.

market in 1986 [547]. Polyolefin carrier films for protective films are manufactured as blown or cast films. Monolayer or coextruded, nonoriented and oriented films have been developed. As raw materials PE and PP are generally used, but other compositions may be formulated also. Economic and environmental considerations forced the use of polyolefins as carrier material for protective films and downgauging of such films. It must be emphasized (as discussed in Chapter 7) that a minimum mechanical strength (expressed as maximum admitted deformability) is required for such films. Excessive deformation of the films degrades their adhesion. As seen from Table 8.24, tensile strength of less than 8–10 N do not provide enough adhesion for these films. This means that common monofilms cannot have a thickness of less than 40 µm and coex (polyolefin) film has a minimum thickness of about 25 mm.

Special protective films with improved deformability like deep drawing films and security (mirror tape) films need sophisticated formulations that give a balance of plastic deformability and dimensional stability (Table 8.25; see also Chapter 11).

b. Polyvinyl Chloride as Nonadhesive Carrier for Protective Films

Polyvinyl chloride was used as the first carrier material for common protective films [548]. Later it was successfully tested for special applications such as deep drawable. Problems related to its recycling and environmental impact have recently arisen, that forced its replacement with more acceptable polymers.

TABLE 8.25
Screening Formulations for the Carrier of Some Special Protective Films

Protective Film	Formulation Components (%)				
	LDPE	MDPE	LLDPE	EBA	EVAc
Deep drawable	20	—	80	—	—
	—	—	70	—	30
	—	—	90	—	10
Mirror tape	—	40	20	40 (3%)	—
	—	40	20	40 (8%)	—

c. Polyolefins as Adhesive Carrier Material for Protective Films

Stretch films were introduced for wrap-around packaging two decades ago. One-side adhesive, oriented (400%) thin films have been used [555]. The low level of crystallinity of VLDPE provides an intrinsic cling to extruded films that are used in coextruded structures for industrial wrapping stretch film. The stretch film is usually a multilayer structure in which at least one of the external layers is constituted by VLDPE or by a blend of ULDPE and LLDPE. For a standard VLDPE the difference between melting and softening range is about 60°C; for ULDPE it is about 80–90°C. This means that such films are very conformable at 20–40°C [382]. While the tackiness of the stretch wrapping film manufactured by the chill roll casting process can also be achieved through the use of special raw materials in conjunction with the very fast rate of cooling on the chill roll, in the case of blown film extrusion it is necessary either to use compounded raw materials, already incorporating the necessary additives, or to feed the additives into the system during extrusion.

Cast (chill roll) films are applied as self-adhesive protective films (without PSA coating). Such films have a thickness of 50–70 μm. These products have to adhere on extruded and cast plastic (PMMA, PC, etc.) plates. Very low-density polyethylene and common LDPE are used as raw materials. Generally LDPE films are corona-treated [549] (see Section II.B.10).

As known for cast plates the application is carried out by room temperature, for extruded plates by 60°C. In some cases the film has to support thermal forming where the protective film-coated plastic plate is heated (ca. 5 min at 180°C). Polyolefin films containing EVAc, EBAc, or PB have been manufactured and tested (see Section III.B.3). For blown films the incorporation of liquid polyisobutylene offers additional technical superiority over EVAc films, with slowly PIB migrating to the film surface (see also Chapter 5 and Chapter 11).

d. Polar Films as Self-Adhesive Protective Material

As discussed in Chapter 5, film-like peelable protective coatings have various, special compositions. For certain compounds peelability is wet debonding. For instance a N-vinylpyrrolidone-based protective formulation becomes emulsified in 10–20 min [550]. Resin dispersions have been proposed for removable protective coatings too [551]. Common formulation may contain methyl methacrylate–ethyl acrylate copolymer (1/1–1/5), styrene–butadiene latex, melamine and carbamide resins, plasticizer, anticorrosion protection additives, and pigments [552].

Removable protective coatings for metals have been manufactured from PVC also [553]. They are applied by spraying, bursting, or dipping, and are removed as film. Self-adhesive PVC coatings have been used as plastisol too [554,555]. Gelled PVC coatings have been used for protecting metals according to Michel [556]. In this way a soft PVC-like coating was manufactured.

Ethylene copolymers with polar monomers are used as self-adhesive carrier materials (see Section III). Ethylene-vinyl acetate copolymers were the first of these materials from this product range. The sealing temperature of an EVAc copolymer film with 12% VAc 130–150°C; an increase in the VAc content lowers it. Such raw materials are recommended for cold sealing also. Copolymers with 15–18% VAc have been suggested for packaging films to modify their seal properties. Commercial EVAc grades contain 10–40% VAc. It should be noted, that their properties are strongly influenced by the molecular weight and side chains of the polymer. Only high molecular weight EVAc polymers exhibit adequate adhesion to plastics [557]. A content of 32% VAc in EVAc leads to a partially crystalline polymer; at the 40% vinyl acetate level a completely amorphous polymer is achieved. Polymers with 15% VAc display PE-like properties. Polymers having 15–30% VAc posses PVC-like performance, whereas polymers with more than 30% VAc are elastomer-like. For self-adhesive protective films, both the mechanical and blocking properties play an important role. The elongation of VAc copolymers increases with the VAc concentration. The main increase is given by up to 15% VAc content. The optimum tensile strength is achieved for a content of 20–30% of VAc. The clarity of EVAc films increases with the VAc content; however many films with less than 15% VAc are manufactured because of their blocking. Blocking is reduced by slip and antiblocking agents and by cooling during manufacturing. In the manufacture of such films the suggested BUR is 4/1. For this ratio the MD/CD shrinkage values are about 50%. During manufacture tensionless film transfer and air flow parallel to the film are recommended. EVAc foams can be manufactured also.

Polymer blends having graft copolymerized PE with carboxylic groups have been produced for SAFs also [558]. For instance, a cold stretchable self-adhesive film is based on an ethylene-α-olefin copolymer (88–97 wt.%) and with polyisobutylene, atactic polypropylene, cis-polybutadiene, and bromobutyl rubber [559]. This polymer has a density of less than 0.940 g/cm^3 and exhibits an adhesive force of at least 65 g (ASTM 3354-74). Polyvinyl acetate films coextruded with LDPE have been recommended for three-layered SAFs that have a core layer of polyethylene mixed with polybutylene. For instance, such a film may have the following buildup: LDPE (9 μm)/ LDPE-polybutylene (37 μm)/EVAc (4 μm). Partially hydrolyzed PVAc (ester number 50–80) has been proposed as peelable protective film for rough surfaces [560]. EVAc with chlorinated PP, has been suggested as removable protective film [561]. Ethylene–ethyl acrylate copolymers have been proposed for removable clear protective films on metals [562].

9. Formulation and Manufacture of Plastic Films for Release Liners

A separate release liner is used for labels, tapes, bituminous building products, and a wide variety of industrial applications [563]. Plastic films can be used as separate release carrier material or as supplemental layer coated on a main carrier material (paper, nonwoven, or film) to ensure a smooth surface. As known from the manufacture of carrier for tapes, generally such a carrier functions as release also. It possesses release performances or it is transformed in an adhesive-repellent material by coating it with or embedding of release substances. The use of plastic films as carrier material for tapes allows their high-speed automathical application due to their flexibility and high mechanical strength. Like paper, plastic liners for tapes can be pleated. With excellent mechanical resistance, such a liner can be delaminated quickly [564]. The siliconized HDPE liner displays high stiffness and good die-cuttability. For PSA-coated fabric and nonwoven having a high degree of extensibility during processing, extensible film-based release materials should prevent wrinkle buildup.

Embossed plastic liners have also been developed. Such liners are used with or without supplemental siliconizing. Plastic-based release liners work as release due to their low adhesivity or reduced contact surface. Low adhesivity is given by the nature of the film raw material (nonpolar polymers) or by the supplemental release coating of the film. Reduced contact surface is ensured by a special, folded, embossed surface of the plastic film. Using such surfaces, polar plastics may function as an abhesive component also. As stated in Ref. [505], contoured, special PVC film displays only one tenth of its full surface as contact area. The shape of the embossing plays an important role also. Release liners with folded, contoured film are used for tapes which are applied on round items, that is, where conformability is very important. Embossed PVC film is recommended as liner for plasters and adhesive films.

The abhesive layer can be coated on the back side of the carrier material, embedded in the carrier material, or applied on a separate carrier material as liner [247]. If coextruded structures are used, the total thickness of the support film is about 15–50 μm, where the thickness of the abhesive layer is about 1.0–5.0 μm. Polyorganodisiloxanes, particularly a dimethylsiloxane with a well-defined viscosity can be applied as the abhesive component. The abhesive substance is added to the layer in an amount from 0.2 to 3.0%.

Paper as carrier material was discussed earlier. Its composition can also include plastics. Glassine, polyethylene-coated craft, and clay-coated craft papers are the most used carrier materials for release liner [563]. Paper board laminated with PE can be used as release liner for "stickies," to give a smooth surface for siliconizing.

According to Ref. [78], the main plastic films applied as release liner for tapes are polyester, polypropylene, polyethylene, and polystyrene. The polyolefins are used for 60% of the plastic release liners. From the range of plastic films HDPE, LDPE, PP, and PET are used as carrier material for release liner. Low-density polyethylene and HDPE are adequate mainly for double-faced tapes and for foam tapes [78]. High-density polyethylene possesses good tear resistance and stiffness. It is suitable for low-cost release liner for industrial applications, for example, bitumen films, tapes, and insulation panels [563]. LDPE offers good tear resistance also and is inexpensive. It is suggested for industrial applications and for graphic products, where high clarity is not a priority or coloured products are required.

Biaxially-oriented polypropylene can be used as release liner, being siliconized by thermal-, UV-, and EB-curable systems [565]. Oriented PP possesses high clarity, better gauge control, glossy surface, and improved stiffness [563]. It is suggested mostly for labels. For instance, a silicone-treated release liner of OPP having a thickness of 50 μm has been proposed for labels [442]. Sandwich printing is usual for plastic labels coated with hot-melts. Such products have a transparent polypropylene or PET carrier and an OPP release liner [566]. Oriented polypropylene is applied for closure tapes for diapers [78].

Polyester displays the same performance characteristics as OPP and better temperature resistance. PET was first introduced as plastic release liner. Later, after its use as face stock material, oriented blown PP films (with a maximum thickness of 75 μm) were tested. Double- and single-sided release can be produced with different release levels. The main advantages of polyester are its smoothness and transparency. As known, only plastic-based liners provide a really glossy no-label look of the adhesive layer [567]. Paper carrier gives textured surfaces due to the paper fibers. Generally plastic films as release liner display the following advantages:

1. They are transparent.
2. They can be produced in different colors.
3. They can be made in specific thickness.
4. They are dimensional stable.
5. They are sealable.
6. They are smooth.
7. They exhibit fiberless tear.
8. They exhibit no blocking after die-cutting.
9. They are recyclable.

The transparency of the films allows light-sensitive web control and regulation. Different colors allow the use of various release liner colors for various product groups. The ratio between stiffness and thickness (specific thickness) is higher for plastic films than for paper. This allows higher processing speeds. Plastic films do not change dimensionally when the atmospheric humidity changes. No elongation and wrinkles appear. No remoisturization is necessary after coating. For bag closure tapes sealability is one of the main requirements. Smooth, glossy release surfaces give smooth and glossy adhesive surface also. A glossy, barrier surface does not need smoothing primers before siliconizing. A clean, fiberless tear or cutting section is required for medical or electronic tapes because of contamination danger. Blocking after guillotine cutting of paper based PSPs is a disadvantage given partially by their compressibility and porosity of the cut section. Plastic films (polyolefins) can be recycled by extrusion. LLDPE causes less grit [568].

10. Postextrusion Modification of the Plastic Carrier Material

Stretching and orientation have different effects on the molecular structure of polymers. In comparison with a nonoriented film the stretched film displays better mechanical properties (tensile strength, stiffness), optical properties, barrier properties, and shrink properties [456].

Monoaxial- or biaxial-stretching is used. Cross-direction mono-oriented films exhibit low cross direction shrinkage and are recommended for labels [466]. Biaxially-oriented films are manufactured by bubble stretching or by cast film tenter frame process. Bubble stretching allows a good balance of the properties in the machine and cross-direction, and the cast film tenter frame process leads to certain anisotropy due to the two-step character of the procedure. The oriented films used are polyethylene, polypropylene, polyester, and cellulose acetate are used [569].

11. Surface Modification of Synthetic Carrier Films

Bond strength and bond permanence are greatly dependent upon the type of surface which is in contact with the adhesive. The main purpose of surface preparation is to ensure that adhesion develops in the joint between two substrates to the extent, that the weakest link is in the adhesive itself and not at the interface with the adherent (evidently, for removable PSA the weakest link apparently has to develop between the adhesive and the substrate). The anchorage of a coating on the carrier surface depends on its polarity and geometry. Its polarity affects its antistatic characteristics also. The modification of the surface polarity is of special importance for certain self-adhesive plastic films where functionalized surface has to ensure instantaneous bonding.

a. Modification of the Surface Polarity

Surface polarity influences the adhesion of the coated layers on the carrier material. Paper, certain metals and plastics are sufficiently polar to bond with PSA. Other plastics like nonpolar polyolefins, special papers or modified metal surfaces do not possess the required bonding affinity. For instance, a common acrylic PSA exhibits on untreated PE surface a peel force of 4 N/cm; on polypropylene its adhesion is only of 0.45 N/cm [549]. For adhesiveless PSPs increased surface polarity provides self-adhesion in the first phase of adhesion buildup. In special case of cover films, that is, of semiself-adhesive films which adhere to the protected surface without pressure and without following buildup of the adhesion, the increased polarity of the surface provides the only adhesiveness of the product [570]. The increase of the surface polarity may have negative side-effects also. As known, such polarity increase improves the surface tension. The unwinding noise of tapes depends on their surface tension.

The surface of nonpolar carrier materials has to be modified. A broad range of methods has been developed to carry out the modification. Coating and treatment methods compete. Flame, chemical or physical/electrical treatment of web surface have been proposed. As known, in some cases primers are used to improve the adhesive anchorage. Bulk additives (or additives in a separate coextruded layer) can work as adhesion improving agents (tackifiers) also. For instance, biaxially-oriented multilayer PP films for adhesive coating have been modified to improve their adhesion to the adhesive coating by mixing the PP with particular resins. For such films the preferred tackifiers (15−25 wt.%) are nonhydrogenated styrene polymer, methylstyrene copolymer, pentadiene polymers, α-pinene or β-pinene polymers, rosin or rosin derivatives, terpene resins, α-methylstyrene−vinyl-toluene copolymers. To improve their adhesion even further corona discharge treatment has been suggested.

The range of materials which can be physically treated is very broad. Generally, PE and PP carrier materials have to be treated before use. PET, PA, Zellglas, and aluminium may be treated in some cases too. Polyethylene has to be pretreated or precoated with a primer [368]. Sources to modify the polyethylene surface include flame, laser, UV radiation as well as discharge form electrical corona or plasma. These physical methods produce a combination of chemical and morphological modifications including crosslinking, oxidation, grafting of active/polar groups, chain scission, ablation, and roughening at the polymer surface.

Corona discharge is the most common method for surface treatment of carrier materials. According to Prinz [571], primers used for UV-printing of plastics can be replaced by corona treatment, and transfer-coated PSA can be corona treated also [572]. The final carrier material should also be treated and both treated materials should be laminated together [573]. Special treaters have been developed for narrow web printing [574].

The energy level needed for treating of various plastic films varies in the order [575]:

$$OPP > PVC \gg HDPE > LDPE > PS > PET \qquad (8.4)$$

In the range of polypropylene films, the required dosage of treatment energy increases as follows [505]:

$$CPP_h < CPP_c < MOPP_c \ll BOPP_c < C_aPP_c < BOPP_h \qquad (8.5)$$

Cast polypropylene homopolymer (CPP_h) needs less treating energy than cast copolymer (CPP_c), mono-oriented copolymer ($MOPP_c$), bioriented copolymer ($BOPP_c$), calendered copolymer film (C_aPP_c), or bioriented homopolymer ($BOPP_h$). Corona treatment is recommended for PP, PS, LDPE, LLDPE, HDPE, EVA, EVOH, PA [301], and paper [302].

Corona treatment generates polar sites, increasing the free energy of the surface and bonding [576]. Depolymerization, oxydation, and crosslinking of the treated surface are the main effects of such treatment [577]. Nitrogen can also be fixed on the surface of corona-treated polyolefins [549]. According to Wilson et al. [303], a treatment energy of $500 \, J/m^2$, brings about $2-3 \times 10^{14}$ CO groups are formed on the polymer surface. These functional groups may react with the postcoated layer. For instance, the anchorage of acrylates crosslinked with trifunctional polyisocyanate is improved by the reaction with the OH groups of the corona-treated surface [578].

The effect of corona treatment is a function of the film nature, age, slip, and length of storage. Such effects are more pronounced for a film treated during manufacture [579]. As stated in Ref. [580], the main test methods of the effectivity of the corona treatment are based on measurement of the contact angle or of the adhesion. The decrease of the polarity given by corona surface treatment (measured as contact angle) depends on the film nature and age. It is about $3-5 \, mN/m$ after the first 2 days for an aged film and $1-3 \, mN/m$ for a "fresh" film. It is also abrasion-sensitive [580]. Ref. [581] discusses the interdependence between the effectivity of the corona treatment, storage time, and concentration of the slip agents. As stated, the shelf life of the corona treatment for PP is longer than that of similarly treated PE [366]. Orientation also influences the corona treatment. The treatment of OPP and BOPP is more difficult requiring high dosage. The energy level needed for the corona treatment of BOPP may be 10 times as high as that needed for PE [376,582]. Generally, corona treatment of PP during film manufacture (before migration of slip) gives better results [376]. Corona discharge activation of elastomer surfaces leads to the new polar surface having a very short shelf life. Reconstruction of the elastomer surface may occur within some minutes after treatment (e.g., 15 min for ethylene propylene or styrene–butadiene copolymers) [583]. Electrically conductive printing inks make corona treatment difficult [584].

To improve the self-adhesion of polyolefin films their functionalization and high pressure/temperature laminating has been developed. Functionalization means corona treatment. High pressure and temperature lamination means lamination in a laminating machine. Such effects are used for other technologies also. As shown in Ref. [585], a polyolefin laminate can be manufactured by laminating together of two corona-treated surfaces at a temperature which is lower than the softening temperature of the films. High temperature and pressure ensure better conformability and contact. Full surface contact and full surface treatment are absolutely necessary. Therefore by corona treatment of very thin films of PE and PP requires the use of a pressure cylinder to eliminate air bubbles [586,587].

Corona treatment requires working with high voltage ($10-25 \, kV$) high frequency ($30-50 \, Hz$) electrical fields [588]. Values of $12 \, kV$ [581], $12-20 \, kV$ [549], and frequencies $1-3.8 \, MHz$ [589],

20 kHz [549], 15–25 kHz [590,591], and 20–40 kHz [592] were also tested. For polypropylene treating energies of 0.15–40 J/cm^2 have been used [549]. According to Prinz [592,593], a greater frequency improves the corona treatment effect. Blocking increases at low frequencies. The so-called streamers (discharge channels) appear above a frequency of 10 kHz.

As stated by Markgraf [594], during the last years additive levels have steadily increased. High slip polyethylene contains 2000 ppm slip now in comparison with 800 ppm for a decade ago. Therefore corona treatment of such films became difficult. Corona treatment can be improved using a hot air jet to blow up the slip agents from the film surface [595]. Slip additives can be volatilized by using high temperature air jet, and better treatment stability is achieved. Indirect corona treatment without a counter electrode has been developed also. With this system it is possible to treat flat materials of any thickness, and the working distance can be increased up to 20 mm [596]. Electrically inhomogeneous materials (containing metals) and contoure items can be treated with "spray corona" systems which use a gaseous agent [597]. According to Gerstenberg [596], the shelf life of PP films treated by indirect corona is longer than that of films treated with direct corona.

Another new method of corona treatment that prevents the fast fading of the surface energy was developed in the 1970s [598]. This process is based on the deposition of an ultrathin chemically active layer, a so-called "corona deposition." Corona deposition is a process that deposits an ultrathin SiO$_x$ layer made from silanes and oxidizing agents. Therefore, the corona discharge is carried out in a controlled (nitrogen) atmosphere with controlled amounts of silanes and oxidizing agents. The surface tensions achieved with this procedure are higher (50–52) than the 40–43 dyn/cm achieved by common corona treatment. This procedure of corona discharge in a gas atmosphere started the development of corona discharge combined with aerosol deposition; the equipments have a power output up to 30 kW, and use a frequency of 2–40 Hz [599,600]. The aerosol deposition can be carried out simultaneously to the corona discharge or after the corona treatment. In the simultaneous system the aerosol prepared in a generator, having particles with a diameter of about 1 μ, is introduced in the chamber of electrodes. After a storage of 10 weeks the aerosol-corona treated film (PP) has a surface tension of about 55 dyn/cm compared with 48 dyn/cm for the film treated with common corona discharge.

Polyethylene for tapes must be chemically pretreated (1892). Etching produces a hydrophylic surface and creates sites for adhesion. Most commercial etching compositions are solutions of chromic acid or chromic acid plus sulfuric acid. Potassium bichromate, sodium hypochlorite, potassium permanganate, zinc chloride, and aluminum, barium, and magnesium chlorides can also be used. For PE different chemical treatment methods have been tested [601]. For instance, treatment with water-based potassium bichromate and sulfuric acid solutions, sodium hypochlorite, potassium permanganate, etc. can be used. According to a patent [602], solutions of metal salts (e.g., Zn, Ba, Al, Sn, and Mg chloride) improve the anchorage on chemically oxidized PE surface. For tapes potassium bichromate, sulfuric acid and zinc chloride treatments have been tested [603]. Polyethylene terephtalate can be chemically treated with acid-potassium bichromate mixtures (5–30 s at 75–80°C) or with sodium hydroxide and acid/potassium bichromate [604]. Recently a new procedure based on sodium hypochlorite and acetic anhydride was developed [605].

Flame treatment is preferred for packaging tapes [606,607]. Flame treatment can offer advantages, where moisture should be driven out [608]. The temperature of the flame is about 1200°C [609]. According to Cada and Peremsky [610], chemical and flame treatments give better adhesion values. According to Dorn and Bischoff [611], corona treatment ensures better (PP/steel) bond strength than flame treatment. For extrusion coating the best adhesion values have been obtained by using a primer [612].

Poly(ethylene-vinyl acetate) foam can be treated with fluorine [613]. For instance, a web-like foam material used as carrier for double-sides tapes with a thickness of 2 mm has been pretreated with a running speed of 5 m/min. In this case the effect of improving of the surface tension is due to the fluorination of the end methylene groups of vinyl acetate [588]. Unlike PE foam, the poly(ethylene-vinyl acetate) foam loses its treatment rapidly (5 days). The method allows the treatment of complex profiles and forms (e.g., bubble packs to be coated with PSA, or EPDM profiles) and of foams more than 2 mm thick. This method does not cause pinholes. It is used for manufacture of barrier layers also [614]. First chlorine was applied as a gaseous reactive together with UV radiation [588]; later a mixture of fluorine and nitrogen was tested. Recently, the use of sulfonation (treatment with an SO_3 atmosphere) was studied.

In corona treatment air contains 100 ppm ozone [615]. Most rigid pollution control regulations call for the elimination of ozone. Because ozone forms during corona treatment, ozone-eliminating systems are required. On the other hand, ozonization is also used for surface pretreatment too. In this case, the ozone concentration attains 12,500 ppm. A combined treatment with ozone and UV light has also been tested [588]. The use of benzophenone with UV radiation helps to remove a hydrogen atom from the polymer and generates radicals which can change the surface polarity [616]. The range of increasing difficulty in removing hydrogen is given as

$$SBR > Kraton \gg EPDM > PP > PE$$

Plasma treatment can be used too. As shown in Ref. [617], a $4-12$-fold improvement of the adhesion on plastic parts was attained (depending on the types of adhesive and plastic) by oxygen and ammoniacal plasma treatment. Sheets or rolls can be plasma treated [618]. The advantages of the procedure are that there are no pinholes, no side effects, low energy, and no static charges. The charge is maintained for a longer time. There are many variants of plasma used for treatment [619]: chemically inert plasma, reactive nonpolymerisable plasma, reactive graft polymerization plasma, and reactive polymerization plasma.

Oxygen plasma treatment allows higher adhesion for medium molecular weight PE than for HMLDPE [620]. The effect of oxygen plasma is twofold. Plasma treatment can be applied as pretreatment for better adhesion, degreasing, roughening, activating, and priming. Active surface groups are generated, that allow good adhesion properties and reduction of the defects gives smoother polymer surface [621]. Plasma treatment affects deeper molecular layers than corona treatment and produces crosslinking [622]. Plasma treatments are subatmospheric processes, working at low pressure (60–150 Pa), between room temperature and 250°C [623]; 0.1–1.5 mbar, 2.45 GHz [624]. According to Ref. [625], the use of oxygen leads to a pronounced decrease of the contact angle with water. For improvement of the surface affinity and adhesive anchorage chemically reactive plasma can be used. If a reactive nonpolymerization plasma treatment is applied (fluorine, oxygen, or nitrogen) multiple effects appear. Low molecular substances from the polymer surface are eliminated and polar groups are formed. According to Dorn and Wahono [597], a polypropylene surface having a contact angle of 90–92° with water before treatment, displays contact angles of 63–85° after plasma treatment, in comparison with corona treatment, where contact angle values of 60–62° have been obtained. According to these authors [597], PP can be plasma treated using oxidative gaseous mixtures only, polyethylene can be plasma treated using nonoxidative gases also. Dorn and Bischoff [611] stated that by treating of polypropylene with oxygen plasma by 1 HPa and 27.12 MHz, the same dependence of bonding strength on the time has been observed as by oxidative chemical treatment with chromesulfuric acid.

All the pretreatments induce various functional groups into the surface and cause molecular modification to varying depths. The pretreatments have also been found to modify the surface topography of the PP in some cases by roughening the surface and causing the creation of new features, such as created by flame. Green et al. [626] stated with surface analysis of pretreated PP, based on functional groups and O:C ratio (using various treatment methods, e.g., corona, flame, fluorination, vacplasma, and airplasma) that vacuum plasma induced oxygen into the surface in higher concentrations and closer to the surface than that of nitrogen. According to

Kaute and Buske [627], up to 30% oxygen has been measured after treatment compared to 10% with corona treatment.

The formation of double bonds, crosslinking and redox reactions are initiated by plasma. Dorn and Bischoff [611], admit the modification of polypropylene tacticity also. The superficial weak boundary layer is eliminated. This was observed by Rasche [628], for polypropylene too, but it has not been found for polyethylene. Plasma treatment with graft polymerization leads to increase of surface adhesion also. First, an inert plasma treatment is carried out. Then, the activated surface which has long-living radicals is reacted with polymerizable monomers. This procedure is really a plasma treatment followed by coating at low temperatures to modify the polymer surface. Plasma treatment has been used to simultaneously improve the wetting characteristics of PET and reduce static charge accumulation via triboelectricity [629]. For instance, the wetting angle with water has been reduced from $75°$ to $60-35°$, by using acrylic acid as reagent, which is coated with a thickness of 15 to 20 Å. Unfortunately, although the electrical conductivity has also been improved by an order of magnitude, it does not fulfill the practical requirements. Actually plasma treatment for film is mainly a batch process [630]. Plasma-treated surfaces lose their surface treatment also. However, according to Ref. [622], the storageability of the treatment effect is better for the plasma process. For instance, a loss of $15-20\%$ of the treatment effect has been found in about a month [631].

Surface oxidation by UV or ozone has an overall effect similar to that of certain plasma and flame treatments. Treatment with benzophenone under UV light produces crosslinking of the surface without increasing its hydrophylicity [632].

Perforation using corona discharge is possible too. Special electrodes allow the manufacture of holes of $2-5$ μm [633] (see also Chapter 10). Unlike glow discharge treatments, this process could be carried out under ambient conditions as a continuous method [634].

Different treatment methods give different surface tension values for various plastics and a different rate of decay of these values during storage. As stated by Armbruster and Osterhold [635], sulfonation provides the highest surface tension values for PP/EPDM, but surface recovery is the highest also. Plasma treatment leads to stable surface tension values. Direct (on-line) measurement of surface tension (adhesion improvement) during corona treatment has been developed. The method is based on the continuous evaluation of the friction between the treated web and a special transducer surface [636].

b. Modification of Antistatic Properties

Static electricity accumulates during web processing and can achieve a level of 40,000 V [637]. Agents improving the electrical and antistatic properties of plastic films can be added to the formulated raw materials during film manufacture or they can be coated on the film. Electrically conductive PE (HDPE) is used for extrusion of tubes, and for film [638]. Antistatic agents like polyethoxy-tiophene can be coated on the film surface [639] (see also Chapter 5). The electrical performances of plastic films can be modified by using radiation-induced crosslinking also. For instance, a polyolefin film having an electrical resistance of 1600×10^{-2} Ω has been manufactured according to [640] via radiation crosslinking. Conductive fillers improve the electrical conductivity and antistatic characteristics. The most used filler to achieve enhanced electrical conductivity is carbon black. Nentwig [641] suggests a level of $15-45\%$ carbon black as filler. There is a logarithmical dependence between electrical conductivity and carbon black concentration. From the technological point of view triboelectricity can be reduced by avoiding separation operations during web transport and processing [642]. Rotary machine parts, forced transport, and use of ionizing bars are the suggested methods. Passive and active charge eliminating devices have also been developed. Ion spray bars used to eliminate static electricity during converting of web-like products work with a tension of $4.5-8$ kV [643]. Web-cleaning devices use electrically charged air jets and vacuum [644]. The aerosol-corona treatment described by Rau [600], allows the treatment of the web with antistatic agents. Thus for a PE film surface resistance of 10^9 Ω can be achieved.

c. Modification of the Blocking Properties

Blocking occurs in both sheet-like and reel-form products. To avoid the bonding of adjacent (contacting) plastic surfaces, antiblocking agents are used (see Chapter 5). Blocking causes tapes or solid-state tape components to resist unwinding. Tapes with an olefinic carrier material are noisy. If the unwinding is carried out at high speed, as in the operation of cutting the spool for rolls with semiautomatic and automatic machines, high noise level is encountered. It is supposed, that the noise level is related to the mechanical properties and surface characteristics of the carrier material, to the adhesive/abhesive properties, and to unrolling conditions. Modification of the adhesive by addition of a mineral oil causes a substantial reduction of the noise level, according to Galli [266]. Further improvement can be achieved by using a release coat or special treatment of the carrier back. Polyvinyl carbamate, polyvinyl behenate, etc. can be used as the release layer (see Section II.D). It should be mentioned that in some cases blocking and sealability are required properties for a carrier material. For special applications blocking may be required. For instance, the use of heat sealable face stock, allows the manufacture of single material type pocket labels [442].

d. Modification of the Surface Geometry

Embossed carrier materials are used for diaper tapes, medical tapes, wall coves, decor films, etc. Embossing can be carried out at either room temperature or high temperature. For instance, PVC wall coverings have been embossed by high temperature (120–140°C) embossing [645] (see also Chapter 10).

C. OTHER CARRIER MATERIALS

Various other materials are used as carriers for PSPs. Web-like, irregular surface items or discontinuous items having various shapes and a complex buildup have also been applied as carrier too. Cotton fabric (cloth) [646,647], glass fiber-reinforced materials [648], and other materials have been developed as carrier. Special materials have been suggested for medical, insulating, and mounting tapes (see Chapter 9 and Chapter 11).

D. MANUFACTURE OF THE RELEASE LINER

1. General Considerations

In principle, an adhesive-coated PSP is built up from a carrier material and an adhesive layer. In some cases another carrier material with an adhesive layer is also required. This component is the release liner. The manufacture of the release liner includes the production of the carrier material and its coating with an adhesive (release) layer. Certain plastic films do not exhibit chemical affinity or contain antiblocking agents (see Chapter 5). Therefore in such cases the uncoated (nonpolar) plastic films can act as release liner. According to Uffner and Weitz [649], antiblocking agents are materials added to an adhesive (or carrier) formulation to prevent the adhesive coating made therefrom from adhering to its backing, when the adhesive coated carrier material is rolled or stacked at ambient or elevated temperatures and relative humidities. (Blocking can be caused by heat or moisture, which may activate latent tack properties of the adhesive composition.) Certain technological additives (e.g., surface active agents in PVC, or fillers) can work as embedded release agents also.

In other cases the abhesive properties of the liner are given by a special coating. As discussed earlier (see Section II), various release formulations are used. Silicones are the most versatile compounds as release. Different silicone systems, 100% solids, solventless and water-based silicones, (low-temperature) radiation curing and hot air curing options are known [261]. Solvent-free silicones have been introduced on the market since the 1960s. Such materials have been coated using offset gravure [650]. Low-temperature cure solventless systems are two- to three-component formulations coated with a coating weight of about 1 g/m^2. Their cure profile shows that at 110°C (where normal systems require several hours of curing) such materials need a curing time of only 15 sec. At 66°C their cure time is less than 1 sec [651]. In comparison, silanol terminated polydimethysiloxane-based silicone release coating (viscosity 80 cP) applied with a coating weight of 0.3 lb/ream, with a gravure roll and rubber transfer roll, is cured 7 sec at 202°F [258]. Because of their high curing rate UV-curing systems allow a 20% increase of the productivity [652]. Such siliconizing lines may run at rates of 1000 m/min. Coating weights of 1.0–2.0 g/m^2 for paper and 0.5–1.0 g/m^2 for film are applied. On the other hand, UV curing needs initiators, and such additives may influence the quality of the product [653]. The subsequent adhesivity of PSA on UV- or EB-cured release papers is about 95%.

Diorganopolysiloxanes have been suggested as water-based release dispersions [276]. Such release coatings have to adhere to polyethylene-coated paper and to display abrasion resistance. The forecast of Fries [189] concerning the forced use of emulsion-based siliconizing procedures, has not been meet by actual technical development. In 1990 about 40% of the release liner production used solvent-based silicones, 40% solventless silicones, and only 20% water-based silicone emulsions [652]. Now solventless silicones have a market segment of about 50%. Various carrier materials (paper, film, fabric, and laminates) are release coated.

2. Manufacture Technology for Release Liners

Almost all label stocks and double-faced pressure-sensitive tapes use a separate release web. Release papers and release films are manufactured. Mechanical or chemical release has been suggested for various products. Coated and uncoated release liner can be used. Coated or embedded release agents or both can be applied. Static and dynamic (postworking) release have been developed. The coating devices suggested for abhesive components depend on their chemical composition and physical state. Silicone coatings are generally used, but wax or other materials can be applied. Solventless, low solvent, solvent-based and water-based silicones are used. These systems are discussed by Weitemeyer et al. [640]. The release formulation is designed to provide variable release level depending on the method of converting and on whether the adhesive is continuously or pattern (zone) coated.

Release papers are generally high density paper (glassine type) materials (30–60 g/m^2) coated with a release layer. The weight of release liner depends upon the method of label conversion and the type of die-cutting and tooling as well as on mechanical performances of the laminate. The general manufacturing scheme for production of such liners involves compounding of the release formulation, coating curing, and remoisturizing when appropriate. Siliconizing of a film may require the use of a primer (top coat). For instance, carboxylated ethylene polymers, crosslinked with zirconium derivatives are used as the top coat for siliconized films. Silica is suggested as antislip. According to Larimore and Sinclair [82], a barrier layer based on polyvinyl chloride, polyvinyl acetate, and polyvinyl alcohol is coated before siliconizing on paper. Vinyltri-acetoxysilane can be used as formulating component of primers for siliconizing paper [655].

Various coating devices have been developed and experimented for siliconizing. For instance, a 100% solids silicone coating system that is a combination of roll- and extruder-coating, has been developed in 1986 [656]. Solventless silicones are coated using a six-cylinder coating head with 0.7–4 g/m^2. According to Moser [657], a Meyer bar can be used for solvent-based siliconizing of a glassine paper. For solvent-free siliconizing a four-cylinder gravure coating has also been proposed. In both cases a coating weight of about 0.5 g/m^2 has been applied. A conventional (42 lb/ream) supercalendered, densified craft paper with silicone release that is highly crosslinked has been suggested for silicone PSAs [258]. A gravure roll with 150 lines per inch (lpi) with a rubber transfer roll is recommended for silicone coating of about 0.3 lb/ream.

Radiation curing offers numerous advantages. For instance, for high temperature siliconizing remoisturing of overdried papers and methods for handling papers after overdrying are important [258,651]. The complete immediate curing of radiation-curable release coating eliminates face-to-back blocking of rolls of the finished product and provides greater assurance of uniform release characteristics for the outside of a roll to its core [658]. EB-curing allows simultaneous crosslinking of both sides of a two-side-coated release liner (transfer tapes). Common release coating formulations have relatively short pot lives. Formulations for EB curing typically have a shelf life of 6 month to 1 yr. Silicone acrylates allow room temperature curing using UV light or EB radiation [659]. Such systems do not need catalyst, and no postcuring occurs; therefore in-line siliconizing, adhesive coating and lamination can be carried out [660,661]. UV-curing allows a running speed of 400 m/min; a coating weight of 1 g/m^2 is applied [662].

Glassine, polyethylene-coated kraft paper, and clay-coated kraft paper, high and low density polyethylene and polyester are the most common liner carrier materials. Various carrier materials are used for release liners for labels and tapes. They differ according to the end-use of the PSP also. Common labels and tamper-evident products need different release liner. The liners for single-side-coated or double-side-coated tapes differ also. For instance, light weight machine-coated paper face stock combined with strong adhesives can be used, and security edge cuts and perforations applied on the press for tamper-evident labels. These products require special liners. Double- and single-sided release can be coated [563]. For instance, a dual-strippable release liner is manufactured by coating the release layer on a kraft paper/polypropylene laminate [663]. This release liner exhibits a release force of 123.5–170 N/m. A pressure-sensitive removable adhesive tape useful for paper products comprises two outer discontinuous layers (based on tackified SIS block copolymer) and a discontinuous middle release layer [664]. This tape can be manufactured with less adhesive than conventional tapes and its top side can be printed and perforated. For medical adhesive tapes a special release coating has been proposed [665]. Such formulation contains silicone rubber, butoxytitanate–stanniumoctoate complex, oligomeric dimethylphenyl-polysiloxane rubber, and triacetoxymethysilane. For food packaging water-based emulsions have been preferred [664].

The extrusion technology for the manufacture of the release liner can use silicone derivatives also. For instance, according to Ref. [286], release agents that do not migrate in the PSA were prepared by reaction of hydrogen siloxanes with unsaturated hydrocarbons. Methyl siloxane has been reacted with α-diolefins. Polyethylene mixed with 3% of the above product, used with a layer thickness of 15 μm gives adhesion values to PSA tape of 50, 90, and 110 g/25 mm at peel rates of 0.3, 3, and 20 m/min and adhesion retention of the tape of 98.3%.

III. MANUFACTURE OF THE FINISHED PRODUCT

Generally, PSPs exhibit pressure-sensitivity due to their coated or built-in pressure-sensitive layer. The classic manufacture of the finished product means buildup of a PSP from its components, that is, the carrier material and the PSA. In practice other product components such as the release liner and cover film can be built in, in one or more layers giving a laminate with a complex structure (see also Chapter 2). As discussed in Chapter 2, the PSP can be manufactured in one step also if a carrier material with built-in pressure sensitivity is used. Such procedure offers primarily economical advantages. Therefore, economical considerations forced the development of extrusion-made products. Table 8.26 summarizes the main characteristics of the manufacture of adhesiveless and adhesive coated PSPs.

From the technological point of view PSPs can be manufactured by using coating, extrusion coating, or coextrusion. Each of these methods has its advantages and disadvantages (Table 8.27). Therefore their applicability for a given product has to be rigorously examined.

TABLE 8.26
Major Raw Materials and Manufacturing Procedures for Adhesiveless PSPs

Raw Materials	Manufacturing Procedure	PSP
Polyethylene	Extrusion	Protective film
Ethylene–vinyl acetate copolymers	Extrusion, coextrusion	Protective film, business forms
Ethylene–olefin copolymers	Extrusion	Protective film, decoration film
Ethylene–acrylate copolymers	Coextrusion	Protective film, tape
Butene–isobutene copolymers	Coextrusion	Protective film, tape
Vinyl acetate copolymers	Extrusion	Business forms
Vinyl chloride copolymers	Extrusion, casting	Tape

In manufacturing PSPs the specific features of the product class and trends in production technology must also be taken into account. The importance of the product components in the buildup of a PSP depends on the product class. Each product class has its own limits concerning the choice of laminate components and manufacture equipment. For instance, label manufacture is strongly related to printing technology. The choice of the label components and manufacturing method is determined by the requirements of printing technology. The manufacture of labels should start with their design and should include origination and pre-press electronic design and artwork, digital artwork and reproduction, plate making, and plate types. Table 8.28 lists the parameters describing the manufacturing versatility of a production technology for the main PSPs.

Technology change refers to the new technologies or material components used in production. Solvent-free, water-based, and 100% solids adhesives are being used more and more, to meet the toxic products emission standards. The printing industry has also undergone changes to meet these standards. Food laws are getting tougher in controlling the effect of migration of various constituents of the package and their interaction with the packed products. Such changes enhance the use of solvent-free and adhesive-free products.

A. MANUFACTURE OF THE FINISHED PRODUCT BY EXTRUSION

As discussed earlier, for special applications it is possible to produce a virtually adhesive-free (noncoated) carrier film based on a common plastomeric polymer. The pressure sensitivity of such materials is achieved via physical treatment and a special chemical composition using viscoelastic raw materials. The manufacture is a one-step procedure (extrusion), which may include

TABLE 8.27
Manufacturing Possibilities of PSPs

	Advantages of the Manufacturing Procedure					
	Free Choice of Product Components		Production Parameter			
			Running Speed			
Manufacturing Procedure	Carrier	Adhesive	Very Good	Good	Fair	On-Line
Coating	+	+	+			±
Extrusion coating	−	−		+		±
Coextrusion	−	−			+	+

TABLE 8.28
Characteristics of Manufacturing Procedure for Major PSPs

| Product Component | Manufacturing Versatility | Product Class | | |
		Label	Tape	Protective Film
Adhesive	Limited use of certain grades possible	+	+	−
	Limited use of certain coating weight values possible	+	+	−
	Crosslinking-free formulation possible	+	+	±
	Primer-free formulation possible	+	±	±
Carrier	Limited carrier grades possible	−	+	+
	Release-free carrier possible	−	−	+
Equipment	Lamination-free production possible	−	±	+
	Limiting of the coating devices types used possible	+	+	−

on-line physical treatment also. Principally plastomer-, tackified plastomer-, and viscoelastomer-based films can be manufactured. Physically treated extruded plastomer films are used as hot laminating films. The application temperature can be reduced by tackifying the plastomer with incompatible tackifiers. Films that can be laminated at room temperature and are competitive with the adhesive-coated PSPs can be manufactured by using viscoelastic polymers such ethylene-vinyl acetate copolymers or butyl acrylate copolymers.

Nonpolar olefin polymers are used as corona-treated pressure-sensitive, hot laminating films. If coextruded they have an adhesive layer designed to adhere to the surface to be protected, while having the top surface layer, which can be printed, provides slip characteristics and protection. Functionalized polar olefins and other polymers have also been developed as self-adhesive films. The main raw materials and manufacturing procedures used for self-adhesive PSPs are listed in Table 8.29.

1. Nonpolar Hot Laminating Films

Nonpolar hot laminating films do not have a special formulation for visco-elasticity. In actuality they are plastomers and do not exhibit pressure-sensitive adhesivity. Their instantaneous adhesivity is provided by the increased contact built up at high temperatures, under pressure, and with physical polarization treatment. However, they have to be discussed together with the self-adhesives films that display room temperature self adhesivity, because their applications and test methods are similar and because a hot laminated formulation may undergo "soft" transformation into a pressure-sensitive one.

TABLE 8.29
Main Raw Materials and Manufacturing Procedures Used for Self-Adhesive PSPs

Raw Material	Manufacture Technology	PSP
Polyethylene	Extrusion	Protective films
Ethylene–olefin copolymers	Extrusion	Protective films
Ethylene–vinyl acetate copolymers	Extrusion, coextrusion	Protective films, decors
Ethylene–acrylate copolymers	Coextrusion	Protective films, tapes, labels
Polyisobutylene copolymers	Coextrusion	Protective films, tapes
Vinyl acetate copolymers	Extrusion	Forms, tapes

a. LDPE-Based Hot Laminating Films

Such films are manufactured from common PE. Their self adhesivity is due to a special physical treatment. Hot laminating films are used for coating coils and for protecting of unvarnished and varnished metallic webs and plastic plates. Because of their one-component formulation and their manufacture as a monolayer blown film, they are the most inexpensive protective film. However, the mechanism of their bonding and the technology of their manufacture are not completely understood. Their production is based mainly on empirical knowledge.

The use of electrical charges to ensure instantaneous adhesion is a general technical procedure. It is also suggested for sheet handling and temporary adhesion. High tension (100,000 V) electrical field is applied to charge the film [666]. Generally, adhesion on a metal surface depends on the energetic state of the metal (wetting out, adsorption), the electric potential (electrical double layer), the morphology and geometry of the metal surface (roughness), and the chemical structure of the metal surface [667]. The same is valid for the plastic film surface to be applied, where other viscoelastic parameters also interfere Physical treatment modifies some of these surface characteristics.

Adhesion of corona-treated surfaces is well known. As stated in Ref. [580], films having a surface treatment at more than 41 mN/m show autoadhesion. Generally, the treatment quality is tested by measuring surface tension. In practice, films that have a certain surface tension are used to obtain a particular debonding resistance (peel). Therefore, for practical use a correlation is admitted between the peel resistance (P) and treatment quality (Q_s) of the film surface are assumed to be related,

$$P = f(Q_s) \tag{8.7}$$

where the surface quality is also evaluated in terms of surface tension (ζ):

$$P = f(\zeta) \tag{8.8}$$

From Table 8.30, for a given polymer (plastic film) and a given adherend surface there is a correlation between the debonding peel resistance and surface tension of a corona-treated polyethylene film. The adhesion of the hot laminated film increases with the surface tension (treatment degree) of the film.

In practice, the surface tension of a polymer film depends on its chemistry, formulation, manufacture, storage, and treatment. Regulation of the treatment has its limits; therefore the peel level is controlled by regulating both, the surface characteristics of the film (i.e., surface tension) and the intrinsic properties of the web to be coated. In this case the resultant peel for a hot

---

TABLE 8.30
Adhesion Characteristics of Hot Laminated Film[a]

Carrier Thickness (μm)	Surface Tension (dyn)	Peel Resistance (90°) (g/30 mm)
120	52	410
120	52	360
150	48	260
150	42	140

[a]An LDPE film has been laminated on a lacquered coil (241°C) with a pressure of 1.5 bar and running speed of 30 m/min.

TABLE 8.31
Peel Buildup of Hot Laminated Films

Treatment Degree (dyn/cm)	Peel Resistance (90°, g/30 mm)	
	Instantaneous	After 1 Month
42	230	435
42	235	370
44	270	295
44	280	520

laminated, corona-treated polyolefin film can be written as a function of the surface tension of the film and surface quality of the adherend (Q_a):

$$P = f(\zeta, Q_a) \tag{8.9}$$

This means, that in practice the desired peel is achieved by regulating the surface tension of the protective film and by modifying the formulation of the lacquer for the coated coil. Unfortunately, surface tension alone is not sufficient to characterize the adhesion properties of the treated plastic film. As shown by the data of Table 8.31, quite different peel values can be obtained for the same initial surface tension of samples of polyolefine protective film on the same adherend after a period of time. Therefore in the correlation (8.9) one should (for a given adherend quality) include another supplemental parameter (β) that takes into account those surface characteristics that are not characterized by the surface tension:

$$P = f(\zeta, \beta, Q_a) \tag{8.10}$$

Bonding depends on the laminating conditions (C_l) also:

$$P = f(\zeta, \beta, Q_a, C_l) \tag{8.11}$$

where C_l takes into account the influence of laminating temperature, pressure, and time. For a given laminating pressure and temperature, the adhesion of a given film is a function of the treatment degree and the characteristics of the surface to be protected. Similar adhesion behavior is known in the packaging industry where self-adhesive heat-sealable films are used. According to Kuik [668], the peel of heat sealable polar ethylene copolymers depends on the corona treatment, coating weight (i.e., layer thickness) and aging. It has been shown, that the adhesion of LDPE on steel (measured as coefficient of friction) increases with the temperature (above 150°F) and attains a maximum at 220–250°F [669]. The weldline strength of LDPE measured in newtons per 15 mm (time 0.5 sec, pressure 0.5 N/mm^2) possesses a value of about 10 at 110°C, but this value increases to 13 at 130°C.

It is known that corona treatment depends on the crystallinity of the surface and the relative humidity of the air. As a result of the treatment various functional groups like (hydroxy, ketone, carboxy, epoxy, ether, esther, etc.) are formed. The energy of positive ions acting in the discharge "streamers" attains about 100 eV, which is sufficient to destroy bonds with energies of about 4 eV [593]. Each of the new polar groups ensures a different surface energy level, that is, adhesion. Theoretically the same "apparent" surface tension can be achieved with different induced functional groups (depending on their concentration). It is not possible to tailor their synthesis. It is evident that peel buildup may be different for surfaces of different chemical compositions.

According to Markgraf [594], the buildup of a laminar flow of ozone is determinant for corona treatment. It is possible to corona treat a substrate and have no indication about the change in surface tension, even though the adhesion is improved. The effects of treatment depend on the nature and age of the film also. HDPE or PP is more difficult to treat than LDPE [628]. According to Ref. [594], LDPE can be processed with the same equipment and parameters, at a treatment energy level of 7.5 W min/m compared with the 25 W min/m for PP.

Slip agents decrease the effect of surface treatment. For instance, a level of 0.12% of eruca-amide as slip requires a 300% increase of the treating energy to obtain the same surface tension [607]. Surface tension has both polar and nonpolar components. Their increase can be quite different, depending on the treatment energy [522].

Another problem is the dependence of the degree of treatment, measured as surface tension, on the level of treatment energy. According to Prinz [571,572], the treatment of polyethylene is a surface phenomenon up to a treatment energy of 10^4 J/m^2. The autoadhesion of PE attains a maximum at 500 J/m^2. The surface tension for this energy level is 60×10^{-3} J/m^2. This means, that for plastic films there is a limit to the treatment energy of about 500 J/m^2; for paper, board, and metallic films this limit is 2000 J/m^2. The required corona energy level (D) depends on the generator power (G_p), web width (W_w), and web speed (s_w) as follows [593]:

$$D = \frac{G_p}{(W_w s_w)} \tag{8.12}$$

The bonding strength increases continuously with treatment time for PP/steel adhesive joints [611]. As seen from Table 8.32, surface tension is dependent on the treatment energy level (generator output). As shown in Table 8.32 and stated by Menges et al. [549], this dependence is almost linear up to a saturation level. The dependence of the adhesion on the treatment energy level varies with the type of adhesive. For instance, for a classic epoxy-based adhesive, a maximum of the adhesion on polypropylene has been found to be a function of the treatment energy level [441]. There is no such maximum for pressure-sensitive tapes. The adhesion between adherend and treated film increases over several month time. The delay in adhesion buildup increases with treatment level. The saturation level of surface tension depends on the type of polymer; for PP it is achieved at an energy level of 2 J/cm^2. It is astonishing, that the surface tension attains its maximum at relatively low levels of treatment energy. The oxygen concentration attains its maximum at 20 J/cm^2 (depending on the experimental conditions). For

TABLE 8.32
Dependence of Surface Tension on Treatment Energy Level[a]

Treating Energy Level (kW)	Surface Tension of Film (dyn/cm^2)
0.5	38
0.7	43
0.9	46
1.1	48
1.3	50
1.5	52
1.7	52
1.9	52

[a]The surface tension of a PE film was measured after 4 days storage.

the same surface energy level per square centimetre under different experimental conditions, the same surface tension can be obtained at quite a different oxygen concentration.

It is evident, that the degree of treatment depends on the constructional characteristics of the machine and treatment conditions. Generally the discharge channels ("streamers") have the diameter of several micrometers and a service life of about 10 nsec. The required processing time of the material to be treated depends on kind if material. For instance, for fabric, it is 0.05–1 sec [670]. According to Gerstenberg [670], the discharge channels (the active treatment zones) have a diameter of 30 μm and a shelf life of only 6–20 nsec.

The manufacturer must take into account the fact that the exact nature of the dependence of adhesion on the treatment degree is not yet fully understood. To attain the desired final bond strength, the laminating conditions (temperature and pressure) and the substrate surface are more important than the chemical of the film. Therefore, hot laminating films are products that need a perfect cooperation between the lacquer manufacturer, film manufacturer, and laminator (coil coater). For the coil coater the parameters influencing the adhesion of the protective film are the following:

- Electrostatic treatment (polarization) of the film
- Storage time of the film (stability of the treatment)
- Type of the lacquer
- Color and gloss of the lacquer
- Lamination temperature
- Period of time between the lacquering oven and lamination (running speed dependent)
- Period of time between laminating and cooling (running speed and equipment buildup dependent)
- Temperature of coil after cooling
- Evolution of adhesion during storage

Taking the above parameters into account, hot laminating is a complex technology. For quality assurance of such products special apparatus and test conditions are required. The suitability of a given corona-treated hot laminating film (tested on a given surface) for use on another slightly different adherend is minimum. Application requires long-term preliminary tests of the adhesion buildup because the adhesion increases over time. It should be mentioned that warm lamination under pressure is used for laminating a PE film on another PE (or PP) film. Stretched and unstreched films can also be warm laminated.

b. Warm Laminating Films Based on VLDPE

The use of polyolefins as self-adhesive PSPs (without adhesive) is based on the flow properties of low modulus polymers. As discussed earlier (see Chapter 3, Section I), such behavior is the result of special chain buildup and amorphous structure. Linear low-density and very low-density polyethylene exhibit self adhesivity. This property is used by their application as shrink film. As mentioned in Ref. [472], if they are used on LDPE, their self-adhesion can lead to excessive bonding. Therefore, new products have been developed that have an antiblocking/Surlyn layer. VLDPE with a density of 0.885 g/cm^3 is self-adhesive and it is difficult to process for blown film with hot-seal adhesivity. Very low-density PE (0.905 g/cm^3) has a low modulus, 87.5 N/mm^2, in comparison with a value of 170 N/mm^2 for common LDPE (ASTM D882), which gives it a low seal initiation temperature [671]. Such very low-density, low-modulus films can be used as warm laminated protective films also.

Butene-based LLDPE may be used as a cling-modifying component in slot-cast stretch films for pallet wrapping, wrapping of paper reels and for bundling [672]. The polymer should be used in blends at a level of 2–30% LLDPE for 20–25 μm monofilms or as a 3–5 μm layer for coextruded films of the same gauge. The formulation with special raw materials in conjunction with the very fast rate of cooling on the chill roll provides enough tackiness for stretch wrapping film. In the case of blown film extrusion it is necessary either to use compounded tacky raw materials that already incorporate the necessary additives, or to feed the additives into the system, during extrusion [673]. The adhesive strength of LLDPE films on stainless steel can be improved by grafting with vinyltrimethoxysilane [674].

Common hot laminating films used since the early 1970s are LDPE-based. Such films are thermoplastic and do not display autoadhesion under room temperature and low pressure. Their application is possible due to their conformability and self-diffusion of a thin polarized, molten, surface layer during their application under high temperature and pressure. As discussed earlier (Chapter 3, Section I), such polymer deformation is a viscous flow. The bonding deformation of elastomers or viscoelastomers is a viscoelastic flow. That means that viscoelastically bonded surfaces ensure instantaneous debonding resistance given by the elastic characteristics and viscosity of the material. Plastically bonded macromolecular compounds exhibit an instantaneous debonding resistance depending only on the (temperature-dependent) viscosity of the material. Mutual interpenetration of the contact surfaces, regulated by the diffusion of the macromolecular compound, remains for "true" PSAs a mostly hydrodynamically and rheologically controlled phenomenon. For plastomers used as adhesives, the chemistry of the contacting surfaces is more important. The surface treatment has to produce an attraction between the contacting surfaces of hot laminated films (based on double layer of electrical charges) and ensure chemical affinity between the polymer and the contacted surface due to the induced polar functional groups.

2. Nonpolar Self-Adhesive Films

In some cases the finished product is a carrier with "true" built-in pressure-sensitivity. As discussed earlier, plastic films may undergo cold flow, and therefore plastic film-based carrier materials of special chemical compositions may work as one-component PSPs, where pressure-sensitivity is to be understood as bonding at high temperature under increased pressure. In this case the manufacture of the finished product by extrusion is actually the processing of plastics. As is known, the main plastic films are manufactured via extrusion. In a similar manner PSPs that have a built-in adhesivity are produced by extrusion. Their manufacturing technology is extrusion technology, characterized by the formulation of the raw materials and the equipment used for their manufacture.

The formulation of one-component (carrier only) PSPs is a problem of plastics processing with some aspects in common with the formulation of carrier films (see Section II). The raw materials and recipes suggested for such a formulation include the manufacture of a solid-state carrier material, with mechanical characteristics and pressure-sensitivity designed to allow its application and deapplication. In some cases the adhesive characteristics of the product are given by the thermoplasts used for the manufacture of the carrier film (together with physical surface activation); in other cases adhesive components have to be built in. In certain cases the product has a monolayer construction; in other cases it consists of a multilayered sandwich. In such cases the reciprocal adhesion of the layers can sometimes be improved by using primers, or barrier layers have to be built in to prevent the migration of the adhesive component to the other side. For many applications no release layer is required, but some products have a built-in release layer.

It is known, that butylene and halogenated butylene derivatives are recommended for use as tackifiers for elastomers. For instance, the tack and green strength of blends of bromobutyl and EPDM have been improved by increasing the level of bromobutyl [675]. In a similar manner butylene derivatives work as tacky additives and lead to adhesive films if embedded in a plastomeric carrier material. For instance, flexible adhesive tapes for temporary protection of fragile surfaces have been manufactured by coextruding a mixture of polyethylene and synthetic resin adhesive (50 μm) and a mixture of butyl rubber and synthetic resin adhesive (15 μm) [676].

a. Self-Adhesive Protective Films Based on Polybutene and Polyiso-butene

Polybutene has been used as tackifier of plastomers in the manufacture of silage wrap films and for stretch and cling films. The pressure-sensitivity of the products is achieved by using polybutylene as a viscous, tackifying component. Because of its incompatibility with the main polymer, the viscous component migrates to the surface of the film and imparts the needed adhesion [677]. Polybutene (3–6%) is used together with LDPE or LLDPE [678]. For LLDPE a level of 3–5% tackifier is generally used [677]. Its concentration depends on the adhesion required, the grade of polybutene, and processing conditions.

Polybutenes are not plastomers; they are viscoelastic compounds. Their processing consists of mixing of a plastomer with a viscoelastic or viscous compound. As discussed earlier, a normal feed-in of such components is not possible. Polybutene (mainly an isobutene–butene copolymer) can be added to the recipe as a masterbatch (10–20%), as tackified pellets (4–6%), or directly (2–8%).

Blending can be accomplished by direct compounding during manufacture of the pellets, or by mixing with PIB masterbatch. The use of a masterbatch is less expensive than compounding. It has the advantage that the masterbatch can be blended with polymers of choice. Unfortunately, a high masterbatch concentration is required. The masterbatch is sticky, and its dosage is difficult to control.

When tackified pellets are used, there is no need for processing know-how. This is an advantage. On the other hand, like the masterbatch, the granules are tacky and difficult to handle. The extrusion of such raw materials is difficult. Block building may occur during storage. There is an upper limit to the PIB level also. Masterbatch and pellets are expensive special products.

There are many variants of direct feed-in. Direct feed-in is the direct addition of solid-state component or in molten status with the aid of a cavity mixer or other injection possibilities. As for material costs, direct injection into a hopper is the least expensive. Unfortunately, screw slip limits injection. Injection at the hopper throat uses a mixture of PB and solid PE. In this case a reduced throughput screw is required.

Direct injection into the polymer melt avoids problems with screw slip, but high pressure injection, which has high capital costs, is needed. For the addition of the components in molten state, extruder injection is proposed. Such equipment should allow the heating up and dosage of PB. A heatable storage tank, a precision speed-controlled pump, a connecting hose, and a lance are needed. A mixture of PB and molten PE is injected into the extruder body. Direct injection into polymer via the center of the screw is also possible [673] by means of a lance that injects the additives through the screw shank into the screw channel [679]. A special technical modality has been developed which uses a cavity mixer to inject the molten materials between the extruder barrel and screen changer die [680]. A low cost process for injecting PIB uses a specially designed screw and growed feed sections [681].

Stretch films with a cling effect on one side, have been manufactured for packaging by using PB and PIB formulations and technologies [682]. Polyisobutene is blended with LLDPE. The adhesive is on the inside and the outer side has slip properties. Such a product can be manufactured as a one-layer film, but a three-layer coex is better. In this case an LLDPE layer provides the stretch properties and another layer ensures the adhesive characteristics. Generally the addition of polybutene to polyethylene film depends on the film manufacture process, type of polyethylene, the buildup of the film, and the application of the film. The parameters that affect the quality of the product include the type of polybutene, type of polyethylene, processing conditions, concentration of the components, and storage of the finished film.

Peel cling (see Chapter 7) increases with molecular weight of the polybutene and its addition level. Lap cling decreases as the molecular weight of the polybutylene increases, and increases with addition level. The density of the PE influences the migration of PB, therefore the choice and dosage of PE is very important. The suggested PB level also depends on the process to manufacture the film. Generally for blown film 2–8% PB, for cast film only 2–4% PB are suggested. According to Ref. [682], 3–5% tackifier is recommended for blown film and cast film 1–3%.

Die gap, frost kine height, melt temperature, and winding tension influence the adhesivity of the film [677]. Among these parameters winding tension and frost kine height are the most important. The storage time and temperature influence the migration of the PB and therefore the self-adhesive properties also. It is suggested to store the finished product under well-defined conditions (heated room) at least 5 days after manufacture. It should be taken into account that (because of the more amorphous nature of the film) PB migrates rapidly in cast film. Curable isobutene polymers with crosslinkable silicone groups have been synthesized also [670].

Self-adhesive protective films on polybutene basis are applied for the protection of plastic plates, in coil coating, and in automotive industry. For the customer SAFs on PB basis have the advantage of better adhesion than EVAc-based films. Therefore the laminating conditions are less important. On the other hand, because of their higher adhesion, their application and handling are more difficult, and blocking may appear. For the manufacturer this is a relatively new product class, more expensive than PE-based hot laminating films. The manufacturer of such films needs special machines and know how. Therefore the production of such SAFs is not recommended for small firms or newcomers.

3. Polar Self-Adhesive Films

Increasing the polarity of polyolefins improves their adhesion. For instance, ethylene-propylene block copolymers are grafted with maleic anhydride [684]. If hot pressed 10 min at 200°C give a peel adhesion of 8.5 kg on stainless steel and 3.8 kg on aluminium. Vinyl acetate and acrylate copolymers of ethylene are used commonly as raw materials for the manufacture of polar self-adhesive films. Such polymers have been proposed as removable coatings also [569]. Polyvinyl alcohol with an ester number of 50–80 has been used as solution. Special compositions have been suggested also (see Chapter 5).

a. Self-Adhesive Protective Films Based on EVAc Copolymers

It is known that EVAc copolymers display polyethylene-like, or adhesive-like properties depending on their vinyl acetate content. In the practice, their processing for SAFs is the processing of a plastomer, which itself can buildup a carrier material. Unlike PIB-based self-adhesive films where the pressure-sensitive properties are given by a low molecular tackifier component (incompatible with the carrier polymer) EVAc copolymers have themselves pressure-sensitive properties and are compatible with the carrier material.

The main application of EVAc-based polar SAF films is for protection of plastic plates and films (PC, PMMA) laminates, profiles, furniture, sanitary items, etc. Such films possess the advantage of deposit-free removability, resistance to environmental stress cracking, and cuttability. Their main disadvantages are that (i) it is difficult to regulate their adhesion (peel), (ii) their adhesion depends on the processing (application) conditions, and (iii) peel may buildup.

Such films replaced paper-based protective films for PC plates (750 μm – 13.3 mm). They have a thickness of 50–100 μm and are applied at a temperature of 60–135°C. They have to exhibit a peel resistance of 140–290 g/25 mm. Films with a thickness of 50 μm are recommended for the protection of plastic films with a thickness of 175 μm to 1.5 mm. Such films are coextruded having an adhesive layer of 12.5 μm with 7.0% EVAc. The adhesion buildup for such films is a complex phenomenon and the formulation alone does not provide a controllable product quality.

Their main advantage for the manufacturer of such films is the possibility of a one step production procedure. On the other hand, the manufacturer depends strongly on the raw material suppliers; who have the manufacturing know how for producing the films. Special extrusion-winding and confectioning conditions are required. To avoid blocking high-speed machines, with finely regulated winding characteristics are necessary. Such EVAc-based film is a mass and end product; no supplemental working steps (added value) are possible.

For protection of PMMA plates ethylene-vinyl acetate copolymers with less than 7% VAc have been suggested. Generally, protective films based on EVAc are manufactured using polymers with a VAc content of 3–28%. An MFI of 1.5–3 is recommended. The VAc content of the polymer used depends on the application temperature of the substrate to be protected and decreases as that temperature increases. The BUR is also a function of the application temperature, decreasing as the application temperature increases. To achieve a homogeneous mixture the output of the extruder should be limited. Polypropylene can be used as carrier material for EVAc-based SAFs also. The self-adhesive layer is built up by coextrusion of an EVAc layer (EVAc with 18% VAc, layer thickness of about 10 μm) or with an LLDPE layer.

Bilayer heat-shrinkable insulating tapes for anticorrosion protection of petroleum and gas pipelines have been manufactured from photochemically cured low density polyethylene with an EVAc copolymer as adhesive sublayer [685]. The linear dimensions of the two-layer insulating tape decreases by 10–50% during curing depending on the degree of curing (application temperature 180°C). The adhesive strength of such tapes decreased by 10–45% after 1 yr. The use of EVAc copolymers has been proposed for carrier-less self-adhesive films (tapes) applied for bonding roofing insulation and laminating dissimilar materials [179]. Such formulations contain EVAc-polyolefin blends, rosin, waxes, and antioxidant. They are processed as hot-melt and cast as 1–4 mm films.

b. SAF Based on Other Compounds

The autoadhesion of soft PVC films depends on the choice of polymer, plasticizer, slip agent and filler [686]. Self-adhesive PVC coatings were developed first as plastisols. They replaced common plastisols which display the disadvantage that they require a prime coat for their anchorage [687]. Self-adhesive laminating of PVC is practiced for manufacture of films. For instance, a 70 μm PVC film was laminated with a 10 μm LDPE film [274].

Self-adhesive, sealable polypropylene with up to 20% terpene resins have been tested also [688]. Chlorinated polyethylene, acrylonitrile-butadiene rubber, liquid chorinated paraffin and fire retardant fillers (e.g., hydrated alumina, $CACO_3$, zinc borate, Sb_2O_3) have been compounded and molded into oil resistant tapes (0.7 mm thick) for electric wires and cables [689].

Such a laminate can be obtained by coextruding LDPE and a thermoplastic elastomer [690].

Peelable protective films (30 μm) comprising methacrylate (e.g., ethylene-methyl methacrylate) copolymers, inorganic filler, and slip have been blow molded. They are useful for protection of rubber articles and show a peel strength of 165 g/25 mm in comparison with 500 g/25 mm for ethylene-ethyl acrylate copolymers [691]. The production of a protective film from polymethyl methacrylate solution has been described [692]. A water-removable protective varnish has been manufactured from a vinyl chloride-vinyl acetate copolymer [693]. The water-washable film (70 μm, 2 m^2) has been peeled off or removed wit a stream of water in 3 min.

Generally the end-use properties of protective films manufactured using different technologies such as plastomer extrusion/corona treatment, adhesive filled plastomer extrusion, modified plastomer extrusion/corona treatment, or the coating of nonadhesive carrier material are different. The application conditions necessary to achieve adhesion, the value of the peel resistance and its buildup over time strongly differ. The regulation of the peel buildup is more difficult for adhesive-less PSPs, because it depends strongly on the manufacturing and application parameters.

4. Manufacturing Technology and Equipment

The manufacturing technology used for PSPs produced by extrusion is the same as that used for extrusion of carrier materials (see Section II). It must be emphasized that for self-adhesive products the flat die (chill roll) procedure is preferable to others.

B. MANUFACTURE OF THE FINISHED PRODUCT BY CARRIER COATING

This chapter is designed to enable to the PSPs producers to coat and laminate PSA and silicone release liners in house rather than having to purchase these expensive constituents of their finished products. The manufacture technology of PSPs via coating/lamination is a part of the classic lamination technology. Lamination is the bonding of two different layers by pressure with an adhesive used as glue or as molten polymer. Laminating procedures include thermal, extrusion, and adhesive lamination.

The adhesive lamination process is influenced by the choice of lamination technology, laminating components (adhesive, substrate), and laminating machines. Dry lamination is a procedure, in which wet adhesive is coated onto a carrier material, dried, and then laminated, generally on-line with another carrier material. PSAs can be used as adhesives for dry laminating also [694]. The criteria for choosing one or the other technique should be based on (1) technical performance, (2) cost, and (3) environmental impact [695].

The classic way to manufacture a PSP is the coating of a carrier material with a PSA. Printers can siliconize and adhesive coat in their own plant and save as much as one third on their manufacture costs [696]. Manufacture of the finished product by carrier coating is a coating technology completed by other conversion steps that are common for web-like products. For some products laminating is also required. Coating of a carrier material supposes an affinity between it and the adhesive. For polar carrier materials this affinity is given. For nonpolar carrier materials it must be realized by means of a primer coating, or surface treatment. Although surface treatment is carried out during carrier manufacture, generally the surface treatment must be refreshed before the carrier is coated.

Water-based PSAs require stand-alone equipment, space for a drying channel, and remoisturizing. Equipment and energy costs are much higher for these adhesives than for HMPSA coating, although production capacity is higher. Hot-melt coating does not need drying ovens; it can be easily adapted to narrow web production. Such coating technology is more flexible. Pattern and strip coating, high finish for no-label look coating, and coating in specific shapes are possible [87]. The manufacture of PSPs via adhesive coating and laminating has some of the general advantages of adhesive lamination [695] such as (1) simplicity of the line, (2) possibility to laminate any pre-formed web, (3) lay flat, when the web tensions are under control (4) minimum make-ready times, and (5) no extra material wastes for caliper adjustment, edge bead, etc.

1. Coating Technology

Coating and printing technology are used for the manufacture of the finished product by coating. Coating is less complex technologically than printing. According to Frecska [697], the main quality criteria in the printing process are (1) accuracy in the reproduction of all image details; (2) accuracy in the placement of the image elements; and (3) uniformity of ink deposit, within the image and from image to image. Generally this last quality criterion is valid only for adhesive coating.

Coating methods are discussed in Ref. [2]. Their special, product-related features are also described here. The main coating methods for liquid materials are based on cast, spraying, roller, blade, and screen printing technique [698], but calendering has been used also. Screen printing is a special procedure that can be applied for adhesive coating and ink printing also presents the following advantages for PSA coating: (1) The contour (shape) of the adhesive layer can be exactly controlled even for complex design; (2) various face stock materials like metals, glass, plastics, etc. can be used; (3) there are no adhesive losses; and (4) coating weight can be exactly controlled within a range of 7–60 μm [699]. For screen printing, solvent-based or water-based adhesives are used. Acrylics or rubber-resin PSA formulations have been introduced. At the beginning of the development of this technology (1970s), only slightly adhesive PSAs were used.

The main aspects of the coating technology refer to direct or transfer coating, mono- or multilayer coating, in-line or off-line coating, coating technology as a function of the type of the adhesive, and of the (adhesive) end-use properties of the laminate. For special products requiring a very smooth adhesive layer, direct coating has the advantage, that air bubbles are avoided.

Processing lines are designed for the individual processor and combine individual units manufactured by various mechanical engineering companies. Different coating devices as direct gravure, offset gravure, reverse roll an knife over roll can be used [700]. Advances in the raw material development caused important changes in the coating technology and forced improvements in the coating devices. Gear-in-die coating devices designed specially for hot-melts and highly viscous materials can coat $20–900 \text{ g/m}^2$ with a coating tolerance of 2 g/m^2 [701]. The coating weight range broadened in recent years. The new coating machines are 3300 mm wide and actual production speed attains 1000 m/min. The coating equipments include over 60 different application methods (grid rolls, nozzles, casting, roll, and blade processes). Multipressure gravure is the practical solution for modification of coating weight without changing the gravure roll [702]. Thirty years ago PSPs were generally coated with 30–100 μm adhesive [313]. Some times ago, the coating weight of PSA coated webs was between 20 to 60 g/m^2 (dry) and the running speed was 3–12 m/min [703]. The average coating weight for (permanent) labels is now about 20–25 g/m^2, for protective films $2–7 \text{ g/m}^2$, and for tapes $20–40 \text{ g/m}^2$. A decade ago, higher (2–40%) prices for finished products in the United States allowed the use of higher release coating weights [704]. Electron beam curing was considered as a rapid growing sector with large perspectives and industrial applications in the mid-1980s. The choice of a manufacturing technology end equipment became important. Past trends called for multipurpose machines, with moderate speeds; now specialized production requires high speeds and wide web widths [705].

a. Coating Technology as a Function of the Chemical Composition

As discussed earlier the chemical composition of the adhesive and its formulation determine both the type of coating and the coating technology required. Thermoplastic elastomers are coated as hot-melts. Acrylics can be coated as solvent-based or water-based formulations but acrylic hot-melts, and radiation-cured compositions are also known.

The main advantage of thermoplastic elastomers is their "thermo-plasticity," their ability to be processed as plastomers. The coating of HMPSAs has to take into account their high specific heat and low thermal conductivity [706]. Styrene block copolymers require three times as much energy to melting than EVAc copolymers.

Radiation-curable systems can have very different compositions and viscosities. As discussed earlier, the coating of UV-curable formulations was developed from the coating of monomer mixtures. Generally a radiation-curable (UV and EB) monomer-based formulation includes a hard monomer system, a soft monomer system, and multifunctional monomers [8]. Such low viscosity formulations require quite a different coating technology than the coating of oligomer- or macromer-based formulations. Macromer formulations start from monomer mixtures containing a hard monomer $(T_g > -25°C)$, a soft monomer $(T_g < -25°C)$, a tackifying monomer $(T_g < 25°C)$, and multifunctional monomers (e.g., pentaerithrytol triacrylate). For instance, a mixture of acrylic acid–butyl acrylate-diethyl fumarate has been polymerized to a macromer, which can be coated as hot-melt and postcrosslinked via EB curing [707].

TABLE 8.33
Primer for Selected PSP Manufacture Procedures

	Primer Coating			
			Optional	
Manufacture Procedure	Required (more than 70%)	Not Required	Different Formulations Possible	Different Technologies Possible
Coating	−	+	+	+
Extrusion coating	−	+	+	−
Coextrusion	+	−	±	−

According to Klein [708], primers are coated with gravure, flexo, or reverse roll coating (see also Chapter 5). Coating weights of $0.01–0.5$ g/m^2 are used. The problem of the primer coating for extruded or adhesive coated PSPs is related to the need of a primer coating and to the technical means available for coating a top layer (Table 8.33). Such a coating is generally required for coextruded products with a complex buildup, but it also needed for the main polyolefin-based adhesive-coated carrier materials. Advances in carrier formulation, carrier treatment, and adhesive coating allow primer-free procedures also, but generally such compositions have to be more crosslinked or they are coated with a lower coating weight (see also Chapter 7).

b. Coating Technology as a Function of the Physical State of the Adhesive

The replacement of the solvent-based adhesives with water-based ones has been a continuing trend in industry for reasons of air quality, safety, and economics. For instance, acrylic adhesives for packaging tapes offer the following advantages: (1) They are water-borne; (2) they exhibit superior instantaneous adhesion to cardboard, good shear resistance, good tack, clarity, UV-stability, temperature resistance; and (3) they are ready-to-coat [709]. However, the transition from solvent-based to water-based has not been simple and is by no means complete. Aqueous adhesives have found limited acceptance being regarded as generally inferior, in either performance or coating rheology or both. An additional deficiency of latex adhesives is their tendency to buildup coagulum, on the metering roll when the dispersion is applied with a reverse roll coater. The coagulum may arise either from poor mechanical stability under shear or from imperfect doctoring, leading to a thin coating on the roll which dries and cannot be redispersed. These particles transfer to the applicator roll and mark the coating [710]. Water-based systems are more efficiently dried with large volumes of air rather than with air of high temperature [711]. It should be noted that the energy consumption for coating and drying increases exponentially with the coating weight [712].

According to Ref. [713], conventional methods for coating of water-based PSAs include reverse gravure with pressurized chamber (vario gravure), metering bar or slot-die; they have reached their system limits. Curtain coating developed for water-based PSAs preserves the advantages of slot-die coating and improves the coating quality and reduces the costs. This method allows a high solids content, low coating weight, high coating speed, high drying speed, lowest drive rating requirement (max. 1 driven roll), no wear and tear of rubber-covered rolls, metering bars, etc., low pressure impact to the substrate and therefore less web breaks, and no interruption of coating when spliced web goes through the coating head. The curtain coating ensures homogeneous, structureless film, very smooth, structure-free surface, clean, coating-free edges, no film splitting effects compared to roll coating, uniform cross profile, easy changing of coating weight (by variation of rotational speed of the pump, wide range of coating weights, wide range of viscosities, easy cleaning with less water, no or little circulating quantity, simple change of width. It is not sensitive to changes in thickness of the carrier [714]. The curtain-coating procedure works with dies. The required coating weight is premetered; this means that only the fluid required is pumped through the dies. Once through the die, the material to be applied forms a sort of curtain which flows evenly over the substrate and adheres to it [715].

Solvent-based crosslinked formulations have been developed for each product class due to their relatively broad raw material basis and the ease with which their degree of crosslinking can be regulated through the choice of raw materials or manufacturing conditions (see Chapter 6). The crosslinking of solvent-based formulations has been used since the beginning of PSA technology. Natural rubber can be used together with polar elastomers and reactive resins (phenolic resins). The resulting crosslinked adhesive has both temperature and solvent resistance [716]. The reaction is carried out by mastication of the components. Crosslinked, solvent-based formulations require special drying conditions to achieve the simultaneous drying and crosslinking of the adhesive.

According to Ref. [717], HMPSAs are used for medical and hygienic products, masking tapes, plotter films, office tapes, labels, and nameplates. Hot-melt coating machines can coat paper, fabric, vinyl acetate, and polyester films [718]. Hot-melt coatings can be defined as thermoplastic coatings that have a minimum viscosity of 9000 cP. Less viscous adhesives are generally applied with roll coaters having 2–4 metering and application rollers.

Coatings of higher viscosities (over 50,000 cps) are processed with a slot orifice coating device that acts like an extruder. It should be mentioned that the data in the literature are very contradictory concerning the limits (viscosity) of the usability of each coating method (Table 8.34).

For hot-melts slot-die coating is used below 7 g/m^2 and above 10 g/m^2. For tapes generally coatings of 5–35 g/m^2 are applied [732]. The introduction of offsetting rollers between the die or application roller and the substrate is an useful technique to lower the coating weight [733]. Slot-die application is normally offered by machinery makers for use up to 120,000 cP (at 200°C), roller coating below 100,000 cP. Hot-melts can be coated without problems up to a viscosity of 100,000 mPa sec on roller coating device [734]. According to Fries [189], coating viscosities for hot-melts can increase up to 17 to 20,000 mPa sec (at 180°C). Rotary screen hot-melt applicators for zone and pattern coating are available also [87].

Hot-melt coaters can use a carrier preheating drum to improve the adhesion of the HMPSA and its surface profile. The most HMPSA coating machines for labels are intended for continuous coating of 18–24 g/m^2. Such machines have an in-line filter with filter cartridge and replaceable coating heads. By inserting various shim plates and Teflon profiles operators can change application width and pattern. Patterned coating can be carried out using special flexible tool [735]. Full-width, stripped, beaded, spot, patterned or zoned coating is possible. For the coating of high gloss adhesive layers on film carrier materials rotating bars can be used to eliminate streaks. Intermittent or continuous operation is possible. On modern machines the coating head automatically swivels away from the substrate if the unite stops. Laminating rolls mounted to the side of the coating roller are either cooled or heated to maintain an accurate finish and coating weight. A drum melt system ("melt on demand") is applied (usually) for label coating.

A modern hot-melt coating machine is able to work with more than 60 application and lamination processes. Application from pan allows dipping for double-sided coating, premetering and double blade processes, parallel run and reverse run processes, gravure printing, Meyer bar, strip lamination, roller coating and metering. Application from a slot (higher viscosity hot-melt adhesives, 2000 mPa sec) allows parallel run application, strip coating, working speed up to 500 m/min [736].

Spraying has also been developed for hot-melts. At first stripes, dots or full surface coating of HMPSAs was possible. By spraying larger surfaces, temperature sensitive materials can be coated with at least 50% economy [737]. This method is used specially in the nonwoven industry [87]. Contact-less coating methods for HMPSA have been developed to allow the use of hot spray technology in the nonwovens market. In this case the adhesive and air are premixed in the nozzle. Controlled fiberization was introduced in 1986 [738]. The fiberization system uses a set

TABLE 8.34
Manufacturing Versatility of Selected Coating Methods for Different SPs

Coating Method	PSP	Adhesive Type	Coating Weight (g/m²)	Viscosity (Pa sec)	Coating Speed (m/min)	Ref.
Slot die[a]	—	HMPSA	10	>120	—	[719]
Slot die	—	Acrylic, HMPSA	20–200	50–500	—	[720]
Slot die	—	—	8–300	—	250	[721]
Slot die	—	HMPSA	7–10	<120	200	[722]
Slot die	—	HMPSA	1–2500	—	—	[723]
Slot die	—	—	—	35–600		[724]
GID[a]	Label, decor film, tape	—	—	5–80	—	[725]
GID	—	—	30	0.2–1000	—	[720]
Roll coater	—	—	20–1000	200–2000	—	[720]
Roll coater	—	—	—	500–2000	—	[724]
Roll coater	—	HMPSA		>30	—	[726]
Roll coater	—	HMPSA		>100	—	[726]
2 rolls[a]	—	—	100–1500	—	—	[719]
3 rolls	—	—		<15	—	[727]
3–4 rolls	—	—		1500	—	[727]
Reverse roll	—	HMPSA		<30	—	[727]
Reverse roll, nip feed	—	Acrylic, WB	—	1–2	—	[100]
Reverse roll	—	—	20	—	120	[728]
Reverse[b,c] roll	Tape/label medical	Rubber-resin, SB	17	3–5	—	[729]
Reverse roll, 3 rolls	Label	—	12–200	0.3–100	200	[730]
Reverse roll, 4 rolls	Tape	—	60–100	0.01–4	300	[730]
Roll blade	Tape	—	20–200	2–10	60	[730]
Roll blade	—	—	30–50	—	150	[728]
Knife over roll	—	Acrylic, WB	—	4–8	—	[100]
Reverse roll[c]	Tape/label medical	Rubber-resin, SB	17	3–5	—	[729]
Rotogravure	—	Acrylic, WB		0.075–0.1	—	[100]
Reverse gravure	—	WB	18–25	—	300	[728]
Wire rod	—	—	10–20	—	—	[728]
Wire rod	—	—	14–30	—	—	[728]
Wire rod	—	Acrylic, WB	—	0.3–0.9	—	[100]
Screen	—	WB and SB	6–7	—	—	[685]
Screen[c]	Tape mounting	Rubber-resin, SB	17	1.5–2.0	—	[731]
Rotary screen	—	—	30	5–10	—	[720]
Calandering	Tape	—	200–1000	200–2000	—	[720]
Extruder	—	—	—	500–5000	—	[720]
Spray	—	HMPSA	3	—	—	[721]

[a]Carrier: paper, film.

[b]Carrier: PET.

[c]Coating weight in lb/3000 ft².

of air jets to orient the bead of adhesive, drawing it down into a fine fiber applied in a helical pattern. Spiralling the adhesive maintains good edge control, even in intermittent applications up to 33 m/min. The spiralling also cools the adhesive permitting the application of hot-melt on heat-sensitive substrates. Another coating process possesses a head that combines elements of slot coating and spray technology. The adhesive is distributed to the desired pattern width through internal slot channeling. As the adhesive is extruded from the slot nozzle, heated air impinges on it from both sides stretching the adhesive and breaking it into very fine fibers. Hot-melt dot and line application guns, allow noncontact application of minute adhesive spots and are suitable for adhesives up to a viscosity of 300 mPa sec at a max frequency of 400/sec [739]. Hot-melt coating systems are described in Refs. [721,727] and [740–742].

In-line siliconizing and HMPSA coating were developed in 1985 [743]. HMPSAs have some major deficiencies: excessive thermoplasticity, poor UV-resistance, and limited plasticizer tolerance [189]. However, McIntire [744] predicts savings of 30% by combining off-line siliconizing and HMPSA. In 1988 the proportion of HMPSAs in Europe in the hot-melts reached about 25% [745]. In the last decade about 15% growth has been observed.

As discussed earlier, the physical state of the adhesive can vary for the same raw material class or adhesive synthesis method also. For instance, 100% acrylics can be coated as low viscosity monomers, medium viscosity oligomers, or high viscosity hot-melts using postcuring technology. In 1988, there was no industrial use of postcuring of HMPSA [746]. Now there are numerous machines designed especially for the manufacture of tapes, that use this technology. UV- [747] and EB-curable [748] formulations have been developed. As is known, EB-curing displays the advantage of high coating weights ($50-80$ g/m^2) [749]. Radiation processing of acrylic hot-melt PSAs has also been developed [750] (see also Chapter 5). The development of new, low cohesion acrylic warm-melts requires the use of a postcrosslinking [60,751]. Therefore, the large scale application of UV-induced postcuring is enhanced by this raw material development. Before crosslinking the new polymer is a highly viscous fluid (at room temperature), and can be processed at $120-140°C$ (its viscosity is about 10 and 20 Pa sec). The UV lamps in the curing channel either use electrodes or are microwave powered. These adhesives are cured as tackified formulations. For tackifying of UV-crosslinkable polycrylates, esterified, highly stabilized rosin is used at a low level (8.5%) and high coating weight (50 g/m^2). For tackified formulations the radiation dosage must be increased. For instance, a 8.5% resin level requires doubling of the radiation dosage.

A special feature of such radiation crosslinking is its anisotropy. The radiation is partially absorbed, partially transmitted, and partially reflected. The maximum radiation level is given at the top of the adhesive layer. That means that the degree of crosslinking is the maximum on this side. Therefore direct and transfer coating give different adhesive characteristics. A uniform, isotropic adhesive layer can be achieved using a UV-transparent face stock (backing) and irradiation from both sides. The manufacture of such UV-crosslinkable acrylates is more expensive than that of acrylic dispersions. However, they are an alternative to solvent-based adhesives where without too much capital investment, a common hot-melt processing line can be equipped with UV curing lamps.

As mentioned earlier [222], the PSA can also be coated as a powder. A dry blend of elastomer and tackifier is obtained as powder and can be coated then sintered by heating of the adherend surface (at 177°C). Coating of soft, tension-sensitive PVC carrier materials require a relaxation station on the coating machine [732].

c. Coating Technology as a Function of the Product Class

The product class determines the type of carrier and its geometry, the type of adhesive and its geometry, the type of release and its geometry, and their buildup. Adhesive geometry includes the coating weight also. Analyzing the changes that occurred between 1980 and 1994 in the domain of pressure-sensitive technology Bedoni and Caprioglio [517] state that line speed and web width increased, adhesive thickness measurement and feed back became frequently the standard, multitechnology combined machines became available and changeover time has was reduced. This is a general consideration. The advances listed above should be evaluated separately for each PSP class. The special buildup of PSPs in each class needs particular coating techniques. These techniques have been developed in various ways.

Rigid, thick, and uneven surface materials are usually coated on sheet fed curtain coaters or spray coaters. In curtain coating a continuous stream of liquefied coating falls onto the moving material. Spray coating deposits an atomized stream of material.

Impregnation can also be used for the manufacture of carrier and finished tape. In this process the coating permeates the fibers or the spaces between the fibers. At saturation the impregnation level reaches the point where the fibers can not hold any more liquids or the spaces between the fibers are completely filled. In the dip process, impregnation and saturation machines apply simultaneous coatings. Impregnation or saturation can be used in to achieve removability (see Chapter 5), dosage of a liquid component (see Medical Tapes), or reinforcing of an adhesive or carrier material. Roll press coating of (temperature-) softened adhesive mass is applied for tapes also. Foaming (simultaneous or postcoating) has been developed for manufacture of special tapes. These methods are not used for common labels. The release technology suggested for labels and other PSPs differ also. The release technology suggested may differ for different labels. The manufacture of release liner via coating involves a complex chemistry and great precision in coating of low coating weights using gravure or flexo anilox rolls. An in-house choice is between solventless thermal, and radiation-cured one. The choice of a manufacturing technology is strongly influenced by the product class (Table 8.35).

Economic considerations forced the in-line manufacture of PSPs. Production equipment has been developed for simultaneous adhesive/abhesive coating and lamination. In-line extrusion and coating of film-based PSPs is also possible. Table 8.36 summarizes the advantages and disadvantages of in-line coating. The relaxation processes in the extruded film are dependent. Such processes may lead to changes in the geometry or surface quality of the films. In-line production of coated films has to take such phenomena into account.

d. Coating Technology as a Function of the Product Buildup

Product buildup includes the nature, number and construction of the PSP laminate. Details concerning the laminate's components and the buildup of the laminate have been discussed in

TABLE 8.35
Coating Methods Used for the Main PSPs

Coating Method	PSP			
	Label	Tape	Protective Film	Other Products
Cast coating	+	+	−	+
Roller coating	++	++	++	+
Blade coating	+	+	++	+
Slot-die	+	++	−	+
Extrusion	−	++	−	−
Screen printing	++	−	−	++
Calendering	−	++	−	−
Spray coating	−	+	−	++

Note: + +, Main coating method.

TABLE 8.36
Advantages and Disadvantages of In-Line Manufacture of PSPs

Advantages		Disadvantages	
Technical	Economic	Technical	Economic
Treatment exists	No supplementary technical (handling and maintenance) operations are required	Time dependent, postmanufacture processes work in the film and change its surface quality and geometry	Coating speed is lower
	Own carrier Global know-how		Mixed technology

Chapter 2. For products having a multiweb structure, multiple coatings are required. For such products the different coating steps can be carried out using different coating methods. For instance, for a label, the PSA, the release agent, the primer, and the lacquer can be coated by using different coating methods. However, if possible the same aqueous or hot-melt technology should be used for the different coating operations (e.g., primer and adhesive coating or release and adhesive coating).

Porous or heat-sensitive face stocks require the use of transfer coating [752,753]. Fabric is coated via transfer procedure. Undulated (crepe) paper is coated with $60-80 \, gm^2$ HMPSA. Generally plastics are difficult to coat because of their lack of porosity and their nonpolar surface. Cratering, adhesive-free places, and coating weight differences are the main defects. The wetting-out of plastic films depends on temperature, vapour saturation of the atmosphere, and electric charges on the surface [754].

The common manufacturing procedure of tapes includes in-line primer coating, release coating, and adhesive coating. As a classic primer, solutions (in toluene) of butadiene-acrylonitrile-styrene have been preferred. For instance, for solvent-sensitive SPVC films primer dispersions based on Hycar Latex and Acronal 500 D (1/1) have been suggested by Reipp [235]. The backside of the tape carrier has been coated via rotogravure with a 0.15% solution of Hoechst Wachs V in gasoline. Gasoline (boiling range $60-100°C$) has been proposed for use as solvent. Blade coating and drying at $30-80°C$ have been recommended.

As mentioned earlier, thick PSA tapes (0.2 to 1.0 mm) have been prepared by photopolymerization of acrylics [24]. For such polymerization a minimum UV light intensity of $0.1 \, mW/cm^2$ is often a practical limit. Multilayered composites having different adhesion and mechanical performances can be manufactured using this technology. As discussed above, with respect to photopolymerisable acrylic warm melts [60], light-initiated polymerisation allows the buildup of anisotropic products. According to the procedure disclosed in Ref. [49], the acrylic ester PSA formulation is a combination of nontertiary acrylic acid esters of alkyl alcohols and ethylenically unsaturated monomers with at least one polar group that may be substantially in monomer form or may be a low molecular weight prepolymer or a mixture of prepolymers. The formulation can be polymerized as a thick layer (up to about 60 mils). The thick layer may be composed of a plurality of layers, each separately photopolymerized. The acrylic esters generally should be selected in major proportion from those that as homopolymers have some PSA properties. As crosslinking agents, diacrylates in a concentration of about 0.005 to 0.5% by weight are preferred. Typically the crosslinking agent is added after the formulation of the prepolymer. The recommended photoinitiator is 2,2-dimethoxy-2-phenylacetophenone. The polymerization chamber has an atmosphere of nitrogen. The length of polymerization zones and the density of lamps in these zones affect the manufacture. The thickness of the layer is a determining parameter with respect to the productivity, a thicker layer requiring a greater degree of exposure. Thick multilayered PSPs are made using this type of manufacturing. When the thick layer is sandwiched between two thinner layers, the thick layer can be used as a carrier. This support carrier layer can be 25–45 mils thick [24]. Glass microbubbles can be incorporated to enhance immediate adhesion to rough and uneven surfaces. Such carrier-less foam-like tapes are prepared by polymerizing in situ with UV radiation. According to [78] web-like products, that is, tapes were first manufactured using UV-initiated polymerization as disclosed in [755], by thickening of the monomers, coating and UV-polymerization. Using this procedure a coating weight of 1000 g/m^2 can be achieved. The curing time is 60 sec.

According to Ref. [22], thick, triple-layered adhesive tape, having filler material in the center layer can also be manufactured by UV-photopolymerization. In this way a carrier-less tape can be produced. As the filler, fumed silica is suggested. The first polymerization zone for an adhesive layer is shorter than the second zone for the polymerization of the carrier. The first prepolymer layer contains 1.0 part photoinitiator and 0.6 part of 1,6-hexane-diol diacrylate (crosslinker). The second prepolymer layer includes 1.0% by weight photoinitiator, 0.3 part of crosslinking, multifunctional monomer, and a polyvinyl acetate polymer used as filler. The second adhesive layer has the same composition as the first one. Line speed of 5 ft/min is suggested. Cooling is provided by the use of cooled nitrogen gas. The adhesive has been dried 1 h at 350°F.

As stated by Schmalz und Neumann [756], for common PSPs UV-curing displays the following disadvantages in comparison with EB- (and thermal) curing: rough paper surfaces are unsuitable, curing is insufficient, and cannot be use for laminates. Because of absorbtion of the radiation by the adhesive layer, UV-curable products must have a limited coating weight.

A multilayered structure can also be obtained by postcrosslinking. This is carried out after coating, by using immersion or printing techniques. According to Larimore and Sinclair [82], a carrier-less PSP (the adhesive layer provides an adequate strength to permit it to be used without a carrier material), with good conformability can be manufactured by applying a special procedure. The product has an adhesive surface and an adhesion-free surface. The tack-free surface is achieved by superficial crosslinking of the adhesive by polyvalent cations (Lewis acid, polyvalent organometallic complex, or salt). The crosslinking penetration depends on the polymer and solvent used. Crosslinking can also be achieved by dipping. Water or water-miscible solvents should be used. For such tapes without backing, strength can be provided by a tissue-like scrim (cellulose or polyamide) embedded in and coextensive with the adhesive layer. The superficially crosslinked adhesive layer may be stretched to fracture the crosslinked portions, exposing the tacky core in order to bond it. Crosslinking in tiny, separated areas can be achieved by printing techniques (halftone printing). Thus microscopic small, tack-free areas can be created, and separate the narrow tacky portions. A coating weight of 81.5 g/m^2 (thickness 0.2 mm) is applied [757].

Decalcomania (transfers) are carrier-less products similar to transfer tapes. For such products the solid-state component is a technological (application) aid only. Such products can be manufactured by using screen printing. First a clear carrier lacquer is printed. This layer provides the mechanical resistance of the product. The image (in mirror print) is coated on this layer, by using screen (or offset) printing. The next layer is a transparent layer followed by a screen printed PSA [758]. In another procedure the PSA is coated on release paper. The PSA-coated liner is laminated together with a printed film [759].

According to Ref. [760], multilayered self-adhesive labels (such as business forms) comprise a carrier label containing silicones, an adhesive layer, a printed message, another carrier layer (e.g., polyester) a release layer (e.g., silicone), a second adhesive layer, a third carrier layer (e.g., polyethylene) and a printed message. Labels of this type may have more than two (adhesive-carrier layer-message layer) units.

2. Manufacture of Labels

In 1982 in Europe as much as 30–40% the of the HMPSAs produced were used for labels [587]. After two decades their market share attained about 50%. However, taking into account the balanced adhesive properties required for labels, except for some low cost products the majority

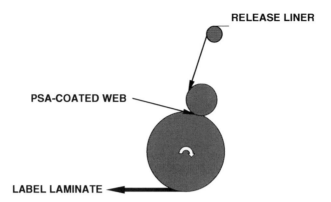

FIGURE 8.8 Schematic presentation of the laminating operation in label manufacture.

of labels are manufactured with water-based or solvent-based adhesives. Therefore, their coating technology uses coating devices specialized for such adhesives. Their choice depends on the coating weight and economic considerations. Label manufacture generally consists of coating of medium thick (15–25 μm) dry adhesive layer having a very smooth image on a nontextured, smooth carrier surface and laminating of the adhesive coated web with the release liner or the final face stock material. According to Ref. [760], the usual width of the coating machines (of 70, 100, 140, and 200 cm) is determined by the 70 × 100 cm sheet format.

Labels use a separate release liner sheet that is normally applied in a single pass. Label stock manufactured on a one-station coater might require two or three passes to complete the process of applying a primer and adhesive and laminating. Two- or three-station coaters would perform all of these operations in a single pass. Such multiple-station coaters deposit three or more coatings on one or more carrier materials in a single machine pass. Roll-fed coaters and sheet-fed coaters are used. Sheet-feed printing presses have the ability to print images of variable sizes on any sheet size. Sheet-feed coaters can apply a top or bottom coating. From the technologic point of view the main characteristic of label production machines is the existence of a laminating unit, which is necessary to buildup the label laminate from the face stock material and the separate solid-state release liner (Figure 8.8). Common tapes and protective films do not need such equipment.

Forms (which, because to their construction, can be considered complex labels) are manufactured by coating and laminating "prefabricated" webs (Figure 8.9). For instance, in the first manufacture step the paper carrier is coated with release agent or adhesive. The web coated with adhesive (1), can be laminated with a film, and the release liner coated with adhesive (2) can be laminated with another release liner to give a temporary laminate. The sandwich structures thus manufactured can be combined to give a pressure-sensitive product. In coating practice adhesives with different bonding strength and release liners with different abhesive characteristics are combined to achieve a partially detachable product.

3. Manufacture of Tapes

Tapes are used in a very broad range of applications. Their performances and buildup are various also (see Chapter 11). Their raw material basis and coating technology vary widely according to their general or special character. Mainly rubber-resin and acrylic adhesives are used for tapes. Both can be applied as hot-melts, solutions, or water-based dispersions with the appropriate equipment.

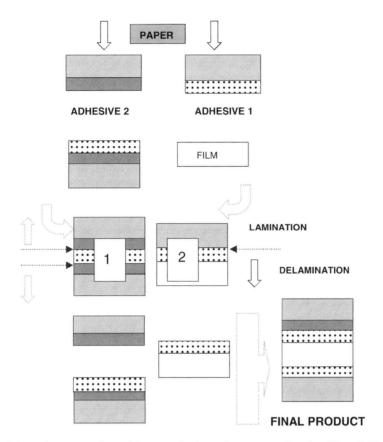

FIGURE 8.9 Schematic presentation of form production using precoated webs. (1) and (2) are temporary laminates.

The coating technology of tapes generally includes multiple coating of the carrier (face and back) with different (primer, adhesive, and abhesive) layers. Crosslinking of the adhesive coated with higher coating weights is a general feature of tape manufacture. The major coating systems include equipments for [761]:

1. In-line production (e.g., release–corona–primer–PSA; corona–primer–PSA).
2. Very thin coatings (e.g., solventless silicones). Such systems have problems with broad webs and high running speeds).
3. For medical products (e.g., in-line and tandem production; solvent-based and water-based PSAs; soft PVC; nonwoven, fabric and paper carriers).
4. Crepe paper coating (such systems are strongly influenced by the amount of pleating).
5. High coating weight on a textile carrier.
6. Wetting out of difficult carrier materials.
7. Labels (e.g., for smooth coated layer, rough carrier surface, nonuniform carrier profil, etc.).
8. Protective films (with a broad carrier web).
9. Rubber solutions with high solids content.

The solvent-based technique is the most proven and popular technology for production of tapes [762]. The development of tapes started with rubber-resin-based formulations. Tapes include a release coating, primer, and adhesive coating [763]. Primer and release coatings can be dried in either vertical or horizontal ovens. Tape manufacturing can require the in-line double-sided coating of the carrier. For instance, a release coating $(0.2 \, g/m^2)$ is applied on the bottom side and a primer $(1 \, g/m^2)$ is coated on the top of the carrier. In this case gravure coating is recommended but alternatively two smooth rollers or reverse gravure can be applied also [764].

Packaging tapes are manufactured with water-based formulations too. For such products medium coating weight $(20-30 \, g/m^2)$ is suggested. Their coating technology is similar to that of common labels. A large portion of tapes of general use (packaging, carpet, office, etc.) are formulated with hot-melts. Such formulations can be coated by slot-die coating. The main advantages of slot-die coating for HMPSAs are that it can be used for processing of high viscous HMPSAs; the coating device is simply constructed; the coating weight is easy regulated; and the device is a closed melting system is, with low oxidation. The disadvantages of slot-die system are that the die can be cleaned only by changing the formulation and problems are caused by dirt [765]. The problem of cleaning is determining factor for HMPSA die-coating; it depends on the construction of the die [766]. The extent to which the coating device is abraded depends mainly on the running speed. Generally highly viscous melt (up to 400,000 mPa sec) can be coated [767]. Hot-melts for tapes can be coated using a porous coating cylinder and a knife that forces the adhesive through the cylinder as in screen printing (the rotary screen printer is a hollow perforated applicator roll), by so-called rotation extrusion [768]. Spray-coating of hot-melts requires less adhesive (up to 50% less) [769].

The solvent content can be reduced by adding high solids or using roll press calendering. High solids content (50%) solvent-based (viscosity of 500 Pa sec) concentrated rubber-resin adhesives have been coated on PVC carrier for tapes [85]. Tapes can also be coated by calendering. In this case if multicomponent crosslinking agents are used, one crosslinking component can be used as the primer coat. The crosslinking agent used for a rubber-based PSA for tapes can be incorporated in an oil and mixed with the adhesive components in a cavity transfer mixer [747]. Such a crosslinking agent can be used in the partially crosslinked adhesive together with a crosslinking activator in the primer. A precrosslinked butyl rubber can be used also instead of partial crosslinking in the Banbury mixer. The degree of partial crosslinking prior to mixing with other components can vary between 35 and 75%. The proportion of partially crosslinked rubber in the composition varies inversely to the percentage of crosslinking. p-Quinone dioxime, p-dinitrosobenzene, Phenolic resins, etc., can be added as the crosslinking agent. A precrosslinking during mixing of the components in Banbury is followed by a supplemental crosslinking (additional crosslinker) in a second mixer (a two-roll mill), where a quantity of p-quinone dioxime is added. For the initial crosslinking, slower acting phenolic resins are used to obviate the lumping problem. The amount of crosslinker for premix is about 3 to about 10 phr. After the second mixing, the resulting adhesive is coated on a carrier, typically by calendering, to form a tape. This method has been used for the manufacture of medical tapes, where the components of the adhesive were calendered and the hot adhesive was coated under pressure on the carrier material [770].

Tapes made with special extensible or pleated carrier materials are made via transfer coating. For instance, a PSA tape for the protection of printed circuits, based on acrylics and phenolic terpene resin with a $100-200 \, g/m^2$ siliconized crepe paper, is made by coating the adhesive mass first on a siliconized polyurethane release liner and transferring it under pressure $(10-30 \, N/cm^2)$ on the final carrier material [333]. Coextruded ethylene/propylene can be stretched for tape carrier also [770]. It is press rolled with the adhesive. Microencapsulated adhesive can also be coated on tapes [771]. In Europe, EB crosslinking was used for tapes first with a 1300 mm width machine using the scanning principle. This equipment uses normal adhesive formulation [772]. It must be emphasized that the majority of tapes are coated with a medium coating weight. This is illustrated by the data of Table 8.37.

TABLE 8.37
Coating Weight Used for Tapes

Type of Tape	Carrier	Adhesive	Coating Weight	Ref.
Packaging	—	—	$5-35 \ \text{g/m}^2$	[773]
Insulating	Aluminium	Silicone	$50 \ \mu\text{m}$	[187]
Packaging	PP	NR	$30 \ \mu\text{m}$	[183]
Special	Paper	EPR	$20 \ \mu\text{m}$	[774]
Packaging	PET	Acrylic	$35 \ \text{g/m}^2$	[775]
Insulating	Aluminium	Silicone	$50 \ \mu\text{m}$	[186]
Insulating	PET	Silicone	$42 \ \mu\text{m}$	[776]
Insulating	Aluminium Paper	—	$40 \ \text{g/m}^2$	—
Removable from paper	—	Acrylic	$5 \ \text{g/m}^2$	[66]
Removable	PE	Acrylic	$10 \ \mu\text{m}$	[664]

a. Manufacture of Packaging Tapes

As stated in Ref. [762], the majority of tapes are packaging tapes. Packaging tapes are manufactured by coating. Recent productivity developments in tape manufacture are related mostly to these products. A decade ago coating machines for tapes ran at $60-220 \ \text{m/min}$ for solvent-based PSAs and $100-250 \ \text{m/min}$ for water-based PSAs [516]. Now they run at about $250-500 \ \text{m/min}$ depending on the type of adhesive. According to Ref. [762], the production speed of such tapes is increased from about $250 \ \text{m/min}$ in 1985 to about $600 \ \text{m/min}$ in 1995 and the coating machine width increased from ca. 1400 mm in 1985 to about 2000 mm in 1995. The average coating speed increased from about 100 to about $300 \ \text{m/min}$. The average coating machine width increased from 1300 to 1500 mm. A new coating machine for tapes having a width of $1200-2400$ mm is run with HMPSA on a $25 \ \mu\text{m}$ BOPP film as carrier material, at a speed of $400-600 \ \text{m/min}$. Hot-melt PSAs, acrylics (mostly water-based) and rubber-resin formulations have been used for such tapes. Advances have also been made in coating devices. An attempt has been made to combine the forced adhesive feed of gear-in-die systems with the web transport and adhesive coating of the roller-based systems. Generally with gravure systems there is the problem, that the gravure cylinder is not filled with the adhesive, because of its viscosity, rheology, speed, foam, etc. In Accugravure systems high pressure is used to fill the cylinder; in the Vario coater a knife is pressed on the cylinder to allow better filling [777]. The Accugravure system ensures the complete filling of the gravure lines under hydraulic pressure. The curtain coater discussed earlier is another improved coating device.

In the United States biaxially-oriented PP tapes became the most important packaging tapes [778]. These products require high tack toward corrugated paper board at low temperatures and high shear. There is a trend on the market to provide customers with printed packaging tapes. The text or graphics is imprinted on the nonadhesive side of the carrier material before the tape is made. The problem with hot-melts is that a release coating has already been applied to the backing material before it is imprinted. The coating is necessary because hot-melts do not release easily from the nonadhesive side of the carrier material.

Suppliers for coating equipment offer the full range of machines and devices for a coating line, for example, coating devices with air knife, reverse roll, roll-blade, blade, gravure roll, slot-die, curtain coater, driers, winding machines, humidification equipment (jetsteamer), etc. [779]. The most important developments in machine design are unwinders with fully automatic reel changeovers at top line speed, with a very short and constant overlapping tail;

preconditioning rolls with tension nips, in-line flame treater, release coating unit with total solvent capture, air flotation dryer, coating device with slot orifice die with automatic adhesive thickness control and rewinder with independently indexing arms, allowing changeover at full speed [762].

Except for the combination of BOPP-water-based adhesive, and that of filament-reinforced BOPP-hot melt adhesive (with higher stiffness and lower contact surface) generally strapping, packaging and masking tapes usually require a primer. Extensible reinforced packaging tapes have been developed that use a fiber-like reinforcing material laminated on or embedded in the tape carrier material [780]. A closure tape based on crosslinked acrylate terpolymer uses a reinforced web and a reinforcing filler. Such an adhesive is useful for bonding the edges of a heat recoverable sheet to one another to secure closure. An amine-formaldehyde condensate and a substituted trihalomethyltriazine are used as additional crosslinking agents [781].

b. Manufacture of Insulating Tapes

Insulating tapes are generally carrier based, adhesive-coated products but they can also be carrier-less (extruded) or adhesiveless (extruded) products are. Special (temperature, chemicals, and humidity-resistant) adhesives with high coating weight are coated on soft or metallic carrier materials. For instance, silicone PSAs that can be cured at low temperature have been developed from gum-like silicones containing an alkenyl group, a tackifying silicone resin, a curing agent, and a platinum catalyst. Such adhesives for heat-resistant aluminum tapes are coated with a thickness of 50 μm and cure in 5 min at 80°C. Heat-resistant insulating tape formulation can include a fluoropolymer, according to Ref. [782]. Such a tape displays a rolling ball tack of 14 cm and peel value of 45 g/20 mm. An aluminum carrier and a tackified CSBR coated with 40 g/m^2 are used as insulating tape in air conditioning.

c. Manufacture of Mounting Tapes

Mounting tapes are generally products coated on both sides. Such tapes may or may not include a carrier material. Generally their manufacture consists of the coating (on both sides) of a carrier material, that is postlaminated with a solid release liner. Mounting tapes with carrier can also be designed as "adhesiveless," that is, self-adhesive. Such products are extruded. Carrier-less mounting tapes are transfer tapes. Transfer tapes are carrier-less mounting tapes that are manufactured by coating, coating and foaming, or extrusion (Table 8.38).

The coater of double-sided tapes performs both transfer and direct coating. The first coating station deposits the adhesive on the release liner (direct coating). In a combining nip the adhesive coated release liner is sandwiched with the carrier material (transfer coating). The duplex structure then proceeds to the second coating station, where a second adhesive layer is applied on the carrier material (direct coating). The manufacturer of foam mounting tapes or other double-faced tapes must often use relatively thick films (6–8 dry mills or higher) of PSA. This is done to ensure that the adhesive film will be able to conform to a surface that is irregular or unpredictable [785]. Generally, high coating weights are suggested. Polished cylinder and wire-wound rod are used for higher coating weights, particularly for paper. They have a coating weight tolerance of 15%. For instance, for mounting tapes a 88 μm thick adhesive layer is coated; for splicing tapes 50 μm adhesive thickness is used [44]. For double-faced tapes different degrees of release can be achieved on the two faces using the same release agent but with polyethylene-coated release paper on one side. Splicing tapes are manufactured by transfer coating [135].

A Belgian patent [76] describes the polymerization by UV light of acrylic monomers direct on the carrier to manufacture a tape. The liquid monomers are applied to the carrier with a doctor blade, roller, or sprayer in a polymerization tunnel from which oxygen has been excluded. This patent uses a temporary carrier that is an endless belt, which does not become incorporated in

TABLE 8.38
Manufacture of Carrier-less Tapes

Code	Chemical Basis	Manufacturing Process	Ref.
Special	Acrylic emulsion, ammonium stearate, tackifier	Mechanically foamed, coated on siliconized PET, impregnated with acrylic adhesive	[783]
Sealing	Acrylic, UV-cured, silica filled	Filled, coated, cured	[784]
Releasable sheet	Acrylic, filled with expandable microbubbles	Filled, coated, cured, blown	[63]
Releasable sheet	Acrylic-urethane filled with expandable microbubbles	Filled, coated, cured, blown	
Transfer	Acrylic copolymer	Filled, coated, cured, blown	[61]
Insulating	Chlorinated polyethylene, acrylonitrile butadiene rubber, liquid chlorinated paraffin	Filled, extruded	[683]

the final product, and the polymerization is completed thermally in an oven. The tapes have PSA on both sides of the carrier.

Curing by UV light is a preferred method in the manufacture transfer tapes. A prepolymer having a viscosity of 0.3–20 Pa sec can be synthesized by UV radiation. The crosslinking agent is added, and the mixture is coated [29], and polymerized by UV to the final product. This composition can be used as a transfer tape. A foam carrier can be applied too. Both surfaces of the carrier may have low adhesion. For a crosslinkable formulation highly volatile vehicles must be used as solvents in order to avoid excessive heating. Such a tape is has a coating weight of 0.5–1.5 mm. A thick coating can be laid down in one pass according to Ref. [27]; if solvents or aqueous adhesive are used multiple coating is necessary to achieve a thickness of more than 0.2 mm. Foam carriers coated on one or both sides can have an overall thickness of 0.1 to 2.0 mm. A knife coater is recommended for coating a photopolymerizable acrylic prepolymer filled with glass microbubbles and used as transfer tape [24]. Sealant tapes are made by extrusion on a release film [50].

According to Harder [72], the main raw materials used for PSA tapes have been acrylonitrile copolymers, butyl rubber, natural rubber, polyacrylates, polyester, polyurethane, styrene–butadiene–styrene (SBS) and styrene–isoprene–styrene (SIS) block copolymers, and silicones. The number of suppliers for acrylic HMPSA is very small. Such materials are generally low viscous prepolymers which must be postcured or polymers having excessive viscosity (see also Chapter 5). For instance, a HMPSA on acrylic basis, obtained by solution polymerization, can be coated at 90–140°C and is crosslinked by using EB and UV-radiation or chemical crosslinking (by NCO-containing systems).

Coating lines can be divided into equipment for adhesives with viscosities of less than 500, 200–2000, and for 500–5000 Pa sec [77]. Most acrylic HMPSAs belong to the lowest viscosity range and can be coated with die systems. Coating weights of 20–200 g/m^2 can be applied. Adhesives belonging to the medium viscosity range, have higher molecular weight, and demand lower coating temperatures and shear stresses. In this case a combination of roll coater and calender coater is used (with two heated rolls). Coating weights between 20 and 100 g/m^2 are recommended. For high coating weights the high viscosity acrylic HMPSAs are coated by using an extruder.

A foam-like pressure-sensitive transfer tape with good shear strength and weather resistance useful in bonding of uneven surfaces has been prepared by forming foamed sheets from agitated acrylic emulsions, impregnating or laminating them with adhesives and drying. A similar film is coated on a PET release film, as a 1.2 mm adhesive sheet, having a density of 0.75 g/cm^3. It displays (in the bonding of acrylic polymer panels) an instantaneous shear strength of 4.5 kg/cm, and shows after weatherometer exposure twice as much resistance as a common polyurethane film [786]. Tapes with microspheres useful for sealing windows, and bonding side moldings onto automobiles contain microspheres and silica [784]. Such tapes are UV-cured on PET. For such products the hydrophobic silica improves the adhesive properties. Similar carrier-less UV-curable tapes are filled with glass microbubbles and silica.

d. Manufacture of Medical Tapes

Medical tapes are carrier-based products, that are generally manufactured by adhesive coating. Because of the absorbtion, dosage, and storage function of such tapes, they are coated on special porous carrier materials with high coating weight of special adhesives. These adhesives do not contain volatiles. Therefore medical tapes were first manufactured by roll pressing or with high solids, later by extrusion and radiation curing. According to Ref. [778], commercial PSAs used for skin application are based on acrylic adhesives, which are easier to remove and cause less skin irritation than rubber–zinc oxide adhesives, which adhere well to the skin but may cause skin irritation because of their higher adhesion level. Acrylic solutions and PUR solutions both are used. Both adhesives exhibit a continuous decrease of G' as a function of temperature. Acrylics used for finger bandage show a more abrupt decrease; their tan δ peak is shifted to higher temperatures (see also Chapter 3).

The first compositions for medical tapes contained natural rubber or polyisobutylene. Later polyvinyl ethers have been applied. Polyvinyl ethers can also be coated by extrusion. As discussed in Ref. [783], polyisobutylene or a blend of polyisobutylene (5–30% w/w) and butyl rubber, a radial SBC block copolymer (3.0 to 20% w/w) mineral oil (8.0–40% w/w), wate-soluble hydrocolloids (15 wt.%, tackifier (7.5–15 wt.%) and a swellable cohesive strengthening agent are used to formulate a medical adhesive. A mixture of *cis*-polybutadiene, an ABA block copolymer, and tackifier resins have also been proposed for medical adhesive. Such an adhesive mixed with an antiinflammatory paste is spread on a film carrier material for medical poultices [787].

Polyvinyl ethers do not produce skin irritation, therefore their use is suggested for medical applications [788]. Polyvinyl ether processed as hot-melt can be used for double-faced tapes with textile carrier [789]. The water vapour porosity of PVE is about the same as that of the skin 259 g/m^2 over 24 h.

Acrylic adhesive compositions for medical tapes which do not leave adhesive residues on skin contain *tert*-butyl acrylate-maleic anhydride-vinyl alcohol copolymers having a molecular weight of 2500 to 3000 and acrylic acid–butyl acrylate copolymers [790]. The adhesives for medical tapes have to meet the following requirements [778]: they should not irritate the skin, they should adhere well but be easily removable, and they should be water resistant. For medical uses the direction of extensibility is very important. Cross-direction elasticity may be given by a special nonwoven material [791]. The wound plaster placed in the direction of elasticity on the carrier is narrower than the carrier. Soft PVC is used as plaster (70–200 μm) with a silicone liner. These films are dimensionally stable up to 200°C [505].

Cotton cloth and hydrophobic textile materials coated with low energy polyfluorocarbon resins are used as carrier materials for surgical tapes, and coated with polyacrylate-based water-soluble PSA. Reusable textile materials as Goretex, polyamide, polyether sulphone/cotton, 100% cotton, or other hydrophobyzed carrier materials are suggested for medical PSPs (operating tapes, labels, and bioelectrodes). Special tapes are used for biolelectrodes because of their low electrical impedance and their conformability to the body. Such products are manufactured by photopolymerization [82]. Compositions include acrylic acid polymerized in a water-soluble polyhydric alcohol and crosslinked with a multifunctional unstarurated monomer. A prepolymer is synthesized which is polymerized to the final product using UV radiation [98]. This polymer contains water, which affords electrical conductivity. Electrical performance can be improved by adding electrolytes to the water. As known, electrically conductive PSAs can be manufactured as gels containing water, NaCl, NaOH [792]. Such gels are useful in detecting voltages in living bodies. Such polymers based on maleic acid derivatives have a specifical resistivity of 5 kΩ (1 Hz).

Radiation-curing has been proposed for medical tapes also. This method is limited only by the coating weight. UV-coating is practicable up to 70 g/m^2 [793]. UV-curable formulations can use difunctional or polyfunctional vinyl ethers for increased adhesive bond strengths after curing [794].

e. Manufacture of Application Tapes

Application tapes are mounting tapes used for placing of signs, letters, decorative elements, etc., die-cutted from plotter films. Such products are special removable tapes with a carrier base. Paper and film carriers are used for application tapes. Special adhesive formulation and special coating methods ensure removability.

In some cases the coating technology used depends on the adhesive surface geometry. As discussed earlier (see Chapter 5), the reduction of the adhesive contact surface enhances removability. This reduction can be achieved by the choosing an appropriate coating device. As known, film-based application tapes have to be conformable but removable. Therefore for such products gravure printing is favored. The pattern-like, striped surface structure of the adhesive ensures less contact surface and better removability. A decrease in the contact surface decreases the buildup of static electricity also. Such a coating can be unwound with less buildup of static charge [795].

As mentioned earlier application tapes are produced as paper- or film-based PSPs. Although used for the same application products based on paper are quite different in buildup from those based on plastic film. Products with a film carrier are more conformable than those with paper. For an excellent lay flat the carrier contains HDPE and is manufactured via coextrusion. To avoid curl from unwinding during application, adhesion should be minimum on the backside. Because of the different surface qualities of plotter films (calendered or cast PVC, cast PE, etc.), which function as the adherend surface for application tapes, a range of application films has been developed, that have different coating weights. The choice of hard acrylic dispersions as base formulation for the application films allows the design of an uncrosslinked formulation for the film-based products. Paper-based products are formulated with NR latex as main component. Both grades of tape can be produced with or without a supplemental release layer.

f. Masking Tapes

Generally the manufacture technology used for masking tapes is the same as that used for protective films. However, there are some special tapes that are produced with a complex technology. For instance, a laminated PSA tape having good chemical, heat and water resistance, that is applied for soldering portions of a printed circuit board, is manufactured using a kraft paper carrier material impregnated with latex, treated with a primer or corona discharge and coated on one side with a rubber-resin adhesive, and on the other side with a polyethylene film and release agent [796].

Thick masking tapes with "bubblepack" character and improved cuttability can be manufactured with classic coating technology, including an expandable filler in the adhesive. According to Ref. [62], a photocrosslinked acrylic–urethane acrylate formulation used for holding semiconductor wafers during cutting and releasing the wafers after cutting, exhibits a peel strength of 1700 g/25 mm without crosslinking and 35 g/25 mm with crosslinking. Adding an expandable

TABLE 8.39
Adhesive Characteristics of Selected PSPs

Adhesive Formulation	Adhesive Characteristics	
	Peel Resistance (N/25 mm)	Rolling Ball Tack (cm)
Unfilled AC PSA	18.0	2.8
Filled (10%) AC PSA, not expanded	17.0	3.5
Filled (10%) AC PSA, expanded (140°C)	8.0	3.2

filler to the formulation (6 to 100 parts adhesive) reduces the peel resistance. Such composition exhibits $1500 \, \text{g}/25 \, \text{mm}$ before crosslinking and $15–20 \, \text{g}/\text{m}^2$ after crosslinking. A releasable foaming adhesive sheet can be prepared by coating blowing agent-containing PSA (30%) for a $30 \, \mu\text{m}$ dry layer [63]. Blowing up of the adhesive layer reduces contact surfaces. Such products exhibit lower bond strength (Table 8.39). However, they are only temporarily removable. Adhesive flow leads to rapid buildup of the peel resistance (Table 8.40). Therefore for adequate removability, both blowing up of the adhesive layer and crosslinking are needed.

4. Manufacture of Protective Films

The choice between adhesiveless and adhesive-coated protective films is determined by the main characteristics of the products. The manufacture of adhesiveless protective films or protective films with embedded adhesive was discussed earlier. The procedures yield removable products with an adhesion of maximum $1.5 \, \text{N}/10 \, \text{mm}$. Such peel resistance values can be achieved with special adhesives and crosslinked formulations and low coating weight. Therefore adhesive coated protective films generally need a crosslinked adhesive, with low coating weight.

Similar low coating weights of a crosslinkable adhesive are used in laminating also. According to Klein [797], coating weights of $0.8–1.5 \, \text{g}/\text{m}^2$ are used for laminating smooth surfaces, $1.2–1.8 \, \text{g}/\text{m}^2$ for printed films, and $3.5–4.5 \, \text{g}/\text{m}^2$ for paper laminates. Therefore, at least theoretically, the coating technology of protective films is related to the well-known procedure used for laminating films. Its choice depends on the type of the adhesive and its coating weight and geometry, on type of carrier and its geometry, and on the end-use of the product.

Hot-melts are limited by their low heat resistance, low penetration into porous substrates and relatively high application temperatures of $150–200°C$. Even at these application temperatures, some hot-melts are highly viscous (which affects their ability to be coated with very low coating

TABLE 8.40
Peel Buildup for Expanded PSP[a]

Dwell Time (day)	Peel Resistance (N/25 mm)
0	10
1	15
2	19
5	PT[b]

[a] A $30 \, \mu\text{m}$ (wet) PSA layer was coated on paper; peel on glass has been measured.

[b] PT, paper tear.

TABLE 8.41
Coating Weight Tolerances for Protective Films

Product Code	Coating Weight Value (g/m^2)		Coating Weight Tolerances (g/m^2)		
			Extreme Tolerance Values		
	Theoretical	Measured	Minus	Plus	Average Tolerance (%)
1	1.6	1.86	0.2	0.8	16
2	1.7	1.77	—	0.2	4
3	1.7	1.85	—	0.2	9
4	2.2	1.90	0.4	—	7
5	2.4	2.57	0.3	0.4	7
6	2.8	3.00	0.1	0.9	7
7	2.9	3.34	—	0.7	15
8	4.0	4.60	—	0.5	15
9	4.5	4.28	0.1	—	5
10	5.7	6.30	—	0.4	11
11	7.0	7.10	—	—	1.5

weights) and can cause damage to sensitive substrates. The applicability of hot-melts is limited by the restricted formulating freedom for removable recipes too.

Aqueous, crosslinked adhesives for protective films can be coated using a gravure cylinder (quad system) 65, 96, and 120 quads/in. For instance, 65 quads/in. give a coating weight of 8 g/m^2 and 120 quads/in. give a coating weight of 4.4 g/m^2. In this case, a doctor blade is used to smooth the adhesive surface. An intermediate cylinder and a pressure cylinder are used also. According to Fust [402], flexo, letterpress, and gravure printing provide a coating weight of 1.0–3.0 g/m^2. Therefore, it would be possible to use them for the manufacture of protective films also. Screen printing works with a coating thickness of 5–10 μm, therefore it is suitable for label, tape or protective film production. Depending on the roughness of the product surface, on the laminating, delaminating and processing conditions chemically different adhesives, with different coating weights are applied with various coating devices. The mean nominal coating weights are between 1.5 and 7.0 g/m^2. For low coating weights, positive tolerances are preferred (Table 8.41).

As discussed in Chapter 7 the peel resistance values required for the removable protection of different surfaces are very different. Generally peel values of 0.3–1.3 N/25 mm are recommended. In practice there are much lower values also. For such products the coating weight is between 0.5 and 1.5 g/m^2 and the range of coating tolerances is very broad and may be as much as ± 0.5 g/m^2.

As is known, the ratio of the viscosities of adhesive compositions based on unmilled and milled rubber is about 5–10, which means that at least theoretically low viscosity rubber-resin solutions may be coated for protective films, although the resin concentration of such recipes is much lower than for common label or tape formulations (Table 8.42). Unfortunately, crosslinking leads to a decided increase in viscosity.

Release coating for protective films and tapes presents a complex problem. It depends on the nature of the adhesive coated and its coating weight. As seen from Table 8.43, different release coatings exhibit different debonding forces and different increases in release force per unit coating weight. The best formulations display a "constant" release force for different coating weights. The carbamate-based release coatings (1 and 2 in Table 8.43) provide low level release

TABLE 8.42
Elastomer/Tackifier Concentration in Removable Products

PSP	Elastomer	Nonresin Tackifier	Plasticizer	Tackifier	Ref.
		Concentration of Components (pts)			
Removable sheet	NR (65–75)	PIB (0–10), PB (0–10)	—	Hydrocarbon resin (25–35)	—
Masking tape	NR (70)	—	—	Terpene resin (50)	[159]
Masking tape	AC (100)	—	—	Terpenephenol resin	[183]
Protective sheet	NR (100)	—	DBP (5–10)	Rosin (100)	—
Protective sheet	NR (100)	—	—	Rosin (55)	—
Protective sheet	NR (100)	—	—	Rosin (60)	—

force for different coating weights. A polyamide-fluoropolymer base formulation gives very low release forces at low coating weight, but high release forces (as high as without coating) at high adhesive coating weight.

5. Main Equipment

Generally the following parameters are determinant concerning the versatility of a coating equipment to produce PSPs belonging to different product classes: the ability of the coating device to process PSAs of different physical states, with different coating geometries, the on-line slitting and cutting possibilities, the existence of lamination equipment, the possibility of in-line flexo printing, and the possibility of in-line release coating. According to Massa [705], the main processes in self-adhesive coating and lamination are web handling, coating, drying and/or curing, and moisturizing.

In the recent decades specially designed new machines have been built, but "second hand" modernized old machines still exist [798]. For such machines modular design is important for flexibility and for the possibility of upgrading or rebuilding in the future.

TABLE 8.43
Dependence of the Peel Resistance on the Back of Protective Film on the Release Agent Used

Coating Weight (g/m^2)	Without Release Coating	Release Coating 1	Release Coating 2	Release Coating 3
	Peel Resistance (N/10 mm)			
2.5	0.20	—	0.15	0.05
4.0	0.25	—	—	—
5.0	0.30	—	0.05	0.05
6.0	0.33	0.20	0.05	—
7.0	0.43	0.15	—	—
9.0	0.65	—	—	0.50
10.0	0.65	0.20	0.27	—
11.0	0.68	0.20	—	—
12.0	0.72	—	—	—
13.0	0.77	0.20	—	0.80

a. Coating Device

Generally an apparatus for applying low and medium viscosity liquid adhesives includes a rotating applicator cylinder, an adhesive tank to supply the adhesive to the roller, and at least one doctor blade operatively associated with the applicator roller. The liquid coating components may interact with the coating cylinder; therefore it is recommended to take the adequate materials [799]. The applicator cylinder may be polished or engraved.

Generally gravure coating is used for laminating (solvent-based, solventless, and water-based systems) and for coating solvent-based, water-based, or hot-melt PSAs [800]. The characteristics of such cylinders are described in Ref. [801]. The main parameter affecting the choice of a coating device is the rheological behavior of the adhesive. It is well known from the practice of laminating, that adhesives with different flow characteristics require different coating devices. For instance, for low viscosity (14–15 sec, DIN Cup No. 4) solvent-based adhesives, smooth reverse rolls are recommended; for the same adhesives with 20 sec viscosity, gravure rolls are suggested. Gravure and reverse gravure require low viscosities, 200–500 cP [802]. Such coaters are well suited for water-based PSAs. Reverse gravure ensures narrow tolerances for the coating weight. On the other hand, for such a coating device it is necessary to change the coating cylinder for different coating weights. The maximum viscosity is limited to about 1000 cP [705]. At high speeds air may be entrained with the cylinder, that is, foaming may appear. Air entrainement may be avoided by using an enclosed supply chamber (Figure 8.10). For highly viscous, cross-linked adhesives gravure coating may lead to a discontinuous or contoured coating layer. Such coating is preferred for certain removable products (e.g., application tapes). For water-based PSAs, the main coating device used is the one-roll reverse coating with metering bars. This is an inexpensive, easily operated, flexible device. Its main disadvantage appears at high speeds where foaming may be a problem. The maximum working speed is about 200 m/min [705].

A gear-in-die, slot orifice die has also been developed. This device has a liquid pumping section built into the die. The pump is designed to cover the complete die length. Advantages include higher solids, accuracy, no foaming, and low shear. An air knife can also be used as coating device [803].

Common application systems for hot-melts include rotogravure, transfer gravure, and smooth roll. The parameters that affect processing of HMPSAs are viscosity, coating, speed and coating weight. Rubber-based conventional hot-melts exhibit a viscosity range situated between 10,000 and 60,000 mPa sec [756]. Rubber-based EB-curable hot-melts have a viscosity range of

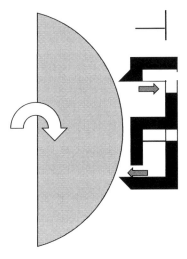

FIGURE 8.10 Enclosed supply chamber.

20,000–100,000 mPa sec; acrylate-based EB- or UV-curable systems have lower viscosities (10,000–30,000 mPa sec at 120–170°C). For low viscosities (100 P), roll coating is suggested, and for higher viscosities (350–5000 P), slot-die [723,724]. Slot-die is used for 10 g/m^2 and 120,000 mPa sec or more [719]. Slot-die coating systems can coat 1–2500 g/m^2. According to Ref. [648], for hot-melts, slot-die application is suggested for use up to 120,000 cP (200°C), and roller coating for 100,000 cP. Coating weights of 1–80 g/m^2 (in special cases down to 7 g/m^2 and above 80 g/m^2) are obtained [804]. Coating trials with acrylic block copolymers allowed a coating speed of 500 ft/min using a die-coater [805]. The slot-die is recommended for labels, plaster, pressure-sensitive foams, and pressure-sensitive textile materials [806]. Slot-die coating devices may run faster (exceeding 400 m/min) than roller systems, have a broader working width (3000 mm or more), produce less oxidation of the PSA, and coat viscosities up to 1×10^6 mPa sec more exactly ($\pm 0.5\%$), but changing the coating material or coating width, startup and cleaning are more complex. Therefore roller systems are suggested for frequent product changes or short production runs. Slot-die is suitable for one product, high speed, 24 h operation.

Roll coaters can coat hot-melts with a viscosities between 500,000 and 1×10^6 or 2×10^6 cP [723]. Roll-coating systems for HMPSA are suggested for 100–1500 g/m^2 [806]. According to Drechsler [807], roller-coating systems for hot-melts are adequate up to viscosities of 15,000 cP. For such viscosities a three-cylinder roller-coating system is suggested. For lower viscosities (70 P) a central driving system is applied. For viscosities over 50 P forced feed of the hot-melt onto the gravure cylinder is recommended [808]. Heated coating cylinders with 20 lines/cm and halfmoon-like cells of 80–100 μm depth are suggested for a coating weight of 12 g/m^2, and 18–34 lines/cm and cells of 40–50 μm depth for a coating weight of 7–8 g/m^2. The temperature of the gravure cylinder should be kept 10–15°C higher than that of the molten hot-melt adhesive [809]. According to Ref. [810], gravure cylinders are suggested for hot-melt coating for viscosities less than 5000 cP. Reverse roll coaters and kiss coater are used up to 20,000 and 30,000 cP, respectively.

Spray-coating technology for hot-melts has been developed from first generation melt-blowing to systems with controlled fiberization. Their main functional criteria are fiber size, fiber density, absorbency control, edge control, product aesthetics, minimal operator involvement and minimal air consumption [811]. As substrates are bonded at thousands of points of contact, fiberization allows the manufacture of light weight products and can also be applied for thermally sensitive carrier materials. Coating weights as low as 2 g/m^2 can be achieved [812].

Coating of HMPSAs supposes the existence of premelters, melt tanks, melt extruders, filters, feeding lines, pumps, and heating systems. Such systems are described in a detailed manner in Refs. [813,814].

Quite different coating weights are required for permanent or removable labels, tapes, labels, or protective films. Different adhesive layer buildups may be necessary for special products. Therefore other, special coating devices are used also. For instance, one special coating device works according to the principle of rotary screen printing [720]; it is recommended for tapes and labels. It can also coat double-faced materials. Screen printing has been recommended for adhesive layers of 7–60 μm [694]. In this case the coating device includes a melt supply, coating slot-die, gear pump, temperature controllers, metering bar, coating of the backup roller, chill roll, and tension control.

As illustrated by Table 8.44, very different products need pattern coating.

Special coating devices were developed for pattern coating, that is, coating of discrete portions of the product. A common reverse gravure system does not allow pattern coating. A special patterned reverse gravure coating is described in Ref. [815], in which a tampon cylinder (which contacts the gravure roll) is profilated. The printing cylinder is used for pattern-coating of hot-melts; dots in excess of 24,000/in.^{2} are coated with a speed of 500 ft/min, and specific zone coat patterns — squares, circles, rectangles, or stripes can be applied. Pattern transfer and lay down are affected by the cylinder mesh sizes, adhesive viscosity, doctor blade pressure and tangent point,

TABLE 8.44
Pattern Coating for PSPs

Product	Application	Adhesive	Coating Method	Ref.
Tape	Removable from paper	HMPSA	Pattern	[664]
Label	Wrinkle resistance for squeeze bottle	HMPSA	Circular dots	[413]
Tape	Diaper closure	HMPSA	Pattern	[690]
Protective paper	Plastic plate protection	NRL	Profilated coating	[251]

temperature, and operating speed [816]. Patterned-coatings can also be applied with a perforated cylinder [817]. Tools of the same thickness as the coating weight and with an elastic cylinder are also used for patterned coating [818].

A removable display poster is manufactured by coating of distinct adhesive and nonadhesive strips on the carrier [819]. The adhesive strips lie on the same plane, the plane being elevated with respect to the product surface. Another procedure described by Gleichenhagen [3], uses a piramidal PSA coating site (30×600 μm base dimensions), coated via screen printing. Rotary screen printing can be used with a coating weight of $1-20$ g/m^2, a running speed of $10-100$ m/min, screen geometry $15-40$ holes/cm, blade $1.5-30$ mm; blade thickness $150-300$ μm, and pressure of $2-6$ N/mm. Patterned hot-melt coating may reduce adhesive costs by 50% [820]. According to a patent for removable PSPs [642], the carrier may be coated alternatively with adhesive and release stripes. A patterned-coating can be replaced by special postcoating reduction of the adhesive surface as discussed in Ref. [821]. The regulation of the adhesive properties can be achieved by partial detackifying after the adhesive is coated on a temporary release liner and transferred onto the final support.

Special coating machines allow sheet-like materials to be coated also. The procedure uses transfer coating of the adhesive from a transfer cylinder or tape onto a primed and release coated sheet-like face stock material [652].

Primers are supplied either as solvent-borne, water-borne, or solid-state products. Equipment can influence the choice of a primer. It determines whether the primer is to be applied in-line or off-line [653]. The need for a prime coat influences the complexity and economics of PSP manufacture.

Different adhesives and coating procedures give different coating images. For crosslinked adhesives whose fluidity is limited in time (because of their gelling) this phenomenon can lead to quite different product build-up with the same application characteristics. As discussed earlier, the same grade of protective film manufactured with different coating devices will have different coating weights (but exhibit the same application characteristics). On the other hand, the same type of coating device but with different characteristics (e.g., number of Meyer bars), may lead to higher coating weights (as usual) and much higher buildup of the coating weight. The very low coating weights as usual for protective films, can coated with a four-cylinder nip-feed coating device (Figure 8.11).

b. Drying

The drying capacity has to be increased for water-based coatings in order to provide the extra energy required to evaporate water compared to other solvents. This is difficult to achieve without extending the length of the dryers, reducing the drying speed, and affecting the printing register, a critical point when coating plastic webs. Drying influences surfactant migration also. According to Pitzler [753], in the transfer coating of acrylic dispersions, the migration to the film surface is increased by drying.

PSA

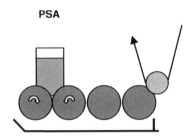

FIGURE 8.11 Nip-feed, four-cylinder coating device for low coating weights.

Drying speed is influenced by the type and formulation of the adhesive. Because of use of multiple-coated layers with different dispersing media, the buildup of the drying equipment may be complex. For instance, for drying in-line-coated double-sided tape (with release and primer) because of the different solvents and coating weights applied, the use of two different drying units is suggested. For drying the release layer a tangential air jet dryer, and for the primer air jet with a longer channel are recommended. The use of solvent-based adhesives imposes the removal of organic solvents during the drying, and special solvent recovery system.

On the other hand, the same drying channel is frequently used to handle both solvent-based and water-based adhesives. For solvent-based coatings the air volume necessary for drying (V_{min}) is given by the following formula [814]:

$$V_{min} = \frac{G_{max}(273° + \vartheta_{max})}{(C_r K_s f \times 293)} \qquad (8.13)$$

where G_{max} is the maximum amount of solvent, ϑ_{max} the maximum temperature of recirculated air, C_r the lower explosion limit, K_s a security parameter, and f the parameter that takes into account the aerodynamic conditions in the channel. In Equation (8.13) G_{max} is given as a function of the coating weight C_w, web width W_w, and web speed W_s.

For water-based adhesives the relative humidity of the air is the most important parameter of drying. Different adhesives need different drying temperature. For instance, the drying temperature recommended for a thermoplastic adhesive is about 200°C, whereas thermosetting adhesives need 250°C, self-crosslinking adhesives require 200°C, and emulsion-based adhesives are dried at 212°C.

IR drying offers the possibility to regulate the drying intensity in the cross direction, and is characterized by low space requirements [822]. IR drying can be used as intermediate drying between coating and printing devices [823]. A combination of IR and air drying gives better results [705]. Drying consumes 70% of the energy of a printing machine [824]. Exhaust and recycling have to be adjusted according to the solvent concentrations. If multiple layers are coated, adequate drying time should be allowed for each layer. In a special system the thin water- or solvent-based wet coat is transferred onto a special heat-storing endless release belt. This system can work with high temperatures without the risk of skin formation and considerably shorter than standard machines.

A classic machine for OPP adhesive tapes for cardboard sealing, designed for a production speed of 350/min, uses a solvent-based PSA with a coating weight of 20 g/m² (solids content 28%), a solvent mixture of hexane (70%) and toluene (30%), and an OPP carrier of about 40 μm.

Electrostatic control is fundamental for safe conversion. The formation of an electrostatic charge is associated with low relative humidity also. Therefore humidifers (humidification systems) are used. For paper processing a temperature of 20–22°C and air at 55–60% relative humidity (RH) are suggested [825]. Deionizing bars or systems (static eliminator bars which neu-tralize the charge with a vacuum hood to remove contaminating particles) are recommended [826].

c. Laminating

Laminating is required for multiweb constructions. These include PSPs that have a separate release liner (e.g., labels and tapes). According to Bulian [695], the most important part of a lami-nating machine is a nip between a rigid roller and an elastic roller, so that enough pressure is achieved to bring the two materials in contact.

6. Auxiliary Equipment

Various kinds of special equipment ensure the web-like processing of PSPs. Winding and printing devices are the most important special parts of the PSP manufacturing equipment. Winders are constituents of web-handling equipment in paper, film, and metal processing industries among others. Printing is a complex industry itself. It is not the aim of this book to discuss in details the problems related to the construction and use of such equipment. Therefore only some special features are discussed here.

a. Winding Equipment

Unwinder web tension control is a main parameter. The winding tension affects the pressure in the roll and the diffusion (migration) of viscous components in the carrier for self-adhesive films (SAF). It depends on the hardness of the film and has to be adjusted on the winder [665].

Many versions of unwinders and rewinders are known: simplex, duplex, nonstop, fully auto-matic, with disc brakes or with dc motors, with different methods of reel core chucking and locking, with constant or variable tension. Flying splice unwind/rewind stands with electronic tension control units situated between the unwind reel stand and first coating station, and between the last coating station and the rewind stand are used. Tension can be controlled by fully regenerative d.c. motor driven draw rollers operating through a low friction dancer roller with potentiometer and tachometer feedback reference [827].

Controlled, constant web tensions are very important for thin films. Generally a two-component regulating system is used. The first component regulates the torque, and the second acts as feed back. Unwind reel diameter is controlled and torque is regulated proportionally. A sensitive dancer roller measures the tension. Fine control of the web tension was first required in the coating of soft PVC films [828]. Similar machines were developed later for thin protective films.

Winding machines with a web tension regulation of $50-200$ g/mm^2 and web tensions of 0.1 to 10 kN are common [208,423]. The maximum winding speed is $20-140$ m/min for blown films, $120-400$ m/min for cast film, and $280-350$ m/min for biaxially oriented film [168] (see also Chapter 10). Web cooling before winding may be required also [829]. Winding is described in details in Refs. [830,831]. The lowest web tension is about 20 N [832]. Generally winding should be tight enough that the film does not shrink later due to crystallization [833]. Web tension should not exceed 10 N for thin plastic films. Common winding machines used for blown film manufacture do not meet these requirements; because of the excentricity of the roll forces of $200-300$ N may appear. A surface- or center rewinder allows production of hard or soft rolls with laminates up to 200 g/m^2.

b. Web Control Equipment

On-line thickness measurement is carried out with sensors that are not in contact with the web. Optoelectronic (laser triangulation, laser scanner, light transmittance), capacitive (condensator), inductive roll sensor, contact, pneumatic (pressure die), radiometric (β-radiation) or acoustical (ultra sons) principles are used. According to Ref. [834], the sensors most often used for measuring film thickness usually fall into one of three categories: caliper, nuclear or infrared. Caliper gauges make a direct, physical measurement of total thickness. Such gauges can contact the product on both sides or may have a relatively thin contact layer of air on both sides. Such devices are not capable of measuring the individual layers of the laminate. Nuclear sensors pass β or γ radiation through the material. They cannot measure the thickness of individual layers either. Infrared sensors use the 1.30–2.70 μm portion of the near-infrared spectrum and work like a spectrometer. The limited number of filters in a sensor restricts the number of polymer layers in a laminate that can be measured simultaneously. Web side control by ultrasound is possible at running speeds as high as 1200 m/min [835].

c. Printing Equipment

The majority of PSPs are printed during the manufacture. Such printing can be carried out one side or both side, in different qualities. Printing during manufacture of the pressure-sensitive laminate may provide a technical aid for postcoating (e.g., labels), or postprocessing of the PSP (e.g., protective films), or serve for advertising. A solvent-based or water-based printing technique may be used. Printing with water-based inks is more difficult because the different properties of the carrier liquid and ink components (see also Chapter 5 and Chapter 10). Drying and cleaning of the gravure cylinder need special operations. The risk of scumming problems is increased, and scum is more difficult to remove. With solvent-based formulations, when scum appears, it is common practice to blow compressed air on the printing cylinder, between the doctor blade and the impression roller.

For labels generally both printing and overprinting (see Chapter 10, Section I.c) are carried out. Narrow web printing is done after confectioning and is more and more often carried out by the end-user. Labels are printed with various printing methods including letterpress, flat-bed, semirotary, rotary, flexography; lay and rotary offset, lithogravure, hot foil, and in-line finishing. Reduced set up times, good production speeds, and the ability to change sizes without changing cylinders are important parameters [399]. The introduction of UV-cured inks allows flexographic printing on nonpaper substrates that used to be printed with letterpress machines (see Chapter 10).

7. Environmental Considerations

Environmental aspects of the manufacture of PSPs include the use of recyclable raw materials and their recycling technology. Recyclable raw materials include common materials and biodegradable products developed to replace the classic materials used for PSPs. Both are discussed in detail in Ref. [836].

REFERENCES

1. I. Benedek and L.J. Heymans, *Pressure Sensitive Adhesives Technology*, Marcel Dekker, Inc., New York, 1997, chap. 5.
2. I. Benedek, *Pressure Sensitive Adhesives and Applications*, Marcel Dekker, Inc., New York, 2000, chap. 8.
3. U.S. Patent 3,691,140, in P. Gleichenhagen, E. Behrend and P. Jauchen, EP 0149135B1, Beiersdorf AG, Hamburg, Germany, 1987.
4. H. Miyasaka, Y. Kitazaki, T. Matsuda, and J. Kobayashi (Nichiban Co., Ltd. Tokio, Japan), Offenlegungsschrift, DE 3544868A1, 1985.
5. *Druck Print*, (4), 18, 1988.
6. G.R. Frazee (Johnson S.C. and Son, Inc.), EP 258753/1986, in *CAS, Emulsion Polymerization*, 19, 2, 1988.
7. G.R. Frazee (Johnson S.C. and Son, Inc.), EP 259842, in *CAS, Colloids (Macromolecular Aspects)*, 19, 2, 1988.

8. W.J. Traynor, C.L. Moore, M.K. Martin, and J.D. Moon (Minnesota Mining and Manuf. Co., St. Paul, MN, USA), U.S. Patent 4,726,882, in *CAS, Adhesives*, 17, 4, 1988.

9. H. Kuroda and M. Taniguchi (Bando Chem. Ind. Ltd.), Japanese Patent 6,343,988/1988, *CAS, Adhesives*, 21, 4, 1988.

10. M. Koehler and J. Ohngemach, *Polym. Paint Colour J.*, 178, 203, 1988.

11. Loctite Co., Japanese Patent 621,795,006/1986, *CAS, Siloxanes and Silanes*, 13, 1, 1988.

12. F. Buehler and W. Gronski, *Makromol. Chem.*, 188, 2995, 1988.

13. M.A. Johnson, *J. Plast. Film Sheet.*, 4(1), 50, 1988.

14. M. Dai, L. Zhang, R. Zhu, and K. Zang, Chinese Patent 8,505,449/1987, in *CAS, Adhesives*, 17, 4, 1988.

15. D. Akihito, O. Tomohisa, and U. Toshishige (Hitachi Chem Co. Ltd), Japanese Patent 6,368,683, in *CAS, Crosslinking Reactions*, 14, 7, 1988.

16. K.H. Schumacher and T. Sanborn, *UV-Curable Acrylic Hot-Melt for Pressure Sensitive Adhesives— Raising Hot-Melts to a New Level of Performance*, in Proceedings of the 24th Annual Meeting of Adhesives Society, Feb. 25–28, 2001, Williamsburg, VA, USA, p. 165.

17. Belgian Patent 675420, in D.K. Fisher and B.J. Briddell, EP 0426198 A2, Adco Product Inc., Michigan Center, MI, USA, 1991.

18. Lehmann et al., U.S. Patent, 3,729,338, in D.K. Fisher and B.J. Briddell, EP 0426198 A2, Adco Product Inc., Michigan Center, MI, USA, 1991.

19. *Adhes. Age*, (7), 53, 1994.

20. E. Fink, *Paper Film and Foil Converter*, (8), 65, 1973.

21. *Adhes. Age*, (11), 44, 1995.

22. D.K. Fisher and B.J. Briddell, EP 0426198 A2, Adco Product Inc., Michigan Center, MI, USA, 1991.

23. Y. Sasaaki, D.L. Holguin, and R. Van Ham (Avery International Co., Pasadena, CA), EP 0252 717 A2, 1988.

24. U.S. Patent 4,223,067, in D.K. Fisher and B.J. Briddell, EP 0426198 A2, Adco Product Inc., Michigan Center, MI, USA, 1991.

25. M. von Bittera, D. Schäpel, U. von Gizycki, and R. Rupp (Bayer AG, Leverkusen, Germany), EP 0147588 B1, 1985.

26. R.R. Charbonneau and G.L. Groff, EP 0106559 B1, Minnesota Mining and Manuf. Co., St. Paul, MN, USA, 1984.

27. U.S. Patent 2,925,174, in R.R. Charbonneau and G.L. Groff, EP 0106559 B1, Minnesota Mining and Manuf. Co., St. Paul, MN, USA, 1984.

28. U.S. Patent 4,286,047, in R.R. Charbonneau and G.L. Groff, EP 0106559 Bi, Minnesota Mining and Manuf. Co., St. Paul, MN, USA, 1984.

29. U.S. Patent 4,181,752, in R.R. Charbonneau and G.L. Groff, EP 0106559 B1, Minnesota Mining and Manuf. Co., St. Paul, MN, USA, 1984.

30. I. Benedek, *Pressure-Sensitive Formulation*, VSP, Utrecht, 2000, chap. 3, sec. 1.2.

31. I. Benedek, *Pressure-Sensitive Adhesives and Applications*, Marcel Dekker, Inc., New York, 1997, chap. 8.

32. I. Benedek, *Pressure-Sensitive Formulation*, VSP, Utrecht, 2000.

33. G. Holden and S. Chin, *Adhes. Age*, (5), 22, 1987.

34. R. Higginson (DRG, UK Ltd.), PCT WO 8803477, in *CAS, Colloids (Macromolecular Aspects)*, 22, 6, 1988.

35. K. Akasoka, S. Sanuki, and T. Matsuyama, Japanese Patent, 62,243,669, Nippon Synth. Chem. Ind. Co., 1987.

36. S. Shinji and Y. Yoshiuki, Japanese Patent 62243670, Nippon Synth. Chem. Ind. Co., 1987, in *CAS, Adhesives*, 13, 4, 1988.

37. J.W. Otter and G.R. Watts, U.S. Patent 5,346,766, Avery International Co., Pasadena, CA, 1994.

38. DE-A-2407494, in P. Gleichenhagen, E. Behrend, and P. Jauchen, EP 0149135 B1, Beiersdorf AG, Hamburg, Germany, 1987.

39. J.L. Walker and P.B. Foreman (National Starch), EP224795/1987 in *CAS, Silicones and Siloxanes*, 2, 3, 1988.

40. I. Benedek, *Pressure-Sensitive Formulation*, VSP, Utrecht, 2000, chap. 3, sec. 2.1

41. M. Avella and E. Martuscelli, *Polymer*, (10), 1734, 1988.

42. R. Stadler, L. de Lucca Freitas, V. Krieger, and S. Klotz, *Polymer*, (9), 1643, 1988.
43. F. Chambon and H.H. Winter, *J. Rheol.*, 31, 683, 1987.
44. D.G. Pierson and J.J. Wilczynski, *Adhes. Age*, (8), 52, 1990.
45. G. Bonneau and M. Baumassy, *New Tackifying Dispersions for Water Based PSA for Labels*, in Proceedings of the 19th Munich Adhesive and Finishing Seminar, Munich, Germany, 1994, p. 82.
46. T.G. Wood, *Adhes. Age*, (7), 19, 1987.
47. P. Dunckley, *Adhäsion*, (11), 19, 1989.
48. H. Colon et al., U.S. Patent 4,331,576, in W.L. Bunelle, K.C. Knutson, and R.M. Hume, EP 0199468 A2/H.B. Fuller Co., St. Paul, MN, USA, 1986.
49. S.L. Scholl, K.C. Knutson, and W.L. Bunelle, EP 0193427 A2, H.B. Fuller Licensing and Financing Inc., St. Paul, MN, USA, 1986.
50. M. Vipin, *Application of Acrylic Rubbers in PSA*, in Technical Proceedings of the Advances in Pressure Sensitive Tape Technology, Tech 12, Itasca, IL, USA, May, 1989, p. 191.
51. U.S. Patent 4,052,368, Morrison, in W.L. Bunelle, K.C. Knutson, and R.-M. Hume, EP 0199468 A2, H.B. Fuller Co., St. Paul, MN, USA, 1986.
52. Japanese Patent A 8287481, in P. Gleichenhagen, E. Behrend, and P. Jauchen, EP 0149135 B1, Beiersdorf AG, Hamburg, Germany, 1987.
53. Vitta Corporation, U.S. Patent 3598679, in *Coating*, (6), 338, 1969.
54. H. Mueller and J. Tuerk, EP 0118726, BASF AG, Ludwigshafen, Germany, 1984.
55. R.G. Jahn, *Adhes. Age*, (12), 35, 1977.
56. R.J. Shuman and B.D. Josephs (Dennison Manuf. Co., Framingham, MA, USA), PCT, WO 88/01636.
57. *Adhäsion*, (4), 172, 1985.
58. J.C. Pasquali, EP 0122847, 1984.
59. I. Benedek, *Pressure-Sensitive Formulation*, VSP, Utrecht, 2000, chap. 3, sec. 2.2.
60. G. Auchter, J. Barwich, G. Rehmer, and H. Jäger, *Adhes. Age*, (7), 20, 1994.
61. W.K. Darwell, P.R. Konsti, J. Klingen, and K.W. Kreckel, EP 257984, Minnesota Mining and Manuf. Co., St. Paul, MN, USA, 1986.
62. E. Kazuyoshi, N. Hiroaki, T. Katsuhisa, K. Yoshita, and T. Saito (FSK Inc.), Japanese Patent 6317981, in *CAS, Adhesives*, 12, 5, 1988.
63. T. Kurono, N. Okashi, and N. Tanaka (Nitto Electric Ind. Co. Ltd.), Japanese Patent 6333487/1988, in *CAS, Adhesives*, 12, 5, 1988.
64. T. Kinoshita, U.S. Patent, 4,645,783, Sanyo Kokkusaku Pulp Co. and Saiden Ind. Co., Tokyo, 1987.
65. Sherwin Williams Company, Cleveland, OH, U.S. Patent 1,519,362, in *Coating*, (1), 24, 1971.
66. Y. Kitazaki, T. Matsuda, and Y. Kobayashi, Japanese Patent 6,226,3273/1987, Nichiban Co. Ind. Ltd., in *CAS, Colloids*, 7, 7, 1988.
67. W.K. Darwell, P.R. Konsti, J. Klingel, and K.W. Kreckel, EP257984, Minnesota Mining and Manuf. Co., St. Paul, MN, USA, in *CAS, Adhesives*, 20, 3, 1988.
68. Minnesota Mining and Manuf. Co., Japanese Patent, 6317091, in *CAS, Adhesives*, 25, 7, 1988.
69. M.C. Bricker and S.T. Gentry, *Adhes. Age*, (7), 30, 1994.
70. French Patent 2,331,607, in F.F. Lau and S.F. Silver, EP 0130087B1, Minnesota Mining and Manuf. Co., St. Paul, MN, USA, 1985.
71. V.L. Hughes and R.W. Looney, EP 0131460, Exxon Research and Engineering Co., Florham Park, NJ, USA, 1985.
72. C. Parodi, S. Giordano, A. Riva, and L. Vitalini, *Styrene-Butadiene Block Copolymers in Hot Melt Adhesives for Sanitary Application*, in Proceedings of the 19th Munich Adhesive and Finishing Seminar, Munich, Germany, 1994, p. 119.
73. P.A. Mancinelli, *New Developments in Acrylic Hot Melt Pressure Sensitive Adhesive Technology*, in Technical Proceedings of the Advances in Pressure Sensitive Tape Technology, Tech 12, Itasca, IL, USA, May, 1989.
74. E.L. Scheinbart and J.E. Callan, *Adhes. Age*, (3), 17, 1973.
75. J.N. Kellen and C.W. Taylor, EP 0246482 A2, Minnesota Mining and Manuf. Co., St. Paul, MN, USA, 1987.

76. Martens et al., U.S. Patent 4,181,752 in J.N. Kellen and C.W. Taylor, EP 0246 A2, Minnesota Mining and Manuf. Co., St. Paul, MN, USA, 1987.
77. C. Harder, *Acrylic Hotmelts Recent Chemical and Technological Developments for an Ecologically Beneficial Production of Adhesive Tapes State and Prospects*, European Tape and Label Conference, Brussels, Belgium, April 28–30, 1993.
78. K. Wabro, R. Milker, and G. Krüger, Haftklebstoffe und Haftklebebänder, Astorplast GmbH, Alfdorf, Germany, 1994, p. 48.
79. C.I. Simionescu, N. Asandei, and I. Benedek, *Rev. Roum. Chim.*, (7), 1081, 1971.
80. C.I. Simionescu, N. Asandei, and I. Benedek, Romanian Patent, 54213, 1969.
81. M. Gerace, *Adhes. Age*, (8), 15, 1983.
82. F.C. Larimore and R.A. Sinclair, EP 0197662A1, Minnesota Mining and Manuf. Co., St. Paul, MN, USA, 1986.
83. H.F. Huber and H. Müller, in Proceedings of the 10th International Conference Radcure '86, December 1–12, 1986.
84. National Starch and Chemical Corp., Japanese Patent, 6,306,076/1988, in *CAS, Adhesives*, 19, 2, 1988.
85. C. Cervelatti, G. Capaldi, and J.L. Erich, EP 273585, Exxon Chemical Patents Inc., 1988.
86. *Coating*, (1), 12, 1988.
87. B. Hunt, *Labels Label. Int.*, (5/6), 34, 1997.
88. T.R. Mecker, *Low Molecular Weight Isoprene Based Polymers-Modifiers for Hot Melts*, in Tappi Hot Melt Symposium, 1984, in *Coating*, (11), 310, 1984.
89. Esso Research and Eng. Co., U.S. Patent 3,351,572, in *Coating*, (7), 210, 1969.
90. D.J. Kim, H.J. Kim, E.C. Kim, and D.H. Yoon, *Performance of SIS-Based Hot-Melt PSAs*, in Proceedings of the 27th Annual Meeting of the Adhesion Society, Wilmington, NC, USA, Feb. 11–18, 2004, p. 385.
91. D.A. Shields, *Hot Melt Pressure Sensitive Adhesives. A Case History*, in Tappi Hot Melt Adhesives and Coating Short Course, Hilton Head, SC, USA, May 2–5, 1982.
92. *Coating*, (7), 187, 1984.
93. C.H. Hill, N.C. Memmo, W.L. Phalen Jr., and R.R. Suchanec (Hercules Inc.) EP 259697/1988, in *CAS, Adhesives*, 14, 5, 1988.
94. Klebstoffvorprodukte, *Prüfung von Haftklebstoffen*, D-EDE/K, Juni/Juli 81, 2, BASF AG, Ludwigshafen, Germany.
95. *Adhes. Age*, (7), 36, 1986.
96. *Adhäsion*, (2), 123, 1974.
97. J.C. Fitsch and A.M. Snow, *Adhes. Age*, (10), 23, 1977.
98. *Coating*, (6), 198, 1996.
99. *Coating*, (7), 232, 1989.
100. F.T. Sanderson, *Adhes. Age*, (11), 26, 1983.
101. Engel, U.S. Patent 514950/1983, in F.C. Larimore and R.A. Sinclair, EP 0197662 A1, Minnesota Mining and Manuf. Co., St. Paul, MN, USA, 1986.
102. G.W.H. Lehmann and H.A.J. Curts, U.S. Patent 4,038,454, Beiersdorf AG, Hamburg, Germany, 1977.
103. E.E. Ewins Jr. and J.R. Ericson, *Tappi J.*, 71(6), 155, 1988.
104. Minnesota Mining and Manuf Co., St. Paul, MN, U.S. Patent 1,594,178, in *Coating*, (8), 240, 1972.
105. E. Djagarowa, W. Rainow, and W. Dimitrow, *Plaste u. Kautschuk*, (1), 28, 1970.
106. Adhesives Tapes Ltd., British Patent, 861,358, in *Coating*, (6), 185, 1969.
107. K.F. Schroeder, *Adhäsion*, (5), 161, 1971.
108. Technical Booklet, BASF, TI-2.2-21 d/November 1979, teil 3, blatt 5.
109. SelbstklebendeDekorationsfolien, Technical Booklet, BASF, TI-2.2-21 d/November 1979, teil 3, b.3,
110. P. Ford, *Eur. Adhes. Seal.*, (6), 22, 1995.
111. Johnson and Johnson, NJ, U.S. Patent 3,161,554, in *Adhäsion*, (6), 277, 1966.
112. Z. Czech, DE 43 03 183 C1, Lohmann GmbH, Neuwied, Germany, 1993.
113. *Coating*, (12), 344, 1984.
114. *Avery Rating Plates and Type Plates Last a Fairly Long Time Prospectus*, Avery Etikettier-Logistik GmbH, Eching b. München, Germany.

115. *Adhes. Age*, (7), 36, 1983.
116. *Finat Labelling News*, (3), 29, 1994.
117. L.M. Schrijver, *Coating*, (3), 70, 1991.
118. R. Milker and Z. Czech, in Proceedings of the 16th Munich Adhesive and Finishing Seminar, Munich, Germany, 1991, p. 136.
119. *Coating*, (1), 29, 1978.
120. G. Fuchs, *Adhäsion*, (3), 24, 1982
121. *Neue Verpackung*, (1), 156, 1991.
122. *Coating*, (1), 24, 1969.
123. *Coating*, (8), 248, 1969.
124. *Coating*, (3), 68, 1985.
125. A. Kenneth, J.R. Stickwell, and J. Walker, EP 0147, Allied Colloids, 1985.
126. I. Benedek, *Adhäsion*, (3), 22, 1987.
127. I. Benedek, *Adhäsion*, (4), 25, 1987.
128. H. Monsey and A. Maletsky, U.S. Patent 4,331,576/1982, Franklin, NJ, in *Adhes. Age*, (12), 53, 1983.
129. H. Colon and A. Maletsky, EP 0057421 A1, NY, USA, 1982.
130. Preliminary Technical Information, PTI No.61/01.1989, GAF, Frechen, Germany.
131. Morrison, U.S. Patent 40,52,368, in W.L. Bunelle, K.C. Knutson, and R.M. Hume, EP 0199468 A2, H.B. Fuller Co., St. Paul, MN, USA, 1986.
132. T.P. Flanagan, EP 0212135B1, National Starch and Chemical Investment Holding Co., Wilmington, DE, USA, 1987.
133. D.K. Ray-Chaudhuri, T.P. Flanagan, and J.E. Schoenberg, German Patent 2,507,683, National Starch and Chem., Co., Bridgewater, 1975.
134. Rahmenrezeptur, GAF, Frechen, Germany, 1995.
135. Z. Czech, *Adhäsion*, (11), 26, 1994.
136. Z. Czech, DE 44 31 053, Lohmann GmbH, Neuwid, Germany, 1994.
137. U.S. Patent 3,661,874, in F.D. Blake, EP 0141504 A1, Minnesota Mining and Manuf., Co., St. Paul, MN, USA, 1985.
138. D.J. Terriault and M.J. Zajaczkowski, U.S. Patent 5352516/Adhesives Research Inc., Glen Rock, PA, USA, 1994.
139. Eastmann Kodak Co., Rochester, NY, U.S. Patent 3,152,940, in *Adhäsion*, (2), 47, 1966.
140. U.S. Patent 3,096,202, in P. Gleichenhagen and I. Wesselkamp, EP 0058382 B1, Beiersdorf AG, Hamburg, Germany, 1982.
141. British Patent 941,276 in P. Gleichenhagen and I. Wesselkamp, EP 0058382 B1, Beiersdorf AG, Hamburg, Germany, 1982.
142. U.S. Patent 3,441,430 in P. Gleichenhagen and I. Wesselkamp, EP 0058382 B1, Beiersdorf AG, Hamburg, Germany 1982.
143. U.S. Patent 3,152,940 in P. Gleichenhagen and I. Wesselkamp, EP 0058382 B1, Beiersdorf AG, Hamburg, Germany, 1982.
144. K. Allan, J.R. Stockwell, and J. Walker, EP 0147067A1, Allied Colloids Ltd., Bradford, UK, 1985.
145. A. Goel, U.S. Patent 4,626,575/1986, Ashland Oil Inc., Ashland, KY, in *Adhes. Age*, (5), 26, 1987.
146. U.S. Patent, Blake, 3,865,770, in F.C. Larimore and R.A. Sinclair, EP 0197662 A1, Minnesota Mining and Manuf. Co., St. Paul, MN, USA, 1986.
147. DDR Patent. 64,111, in *Coating*, (1), 24, 1969.
148. E.H. Andrews, T.A. Khan, and H.A. Majid, *J. Mater Sci.*, 20, 3121, 1985.
149. *Coating*, (6), 184, 1969.
150. D. Allen Jr. and E. Flam, U.S. Patent 4,650,817, Bard Inc., Murray Hill, NJ, 1987.
151. J.R. Pennace, C. Ciuchta, D. Constantin, and T. Loftus, WO8703477 A, Flexcon Co. Inc., Spencer, MA, USA, in *Adhes. Age*, (5), 24, 1987.
152. U.S. Patent 475,373, in S.E. Krampe and C.L. Moore, EP 0202831 A2, Minnesota Mining and Manuf. Co., St. Paul, MN, USA, 1986.
153. U.S. Patent 3,532,652, in S.E. Krampe and C.L. Moore, EP 0202831 A2, Minnesota Mining and Manuf. Co., St. Paul, MN, USA, 1986.
154. Pittsburgh Plate Glass Co., U.S. Patent, 3,355,412, in *Coating*, (7), 274, 1969.
155. R.L. Sun and J.F. Kenney, U.S. Patent 4,762,688, Johnson and Johnson Products Inc., in *CAS, Hot Melt Adhesives*, 26, 1, 1988.

156. D. Allen Jr., E. Flam, and C.R. Bard, U.S. Patent 4,650,817/1987, in *Adhes. Age*, (5), 24, 1988.
157. Morstik 103 Adhesive, Morton Thiokol Inc., Adhesives and Coatings, MTD-MS103-12/1988.
158. *Coating*, (1), 23, 1969.
159. *Finat Labelling News*, (3), 29, 1994.
160. A. Dobmann and A.G. Viehofer, in Proceedings of the 19th Munich Adhesive and Finishing Seminar, Munich, Germany, 1994, p. 168.
161. S. Toshiki, T. Yasuo, I. Hidaharu, and M. Takumi, Japanese Patent 6,366,254, Japan Synthetic Rubber Co. Ltd., 1988.
162. *Coating*, (6), 175, 1972.
163. Bakelite Xylonite Ltd., British Patent, 1,081,291, in *Coating*, (4), 114, 1969.
164. E.L. Scheinbart and J.E. Callan, *Adhes. Age*, (3), 17, 1973.
165. T.K. Bhaumik, A.K. Bhowmick, and B.R. Gupta, *Plast. Rubber Process. Appl.*, 7, 43, 1987.
166. R.J. Nichols and F. Kheradi, *The Interrelationship of Machine Design and Processing of Permanent Hot Melt Adhesives*, Tappi Hot Melt Adhesives and Coating Short Course, Hilton Head, SC, May 2–5, 1982.
167. Sealock, L1233 Adhesive, Product Data, Andover, Hampshire, UK.
168. F. Hufendiek, *Etiketten-Labels*, (1), 9, 1995.
169. *Adhäsion*, (3), 19, 1984.
170. M. Fairley, *Labels Label.*, (7/8), 34, 1995.
171. *Convert. Today*, (11), 13, 1990.
172. *Adhes. Age*, (1), 6, 1985.
173. *Adhäsion*, (11), 9, 1988.
174. Minnesota Mining and Manuf. Co. St. Paul, MN, U.S. Patent 3,129,816, 1987.
175. Kimberley Clark, U.S. Patent 799,429, in *Coating*, (1), 9, 1970.
176. G. Grzywinski and E.J. Foley (Scott Paper Co.), *Adhes. Age*, (12), 281, 1988.
177. K. Nakamura, Y. Miki, and Y. Nanzaki, Japanese Patent 6,386,787 Nitto Electric Industrial Co. Ltd., in *CAS, Adhesives*, 19, 5, 1988.
178. H. Suchy, J. Hezina, and J. Matejka, Czech CS 247802/1987, in *CAS, Adhesives*, 19, 4, 1988.
179. K. Mitsui, PCT, WO8802767, Mitsui Petrochemical Ind. Ltd., in *CAS, Adhesives*, 22, 4, 1988.
180. I. Yorinobu, W. Yasuhhisa, and T. Hiroshi, Japanese Patent 6,386,777 Japan Synthetic Rubber Co., in *CAS, Adhesives*, 24, 3, 1988.
181. R.M. Enanoza, EP 259968/1986, Minnesota Mining and Manuf. Co., in *CAS, Adhesives*, 22, 3, 1988.
182. T. Sugiyama, N. Miyaji, I. Yoshihide, and T. Tange, Japanese Patent 62,199,672/1987, Nippon Carbide Ind. Co. Inc., in *CAS, Adhesives*, 14, 3, 1988.
183. T. Hiroyoshi, H. Kuribayashi, and E. Usuda, EP 254002/Sumitomo Chem Co. Ltd., in *CAS, Adhesives*, 14, 3, 1988.
184. T. Moldvai and N. Piatkowski, Romanian Patent RO 93124/1987, Inst. Cerc. Pielarie si Incaltaminte in *CAS, Adhesives*, 19, 3, 1988.
185. T. Kishi, Japanese Patent 6,369,879, Sekisui Chem. Co. Ltd., in *CAS, Adhesives*, 19, 6, 1988.
186. H. Yaguchi, H. Fukuda, T. Masayuki, and T. Ohashi, Japanese Patent 6,327,583, Bridgestone Corp., in *CAS, Adhesives*, 19, 6, 1988.
187. I. Murakami, Y. Hamada, and O. Takuman, EP 253601, Toray Silicon Co. Ltd., in *CAS, Adhesives*, 14, 3, 1988.
188. *Coating*, (11), 341, 1972.
189. J.A. Fries, *Hot Melt Pressure Sensitive Paper Label Application, An Overview of the Adhesives Market*, in Hot Melts—The Future is Now, Tappi Symposium, Toronto, Canada, June 2–4, 1980.
190. U.S. Patent 4,163,077, in F.F. Lau and S.F. Silver, EP 0130087B1, Minnesota Mining and Manuf. Co., St. Paul, MN, USA, 1985.
191. L.A. Sobieski and T.J. Tagney, *Adhes. Age*, (12), 23, 1988.
192. A.B. Wechsung, *Coating*, (9), 268, 1972.
193. *Adhes. Age*, (1), 6, 1985.
194. Mobil-Bicor, Mobil Platics, Virton, Belgium, (5), 4, 1984.
195. P. Penczek and B. Kujawa-Penczek, *Coating*, (6), 232, 1991.
196. A.J. Risser, U.S. Patent 3,759,780, Cities Cervice Co., 1973.

197. H.K. Porter Co. Inc., Pittsburgh, PA, U.S. Patent 3,149,997, in *Adhäsion*, (3), 79, 1966.
198. G.L. Burroway and G.W. Feeney, *Adhes. Age*, (7), 17, 1974.
199. A.N. Anisimov et al., USSR Patent, 249,525, in *Coating*, (1), 24, 1969.
200. Hyvis/Napvis Polybutenes, PB 301, BP Chemicals.
201. *Coating*, (3), 12, 1985.
202. J. Verseau, *Coating*, (11), 309, 1971.
203. *Eur. Adhesives and Sealants*, (6), 36, 1995.
204. U.S. Patent 1,569,888, Beiersdorf AG, Hamburg, in *Coating*, (8), 20, 1972.
205. E. Djagarowa, *Plaste u. Kaut.*, (19), 748, 1969.
206. *Coating*, (6), 185, 1969.
207. *Adhes. Age*, (4), 6, 1983.
208. *Coating*, (6), 184, 1969.
209. G. Chand, British Patent 723,226, *Coating*, (6), 184, 1969.
210. G. Chand, British Patent 790,087, *Coating*, (6), 184, 1969.
211. Johnson & Johnson, U.S. Patent 2,882,179, *Coating*, (6), 184, 1969.
212. Adhesives Tapes Ltd., British Patent 861,358, in *Coating*, (6), 185, 1969.
213. Johnson and Johnson, British Patent 798,471, in *Coating*, (6), 185, 1969.
214. B.B. Blackford, British Patent 8,864,365, in *Coating*, (6), 185, 1969.
215. G.F. Vesley, A.H. Paulson, and E.C. Barber, EP 0202 938 A2/1986.
216. A.F. Carr, *Coating*, (11), 334, 1973.
217. J. Grabemann and R. Hauber, DE 42 31 607, Hans Neschen GmbH & Co KG, Bückeburg, Germany 1994.
218. W.E. Lenney, Canadian Patent 1,225,176, Air Products and Chemical Inc., USA, 1987.
219. U.S. Patent 3,257,478, in W.E. Lenney, Canadian Patent 1,225,176, Air Products and Chemical Inc., USA, 1987.
220. U.S. Patent 3697618, W.E. Lenney, Canadian Patent, 1,225,176 Air Products and Chemical Inc., USA, 1987.
221. U.S. Patent 3,998,997, W.E. Lenney, Canadian Patent 1,225,176, Air Products and Chemical Inc., USA, 1987.
222. F. Korpmann, N.J. Bridgewater, and G. Perry, New Brunswick, NJ, U.S. Patent 4,325,770, 1982.
223. Y. Kitazaki, T. Matsuda, and Y. Kobayashi (Nichiban Co., Ltd.), Japanese Patent 62,263,273/1987, *CAS, Colloids*, 7, 7, 1988.
224. H. Kenjiro and S. Kotaro (Nitto Electric Ind. Co. Ltd.), Japanese Patent 63,118,383/1988, in *CAS, Adhesives*, 24, 5, 1988.
225. I. Benedek, *Eur. Adhes. Sealants*, (2), 25, 1996.
226. H. Satoh and M. Makino, Ger. Offen. DE 3740222, Nippon Oil Co. Ltd., in *CAS, Adhesives*, 22, 4, 1988.
227. I. Yorinobu, W. Yasuhisa, and T. Hiroshi, Japanese Patent 6,386,777/Japan Synthetic Rubber Co., in *CAS, Adhesives*, 24, 3, 1988.
228. K. Akasaka, S. Sanuki, and T. Matsuyama, Japanese Patent 62,243,669/1986, Nippon Synthetic Chem. Ind. Co. Ltd., in *CAS, Adhesives*, 11, 5, 1988.
229. H. Nakagawa, A. Baba, S. Furukawa, and S. Fukuchi, Japanese Patent 62,292,877/1987, Nippon Shokubai Kagaku Kogyo Co. Ltd., *CAS, Adhesives*, 11, 6, 1988.
230. V. Stanislawczyk, EP 264903, Goodrich B.F. Co., *CAS, Colloids (Macromolecular Aspects)*, 17, 7, 1988.
231. H. Miyasaka, Y. Kitazaki, T. Matsusa, and J. Kobayashi, Japanese Patent 62,263,273/1987, Nichiban Co. Ltd., in *CAS, Colloids (Macromolecular Aspects)*, 7, 7, 1988.
232. T. Kishi, Japanese Patent 6,369,879, Sekisui Chem. Co. Ltd., in *CAS, Adhesives*, 19, 6, 1988.
233. K. Maeda and Y. Kitazaki, Japanese Patent 6,348,380, Nichiban Co. Ltd., in *CAS, Adhesives*, 19, 3, 1988.
234. T. Tatsuno, K. Matsui, M. Takahashi, and M. Wakimoto, Japanese Patent 6,228,597/1987, Kansai Paint Co. Ltd., *CAS, Adhesives*, 14, 3, 1988.
235. H. Reipp, *Adhes. Age*, (3), 17, 1972, in *Coating*, (6), 187, 1972.
236. *Adhäsion*, (1/2), 44, 1987.
237. N.W. Malek (Beiersdorf AG, Hamburg, Germany), EP 0095093, 1983.

238. Kjin Co., Japanese Patent 28520, *Coating*, (2), 71, 1972.
239. Mystik Adhesives Prod. Inc., U.S. Patent 2,878,142/1976, 1976.
240. I. Benedek, *Pressure-Sensitive Formulation*, VSP, Utrecht, 2000, chap. 5.
241. *Coating*, (6), 185, (1969).
242. H.H. Hub, *Composition, Characteristics and Application of Polar Ethylene Copolymers*, Polyethylene '93, Oct. 4, 1993, Session III, Maack Business Service, Zürich, Switzerland.
243. *Coating*, (12), 450, 1986.
244. Dow, Primacor Polymers, CH 272-047-E-991, Specification, Dow Chemicals, Horgen, Switzerland.
245. A. Ridgeway and L.K. Mengenhagen, *EAA/Polybutylene Blend for Packaging Applications*, Tappi Proceedings, Polymers, Laminations & Coatings Conference, 1992, p. 52.
246. *Tappi J.*, (5), 103, 1988.
247. R.N. Henkel, *Paper Film and Foil Converting*, (12), 68, 1968.
248. I. Benedek, Ebert Folien AG, Wiesbaden, Germany, Deutsches Gebrauchsmuster, G 9113755.1, 1991.
249. *Verpackungs Rundschau*, (5), 556, 1988.
250. D.H. Teesdale, *Coating*, (6), 246, 1983.
251. Rohm & Haas Co., Philadelphia, PA, U.S. Patent 3,152,921, in *Adhäsion*, (2), 80, 1966.
252. Vorwerk & Sohn, DBP 1078285, in *Coating*, (6), 184, 1969.
253. Johnson & Johnson, U.S. Patent 2,973,859, in *Coating*, (6), 184, 1969.
254. *Allg. Papier Rundschau*, (48), 1360, 1985.
255. M.J. Owen, *J. Coat. Technol.*, 53 (679), 49, 1981.
256. H. Toepsch, *Wochenblatt für Papierfabrikation*, (11/12), 320, 1971.
257. *Coating*, (1) 12, 1984.
258. P. John and B. Sharon (Flexcon Co., Inc.), PCT Inc., Int. Appli., WO 8703537/1987, in *CAS, Silicones*, (2), 3, 1988.
259. J. Pennace and G.E. Kersey (Flexcon Co. Inc., Spencer, MA, USA), PCT, WO 87/035537, 1987.
260. P.L. Brown and D.L. Stickles (Dow Corning Co.) EP 251483, 1988, in *CAS, Siloxanes and Silicones*, 13, 5, 1988.
261. G.A.H. Kupfer, *Siliconizing with Radiation Curing Systems*, Tappi Hot Melt Adhesives and Coatings, Short Course, Hilton Head, SC, USA, June, 1982.
262. H.J. Timpe, R. Wagner, and U. Müller, *Adhäsion*, (2), 28, 1985.
263. R. Eckberg, *Silicone Release Coatings: Testing Pressure-Sensitive Adhesion*, in Proceedings of the 23th Annual Meeting of the Adhesion Society, Wilmington, NC, USA, Feb. 11–18, 2004, p. 307.
264. V.F. Andreev et al., SSSR Patent 263,409, in *Coating*, (5), 130, 1971.
265. T. Shibano, I. Kimura, H. Nomoto, and C. Maruchi, U.S. Patent 4,624,893, 1986, Sanyo Kokusaku Pulp Co., Ltd., Tokyo, Japan, in *Adhes. Age*, (5), 24, 1987.
266. G. Galli, Italy, EP 0191191 A1, Manuli Autoadesivi Spa., Cologno Monzese, 1986.
267. Italian Patent, 21842 A/82, in G. Galli, EP 0191191A1, Manuli Autoadesivi Spa., Cologno Monzese, Italy, 1986.
268. *Adhes. Age*, (8), 42, 1987.
269. Norton Co., U.S. Patent, 1,123,014.
270. R.A. Bafford and G.E. Faircloth, *Adhes. Age*, (12), 353, 1987.
271. Release Coat K, Kalle Folien, Data Sheet, Wiesbaden, Germany.
272. N. Yataba and Y. Miki, Japanese Patent 6,333,485, Nitto Electric Co. Ltd., in *CAS, Hot Melt Adhesives*, 13, 1, 1988.
273. Acheson Ind. Inc., Port Huron, MI, DBP 1278652, in *Coating*, (7), 274, 1969.
274. H. Kenjiro, O. Kazuto, A. Nagai, and K. Kiyohiro, Japanese Patent 63,137,841, Nitto Electric Ind. Co. Ltd., in *CAS, Adhesives*, 24, 5, 1988.
275. T. Kikuta and H. Hori, Japanese Patent 6,351,478, Nippon Shokubai Kagaku, Kogyo Co. Ltd., in *CAS, Coating, Inks and Related Products*, 17, 11, 1988.
276. T. Kikuta and H. Hori, Japanese Patent 6,333,482, Nippon Shokubai Kagaku, Kogyo Co. Ltd., in *CAS, Coating, Inks and Related Products*, 17, 11, 1988.
277. T. Kikuta and H. Hori, Japanese Patent 6,322,812, Nippon Shokubai Kagaku, Kogyo Co. Ltd., in *CAS, Emulsion Polymerization*, 18, 2, 1988.

278. K. Nitzl and L. Birk, *Coating*, (12), 344, 1984.

279. *Coating*, (12), 345, 1984.

280. A. Fau, EP 0169098B1, Rhone Poulenc Chimie, Courbevoie, France, 1986.

281. K. Brack, U.S. Patent 4288479, Design Coat Co., in *Adhes. Age*, (12), 58, 1981.

282. G. Camerini (Coverplast Italiana SpA), EP 248771, 1986.

283. V.D. Babayants, V.V. Kolesnitschenko, and S.G. Sannikov, *Lakokras. Mater. Ikh. Primen.*, (3), 45, 1988, in *CAS, Coatings, Inks & Related Products*, 17, 9, 1988.

284. L. Bothorel, *Emballages*, (278), 372, 1972.

285. M. Omote, I. Sakai, and T. Matsumoto, Japanese Patent, 6,386,788, Nitto Electric. Ind. Co., in *CAS*, 19, 4, 1988.

286. British Patent 1,123,014, in *Coating*, (7), 274, 1969.

287. *Coating*, (6), 184, 1969.

288. TNII Bumagi, SSSR Patent 300,561, in *Coating*, (7), 368, 1969.

289. Y. Naoto, Japanese Patent 62,218,467/1987, Toyo Ink. MFG. Co. Ltd., in *CAS, Siloxanes & Silicones*, 2, 4, 1988.

290. S. Ohara and R. Kitamura, EP 254050, Goyo Paper Working Co. Ltd., in *CAS, Adhesives*, 13, 11, 1988.

291. H. Komatsu, Japanese Patent 62,246,973, 1987, in *CAS, Colloids (Macromolecular Aspects)*, 4, 10, 1988.

292. *Adhes. Age*, (1), 6, 1985.

293. *Etiketten-Labels*, (1), 49, 1996.

294. *Papier und Kunststoff Verarbeiter*, (9), 57, 1988.

295. S. Heimlich, *Papier und Kunststoff Verarbeiter*, (11), 26, 1973.

296. *Brit. Plastics*, (10), 69, 2001.

297. *Allg. Papier Rundschau*, (5), 100, 1988.

298. O. Huber, *Wochenblatt Papierfabrikation*, (17), 657, 1973.

299. R. Stockmeyer, *Deutscher Drucker*, (13) 42, 1988.

300. A.T. Franklin, *Print. Trades J.*, (1044), 46, 1974.

301. T. Arai, T. Yamasaki, K. Suzuki, T. Ogura, and Y. Salai, *Tappi J.*, (5), 47, 1988.

302. V.B. Topliss, *Finat News*, (2), 45, 1989.

303. J.E. Wilson, R.A. Ravier, D.A. Ewaniuk, and R.S. McDaniel, PTSC XVII Technical Seminar, 4 May, Woodfield Shaumburg, IL, USA.

304. E. Chicherio, *Coating*, (10), 285, 1982.

305. *Etiketten-Labels*, (5), 91, 1995.

306. *Etiketten-Labels*, (5), 22, 1995.

307. E.B. Strahler, *Coating*, (5), 163, 1996.

308. M. Bateson, *Finat News*, (3), 29, 1989.

309. D. Lacave, *Labels and Labelling*, (3/4), 54, 1994.

310. R.H. Feldkamp, *Allg. Papier Rundschau*, 14, 367, 1986.

311. *Papier Kunststoffverarbeiter*, (9), 57, 1988.

312. *Coating*, (5), 212, 1990.

313. C. Bayer, *Adhäsion*, (9), 349, 1965.

314. *Etiketten-Labels*, (3), 9, 1995.

315. *Etiketten-Labels*, (5), 24, 1995.

316. *Etiketten-Labels*, (3), 32, 1995.

317. *Druck Print*, (10), 32, 1987.

318. *Etiketten-Labels*, (3), 10, 1995.

319. J. Paris, *Papier Kunststoffverarbeitung*, (9), 57, 1988.

320. *Paper, Film and Foil Converter*, (8), 28, 1973.

321. M. Fairley, *Labels and Labelling International*, (5/6), 28, 1997.

322. Mactac, Spitzentechnologie für Selbstklebende Materialien, MACtac Europe S.A., Soignies, Belgium.

323. H. Senn, *Allg. Papier Rundschau*, (16), 170, 1986.

324. L.C. Fehrmann, *Adhes. Age*, (6), 48, 1970.

325. *Labels and Labelling* (3/4), 44, 1994.

326. J. Young, *Tappi J.*, (5), 78, 1988.
327. *Coating*, (8), 242, 1972.
328. *Allg. Papier Rundschau*, 18, 405, 1987.
329. B.H. Gregory, *Extrusion Coating Advances – Resins, Processing, Applications, Markets*, Polyethylene '93, The Global Challenge for Polyethylene in Film Lamination, Extrusion, Coating Markets, Oct. 4, 1993, Maack Business Services, Zürich, Switzerland.
330. Raflatac Synthetic Materials Available as Rollstock, Booklet, Raflatac Oy, Tampere, Finland.
331. *Paper, Film and Foil Converter*, (9), 32, 1989.
332. E. Park, *Paper Technology*, (8), 15, 1989.
333. D.M. Stott, *Surface Coatings*, (11), 296, 1969.
334. Johnson and Johnson, U.S. Patent 3,403,018, in *Coating*, (1), 24, 1969.
335. Domtar Ltd., Canadian Patent 857,847, *Coating*, (2), 38, 1972.
336. Ludlow Corporation, U.S. Patent 1,157,154, in *Coating*, (12), 353, 1970.
337. K. Hamada, U. Uchiyama, and S. Takemura, Japanese Patent 13,765/69, in *Coating*, (12), 353, 1970.
338. I.P. Rothernberg, U.S. Patent 4,650,704, Stik-Trim Industries Inc., New York, NY, USA, 1987.
339. B. Reinhardt, M. Hottenträger, B. Gather, B. Hartmann, and M.D. Lecher, Proceedings of the 19th Munich Adhesive and Finishing Seminar, Munich, Germany, 1994, p. 48.
340. C.M. Brooke, *Finat News*, (3), 34, 1987.
341. Arhoco Inc., U.S. Patent 3,509,991, in *Coating*, (5), 130, 1971.
342. *Adhes. Age*, (1), 6, 1985.
343. *Papier und Kunststoff Verarbeiter*, (1), 20, 1996.
344. Maratech International, Ltd. Labelexpo Asia-News, 1998, Labels and Labelling Asia, Cowise International, Abbey House, UK.
345. J.R. Allen, *New Silicone Coating Base Papers*, in Tappi Hot Melt Adhesives and Coating Short Course, Hilton Head, SC, May 2–5, 1982.
346. J.R. DeFife, *Labels and Labelling*, (3/4), 14, 1994.
347. *Packaging Today*, (12), 9, 1995.
348. *Papier und Kunststoff Verarbeiter*, (9), 57, 1988.
349. *Adhes. Age*, (10), 125, 1986.
350. S. Moser, *Allg. Papier Rundschau*, (16), 452, 1986.
351. L. Placzek, *Coating*, (1), 2, 1986.
352. I. Benedek and L.J. Heymans, *Pressure Sensitive Adhesives Technology*, Marcel Dekker, Inc., New York, 1997, chap. 7.
353. R. Hinterwaldner, *Adhäsion*, (6), 11, 1984.
354. *Finat News*, 4, 10, 1996.
355. *Label Buyer International*, Spring 97, p. 26.
356. *Neue Verpackung*, (1), 156, 1991.
357. *Etiketten-Labels*, (5), 136, 1995.
358. *Films and Sheets Portfolio for Printing and Advertising*, Klöckner Pentaplast, GmbH & Co KG, Technical Booklet, Montabaur, Germany, Ed.03.03.
359. J. Michel, *Verpackungs Rundschau*, (12), 1425, 1985.
360. *Converting Today*, (11), 8, 1990.
361. *Labels and Labelling*, (3/4), 82, 1994.
362. R. Kasoff, *Paper, Film and Foil Converter*, (9), 85, 1989.
363. *Coating*, (3), 65, 1974.
364. M.H. Mishne, *Screen Printing*, (5), 60, 1986.
365. *Etiketten-Labels*, (3), 9, 1995.
366. W.J. Busby, *Processing Problems with PE Films*. Polyethylene '93, p. 3, Session VI, 3-3, Maack Business Service, Zürich, Switzerland.
367. *Allg. Papier Rundschau*, 40, 1102, 1988.
368. *Druckwelt*, (14/15), 27, 1988.
369. *New Values in Packaging*, Mobil Plastics Europe, (6), 2, 1997.
370. McCauley, *Tappi J.*, (6), 159, 1985.
371. H. Roder, *Adhäsion*, (6), 16, 1983.
372. *Handling*, (5/6), 12, 1994.
373. Dowlex, *Polyethylene Resin*, Data Sheet, 2740E.

374. K. Suenaga, Japanese Patent 6,386,784, Nitto Electric Industrial Co., 1988, in *CAS*, 21, 5, 1988.

375. Lonza, *Cast Plastic Films for Technical Applications*, Buhr, Lonza Werke GmbH, Weil am Rhein, Germany.

376. P. Kriston, *Müanyag Fóliák*, Müszaki Könyvkiadó, Budapest, 1976, p. 45.

377. Mitsubishi Rayon Co., French Patent, 2,020,683, in *Coating*, (8), 240, 1972.

378. *Kunststoffe*, 75 (10), XXIV, 1985.

379. F. Altendorfer and A. Wolfsberger, *Kunststoffe*, (6), 691, 1990.

380. F. Hensen, *Papier und Kunststoff Verarbeiter*, (11), 32, 1988.

381. BASF Kunststoffe, Lupolen, B 581, d/12.92 p. 38, BASF AG, Ludwigshafen, Germany.

382. M.A. Barbero and A. Amico, *A New Performance ULDPE/VLDPE from High Pressure Technology-Potential Applications*, Polyethylene '93, Oct. 4, 1993, Session III, p. 6, Maack Business Service, Zürich, Switzerland.

383. A. Stroeks, *New Developments in Polyolefins (LDPE, LLDPE, VLDPE) for Flexible Packaging*. Specialty Plastics Conference '87, Polyethylene and Copolymer Resin and Packaging Market, Dec. 1, 1987, Maack Business Service, Zürich, Switzerland.

384. *Converting Today*, (1), 16, 1992.

385. W.W. Bode, *Tappi J.*, (6), 133, 1988.

386. *Neue Verpackung*, (5), 56, 1991.

387. *Trouble Shooting Guide for Processing Films*, Conversion Industry Reference Report, MBS No. 903, 1994, Maack Business Service, Zürich, Switzerland.

388. Dow, LDPE, Data Sheet, 2583-F-586.

389. *Neue Verpackung*, (8), 66, 1993.

390. W. Michael and R. Harms, *Papier und Kunststoff Verarbeiter*, (10), 7, 1990.

391. D. Djordjevic, *Tailoring Films by the Coextrusion Casting and Coating Process*, Speciality Plastics Conference '87, Polyethylene and Copolymer Resin and Packaging Markets, Dec. 1, 1987, Maack Business Service, Zürich, Switzerland.

392. H. Groß, *Kunststoffe*, (11), 1548, 1994.

393. B. Kunze, S. Sommer, and G. Düsdorf, *Kunststoffe*, 84 (10), 1337, 1994.

394. A. La Mantia, and T. Manlio, *Acta Polymer.*, (11/12), 696, 1986.

395. N. Verse, *Papier und Kunststoff Verarbeiter*, (10), 23, 1990.

396. E.B. Parker, *Polyethylene Polar Copolymers and their Application*, Speciality Plastics Conference '87, Dec. 1, 1987, Maack Business Service, Zürich, Switzerland, p. 259.

397. *Finat News*, (4), 27, 1995.

398. W. Feistkorn, *Coating*, (9), 310, 1995.

399. Dow Chemicals, Horgen, Switzerland, Dowlex, *Films for Lamination*, CH-254-052-E-288, Data Sheet, 1995.

400. *Coating*, (2), 35, 1984.

401. *British Plastics*, (10), 52, 2001.

402. K. Fust, *Coating*, (2), 66, 1988.

403. *Finat Labelling News*, (3), 12, 1996.

404. W.A. Fraser, *Novel Processing Aid Technology for Extrusion Grade Polyolefins*, Maack Speciality Plastics Conference '87, Polyethylene and Copolymer Resin and Packaging Markets, Dec. 1, 1987, Maack Business Service, Zürich, Switzerland.

405. *HamLet*, Polyethylene '93, Oct. 4, 1993, Session III, p. 56, Maack Business Service, Zürich, Switzerland.

406. *Kaut. Gummi, Kunststoffe*, (6), 564, 1985, in *Coating*, (1), 24, 1987.

407. *Allg. Papier Rundschau*, 42, 1499, 1986.

408. *Polyethylene Film Market Applications*, Western Europe, Conversion Industry Reference Report, MBS 904, 1994, Maack Business Service, Zürich, Switzerland.

409. E. Beier, *Technica*, 21, 35, 1995.

410. Klebeband Forum, No.27, Hoechst Films, Hoechst AG, October, 1989.

411. *Papier und Kunststoff Verarbeiter*, (2), 18, 1996.

412. *Coating*, (3), 64, 1974.

413. G.L. Duncan, U.S. Patent 4,720,416, Mobil Oil Co., 1988.

414. K. Taubert, *Adhäsion*, (10), 379, 1970.

415. G. Meinel, *Papier und Kunststoff Verarbeiter*, (19), 26, 1985.
416. P. Dippel, *Adhäsion*, (4), 44, 1988.
417. *Kaut. Gummi, Kunststoffe*, 39 (9), 778, 1986.
418. *Coating*, (11), 272, 1985.
419. Verpackungs-Rundschau, Interpack 96 Special, E25.
420. *Etiketten-Labels*, (35), 21, 1995.
421. *Printing Trycite*, Dow Chemical USA, Designed Products Department, Midland, MI.
422. U. Reichert, *Kunststoffe*, 80(10), 1092, 1990.
423. G.M. Miles, *Papier und Kunststoff Verarbeiter*, (2), 64, 1988.
424. *Labels and Labelling*, (7/8), 70, 1988.
425. *British Plastics & Rubber*, (10), 56, 1998.
426. C.D. Weiske, *Kunststoffe*, (8), 518, 71.
427. *Neue Verpackung*, (3), 60, 1997.
428. B. Wright, *Adhes. Age*, (12), 25, 1971.
429. *Converting Today*, (11), 9, 1991.
430. *Packaging Today*, (12), 31, 1995.
431. *Adhäsion*, (8), 296, 1973.
432. *Adhäsion*, (1), 15, 1974.
433. F. Haag and E. Rohde, *Kaut. Gummi, Kunststoffe*, 39(12), 1216, 1986.
434. *Coating*, (6), 154, 1974.
435. *British Plastics*, (10), 63, 2001.
436. *Chemie Ingenier Technik Plus*, (6), 47, 2004.
437. S. Sacharow, *Adhäsion*, (6), 268, 1966.
438. J.S. Cheng Shiang, U.S. Patent 4,738,810, Dow Chemical Co., in *CAS, Adhesives*, 17, 6, 1988.
439. Gilman Brothers Co., *Plastics Technology*, (7), 72, 1968.
440. J.S. Razzano and R.B. Bush, U.S. Patent 4,728,567, General Electric Co., in *CAS, Siloxanes and Silicones*, 17, 4, 1988.
441. R.G. Nelb and K.G. Saunders, U.S. Patent 4,738,990, Dow Chem. Co., in *CAS, Siloxanes and Silicones*, 17, 5, 1988.
442. L. Waeyenbergh, in Proceedings of the 19th Munich Adhesive and Finishing Seminar, Munich, Germany, 1994, p. 139.
443. Packlabel News, Packlabel Europe '97, Ausgabe 2.
444. *Packaging Today*, (12), 28, 1995.
445. R. Hummel, Basismaterial für Haftetiketten, in Proceedings of the 19th Munich Adhesive and Finishing Seminar, Munich, Germany, 1994, p. 58.
446. *Coating*, (4), 139, 1997.
447. *Label Management for Wines, Spirits and Preserves*, Technical Booklet, Brigl and Bergmeister GmbH, Niklasdorf, Austria, 2004.
448. *Druck/Print*, (9), 65, 1986.
449. M. Fairley, *Labels and Labelling International*, (5/6), 76, 1997.
450. *British Plastics*, (10), 64, 2001.
451. P. Hammerschmidt, *Allg. Papier Rundschau*, 7, 190, 1986.
452. R. Nurse, *HDPE Applications, PE Developments*, Polyethylene '93, Session 3, Oct. 4, 1993, Maack Business Service, Zürich, Switzerland, p. 1.
453. *Bimodale HDPE*, Polyethylene '93, Oct. 4, 1993, Session III, Maack Business Service, Zürich, Switzerland, p. 77.
454. G. Bolder and M. Meier, *Kaut. Gummi, Kunststoffe*, (8), 715, 1986.
455. *Packaging Today*, (12), 28, 1995.
456. Dowlex. For Oriented Films. CH -254032-E-787, Data Sheet, Dow Chemicals, Horgen, Switzerland.
457. Dowlex. For Lamination Films. CH-254-052-E-288, Data Sheet, Dow Chemicals, Horgen, Switzerland.
458. *Etiketten-Labels*, (5), 30, 1995.
459. Valeron Film, Van Leer Flexible Packaging,Van Leer Flexibles, Essen, Belgium.
460. *Adhäsion*, (1), 14, 1974.
461. *Etiketten-Labels*, (5), 6, 1995.

462. *Converting Today*, (11), 23, 1990.

463. *Mobil-OPP Art*, (3), 1, 1993.

464. *Etiketten-Labels*, (5), 16, 1995.

465. *Kunststoffe*, 83 (10), 737, 1993.

466. E. Pilipponen, *Allg. Papier Rundschau*, 40, 1100, 1988.

467. R. Hummel, *Adhäsion*, (1–2), 49, 1972.

468. *Adhäsion*, (9), 19, 1984.

469. B. Martens, *Coating*, (6), 187, 1992.

470. Y. Yuzo, K. Kyotaka, N. Kanji, and K. Hironori, Japanese Patent 62,180,780/1987, Nippon Foil Manuf. Co. Ltd., in *CAS, Colloids*, 3, 13, 1988.

471. Sengewald Report, Halle, Germany, (1), 2, 1984.

472. *Coating*, (5), 173, 1996.

473. *Kunststoffe*, 83 (10), 725, 1993.

474. *Verpackung Rundschau*, (9), 994, 1983.

475. *Paper, Films and Foil Converter*, (9), 28, 1989.

476. *Finat Labelling News*, (3), 29, 1994.

477. *Coating*, (1), 22, 1985.

478. D. Boettger, *Labels and Labelling*, (3/4), 58, 1994.

479. U.S. Patent 3,321,451, in S.E. Krampe, and C.L. Moore, EP 0202831 A2, Minnesota Mining and Manuf. Co., St. Paul, MN, USA, 1986.

480. J. Verseau, *Coating*, (7), 181, 1971.

481. W. Witke, *Coating*, (12), 321, 1985.

482. W. Witke, *Coating*, (8), 278, 1988.

483. W. Witke, *Coating*, (9), 340, 1988.

484. *Coating*, (6), 187, 1972.

485. Mystik Tape Inc., IL, U.S. Patent 3,161,533, in *Adhäsion*, (6), 277, 1966.

486. H. Becker, *Adhäsion*, (3), 79, 1971.

487. *Labels and Labelling International*, (5/6), 18, 1997.

488. Toshio Nakajima, Ken Oda, Kazumi Azuma, and Kazuhide Fujita, Japanese Patent 6,327,579, Nitto Electric Ind. Co., 1988, in *CAS, Siloxanes and Silicones*, 17, 5, 1988.

489. *Adhes. Age*, (9), 82, 1984.

490. DBP 1079252, in *Coating*, (6), 185, 1969.

491. U. Füssel, *Kunststoffe*, 81 (10), 915, 1991.

492. Scholl Manuf. Co., British Patent 925,810, in *Coating*, (6), 184, 1969.

493. Bunyan, British Patent 815,121, in *Coating*, (6), 184, 1969.

494. *Kunststoff Information*, (1083), 4, 1991.

495. R. Bulet and J. Michel (Stamicarbon B.V.), Dutch Patent, 8601984, in *CAS, Adhesives*, 20, 3, 1988.

496. Minnesota Mining and Manuf. Co., St. Paul, MN, USA, EP 0306232B1, 1993.

497. S. Aoyanagi, H. Suzuki, and S. Takeda, Japanese Patent 6,315,872, Kanzaki Paper Manuf. Co. Ltd., in *CAS, Adhesives*, 14, 4, 1988.

498. German Offenlegungsschrift, 3.216.603 in EP 4.673.611.

499. Vorwerk & Sohn GmbH & Co. KG, Wuppertal, DE 4040 917, 1992.

500. G. Crass and A. Bursch (Hoechst AG, Frankfurt am Main, Germany), U.S. Patent 4,673,611, 1987.

501. R.B. Lipson, U.S. Patent 5,468,533, Kwik Paint Products, in *Adhes. Age*, (5), 12, 1996.

502. M.J. Huber, U.S. Patent 546,692/1995, Quality Manuf. Inc., in *Adhes. Age*, (5), 12, 1996.

503. T. Horie, T. Kino, and T. Nagaresugi, Japanese Patent 6,823,834, Japan Styrene Paper Co., in *CAS, Siloxanes and Silicones*, 11, 3, 1988.

504. K. Shiraishi, Japanese Patent 63,117,068, in *CAS, Colloids (Macromolecular Aspects)*, 22, 4, 1988.

505. Klebeband Träger u. Abdeckfolie, Datenblatt, Ausgabe 07/92, Hoechst AG, Wiesbaden, Germany.

506. *Allg. Papier Rundschau*, (18), 572, 1987.

507. Allmänna Svenska Elektriska AB, DBP 1276771, in *Coating*, (6), 184, 1969.

508. Kalle Folien, Siegel Band, Hoechst., Mi 1984, 38T 5.84 LVI, Wiesbaden, Germany.

509. Pritchett and Gold, British Patent 821,959, in *Coating*, (6), 184, 1969.

510. Blackford-Gross, British Patent 829,715, in *Coating*, (6), 184, 1969.

511. Scholl Manuf. Co., British Patent 871,504, in *Coating*, (6), 184, 1969.

512. Scholl Manuf. Co., U.S. Patent 2,953,130, in *Coating*, (6), 184, 1969.
513. American White Cross Laboratories Inc., Canadian Patent, 647,454, in *Coating*, (6), 184, 1969.
514. Johnson & Johnson, U.S. Patent 2,882,179, in *Coating*, (6), 184, 1969.
515. *Adhes. Age*, (10), 125, 1986.
516. P. Hammerschmidt, *Coating*, (4), 194, 1986.
517. D. Bedoni and G. Caprioglio, *Modern Equipment for Label and Tape Converting*, in 19. Münchener Klebstoff und Veredelungsseminar, 1994, p. 37.
518. *Adhäsion*, (11), 37, 1994.
519. *Adhäsion*, (1/2), 27, 1987.
520. R. Shibata, H. Miyagawa, Japanese Patent 6,386,785, Hitachi Condenser Co. Ltd, in *CAS, Adhesives*, 20, 4, 1988.
521. J.E. Riedel and P.G. Cheney, U.S. Patent, 4,292,360, Minnesota Mining and Manuf. Co., St. Paul, MN, USA, in *Adhes. Age*, (12), 58, 1981.
522. R. Milker, *Coating*, (3), 60, 1984.
523. B.F. Goodrich, U.S. Patent 3,788,878, in *Coating*, (10), 47, 1974.
524. Scholl Manuf. Co., U.S. Patent 3,039,459, in *Coating*, (6), 184, 1969.
525. Johnson and Johnson, U.S. Patent 3,077,882, in *Coating*, (6), 184, 1969.
526. Lohmann KG, U.S. Patent 3,086,531, in *Coating*, (6), 185, 1969.
527. *Adhes. Age*, (12), 67, 1984.
528. Westinghouse Electric Co., U.S. Patent 3,772,064, in *Coating*, (19), 47, 1974.
529. *Adhes. Age*, (3), 8, 1987.
530. Monsanto Co., U.S. Patent 3,752,733, in *Coating*, (19), 47, 1974.
531. Kendall Co., U.S. Patent 3,723,236, in *Coating*, (6), 154, 1974.
532. *Adhäsion*, (9), 254, 1976.
533. J.R. Wagner, *Tappi J.*, (4), 115, 1988.
534. J.M. Oelkers and E.J. Sweeney, *Tappi J.*, (8), 69, 1988.
535. L.E. Grunewald and D.J. Classen, EP 256662/1986, Minnesota Mining and Manuf. Co., St. Paul, MN, USA, in *CAS, Adhesives*, 15, 3, 1988.
536. K. Sakai and K. Inoue, Japanese Patent 6,348,379, Marubeni Co., in *CAS, Adhesives*, 15, 3, 1988.
537. Scotchmount, Doppelseitige Klebebänder mit Schaumstoff Träger, Brochure, 3M Deutschland GmbH, Neuss, Germany.
538. *Adhes. Age*, (1), 6, 1985.
539. A.D. Little, Inc., U.S. Patent 3,039,893, in *Coating*, (6), 185, 1969.
540. H. Meichner, Rehau, Germany, Gebrauchsmuster, GM 8607368, 1986.
541. S.B. Driscoll, L.N. Venkateshwaran, C.J. Rosis, and L.C. Whitney, Soc. Plast. Eng. Annu., Tech., Conf., Technical Paper, 1,450(1985), in *Kaut. Gummi, Kunststoffe*, 39(2), 161, 1986.
542. A.S. Wood, *Moder. Plast.*, 14(3), 36, 1984.
543. Y. Tanaka, K. Mai, H. Takegawa, S. Watanabe, and A. Midorikawa, Japanese Patent 62,177,043/1987, Dainippon Ink and Chemical Inc., in *CAS, Emulsion Polymerization*, 10, 2, 1988.
544. G.D. Bennett, U.S. Patent 925,735/1987, Simmonds Precision, New York, in *Adhes. Age*, (5), 28, 1988.
545. W. Geier, *Allg. Papier Rundschau*, 20, 618, 1987.
546. W.R. Grace and Co., French Patent 1,391,908, in *Coating*, (11), 336, 1972.
547. *Allg. Papier Rundschau*, 41, 138, 1986.
548. *Verpackungs Rundschau*, (7), 33, 1986.
549. G. Menges, W. Michaeli, R. Ludwig, and K. Scholl, *Kunststoffe*, 80(11), 1245, 1990.
550. R.L. Francisco, U.S. Patent 4,732,695, Texo Co., in *CAS, Coatings, Ink & Related Products*, 13, 9, 1988.
551. *Adhäsion*, (2), 45, 1974.
552. U. Zorll, *Adhäsion*, (90), 236, 1975.
553. Takdust Products Co., Rubber and Plastics Age, (London), (6), 559, 1968.
554. *Coating*, (8), 244, 1969.
555. H. Tsukamoto, *Japan Plastics Age*, (6), 56, 1969, in *Coating* (1), 18, 1971.
556. M. Michel, *Adhäsion*, (4), 154, 1966.
557. *Adhäsion*, (3), 83, 1974.
558. Chemplex Co., Rolling Meadows, IL, DE-PS 33 13 607/1983, in *Coating*, (5), 191, 1988.

559. A. Haas (Société Chimique des Charbonnage-CdF Chimie, France), U.S. Patent 4,624,991/1986, in *Adhes. Age*, (5), 26, 1987.

560. Les Colloides Industriels Francais, DBP 1289219, in *Coating*, (9), 272, 1969.

561. Nitto Electr. Ind. Co. Ltd., Japanese Patent 24,229/70, in *Coating*, (2), 38, 1972.

562. R.J. Litz, *Adhes. Age*, (8), 38, 1973.

563. *Labels and Labelling*, (3/4), 64, 1994.

564. *Adhäsion*, (9), 19, 1984.

565. *Etiketten-Labels*, (5), 30, 1995.

566. *Etiketten-Labels*, (5), 25, 1995.

567. *Etiketten-Labels*, (3), 53, 1995.

568. *Neue Verpackung*, (2), 61, 1991.

569. Eastman Kodak Co., U.S. Patent 3,718,728, in *Coating*, (6), 154, 1974.

570. I. Benedek, E. Frank, and G. Nicolaus (Poli-Film Verwaltungs, GmbH, Wipperfürth, Germany), DE 4433626A1, 1994.

571. E. Prinz, *Coating*, (2), 56, 1978.

572. E. Prinz, *Coating*, (10), 269, 1979.

573. *Coating*, (10), 270, 1979.

574. *Paper, Film, and Foil Converter*, (9), 32, 1989.

575. Softal Electronic, Erhöhung der Oberflächeneenergie von Kunststoffolien durch Softalisierung, Softal, Electronic, GmbH, Hamburg, Germany.

576. R.M. Podhajny, *Converting and Packaging*, (3), 21, 1986.

577. T.J. Blong and D.F. Klein, *The Influence of Processing Additives on Optical, Surface and Mechanical Properties of LLDPE Blown Film*, Polyethylene '93, Oct. 4, 1993, Maack Business Service, Zürich, Switzerland.

578. H.J. Fricke and L. Maempel, Kleben & Dichten, *Adhäsion*, (11), 14, 1994.

579. K.W. Gerstenberg, *Corona Treatment for Wetting and Adhesion on Printed Materials*, European Tape and Label Conference, Brussels, April 28, 1993.

580. *Deutscher Drucker*, (12), 26, 1988.

581. *Adhäsion*, 23, 136, 1979.

582. *Allg. Papier Rundschau*, 29, 798, 1988.

583. D.F. Lawson, *Rubber Chem. Technol.*, 60(1), 102, 1987.

584. *Coating*, (2), 54, 1987.

585. Milprint Overseas Corporation, Milwaukee, WI, U.S. Patent 1,504,556, in *Coating*, (8), 240, 1972.

586. *Coating*, (9), 340, 1995.

587. *Coating*, (7), 33, 1984.

588. *Adhäsion*, (6), 255, 1966.

589. *Papier und Kunststoff Verarbeiter*, (11), 44, 1986.

590. R. Milker and A. Koch, *Coating*, (1), 8, 1988.

591. *Coating*, (1), 20, 1978.

592. E. Prinz, *Coating*, (10), 360, 1979.

593. Softal Electronic, Report, Nr.102, Softal Electronic GmbH, Hamburg, Germany.

594. D.A. Markgraf, *Covering and Packaging*, (3), 18, 1986.

595. K.W. Gerstenberg, *Coating*, (8), 260, 1983.

596. K.W. Gerstenberg, *Coating*, (5), 172, 1991.

597. L. Dorn and W. Wahono, *Kunststoffe*, 81(9), 764, 1991.

598. J. Nentwig, *Papier und Kunststoff Verarbeiter*, (2), 50, 1996.

599. Ahlbrandt System, Technical Booklet, Ahlbrandt System GmbH, Lauterbach/Hessen, Germany, 2004.

600. A. Rau, *Coating*, (5), 185, 2004.

601. E. Djagarowa, *Plaste und Kautschuk*, (9), 678, 1969.

602. Japanese Patent 3571/1964, in *Coating* (1), 19, 1969.

603. *Coating*, (6), 174, 1972.

604. A.M. Slaff, *Coating*, (7), 198, 1973.

605. C.L. Aronson, L.G. Beholz, B. Burland, and J. Perez, *Investigation of a New Mechanism for Rendering High Density Polyethylene Adhesive*, in Proceedings of the 24th Annual Meeting of Adhesive Society, Williamsburg, VA, USA, Feb. 25–28, 2001, p. 294.

606. Midland Silicones Ltd., London, DBP 1293374, in *Coating*, (1), 6, 1970.
607. *Adhäsion*, (7), 198, 1974.
608. *Converting Today*, (1), 13, 1991.
609. W. Möhl, *Kunststoffe*, 81(7), 576, 1991.
610. O. Cada and P. Peremsky, *Adhäsion*, (5), 19, 1986.
611. L. Dorn and R. Bischoff, *Maschinenmarkt*, (43), 64, 1987.
612. *Das Papier*, (10A), 1985, in H. Klein, *Coating*, (12), 431, 1986.
613. R. Milker, *Coating*, (11), 294, 1985.
614. *Neue Verpackung*, (2), 60, 1991.
615. *Papier und Kunststoff Verarbeiter*, (7), 49, 1986.
616. R.A. Bragole, *Adhes. Age*, (4), 24, 1974.
617. S.L. Kaplan and P.W. Rose, *Plast. Eng.*, 44(5), 77, 1988.
618. *Adhäsion*, (7–8)39, 1996.
619. *Papier und Kunststoff Verarbeiter*, (6), 10, 1988.
620. *Adhäsion*, (1), 16, 1987.
621. Ch. Bichler, M. Bischoff, H.C. Langowski, and U. Moosheimer, VR Interpack '96, Special, E37.
622. *Deutsche Papierwirtschaft*, (2), IV, 1988.
623. R. Mannel, *Papier und Kunststoff Verarbeiter*, (10), 48, 1988.
624. W. Möhl, *Kunststoffe*, 81(7), 576, 1991.
625. *Caoutch. Plast.*, 64(674), 101, 1987.
626. M.D. Green, F.J. Guild, and R.D. Adams, *Analysis of Functional Surface Modification of Pre-treated Polypropylene, using Multi-Modal XPS and AFM*, in Proceedings of the 24th Annual Meeting of Adhesives Society, Williamsburg, VA, USA, Feb. 25–28, 2001, p. 254.
627. D.A. Kaute, and C. Buske, *High Performance, Truly Environmentally Friendly and Cost Effective Bonding Solutions with Atmospheric Pressure Plasma*, in Proceedings of the 24th Annual Meeting of Adhesive Society, Williamsburg, VA, USA, Feb. 25–28, 2001, p. 310.
628. M. Rasche, *Adhäsion*, (3), 25, 1986.
629. Y. Nishiyama, S.Y. Mo, and K.S. Bae, *Nippon Kagaku Kaishi*, 1118, 1985, in *Adhäsion*, (3), 32, 1986.
630. *Converting Today*, (1), 13, 1991.
631. A. Kruse, K.D. Vissing, A. Baalmann, and M. Hennecek, *Kunststoffe*, 83 (7), 522, 1993.
632. U. Zorll, *Adhäsion*, (7), 222, 1978.
633. *Coating*, (9), 279, 1969.
634. *European Adhesives*, (6), 22, 1995.
635. K. Armbruster and M. Osterhold, *Kunststoffe*, 80(11), 1241, 1990.
636. *Papier und Kunststoff Verarbeiter*, (6), 37, 1995.
637. *Coating*, (6), 163, 1974.
638. *Kunststoffe*, 84(11), 1569, 1994.
639. K.H. Kochem, *Kunststoffe*, 82(7), 578, 1992.
640. *Adhäsion*, (5), 211, 1967.
641. J. Nentwig, *Kunststoff J.*, (19), 8, 1991.
642. Trespaphan Information, (2), Hoechst Folien, Business Unit Trespaphan, Hoechst AG, Wiesbaden, Germany, September, 1991.
643. *Allg. Papier Rundschau*, (18), 686, 1986.
644. H. Klein, *Coating*, (12), 431, 1986.
645. *Coating*, (1), 28, 1978.
646. *Coating*, (1), 29, 1978.
647. Minnesota Mining and Manuf., St. Paul, MN, U.S. Patent 2,814,601, in *Coating*, (1), 24, 1987.
648. Minnesota Mining and Manuf., Co., St. Paul, MN, U.S. Patent 2,882,183, in *Coating*, (1), 24, 1987.
649. M.W. Uffner and P. Weitz, U.S. Patent 3,345,320/General Aniline and Film Co., New York, NY, 1967.
650. G.L. Booth, *100% Solids Silicone Coating*, in Tappi Hot Melt Adhesives and Coating Short Course, Hilton Head, SC, USA, May 2–5, 1982.
651. M.D. Fey, *Low Temperature Cure Solventless Silicone Paper Coatings*, in Tappi Hot Melt Adhesives and Coating Short Course, Hilton Head, SC, USA, May 2–5, 1982.

652. *Etiketten-Labels*, (35), 18, 1995.
653. C. Weitemeyer, J. Jachmann, D. Allstadt, and H. Brus, *TEGO Silicone Acrylates RC for Release Coatings*, in Tappi Hot Melt Symposium, Hilton Head SC, USA, June 16–19, 1985.
654. *Papier und Kunststoff Verarbeiter*, (9), 50, 1990.
655. H.J. Northrup, U.S. Patent 3,691,206, in *Coating*, (6), 154, 1974.
656. T. Bald, *A 100% Solid Silicon Coating System*, PSTC Technical Seminar, Itasca, IL, USA, May 7–9, 1986.
657. S. Moser, *Allg. Papier Rundschau*, (16), 452, 1986.
658. *Tappi J.*, (9), 277, 1987.
659. *Adhäsion*, (4), 23, 1983.
660. P. Lersch, T. Ebbrecht, and D. Wewers, in *Coating*, (2), 44, 1993.
661. *Kunststoffe*, (2), 85, 1992.
662. B. Hunt, *Labels and Labelling*, (2), 29, 1997.
663. K.B. Kasper and D.R. Williams, EP 252712 Schoeller Technical Papers Inc., in *CAS, Siloxanes and Silicones*, 17, 7, 1988.
664. K. Palli, M. Tirkkonnen, and T. Valonen, Finnish Patent, 74723, Yhtyneet Paperitehtaat Oy, in *CAS, Adhesives*, 14, 4, 1988.
665. T.N. Skuratovskaya, O.V. Annikov, N.Z. Kvasko, V.I. Stolyarov, M.D. Podvolotskaya, Yu.A. Yushelevski, and A.S. Filenko, USSR Patent 1,395,721, in *CAS, Crosslinking Reactions*, 22, 13, 1988.
666. *Coating*, (1), 25, 1978.
667. *Adhäsion*, (11), 15, 1986.
668. J.F. Kuik, *Papier und Kunststoff Verarbeiter*, (10), 26, 1990.
669. Polyethylene '89, Maack Business Service, Zürich, Switzerland, p. 53.
670. W. Gerstenberg, *Coating*, (9), 304, 1992.
671. Dow, XZ87131.32, Experimental Data Sheet, Dow Chemicals, Horgen, Switzerland.
672. Neste Chemicals, Technical Information, Polyethylene, F0133, 1991.
673. Battenfeld, Press-Info, 2713-GB, Feb., 1988, p. 2.
674. C. Momose, K. Nakakawara, and M. Matsui, Japanese Patent 6,351,488, Mitsubishi Densen Kogyo K.K. 1988, in *CAS, Adhesives*, 20, 3, 1988.
675. T.K. Bhaumik, B.R. Gupta, and A.K. Bhowmik, *J. Adhes.*, 24(2/4), 183, 1987.
676. J.P. Croquelois and P. Phandard, French Patent 22,600,981, Papeteries Elce, S.A. 1988.
677. G. Panagopoulos, S.E. Pirtle, and W.A. Khan, *Film Extrusion*, 1991, p. 103.
678. *Papier und Kunststoff Verarbeiter*, (9), 23, 1987.
679. *British Plastics and Rubber*, (5), 24, 1988.
680. Rapra Technology Limited, The Extrusion of Tacky Polyethylene Film, RTL/1705, 1990.
681. *Converting Today*, (1), 15, 1992.
682. *Deutsche Papierwirtschaft*, (3), 133, 1987.
683. K. Noda and K. Isayama, EP 252372, Kanegafuchi Chem. Ind. Co. Ltd., in *CAS, Siloxanes & Silicones*, 12, 4, 1988.
684. I. Hitoshi, T. Tokihiro, and K. Kiyonori, Japanese Patent, 62,270,642/1987, Tokuyama Soda Co. Ltd. in *CAS, Adhesives*, 13, 1, 1988.
685. V.M. Ryabov, O.I. Chernikov, and M.F. Nosova, *Plast. Massy*, (7), 58, 1988.
686. M.W. Janczak, *Polymer*, (9), 381, 1964.
687. *Coating*, (8), 4, 1969.
688. *Neue Verpackung*, (2), 61, 1991.
689. M. Toshio, Japanese Patent, Furukawa Electric Co Ltd., in *CAS, Adhesives*, 24, 6, 1988.
690. Y. Torimae, Japanese Patent 63,165,474, Kao Corp., in *CAS, Adhesives*, 24, 5, 1988.
691. K. Yamada, K. Miyazaki, Y. Owatari, Y. Egami, and T. Honma, EP 257803, Sumitomo Chem. Co., Ltd., in *CAS, Coatings Inks & Related Compounds*, 17, 6, 1988.
692. A.E. Khoklovkin, N.B. Vladimirokaya, and E.S. Bushkova, Sovrem. Lakokrasokh. Mater. i Tekhnol. ikh Primeneniya, Mater. Semin. N., in *CAS, Coatings Inks & Related Products*, 18, 2, 1988.
693. M. Chladek, J. Jilek, and F. Benc, Czech CS 246986/1987, in *CAS, Coatings, Inks & Related Products*, 18, 13, 1988.
694. W.E. Havercroft, *Paper, Film and Foil Converter*, (10), 52, 1973.

695. G.F. Bulian, *Extrusion Coating and Adhesive Laminating: Two Techniques for the Converter*, Polyethylene '93, 4/6, 1993, Maack Business Service, Zürich, Switzerland.

696. *Coating and Laminating in House*, Cowise Management and Training Conference, Amsterdam, 1997.

697. T. Frecska, *Screen Printing*, (2), 64, 1987.

698. G. Perner, *Coating*, (6), 237, 1991.

699. *Quelques Produits dont vous ne Pouvez vous Passe*r, Etilux, Booklet, S.A Etilux N.V., Brussel, Belgium.

700. *Adhes. Age*, (8), 34, 1983.

701. *Coating*, (4), 139, 1997.

702. Kroenert Technical Booklet, Maschinenfabrik Max Kroenert GmbH & Co, Hamburg, Germany, May 2004, p. 7.

703. *Adhäsion*, (9), 351, 1965.

704. R. Hinterwaldner, *Adhäsion*, (6), 11, 1984.

705. C. Massa, *Pressure Sensitive Tapes and Laminates Marketing and Production Techniques*, in Proceedings of the 19th Munich Adhesive and Finishing Seminar, Munich, Germany, 1994, p. 35.

706. R. Schieber, *Adhäsion*, (5), 21, 1982.

707. B.K. Bordoloy, Y. Ozari, S.S. Plamthottam, and R. Van Ham, EP 263686, Avery Internat. Co., 1988, in *CAS, Radiation Curing*, 19, 1, 1988.

708. H. Klein, *Coating*, (12), 430, 1986.

709. Rohm & Haas Co., Philadelphia, Polymers Resins and Monomers, Adhesives, Rhoplex PS-83D, 1982, p. 2

710. U.S. Patent, 0,212,358, p. 2.

711. *Adhes. Age*, (3), 41, 1985.

712. E. Bradatch, *Papier und Kunststoff Verarbeiter*, (6), 33, 1987.

713. BMB Curtain Coating, BMB HSC Coater, Kroenert Technical Booklet, 01-03-D, 2003, Maschinenfabrik Max Kroenert GmbH & Co., Hamburg, Germany.

714. Das Technikum, Kroenert Group, Technical Booklet, Maschinenfabrik Max Kroenert GmbH & Co., Hamburg, Germany, 2004.

715. *Coating*, (5), 162, 2004.

716. P. Beiersdorf & Co. AG, Hamburg, U.S. Patent 1,569,882, in *Coating*, (11), 336, 1972.

717. M. Guder and J. Auber, *Die Verarbeitung von HMPSA im industriellen Bereich unter Technischen und Ökonomischen Aspekten*, in Proceedings of the 19th Munich Adhesive and Finishing Seminar, Munich, Germany, 1994, p. 232.

718. *Paper, Film and Foil Converter*, (8), 28, 1993.

719. W. Schaezle, *Adhäsion*, (9), 5, 1983.

720. *Druck Print*, (8), 24, 1986.

721. H.J. Claasen, *Allg. Papier Rundschau*, (50–51), 1741, 1986.

722. *Coating*, (11), 291, 1982.

723. H.J. Einfeldt and H.J. Meissner, *Schmelzklebstoff Auftragssysteme*, in Proceedings of the 16th Munich Adhesive and Finishing Seminar, Munich, Germany, 1991, p. 230.

724. H. Klein, *Coating*, (6), 210, 1986.

725. *Coating*, (1), 12, 1988.

726. *Coating*, (12), 344, 1984.

727. G. Drechsler, *Coating*, (6), 153, 1971.

728. J. Türk, *Papier und Kunststoff Verarbeiter*, (10), 22, 1985.

729. Morstik 103 Adhesive, Morton Thiokol Inc., *Adhesives and Coatings*, MTD-MS103-12, 1988.

730. Tappi, Hot Melt Symposium, 1988, Hilton Head, SC, in *Adhäsion*, (4), 234, 1986.

731. Morton Thiokol Inc., Adhesives and Coatings, Morstik 16 Adhesive, MTD-MS106-12, 1988.

732. *Coating*, (11), 394, 1990.

733. *Coating*, (11), 291, 1982.

734. G.W. Drechsler, 11. Münchener Klebstoff u. Veredelungsseminar, October 20–22, 1986, in *Coating*, (3), 97, 1987.

735. Planatolwerk, W. Hesselmann, Rosenheim, DBP 1288070, in *Coating*, (9), 272, 1969.

736. PAK 600, Kroenert Technical Booklet, Maschinenfabrik Max Kroenert GmbH & Co, Hamburg, Germany, 2004.

737. *Adhäsion*, (1–2), 36, 1985.
738. J. Weidauer, *European Adhesives and Sealants*, (5), 26, 1995.
739. Leimaufrags-Systeme, Technical Booklet, hhs, Krefeld, Germany.
740. H. Klein, *Adhäsion*, (7), 248, 1973.
741. F.M. Fischer, *Coating*, (3), 90, 1973.
742. V. Bohlmann, *Papier und Kunststoff Verarbeiter*, (2), 12, 1979.
743. W. Grebe, *Papier und Kunststoff Verarbeiter*, (2), 46, 1985.
744. F.S. McIntire, *The Future of Inline UV Silicone and Hot Melt Pressure Sensitive Adhesive Coatings for Label and Tape Products*, Hot Melt Adhesives and Coating Short Course of Technical Association of the Pulp and Paper Industry, Head, SC, June, 1982, in *Adhes. Age*, (8), 35, 1983.
745. *Coating*, (5), 176, 1988.
746. E.G. Huddleston, U.S. Patent 4,692,352, The Kendall Co., Boston, MA, 1987.
747. M. Dollinger, UV Strahlungstechnik für die Verarbeitung, von UV Härtenden Klebstoffen, Möglichkeiten, Neue Entwicklungen, 11 Klebetechnik Seminar, Jan. 25, 1989, Rosenheim, Germany.
748. P. Holl, Elektronenstrahlvernetzung/Vulkanisation von Klebstoffen für Flexible Materialien, 11 Klebetechnik Seminar, Jan. 25, 1989, Rosenheim, Germany.
749. U. Schwab, *Coating*, (5), 171, 1996.
750. E. Smit, Pressure Sensitive Adhesives and Adhesive Coating, Cowise Management and Training Service, Amsterdam, 1996.
751. H. Braun, UV-Curable Acrylic Based Hot Melt Pressure Sensitive Adhesives, in *Coating*, (12), 477, 1995.
752. A. Dobmann and J. Planje, *Papier und Kunststoff Verarbeiter*, (1), 38, 1986.
753. G. Pitzler, *Coating*, (6), 218, 1996.
754. H. Kamusewitz, *Die thermodynamische Interpretation der Adhäsion unter besonderer Beachtung der Folgerungen aus der Theorie von Girifalco und Good*, Dissertation, Halle, 1988, in G. Pitzler, *Coating*, (6), 218, 1996.
755. Novacel, France, DE 1594193, in K. Wabro, R. Milker, and G. Krüger, Haftklebstoffe und Haftklebebänder, Astorplast GmbH, Alfdorf, Germany, 1994, p. 48.
756. A. Schmalz and W. Neumann, *Coating Lines for the Production of Self Adhesive Tapes and Labels Using Radiation Curable Hot Melt Systems*, in European Tape and Label Conference, Brussels, April 28, 1993.
757. H.J. Voss, *Coating*, (5), 150, 1987.
758. H. Hadert, *Coating* (1), 11, 1969.
759. Johnson & Johnson, Canadian Patent 583,367, in *Hans Hadert, Coating* (1), 11, 1969.
760. U.P. Seidl, Ger. Offen, DE 3625904, Schreiner Etiketten und Selbstklebetechnik GmbH und Co., 1988, in *CAS, Siloxanes and Silicones*, 16, 3, 1988.
761. H. Klein, *Coating*, (10), 372, 1988.
762. D. Percivalle, *High Speed Machines for the Production of Self Adhesive Tapes and Labels*, European Tape and Label Conference, Brussels, April 28, 1993, p. 133.
763. H. Klein, *Adhäsion*, (5), 190, 1973.
764. H. Klein, *Coating*, (3), 4, 1993.
765. C.S. Watson, *Coating*, (11), 399, 1987.
766. H. Klein, *Coating*, (11), 390, 1986.
767. H.G. Reinhardt, *Application Methods for Highly Viscous Melts*, in Proceedings of the 16th Munich Adhesive and Finishing Seminar, Munich, Germany, 1991, p. 15.
768. B.C. van Oosten, *Allg. Papier Rundschau*, 16, 382, 1988.
769. H.W. Jakobs, *Allg. Papier Rundschau*, 16, 380, 1988.
770. *Coating*, (6), 184, 1969.
771. S. Aoyanagi, H. Suzuki, and S. Takeda, Japanese Patent 6,315,872, Kanzaki Paper Manuf. Co. Ltd., in *CAS, Adhesives*, 14, 4, 1988, 108:222810 b.
772. *Adhäsion*, (4), 4, 1985.
773. *Coating*, (11), 393, 1990.
774. Y. Mizutani, T. Noguchi, H. Kuroki, and T. Imahama, Japanese Patent 63,265,369/1987, Tosoh Corp., in *CAS, Hot-Melt Adhesives*, 11, 1, 1988.

775. G. Wouters, EP 251726, Exxon Chemical Patents Inc., in *CAS, Emulsion Polymerization*, 18, 2, 1988.

776. B. Copley and K. Melancon, U.S. Patent 878,816/1986, Minnesota Mining and Manuf. Co., in *CAS, Siloxanes & Silicones*, 14, 2, 1988.

777. 14th Munich Adhesive and Finishing Seminar, Munich, Germany, 1989, p. 9.

778. D. Satas, *Adhes. Age*, (8), 30, 1988.

779. *BMB Veredelungsanlagen für Papier, Karton, Kunststoff-und Aluminiumfolie*, Technical Booklet, Bachofen Meier AG, Bülach, Switzerland, 2004.

780. Universum Verpackung GmbH, Rodenkirchen, U.S. Patent 1,297,790, in *Coating*, (1), 274, 1969.

781. T.J. Bonk, Tsung I. Cheng, P.M. Olsom, and D.E. Weiss, PCT, WO 87/00189, 1987.

782. A. Saburo, Japanese Patent 63,117,085, Central Glass Co. Ltd., 1988.

783. A. Nagasuka and M. Kobari, Japanese Patent 6,389,585, Sekisui Chem. Co., Ltd., in *CAS, Adhesives*, 19, 6, 1988.

784. Minnesota Mining and Manuf. Co. Ltd., Japanese Patent 63,175,091, in *CAS, Adhesives*, 25, 7, 1988.

785. Y. Yamada and O. Takuman, EP 253601, Toray Silicone Co., Ltd., 1988.

786. *Adhäsion*, (6), 14, 1985.

787. E.R. Squibb and Sons, Princeton, NJ, U.S. Patent 4,551,490.

788. T. Kishi, Japanese Patent, 6,369,879, Sekisui Chem. Co. Ltd., in *CAS, Adhesives* 19, 6, 1988.

789. H.W.J. Müller, *Adhäsion*, (5), 208, 1981.

790. I.A. Gritskova, L.P. Raskina, S.A. Voronov, G.K. Channova, D.N. Avdeev, E.B. Malyukova, T.K. Vydrina, N.D. Sandomirskaya, and O.M. Solozhentseva, *Otkrytiya, Izobret.*, (42), 87, 1987, in *CAS, Coating & Inks and Related Products*, 11, 11, 1988.

791. Beiersdorf A.G., Hamburg, Germany, DBP 1,667,940, in *Coating*, (12), 363, 1973.

792. K. Takashimizu and A. Suzuki, Japanese Patent 6,392,683/Advance Co. Ltd., in *CAS, Adhesives*, 19, 5, 1988.

793. *Coating*, (6), 198, 1993.

794. *Druck Print*, (4), 19, 1988.

795. Application Tape Folie, Technische Anforderungen, Tacfol Klebfolien GmbH, Germany.

796. R. Higginson, PCT/WO 88 03477, DRG, U.K., in *CAS, Colloids (Macromol. Aspects)*, 22, 6, 1988.

797. H. Klein, *Coating*, (7), 270, 1988.

798. G.W. Drechsler, *Coating*, (5), 167, 1996.

799. *Kaut. Gummi, Kunststoffe*, 37(3), 252, 1984.

800. G.W. Drechsler, *Coating*, (3), 62, 1984.

801. W.K. Behrendt and R. Schock, *Maschinenmarkt*, 88(81), 1648, 1982.

802. B.W. McMinn, W.S. Snow, and D.T. Bowman, *Adhes. Age*, (11), 37, 1995.

803. H. Klein, *Coating*, (6), 198, 1996.

804. A. Hecht Beaulieu, D.R. Gehman, and W.J. Sparks, *Tappi J.*, (9), 102, 1984.

805. D.H. Treesdale, *Coating*, (11), 290, 1982.

806. M. Guder and J. Auber, *Die Verarbeitung von HMPSA im Industriellen Bereich unter Technischen und Ökonomischen Aspekten*, in Proceedings of the 19th Munich Adhesive and Finishing Seminar, Munich, Germany, 1994, p. 232.

807. G. Drechsler, *Coating*, (6), 153, 1971.

808. F.M. Fischer, *Coating*, (3), 90, 1973.

809. L.W. Fuller GmbH, Lüneburg, Wachse, Hotmelts, Klebstoffe, 1977.

810. H. Hadert, *Coating*, (7), 203, 1970.

811. J. Raterman, *Evolution of Pressure Sensitive Adhesive Spray Technology*, in Tappi Hot Melt Symposium 85, Hilton Head, SC, USA, June 16–19, 1985.

812. *European Adhesives and Sealants*, (6), 24, 1966.

813. W. Schaezle, *Coating*, (7), 450, 1981.

814. H. Klein, *Coating*, (3), 74, 1993.

815. K.H. Honsel, Bielefeld, Germany, DBP, 2110491, in *Coating* (12), 363, 1973.

816. *Adhes. Age*, (5), 34, 1987.

817. *Coating*, (3), 172, 1969.

818. K. Ochi, S. Uozu, and K. Toyama, EP 0070524, Nippon Carbide Kogyo K.K. Tokyo, Japan, 1981.

819. M. Hasegawa, Tokyo, U.S. Patent 4,460,634/1984.

820. G.E. Davis, *Precision Controlled Application of Continuous Patterned Hot Melt Adhesive*, in Tappi Hot Melt Symposium 85, Hilton Head, SC, USA, June 15–19, 1985.
821. Japanese Patent A 5,695,972, in P. Gleichenhagen, E. Behrend, and P. Jauchen, EP 0149135 B1, Beiersdorf AG, Hamburg, Germany, 1987.
822. *Der Siebdruck*, 33 (3), 43, 1987.
823. *Druckwelt*, (6), 50, 1988.
824. H. Mürmann, *Papier und Kunststoff Verarbeiter*, (11), 26, 1987.
825. E. Mandershausen, *Coating*, (10), 271, 1974.
826. *Converting Today*, (4), 22, 1991.
827. *Converting Today*, (11), 43, 1990.
828. *Adhäsion*, (1/2), 18, 1987.
829. H. Klein, *Coating*, (19), 346, 1992.
830. U. Temper, *Allg. Papier Rundschau*, 42, 1156, 1988.
831. H. Klein, *Coating*, (5), 177, 1996.
832. *Coating*, (1), 15, 1978.
833. A. Erdmann, *Papier und Kunststoff Verarbeiter*, (3), 46, 1997.
834. S.I. Shapiro, *Tappi J.*, (5), 97, 1988.
835. *Kaut. Gummi, Kunststoffe*, 37(1), 57, 1984.
836. I. Benedek, *Pressure Sensitive Adhesives and Applications*, Marcel Dekker, Inc., New York, 2000, chap. 9.

9 Molecular Design of Hydrophilic Pressure-Sensitive Adhesives for Medical Applications

Mikhail M. Feldstein, Nicolai A. Platé, and Gary W. Cleary

CONTENTS

As discussed in Chapter 8, a major part of medical applications require the use of pressure-sensitive products (PSPs). Some of them are classic constructions with or without a carrier, and a pressure-sensitive adhesive (PSA), that are manufactured using common procedures; the most promising of them include the new class of hydrophilic PSAs. Their design and manufacture is sophisticated than that of common pressure-sensitive products. On the basis of the fundamental features of pressure-sensitive formulation described in Chapters 4, 5, and 7, this chapter discusses the design of hydrophylic PSAs for medical applications.

I. PERFORMANCE PROPERTIES OF PRESSURE-SENSITIVE ADHESIVES FOR MEDICAL APPLICATION

PSAs have diverse applications in numerous areas of industry and health care, in which they are critical components of drug delivery systems, serving both as the means of device attachment to patient skin or mucosal tissues and as a reservoir vehicle controlling the rate of drug delivery [1]. The following four major performance properties of PSAs constitute a basis for their application in drug-loaded adhesive patches:

1. *Adhesive:* High tack coupled with an optimum slip-stick transition point
2. *Transport:* Drug release kinetics controlled in terms of transdermal delivery rate and the functional lifetime of device
3. *Reservoir:* Drug compatibility and ability to be stored in a stable form tailored to incorporated drug of interest
4. *Biological:* No toxicity, skin irritation, and sensitization

As PSA is a complex phenomenon, and the adhesive and transport properties are interrelated as has been established by Equation (4.8), the above-mentioned performance properties are difficult to combine in a single system. It is therefore no wonder that there are only few major types of medical-grade PSAs: natural and synthetic rubbers, polyisobutylene, polydimethyl siloxane, and acrylates [2,3] (see also Chapter 8). It is pertinent to note that currently available medical-grade adhesives are hydrophobic polymers [1].

For many pharmaceuticals, the solubility of active agent in the reservoir of transdermal drug delivery device is of decisive importance. With higher solubility, it is possible to increase the rate of transdermal delivery. Because many therapeutic agents are ionogenic organic substances having a higher solubility in hydrophilic media than in lipophilic vehicles, adhesive reservoirs based on the hydrophilic polymers would be more versatile than those based on hydrophobic polymers [4]. Other general advantages of hydrophilic adhesives are reviewed in Ref. [5].

In Chapter 4 a typical hydrophilic PSA has been described, which is based on the blends of high molecular weight (MW) poly(*N*-vinyl pyrrolidone) (PVP) with oligomeric poly(ethylene glycol) (PEG). The PVP–PEG adhesive can serve as a universal diffusion matrix for transdermal delivery of drugs spanning a wide range of chemical structures, physicochemical properties, and therapeutic categories [4,6,7]. Transdermal therapeutic systems with six various drugs incorporated into the PVP–PEG universal adhesive matrix are currently approved for clinical practice. As shown in Chapter 4, the essential advantage of the hydrophilic PVP–PEG adhesive when compared with hydrophobic PSAs is an increase in adhesion in the course of water or moisture absorption. However, solubility in large amounts of water hinders the application of the PVP–PEG adhesive to the biological substrates with very high level of hydration, such as oral mucosa or dental tissue.

Thus, the development of water insoluble but swellable hydrophilic PSAs is an actual challenge. The first few families of such PSAs have been recently created and known as Corplex™ adhesives, which are commercially available from Corium International, Inc., Redwood City, CA [8–10].

II. KEY FACTORS UNDERLYING THE DESIGN OF HYDROPHILIC PRESSURE-SENSITIVE ADHESIVES

All the PSAs are in viscoelastic state at the temperature of their usage. Normally, the glass transition temperature T_g of the PSAs is 50–70°C lower than the temperature of their application [11]. Traditional way of formulation of hydrophobic PSA is to select an elastomer with appropriate T_g value and optimize the adhesion by mixing the elastomer with suitable plasticizers and tackifiers (see Chapter 8). The former leads to the reduction of T_g, whereas the tackifiers cause the increase of T_g [12]. However, the PVP–PEG hydrophilic adhesive is formulated based on a high molecular weight amorphous polymer with T_g at about 150°C above the temperature of application. By adding liquid short-chain PEG with $M_w = 400$ g/mol, in miscible PVP–PEG blends the T_g of the composition is reduced to the value of -55°C, which is typical for PSAs (see Chapter 4 and [13]).

It is well known that crosslinked hydrophilic polymers are capable of swelling by absorption of large amounts of water, but they are essentially insoluble in water. Consequently, to make the PVP–PEG adhesive swellable but insoluble in water, the long PVP chains are to be crosslinked, forming a supramolecular network structure.

As demonstrated in Chapter 4, at a molecular level the PSA results from a specific balance between high energy of intermolecular cohesion and a large free volume. Factors appealing to enhance the cohesion and increase the free volume are as follows:

Factors Enhancing Cohesion	Factors Increasing Free Volume
Covalent crosslinking	Crosslink length and flexibility
Noncovalent crosslinking	Adding plasticizers
Ionic bonding	Adding tackifiers
Electrostatic bonding	
Hydrogen bonding	
Increase in network density	

Let us consider the significance of these factors for the molecular design of hydrophilic PSAs.

III. HYDROPHILIC ADHESIVES: HYDROGEN BONDS VS. COVALENT CROSSLINKING

A strong favorable molecular interaction between polymer chains leads normally to the decrease in free volume. The development of new PSAs requires the molecular structures that reconcile the cohesion and free volume. Those are usually conflicting properties and their combination within the same material is not easy to provide. Coupling the strong cohesion with large free volume may be achieved by crosslinking the chains of high molecular weight polymer through long and flexible connecting chains as it occurs, for example, in the PVP–PEG blends described in Chapter 4. As a result of the hydrogen bonds formed between OH groups at the ends of PEG chains and complementary carbonyl groups in the monomeric units of PVP macromolecules, the PEG acts as a noncovalent crosslinker of longer PVP chains and produces H-bonded network that assures the increase in cohesion. At the same time, as a result of appreciable length and flexibility of the PEG crosslinks, the intermolecular association between PVP macromolecules is followed by the increase in free volume. But question is pertinent: how does a type of intermolecular crosslinking affect the adhesive performance of cured material? In other words, what is the role of H-bonding in the adhesive capability of the PVP–PEG system? In order to answer this question we have synthesized the covalent replica of the PVP–PEG H-bonded complex [14].

Stoichiometric PVP–PEG complex that provides the best adhesion contains 19–20 PEG chains per 100 monomeric units of PVP [13]. To gain a molecular insight into the contribution of hydrogen bonds to the adhesion of PVP–PEG adhesives we have attempted to replace fully or partly the H-bonds in the PVP–PEG mixture for covalent bonds, leaving the same structure and large free volume in covalent replica of the H-bonded network complex. With this purpose we have synthesized copolymers of vinyl pyrrolidone (VP) with PEG-400 diacrylate (PEGDA) and PEG-400 monomethacrylate (PEGMMA). The proposed structures of PVP–PEG complex formed by mixing PVP and PEG-400 along with the structures of vinyl pyrrolidone (VP)–PEGDA and VP–PEGMMA copolymers are shown in Figure 9.1.

The VP–PEGDA crosslinked copolymer (Figure 9.1c) represents a covalent-bonded replica of the H-bonded PVP–PEG stoichiometric complex shown in Figure 9.1a. The VP–PEGDA copolymer containing about 25 PEGDA crosslinks per 100 VP units has a network density approaching the density of H-bond network in the stoichiometric complex within PVP–PEG mixture [13]. The major difference is that both hydrogen bonds in the PVP–PEG complex are replaced by two covalent bonds in the VP–PEGDA crosslinked copolymer. However, in the comb-like VP–PEGMMA copolymer, only one of the hydrogen bonds is replaced by a covalent bond, leaving the other hydroxyl group at opposite end of the PEG side-chain accessible for forming hydrogen bond with the carbonyl group in PVP (Figure 9.1b). The VP–PEGMMA comb-like polymer (one hydrogen-bond and one covalent bond) has a structure intermediate between the hydrogen-bonded PVP–PEG stoichiometric complex (two hydrogen bonds) and its covalent bonded replica in VP–PEGDA copolymer (two covalent bonds). The VP–PEGMMA and VP–PEGDA copolymers spanning entire range of monomeric compositions have been prepared. The adhesion of the copolymers has been measured with 180° Peel Test, whereas the relaxation properties have been evaluated as described in [14] and Chapter 4, Section III.B.

As follows from the results presented in Figure 9.2, complete replacement of H-bonded crosslinks by covalent bonds in the VP–PEGDA crosslinked copolymer leads to dramatic drop in the longer retardation time and adhesion. If only 50% of the H-bonds are replaced by the covalent crosslinks as in the comb-like VP–PEGMMA copolymer, the longer retardation time is increased compared with VP–PEGDA copolymer and, consequently, the adhesion is also improved. In this way, obtained results confirm unequivocally the significance of longer retardation time that relates to large-scale structure rearrangement component of the relaxation process for the PSA. This is in full agreement with Equation (4.8) which predicts the importance of longer relaxation time and high coefficient of self-diffusion for proper adhesion.

With the rise in PEGDA and PEGMMA content, the longer retardation time tends to decrease and for PEGDA homopolymer only shorter retardation time has been found to occur. In dry state the VP copolymers with PEGDA and PEGMMA possess a low or no adhesion. The lack of adhesion is mainly due to insufficient molecular mobility of covalently crosslinked polymer embedded by the longer retardation time and relevant modulus. The contribution of molecular mobility can be nevertheless significantly enhanced if the VP–PEGDA or VP–PEGMMA copolymers are swollen in water, because in the swollen state both the self-diffusion coefficient of polymer segments and the retardation time is significantly increased (see Equation (4.8)). And really, the adhesion of the copolymers has been shown to increase appreciably if the amount of absorbed water achieves 15–40%.

Thus, the contribution of hydrogen-bonds is of crucial importance for the development of hydrophilic PSAs. Transient character and lability of H-bonds allows their continuous breakdown and reformation in the course of material straining and relaxation, providing a tool for the dissipation of a large amount of energy during adhesive bond formation or failure.

Hydrogen-bonded adhesives can be easily prepared by mixing appropriate complementary polymeric components. The advantages of blended adhesives over synthesized through covalent

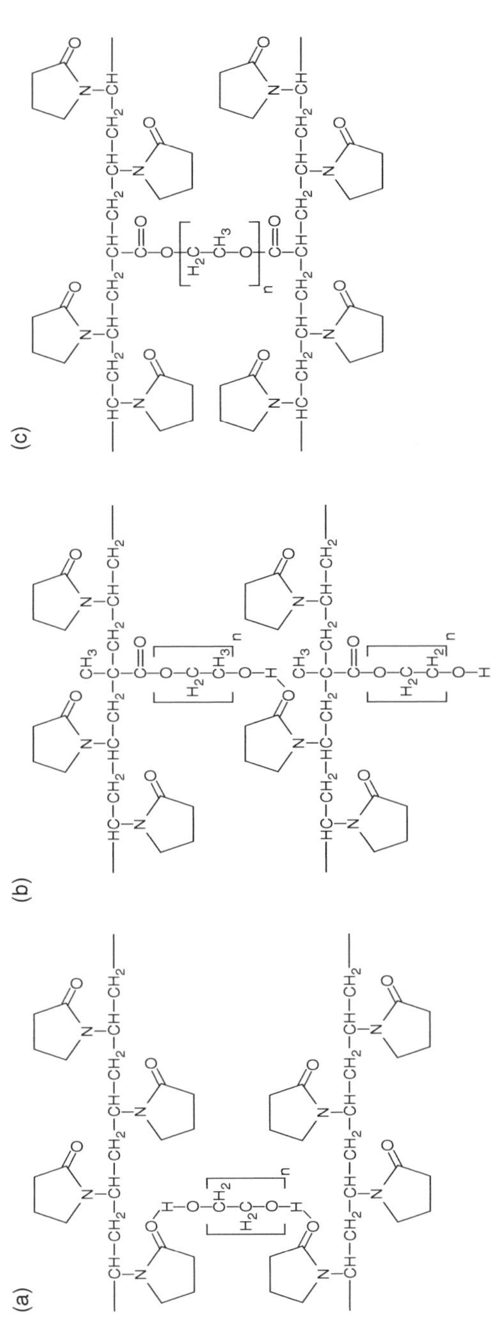

FIGURE 9.1 Chemical structures. (a) Stoichiometric crosslinked H-bonded complex formed in PVP–PEG blend (two H-bonds per PEG molecule and no covalent bond between PVP and PEG); (b) The VP–PEGMMA comb-like copolymer (one H-bond and one covalent bond); (c) The VP–PEGDA copolymer (no H-bonds and two covalent bonds).

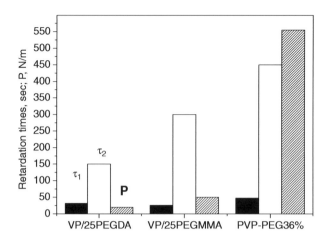

FIGURE 9.2 Effect of the replacement of hydrogen bonds by the covalent crosslinking of PVP macromolecules through comparatively shorter and flexible PEG-400 chains on the retardation times, τ, and peel adhesion, P. The VP–PEGDA: no H-bonds, two covalent bonds per PEG chain; VP–PEGMMA: one H-bond and one covalent bond; PVP–PEG: two H-bonds and no covalent bond.

crosslinking are easier production, lower cost, and capability to obtain materials, which provide film-forming properties. In turn, the latter property is of crucial significance for the preparation of diverse drug delivery systems designed for skin or mucosal applications. For this reason, from this point onwards our focus is on the development of film-forming blended adhesives.

IV. PRINCIPLES OF FORMULATING THE ADHESIVE HYDROPHILIC POLYMER BLENDS

Blending the polymers provides a convenient way to obtain composite materials with specifically tailored properties, as the properties of the blend are typically intermediate between those of the unblended components when the components are immiscible or partly miscible. In order to make the composite insoluble in water, water-soluble materials are usually mixed with water-insoluble components. When this is done, however, a phase separation can often occur that does not favor adhesion. Moreover, the insolubility of blend components may hamper the procedure of blend preparation, which often involves the dissolution of all the components in a common solvent, followed by casting the solution and drying. Preparation of polymer composite materials whose properties are new and untypical of parent components requires a high skill of a material designer. This challenge may be resolved if the blend components are capable of a strong favorable interaction with each other. More often, such interaction is hydrogen bonding, electrostatic bonding or ionic bonding. In this instance, mixing of two or more soluble polymers can give their ladder-like complex that is schematically shown in Figure 9.3. Such complex is swellable, but either insoluble

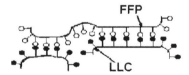

FIGURE 9.3 Schematic representation of a ladder-like complex formed between a film-forming hydrophilic polymer (FFP) and a complementary proton-donating polymer named the ladder-like noncovalent crosslinker (LLC).

or only partly soluble. In order to resolve these problems, this work is directed to a method of obtaining water-insoluble, film-forming compositions by blending soluble polymers, more specifically by blending hydrophilic polymers with complementary macromolecules that are capable of hydrogen bonding, electrostatic bonding or ionic bonding.

The complementarity of polymer components is an important aspect of this approach. A list of exemplary functional groups and the types of bonding for the film-forming polymer (FFP) and the ladder-like noncovalent crosslinker (LLC) is presented in Table 9.1.

A distinctive feature of hydrogen bonding between proton donating and proton accepting complementary groups is that both the reactive groups and the product of their interaction bear no electric charge. Electrostatic bonding is the interaction of proton donating and proton accepting groups, which are initially uncharged, but their interaction causes the formation of ionic bond. And finally, ionic bonding is the interaction of oppositely charged (cationic and anionic) groups with the formation of ionic (salt) bond.

Before describing the major principles of the design of hydrophilic adhesives in detail, it is useful to outline briefly relevant definitions. The definitions of "hydrophobic" and "hydrophilic" polymers are based on the amount of water vapor absorbed by polymers at 100% relative humidity. According to this classification, hydrophobic polymers absorb only up to 1 wt.% of water at 100% relative humidity, while moderately hydrophilic polymers absorb 1–10 wt.% of water. Hydrophilic polymers are capable of absorbing more than 10 wt.% of water, and hygroscopic polymers absorb more than 20 wt.% of water. A "water-swellable" polymer is one that absorbs an amount of water greater than at least 50 wt.% of its own weight, upon immersion in an aqueous medium. The term "crosslinked" herein refers to a composition containing intramolecular and intermolecular crosslinks, whether arising through covalent or noncovalent bonding. Noncovalent bonding, that is in focus of our further discussion, includes both hydrogen bonding, electrostatic bonding, and ionic bonding.

The term "hydrogel" is used in the conventional sense to refer to water-swellable polymeric matrices that can absorb a substantial amount of water to form elastic gels, where the "matrices" are three-dimensional networks of macromolecules held together by covalent or noncovalent crosslinks. Again, in following we consider mainly noncovalent type of crosslinking. Upon placement in an aqueous environment, dry hydrogels swell to the extent allowed by the degree of crosslinking.

The term "complex" or "interpolymer complex" refers to the associates of macromolecules of two and more complementary polymers formed owing to favorable interactions between the reactive groups in their macromolecules. When the complementary polymers contain reactive functional groups in the repeating units of their backbones, the resulting complex has a so-called "ladder-like" structure. However, due to entropic reasons functional groups react not separately but in a cooperative manner, forming the sequences of relatively short and tough intermacromolecular bonds; the

TABLE 9.1
Complementary Groups and Bonding Types

Complementary Groups		Type of Bonding
$-COOH, -PhOH, -SO_3H$	$-NH_2, -NHR, -NR_2$	Electrostatic
	$-OH, -C-O-C-, -CONH_2, -CONHR, -CONR_2$	Hydrogen
$-COO^-$	$-NH_3^+, -NH_2R^+, -NHR_2^+ -NR_3^+$	Ionic
$-OH$	$-CONH_2, -CONHR, -CONR_2$	Hydrogen

Note: R and Ph represent alkyl or phenyl radicals, respectively.

schematic structure of this complex shown in Figure 9.3 resembles a ladder. The ladder-like type of interpolymer complexes was first described by Kabanov and Zezin [15].

The PVP–PEG complex described in Chapter 4 combines a high cohesive toughness (due to PVP–PEG H-bonding) with a large free volume (resulting from considerable length and flexibility of PEG chains). In order to emphasize enhanced free volume in the PVP–PEG blend, this type of complex structure is defined as a "carcass-like" structure. The carcass-like structure of the complex, shown schematically in Chapter 4 (Figure 4.2) results from the location of reactive functional groups at both ends of PEG short chains. The term "carcass-like" defines the complex or the mechanism of complexation leading to the association of complementary macromolecules and oligomeric chains, wherein specific interaction occurs between the complementary functional groups in the repeating units of the backbone of a longer chain polymer component and the reactive groups at the both ends of a shorter oligomeric chain. In contrast to the ladder-like complex, the term "carcass-like" appealed to emphasize that the crosslinks between the longer polymer chains are of appreciable length and that the density of network formed by this type of bonding is comparatively low.

While the formation of the carcass-like complex leads to enhanced cohesive strength and free volume (which determines the adhesive properties of PVP–PEG blends), the formation of the ladder-like complex is accompanied by the loss of blend solubility and the increase of cohesive strength coupled with the decrease in free volume. For this reason, the structure of the ladder-like complex provides no adhesion.

Both the ladder-like and carcass-like complexes are crosslinked due to specific interaction between reactive groups in complementary macromolecules and represent "networks." Figure 9.4 demonstrates a schematic representation of interpolymer complex combining the carcass-like and the ladder-like types of crosslinking. In the context of present work the term "network" is synonymous of the complex, but refers to the supramolecular structure of the interpolymer complex.

Every component of hydrophilic adhesive polymer blends is destined for a special mission to provide the tailored performance properties of the composition. Thus, the term "film-forming" polymer (FFP) relates to a hydrophilic polymer that presents in the composition in a greater amount. The terms "noncovalent" crosslinker, "ladder-like" crosslinker (LLC), or "carcass-like" crosslinker (CLC) refer to the polymers that are complementary with respect to the FFP but the amount of which in blend is less compared with that of the FFP. The term "plasticizer" is used in conventional meaning and designates a low molecular weight or oligomeric substance that is miscible with polymer components and is capable to decrease the T_g and elasticity modulus of the composition. The terms "tackifier" and "tackifying agent" relate to the low molecular weight compounds, which enhance the tackiness of the composition. In this sense, these terms are equivalent to "the adhesion promoters."

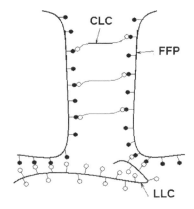

FIGURE 9.4 Schematic view of interpolymer complex combining the carcass-like and ladder-like types of noncovalent crosslinking.

Our purpose is to obtain water-insoluble, water-swellable hydrophilic adhesive polymer blends (adhesive hydrogels) that are capable to form homogeneous films either upon casting a solution to a backing layer followed by drying, or under external pressure or by means of extrusion. The film-forming capability requires that the blend has to be free of covalent crosslinks.

Another approach towards formulating of adhesive polymer blends is illustrated in Figure 9.5 that exhibits a scheme of the ladder-like complex mixed with a plasticizer and tackifier.

For different particular needs it is very useful to employ the copolymers of hydrophilic and hydrophobic monomers as both the FFPs and the LLCs. Although the formation of the ladder-like complex proceeds only through specific bonding between complementary groups in the hydrophilic repeating units, the hydrophobic monomer units in polymer backbones dilute the density of noncovalent crosslinks and serve to decrease the hydrophilicity and swelling capability of adhesive composite.

Employment of the copolymers of hydrophilic and hydrophobic monomers makes polymer blends miscible with hydrophobic plasticizers and tackifiers. The purpose of tackifiers in adhesive blends composed of the copolymers of hydrophilic and hydrophobic monomers is to improve tack. Mechanism underlying tack improvement results from a large size and hydrophobic character of tackifier molecules. Being mixed with interpolymer complex, tackifier increases free volume causing only comparatively slight impact upon energy of cohesion. Examples of suitable tackifiers herein are those that are conventionally used with PSAs, for example, rosins, rosin esters, polyterpenes, and hydrogenated aromatic resins.

The compounds referred to herein as "oligomers" are polymers having a molecular weight below about 800 Da. As shown in Chapter 4, oligomers of polyalkylene oxide such as polyethylene (PE) and polypropylene (PP) oxide couple the functions of the carcass-like crosslinker (due to the location of reactive OH-groups at the ends of their short chains) and the plasticizer (owing to their low T_g values).

The approach toward molecular design of adhesive hydrophilic polymer blends, based on the structures outlined in Figure 9.4 and Figure 9.5, is illustrated in Sections V and VI.

V. HYDROPHILIC ADHESIVES PRODUCED BY COMBINATION OF LADDER-LIKE AND CARCASS-LIKE TYPES OF NONCOVALENT CROSSLINKING

Most typical and illustrative example of this series of hydrophilic adhesives is obtained by mixing the PVP–PEG adhesive (considered in Chapter 4) with a copolymer containing carboxyl groups, which is complementary with respect to the PVP [8,10,16,17]. In this blend the PVP serves as

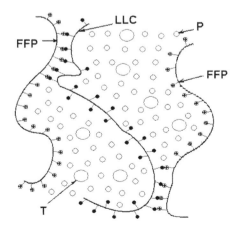

FIGURE 9.5 Schematic representation of the structure of the ladder-like interpolymer complex formed in blends of a FFP with a LLC, which also contains a plasticizer (P) and tackifier (T).

the FFP, PEG is the CLC and the LLC is complementary copolymer of a hydrophilic monomer bearing carboxyl group with a hydrophobic monomer.

A. PHASE STATE AND MECHANISM OF INTERACTION

It is known that the amount of PEG bound into a complex with PVP can be easily evaluated on the endotherms of PEG fusion in differential scanning calorimetry (DSC) thermograms, because only unbound PEG is available for crystallization [18]. As DSC traces of binary PEG blends with PVP and LLC demonstrate, the PEG binding capacity with PVP approaches 50%, whereas that of LLC is as high as 80%. According to both DSC and FTIR data, the strength of interaction diminishes in a row PVP–LLC > LLC–PEG > PVP–PEG.

As the LLC is added to PVP–PEG blends wherein the carcass-like complex is formed, endotherm of PEG melting, which is coupled with symmetric exotherm of its cold-crystallization, appears at 11.5% concentration of LLC and vanishes again at 35.7% of LLC (Figure 9.6).

Because these transitions reveal the existence of PEG, which is free of binding with polymer into a complex within amorphous phase, we can characterize the PVP–PEG–LLC blends as a system with strong favorable and competing interactions between components. At low concentrations of the LLC, all PEG is associated with PVP within amorphous phase and no evidences of the presence of PEG crystalline phase is observed at low temperatures. The increase of the content of LLC leads to appearance of crystallizing PEG due to the replacement of PVP–PEG complex by PVP–LLC one. At higher concentrations of the LLC, all PEG becomes associated with LLC. Existence of only single glass transition in the blends underloaded with LLC and two T_g's featuring for the blend with 35.7% LLC is a sign of partial miscibility of the LLC with single-phase PVP–PEG blends (Figure 9.6).

B. MECHANICAL PROPERTIES

Tensile properties of the Corplex hydrogels are typical of those for slightly cured rubbers (Figure 9.7).

Adding the ladder-like noncovalent crosslinker to the PVP–PEG adhesives causes an appreciable gain in mechanical strength and the loss of compliance or malleability. The ultimate tensile stress goes through a maximum at 8% concentration of the LLC, while the maximum elongation at break decreases smoothly with the rise of LLC concentration for single-phase blends. Two-phase

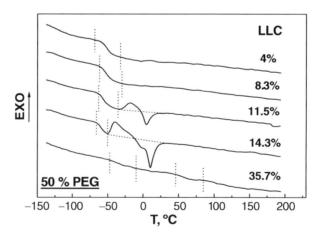

FIGURE 9.6 DSC heating thermograms of PVP blends with 50 wt.% of PEG-400, containing different amounts of LLC [8].

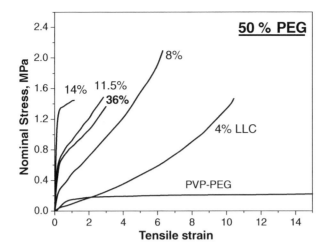

FIGURE 9.7 Effect of the concentration of LLC on tensile stress–strain curves of adhesive hydrogels based on interpolymer complex involving both the carcass-like (PVP–PEG) and ladder-like (PVP–LLC) types of noncovalent crosslinking [10].

compositions exemplified by PVP–PEG blend with 36% of the LLC show the increase of compliance, accompanied with the loss of cohesive strength.

As described in Chapter 4, the ultimate tensile stress at the break of polymer material is a direct measure of the cohesive strength of a material. At the same time, as shown in Chapter 4 (Figure 4.8), the maximum elongation at the break of polymer material under uniaxial drawing is indirect measure of free volume within the polymer of a given molecular weight. The significance of data presented in Figure 9.7 is that the LLC increases the energy of cohesive interaction and decreases appreciably the free volume. As the ratio between the cohesive strength and the free volume is a factor accounting for adhesive capability of material, and taking into consideration that the carcass-like PVP–PEG complex is tacky, it is logical to expect that adding the LLC to the carcass-like complex would upset the specific balance between cohesion and free volume and deteriorate the adhesion.

The CLC (PEG) is a good plasticizer for the PVP blends with the LLC. As is evident from the data in Figure 9.8, the rise in PEG content promotes the ductility of PVP–PEG–LLC blends by increasing the free volume.

C. Effect of Swelling in Water on Adhesion

As predicted from the results of tensile test shown in Figure 9.7, in dry state the PVP blends with LLC and comparatively small amounts of plasticizer (PEG) are generally nontacky. According to the data presented in Figure 9.6, the T_g value of the ternary PVP–PEG–LLC blend is around $-45°C$. It is only 10°C higher than that for binary PVP–PEG composition that provides good adhesion (Chapter 4). Adding plasticizers can reduce the T_g. Water is a good plasticizer for hydrophilic PVP–PEG–LLC blends. Consequently, we can expect the appearance of adhesion upon hydration of the PVP–PEG–LLC blend, when the T_g will decrease in such extent that it will meet the value providing a proper adhesion ($-55°C$). And actually as the data of Probe Tack Test in Figure 9.9 have shown, the blend of PVP–PEG carcass-like complex with the LLC becomes increasingly tacky in the course of water sorption.

The stress–strain curves in Figure 9.9 are informative on the fundamental properties of material. The peak relates to the nucleation of cavities and the loss in integrity of adhesive material in the plain normal to the direction of detaching force [12]. The maximum stress is a measure of

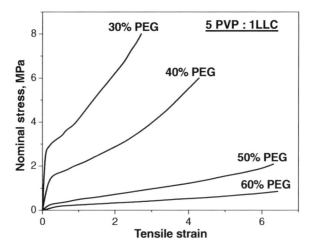

FIGURE 9.8 Effect of PEG concentration on tensile stress–strain curves of adhesive hydrogels based on interpolymer complex involving both the carcass-like (PVP–PEG) and ladder-like (PVP–LLC) types of noncovalent crosslinking.

tack. Initial slope of the stress–strain curve relates to the cohesive strength expressed in terms of the elastic modulus. As the stress decreases gradually with elongation, the growth of cavities and their evolution into elongated microfoam structure occurs. The presence of a plateau on the stress–strain curve is a result of orientation of polymer chains in the direction of traction, fibrillation, and elongational flow of adhesive. Finally, the fibrils either break cohesively or detach from the probe surface, and complete debonding occurs. The length of the plateau and the value of maximum elongation are indirect characteristics of material fluidity and free volume. The most accurate measure of adhesion is the area under the stress–strain curve that defines the total amount of energy dissipated during debonding [11].

Below 10% degree of polymer hydration, the Corplex PVP–PEG adhesive with 12% LLC exhibits no tack. As seen from the data in Figure 9.9, the blend containing 11 wt.% of water has a low tack. Following uptake of sorbed water to 15–17% leads to appreciable increase in tack.

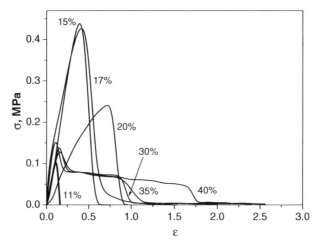

FIGURE 9.9 Effect of amount of sorbed water (wt.%) on the stress–strain curves during the probe tack test of Corplex SA-1212 adhesive (Kireeva and Novikov).

A sharp peak of stress and debonding at comparatively low strains, which are typical for cohesively strong polymer blends containing 11–17% of water, imply the cavitation without fibrillation [19] and macroscopically adhesive mode of fracture when the foot of the wall between the cavities detaches from probe surface [11]. The drop in maximum stress and the growth of material ductility accompanies the further increase in polymer hydration up to 20%. It is interesting to note that absorbed water behaves as a typical plasticizer of hydrophilic blend upon 17% degree of hydration, whereas at lower levels of hydration the water improves the tack most likely by facilitating the formation of a perfect contact between polymer material and probe surface.

Above 20% hydration level, further increase in water uptake results in a rapid change of the mode of polymer deformation evidenced by the appearance of a plateau on the stress–strain curves (Figure 9.9). The maximum stress becomes lower and shifts to the smaller strains, while the appearance of plateau leads to larger area under curve and greater the work of debonding. The higher the hydration, the higher the strain values at break, while the plateau stress is practically unaffected by hydration. As the result of such behavior, the work of adhesive debonding remains high under degrees of hydration ranging between 15 and 40% (Figure 9.10). With the increase in hydration above 40% level the debonding energy tends to be constant, whereas the stress–strain curves become less reproducible due to nonuniform distribution of sol and gel fractions within a swollen hydrogel. The effect of water sorption on adhesive behavior of the Corplex hydrogel characterizes this family of materials as typical bioadhesives. The term "bioadhesive" means a hydrogel that exhibits a pressure-sensitive character of adhesion toward highly hydrated biological surfaces such as mucosal tissues. Existing bioadhesives are generally nontacky in the dry state, but adhere to wet substrates. The adhesive strength of such bioadhesives, however, is usually much lower than that of the PSAs. As shown in Section VII, the Corplex adhesive hydrogels share the properties of classical pressure-sensitive adhesives and bioadhesives, coupling a high adhesive strength typical of PSAs with capability to improve adhesion under hydration that is featured for bioadhesives.

D. Concentration of the Ladder-Like Crosslinking as a Factor Accounting for Adhesive Performance

In the PVP–PEG–LLC blends, the initial tack may be provided by the decrease in crosslinking density that can be easily achieved by the decrease in the content of the LLC. As Figure 9.11

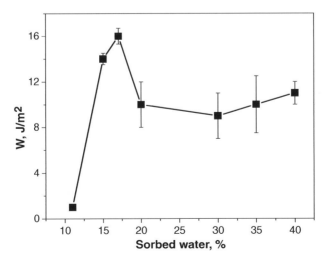

FIGURE 9.10 Effect of sorbed water (wt.%) on the work of adhesive debonding for Corplex PVP–PEG–LLC adhesive (Kireeva and Novikov).

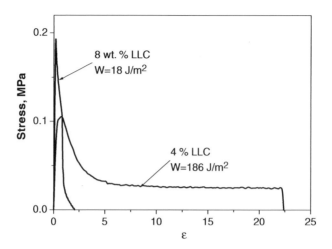

FIGURE 9.11 Effect of the concentration of LLC on probe tack stress–strain curves of the PVP blends with 50 wt.% of PEG-400 and 20% of absorbed water. The contents of LLC and the values of the work of debonding are also shown (Kireeva and Novikov).

demonstrates, at lower concentrations of the LLC in triple blends the cohesive energy is so reduced and free volume increased that the work of debonding grows dramatically achieving the value exceeding the average strength of adhesive bonds found for many conventional PSAs. At the same time, the deformation type changes appreciably, manifesting facilitated elongational flow of cohesively strong fibrils under detaching stress. Although the value of maximum elongation under adhesive joint failure is very high, the mode of debonding is nevertheless adhesive as seen by an abrupt drop of plateau stress in the point of probe detachment.

E. IMPACT OF IONIZATION OF LADDER-LIKE CROSSLINKING UPON ADHESION

Specific balance between the energy of intermolecular cohesion and free volume, which is a factor accounting for pressure-sensitive adhesion (see Chapter 4), can be easily manipulated within the blends involving polyelectrolytes by varying pH and the degree of ionization of ionogenic groups. Figure 9.12 displays the impact of the ionization of carboxyl groups in the LLC on adhesive properties of its blend with PVP as FFP and PEG, which serves both as CLC and plasticizer in the Corplex SA-1212 adhesive hydrogel.

Partial ionization of reactive groups in the LLC has a twofold effect on network formed between PVP and carboxyl-containing polymers. On the one hand, carboxylate ions are unable to form hydrogen-bonds with carbonyls in PVP units and with the increase in ionization degree the network density and cohesion energy go down. On the other hand, introducing of negative charges into the chain of the LLC causes the charge repulsion that leads immediately to expanding of polymer chains and to the increase in their apparent rigidity. As a result, a free volume increases. The changes in cohesion energy and free volume affect the adhesion. Obviously the stress–strain curves in Figure 9.12, partial ionization of the LLC in the blends with PVP–PEG carcass-like complex improves the adhesion appreciably but does not change the mechanism of adhesive deformation in the course of debonding process.

F. SWELLING AND DISSOLUTION PROPERTIES

Due to the LLC supramolecular structure shown schematically in Figure 9.4 and Figure 9.5, the Corplex adhesive hydrogels are insoluble in water. Solubility in water is evaluated in terms of the amount of sol fraction, whereas capability to swell and hold absorbed water is traditionally

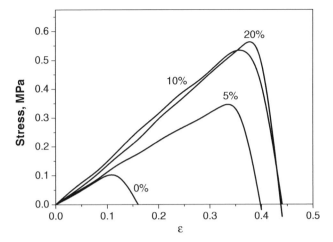

FIGURE 9.12 Effect of ionization of carboxyl groups in the LLC on the stress–strain curves of the Corplex PVP–PEG–LLC adhesive hydrogel containing 12 wt.% of sorbed water (Kireeva and Novikov). The degree of ionization (%) is also shown.

characterized in terms of swell ratio. The swell ratio is a hydrogel weight in a swollen state divided by the dry weight of gel fraction.

Figure 9.13 shows the effect of the concentration of the LLC on the swell ratio and sol fraction of adhesive hydrogels. The content of the LLC defines the density of the ladder-like network. The higher the concentration of noncovalent LLC, the lower the values of both swell ratio and the content of soluble fraction. The swell ratio is significantly affected by the change in pH value. With the increase in pH from 4.6 to 5.6 a partial ionization of the LLC and the decrease in the ladder-like network density occur. As a result, the values of swell ratio at pH 5.6 are about 25–30% higher than those at pH 4.6, while the content of uncrosslinked soluble fraction is insignificantly affected by pH. Ionization of LLC in the complex with PVP and PEG up to 5% affects comparatively slightly the values of sol fraction and swell ratio, however at 10% ionization degree the blend becomes soluble.

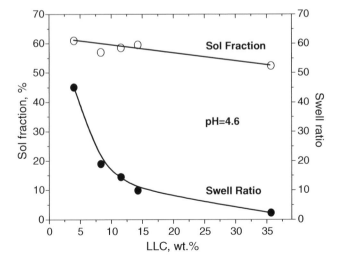

FIGURE 9.13 Swell ratio and sol fraction (%) of adhesive hydrogels based on the combination of the ladder-like and carcass-like interpolymer complexes as a function of the content of the LLC in blends, pH 4.6 [10].

From the data shown in Figure 9.14, sol fraction is predominantly composed of PEG that is relatively loosely bound into complex, while the total dissolution of PVP and LLC in water (that relates to the intercept on *y*-axis) is about 20–25%.

Mechanism of swelling the Corplex PVP–PEG–LLC adhesive in water is illustrated in Figure 9.15 by the optical wedge microinterferogram of the contact area between the water and the adhesive. In an Optical Wedge Microinterferometry experiment, information about the miscibility and interdiffusion of two media with different refractive indices is extracted from interference patterns spanning a contact interface between the two media. When a medium of constant refractive index is introduced into the wedge-shaped cell, the interference pattern consists of alternating bright and dark fringes parallel to the wedge vertex (left and right extremes of Figure 9.15). The number of fringes crossing the unit distance is proportional to the refractive index of the medium [20].

When two components with different refractive indices are brought together in the wedge-shaped cell, with interface running perpendicular to the wedge vertex, the interference pattern at the moment of contact comprises two fields with parallel fringes separated by a sharp dark zone (optical boundary) running along the contact interface. The fields with parallel fringes correspond to unblended components, whereas the dark zone relates to a region with a sharp jump of refractive index. If the two media are miscible, interdiffusion commences between these components across the contact interface. Because the pure components have different refractive indices, a gradient in refractive index is developed along the direction of mass transfer and distortion of interference fringes occurs (Figure 9.15). The distribution of refractive index that corresponds to the concentration distribution over the interdiffusion zone can be easily obtained from the intersection points of the distorted fringes with a line running parallel to the wedge vertex [19].

The region on the left-hand side in Figure 9.15, wherein the curvature of fringes vanishes and fringes become parallel, corresponds to nonswollen adhesive that is supported by impermeable film. The right-hand side in Figure 9.15 corresponds to the water phase. The curvature and density of the distorted interference fringes increase from the area occupied by neat adhesive towards the region of bulky water, indicating a refractive index gradient and hence water concentration gradient in this direction. Phase boundary separates the areas of swollen hydrogel and solution. This observation is typical of crosslinked hydrogels with limited solubility in water.

The total amount of water associated with the PVP–PEG–LLC adhesive through specific interaction (i.e., solubility of water in the adhesive) has been evaluated using DSC technique

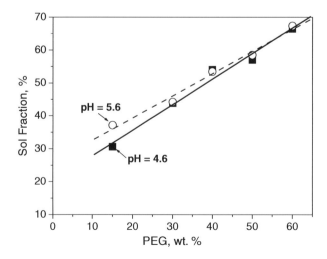

FIGURE 9.14 Dependence of the amount of sol fraction (%) on PEG content for PVP–PEG with LLC. [PVP]:[LLC] = 5:1 (Kireeva).

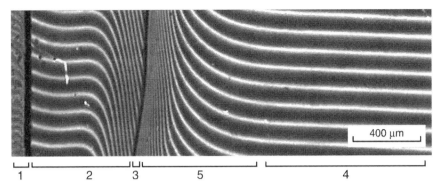

FIGURE 9.15 Microinterferogram of the contact interface between Corplex PVP–PEG–LLC adhesive and water in 5 min upon contact. 1, Impermeable backing film–adhesive interface; 2, Corplex SA-1212 adhesive; 3, adhesive/water interface; 4, pure water; 5, dissolving fraction of adhesive (Khasbiullin, Bairamov, and Chalykh).

by measuring the fraction of water that is capable of crystallization in swollen hydrogel at low temperatures. Crystallizing water can be treated as free water, whereas the difference between total amount of absorbed water and the free water is defined as the water bound with polymer material in amorphous phase. The fraction of water dissolved within the Corplex PVP–PEG–LLC adhesive through specific interaction has been found to be around 34%. This critical degree of hydration has to be taken into account when we attempt to treat the effect of water absorption on adhesive behavior illustrated in Figure 9.9. As seen from Figure 9.9, the change of mode of adhesive detachment from deformation without fibrillation to the elongational flow of the adhesive material in extending fibrils (embedded by a plateau on the stress–strain curves) takes a place between 20 and 30% of hydration. Considering that water sorption tends to decrease appreciably under mechanical stress applied to absorbing material [21,22], there are strong grounds for believing that the appearance of free water is a key factor responsible for the emergence of a plateau zone on the stress–strain curves in Figure 9.9 with the rise in hydration due to the decrease of cohesive strength and onset of yielding.

Within the frameworks of present approach, the same materials may be used as either the FFPs or as the noncovalent LLCs. Actually, both represent the same class of polymers, which bear reactive groups capable of hydrogen, electrostatic bonding or ionic bonding in the repeating units of polymer backbones. The functions of complementary polymers and their role in the composition will be determined by the amount of material present in the composition. The material present in the greatest quantity acts as the FFP. In this way, the difference between the FFP and its LLC is an issue of their concentration. Thus, it is not critical what polymer serves as the major FFP, and what serves as the noncovalent LLC. The change in concentrations of the LLC and FFP in such a way that the carboxyl-containing copolymer with hydrophobic monomer becomes predominant component of the system and FFP, while more hydrophilic PVP serves as the LLC, allows us to decrease essentially the hydrophilicity of the adhesives. For example, while typical values of swell ratio for PVP–PEG–LLC hydrogels in dependence on crosslinking density range between 12 and 60, for the blends overloaded with the carboxyl-containing copolymer with hydrophobic monomer, the values of swell ratio vary from 2.2 to 13.6.

VI. AMPHIPHILIC ADHESIVES PRODUCED BY LADDER-LIKE CROSSLINKING AND PLASTICIZATION OF FILM-FORMING POLYMERS

While in the Section V we have considered interpolymer complexes formed owing to hydrogen-bonding between comparatively weak polybase (PVP) and polyacid, in this section our focus is on the complexes of much stronger polybases such as polyamines. As follows from Table 9.1, amino groups are involved into electrostatic bonding with the carboxyl groups of polyacids, giving ionic (salt) bonds due to reaction of neutralization:

$$
\begin{pmatrix} & CH_3 \\ & | \\ -N & + \ HOOC \\ & | \\ & CH_3 \end{pmatrix} \longrightarrow \begin{pmatrix} & CH_3 \\ & | \oplus \quad \ominus \\ -NH \ \cdots \ OOC \\ & | \\ & CH_3 \end{pmatrix}
$$

This type of bonding is much stronger compared with hydrogen-bonds. As a FFP we select for this purpose a copolymer of dimethylaminoethyl methacrylate (DMAEMA, 50 mol%) with a hydrophobic monomer (alkylmethacrylate). The LLC in this case is a copolymer of methacrylic acid (MAA, 50 mol%) with ethylacrylate.

PSAs obtained by blending the copolymers of DMAEMA and MAA followed by the blend plasticization were first described by Assmus et al. [23]. Here we advance and elaborate this approach by considering it as a constituent part of more general concept of interpolymer complexation (see scheme presented in Figure 9.5) and gaining an insight into unexplored molecular mechanisms allowing us to exert control over the adhesion.

Because both FFP and LLC contain in their backbones 50% of relatively hydrophobic monomer units, the blends prepared from these polymers are miscible with moderately nonpolar plasticizers such as alkyl citrate and citrate esters (e.g., triethyl citrate [TEC], tributyl citrate, acetyl triethyl citrate) [22]. By this means, in contrast to much more hydrophilic adhesive blends described in Section V, the adhesive composites disclosed below possess ideal hydrophilic–hydrophobic balance being compatible with both polar and lipophylic substances and may be defined as amphiphilic adhesives.

A. Phase State and Mechanism of Interaction

Fourier-transformed Infrared Spectroscopy (FTIR) spectroscopy has established electrostatic interaction between tertiary amino groups of DMAEMA and MAA carboxyls. Figure 9.16 demonstrates the effect of plasticizer concentration (TEC) on the T_g of the blend, measured with DSC. Existence of a single glass transition in the blend is a sign of TEC miscibility with interpolymer ladder-like

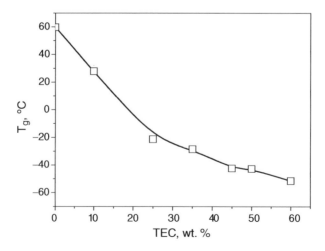

FIGURE 9.16 Glass transition temperature (T_g) of the blend of DMAEMA and MAA copolymers (10:1) as a function of plasticizer concentration (TEC) (Shandryuk and Kiseleva).

complex between DMAEMA and MAA copolymers. Taking into consideration that T_g value in the blend achieves the range from -40 to $-50°C$ at 45–50% concentration of the plasticizer, there is good reason to expect that the blend reveals best adhesion within this composition limits.

B. Effect of Ladder-Like Crosslinking on Tensile Properties

As shown in Chapter 4 and in Ref. [22], the characteristics of tensile stress–strain curves make possible the evaluation of cohesive strength in terms of ultimate tensile stress under fracture of adhesive film, whereas free volume may be assessed qualitatively in terms of maximum elongation under rupture. The area under stress–strain curve represents the work of viscoelastic polymer deformation up to break, and this value correlates to the work of adhesive debonding (see Figure 4.7).

As Figure 9.17 has shown, mixing the FFP with LLC in a ratio of [FFP]:[LLC] = 10:1 leads to dramatic increase of cohesive strength (the value of ultimate stress increases by 6.6 times), whereas the free volume drops appreciably (the value of maximum elongation decreases by a factor of 4.3).

C. Adhesive Behavior

1. Effect of Plasticizer

As is obvious from the probe tack data shown in Figure 9.18 and Figure 9.19, in full agreement with T_g prediction (see Section VI.A), the blends of DMAEMA copolymer with the copolymer of MAA used as a ladder-like electrostatic crosslinker, being taken in a ratio of 10:1, reveal the best adhesion within the range of plasticizer concentrations of 45–50 wt.%.

According to Equation (4.10), the T_g value establishes specific balance between energy of cohesion and free volume, which is a factor controlling adhesion at most fundamental molecular level. As Figure 9.18 has shown, at comparatively low plasticizer concentration (25 wt.%) the blend of DMAEMA and MAA copolymers exhibits low tack and adhesive mechanism of debonding without fibrillation that is featured for cohesively strong, crosslinked or hard adhesives. With the rise of plasticizer content, the peak stress grows rapidly achieving the maximum at 35–45 wt.% of TEC. Respectively, the maximum elongation at probe detachment increases also implying that the blend imparts a ductility needed to dissipate more energy. However, if a peak

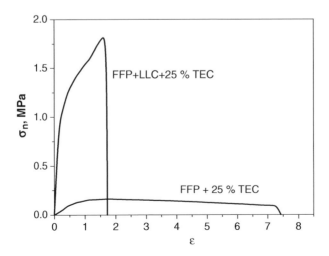

FIGURE 9.17 Nominal stress–strain curves for uniaxial drawing for the mixture of FFP with 25 wt.% of TEC and for the ladder-like interpolymer complex ([FFP]:[LLC] = 10:1) plasticized with the same amount of TEC. Drawing rate is 20 mm/min (Kiseleva and Novikov).

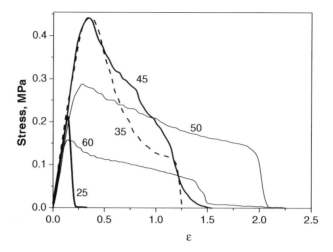

FIGURE 9.18 Impact of TEC concentration on probe tack stress–strain curves of the blends of DMAEMA and MAA copolymers (10:1). The TEC concentrations are also indicated (Kiseleva and Novikov).

value of stress passes through a maximum at 35–45 wt.% of plasticizer concentration, the total amount of dissipated energy has maximum at 45–50 wt.% of TEC, when fibrillation process is much more elaborated and the blend demonstrates appreciable elongational flow. Following increase in the plasticizer concentration leads to cohesively weaker compositions for which the contribution of free volume dominates over the cohesive strength.

2. Effect of Ladder-Like Crosslinking

Technology of polymer blends enables easy manipulating the specific balance between the cohesive strength and fluidity of adhesive composite by the increase in the content of LLC. As follows from the stress–strain curves presented in Figure 9.20, binary blend of the FFP (DMAEMA copolymer) with 35 wt.% of plasticizer TEC that contains no crosslinker is highly tacky fluid and debonds

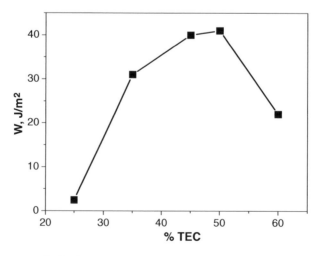

FIGURE 9.19 The energy of adhesive debonding as a function of TEC concentration in the ladder-like interpolymer electrostatic complex formed by DMAEMA copolymer as a FFP and MAA copolymer as a LLC. [DMAEMA]:[MAA] = 10:1 (Kiseleva and Novikov).

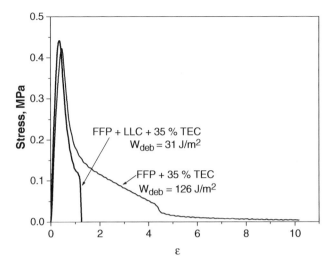

FIGURE 9.20 Effect of electrostatic crosslinking of film-forming polybase (DMAEMA) by polyacid (MAA copolymer) on probe tack stress–strain curves (Kiseleva and Novikov).

cohesively at high values of relative elongation leaving the remainder of the adhesive at the surface of probe.

Mixing the film-forming polybase with complementary LLC (polyacid) in a ratio of [FFP]:[LLC] = 10:1 leads to immediate change of debonding mechanism from cohesive to adhesive. The implication of the data in Figure 9.20 at a molecular level is that the LLC causes the rapid decline of free volume that prevents creep, having only comparatively negligible effect on the value of maximum stress. The latter value characterizes the level of stress under which a process of cavitation or the fracture of adhesive material occurs. Referring back to the tensile test data presented in Figure 9.17, it becomes obvious that the value of maximum stress cannot be regarded as universal measure of cohesive strength because this value has different sense under different mechanisms of debonding. In the blends involving the LLC the cohesive strength is higher and at comparable level of maximum stress the processes of cavitation, fibrillation, and elongational flow of adhesive material are inhibited.

3. Effect of Tackifier

Owing to optimum hydrophilic–hydrophobic balance, the amphiphilic adhesives considered here are miscible with tackifiers, which are extensively used in adhesive technology to improve tack. Figure 9.21 demonstrates the probe tack stress–strain curves for two grades of amphiphilic adhesives based on the ladder-like interpolymer complex formed by DMAEMA and MAA copolymers, plasticized with TEC and additionally containing a typical tackifier (rosin). Adding the tackifier modifies essentially the adhesive performance of blended adhesive. While plasticizers contribute mainly to the increase of material ductility, the tackifier enhances appreciably its cohesive strength.

4. Effects of Ionic Interaction

Similar to the properties of interpolymer complexes involving the combination of carcass-like and ladder-like types of crosslinking (see Section V), the performance of amphiphilic adhesives is pH-sensitive. However, while the hydrophilic adhesives described in Section V involve only one polyelectrolyte (carboxyl-containing polyacid), the specific feature of amphiphilic adhesives is that they are composed of two complementary polyelectrolytes: polyacid and polybase. Accordingly, the

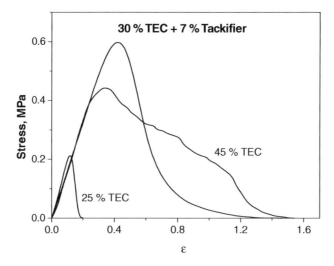

FIGURE 9.21 Comparative effects of plasticizer (TEC) and tackifier (glycerol ester of tall oil rosin) on probe tack stress–strain curves of amphiphilic adhesives based on the ladder-like electrostatic complex of DMAEMA and MAA copolymers (10:1) (Kiseleva and Novikov).

adhesion of amphiphilic adhesives can be affected by partial ionization of both polyacid and poly-base macromolecules.

As seen from the stress–strain curves in Figure 9.22, the ionization of the film-forming poly-base by pretreatment with HCl solution enhances the modulus and the tack by the increase of initial slope and maximum stress. Maximum elongation in the point of debonding (that is indirect measure of free volume) also increases. Correspondingly, the area under curve increases progressively with ionization degree indicating the increase in detaching energy. The mode of deformation of comparatively hard adhesive containing the lack in plasticizer (25% TEC) is invariant and featured for the adhesives detaching without fibrillation. Similar effect of ionization on tack has been earlier reported for interpolymer complexes underlying carcass-like and ladder-like mechanisms of polymer chain crosslinking (see Section V).

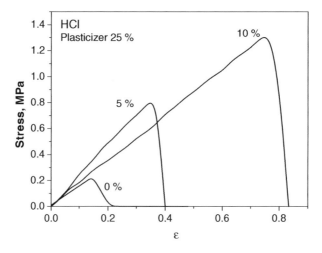

FIGURE 9.22 Effect of partial ionization of FFP by HCl solution on the tack of amphiphilic adhesive containing 25 wt.% of plasticizer TEC (Kiseleva and Novikov).

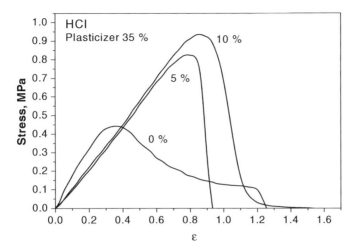

FIGURE 9.23 Effect of partial ionization of film-forming polymer by HCl solution on the tack of amphiphilic adhesive containing 35 wt.% of plasticizer TEC (Kiseleva and Novikov).

For comparatively ductile adhesives (exemplified here by the composition containing 35 wt.% of plasticizer), which reveal fibrillation (a plateau on the stress–strain curves), partial ionization of film-forming polybase by HCl solution enhances the cohesive strength dramatically and the adhesive debonds without fibrillation (Figure 9.23). The maximum elongation in the point of debonding first decreases with 5% ionization and then increases again (at 10% ionization), implying that under comparatively small degree of polymer chain ionization the enhancement of cohesive strength is a predominant factor, whereas further increase in the ionization degree is accompanied by the formation of large free volume. The enhancement of cohesive strength tends to a maximum above 10% of the ionization of FFP (Figure 9.23).

It is very intriguing to compare the effects of ionization of the film-forming polybase and the LLC (polyacid) on tack. As is obvious from the data shown in Figure 9.24, partial ionization by NaOH solution of the LLC bearing carboxyl groups in the ladder-like polycomplex follows the pattern provided by ionization of this component in the hydrophilic adhesives involving the

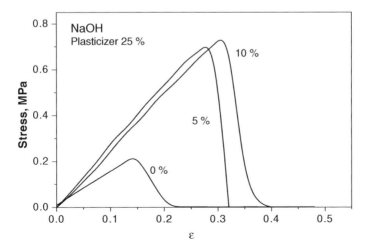

FIGURE 9.24 Effect of partial ionization of LLC by NaOH solution on the tack of amphiphilic adhesive containing 25 wt.% of plasticizer TEC (Kiseleva and Novikov).

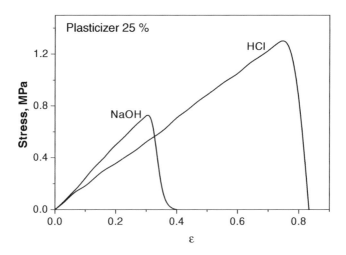

FIGURE 9.25 Comparative effects of partial ionization of film-forming polymer (by HCl) and LLC (by NaOH) on the probe tack stress–strain curves for amphiphilic adhesive containing 25 wt.% of plasticizer TEC (Kiseleva and Novikov).

combination of the ladder-like and carcass-like types of crosslinking (compare Figure 9.24 and Figure 9.12).

By comparing the stress–strain curves presented in Figure 9.24 and Figure 9.22, we see that qualitatively the mechanisms of tack enhancement by ionization of the LLC and the FFP are similar. However, as follows from the data shown in Figure 9.25, in quantitative terms the effect of ionization of the film-forming polybase on adhesion is much stronger than that observed for the ladder-like crosslinking polyacid. This fact is quite explicable because the FFP is a major component of a system, which is incorporated into the blend in a greater amount.

If both the FFP and the LLC are preliminary ionized by treating, respectively, with HCl and NaOH, then the ionic bonding between cationic groups of DMAEMA copolymer and anionic groups at MAA copolymer contributes to the adhesive behavior of the interpolymer complex along with hydrogen bonds formed between uncharged groups. As follows from the data shown in Figure 9.26, in this case the adhesive properties of the complex are intermediate between those featured for the complex involving partial ionization of either the film-forming polymer or the LLC.

Effects of macromolecular ionization on the tack of adhesive composites involving polyelectrolytes have never been earlier reported in literature and need in further characterization.

D. SWELLING AND DISSOLUTION

Figure 9.27 displays the effects of LLC and plasticizer concentrations on the content of soluble fraction in adhesive blends based on the interpolymer complex formed between polybase as a FFP and polyacid as the LLC. The amount of soluble fraction varies in direct proportion to the content of plasticizer, implying that sol fraction is mainly composed of the plasticizer. At the same time, the increase in the content of LLC and the density of electrostatic crosslinking does not result in the reduction of sol fraction as has been earlier observed for the hydrophilic adhesives based on PVP–PEG–LLC complex (Figure 9.13).

The lack in direct proportionality between network density and the content of sol fraction signifies most likely that the crosslinked (gel) fraction is insoluble in water. This supposition seems to be quite probable tacking into account that the ternary blends of DMAEMA and MAA copolymers with TEC are much less hydrophilic when compared with PVP–PEG–LLC blends.

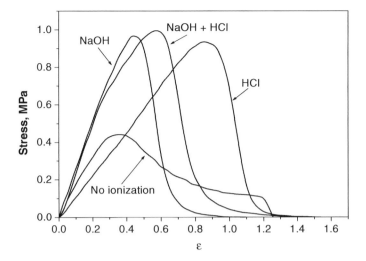

FIGURE 9.26 Probe tack stress–strain curves for the amphiphilic adhesive containing 35 wt.% of plasticizer TEC under 10% ionization of FFP and LLC and for the complex formed between partially ionized polymer components at 10% degree of ionization (Kiseleva and Novikov).

Figure 9.28 demonstrates the impact of network density (expressed in terms of FFP:LLC ratio) and the concentration of plasticizer (TEC) upon swell ratio for the blends involving the ladder-like interpolymer complex formed between DMAEMA and MAA copolymers. Following from these data, the swell ratio is increasing function of both plasticizer concentration and the [FFP]:[LLC] ratio.

Partial ionization of the LLC in interpolymer complex with FFP does not affect the swelling and dissolution of the adhesive. However, the 10% ionization of the film-forming polymer with HCl solution results in appreciable increase of swell ratio from 3.5 to 22.5, while the amount of soluble fraction is practically unaffected by the degree of ionization.

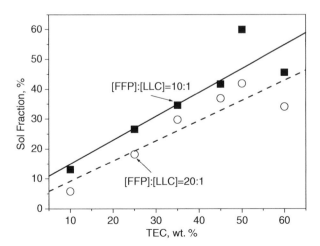

FIGURE 9.27 Amount of soluble fraction in the blends based on ladder-like interpolymer network complex as a function of concentration of plasticizer in the blends at different ratios of FFP to LLC (Kiseleva).

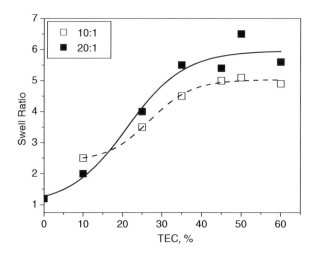

FIGURE 9.28 Effects of crosslinking density (FFP:LLC = 10:1 or 20:1) and the content of plasticizer on the value of swell ratio for the ladder-like interpolymer complex plasticized with TEC (Kiseleva).

If we interchange the polybase and polyacid in the ladder-like complex in such a way that the polyacid serves now as the FFP and the polybase is the LLC, we obtain adhesive materials wherein the treatment with NaOH has a greater effect on adhesion and water sorption.

The value of swell ratio is a direct measure of material hydrophilicity and absorption capability. The greater the swell ratio, the higher the hydrophilicity. By comparing the data shown in Figure 9.28 and Figure 9.13 we see that the swell ratio for amphiphilic adhesives considered here ranges between 1.26 and 6.5, whereas the swell ratio for hydrophilic PVP–PEG–LLC adhesives described in Section V varies from 2.5 to 45.3. In this way, the technology of blended adhesives based on interpolymer complexes covers the entire range of the hydrophilicity of adhesives. The value of swell ratio serves as a basis for the classification of Corplex adhesives. The adhesive composites providing the values of swell ratio from 10 and higher (more than 1000% of absorbed water) are super absorbents (SA) of moisture and represent the SA series (Corplex SA adhesives). Adhesive blends characterized with the swell ratio ranged between 2 and 10 (100–1000% of absorbed water) have been defined as moderately absorbing (MA) adhesives (Corplex MA series). And lastly, little absorbing (LA) materials feature the values of swell ratio lying within the range from 1 to 2 (from 0 to 100% of absorbed water) making up the Corplex LA series of adhesives.

The use of hydrophilic polymers, which are suitable for the preparation of adhesive blends, is not confined to the polymers described earlier. In essence, all the polymers bearing complementary groups listed in Table 9.1 are employable as the parent components for the Corplex technology of adhesive polymer blends. Versatility of the Corplex grades allows easy manipulation of the entire spectrum of performance properties of these hydrophilic and amphiphilic adhesives.

VII. CORPLEX ADHESIVES COMPARED WITH CONVENTIONAL PRESSURE-SENSITIVE ADHESIVES AND BIOADHESIVES

The Corplex technology of adhesive polymer composites provides the fundamental principles needed for the molecular design of novel PSA materials of diverse hydrophilicity degrees with tailored performance properties. Table 9.2 illustrates the similarity and differences in physical properties of three different series of the Corplex adhesives: water-soluble adhesives exemplified by the PVP–PEG blends described in Chapter 4, hydrophilic adhesives and amphiphilic ones outlined,

TABLE 9.2
Corplex Adhesives Compared to Conventional PSAs and Bioadhesives

Attribute	PSA	Bioadhesives	Water Soluble	Hydrophilic	Amphiphilic
Peel resistance (N/m)	300–600	—	370–550	—	140–710[a]
In dry state	None	None	50–70	10–30	—
In hydrated state	—	10–60	300–550	100–300[b]	—
Solubility in water	Insoluble	Insoluble, Swellable	Soluble	Insoluble, Swellable	Insoluble, Swellable
Water sorption capability[c] (%)	<1	98	Nonlimited	96[d]	17–85[e]
Film forming capability	Yes	No	Yes	Yes	Yes
Modulus of elasticity (Pa × 10^5)	1.0–5.0	0.09–0.9[f]	1.3–5.0[g]	0.4–40[h]	1.0–7.3[i]
Maximum elongation (%)	22	>30	22[g]	2.7[h]	1.71[i]
Ultimate tensile strength (MPa)	16	0.01	12[g]	30.4[h]	5[i]
Logarithm yield stress (MPa)[j]	4.1	2.6	3.7–4.9	5.0	Not available

[a]See also Figure 9.18–Figure 9.25.

[b]See also Figure 9.9 and Figure 9.10.

[c]In wt.% of absorbed water in swollen hydrogel.

[d]See Figure 9.13 and Figure 9.14.

[e]See Figure 9.27 and Figure 9.28.

[f]See Ref. [24].

[g]See Chapter 4 (Figure 4.12) and Ref. [23]

[h]See Figure 9.7 and Figure 9.8.

[i]See Figure 9.17.

[j]See Ref. [10].

respectively, by the Sections V and VI. In turn, the table compares the properties from the Corplex adhesives with those futured for conventional PSAs (DURO-TAK® 34-4230, a hydrophobic PSA based on styrene–isoprene–styrene (SIS) block copolymer) and classical bioadhesives (Carbopol® and Noveon®, covalently crosslinked acrylic (AC) acid). As follows from the data listed in Table 9.2, adhesive properties of the Corplex adhesives bridge the gap between conventional PSAs and typical bioadhesives, whereas in terms of other performance properties the Corplex adhesives have opened up new avenues for creation of numerous products on their basis.

Figure 9.29 compares the peel adhesion towards dry and moistened human forearm skin *in vivo* for conventional acrylic PSA and three grades of adhesives based on interpolymer complexes. According to these data, the adhesive properties of polymer composites described in Chapter 4 and Sections V and VI of this chapter share the properties of PSAs and bioadhesives by combining high adhesion featured for conventional PSAs with capability to adhere to moistened skin and biological tissues typical of bioadhesives.

Stress–strain curves obtained in the course of Probe Tack Test (Figure 9.30) are much more informative on the mechanisms of adhesive debonding than the peel force traces presented in Figure 9.29. In this figure, the adhesive behaviors of water-soluble PVP–PEG adhesives (described in Chapter 4), PVP–PEG–LLC adhesive hydrogels (Section V) and the amphiphilic FFP–LLC adhesives plasticized by TEC and filled with tackifier (T) rosin (Section VI) have been compared with the properties of two different grades of conventional PSAs.

Being expressed in terms of maximum stress under debonding, the tack of adhesives based on interpolymer complexes is comparable with that typical of conventional PSAs. However, the distinctive feature of the adhesive blends shown in Figure 9.30 is the lower values of maximum elongation that results from noncovalent crosslinking of the chains of FFP. Because the carcass-like crosslinking is significantly looser than the ladder-like crosslinking, it is no wonder that the water-soluble PVP–PEG adhesive demonstrates higher stretching at probe detachment than the adhesives involving the ladder-like type of crosslinking. In this connection it is pertinent to note that the main tools to increase fluidity and maximum elongation of the adhesives provided by the ladder-like crosslinking it is the dilution of network density due to mixing with plasticizers (Figure 9.8 and Figure 9.18), in the course of swelling in water (Figure 9.9) and also the decrease in concentration of the LLC (Figure 9.7, Figure 9.11, and Figure 9.20).

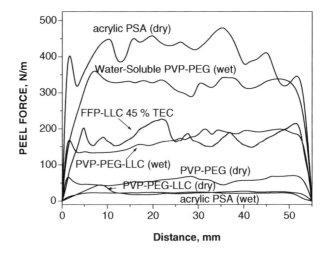

FIGURE 9.29 Peel force traces towards dry and wet human skin for Gelva acrylic PSA, water-soluble adhesive based on carcass-like PVP–PEG complex, hydrophilic PVP–PEG–LLC adhesive and amphiphilic adhesive based on the ladder-like FFP–LLC complex containing 45% of plasticizer triethylcytrate (TEC) (Novikov, Kireeva, and Kiseleva).

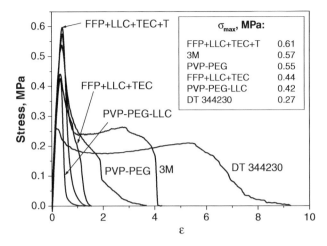

FIGURE 9.30 Probe tack stress–strain curves for water-soluble PVP–PEG (36%) adhesive, amphiphilic adhesives FFP–LLC–TEC (35%) and FFP–LLC–TEC (30%) + 7% of tackifier (T), hydrophilic PVP–PEG–LLC adhesive at 17% of absorbed water in comparison with two grades of conventional PSAs: SIS-based DURO-TAK®, 34-4230 and acrylic PSA manufactured by 3M (Novikov).

VIII. PHARMACEUTICAL APPLICATION OF THE CORPLEX ADHESIVES

A new class of water-soluble and insoluble adhesive hydrogels, described herein, can be used for the application in various pharmaceutical or consumer products. Figure 9.31 compares *in vivo* release profiles of hydrogen peroxide from Crest™ teeth whitening strips (Procter & Gamble) based on conventional bioadhesive Carbopol® and on the Corplex PVP–PEG–LLC hydrogel in the course of their application to human teeth. The data show that the Corplex SA hydrogel stays longer on teeth surface and provides prolonged release of the active agent than Crest Whitening Strips.

Taking into consideration their adhesive properties presented in Figure 9.29 and Figure 9.30, the water-soluble PVP–PEG adhesive described in Chapter 4 and amphiphilic adhesives outlined by Section VI of this chapter are most suitable candidates as skin-contact adhesives in transdermal

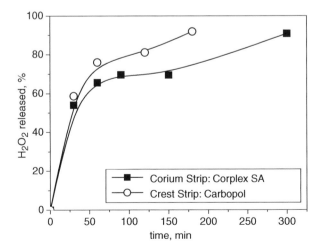

FIGURE 9.31 *In vivo* release of hydrogen peroxide from Teeth Whitening Strips based on Carbopol bioadhesive (Crest™) and on the Corplex SA PVP–PEG–LLC hydrogel film.

systems, topical dermal patches and as electroconductive adhesives. The application of water-soluble PVP–PEG adhesives in the form of diffusion matrices designed for transdermal delivery of various drugs has been earlier described in relevant publications [4,6,7,25]. Adhesive PVP–PEG–LLC hydrogels based on interpolymer complex combining carcass-like and ladder-like mechanisms of noncovalent crosslinking are most appropriate as materials for wound dressings owing to their high absorbing capacity and the capability to form a good adhesive contact to the bottoms of exudating ulcers, burns, and wounds. Owing to high initial tack value, the PVP–PEG blends with comparatively low contents of LLC (Figure 9.11) can be also employed as skin-contact PSAs.

The behavior of the Corplex adhesives is typical of covalently crosslinked polymers. In contrast to covalently crosslinked systems, however, the Corplex adhesives can be easily prepared using a simpler blending process, and, furthermore, provide film-forming properties that are unattainable using covalently crosslinked polymers. Applicability of the approach described in this chapter to a wide range of complementary hydrophilic polymers gives grounds to consider the Corplex technology for a rational design of novel PSAs as a universal tool promising numerous adhesive materials with tailored performance properties.

REFERENCES

1. T.K. Ghosh, W.R. Pfister, and S.I. Yum, *Transdermal and Topical Drug Delivery Systems*, Interpharm Press Inc., Buffalo Grove, IL, USA, 1997.
2. I. Benedek, *Pressure-Sensitive Formulation*, VSP, Utrecht, 2000, chap. 5.
3. I. Benedek, *Pressure-Sensitive Adhesives and Applications*, Marcel Dekker, New York, 2004, chap. 8.
4. M.M. Feldstein and N.A. Platé, in *NBC (Nuclear, Biological and Chemical) Risks—Current Capabilities and Future Perspectives for Protection*, T. Sohn and V.A. Voicu, Eds., Kluwer Academic Publishers, NATO Science Series: 1. Disarmament Technologies, Vol. 25, Dordrecht, 1999, p. 441
5. A.A. Chalykh, A.E. Chalykh, M.B. Novikov, and M.M. Feldstein, *J. Adhesion*, 78, 667, 2002.
6. M.M. Feldstein, V.N. Tohmakhchi, L.B. Malkhazov, A.E. Vasiliev, and N.A. Platé, *Int. J. Pharm.*, 131, 229, 1996.
7. M.M. Feldstein, I.M. Raigorodskii, A.L. Iordanskii, and J. Hadgraft, *J. Control. Release*, 52, 25, 1998.
8. G.W. Cleary, M.M. Feldstein, and E. Beskar, *CORPLEX^{TM}—A Versatile Drug Delivery Platform for Dry and Wet Dermal and Mucosal Surfaces*, Pharmatech, Business Briefing, 2003, p. 1.
9. M.M. Feldstein, N.A. Platé, A.E. Chalykh, and G.W. Cleary, PCT Application WO 02/04570 A3 (2002), U.S. Patent 6,576,712 B2, 2003.
10. G.W. Cleary, P. Singh, and M.M. Feldstein, Proceedings of the 27th Annual Meeting Adhesion Society, Wilmington, NC, USA, 2004, p. 376.
11. C. Creton and P. Fabre, in *The Mechanics of Adhesion*, D.A. Dillard and A.V. Pocius, Eds., Elsevier, Amsterdam, 2002, chap. 14.
12. D.J. Yarusso, in *The Mechanics of Adhesion*, D.A. Dillard and A.V. Pocius, Eds., Elsevier, Amsterdam, 2002, chap. 13, p. 529.
13. M.M. Feldstein, A. Roos, C. Chevallier, C. Creton, and E.D. Dormidontova, *Polymer*, 44(6), 1819, 2003.
14. M.M. Feldstein, M.B. Novikov, I.L.Valuev, P. Singh, A.L. Faasse, and G.W. Cleary, Transactions 31st Annual Meeting Control. Release Soc., Honolulu, USA, 2004 #441.
15. V.A. Kabanov and A.B. Zezin, *Makromol. Chem.*, Suppl. 6, 259, 1984.
16. G.W. Cleary, S. Parandoosh, M.M. Feldstein, and A.E. Chalykh, PCT Application WO 02/087645 (2002), U.S. Patent Application 2003/0170308, 2003.
17. P. Singh, G.W. Cleary, S. Mudumba, M.M. Feldstein, and D.F. Bairamov, U.S. Patent Application US 2003/0152528, 2003.
18. M.M. Feldstein, S.A. Kuptsov, G.A. Shandryuk, and N.A. Platé, *Polymer*, 42(3), 981, 2001.
19. A. Zosel, Proceedings of the 5th European Conference on Adhesion (EURADH'2000), Lyon, France, 149, 2000.

20. D.F. Bairamov, A.E. Chalykh, M.M. Feldstein, R.A. Siegel, and N.A. Platé, *J. Appl. Polym. Sci.*, 85, 1128, 2002.
21. S.C. Cutié, P.B. Smith, R.E. Reim, and A.T. Graham, in *Modern Superabsorbent Polymer Technology*, F.L. Buchholz and A.T. Garaham, Eds., Wiley-VCH, New York, 1998, chap. 4.
22. M. Assmus, T. Beckert, G. Bergmann, S. Kaehler, and H.U. Peterelt, U.S. Patent 6,063,399, 2000.
23. M.B. Novikov, A. Roos, C. Creton, and M.M. Feldstein, *Polymer*, 44(12), 3559, 2003.
24. T.A. Borodulina, M.M. Feldstein, S.V. Kotomin, V.G. Kulichikhin, and G.W. Cleary, Proceedings of the 24th Annual Meeting Adhesion Society, 147, 2001.
25. M.M. Feldstein, *Polym. Sci., Ser. A*, 46(11), 1265–1291, 2004.

10 Converting Properties of Pressure-Sensitive Products

István Benedek

CONTENTS

The uses of pressure-sensitive products (PSPs) are made possible by their adhesive and mechanical characteristics. As discussed throughout the preceding chapters, PSPs are manufactured generally as web-like products. They are applied as web-like or finite products whose dimensions or surface characteristics differ from those with which they manufactured. Therefore, before application they have to be finished. In this case, finishing means the transformation of the continuous, web-like product that has the optimal geometry for manufacture in a product that has the optimal characteristics for use. This production step is called conversion. Benedek [1] has discussed convertibility as the sum of the convertibility of the adhesive and that of the laminate. Convertibility of the adhesive has been described as its coatability. For some PSPs, coating is included in their conversion. Certain PSPs are not coated. For some products, converting includes cutting (slitting), laminating or delaminating, die-cutting, and other operations (called confectioning). Other PSPs are not laminated, not cut, but torn. Numerous PSPs are used for applications where the adhesive and mechanical performances have to be complemented with other properties. Therefore, these products have certain special performances called conversion properties.

For PSPs manufactured by adhesive coating, converting properties are also influenced by the adhesive. Although the characteristics of the solid-state components of the laminate (face stock and release liner) and the manufacture technology needed for the laminate influence the conversion properties also, in this process the requirements of coaters and laminators primarily concern the adhesive. Good die-cutting and stripping characteristics and good convertibility of the adhesive are required [2]. The adhesive has to exhibit good resistance to gum balls, edge flow, and face bleed and to meet the composition requirements of FDA 21 CFR 175.105, for indirect contact with food products.

I. COATING PROPERTIES OF PRESSURE-SENSITIVE PRODUCTS

In manufacturing a PSP by way of coating, the solid-state components of the laminate must be coated with adhesive and, in some cases, with a release layer (e.g., labels, tapes, and certain protective films). The release layer can be manufactured as a separate solid-state component also. In some cases, the adhesive is coated on the release liner (transfer coating). Therefore, the coating properties of the carrier material for the liner, as well as those of the abhesive-coated release liner are important. For both PSP classes (adhesive-coated and extrusion manufactured products), there is a need for postcoating (e.g., lacquering) also. Therefore, coating properties generally play a special role in the manufacture of PSPs.

It should be emphasized that coating properties are the sum of the coating performance characteristics of the solid-state component to be coated (face stock or release liner), the coated mass (adhesive, abhesive, lacquer, and printing ink), and the coating technology. Unfortunately, the coater can regulate only some of these parameters. Coating of PSPs was discussed in detail in Chapter 8.

A. ADHESIVE COATING

The coating of the adhesive on a carrier material depends on the rheology of the liquid adhesive (dispersed/dissolved, or molten), on the characteristics of the carrier material, and on the coating technology. For a given coating technology, the coating properties depend on the rheology of the adhesive and the surface characteristics of the carrier material. For a dispersed or dissolved adhesive the coating properties depend on the components of the dispersed system and its wetting characteristics. The surface of the carrier material influences the coating through its texture (structure), porosity, surface tension, and chemical affinity.

As is known, there are certain tape manufacturing procedures in which a soft, solvent-free adhesive is calendered onto or into a fiber-like, textile material (roll press coating). In such cases, the penetration of the adhesive ensures its anchorage. Penetration as an adhesion-improving factor should be taken into account in the manufacture of removable products also.

Chemical affinity is a parameter that ensures the anchorage of many solvent-based adhesives on plastic films, or of functionalized (carboxylated) latices on paper. It is the reason for precoating a carrier with a reactive (e.g., polyisocyanates, reactive resin) primer also. Chemical affinity is also a factor in the regulation of the bonding force of polar self-adhesive films (SAFs).

Wetting-out depends on the surface tension of the carrier material, and the surface tension and rheology of the adhesive. Surface treatment of the carrier and appropriate choice of the surface-active agents for a formulation allow control of dynamic wetting-out. In adhesive coating the manufacturer has the possibility to regulate the coating properties by proper formulation of the adhesive, by the choice of the carrier material, by the use of a primer, and by the choice of coating device or coating technology.

1. Solvent-Based Coating

In coating solvent-based adhesives, highly viscous, low surface tension liquids are used; therefore, wetting-out is easily controlled. In this case the main problem is related to the chemical sensitivity of many plastic carrier materials to organic solvents. Another special problem is change in viscosity of crosslinked systems.

2. Water-Based Coating

A variety of water-based (WB) products such as cold sealing adhesives, adhesives for lamination, primers, and finishing coats are routinely used in film conversion. Blisters, grit, ribbing, haze, and foam are always found with such pressure-sensitive adhesives (PSAs). Ideally, water-based coatings should be smooth and dry quickly, stick under stress and heat, and resist humidity.

B. ADHESIVE COATING

Siliconizing is a well-known technology that uses solvent-based, and water-based or solventless thermally cured silicones or radiation cured 100% solids. Therefore, the silicone coating specialist must have the skill to operate in each of the related fields. Surfaces that are difficult to coat (such as plastics or coated papers) appeared only in the last decade. On the contrary, there are many mixed systems, where "solventless" formulations contain a certain level of solvent to regulate the viscosity. Adhesive formulation know-how and coating technology know-how can be accumulated by the manufacturer; the skills related to silicone formulation and coating belong more to the suppliers of such materials.

Special problems appear by coating of carbamate-based release liners. There are only a limited number of solvents that can be used. Carbamates have very low solubility even in hot solvents, and the release solution has to be maintained and coated at elevated temperatures on chemically and thermally sensitive carrier material. The anchorage of such release layers is low.

In some special cases polyethylene (PE) is used as release layer. Extrusion-coating is the most important method to manufacture PE-coated papers and fabrics. It applies a low amount of melt $(4-5 \text{ g/m}^2)$ [3].

C. PRINTING

Adhesives are coated on the carrier material with the use of special coating devices. Other components (inks, antistatic agents, primers, etc.) are coated with common printing techniques. In some cases, the adhesive or the release agent can be coated with these printing techniques also.

Most PSPs such as labels or certain tapes are information carriers. The information has to be printed on them before they are end-used. Generally, printing occurs during the manufacture of the web-like, nonconfectioned product. For some products (mainly labels and tapes) a postprinting of the confectioned (cut, die-cut product) can be carried out also, and, in some cases, printing (writing, stamping, etc.) can be done during their end-use. Therefore, generally one can speak of printing and postprinting. Some time ago, the basic printing methods for printing of the PSP during the manufacture of the web-like product and for postprinting (in the prelaminated, confectioned, or postlaminated state) were the same. Differences were given by the web width and the combination of the special materials and printing/confectioning technologies. Some years ago, special nonimpact printing methods and computerized design and printing of PSPs were developed. Thus, the last step in conversion or printing is transferred to the end-user (or "almost" end-user). Therefore, now (and in the future) printing and postprinting technologies may not be the same. As a consequence, the surface quality (printability) required for a certain printing method used in the manufacture of the web-like product (e.g., flexo printing of a plastic film) could differ from the surface quality required by postprinting of the confectioned product (e.g., screen printing or laser printing).

1. General Printing Considerations

Printability includes quite different performance characteristics such as lay flat, dimensional stability (under various mechanical and environmental stresses), good anchorage and adequate machinability [4]. When choosing a printing base certain parameters have to be taken into account [5]: required quality standards, performance ratio, and subjective assessment of the finished product. Quality includes printability, roughness on front and reverse side, gloss, opacity, density, thickness, tensile strength, and bending stiffness.

The importance of printability is quite different for various PSP product classes. For labels it is determinant. Printing of labels is most important for extended text labels. Aesthetic products such as labels or certain tapes and those carrying information have to be printed with high quality image or text. Resolution is one of the main characteristics of a printing procedure used as a criterion by the choice of printing method. According to Frecska [6], resolution is the minimum distance between two objects giving a clearly separated image at a given magnification for a given optical system. The resolutions of the main printing methods are listed in Table 10.1.

Generally, there are reciprocal influences between the printing ink and the paper or film [18]. Printing properties depend on the substrate to be printed. Printing quality depends on the surface finishing, roughness, and wettability [19]. For direct thermally printed papers the choice of the paper (nontopcoated or topcoated) can be dictated by the method of preprinting [20]. Nontopcoated grades have less protection against solvents or monomers. Therefore, water-based or low solvent flexo printing are best suited for this grade. Topcoated papers can absorb out the monomers in ultraviolet (UV)-curable inks. The nonabsorbancy of films has to be compensated for by the proper choice of ink and drying conditions. Inks for normal and thermal papers and for nonabsorbent carrier materials such as aluminum, polyvinyl chloride (PVC), and polypropylene (PP) have been developed. Good printability [21] requires: sufficient smoothness (low surface roughness); more or less surface absorbancy, depending on the printing ink; sufficient surface energy; sufficient resistance to solvents in printing inks; sufficient resistance to heat produced during ink drying or hot foil stamping, and absence of separating agents and other impurities from the imprintable side.

The chemical composition of the film substrate affects the bond. Additives (compatible or migratory) in the polymer can also influence the bond. Slip agents can contaminate the surface.

TABLE 10.1
Performance Characteristics of the Main Printing Methods Used for PSPs

Printing Method	Resolution (dpi)	Other Characteristics	Reference
Offset	2400 × 2400	—	[7]
Mask	400 × 400	—	[8]
Thermotransfer	300 × 300	Low chemical and abrasion resistance of the image	[9–11]
	300 × 400		[12]
DCI	1200 × 1200	—	[13]
Laser	600 × 600	Low chemical resistance of the image, temperature-resistant materials required	[14]
Ink jet	150 × 150	Porous face stock and long drying time required	[15]
	300 × 360		[16]
Dot matrix	150 × 150	Good chemical, weathering, and abrasion resistance, good anchorage on plastics, noisy	[14,17]

Surface-active agents create an adhesively weak layer. Lower molecular weight (MW) oligomeric fractions from the polymer can cause problems. Excessive additives on the film surface (e.g., slip and processing aids) can cause printing problems. For instance, linear low density PE (LLDPE) sometimes contains more low molecular weight species and will require more antiblock additive to prevent blocking [22]. Residual oil on the film surface is dangerous. The smoothness of the surface can also affect the bond. The rough and porous surface of paper provides some mechanical adhesion, while the smooth surface of film does not. Surface energy determines wetting and also influences bonding.

As is known from the printing practice, different plastic films display different degrees of printability. Acetate films display good printability; they can be used as label or tape carrier materials. A 50 μm cellulose acetate film needs no pretreatment for normal printing. It is computer-printable also [23]. For polystyrene (PS) films flexography, roll feed lithography, and gravure printing are recommended. Seal tapes used for closing are based on PVC. They can be printed by using sandwich printing (flexo for the top side and gravure printing for the back side between the film and the adhesive) [24]. For PVC and vinyl chloride copolymers solvent-based gravure and flexo printing inks have been suggested [25]. Predecoration of packages involves graphic application before a container is filled with the product. This is used by bottle manufacturers and contract packagers [26]. Predecoration methods include: application of label in the package mold, direct screen printing on container, hot stamping, heat transfer decoration, shrink sleeve system, and pressure-sensitive systems. Decoration can also be applied in-line with filling. In choosing a carrier material and a printing technology, the composition of the printing ink should also be taken into account. Printing inks generally consist of two basic components: pigment and vehicle.

Gravure, flexo, and screen processes use low viscosity inks and simple inking systems. Lithographic and letterpress systems apply printing ink with higher viscosity and use a complex metering system. Vehicle systems contain film forming parts (resins and oils), volatile materials (solvents, water, viscosity regulators), and additives (plasticizers, dispersing agents, drying agents). Different printing processes need different inks, of various compositions and rheology. Flexographic inks are low viscosity recipes based on water or alcohol, with nitrocellulose, PVC, or polyurethane (PUR) as base polymeric components [27]. Lithographic inks are water-soluble high solids compositions. Gravure inks are solvent-based high viscosity compositions.

Different printing procedures use different solvents, drying temperatures and cause different stresses in the web. Therefore, printability is a function of the carrier thickness also. Gravure printing is suggested for PE films that are at least 25 μm thick and for PP films with a minimum thickness of 12 μm.

Printability depends on the surface wettability, that is, degree of treatment. As is known, the type of film also influences pretreatability. The loss of treatment effect also depend on the nature of the film (see also Chapter 8). Blown films maintain their degree of treatment for a longer period of time than cast films. Triboelectricity affects printability also. As is known from printing and web handling, polyester displays a high level of triboelectricity.

2. Special Printing Considerations

Special printing parameters should be taken into account depending on what carrier material is used, the product class, and the type of ink. Most PSPs are plastic-based. Plastics printing is a special domain. The printing of nonpolar plastics with water-based inks, as a consequence of technological trends causes additional problems. Curling and shrinkage are some of the problems that appear during printing, partially as a consequence of the printing process (see later, Printing-Related Performance Characteristics of the Carrier Material).

a. Printing of Plastics

The printability of plastics is described by Verseau [25]. Electrostatic charges, lack of porosity, high surface tension, thermoplasticity, sensitivity to solvents, and environmental stress cracking are some of the main disadvantages of printing on plastic films [28]. Before coating with varnishes or lacquers, a plastic surface has to be pretreated with antistatic agents to deionise the air [29]. It is recommended that static charge elimination systems be placed after the printing cylinders to help dissipate charges.

b. Nonpolar Carrier Materials Are Difficult to Coat

In some cases, a top coating is applied before printing. As is known, PP is an alternative for polyethylene terephtalate (PET) for "no-label look" labels [30]. For such products a 60 μm PP film is used. This is coated with a lacquer to allow printing via screen printing (solvent-based and UV-cured), flexo printing (water-based, solvent-based, and UV-cured), and UV-cured letterpress printing. According to Waeyenbergh [31], good ink anchorage is achieved for oriented polypropylene (OPP) by the choice of suitable ink and lacquer, adequate drying capacity, film surface finish (pretreatment level or primer), and corona treatment on the printing press. The product class should be taken into account also concerning the special requirements for different products (see later).

c. Technological Trend

There is a growing interest in printing technologies such as gravure and flexo that use water-based inks. There are differences between the characteristics of water-based inks and those of solvent-based inks. Apart from the fact that different resins and pigments are used in their formulation, major differences arise from the physical and chemical properties of water and organic solvents (see also Chapter 5). These differences concern the boiling temperature, evaporation rate, evaporation heat, and surface tension. Water has a higher boiling point than ethyl acetate or ethyl alcohol. Its evaporation rate is higher also. On the contrary, water needs more evaporation heat.

It is well known that the surface tension of water is much higher than that of most other solvents, and, therefore, it displays lower wettability. Water-based gravure ink requires cylinders with small cell volumes. The same approach has been confirmed for flexo printing by the use of ceramic anilox with line screens [32].

3. Printing Methods

There are a number of different printing methods. The main procedures are impact (flexo, gravure, screen, etc.) and nonimpact (ink-jet, laser, thermotransfer, etc.) printing. The requirements concerning the carrier material for these procedures are quite different. For instance, nonimpact printing procedures need paper that has a porous surface [33]. The use of such printing methods is also related to their versatility to broad or narrow web printing also. Table 10.2 lists the main printing methods used for labels, tapes, and protective films.

4. Printing of Labels

As discussed earlier, label manufacturing includes printing and label overprinting. Label overpinting is really a converting step required to fix supplementary, variable data on the label [17]. For label converters the stock must meet printing and die-cutting specifications. Because the printing of labels is the most exigent domain of PSP printing, a short description of the main features of different printing techniques used for labels follows. Generally, the choice of the best printing method depends on run length, quality of printing required, type of equipment available, and sheet or web form. The type of the adhesive and the end-use of the printed item influence the printing method also. As an example, sandwich printing is usual for plastic labels coated with hot-melts. These have a transparent PP or PET carrier and an OPP release liner [34]. Rating plates can also be sandwich printed, with the reverse side of a transparent material printed with mirror lettering [35]. According to Fust [36], the main printing methods used for labels are flexography, letterpress, offset, gravure, screen, dry/hot stamping, and magnetography.

The most important method is letterpress printing and the second is flexographic printing. In Europe gravure and offset printing are too expensive; therefore, letterpress, flexography, and screen printing are considered by Fust [36] to be the printing methods of the future. Screen printing should be used for more expensive (quality) labels than flexo- or letterpress printing. The choice of a printing procedure is influenced by printing quality, product range, dimensions and form, and costs [37].

TABLE 10.2
Main Printing Methods Used for PSPs

	Printing Method	
PSP	**Broad Web**	**Narrow Web**
Label	Gravure	Flexo
	Flexo	Thermal
	Offset	Dot matrix
	Letterpress	Ink jet
	Screen	Laser
	Tampon	Digital offset
	Mask	
	Hot stamping	
	Cold foil	
Tape	Flexo	Flexo
		Thermal
		Dot matrix
		Ink jet
Protective film	Flexo	—

a. Gravure Printing

In gravure printing the image is etched into the metal cylinder. The wells are filled with the ink and the excess of ink is wiped off with a doctor blade. In this procedure the pressure of the knife on the cylinder attains $12 \, N/mm^2$, that is, 2100 atm [38]. Gravure printing presses for labels have to secure excellent printing quality and productivity, as well as the possibility to perform all the other converting operations required for labels such as lamination and die-cutting, in-line with the printing. According to Schibalski [39], the hot sealing adhesive of in-mold labels has to be printed on varnishing or gravure printing machines. In-mold labels have gravure printing, sheet-set offset gravure, narrow-web-UV, special inks, and special die-cutting. There is a general trend in the packaging industry to replace gravure printing by flexo- and offset printing [40]. The printing cylinders are cleaned with special burst cylinders.

b. Flexo Printing

Flexography is a form of web-fed (letterpress) printing. This technology uses flexible (rubber-like) printing plates mounted on a printing cylinder. The print pattern is raised, and the ink is applied only to these raised portions. In-line, stack, or central impression flexo are the main machine configurations [41]. Flexo printing of films with one-cylinder printing machines needs high cylinder precision and web regulation [42]. Flexo for label printing has had a growth rate of 6% per year [43,44].

The classical flexo printing uses four cylinders in the printing device. An ink transport cylinder transfers the ink to the gravure cylinder. The gravure cylinder transports the ink to a plate cylinder that has a soft printing plate (stereotype). A pressure cylinder ensures contact with the web [45]. The first generation of flexo printing machines used solvent-based low viscosity printing inks, giving low quality printing [46]. For instance, flexo printing ink with a viscosity of 21–23 sec (Zahncup nr. 2) was suggested for PP [47]. In the 1980s a new generation of rotary flexo printing machines was developed. Water-based and UV-cured printing inks give a seal-resistant (over 200°C) image.

For labels a common flexo rotary press has one to three printing devices, die-cutting device, a winder for labels and waste material, and a lamination device [48]. Common flexo printing inks for labels are based on alcohol or water with a maximum of 3% solvent. Flexo presses for overprinting labels can print in eight colors, have a 420–620 mm web width, and run at a speed of 150–200 m/min, depending on the carrier material [49]. Narrow web (420–620 mm) flexo printing machines can use webs with a thickness of 20–300 μm [50]. In the United States, flexo printing is the main label-printing procedure [51]. Flexo postprinting machines are suitable for producing screen rulings of up to 60 line/cm [52]. UV drying was developed for flexo printing also. Switching to UV inks from solvent-based inks increases the printing speed by up to 20% [53].

c. Screen Printing

Screen printing is the least technologically advanced printing method, but it can be used for the greatest number of types of substrates [54]. It can be combined with other printing methods such as letterpress and hot-foil stamping [36], and it is recommended for plastics [55]. The main market segment for screen printing include food, pharmaceuticals, cosmetics, toiletries, and household.

Rotational screen printing machines have been in use since 1986; they run at 50 m/min [56]. According to Ref. [57], rotary screen printing was developed earlier, in the 1970s. Today it allows 30 m/min; tolerances lower than 50 μm, and printed areas of $5 \, m^2$. There are special machines developed for labels [58]. For this procedure the drying time is short and energy costs are low, but there may be register problems [59]. PVC films having a minimum thickness of 0.1 mm and hard foams with a thickness of maximum 20 mm are printable via screen printing [60].

Reel-to-reel screen printing lines allow labels to be printed with a production capacity of $10-20$ m^2/min and a web width of $150-750$ mm. Posters, labels, membrane switches, and transfers can be manufactured. Laminators, embossing units, die-cutters, sheet-fed dryers (sheet feed from roll) can be attached also. The image can be transferred onto surfaces with different levels [61]. Weather-resistant decals are printed via screen printing [62]. Screen-printed polycarbonate (PC) labels are used for in-mold labeling (IML) [60]. Polyester with a maximum thickness of 350 μm is printed via screen printing. Printing screen should be cleaned shortly before use with special chemicals [63]. Special chemicals and pastes are used for cleaning of screens [64].

d. Offset Lithography

In offset lithography the image and nonimage area are in the same plane. The image area is oil-receptive and hydrophobic. The inks adhere to this portion. The nonimage area is hydrophilic. Special offset printing methods use a water-free offset printing plate. The printed image is built up on the plate by erosion [65]. There are two basic types of lithography, direct and offset. In the offset lithography (used commonly) the image is transferred with a blanketed rubber offset cylinder. In roll offset printing during drying the humidity content of the paper may decrease from $5-7$ to $1-2\%$, which causes dimensional changes [66]. Excessive running rates during roll offset printing can cause cylinders to overheat and deform (bomb) due to the high dilatation coefficient of plastic sleeves. This deformation produce printing defects [67]. Offset printing machines (sheet) can be used for lacquering also [68]. Offset printing allows a resolution of 2400×2400 dpi [7]. In the mid-1990s in Europe about 50 roll offset machines were improved with electron beam (EB) curing systems [69]. Narrow web label postprinting machines use impact printing methods: offset, flexo, or dry offset [51].

e. Letterpress Printing

Letterpress printing is largely used for labels. Label printing machines run at 100 m/min [70]. Rotational letterpress machines have been developed. They are complex pieces of equipment with numerous cylinders and stable construction [46], so they are more expensive than flexo printing machines. Letterpress-printed labels are printed with 60 lines/cm, at a running speed of 60 m/min [71].

UV drying is also applied [54]. For UV-curing letterpress machines, $15-20$ cylinders are used to transport the highly viscous UV-curing printing ink to the hard printing plate. Web contact is ensured by a soft, flexible pressure cylinder. The so called short roller train coating devices have also been developed; these constructions have only three cylinders. A special gravure cylinder transfers the ink to the soft plated cliché cylinder [71].

Letterpress printing is the most common method for printing labels in Japan [72]. In Europe in the 1980s, 70% of labels were printed by letterpress, 20% by flexo, 5% by offset, and ca. 3% in screen printing. Flexographic printing is also growing, but the main printing method is actually letterpress printing [51]. Gravure, flexographic, offset, screen, nonimpact, and digital are the printing methods most used for labels.

f. Tampon Printing

This is an indirect gravure (deep) printing method. The image is transferred from the (steel) gravure plate with the aid of an oval or circular rubber pad [57]. Tampon printing has been in use since the mid-1970s. It allows on-the-spot printing of nonuniform, nongeometrical surfaces, with small details at high running speed. It is less expensive than screen printing or hot stamping [73,74].

g. Mask Printing

Mask printing can be used for direct imaging also. Here the printing tool is a perforated carrier. The ink penetrates into the perforations [75]. The image is given by built-up perforations. The mask, or master, is manufactured by using a thermal procedure. A PET film is perforated with the aid of a computer-controlled thermal head, giving a resolution of 400 dpi [8]. The master is used for a scanner and an image processor for printing or for modifying the image before printing.

h. Hot Stamping

This method is a lithographic printing procedure that uses predried ink. It may be considered as a lamination method also [76]. Hot stamping is a dry lamination. The printing tool is electrically heated. The transfer temperature is about 120–160°C. During hot stamping the image is transferred with a pressure of 100 kN onto a surface of 150 × 250 mm [77]. The dry transfer of the image has the advantages [78] that (i) no drying is needed and (ii) the image can be modified of as it is being transferred.

Hot stamping allows a protective lamination too and it is used often for labels. It has the supplemental advantages that (i) there are no problems with pretreatment; (ii) as no solvent is used there is no problem with solvent sensitivity; and (iii) a metallic effect can be achieved; normally such an effect is achieved only by metallizing [79]. Hot stamping does not pollute [80]. No registry problems appear. This is an one-shot operation; no intermediate drying steps are necessary; common inks can be combined with partially metallized inks in the same image; different surface structures can be combined, and fine color nuances are obtained. The substrate must not have a geometric form but various substrates can be used. The ink for this procedure is less expensive than the transfer film [57].

There are two different hot stamping printing machines: one with a stamping cylinder (for circular or flat printing) and the other with a flat device. The circular/flat principle ensures a continuous laminating on a large surface without air bubbles and overheating [80]. There are different types of hot stamping: plan, with structure, and with relief [81]. Hot stamping can be considered as a special case of embossing also. Embossing is a converting operation used for self-adhesive wall coverings too [82].

The first German patent concerning hot stamping appeared in 1892 (granted to Ernst Oeser). Today, materials of this type are used in many variants of color and processing method [80]. Hot stamping-die-cutting machines have been developed. These systems offer the possibility to produce decorative labels, resistant type-plates, and informational labels. Depending on specified requirements, the machine can be equipped with various work units (stamping station, numbering station, laminating station, and die-cutting station). Longitudinal and cross-cutting devices are used. The machine has a head for printing and a head for die-cutting [81]. Generally, material is processed in reels [79]. Hot foil labels are printed at a rate of more than 10,000 impressions per hour.

Hot stamping can be used for in-mold labeling also [80]. Because of the high temperature and pressure used in this procedure, (in-mold labeling is carried out with a meltable adhesive), it may be considered as a case of hot stamping.

Hot stamping of holograms is a very complex procedure (having a running speed of 70–90 steps/min) that requires images to be exactly positioned. Hologram hot stamping was developed in 1986 [79]. Hot stamping of holograms is adequate for the transfer of holograms, having a replica of the image on the stamp. It needs a very exact placement of the image. Single-pass holograms have been developed, that allow holograms to be created in-line on any narrow-web press. Such process uses a patented plastic imaging foil with a metal layer 20–100 μm thick that adheres to the labelstock during the stamping process through a lacquer and a heat activated adhesive. Hot stamping and embossing the foil directly on the machine eliminate the security risks [83].

Holography was discovered by Dénes Gábor in 1948, but the first holograms that could be seen in white light were not developed until 1963. The retrievable storage of holograms was achieved in 1980 by embossing, and the transfer of the multiple embossed copies of holograms was developed in 1984 [84]. The main steps of the printing of holograms are (i) manufacture (mastering) of the hologram; (ii) replication of the master hologram; and (iii) transfer of the replica. The master hologram is used to prepare the shims, the embossing tools for hologram transfer. With a specially rotational embossing system the holograms can be transferred using a nickel embossing tool on a PET film for hot stamping. The carrier film should display tear resistance. Therefore a 19–23 μm PET is recommended as the transfer film for hot stamping. This is a metallized film coated with a special lacquer that can be embossed. The hologram on a special fotoresist is coated with a nickel layer; this is the master embossing tool. Two methods are used to transfer the holograms: the hot stamp method and PSA lamination. Thus, PSA labels can be "printed" with holograms using hot stamping or a PSA hologram. Diffraction films are made using the same procedure. They contain two-dimensional images only. Pressure-sensitive holograms can be laminated, cut, and overprinted. For pressure-sensitive holograms a 50 μm PET has been used. In 1984 a special method was developed to permanently bond holograms onto a substrate, for use as identification [84]. Hot foil hologram presses can apply up to four holograms in one pass upon a web up to 250 mm wide [85]. Label manufacturing machines are completed with hologram-dispensing units allowing the application 70–90 pieces per minute [76]. For advertising and promoting, pressure-sensitive labels with holograms play a decisive role. They are die-cut on the roll, have various carrier materials, and are used for direct mailing. If they have a sandwich build-up, using a transparent cover film, the label can be separated into two parts, one carrying the technical (hidden message) information [86].

Transfer films are hot stamping films applied to large substrate surfaces. They have PET or PP (coex or oriented) carriers and a flexible and conformable ink composition [87]. Heat transfer decoration uses a preprinted paper carrier web that transfers the graphic image onto a treated container via a thermal applicator [88].

i. Cold Foil Transfer

The cold foil transfer may replace hot stamping in certain applications. In these cases a special adhesive is used to fix the image on the carrier. The adhesive is cured with an UV curing system having a time window for lamination. A special cold laminating head transfers the image on the substrate [89]. Laminating of photos on labels can also be carried out [90].

j. Label Overprinting Methods

These procedures include mechanical printing; dot matrix, electronic and computerized systems; direct thermal and thermal transfer; ink jet, laser; ion deposition; magnetography; and digital color printing. Direct thermal printing is the largest segment, followed by methods that use a toner (laser, ion deposition, EB, and magnetography) [91]. The nonimpact techniques of direct thermal and thermal transfer printing are the most reliable [92]. Ion deposition has been used in the label industry since 1983; it prints at a speed of 90 m/min [93]. Ion deposition works like a copier, but uses an increased pressure for toner absorbance [94].

Printing methods can be classified as impact and nonimpact printing methods. Nonimpact printing methods allow low noise level, high speed, and high quality printing. The computer label business is one of the fastest growing segments of the label market [95]. Variable data bar-coded labels (routing labels for mailing, inventory control labels, document labels, individual part and product labels, supermarket shelf labels, health sector labels) are the most important sector of computerized nonimpact printing. The main nonimpact printing methods are [96] electro-static and electrosensitive procedures, magnetography, ink-jet printing, and thermography.

Universal product codes (bar codes) are a major reason for the industry's growth, as their use has increased at the rate of 25–50% per year [92]. Digitized variable image printing (VIP) allows customers' data supplied on various media to be downloaded, manipulated electronically, and printed.

Thermal printing which uses heat to color the face stock, comprises direct thermal printing and transfer thermal printing. Direct thermal (thermosensitive or thermochemical) printing applies a heat-sensitive printing material. This undergoes a color change when heated. This method of printing was developed in Japan, for facsimile paper, in the early 1970s. In thermal transfer printing there is a thermal print head (which is rapidly heated and cooled again) and an ink-transfer ribbon. This method does not use a heat-sensitive label stock. Ink ribbons are based on a wax/color composition. During printing, the ink melted on impact and for a very short time (milliseconds), is transferred onto the substrate [96].

Thermotransfer printing for labels is used for different substrates (paper up to 300 g/m^2, PET, PVC, fabric, etc.). According to Hufendiek [17], transfer thermal printing allows "high density" bar codes. It was developed for paper labels. Because the ink does not penetrate in the paper, its chemical resistance and abrasion resistance are relatively low. Thermal transfer printers with a near edge-type printing head allow a resolution of 300 dpi (12 dots/mm), a running speed of 300 mm/sec with a 10 mm print-free area, and a running speed of 250 mm/sec [9,10]. Materials weighing up to 350 g/m^2 can be printed. Fanfold material can also be processed. Single labels, strip labels, and label strips can be printed. Labels having a width of 30.2 to 164 mm cover almost all areas of application. Within recent years the technical performance characteristics of thermotransfer printing machines have been improved substantially. Common thermotransfer printers have running speeds of 128 mm/sec [97]; high speed, high resolution thermal printers are capable of printing at speeds of up to 50 mm/sec, with excellent abrasion resistance, and bar codes in either picket or ladder form are produced [15,98]. According to Ref. [99], thermo transfer printing presses work with 150–600 mm/sec and gives a resolution of 300 dpi [7,99]. Depending on the need for smudge or scuff resistance, wax-based resin/solvent, or multicoated ribbons are used [98]. Compatibility guides for thermal transfer ribbons and pressure-sensitive films summarize the test results for ribbon and substrate combinations, such as speed, burn setting, print quality, smudge, scratch, and chemical resistance [100].

Presses for printing of small series of labels (20,000 pieces) are recommended for color thermal transfer printing. They run at a speed of 100 mm/sec, have print dimensions of 240 × 374 mm, and can cut and perforate. Such presses produce tractor punching and transverse cut. A resolution of 300/400 dpi is achieved [101]. Thermal transfer printing is used for price-weight labels and computer-printed tamper-evident labels [102]. For instance, a direct thermally printable label described in Ref. [103], has a five-layered construction. The paper carrier is coated with a top layer which includes a heat-sensitive ink embedded in a lacquer. Some labels also have a protective overlayer. Thermal transfer printers are used for rating plates. Their resolution is 7.6–11.4 dots/mm and their printing speed can reach 175 mm/sec [24].

Electrostatic printing is based on the same principle as xerography and electrography. Both give an invisible, latent image. Xerography uses an intermediate image carrier. In electrography the image built up by electric charges is formed directly on the surface to be printed [104]. The main electrostatic procedures are electrophotography and electrography. Electrophotography builds up the image using light and an electrically charged substrate. The image is made visible with the aid of a toner. In electrography the image is built up directly on the substrate using charges. Laser, light emitting diode (LED) magnetographic, and EB devices generate a latent image onto a photoconductive drum. Direct charge imaging (DCI) is a belt type of imaging system. It is a toner-based printing method, where imaging is given by a continuous and seamless dielectric belt, rather than a photoconductive drum [13]. This belt carries the latent image. Fusing is accomplished with heat. This method theoretically allows speed of 300 m/min with 1200 × 1200 dpi resolution. Variable information printing allows the manufacture of personalized labels in very small quantities (e.g., 100–500) [105].

Electrophotography is laser printing and uses toner to fix the image. With laser beam printing, the image is built up by a raster image processor and thermally fixed by laser beam [106]. A laser beam writes the image from the computer onto a drum [7]. The laser beam can be applied to burn out the image in a two-layered laminate also. With laser printing based on electrical conductivity, the image is given by selective discharge of the image carrier material via laser beam and thermal fixing of the image. The use of temperature for fixing, limits the choice of carrier materials. High temperature-resistant carrier materials such as PET can be used [17]. The chemical resistance of the printed image is low. Laser printing is used for printing over-laminated labels also [107].

Laser-printable labels (sheet) have a warranty of 4 yr and may be used between -20 and $+120°C$ [108]. Laser printing allows a resolution of 600×600 dpi [7], and requires temperature-resistant laminate components. Temperature-resistant, high melting point tackifiers should be used for the adhesive formulation. The printing equipment should ensure adequate fixing of the printed image (at a minimum possible temperature) using a special fixing device (Figure 10.1).

An EB printing system is used for high speed in-line application of variable data [109]. With an EB printing system (VIP) the numbers, text, graphics, etc. can be printed by using a computer program. They can be oriented in any direction. The resolution of the method is 300×300 dpi. The machine is based on a cold pressure fusing technique, that presses the dry toner into the substrate. A fastening is recommended to ensure adhesion without smudging or flaking. Varnishing, UV-cured lacquering, or lamination is suggested. The electro-sensitive method uses an electrolyte as a base substrate. There are electro-erosion printers also; they work on a metallized substrate. Magnetography applies magnetic fields for image conservation. EB printers run at a speed of 100 m/min [51]. Modern laser printing machines print 2500 characters/sec. A laser cartridge generates a charged printing "form" onto a rotary light-sensitive cylinder, which attracts the toner [110]. The toner develops the image, which is fixed by fusing, or pressure. Such methods completely reimage the cylinder after every impression. Desktop laser printers allow a printing speed of 4–20 pages (A4)/min, printers for big computers achieve 75 pages/min [111].

An ink jet (or bubble jet) printer has small (0.03–0.06 mm), ink-filled channels, that are heated, with the heating controlled by computer. The molten ink bubbles are directed via electrical charges (and an electric field) to build-up the image. Because of the fluid state of the ink, this procedure can be used for rough surfaces also. Unfortunately, it needs a porous substrate surface (or top coat) and a long drying period. The resolution of the image is low [17]. Bubble jet printers give digital printing with a speed of 150 mm/sec and a resolution of 150×150 dpi [90] to 360 dpi [16]. Ultrahigh speed inkjet printing systems have been designed to print at speeds of up to 300 m/min [112]. Ink jet allows printing of 1800 characters/sec [113]. A one hundred percent retrace-ability to the individual production steps and back to the raw material is demanded in pharmaceuticals and toiletry. The continuous inkjet printing process (CIJ) helps in the attainment of this. CIJ is in a position to be able to print smear free codes on difficult nonporous surfaces such as glass and is difficult in small places [114].

Dot matrix printing is an older printing method, that may be used for "low density" bar codes only. Its resolution is limited at 150×150 dpi, and it uses a colored ribbon [7]. It is an impact method that mechanically embosses the ink layer into the carrier. Because of its liquid state, the ink can rapidly bond chemically with low energy plastic surfaces also [17]. Very good chemical, weathering, and abrasion resistance is given. Unfortunately, this procedure is very noisy and the resolution of the image is low.

Digital printing is used in many cases where parts of the printing equipment are computer controlled [115]. Digital printing is actually the digital buildup of the printing image onto a printing

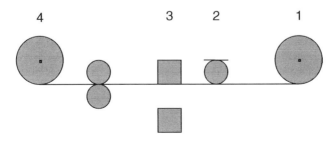

FIGURE 10.1 Schematic buildup of a laser printing line. 1, Winder; 2, laser printing unit; 3, fixing unit; 4, rewinder.

tool with the aid of a computer. This is the computer-to-machine or direct imaging technique. Currently, direct imaging is applied in electrophotography or offset printing.

Until recently, digital printing was carried out primarily on paper and was available only in black. Now the options range from black to as many as six colors on paper and film. Available technologies include dot matrix, color laser, thermal digital printers, color copiers. LED systems, magnetography, EB, electrography, digital offset, and ink jet will influence the label market. Computerized digital label converting equipment can print and die-cut labels [116].

Digital offset technology was developed to satisfy the demand to increase the number of colors used for printing labels. Digital offset is a combination of offset printing and xerography. The print image is digitally built up and transferred to an image cylinder. The electrostatically charged printing ink is transferred from this cylinder to the rubber image transfer blanket and from there to the carrier material. Recently, an one shot color technology was developed. The six colors (images) are transferred to an intermediate image transfer blanket and from this blanket, in one step, to the carrier. The advantages of the system are

1. There are no register problems.
2. The printing quality is regulated in the preprinting step (therefore, it is possible to use less dimensionally stable materials).
3. No drying system is necessary.
4. No change-over in printing is necessary if new products are needed.
5. No printing plate (form) manufacture is required.
6. There are no machine parts that depend on product geometry.

This procedure is recommended for production runs of less than 100,000 [117]. This is a flexible system that allows (at least theoretically) a new image for every cylinder rotation and a printing speed of 100 sheets/h. The face stock material used for digital printing has to have an electrostatically receptive surface, heat-resistant adhesive, and resistance to humidity changes. Running speeds of more than 10,000 impressions per hour are claimed; the common construction is a machine with three heads, two for printing, and one for cutting [118].

k. Combined Printing

For printing during product manufacture (e.g., protective films, or tapes) in-line presses are used, where the inking and impression cylinders are in-line with the web. Printing presses with UV-flexographic and rotary screen or digital offset printing are used [119]. UV-flexo allows printing on nonpaper substrates that was originally accomplished with rotary letterpress machines. The printing machine has a flat bed and rotating die-cutting and UV-lamination equipment. For label printing by flexography, letterpress or screen printing, UV-curable inks have been developed [120]. Special, so-called cold UV systems have been introduced that work with only a low level of thermal infrared (IR) radiation [121]. For PVC decorative films water-based gravure printing is used [122]. Films used for name plates should be printable via flexo, letterpress and computer printing methods. Such films must have a matte, metallic appearance, similar to that of anodized aluminum. The matte surface has to absorb ink, and maintain sharp graphic contrast, good color intensity and excellent legibility. It has to be printable with a range of common or special computer printers. Cationic printing inks have been tested industrially in Europe since 1993 [123]. Free radicalic UV systems contain acrylics (ACs) as base macromolecular compound. Cationic systems include low viscosity epoxides. Both are 100% solids in comparison with common, solvent-based inks, which have 30% solids.

Important growth in the label industry has come from blank labels or nearly blank labels [124], for which more printing is done at the point of use. Therefore, combined printing methods have been among the main processes used in Europe (letterpress, flexo, offset, and digital) [125].

I. Lacquering

The standard pretreatments carried out by foil manufacturers are not sufficient to meet the extremely high requirements for the printability in the field of computer printing. Although PVC foils can be quite easily printed, OPP foils may require considerable effort to satisfy the special requirements. The possibilities for chemical print treatment are often limited, if the result is to be highly transparent. Certain pretreatments can only be achieved with opacity. Therefore, lacquering of the face stock material is a necessity in many cases. Synthetic carrier materials with a surface coating can be printed with the usual paper printing inks in all normal printing processes; they can be used in copiers, laser printers, and color ink jet printers [126].

There are a number of ways to produce a white label based on a clear, colorless carrier material. A white pigment can be added to the adhesive, or a white top coat can be applied to the clear film. High opacity ensures accurate scanning [127]. High reflectivity given by a glossy surface offers high print contrast for bar codes, which is necessary for scanning accuracy and high first-read rates. Dimensional stability and abrasion resistance are important in maintaining bar code scanner and visual readability. A vapor-deposited aluminum layer is sensitive to mechanical stress (wear) and oxidation and must not remain unprotected on the front of the label. Therefore, it is covered with a top coat [128]. Such a primer offers excellent conditions for standard print processes. Full surface lacquering of paper labels followed by UV curing improves their gloss [46]. Linerless labels are supplied as a continuous tape-like, monoweb material. A special coating on the top surface of the label prevents blocking of the adhesive layer [129].

Varnishing can be done on-line or off-line with dispersion varnishes or with EB-curable varnishes. There is a wide variety of dispersion, ink unit, oilprint, functional, and NC varnishes. UV, cationic, and calendered varnishes can be added. Varnishes can be applied by using water pans, varnishing units, varnishing modules, or coating units (tower). In the case of varnishing within the offset press, the varnish is transferred to the substrate directly from the printing plate. Direct varnishing ensures that thicker varnish and ink coatings can be deposited, improving the mechanical properties and gloss of the product. According to Hummel [21], varnishability is a quality parameter for semifinished products for pressure-sensitive labels. Common laquering methods are described by Hadert [130].

According to Ref. [131], gravure printing gives the best results, three- or four-cylinder coating devices with a dosage based on cylinder rotation speed allow coating of up to 4 g/m^2 (wet) lacquer. Lacquering is carried out by using 60-line gravure printing, to a gravure depth of $45-50$ μm [132]. UV-cured lacquer for labels has a coating weight of $2-4$ g/m^2. The coating device is a gravure cylinder ($80-120$ lines/cm, $12-20$ μm deep) [133].

5. Printing of Tapes

Meinel [134] states that requirements for printable tapes include very different performance characteristics: the carrier surface must be abhesive to ensure release properties (i.e., unwind resistance) but at the same time it must exhibit good affinity to the printing inks and anchorage.

Tapes can be printed via gravure, flexo, offset, or screen printing procedures, like labels [135,136]. Tapes with OPP carrier have been printed with solvent-based printing inks. Water-based technology was developed toward the end of the 1980s [137]. Sealing tapes based on PVC carrier are printed by sandwich printing (flexo for the top side and gravure printing for the backside between the film and the adhesive) [24].Vinyl marker tapes must be capable of being printed on a standard printer with vinyl ink [138]. Unfortunately, while the printability of PE tapes, can be evaluated by testing the surface tension (wettability), the printability of PVC tapes cannot [134].

The most used printing method for tapes is flexographic printing [139]. Chromed steel gravure cylinders having $60-100$ engraved pyramidal per square centimeter points are used together with steel or plastic (polyamide (PA)) blades. A decade ago, the majority of tapes (70%) were only one color. Now the majority of tapes are printed with multiple colors. Common three-color printing machines for tapes have a width of 100–300 mm and run up to 200 m/min. Tapes are generally flexo printed [140].

Seng [123,141] discussed the printing of PP tapes using UV-cured systems (flexo printing inks). As is known, radiation drying of printing inks is based on microwaves and IR radiation. Radiation curing uses UV light, X-ray radiation, and EBs that are capable to break the C—C, C—H or C—O links (i.e., have an energy of more than 3.6–4.3 eV). Ionic (cationic) and free radical polymerization can be used. Cationic initiators are onium salts that give an acid by photolysis (see also Chapter 5 and Chaper 6). The UV light absorbtion domain for most proposed cationic systems is 275–450 nm. Good results are obtained with inks based on cycloaliphatic epoxides (chain opening polymerization via oxirane rings). Ionic UV systems possess the advantages that they are not inhibited by oxygen and they result in less shrinkage and less toxicity. Unfortunately, such systems are less reactive, react more slowly (curing is not finished after radiation; postcuring is necessary), and exhibit lower penetration. Therefore, double-cure (dual cure) systems are proposed. A peroxidic initiation is followed by UV curing.

6. Printing of Protective Films

Generally, protective films do not carry informations. They are printed for opacification, as a technological aid or for publicity. Generally, flexo printing is used. In special cases gravure or screen printing is applied. Taking into account that certain protective films (e.g., films for plastic plates) are postprocessed, printing inks used for protective films have to support the elevated processing temperatures. For instance, inks used for protective films for PC plates have to be stable up to 160°C. The ink must be compatible with the protective lacquer on the mask and should not penetrate through the masking during cold or hot line bending of the sheet and during drape forming.

7. Printing-Related Performance Characteristics of the Carrier Material

Some special performance characteristics of the carrier material allow evaluation of its printability. These are related to the dimensional stability of the material. Shrinkage, lay flat, smoothness, and stiffness affect printability. It is evident that direct printability tests are also necessary to evaluate printability. Label printability can be evaluated by evaluation of the print quality (visible appearance); color tone of material, and curling of printed material [66]. Actually, print quality evaluation is done automatically at productivity rates as high as 100,000 labels per hour [96].

a. Shrinkage

Dimension stability includes the stability of the geometrical dimensions and shape (form) of a PSP. Shrinkage is a phenomenon that change the original dimensions of the PSP with or without changing its original shape. This is the result of built-in and "processed-in" tensions (Figure 10.2).

The carrier material (e.g., plastic film) is tensioned during manufacture and conversion. Such material suffers detensioning (relaxation) as a function of time and temperature. This is the manufacture-induced component (S_M) of the shrinkage. For instance, the shrinkage of extruded PVC is partially the result of such tensions arising during extrusion of the film (see also Chapter 8). Cast or calendered films are more dimensionally stable. On the contrary (independently of the manufacture procedure), each material is influenced by the environmental factors: the temperature and substances that migrate from the atmosphere into the carrier or from the carrier into the environment. In the case of paper, there is an equilibrum with the humidity of the air. In the case of plastics, solvents can interact with the film and in special cases (e.g., cellulose derivatives, polyamide, PC) water interacts with the film also. This is the environmental component (S_E) of the shrinkage. Other influences result from coating of the carrier with adhesives or printing inks (S_C). For instance, the film can be printed by means of screen printing which uses solvents. The solvents can migrate into

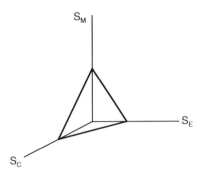

FIGURE 10.2 The main components of shrinkage. S_M, manufacturing induced component; S_E, environmental component; S_C, coating-related component.

the adhesive and dilute it. Certain solvents can penetrate into the film also. Therefore, screen printing of PVC increases its shrinkage. In a similar manner, the special elastic modulus of paper (E_M^*) is related to the humidity elongation coefficient (β) in machine (M) and cross (C) direction according to the correlation [142]:

$$E_M^* \beta_M = E_c^* \beta_C = \text{constant} \tag{10.1}$$

where β is a function of the relative humidity of the air (H_a), its variation (ΔH_a), and the variation of the length of the sample (Δ_l):

$$\beta = \frac{\Delta_l \times 100}{1 \times \Delta H_a} \tag{10.2}$$

It should be mentioned that shrinkage, that is, dimensional stability is also a function of temperature resistance. Generally, the processing temperature range is given by the film supplier for the common carrier materials. Shrinkage is due to tensions in the film by shrinkage-resistant adhesives (e.g., crosslinked acrylics), solvents used in printing, and plasticizers influence it [143]. The importance of shrinkage differs for different PSP classes. Acceptable shrinkage values vary depending on the postconversion steps or end-use of the product (Table 10.3). Because of the high quality printing of labels and the mutual interaction between label and printing, the control of shrinkage plays a weighty role in their quality. In the domain of tapes there are applications where shrinkage is required.

TABLE 10.3
Shrinkage Test Conditions and Values for Selected PSPs

Product	Carrier	Comments	Temperature (°C)	Time (min)	Shrinkage Value (%)	Reference
Closure tape	OPP	15×15 cm^2 sample	125	10	1.0	—
Protection film	Polyolefin	—	120	5	1.1	—
Plotter film	PVC	Bonded on aluminum	70	2880	0.25–2.0	[144]
Medical plaster	PVC	DIN 53374	—	15	<7	[145]
	Polyolefin	—	125	15	26	
Label	—	DIN 40634	—	—	1.1	[21]

Shrinkage of the carrier can be used for special PSPs like heat-shrinkable insulating tapes; during application, the dimensions of the tape decrease by up to 10–50%. As an example, bilayered heat-shrinkable insulating tapes for anticorrosion protection of petroleum and gas pipelines have been manufactured from photochemically cured low density polyethylene (LDPE) with an EVAc copolymer as adhesive sublayer [146]. The linear dimensions of the two-layered insulating tape decrease depending on the curing degree (application temperature 180°C). The application conditions and physico-mechanical properties of the coating were determined for heat-shrinkable tapes with 5% shrinkage, at a curing degree of 30% and shrinkage test force of 0.07 MPa [147]. Shrinkage of self-adhesive EVAc-based protective films is used as a quality criterion. Shrinkage after crosslinking is discussed in detail in Chapter 6, Section III. Test methods for shrinkage are described in Ref. [148].

b. Lay Flat

Lay flat is the stability of the shape of a PSP. For laminated products it is the sum of the geometrical stabilities of the components. Inadequate lay flat is generally manifested as curl of the solid-state components. Like shrinkage and most other dimensional changes during processing and application, curl is the result of the tensions or their relaxation in a carrier material. Such tensions may appear as the result of the composite structure of the laminate or laminate components. Paper itself is a composite built up from fibers. Its humidity balance produces changes in the fiber diameter (across the original machine direction [MD]) [149]. These changes may be different in the middle and external paper layers. Therefore, paper curl can be caused by humidity changes also. In this case, the stress σ in the paper layer is the result of deformation by mechanical forces ε, and by humidity (e) [142]:

$$\sigma = E(\varepsilon - e) \tag{10.3}$$

where E is the modulus. It is well known, that to attain a better humidity balance, carbamide, glycerine, or carbamide-calcium nitrate are added to paper [149].

Some films have an anisotropic buildup. Edge curl (corner drag) of printed films depend on the film thickness, thickness of the printing ink layer, differences in the elasticity of the printed and carrier layers, and the diffusion of low molecular substances in the carrier film [150]. In particular, the printing ink film is hardened more at a high drying speed and is less extensible by room temperature than the PVC film. This can lead to curl. "Frozen" processing tensions in the film may lead to curl also. When plastic is melted, forced through a slit in a die, subjected to pressure by finish rolls, and cooled rapidly, it develops strains. When the sheet is later subjected to heat, or solvents, it relaxes and warps, buckles, bows, or "dishes". The larger the piece and the heavier the gauge, the greater is the potential for problems. The shear rate during the processing of raw materials is quite different for calendering (e.g., $10-10^2$ s^{-1}) and extrusion (10^2-10^3 s^{-1}); therefore, the tensions in the materials are different also [151]. Curl of the film edges may be due to, or amplified by screen printing. Drying of labels printed by screen printing may produce tensions in the material depending on the printing ink used. These tensions cause curling. Ideally, printing inks have to allow rapid drying and optimum screen opening. Very elastic acrylic inks having alcohol as solvent may be processed with low curling. If a printing ink on PVC basis (copolymer) is used for PVC, low curling occurs. High gloss printing inks on an acrylic basis give low curling also. Unfortunately, mixed compositions that allow high speed printing and open screen produce high curling for PVC. Curling depends on coating weight and machining conditions also. Nip pressure and web tension should have low values [23].

Curling also depends on the type of adhesive. Because of the lack of sensitivity to atmospheric humidity, polyvinyl methyl ether has been added to formulations for paper tapes to avoid curl. As proposed by Sigii et al. [152], a water-based acrylic PSA with good cohesive strength that prevents adherents from recovering from bending is based on ethyl hexyl acrylate, ethyl acrylate, and acrylic acid.

Curling is also a function of the laminate buildup. The use of a stiff release liner (e.g., 80 lb kraft liner) for labels helps to avoid curling [153]. The liner plays an important role in the functionality and cost of most PSPs. Tear strength, dimensional stability, lay flatness, and surface characteristics are the most important features of the release liner [154]. Tear strength is important for label converting and dispensing. Good gauge control and liner hardness (lack of compressibility) are the key parameters. Dimensional stability, the ability to maintain the original dimensions when exposed to high temperature and stresses, is important for print-to-print and print-to-die registration. If the liner stretches under heat and tension, the labels may be distorted. Liner stretch can also affect label dispensing. Lay flat is very important for sheet labels or pin-perforated and folded labels. Labels that are butt-cut, laser-printed, and fanfolded must lie flat to allow them to be properly fed and stacked. Liner lay flat is affected by resistance to humidity, dimensional stability at different temperatures, and the stiffness (thickness) of the liner. The use of radiation-cured or low temperature thermally-cured silicones allow a better humidity balance of carrier paper and better lay flat.

When unwound from the roll, tapes should have no tendency to curl [155]. It should be mentioned that curl of tapes may be due to inadequate carrier design (two-layered coex films with different mechanical properties), processing, and transport tensions or printing. Shrinkage and lay flat are also very important quality characteristics for heat-laminated protective films. Lay flat influences the control of delamination, the separation of the protective film from the protected item.

Films made with a MD orientation display outstanding stiffness in the MD and flexibility in the cross-direction (CD). Because an isotropic material is necessary for labels, ideally the films should be biaxially oriented. Lay flat in printing is complemented by on-pack wrinkle-free squeezability [124]. In other words, lay flat and conformability are contradictory requirements. Migration of plasticizer from the substrate, insufficient adhesion due to slip and release material layers on leather and plastics, insufficient adhesion on round adherends, and damage due to UV light have been the main problems with labels for many decades [156]. Paper labels should be laid out in the long grain direction to provide maximum conformability [157].

c. Smoothness

Liner and face stock smoothness affect several performance features. The adhesive surface is a replica of the face stock or release surface. If it is rough, the level of initial adhesion will be reduced. If a rough label is laminated with a clear face stock material, its clarity will be negatively influenced. If the liner is too smooth, air entrapment during dispensing can become a problem. The roughness of the label back can affect web tracking on the printing press. Smooth liners can weave, causing cross-web print registration. Also in laser printing the roughness of the liner plays a critical role in sheet feed ability and tracking through the printer. Smoothness may influence telescoping also. Mounting tapes used in fashion and textile industry are protected against telescoping during storage with a woven or nonwoven supplementary layer [158].

d. Stiffness

Stiffness influences the conformability of the PSP in the printing press and wrinkle buildup. It affects die-cutting and dispensing also. For instance, glassine liners are harder, densified; they are used for high-speed dispensing (see also Chapter 3).

e. Elongation in Printing

During printing, tensions may build up in the material and cause dimensional changes. They depend on the forces applied in the machine and on the temperature in the machine, and they are also function of the material characteristics. There is a correlation between machining conditions and dimensional changes in different films. There is a correlation between tensile strength during printing, tensile strength during print drying, and elongation during printing for various films. For instance, OPP exhibits a higher elongation at 80°C than polyamide at the same temperature [159]. There is a correlation between the gravure roll inserting tensile strength and elongation during printing. The tensile strength of the material influences its elongation during lamination also. Lamination temperature also affects elongation.

f. Wrinkle Buildup

Wrinkle buildup during printing is mainly due to overdrying of the paper carrier. The humidity content of paper may decrease to $1-2\%$, changing its dimensions. This phenomenon depends on the stiffness of the paper. For papers of lower weight ($40-80$ g/m^2) the buildup of wrinkles is more accentuated. Such papers wrinkle with shorter wavelenghts than papers of 100 g/m^2. Wrinkle buildup depends on machine construction also. As stated in Ref. [160], tension fluctuations can cause wrinkles. Rewinders with independent indexing arms allow reel changeovers at full speed without any tension flutters.

8. Conversion of Labels

Conversion of labels is the production of narrow rolls (or sheets) of finished labels, printed, die-cut, slit, and rewound. A revarnishing may be necessary also, requiring a separate drying unit (e.g., UV dryer) [161]. The conversion of label material includes the operations necessary to transform the coated laminated web into labels by cutting, die-cutting, printing, perforating, postprinting, and lacquering. MD or CD cutting (slitting) of the web are common operations for labels, tapes, and protective films. Die-cutting is generally the operation that transforms a PS web into a label (or business form) laminate, but it is used for special tapes also. Printing of the full width web is a common operation for tapes and protective films also. Postprinting is specific for labels. Postlacquering is required for special label applications only. Narrow web printing offers the advantage that it can be used for other operations also [110]. Users can modify or attach a variety of nonstandard devices to manufacture different products. In-line sheeter, nonimpact variable information printer, or web tinter can be attached. Flexographic varnishing units, plough folders, spot and pattern gluing modules, self-sealing adhesive modules, and card and label applicators can be associated with the main printer. Such equipment can be used to produce forms also. Electronic web tension devices, web guide sensors, and reinserters of preprinted web are used also. In-line sheeters allow form producers to compete by extending cut-sheet laser and ink jet market. Batcher stackers (with chipboard inserters) and wrapping machines complete the equipment. Certain form printing machines offer custom-built commercial jaw folders to produce complete printed parts. Certain security labels have a complex construction. Their conversion includes a range of operations. Such labels are used for ware hanging, advertising or product information text, bar codes, and electronic security elements [92].

II. CONFECTIONING PROPERTIES

Confectioning requires roll and material handling equipment; slitter, sheeter, and surface treating apparatus; die-cutting and embossing equipment; and windowing and tag machines. Electronic confectioning machines for labels cut, die-cut, punch and perforate the paper with a running speed of up to 200 m/min. The dimensions of the processed materials and their characteristics can vary widely. Thin films of $15-20$ μm and cardboard of 300 g/m^2 can be processed [162].

Special preprocessing and postprocessing operations are related to printing [95]. Thus, humidification/dehumidification, ionization/deionization, decurling, coating/varnishing, cleaning, line/hole punching and preprinting can be carried out before printing. Varnishing, overprinting, laminating, die-cutting/waste removal, cross perforating, slitting, and quality control should be carried out after printing.

A. CUTTING

The general aspects of cuttability are discussed in a detailed manner in Ref. [1]. Cutting is the operation carried out to transform the full width web into end-use width reel material by slitting, or into sheet material by transverse cutting. Simultaneous cutting in both directions is required for plotter film (pattern paper) cutting. Die-cutting is a partial cutting of the laminate, into laminate sections. The importance of cutting and die-cutting differ for various PSP classes (Table 10.4).

Products supplied and used as continuous, web-like materials have to be cut into narrow webs. Here, slitting is the main cutting operation. These products include protective films and the main tapes. In-line slitters remove the edges of (trim) the material and also divide (slit) the web laterally into two or more narrow widths (ribbons) that are formed into individual rolls on the rewinder. Off-line slitters receive full-width rolls and form rolls with small diameters. Off-line slitters are normally used when the finished rolls have smaller diameters than the unwinding roll and when there are many narrow ribbons. The jumbo size coated rolls are transferred from the rewinder to special slitter-rewinder that produce small retail or consumer size rolls of tape. Narrow width rolls range from 12 to 25 mm and contain 3–30 m of tape per roll. Sealing tapes for laminated glass are supplied preslit [163]. Common slitting machines (600/800/100) have a change-over time of 40 sec for finished rolls [164].

Other products are finite elements cut out from the continuous web. Here, cutting is carried out throughout the whole section of the web (e.g., transformation of reel material into sheet material) or through the face stock material only (die-cutting) to preserve the web-like character of the product, that is, to ensure automatic handling, conversion, and end-use. New slitter cross cutting machines are claimed to provide absolute flatness without curl of the extruded material. Cross cutting is by guillotine or pendulum method [165].

Cuttability depends on the plasticity and elasticity of the material. Both, the solid-state components of the PSP and the adhesive influence cuttability (Figure 10.3). For instance, linear low density PE is neither foldable nor stiff. It is difficult to cut or die-cut it and to punch it because of its high elasticity. Its extensibility is a disadvantage for the converter [166]. In this case, special cutting tools with a higher pressure should be used. There is a trend to multilayer films (5–9 layers); such films are more difficult to be cut [167].

The influence of the solid-state components and that of the adhesive on cuttability has been discussed in a detailed manner in Ref. [1]. As stated there, the mechanical performance characteristics (e.g., flexural resistance) and geometry of the solid-state components are decisive concerning the cuttability of pressure-sensitive laminates (Table 10.5).

TABLE 10.4
Main Characteristics of the Cutting Process

			Cutting Characteristics	
PSP	Product Form	Cutting Operation	Cutting Line	Cut Material
Label	Reel	Slitting	Linear	Laminate
		Guillotining	Linear	Laminate
		Die-cutting	In plane	Laminate component
	Sheet	Guillotining	Linear	Laminate
		Die-cutting	In plane	Laminate component
Tape	Reel	Slitting	Linear	Laminate component
		Die-cutting	In plane	Laminate component
Protective film	Reel	Die-cutting	In plane	Laminate component
Plotter film	Reel	Slitting	Linear	Laminate component
		Die-cutting	In plane	Laminate

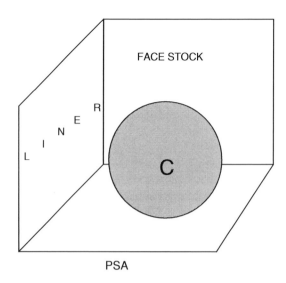

FIGURE 10.3 Product buildup parameters influencing cuttability (C).

As discussed by Medina and DiStefano [168], formulating modalities that improve the high frequency modulus of the adhesive will facilitate conversion (see Chapter 3, Section I). This is also confirmed by the data given in Figure 10.4, which illustrate the dependence of the guillotine-cuttability on the tack of the adhesive. Soft, high tack adhesives exhibit low cuttability. Waste stripping properties of an adhesive are very important. Such characteristics depend on the adhesive properties too. The slitting properties of the tapes strongly depend on their rewindability. Rewindability (considered a function of the peel resistance on the carrier back side of tapes) is also a function of the adhesive performance. However, in the case of die-cuttability, the deformability and tear characteristics of the solid-state components and the global rigidity of the laminate play a major role.

A special case of cutting is processing of plotter films. For such films, cuttability is really an end-use performance characteristic. Texts and logos are cut with a speed of up to 400 mm/min. Roll and sheet cutters have been developed that not only have a cutting mode, but can also be used for drawing [169]. Plotter-cutting is very high speed cutting. For instance, tabletop plotter cutting machines work with a speed of 800 mm/sec and can process films with a width of 50–634 mm [170].

TABLE 10.5
Dependence of Cuttability on the Flexural Resistance of the Laminate

Product Code	Flexural Resistance (mN/10 mm)	Cuttability	Comments
1	226	Inadequate	Permanent, RR, HMPSA
2	331	Fair	Permanent, WBPSA
3	336	Inadequate	Permanent, WBPSA
4	384	Inadequate	Removable, HMPSA
5	483	Fair	Permanent, WBPSA
6	607	Very good	Permanent, RR, crosslinked HMPSA

Note: Paper-based laminates were tested by guillotine cutting.

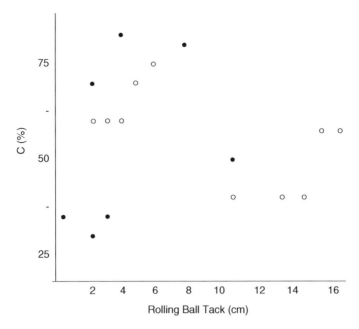

FIGURE 10.4 Dependence of the cuttability (C) on the rolling ball tack of permanent paper label laminates. ●, Solvent-based rubber-resin adhesive; ○, water-based acrylic adhesive.

B. Die-Cutting

Converting involves processes such as die-cutting of roll stock and guillotining of sheet stock. Die-cutting is the technical operation allowing the transformation of the web-like face stock material into discrete items, that is, labels. Therefore, die-cutting has great importance for labels. For special medical tapes, where discontinuous constructions are used to allow breathing and diffusion, die-cutting is also important [102]. The various types of cutting include cutting through, cutting out of the top layer, perforation, engraving, and marking [171].

For labels or label-like products the cutting equipment is generally integrated in the converting/printing line. For instance, type plate printing units possess an automatic cutter [23]. Operating on-line finishing equipment supposes the existence of die-cutter and matrix stripper, re-reeler, sheeter, etc.

Die-cutting requires that the label be designed and its shape and distribution of label area on the web be calculated to allow web (and waste matrix) transport/removal after die-cutting. The unusable areas surrounding the cut labels are stripped from the release paper and wound into a roll. The cut labels remain attached to the release paper and are also wound into rolls.

Flat bed cutters, rotary cutters, and wrap-around dies are recommended for labels. Die-cutting machines can be either nonrotary or rotary. Nonrotary mechanical die-cutting machines have a maximum speed of 350 tacts/min [172]. Flat bed dies from 7–12 mm in line height, using laser technology or grids, rotary dies with a die plate 0.41–1 mm thick, and magnetic or nonmagnetic, milled or etched versions have been developed. The development of form-label combinations where a die-cut siliconized patch transports one or more butt-, or kiss-cut paper labels forced the development of rotary die-cutting technology coupled with matrix waste removal.

Stanton Avery developed the rotating die-cutting machine [173]. Such machines include a geared die cylinder with bearers and an anvil roller assembly. Flexible dies that wrap around a stainless cylinder containing magnetic inserts are less expensive. Flexible steel strip cutting knifes have been known since the 1930s. Rotary die-cutting machines can cut card board as thick as 200–500 μm [174]. The clearance and tolerances of rotative die-cutting machines depend on the nature of the material to be cut [175]. The critical cutting range, that is, usable clearance, for common materials is 0.010–0.015 mm. For cutting through PE films or nonwovens it is about 0.005 mm. Cast PE exhibits better die cuttability and precision registry [176].

Steel strips have been used as cutting tool in the card board industry since the 1930s [177]. Magnetic cutting plates are used for a cutting height of 0.42–1.00 mm. A cutting angle of 60–100° is used [178]. A dual-knife sheeter for high speed work is a rotary dual-knife cutter with a delivery and stacking system. Metal-to-metal contact occurs between the upper and lower knives. For such machines sheet length and squareness accuracy of ±0.2 in. can be maintained continuously [179].

As discussed in Ref. [1], cutting and die cutting depend on the solid-state components of the laminate (face stock and release liner). Their quality and combination are very important. A good die-cutting is provided by a uniform caliper densified paper liner [175]. Cuttability tests have been carried out by Hartmann et al. [180] using PE, PE/PP, PET, and PET/SiO$_2$ and paper (80 g/m^2) as liner and face stock material to test cutting knife wear. The influence of the face stock material and release liner was also studied with rotary die-cutting [181]. The investigations show the importance of an appropriate release liner for each face stock material, depending on its cutting and tear mechanism. The versatility of the release liner as a cutting basis depends on its chemical nature, mechanical properties, and dimensional tolerances.

The thickness and density of the paper used as face stock material are less important [180,181]. Paper can be compressed and split. Therefore, in paper cutting the deformation of the cutting tool is minimum and cutting forces are minimal. Adjustment of the cutting tool is easier and stable. Cutting of PE includes splitting and separating of the split polymer into two separate phases. Because of the plastic flow of the material, there is a need for a hard bottom surface, with low dimensional tolerances and no compressibility. In this case, cuttability (its quality and speed) depends on the density of the release paper used. Kraft paper is not recommended as a release backing for PE labels. The best results are obtained with high density glassine paper and PET. It is evident that material combinations for face stock and release liner lead to the best die-cuttability results. The cutting of PET is similar to that of paper. It is a crack or slitting dependent process, initiated at first contact points between the knife and the PET face stock materials (zip effect). Tear resistance influences cuttability. Therefore, release papers with broad tolerances give the best cuttability results. Polyester needs a hard but low quality glassine paper as release (backing) material. Therefore, one can state that the cuttability of different face stock materials can be estimated as follows:

$$\text{Paper} > \text{PET} > \text{PE} \tag{10.4}$$

and the versatility of different release liner materials may be estimated as

$$\text{PET} > \text{glassine paper} > \text{kraft paper} \tag{10.5}$$

Voluminous and fragile (inelastic) solid-state laminate components generally improve cuttability. Therefore postlamination blowup of the laminate improves cuttability. According to Ref. [182], a photocrosslinked acrylic-urethane acrylate formulation used for holding semiconductor wafers during cutting and releasing the wafers after cutting, with enhanced cuttability, can be prepared by crosslinking and blowing up of the adhesive. Such releasable foaming adhesive sheet can be prepared by coating blowing agent (e.g., of microcapsular type) containing PSA (30%) as a 30 μm dry layer [183]. Such adhesive is removable and improves the cuttability.

Cuttability of label films can be evaluated by measuring the cutting forces in both, the MD and CD. For optimum die-cuttability the MD/CD ratio of the measured values has to be about one. As illustrated by the data of Figure 10.5 the manufacturing parameters of the plastic film strongly

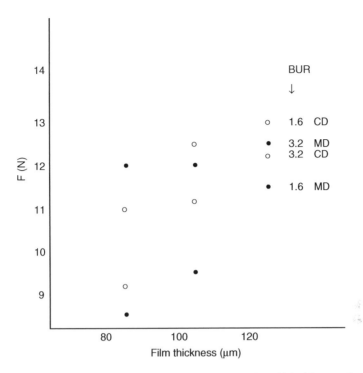

FIGURE 10.5 Dependence of the die-cuttability of a blown, polyolefin-based label face stock material on the film thickness and BUR. Symbols as in Figure 10.4.

influences its die-cuttability. The choice of blow up ratio (BUR) together with the thickness of the film (for a given formulation) allows the regulation of the die-cuttability.

As stated in Ref. [184], current laser technology offers a limited power/cost ratio; therefore, in-line laser cutting is not viable. Several laminates show undesirable side effects (discoloration of the carrier, and differential cutting of paper and liner) because of the thermal effects of the laser beam. More perspectives are given to laser marking, where special foils accept laser marking. In this case, the minimum width that can be engraved is about 1 mm. Thermal die-cutting was developed for pressure-sensitive decals. According to Ref. [185], a 0.004 in. cutting line heated at 300°C will cut vinyl but not paper. Blank self-adhesive labels, precut on silicone backing paper are manufactured also.

C. Perforating, Embossing, and Folding

Perforating and microperforating are common operations in the conversion of labels and tapes. Perforating is used mainly in order to weaken the carrier material to achieve detachability. Microperforating is performed to achieve porosity. Label die-cutting, liner perforation, and pin-feed hole cutting (for labels printed on dot matrix printer) are also necessary. Slits or perforations in the face stock improve the tamper-evident properties of labels [186]. For computer use, pin-feed hole punching is necessary. It can be accomplished through the use of male/female dies. Commonly, a standard rotary die is used to die-cut through the liner. The labels can be either rolled or fanfolded. Special acrylic adhesives with excellent adhesion at high temperature are suggested for PSPs subjected to punching process [187]. Some price labels require perforation for transport [188]. Special pharmaceutical labels can also have take-off (detachable) parts to allow a multiple information transfer. Parts of the label remain on the drug, or on the patient [189]. The laminated structure is perforated to allow multiple detachment (i.e., multiplication) of the labels [190]. An adhesive tape used for holding together printer paper is perforated along the centerline in the long direction [191]. Electrical insulating tapes used for taping may also be perforated [192]. Common mechanical perforation machines have 25,40,50,70 teeths/in. [193].

Microperforating can be used to transform some carrier materials in a porous web [194]. Microporous webs are needed for medical tapes, condensed water-free packages, etc. Various methods are used for perforation: hot needles, liquid or gas jet, ultrasons, high frequency laser, and electrostatic microperforation. Hot needles are used for PE and PP films and nonwovens. The hole diameters are 200–500 μm. Webs having a width of 1500 mm can be processed with a speed of 10–30 m/min. Gas and liquid jets are used for perforating soft webs (foam, nonwovens, board). Thermoplast-like polyolefins and PVC can be cut with ultrasons. High frequency cutting is limited to thicknesses of 5–30 μm. Laser perforating is used for narrow webs. It allows running speeds of 600 m/min and is used mainly for fine paper. The hole diameters are 50–200 μm.

Electrostatic microperforation can be used only for electrically nonconductive materials. Webs with a width of 50–1000 mm and weights of 5–100 g/m^2 are processed, at a running speed of 300 m/min. Paper, plastics (LDPE, PET, PP, and PVC), laminates (PET/PE), nonwovens, and other materials can be processed. The diameter of the holes is 2–70 μm, the pore number of 1.6 Mio/m^2. The interpore distance is of 1.0 mm. The reciprocal distance of the stripes is 0.25 mm. The procedure uses electrodes with high voltage and high frequency (500–5000 Hz). Microperforation (diameters of 2–70 μm) and macroperforation (diameter of 50–500 μm) can be realized [194]. Special electrodes allow the manufacture of perforations having a diameter of 2–5 μm using corona treatment [195]. Flame perforation is also used; this procedure allows the manufacture of hundreds of holes per square inch, with a hole diameter of 0.4 mm [196].

Embossing is generally carried out during the manufacture of plastic carrier films (see also Chapter 8 and Chapter 11). Embossing of metal films, metallized PE, and paper is carried out more easily and runs at much higher speeds (up to 400 m/min) than that of PE [197]. Microembossed films are used for image holograms, two- and three-dimensional background designs, decorative laminates, and labels [198]. Embossed paper or film provides mechanical release.

Folding is used mainly for computer-printable labels and tapes. A special application apparatus is suggested for folded tapes. Folding along the longitudinal axis is proposed in order to prevent back rolling of the tapes [199].

D. WINDING PROPERTIES

Winding properties influence the manufacture of carrier material (see Manufacture of Plastic Carrier), coating, laminating, and converting (see Chapter 6, Section III) and the end-use of PSPs. Winding properties play a special role for PSPs applied as webs (tapes and protective films). Such characteristics are primary conversion parameters for tapes because of their high degree of confectioning (slitting). Table 10.6 illustrates the common values of the unwinding resistance for tapes. As can be seen from Table 10.6, the unwinding force of common tapes is situated as order of magnitude in the debonding force (peel resistance) domain of removable labels or protective films.

The stresses which appear during winding depend on the running speed of the web. Their order of magnitude and their degree of influence differ during film manufacture, coating, and confectioning. In the case of film manufacture, the maximum winding speed is situated at 20–140 m/min for blown films, 120–400 m/min for cast film, and 280–350 m/min for biaxially oriented film [204]. Common coating speed values are 100–200 m/min [92]. The rewinding speed for narrow web is generally lower. For instance, one tape specification [205] states that when unwound at a rate of 7.3 m/min the adhesive of the tape shall not transfer to the adiacent layer. During unwinding a tape may tear or split [206]. As shown in Ref. [134], the unwinding resistance (R_w) is a function

TABLE 10.6
Unwind Resistance of Different Tapes

Grade	Carrier Material	Unwind Resistance		Method	Reference
		Value			
Flame resistant	PVC	11.1–18.1	N/32 mm	OCT-906	[200]
Closure	PE	1.4	N/10 mm	BDF A 07	—
Special	PP	1.5	N/50 mm	JIS/Z 0237	[201]
Special	PVC	0.4	N/20 mm	—	[202]
Packaging	PE	0.02	N/10 mm	—	[203]

of the temperature (T_w) and unwinding speed (v_w):

$$R_w = f(T_w, v_w) \tag{10.6}$$

In a similar manner, tape tear number (N_t) depends on the unwinding speed and on the temperature according to the correlation [206]:

$$N_t = f(v_w, T_w) \tag{10.7}$$

It is well known from the unwinding and slitting of tapes, that the reel should be tempered before confectioning. The adhesion on the back of the tape (blocking) depends on the temperature. The resistance to unwinding can be considered as equivalent to peel resistance. Therefore, the unwinding force is a function of unwinding speed and temperature. Generally the unwinding force depends on a number of parameters:

$$F_w = f(F_a, E_a, E_f, h_a, h_f, w_t) \tag{10.8}$$

where F_a is the adhesion force; E_a and E_f are the moduli of elasticity of the adhesive and of the film, h_a and h_f are the thickness of the adhesive and film, and w_t is the width of the tape. As is known, the adhesive is a viscoelastic substance, therefore, the modulus of elasticity of the adhesive depends on the time and temperature, that is, on the unwinding speed and temperature. The modulus of elasticity of the film carrier material also depends on the temperature. In this case, the adhesion force is the peel resistance of the PSA from the back of the tape. The peel resistance from the abhesive layer protecting the active face of the PSP is also measured for labels. For labels, the release layer is a separate component, and delamination occurs during end-use; it is an end-use characteristic. For tapes and protective films delamination occurs during rewinding. It is both a converting and end-use performance characteristic.

Maschke [206] studied the dependence of the number of tear cases on the unwinding temperature and unwinding speed. Unwinding temperatures of 22 and 28°C were chosen, and the number of cases of tape tearing per 10,000 m^2 of processed tape surface was registered. As shown by Maschke [206] increasing the unwinding speed from 120 to 220 m/min (by 22°C), increases the tear number almost 100 times. Increasing the unwind temperature from 22 to 28°C, decreases the tear number by only 10 times. Therefore, in this case in order to balance the negative influence of the speed increase, a temperature increase of 12–15°K would be necessary. As found by examination of the tear surface, tape tear is due mainly to bubbles and bubble ranges ("chains"), and small (diameter less than 2 mm) holes. Bubbles appear on the tape carrier on places where the film is too thin.

Telescoping of a slit roll is related to the unwinding quality. For narrow webs it is more critical. Its maximum acceptable value is generally given as a product specification. For instance, according to Ref. [205], for a 32.0 mm PVC tape, telescoping shall be limited to 6.35 mm. Some basic operations of web-converting are summarized in Ref. [207].

III. DISPENSING AND LABELING

The labeling process is influenced by the label and by the product to be labeled [208]. The main label-dependent parameters of labeling are adhesive type, coating weight, release degree, labeling system, labeling speed, label geometry, labeling tolerance, label material, label printing, and converting. The product to be labeled influences the instantaneous and final bond strength, by its surface temperature, surface structure and geometry, and by its content.

Many pressure-sensitive labels are still affixed by hand, but a growing percentage are applied by automated equipment. Adhesives used for products with automatic labeling should have a level of quick stick that will provide for secure adhesion with minimal application pressure [155]. Conventional dispensers fo paper-based pressure-sensitive materials do not require a special applicator system, because of the inherent stickyness of paper. For conformable and squeezable labels special systems are needed. Tight label placement tolerances of 0.015 in. are usual for application systems.

The label parameters required by its application differ for wet and pressure-sensitive labels. According to Ref. [209], the main labeling parameters of wet labels are (i) the quality of the carrier paper, (ii) the nature of the adhesive, (iii) the surface to be bonded, and (iv) the labeling equipment.

The main features of labels with good labeling performances are (i) lay flat, (ii) narrow die-cut tolerances, (iii) good mechanical properties (tear resistance in wet and dry states), and (iv) low stiffness. Although some of these statements are valid for PSA labels also, other requirements are quite different. Because of the differences in labeling systems used for wet and PSA labels, different carrier characteristics are required for each class. Labeling speed for common pressure-sensitive labels is lower than for wet labels. For instance, labeling speeds of up to 150 packs/min can be achieved with a print-and-apply system for food packaging [210].

Label application machines (labelers) of quite different constructions and capabilities have been developed. Suppliers of such machines offer hardware and consumables. A hand labeling device contains a storage roll, a pressure roll, and a cutting device [211]. Generally, a hand-held labeler used to print and apply PSA labels releasably laminated on a carrier web, comprises a housing having a handle, a label roll printing device, delaminating device, applying device, and some means for advancing the carrier web. A hand-operated printing and labeling machine may possess a carrier band on which the labels are temporarily positioned. The machine has a conveyor system that draws the label carrier head around a label detaching edge. Pneumatic (touch blow) label transport and application has been known for many years [212].

Labeling speed depends on the dimensions of the label. The maximum labeling capacity is a function of the length of the label but is strongly influenced by the construction of the labeling machine. A common labeling system allows the application of 50 labels/min with label dimensions of 120×1450 mm. The same equipment ensures a labeling speed of 30 labels/min for labels of 13×350 mm. Labels having dimensions of 230×999 mm can be dispensed at a rate of 600 labels/min [213]. Labeling speed is a function of label stiffness also. Biaxially oriented PP films are applied in cosmetics and bottle labeling [214]. For this use, films with a thickness of 30–80 μm are recommended. Bottle labels need high scratch resistance, chemical resistance, mechanical resistance during filling, and resistance against pasteurization and washing [215]. Control system checks up to 800 units/min and separates mislabeled bottles [216]. Label applicators in the pharmaceutical industry affix wrap-around labels on ampoules at a speed of 30,000 items per hour [217]. For high speed labeling, films having a thickness of 50 μm are suggested [218].

Tamper-evident constructions have to stay put as the waste matrix is removed, yet break up or tear when removed from the substrate [219].

Labeling speed depends on the labeled item also. For instance, a hand-held labeling device is used for marking wires and cables [220]. Hang tabs can be laminated manually, or by an automatic dispenser or automatic machine application. Special machines for labeling plastic containers and cups have a productivity of 300 cups/min [221]. Nonoriented transparent PP is used in a thickness of 60 μm for no-label look labels [218]. A high speed labeling unit allows the labeling of up to 350 bottles/min; the system equipped with servo drive technology, including four automatic labeling units allows a dispensing speed of up to 120 m/min [222]. No-label look labeling of glass bottles is carried out with a productivity of 450 pieces/min. Labeling speed of 80 m/min is attained [223]. Labeling machines with tandem labeling provide better productivity. These machines can simultaneously apply two labels at a running speed of 250 containers/min [224]. The build-up of a machine for two-sides labeling of bottles is shown in Ref. [225]. Round-around labeling allows the labeling of the front of backside and the use of multilayer labels, for example, inforoll-labels; it allows the dispensing of up to 400 labels/min [226]. Round-around labeling machines with 3-rolls system, allows labeling with two labels, on the front and backside, or the application of labels with overlapping [227]. Now, in Europe about 2 mrd tubes/yr are used [228]. Tube labeling machines allow the labeling of up to 140 tubes/min [229]. The build-up of a labeling machine for tubes is shown in detail in Ref. [230]. Wrap-around labeling is recommended if the items to be labeled are not rigid enough, if a horizontal transport of the product is required and high quality positioning of the labels is needed [231].

So-called touch-blow labels use no mechanical contact in label application. The label is blown onto the substrate. This method is used for labeling mechanically sensitive items and deformable surfaces [232]. Labeling machines for modern industrial production lines dispense 800 labels/min [233]. Microprocessor controlled noncontact pressure-sensitive labelers achieve a labeling speed of 30,000 labels/h, and a label placement accuracy of 1 mm [234]. The new labeling systems are complete packages incorporating the own label material, and label handling modules, and are complementary to the relevant production lines [235]. Line scan photo-sensor technology checks for label placement and defects [236]. Technical specifications for labeling machines include label dimensions, label unwind outside diameter, label core diameter, backing paper rewind, label stop accuracy (e.g., +0.05 mm), and dispensing speed.

Tape application machines are known also. These devices were first developed for hot-laminated tapes [237]. They are used mainly for heavy insulating tapes. Transfer tapes are also often die-cut. Such tapes cannot be cleanly dispensed from a common adhesive transfer gun, which has no cutting blade. They tend to elongate during dispensing and leave excess adhesive both at the broken edge of the transferred strip and at the orifice of the gun [238].

REFERENCES

1. I. Benedek, *Pressure Sensitive Adhesives and Applications*, Marcel Dekker, Inc., New York, 2004, chap. 7.
2. *Adhes. Age*, (3), 8, 1987.
3. *Modern Plast. Internat.*, Show Daily, 23 & 24 Oct., 2004, p. 21.
4. H. Haas, *Coating*, (3), 102, 1987.
5. A.K. Das Gupta, *A Label Paper Satisfying the Highest Standard*, 2nd International Cham Tenero Meeting for Pressure Sensitive Materials, Locarno, Switzerland, 1990.
6. T. Frecska, *Screen Printing*, (2), 120, 1988.
7. K. Hanser, *Druck Print*, (1), 15, 1989.
8. R. Hummel, Prägefoliendruck auf Formkörpern, Präegefoliendruck auf Kunststoff Folien und Laminaten, Seminar Bedrucken von Kunststoffen Fachhochschule Stuttgart, *Coating*, (11), 311, 1985.

9. *Packlabel News*, Packlabel Europe '97, Ausgabe 2.
10. TDI, *Label Printer*, Avery Dennison Deutschland GmbH, Eching, 1998.
11. *Verpackungs-Berater*, (5), 26, 1996.
12. *Etiketten-Labels*, (5), 133, 1995.
13. *Intern. Forms*, (3), 16, 1997.
14. K. Hanser, *Druck Print*, (1), 15, 1989.
15. Sato Europe GmbH, Hilden, General Printer Specification, M8485S.
16. *Etiketten-Labels*, (5), 9, 1995.
17. F. Hufendiek, *Etiketten-Labels*, (1), 9, 1995.
18. *Adhäsion*, (10), 6, 1985.
19. J.C. Pommice, J. Poustis, and F. Lalanne, *Paper Technol.*, (8), 22, 1989.
20. E. Park, *Paper Technol.*, (8), 15, 1989.
21. R. Hummel, *Basic Material for Self-Adhesive Labels*, 19th Munich Adhesive and Finishing Seminar, Munich, Germany, 1994, p. 58.
22. W.J. Busby, *Processing Problems with Polyethylene Films*. PE '93, p. 3, Session VI, 3-3 Maack Business Service, Zürich, Switzerland.
23. *Convert. Today*, (11), 9, 1991.
24. Kalle Folien, Siegel Band, Hoechst., Mi 1984, 38T 5.84 LVI, Hoechst A.G., Wiesbaden, Germany.
25. J. Verseau, *Coating*, (7), 189, 1971.
26. J.M. Casey, *Tappi J.*, (6), 151, 1988.
27. R. Ammann, *Verpackungs-Rundsch.*, (5), 24, 2004.
28. E.C. Schütze, Kunststoffe als Bedruckstoffe, Seminar Bedrucken von Kunststoffen, Fachhochschule Stuttgart, *Coating*, (11), 311, 1985.
29. *Allg. Papier-Rundsch.*, 42, 1498, 1986.
30. *Etiketten-Labels*, (5), 16, 1995.
31. L. Waeyenbergh, 19th Munich Adhesive and Finishing Seminar, Munich, Germany, 1994, p. 138.
32. *Mobil-OPP Art*, (2), 2, 1993.
33. *Papier Kunstst. Verarb.*, (9), 57, 1988.
34. *Etiketten-Labels*, (5), 25, 1995.
35. *Avery Rating Plates and Type Plates Last a Fairly Long Time*, Booklet, Avery Etikettier-Logistik GmbH, Eching b. München, Germany.
36. K. Fust, *Coating*, (2), 65, 1988.
37. C. Hars, *Verpackungs-Rundsch.*, (5), 526, 1988.
38. G.M. Milles, *Papier Kunstst. Verarb.*, (11), 41, 1986.
39. W. Schibalski, *Coating*, (3), 87, 1985.
40. *Coating*, (7), 245, 1993.
41. Printing Trycite Plastic Films, Dow Chemical USA, Designed Products Department, Midland, MI, 1995, p. 3.
42. B. Hoffmann, *Qualitätssteigerung im Flexodruck durch Moderne Maschinen Konzeptionen*, Seminar Bedrucken of Kunststoffen, Fachhochschule Stuttgart, in *Coating*, (11), 311, 1985.
43. *Etiketten-Labels*, (3), 10, 1995.
44. R. Neumann, *Coating*, (6), 194, 1987.
45. *Etiketten-Labels*, (5), 95, 1997.
46. B. Ellegard, *Etiketten-Labels*, (3), 47, 1995.
47. R.J. Ridgeway, *Paper Film and Foil Conv.*, (12), 41, 1968.
48. *Etiketten-Labels*, (5), 91, 1995.
49. K.W. Holstein, *Neue Verpackung*, (4), 59, 1991.
50. *Etiketten-Labels*, (5), 22, 1995.
51. K. Ehrlitzer, *Etiketten-Labels*, (3), 34, 1995.
52. *Papier Kunstst. Verarb.*, (10), 54, 1996.
53. J. Allen, *FlexoTech*, 3(9/10), 34, 1995.
54. *Druckwelt*, 17(9), 31, 1986.
55. *Labels, Label.*, (7/8), 67, 1988.
56. *Etiketten-Labels*, (5), 91, 1995.
57. Der Siebdruck, 40, Drupa 94, 5, Sonderausgabe, Halle 3, Düsseldorf.

58. *Pack Report*, (11), 48, 1983.
59. M. Emrich, *Druck Print*, (1), 34, 1986.
60. P. Dippel, *Coating*, (4), 44, 1988.
61. R. Davis, *Fassson Facts International*, (1), 2, 1969.
62. *Der Polygraf*, (23), 2260, 1987.
63. *Coating*, (4), 115, 1987.
64. *Druckwelt*, 17(9), 30, 1986.
65. *Coating*, (12), 494, 1995.
66. *Druckwelt*, (12), 34, 1988.
67. K. Hanke, *Deutscher Drucker*, (8), 16, 1987.
68. A. Jentzsch, *Druckwelt*, (18), 44, 1988.
69. *Papier Kunststf. Verarb.*, (6), 37, 1995.
70. H. Roman, Bedrucken von Folien im Buchdruckverfahren auf Etikettendruckmaschinen, in *Coating*, (11), 312, 1985.
71. *Pack Report*, (11), 48, 1983.
72. H. Mathes, *Etiketten-Labels*, (1), 21, 1997.
73. W. Kaiser, *Der Siebdruck*, (8), 46, 1988.
74. W. Kaiser, *Der Siebdruck*, (9), 48, 1988.
75. *Coating*, (12), 494, 1995.
76. H.J. Teichmann, *Papier Kunstst. Verarb.*, (6), 21, 1995.
77. *Druckspiegel*, (6), 160, 1996.
78. Seminar Bedrucken von Kunststoffen, Fachhochschule Stuttgart, in *Coating*, (11), 311, 1985.
79. R. Hummel, *Coating*, (19), 358, 1992.
80. H.J. Teichmann, *Papier Kunstst. Verarb.*, (11), 10, 1994.
81. H. Klein, *Coating*, (4), 102, 1986.
82. *Convert. Today*, (10), 41, 1991.
83. B. Hunt, *Labels, Label.*, (2), 47, 1997.
84. K. Unbehaun, *Deutscher Drucker*, (6), 43, 1988.
85. *Labels, Label.*, (3/4), 11, 1994.
86. H. Pfeiffer, *Druckspiegel*, 51 (6), 122, 1996.
87. *Coating*, (12), 452, 1986.
88. *Technologien zur Fälschungssicheren Produkt-Authetifizierung*, Interpack News, 25–27, Apr., 2005, p. 2.
89. *Cold Foil Transfer*, Booklet, Arpeco Engineering, Missisauga, ON, Canada, 1999.
90. *Etiketten-Labels*, (5), 26, 1995.
91. *Etiketten-Labels*, (1), 38, 1996.
92. J. Young, *Tappi J.*, (5), 78, 1988.
93. *Etiketten-Labels*, (1), 30, 1995.
94. A. Prittie, *Finat News*, (3), 35, 1988.
95. Folex Symposium, Kommunikationstechnik in der näheren Zukunft, Nov. 27, Gravenbruch, in *Offsetpraxis*, (2), 50, 1988.
96. *Etiketten-Labels*, (5), 142, 1995.
97. *Etiketten-Labels*, (5), 6, 1995.
98. *Labels, Label.*, (3/4), 11, 1994.
99. *Verpackungs-Berater*, (5), 26, 1996.
100. *Labels, Label.*, (2), 29, 1997.
101. *Etiketten-Labels*, (5), 133, 1995.
102. *Etiketten-Labels*, (5), 136, 1995.
103. *Allg. Papier Rundsch.*, (48), 1360, 1985.
104. *Coating*, (5), 117, 1974.
105. *Labels, Label.*, (2), 38, 1997.
106. *Polygraf*, 23, 2237, 1987.
107. Aplitape, October, 1995, South Plainfield, NJ, USA.
108. *Verpackung*, (3), 10, 1995.
109. *Etiketten-Labels*, (5), 6, 1995.
110. B. Hunt, *Intern. Forms*, (3), 22, 1997.

111. Taktik, Samuel Jones and Co. Ltd., Laserdruckern, Herzogenrath, Germany, 2000.
112. *Labels, Label.*, (3), 34, 1994.
113. *Papier Kunstst. Verarb.*, (10), 62, 1987.
114. *Pack Report*, (3), 30, 1997.
115. *Coating*, (12), 494, 1995.
116. *Convert. Today*, (10), 41, 1991.
117. *Etiketten-Labels*, (3), 40, 1995.
118. FINAT, *Labelling News*, (1), 15 1995.
119. *Etiketten-Labels*, (3), 30, 1995.
120. *Etiketten-Labels*, (35), 28, 1995.
121. *Etiketten-Labels*, (5), 22, 1995.
122. *Allg. Papier Rundsch.*, (18), 686, 1986.
123. H.P. Seng, *Coating*, (9), 231, 1993.
124. B. Hunt, *Labels, Label. Int.*, (5/6), 34, 1997.
125. *Etiketten-Labels*, (1), 49, 1996.
126. M. Fairley, *Labels, Label. Int.*, (5/6), 28, 1997.
127. R. Kasoff, *Paper, Film and Foil Convert*, (9), 85, 1989.
128. Folien, Gesamtkatalog '96, Geonit Siebdrucktechnik GmbH, Erfurt, Germany, p. 23.
129. United Barcode Industries, Denmark, Drucker Zubehör, Technical Booklet, 1999.
130. H. Hadert, *Coating*, (6), 175, 1969.
131. *Papier Kunstst. Verarb.*, (5), 98, 1986.
132. R. Mannel, *Papier Kunstst. Verarb.*, (2), 17, 1996.
133. J. Wahl, *Etiketten-Labels*, (3), 44, 1995.
134. G. Meinel, *Papier Kunstst. Verarb.*, (10), 28, 1985.
135. J. Dietz, *Coating*, (11), 296, 1982.
136. *Coating*, (1), 22, 1969.
137. R.M. Podhajny, *Convert. Packaging*, (3), 21, 1986.
138. Packard Electric, Engineering Specification, ES-M-2379.
139. P.M. Guella, *Papier Kunstst. Verarb.*, (10), 39, 1985.
140. *Verpackungs Rundsch.*, (3), 334, 1987.
141. H.P. Seng, *Coating*, (9), 324, 1993.
142. C. Fellers, L. Salmen, and M. Htun, *Allg. Papier Rundsch.*, 35, 1124, 1986.
143. R.R. Lowman, *Finat News*, (3), 24, 1987.
144. Selbstklebefolien maßgeschnitten, Grafityp, Houthalen, Belgium, October, 1992.
145. Klebeband Träger u. Abdeckfolie, Datenblatt, Ausgabe 07/92, Hoechst AG, Wiesbaden, Germany.
146. V.M. Ryabov, O.I. Chernikov, and M.F. Nosov, *Plast. Massy*, (7), 58, 1988.
147. P. Zimmermann, *Verpackungs Rundsch.*, (3), 298, 1972.
148. I. Benedek, *Pressure Sensitive Adhesives and Applications*, Marcel Dekker, Inc., New York, 2004, chap. 10.
149. Beschichtung von Kunststoffen mit Ixan WA, Solvay & Cie, Bruxelles, Belgium, p. 29.
150. *Siebdruck*, (1), 14, 1988.
151. J. Metall, *Kaut. Gummi, Kunstst.*, (3), 228, 1987.
152. T. Sigii, Y. Moroishi, and K. Noda (Nitto Electric Ind. Co. Ltd.), Japanese Patent, 6386 778/18.04.1998, in *CAS, Adhesives*, 19, 2, 1988.
153. *Adhes. Age*, (10), 125, 1986.
154. J.R. DeFife, *Labels, Label.*, (3/4), 14, 1994.
155. Packard Electric, ES-M-2147, Engineering Specification.
156. C. Bayer, *Adhäsion*, (9), 349, 1965.
157. A.W. Norman, *Adhes. Age*, (4), 35, 1974.
158. Kendall Co., Canadian Patent, 853145, 1987.
159. Technical Information 2003, Printing on EVAL Films, Kuraray Co. Ltd., 1992.
160. D. Bedoni and G. Caprioglio, *Modern Equipment for Label and Tape Converting*, in 19th Munich Adhesive and Finishing Seminar, Munich, Germany, 1994, p. 37.
161. *Convert. Today*, (10), 41, 1995.
162. *Etiketten-Labels*, (5), 91, 1995.

163. *Eur. Adhes. J.*, (6), 23, 1995.

164. *Papier Kunstst. Verarb.*, (1), 25, 1996.

165. *Brit. Plastics Rubber* (10), 56, 1998.

166. J. Michel, *Verpackungs Rundsch.*, (12), 1425, 1985.

167. Products and more, *Plastverarb.*, p. 19, K2004, 20–27 Oct., 2004.

168. S.W. Medina and F.V. DiStefano, *Adhes. Age*, (2), 18, 1989.

169. CSR, Turbo, Computerized Signmaking Robot, N.V. Grafityp Houthalen, Belgium, 1997.

170. Grapihtec, RFC3100-60, Tischschneideplotter, Multiplot Grafiksysteme, Bad Emstal, Germany, 1997.

171. G. Strasser, *Etiketten-Labels*, (3), 24, 1996.

172. K.W. Holstein, *Neue Verpackung*, (4), 59, 1991.

173. *Etiketten-Labels*, (3), 8, 1995.

174. *Etiketten-Labels*, (1), 44, 1996.

175. M. Bringmann, *Etiketten-Labels*, (5), 72, 1995.

176. *Packlabel News*, 8–10 Apr., 1997, Frankfurt, Germany.

177. *Allg. Papier Rundsch.*, 31, 1014, 1986.

178. *Etiketten-Labels*, (5), 32, 104.

179. *Paper Film Foil Convert.*, (9), 30, 1989.

180. B. Hartmann, F. Zörgiebel, and R. Wilken, *Etiketten-Labels*, (3), 4, 1996.

181. *Etiketten-Labels*, (3), 14, 1996.

182. E. Kazuyoshi, N. Hiroaki, T. Katsuhisa, K. Yoshita, and T. Saito (FSK Inc.), Japanese Patent, 6317981/25.01.1988, in *CAS, Adhesives*, 12, 5, 1988.

183. T. Kurono, N. Okashi, and N. Tanaka (Nitto Electric Ind. Co. Ltd.), Japanese Patent, 6333487/ 13.02.1988, in *CAS, Adhesives*, 12, 5, 1988.

184. *Finat News*, (4), 10, 1996.

185. *Screen Printing*, 22, 204, 1987.

186. D. Lacave, *Labels, Label.*, (3/4), 55, 1994.

187. Sekisui Chemical Ind. Co. Ltd., Japanese Patent, 072785513/24.10.1995, in *Adhes. Age*, (5), 12, 1996.

188. *Papier Kunstst, Verarb.*, (12), 44, 1986.

189. *Neue Verpackung*, (1), 156, 1991.

190. LOS Lager Organisations System GmbH, Mannheim, Germany, Gebrauchsmuster, G 87 06 322.0/ 02.05.1987.

191. H. Nakahata, Japanese Patent, 07316510/05.12.1995, in *Adhes. Age*, (5) 11, 1996.

192. Packard Electric, Engineering Specification, ES-M-1881.

193. *Etiketten-Labels*, (1), 50, 1996.

194. W. Grosse, *Coating*, (6), 201, 1993.

195. *Coating*, (9), 279, 1969.

196. *Coating*, (12), 492, 1995.

197. *Coating*, (5), 185, 1988.

198. *Convert. Today*, (8), 13, 1994.

199. Better Packages Inc., U.S. Patent, 3510037, in *Coating*, (5), 130, 1971.

200. Packard Electric, Engineering Specification, ES M-2147.

201. K. Nakamura, Y. Miki, and Y. Nanzaki (Nitto Electric Ind. Co. Ltd.), Japanese Patent, 6386787/ 18.04.1988, in *CAS, Adhesives*, 19, 5, 1988.

202. H. Kenjiro and O. Kazuto (Nitto Electric Ind. Co. Ltd.), Japanese Patent, 63137841/09.01.1988, in *CAS, Adhesives*, 24, 5, 1988.

203. Tapes, Data Sheet, Four Pillars, Taiwan, 1999.

204. F. Hensen, *Kunststoffe*, 84(10), 1325, 1994.

205. Packard Electric, EM-S-2147, Engineering Specification.

206. K. Maschke, Hoechst Folien, 4. Klebeband Forum, Frankfurt Main, Nov. 1990.

207. *Coating*, (5), 162, 2004.

208. *Pack Report*, (11), 48, 1983.

209. *Packung Transport*, (6), 20, 1983.

210. *Labels, Label.*, (2), 24, 1997.

211. Alois Stöckerl GmbH, München, Germany, DBP 1183007, in *Adhäsion*, (2), 82, 1966.
212. Thatcher Glass Manufacturing Co., Inc., New York, NY, U.S. Patent, 3152940, in *Adhäsion*, (2), 80, 1966.
213. *Folio, Avery Dennison Info J.*, p. 2, 1997.
214. *Etiketten-Labels*, (5), 30, 1995.
215. Hochwertige Getränkeausstatung mit Selbstklebeetiketten, Technical Booklet, Pago AG, Buchs, Switzerland, 10, 2001.
216. *Package Print Design Internat.*, (1/2), 26, 1997.
217. *Verpackungs-Rundschau Special*, 53, 12, 2002.
218. *Etiketten-Labels*, (5), 16, 1995.
219. Pago System 540, *High Speed in Line System for Top Quality Decorative Labelling*, Technical Booklet, Pago AG, Buchs, Switzerland, 10, 2001.
220. B. Hunt, *Labels, Label.*, (2), 29, 1997.
221. *Labels, Label.*, (3/4), 11, 1994.
222. *Pack J.*, May 14, 1996, p. 3.
223. *Nahrungsmittel + Genußmittel Verpacken*, (3), 2, 1977.
224. *Pack Report*, (6), 40, 1991.
225. Pago System 520, Technical Booklet, Pago AG, Buchs, Switzerland, 10, 2001.
226. Pago System 225, Rundum Etikettierung im Drehsystern, Technical Booklet, Pago AG, Buchs, Switzerland, 10, 2001.
227. Pago System 188, *Round-Around Labelling with 3-Rolls-System*, Technical Booklet, Pago AG, Buchs, Switzerland, 10, 2001.
228. A. Homola, *Creativ Verpacken*, (2), 5, 2001.
229. Pago System 270DT, Labelling of Tubes, Technical Booklet, Pago AG, Buchs, Switzerland, 10, 2001.
230. Tubendekoration mit Selbstklebeetiketten, Technical Booklet, Pago AG, Buchs, Switzerland, 10, 2001.
231. Pago System 233, Rundum Etikettiermaschine mit horizontalem Kettenfördersystem, Technical Booklet, Pago AG, Buchs, Switzerland, 10, 2001.
232. *Etiketten-Labels*, (5), 91, 1995.
233. *Verpackungs-Berater*, (3), 21, 1997.
234. *Willett News*, 1997, Willett Sarl, Roissy, France.
235. B. Topliss, *Finat News*, (2), 45, 1989.
236. *Labels, Label. Internat.*, (5/6), 5, 1997.
237. *Adhäsion*, 10(2), 82, 1966.
238. C.L. Vernon and E.A. Stanek, EP 0,147,093A2, Minnesota Mining and Manuf. Co., St. Paul, MN, USA, 1985.

11 End-Uses of Pressure-Sensitive Products

István Benedek

CONTENTS

The end-use properties together with the adhesive properties are the most important performance characteristics of pressure-sensitive products (PSPs) [1–3]. The number of applications for PSP is growing. Semifinished products (such as sheets, tubes, films, and profiles) and finished ones (bottles, containers, etc.), natural and industrial products (fruits, machine parts, etc.) are made pressure-sensitive. Security products, envelope sealing strips, and materials such as post-it notes are examples [4]. The proportions of the global volume represented by the various PSPs in the produce is continuously changing. In 1983, in Europe, the main PSP applications were tapes, labels, and postage [5]. In 1987, labels, tapes, hygienic products, decorative tapes, building protection products, envelopes, medical products, and tiles were considered the most important application domains [4]. According to Jordan [6], the main application fields for PSPs are medical and hygienic uses (plasters, tapes, dosage systems), masking tapes, plotter films, common tapes (office tapes, marking tapes, packaging tapes) labels, protective films, insulation tapes, decoration foils, and carpet and double-faced tapes. A study carried out by the Skeist Laboratories [5] shows that in 1983 market expansion was expected in graphics, mounting, cushioning, and insulating products. Now low temperature resistance for deep freeze price labeling, structural pressure-sensitive adhesives (PSAs) for holding and mounting, high-shear/peel adhesives for packaging tapes, adhesives resistant to perspiration and body liquids for personal care products, and heat-resistant PSPs for electrical and industrial uses are considered as the most important. According to Ref. [7], in 1996 foodstuffs (15%) cosmetics (13%), pharmaceuticals (7%), industrial (10%), and variable image printed products (40%) have been the main application domains for labels. Recently, the variable image printed products display the highest growth rate. Table 11.1 lists the main converting and end-use properties of PSPs.

As illustrated by Table 11.1, the end-use properties of PSPs have to fulfill the different requirements of a number of application fields. Some of them are general requirements related to the application technology of a product class or adherend or to the processing of labeled, taped, protected, etc. products. These are special criteria related to the application field. There are products manufactured with different technologies that are proposed for the same application (e.g., certain adhesive-coated and adhesiveless protective films). In contrast, some classic products (see Chapter 2) have the same buildup but quite different application.

I. GENERAL CONSIDERATIONS

Generally, end-use properties must be evaluated as "instantaneous" application characteristics and storage- or aging-dependent characteristics.

Both depend on the product buildup and application conditions. It is evident that labels, tapes, or protective films applied from rolls have to meet some general requirements related to the web-like character (e.g., ease of unwinding). Labels, tapes, or protective films, which are marked by hand during or after application (e.g., writable items), have similar surface characteristics.

TABLE 11.1
The Main Conversion and End-Use Properties of PSPs

PSP	Converting Properties	End-Use Properties
Label	Slitting ability	Printability
	Cutting ability	Labeling ability
	Die-cutting ability	Delabeling ability
	Printability	
Tape	Slitting ability	Tearability
	Die-cutting ability	Cuttability
	Printability	Special electrical characteristics
		Special thermal characteristics
		Special dosage characteristics
Protective film	Slitting ability	Laminating ability
	Printability	Mechanical and thermal processability
		Delaminating ability

Labels and tapes applied on the same type of packaging material or item (e.g., bags) have to be recyclable with the item.

In contrast, there are special PSA applications, for example, in the automotive and electronics industries for which specific bond strength, or environmental or physiological behavior etc., may be required. The achievement of specific bond strengths in application to any substrate depends on the adhesive formulation used, the substrate itself, and the environmental circumstances of the application. Two areas of prime concern involve adhesion to nonpolar surfaces and to flexible packaging that can change form [8]. Pressure-sensitive laminating can also be used for over-laminating. Some years ago, thermal laminating was the main encapsulation laminating method. Pressure-sensitive laminating equipment costs considerably less than thermal laminating equipment [9].

As can be seen from Table 11.2, generally the application and deapplication conditions for different types of PSPs are quite different. Labels are laminated under low pressure at room temperature. Their bonding speed is high. Tapes and protection films are laminated under high pressure (tension) at room or elevated temperatures. The application speed is lowest for tapes. Delamination of labels is a low speed process. Delamination of tapes and protective films occurs at high speed. Removable, repositionable, and readherable products require different application conditions and bonding characteristics.

TABLE 11.2
Application and Deapplication Conditions for Major PSPs

Pressure-Sensitive Product	Application Conditions	Deapplication Conditions
Label	Impact or nonimpact labeling	Mechanical, chemical, or thermal deapplication
	High speed application	
	Room temperature or low temperature application	Partial deapplication possible
	Very low application pressure	
	Repeated application possible	
	Manual or automatic application	
	Converting during application	
	Converting after application	
Tape	Impact lamination	Mechanical deapplication
	Room temperature, low temperature, or high temperature application	
	Low or high pressure application	
	High stress application	
	Low or high speed application	
	Converting during application	
	Hand or automatic application	
Protective film	Impact lamination	Mechanical delamination
	High temperature and high pressure lamination	High stresses during delamination
		Delamination of more importance than lamination
		High temperature, high pressure processibility after lamination

II. LABELS AND THEIR APPLICATION

PSAs have been used since the late 19th century mainly for medical tapes and dressings. Industrial tapes were introduced into the market in the 1920s and 1930s and self-adhesive labels in 1935. Various end-uses are known for labels, but a label is always an information medium. Some of the end-uses are general, classic application fields where PSA labels replaced wet adhesive labels or other products designed to place information or aesthetic, decorative elements on a product surface. Others are new domains where only a PSP can meet practical requirements.

A. APPLICATION CONDITIONS

The conditions under which PSPs are applied differ according to the PSP classes and special uses. The application conditions for labels depend on the product to be labeled, the labeling method, and the type of label. Among the product's characteristics, its surface has special importance in application. Application conditions include the application temperature, environment, and speed of application. As mentioned earlier, they may also include readherability.

1. Influence of the Product Surface on Label End-Use

Substrate surfaces used in laminating (labeling) vary widely. Their diversity is due to the raw materials used for their manufacture and the processing technology used. Special application fields need special application (adherend) surfaces. Therefore, the label designer has to have information about the adherend surface in order to tailor the product to serve its intended purpose.

According to Fust [10], at the end of the 1980s, the main label markets were food, cosmetics, computers, pharmaceuticals, sanitary products, and beverages. Food, pharmaceuticals, and cosmetics accounted for 36% of the label market at that time. Heger [11] stated that in 1965, less than 5% of packaging materials were plastics, but by 1987, more than 40% were plastic based. In 1988, in Germany, 44% of packaging materials were paper based [12]. Now less than 30% are.

The variety of adherend surfaces in a given application field is illustrated by the automotive industry which uses various unmodified and modified plastics. Composite materials and alloys, polycarbonate (PCMA), and polyurethane (PUR) foam are used as interior coverings [13]. Polyvinyl chloride (PVC) is no longer used. Thus, in the U.S.A. in 1988, about 80% of interior automotive PVC parts has been replaced with other materials [14]. Most electroplated plastics in the U.S.A. are used for decorative products for automobiles and appliances. The main manufacturing processes for metallized finished plastics are electroplating and electroless plating, evaporative vacuum metallizing, vacuum sputter coating, and hot stamping. The surface quality of metallized items depends on the metallizing procedure used.

Because of the increased use of postimproved surface qualities, that is, items whose final surface is quite different from that of the bulk material, it is difficult to find the correct adhesion level of a PSP without preliminary testing. Injection molded, blown, or thermoformed plastic items may have an outer layer of release material such as silicone [15]. Polystyrene (PS) sheets or thermoformed PS parts (e.g., refrigerator shelves) are covered with clear OPS. In households, glass surfaces and glass-like surfaces are the most common labeling substrates. In some cases, glass surface is coated with TiO_2, or SnO and PVA [16]. Epoxy coatings and acrylics (ACs) can also be used [17]. Colored bottles may have a greater amount of mold release agent applied in their manufacture. Light weight single-use bottles are chemically hardened and postsmoothened using a surfactant, polyethylene (PE), or wax [18]. Low friction coatings for glass bottles are based on PVA and plasticizer [19]. In other cases, the surface quality of the glass is changed by condensed humidity. Therefore, wet tack is necessary to ensure the bonding of labels to condensation-coated surfaces. Injection-molded containers have mold release agents on their surface.

Composite materials also replace other common surfaces. Such composite structures are manufactured by laminating, varnishing, or common processing of plastics. To improve the water resistance of paper or cardboard, chlorinated rubber and paraffin are used [20]. The water resistance of paper can be improved using ethylene–acrylic acid (EAA) copolymers [21]. Paper can be coated with PE dispersion also. Dispersed PE powder (particles of $5-12\ \mu m$ diameter) are used to achieve a thin coating [22]. PE laminates dominate the market of plastic-coated papers [23]. It is well known that such laminates (milk and juice boxes) are difficult to label [24]. Industrial containers and drums made of high density polyethylene (HDPE) or impregnated cardboard are also difficult to label. Ultraviolet (UV)-cured epoxides are used for coating cans [25]. PS is used for the surface coating of imitation woods [26]. Various materials such as paper, paperboard, aluminum, glassine, and plastic films are extrusion-coated with PE [27]. Papers with extrusion-coated PE display different surface properties in the machine and cross directions. Extrusion coating orients the PE material [28]. When blow molded items are rapidly cooled during injection, their surface may have a different structure than that of the bulk material [29].

Flexible packaging industry represents a difficult domain of label application because of the various combinations of packaging materials, their different thicknesses and shapes, and their end-use conditions. According to Bulian [30], the main types of materials used by the flexible packaging industry in the form of thin sheet suitable to "wrap" a product are paper, aluminum foil, PE, polyester, polypropylene (PP), PVC, polyamide (PA), PS, polycarbonate (PC), and PAN. For the packaging of fatty or moist products, $6-15\ \mu m$ thick HDPE film is used [31]. In the packaging industry, there is a trend to replace PS with PP [32]. Blow-molded containers have been manufactured from PAN and PC Polyethylene terephthalate (PET)/PC; PP/ethylene vinyl hydroxide (EVOH), PET/EVOH/PET; PC/EVOH/PC combinations [33].

In-mold labeling (IML) allows a plastic item to be simultaneously manufactured by injection molding and labeled with a special label placed in the mold. No supplemental postlabeling or post-printing of the finished item is necessary. The label is made from the same polymer as the labeled item and both are recycled together. However, the material structure may change due to the cooling or crystallizing conditions. In-mold labeling is used for injection molded items, but it can also be used with thermoformed plastics [34]. It is evident that in this case the surface properties of the adherend are quite different.

According to Maack [35], the complexity of packaging materials and the variety of packaging surfaces have been increased by recycling and downgauging. Downgauged or recycled plastic materials require special adhesive properties. Recycling of postconsumer PET containers is well established. Wide-mouthed glass jars, for example, for coffee or milk powder have been substituted by coinjected PET barrier containers. In many cases, glass bottles have been replaced with HDPE bottles. Puncture-, impact-, and tear-resistant shrink and stretch films, based on linear low density polyethylene (LLDPE), very low density polyethylene (VLDPE), and LLDPE/low density polythylene (LDPE) blends and coextrudates, have replaced craft paper wrapping. Cling films $(13-15\ \mu m)$ have replaced PVC as wrap, household, and catering film (LLDPE blends with ethylene–vinylacetate (EVAc) or PP). Thinnest films $(15-20\ \mu m)$ are manufactured for rack and counterbags. Heavy duty bags produced from coextruded blown LLDPE/HDPE/LLDPE or similar coextrudates have been introduced to replace multi-ply paper bags. Such materials may have an antislip melt-fractured surface. The presence of certain fillers, pigments, and lubricants in these recycled materials may cause gel formation. Many hygienic plastic applications (diapers, garments, drapes, sanitary towels, etc.) use highly filled materials. Some companies develop products with degradable layers to facilitate accelerated decay, if disposal is not by incineration. In medical packaging, injection- and coinjection-molded PET, PP/elastomer, and HDPE/elastomer blends find increasing use as well as coextruded PP/HDPE, HDPE/LMDPE, HIPS/HDPE, and other compositions. Labels for electronic parts have to adhere to paper, board, wood, steel, and glass [36]. Coextruded and coinjected packaging materials can have different surface qualities for the same main carrier. New packaging techniques (aseptic, blow-fill seal, boil-in bag, retort plastics, barrier containers, etc.) need new materials. Labels have to adhere (or readhere) to such surfaces and have be recyclable.

2. Influence of the Product Form on Label End-Use

The product form and flexibility or rigidity influence the application and labeling conditions also. As is known, application on flexibile surfaces is difficult. The flexibility of the packaging materials is increasing as a consequence of downgauging. New vinyl alloys allow blow molders to create cosmetic bottles and tubes with a frosty squeeze effect; this means the product has a satin look, a soft touch, and the ability to recover its shape after compression. Such products are free of phthalates [37].

New packaging films may be manufactured with a reduced wall thickness. As an example, biaxially oriented EVOH film with a wall thickness less than 10 µm can be manufactured with sufficient strength for many packaging applications [38]. In some cases, the label has to be used for unicates having various forms. The adhesive for name plates has to bond with difficult surfaces, high or low surface energy plastics, textured or curved surfaces, and cold surfaces. No-label look transparent labels need a plasticizer-resistant PSA. As shown by Fries [39], such PSAs have to display improved holding power with a plasticizer content of 15%. A common label grade PSA shows a holding power of less than 1 h (with a 2 kg/in.^{2} load) and a plasticizer-resistant grade more than 3 h.

3. Other Product Characteristics

Pressure-sensitive labels are also applied directly to foods such as fruits and vegetables. For these, FDA 175.125, Part (b) applies (pressure-sensitives, raw fruits, and vegetables). The regulation does not distinguish between produce whose skin is normally eaten and produce whose skin is discarded. For adhesives with indirect contact with food, 3-Jod-2-propenylbutyl carbamate (IPBC) has been suggested as stabilizer. This product satisfies the 21CFR 171.1 (b) and is on the list of adhesive additives [40].

4. Application Climate/Aging

Roll storage aging is the resistance to adhesive properties changes upon long-term contact with the face material and release liner. Migration of plasticizers, surface active agents, and antioxidants affect adhesives, printing inks, face stock, etc. Environmental resistance may be evaluated by retention of bond strength and other properties upon exposure to water, humidity, temperature changes, sunlight, etc. The surface temperature of the adherend is another factor influencing its adhesive properties. Most general-purpose adhesives are formulated for tack at room temperature. If the adherend temperature is lower, a higher degree of adhesive cold flow is required to provide adequate wet out. Some products are labeled at room temperature and subjected to lower temperatures later in their life cycle. For tackified formulations, peel reduction at 0°C may attain 300% [41]. Owing to the increase in the modulus of the adhesive at low temperatures, a loss of wetting and jerky debonding may be observed for certain adhesive formulations. According to Dunckley [42], freezer labels should display the same viscoelastic properties at −40°C as at +20°C. At 20°C, a storage modulus value of 10^5 has been suggested for common labels. For freezer labels, a value of 10^4 is recommended. Cable tape with good temperature resistance is made with butadiene rubber (more than 90% *cis*-butadiene content), an styrene-block copolymer (SBC) (styrene–isoprene–styrene [SIS] or styrene–butadiene–styrene [SBS]), and tackifier. This blend exhibits a T_g of −104°C. The peel resistance at −10°C is 790 g/15 mm, and at +40°C, it is 370 g/15 mm. The creep is measured at −5 and +40°C also. The rolling ball (RB) tack changes from 18 to 28 between 3 and −5°C [43].

Generally, there is a difference between labeling and service temperature. Suggested labeling temperatures are between −10 and +10°C [44]. Temperature-resistant labels are used between +40 and +126°C [45]. Fuel and heat make special demands on plastics and labels used in automobiles, 500 h tests at the temperatures typically occurring in the engine compartment (up to 120°C) are carried out [46]. Computer labels are designed for a service temperature of −20 to +90°C [47]. Solventless rubber-based adhesive for filled HDPE milk and fruit juice containers have to operate between $-25°C$ and $+70°C$ with a minimum surface application temperature of $-5°C$. According to Dobmann and Planje [48], office labels have to work at $+23°C$, labels for refrigerated meals at $+40°C$, and labels that are applied to frozen foods at $-18°C$. Bag closure labels must be able to withstand outdoor temperatures and moisture [49]. Labels for butt splicing have to adhere on PP, PE, PA, and coated papers [33]. Such labels have high temperature stability up to $117°C$. For high temperature splicing tapes, shear at $250°C$ has been tested [50]. For ice-proof gums, water resistance for more than 72 h is required [51]. Labels used on PA substrates, that absorb up to 30% water, require moisture-independent peel resistance [50]. Blood bag labels have to be water resistant and be able to withstand very low temperatures, so they are overlaminated [52]. Logging tags have to withstand long exposure to salt water, therefore, they are printed on a special synthetic face stock material [53]. Surgical tapes are used for fastening cover materials during surgery; the adhesive has to be insensitive to moisture and insoluble in cold water, and display adequate skin adhesion and skin tolerance. The water solubility of the product is reached at $65°C$ and at a pH value of more than 9 [54]. Aging on the end-use substrate is a very important practical evaluation criterion. Therefore, shrinkage should be tested on different surfaces such as 2 mil cast vinyl and 4 mil calendered vinyl [50]. Plasticizer resistance is important for exterior graphics on vinyl, semipermanent vinyl labels, and automotive decals, the target values are between 0.3 and 0.7%.

B. APPLICATION METHOD OF LABELS

Labels are applied manually or by labeling machines. Generally, labeling quality depends on the machine, label components, and substrate. According to Grebe [55], labeling is influenced by the labeling system, substrate surface, label material, adhesive, and transport and storage conditions (see also Chapter 10).

The application of pressure-sensitive labels is a cold laminating process. It involves the use of adhesive-backed carrier on a release liner. Equipment for this method varies from hand-fed units to sophisticated equipment with automatic feeders and cutoffs. Sometimes, the labeling device is a part of bag-making equipment, corrugated and folding carton converting machinery, envelope machinery, or engraving equipment. The release from liner should be adequate to prevent predispensing of labels but not so tight as to cause the web to break or the liner to tear. Tear-sensitive labels with low mechanical resistance are dispensed by machine also. Pressure-sensitive labels for tamper-evident application have to be applied automatically [13].

In the label industry, a difference exists between roll and sheet laminates. Requirements are generally less critical for roll applications. Sheet application needs PSA with a sufficient anchorage to prevent pulling out but sufficiently low tack level to prevent gumming of the guillotine. Labeling machines for roll and sheet labels have been developed [56]. Labeling speeds as high as 240 labels per minute are achieved. Application of small roll labels needs almost no mechanical lamination. As is known from the PSA testing practice, a loop tack test simulates the real application conditions of labels. They are blown onto a surface using air pressure [57].

A special case of label application is in-mold labeling where the item to be labeled is applied (melt around) to the solid-state static label. According to Teichmann [58], hot stamping can be used for in-mold labeling also. Microprocessor-controlled PSA label application offers variable speed settings to match product requirements and may have a special dispensing edge and a collapsible backing paper rewind. Where a high degree of conformability is desired, flexible label stock such as saturated paper should be used rather than a high gloss paper. The layout should be in the long grain direction to provide maximum conformability [59].

C. POSTMODIFICATION OF LABELS

As discussed earlier (see Chapter 10), some conversion operations are transferred from the manufacture in the end-use phase of labels. Postdesign, postprinting (writing), and postcutting are carried out by the end-user. Therefore, end-use properties include label versatility for these operations also. Writability, printability, computer printability, laser printability, and cuttability are special characteristics of labels in development. For instance, promotional form labels can have many pages, and their front pages can have a different print quality [60]. Laser beams can be used for cutting, punching, sealing, and printing labels. The label material (adhesive and carrier) has to fulfill the related requirements. The use of lasers in paper processing is examined in Ref. [61].

D. MAIN TYPES OF LABELS

The diversity of labels with respect to construction and end-use is increasing. Label markets and applications include food, toiletries, cosmetics, pharmaceutical, computer and industrial goods, and others. The computer label business is one of the fastest growing segments of the label market [53]. Electronic data processing (simple and high quality) labels were the main market segment in the U.S.A. in 1986. Variable data bar-coded labels were introduced. Routing labels (for mail and goods transport), inventory control labels, document labels (for libraries and documentation centers), individual part and product labels (e.g., for the automotive, aerospace, and electrical and electronic industries), supermarket shelf labels, and health sector labels (blood bags, drugs, medical equipment, and surgical utensils) are the most important sector of nonimpact printing. Labels for variable data and short-run jobs like price labels, short-run product labels, labels and tags, for bulk packaging, tickets, baggage tags, and other transport documents have been introduced. The film label market has grown at a rate of more than 30% per year [62]. Film labels provide a no-label look. Such labels are needed for cosmetics and office use also [63], that is, in applications that require that the label itself virtually disappear when applied onto the background. New labeling methods like sleeve labeling and in-mold labeling have been developed. Extended text labels were introduced. Magnetically encoded mailing labels are used by the German postal service for the automatic identification and tracking of packages. These contain a bar code and magnetic stripe to allow optical and magnetic processing [53]. It is not the aim of this book to list or describe all types of labels or their end-use. However, certain products will be discussed to clarify some technical aspects related to their manufacture.

According to a statistical evaluation by Thorne [64], in 1988, PSA labels were used mainly for information on pack (83%), on-case information (71%), design (32%), off-pack promotions (20%), bar codes (20%), and security (15%). According to Linder [65], the market segments that are the most important for labels are food, EDP/computer, cosmetics/toiletry, and price/weight.

Labels can be classified according to their main performance characteristics related to application conditions (temperature-resistant labels, water-resistant/water-soluble labels, etc.) or special application field (price-weight labeling, bottle labeling, etc.). In their application field they may have different end-uses — automotive labels, temperature labels in the category of technological labels, tamper-evident labels, or nondestructible labels in the category of antitheft labels, etc. Some products (e.g., temperature-resistant, temperature-display, freezer labels) have to support special climatic or temperature condition.

1. Temperature-Resistant Labels

In almost every label application field, there are products that have to be applied on substrate surfaces at extreme temperatures. According to Dunckley [42], temperature-resistant labels are used generally between +40 and +260°C.

2. Temperature Display Labels

Reel labels are used between 40 and 260°C and display the measured (substrate) temperature values with an exactness of ±1% [66,67].

3. Freezer Labels

As discussed earlier, low temperature labeling needs labels having a special PSA composition that ensures adhesive flow for bonding at low temperatures. Plastic film (PE, PP, PVC, PA, ionomer/ PA, coated cellophane) wrapped products (refrigerated meat) require a special, low temperature bonding PSA that has a low modulus in the temperature range between -50 and $+150°F$ [68]. As discussed in Chapter 3, Chapter 7, and Chapter 8, the formulation of such adhesives needs special know how. According to Bonneau and Baumassy [41], one of the main requirements for PSAs used for labels is the retention of the adhesive properties over a wide temperature range. According to Dunckley [42], freezer labels should display the same viscoelastic properties between -40 and $+20°C$. For common labels at $20°C$, a storage modulus value of $10^{4.5}$ is suggested and for freezer labels, a value of 10^4. At low temperatures, tack and peel may be too low. It was found by practical tests that at $0°C$, the peel of an untackified acrylic is almost 25% less than at $23°C$ [41]. Therefore, by formulation screening, $180°$ peel at $0°C$ should be also tested. For rubber-resin-based hot-melt formulations, low molecular weight (MW) polyisoprenes with a low T_g (-65 to $-72°C$) have been used as viscosity regulators [69]. They replace common plasticizers and also improve the migration-resistance and low-temperature adhesion.

4. Water-Resistant Labels

Textile, automotive, and bottle labels (see subsequently) need water resistance. Labels used for marking textiles or clothing should resist 50 laundry cycles and 10 drycleaning cycles [70]. Water resistance was discussed in detail in Ref. [71].

5. Price-Weight Labels

Many applications for labels such as bottles, flexible packaging, and textiles are special cases of price-weight labeling. Price-weight labeling is one of the most dynamic application fields. Price-marking and price-weight applications represent about 15% of European label consumption [72]. In the U.S.A., the 1988 forecast for the volume increase of thermal printed UPC (Universal Product Code) labels by 1990 was 20% [73]. The application requirements for products from this range vary. Simple and complex labels have been developed for different classes of products. Even the simplest products have to meet various criteria. In principle, the "same" price-weight labels should be usable on unpacked items (i.e., banana label), items packed in film [74], or items with a plastic plate label substrate (hang tags) [75]. Some price labels require perforation for transport [76]. According to Ref. [77], removable price labels use almost exclusively paper carrier coated with rubber-resin adhesive. Natural and synthetic rubber and polyisobutylene have been the main formulation components with soft resins and plasticizers added as tackifier. For these solvent-based adhesives, gasoline, toluene, acetone, or ethyl acetate have been used as solvent. The rubber has been calendered and masticated. Legging and migration are characteristics of these formulations. The adhesion buildup on soft PVC and varnished substrates attains very high level after long storage times. By debonding, no clear adhesions failure has been possible.

6. Bottle Labels

Although glass possesses a polar surface, the labeling of glass bottles with pressure-sensitive labels has developed only in recent decades. Economic and technical problems made this development difficult. As mentioned earlier, a chemically treated glass surface, condensed water on the bottle, water resistant of labels, and water-soluble adhesives call for specially tailored adhesive properties. The literature data concerning bottle labeling are summarized by papiertechnische stifting [78]. Pressure-sensitive labels are also used for wine and champagne bottles [79]. In 1977, PET was introduced for carbonated soft drinks [80], and glass bottle labeling was replaced at least partially with plastic bottle labeling. In-mold labeling is a special case of plastic bottle labeling where first the label is manufactured, then the plastic bottle. A combination label and folded information booklet (prospectus) is used for bottle labeling of pharmaceutical products [81]. Thus, no supplementary instructions are needed. Organization and logistic problems for food and beverages are solved by using bottle labeling [82].

A special case of application is wine bottle labeling where different water-based (WB) adhesives are used on a large range of (mainly) paper face stock materials. Such labels have been developed since 1985 [83]. Their design has to meet certain requirements (see also Chapter 8). These products include permanent labels with cold water resistance (ice water) and permanent labels that are removable with hot water. In all these categories, items have to possess good adhesion on moist, condensation-covered surfaces [84]. Special water-soluble compositions are required for bottle labeling. The adhesive must display good adhesion on both wet and dry (polar or nonpolar) surfaces and removability with hot or cold water or with commercial detergents and alkalies [85].

7. Bag Closure Labels

These products are applied to polythylene bags that contain stacks and buckets that are stored outdoors. The labels are applied to the packages at the end of the production process. They must be able to withstand outdoor temperature and moisture [49]. Acrylic adhesives are proposed for such labels. Bag reclosure labels (tapes) are also used, mainly for tobacco, coffee, textiles, and dry and wet tissues (see End-Use of Tapes, p. 553).

8. Nameplates

Equipments, apparatus, electric appliances, and other durable goods have always been given serial numbers. Traditionally, this has been done by using name plates (rating plates, type plates, inventory labels) to identify the manufacturer and state conditions (voltage, amperage, year of fabrication, etc.). They have to provide information over a period of years on serial numbers, date of manufacture, article number, and performance data. Such plates are used in the most varied industrial and commercial domains. They have to be light, chemically resistant, and weatherproof and be able to withstand abrasion. Generally they have to be printed on the spot.

These data once had to be stamped by hand, but stamping was expensive and time consuming. Up to the mid-1960s, these plates were mounted with screws or rivets. The application of adhesives allowed (required) the use of thinner metal plates and faster mounting. The data could be processed with high impact typewriters. At first, only paper and cellulose acetate could be computer printed and they were not suitable for name plates. The development of synthetic films, especially computer-imprintable PET, provided a solution to this problem. Metallized 90 μm PET has been used.

To be printed with flexographic, letterpress, thermal, laser, or computer printing methods, a film must have a matte, metallic appearance, similar to that of anodized aluminum. It has to display the handling characteristics of hard aluminum. No curl should occur after removal. A heat resistance of 150°C is needed. The adhesives used must have adequate aging resistance. The adhesive for nameplates has to bond on difficult surfaces, plastics with high or low surface energy, textured or curved surfaces, and cold surfaces. The liner has to exhibit trouble-free release; dimensional stability; and good die cutting, perforating, prepunching, and fanfolding. The matte surface has to absorb ink and maintain sharp graphic contrast, good color intensity, and excellent legibility. It has to be printable with a range of common or special computer printers and demonstrate abrasion, chemical, and environmental resistance. Polyester and vinyl film and special heat-resistant materials have been suggested for this purpose.

Laser-printable labels (sheet) are warranted for 4 yr and may be used between -20 and $+120°C$ [86]. Laser-engraved acrylic films for name plates are tamper-proof. Such labels are ideal for applications where either contact-free data recording is needed or direct laser inscription of work pieces is not possible [87]. It must be emphasized that formulation of "laser-resistant" adhesives, that is, products that exhibit enhanced temperature resistance, requires special tackifier resins (see also Chapter 8).

For printing such products, letterpress or screen-printed presetting is also used. The printer should have a comprehensive library of types and standard bar codes, electrical symbols, etc. If thermal transfer printing is used, the resolution is 7.6–11.4 dots/mm and the printing speed reaches 175 mm/sec. Rating prates can be printed in sandwich printing also, where the reverse side of a transparent material is printed with mirror lettering [88]. Easy-to-print destructible roll label materials can be used as tamper-evident products. The internal strength of the label stock totally disappears once applied and, if any attempt is made to remove the film, the label breaks into tiny fragments. This is a faster and more cost effective system compared with costly to produce laser etch labels and name plates [89].

9. Instruction Labels

Such labels are removable. They have a PET or PVC carrier and are used in clean rooms and on delicate electronic parts [90]. Such labels for technological use are used in various fields such as the automotive [91] and textile industries. Iron-on labels used in the textile industry have to resist temperatures up to 180°C (10–20 sec) during application and be chemical and water resistant [92]. Tamper-evident polyester label stocks provide an alternative to PVC and acetate films [93] (see Antitheft Labels, p. 828). Various carrier materials can be used: acetate for envelopes, PVC for warning labels for electronic equipment, paper for sterile needle cartridges (see also Medical Labels), etc. [94]. Topcoating or treatment of PET makes it printable via different methods including impact, thermal transfer, and flexo-printing. Applications include rating plates for use at high temperatures and exposure to cleaning solutions. Automotive parts labels have to withstand extreme abrasion, extreme temperatures, and exposure to chemicals. They must carry bar codes that remain readable for long periods. Pharmaceutical labels can have a sandwich structure, that is, a secondary removable label attached to the main label. The secondary label is printable via thermal printing. The laminated structure also possesses perforations to allow multiple detachment (i.e., multiplication) of the labels and instructions [95].

10. Medical Labels

As discussed earlier, medical tapes were the first PSPs introduced in the market (see Chapter 1 and Chapter 2). These products have special requirements imposed by their application under sterile conditions and on difficult substrates (see also Medical Tapes). The main requirements for medical tapes are valid for medical labels also. Table 11.3 lists the main functional requirements for medical PSPs.

Medical labels are used for wound care or in the operating rooms. Surgical labels have to adhere well on the skin and not lose adhesiveness in hot or humid environment. No adhesive transfer should occur during removal of the adhesive from the skin. When the label is removed, no residue is left on the product to attract microorganisms [90]. For certain health care applications, pressure-sensitive labels are exposed to sterilization or autoclaving. Gas-sterilizable packages have to be resistant to ethylene oxide. Polyacrylate-based water-soluble PSAs are suggested for medical PSPs (operating tapes, labels, and bioelectrodes) [54]. A fluid permeable adhesive useful for transdermal therapeutic devices applied to human skin for periods up to 24 h is based on an acrylic, urethane or elastomer PSA mixed with a crosslinked polysiloxane. The therapeutic agent passes through the adhesive into the skin. Uninterrupted liquid flow through the adhesive has to occur over a prolonged period (6–36 h) and at constant rate [104]. The skin adhesive has to present an enhanced level of initial adhesion when applied to skin but resist-adhesion buildup over the time. An acrylate-based skin adhesive should have a creep compliance value of at least 1.2×10^{-3} [99]. The earliest medical tapes have been based on mixtures of natural rubber (NR) plasticized and tackified with wood resin derivatives and turpentine and pigmented with zinc

TABLE 11.3
End-Use Requirements for Medical PSPs

Requirement	Product	Reference
Physiologically compatible	Label, tape, electrode	[96]
Body fluid nondegradable	Label, tape, electrode	[96]
No skin irritation	Label, tape, electrode	[97,98]
Skin adhesion (95%)	Label, tape, electrode	[97,98]
Low adhesive transfer to skin (cohesion)	Label, tape, electrode	[90,96–98]
Conformability	Label, tape, electrode	[96, 99]
Moisture insensitivity	Label, tape, electrode	[100,101]
Cold water insoluble, hot water (65°C) soluble adhesive	Label, tape, electrode	[96,101]
Breathability (50 sec/100 cm^3 per in.^{2})	Label, tape	[101,102]
Vapor transmittance (500 g/m^2, 24 h, 38°C)	Label, tape, electrode	[103]
Liquid transmittance (6–36 h)	Label, tape	[104]
Light resistance	Label, tape, electrode	[103]
Heat resistance (350°F)	Label, tape, electrode	[101,103]
Detachable	Label	[105]
Ethylene oxide resistance	Label	[54,106]
Gamma radiation resistance	Label, tape	[106]
Electrical conductivity	Bioelectrode	[54]

oxide. Later, acrylics were suggested for medical tapes [107]. The irritation caused by the removal of the tape was overcome by including in the adhesive certain amine salts. Adhesive balance, stretchiness, and elasticity are required for medical PSAs. Typical examples of conformable carrier materials used for medical labels and tapes include nonwoven fabric, woven fabric, and medium to low tensile modulus plastic films (PE, PVC, PUR, PET, and ethyl cellulose). The preferred carrier materials are those that permit transpiration and perspiration or the passage of tissue or wound exudate. They should have a moisture vapor transmission of at least 500 g/m^2 over 24 h at 38°C (according to ASTM E 96-80), with a humidity differential of at least about 1000 g/m^2. The theoretical aspects of the design of raw materials for medical PSPs were discussed in Chapter 9.

11. Antitheft Labels

Pressure-sensitive tamper-evident labels are widely used for tamper-proof packaging and sealing [108,109]. This market includes labels for traceability and identification, shock damage or temperature change, and security [110]. Generally, permanent labels are tamper-proof. They tear when any attempt is made to remove them. However, for special requirements, tamper-evident seals, security tags, tactile labels, special backgrounds, special (e.g., magnetic) inks, controlled front or reverse delamination, security threads and numbering, holograms, etc. have been used. Hologram labels were developed which need low light [111]. Papers with watermarks, papers that react to UV light or heat, and iridescent papers are used. The choice of an adequate technical solution for tamper evidence is a complex problem.

Security labels for the pharmaceutical industry are used to protect product integrity. Changes in the product could involve indicators about time and temperature, light sensibility, moisture, impact sensibility, contamination, loss of vacuum, etc. Time–temperature indicators change color to reveal an underlying graphic when exposure to a specific temperature exceeds specified time limit [110]. A special ink can be printed flexographically on the label surface. Using a simple testing kit, it can be validated within 2–3 min [112]. Security face stock materials can work by using color change in a peel-away, reverse peel or peel-apart closure version [113]. The peel-away label is a standard one placed across the opening to be closed. Removing it activates the color loss and exposes a hidden message that indicates tampering. Unlike many fragile security materials, it leaves no deposit on the adherend. Reverse-peel labels are used on transparent surfaces as an uniformly colored overlayer. When removed, the label splits apart, losing its color, to reveal the hidden message. With peel-apart closures, the hidden message is incorporated in a laminated pouch or envelope with a transparent overlay. Opening activates the color loss and exposes a hidden message. A pharmaceutical label, which cannot be removed with steam or water from a container once the label is applied, is builtup as a combination of an image-producing self-contained carbonless label overlaminated with an opaque carrier material [114]. According to FDA recommendations, tamper-proof packaging can be achieved by using a packaged product, which is wrapped in a transparent, boldly labeled printed foil, packagings sealed with foil or adhesive tape, or shrink film [115].

In the healthcare, the goal is zero-defect product, that is, zero-defect packaging and documentation. In such products, resins sensitive to visible light (e.g., laser beams) may be used [116]. Safeguards against theft, or improper use, childproofing, and user friendliness are also necessary. Product integrity and safeguard against mishandling and abuse have to be ensured [117]. As a technical solution, nonremovable and nontransferable constructions have been manufactured as antitheft labels. Destructible, computer-imprintable vinyl film has been used for this purpose. The vinyl film fractures when removal is attended [118]. Such a product exhibits tamper-evident performance on chipboard cartons, glass, painted metal, PE, and PP. Security cuts enhance performance. Excellent conformability, good printability, and destructibility are required. Bonding in less than 24 h is necessary [119].

Fragile pressure-sensitive materials either are very thin or have low internal strength. They are coated with a very aggressive PSA [93]. The construction includes a matte top coated transparent film. This film is coated on its back side with a release layer that has moderate adhesion to the film. This release should be printable. A primer is used to achieve good anchorage of the printing ink (graphics in mirror image), to ensure effective splitting during removal. The adhesive film should be slightly narrower than the web width to prevent possible adhesive contamination of the laminating roll. During manufacture, nip pressure should be light and web tension should be adjusted so as to avoid label curling. Adequate die cutting is provided by a uniformly densified paper liner. Making slits or perforations in the face stock materials improves tamper-evident properties [93]. A decade ago, a special method was developed to pemanently bond holograms onto a substrate, and such products can also be used as identification [120].

Generally, these tamper-evident products are made by ordinary direct- or transfer-coating procedure of a face stock material such as film or paper. To achieve tamper-evident properties, special face stock materials are used; currently a polymer film is used, in several cases with a sandwich structure. Such a multilayer construction is used for security tapes also [121].

Unfortunately, the manufacture and coating of special carrier material involve technical and economical problems because of the low tear resistance of these materials. A quite different construction can also be manufactured. One tamper-proof label is based on chrome polyester [118]. On removal the word "Void" appears on both the marked item and the label, preventing their reuse. Another tamper-evident construction includes a release liner and a composite layer coated on the release liner. The upper part of the composite layer is nontacky. It is a film-like layer with low tear resistance. The bottom layer anchored on the face stock layer is a common PSA. The film-forming polymer (FFP) should be manufactured (cast) by the same coating procedure used for the PSA layer, but first the liner should be coated with an aqueous dispersion. This dispersion made face stock layer of the tamper-evident PSA layer is made by using a low flexibility, low tack, medium cohesion polymer on the basis of acrylate, styrene–acrylate, EVAc, with or without filler. To prepare labels that can be dispensed with labeling gun, $6-250\,g/m^2$ coating weights are used.

Advances in radiofrequency identification (RFID) require the use of special labels including the RFID chip. Their printing needs a raster of 1000 l/cm. Unlike common barcodes, such labels are rewritable [122].

12. In-Mold Labels

This special category of labels is applied on injection molded, thermoformed or blow molded containers or soft spreads. These labels are used for toiletries and chemicals. As carrier material, paper and synthetic paper, PP and metallized films, and papers are applied. Label insertion and molding techniques need special know how. In-mold labeling allows the manufacture of the plastic item by injection molding (or other thermoforming procedures) and its labeling in the mold with a special label placed in the mold before the plastic is added. No supplemental postlabeling or postprinting of the finished item is necessary [32]. In-mold labels are not pressure-sensitive. The PSP manufacturer or user has to be informed about this technology because it is competitive with pressure-sensitive technology, and the manufacture of in-mold adhesives is a hot-melt formulation. For instance, for such labels, paper of $100-120$ g/m^2 is printed, overlacquered, and adhesive coated ($15-20$ g/m^2) with a hot melt [123]. Now film insert molding can add a soft touch-soft feel surface using a flexible thermoplastic polyurethane (TPU) film. Aliphatic TPU films can be colored in bright colors [124].

13. Other Labels

Decals used for clear overlays have to possess an adhesive that is water white before and after drying [116]. For advertising decals weather resistance is required. These products are applied on buses, trucks, and trains and have to possess chemical resistance to washing agents also [125]. Baggage tags for airlines, promotion labels, tickets, and address labels are other label applications that have been developed [126].

III. TAPES AND THEIR APPLICATION

Tapes are web-like PSPs used to provide structural integrity, dimensional stability, shape retention, bonding, or another particular property (thermal, sound, chemical, or electrical insulation, etc.). In comparison with labels (where the role of the nonadhesive, coated-carrier surface is more important for long-term end-use, and the adhesive has to allow primarily only such instantaneous adhesion that ensures the contact between label and labeled item), for tapes, the mechanical resistance of the carrier is the most important parameter, and the adhesive acts as an assembly tool between carrier and the taped product. Tapes function as an adhesive in film form, a prefabricated glue. As a glue in web form, on a solid-state carrier, they exhibit an exactly regulated thickness and allow fast processing. Taping requires only low energy use. No clean up is required and no polluting is produced.

End-use of tapes differs in various regions or countries. In the U.S.A. (1989), the most important tape qualities have been packaging, industrial, healthcare, and consumer tapes. In Japan (1987), the main application fields have been packaging, masking, electrical, application, stationery, and protective tapes [127]. In Europe (1987), the leading tape applications have been packaging, masking, anticorrosion, insulating, double-faced, hygienic, and reinforced products [128]. According to Roder [129], in the 1980s, the main end-uses for double-faced tapes were splicing, laminating, and mounting. The following end-use fields have been considered the most important sectors: packaging/cardboard (70%); industrial packaging (20%), and household/office (10%). In the last decade, industrial packaging became the most important end-use for tapes.

A. APPLICATION CONDITIONS

The tape market area includes packaging, electrical, industrial, surgical, masking, and consumer goods [130]. Most tapes are web-like fixing elements applied manually or with special winding machines. According to Schroeder [131], the winding up of tapes ensures the pressure necessary to adhere by lamination. It should be noted that sometimes ago, to achieve better wetting out and increase the contact surface, tapes were also applied by pressure and solvents. Product surface, geometry, and application climate influence the end-use of tapes. It is evident that the application conditions for tackified adhesive-coated packaging tapes must differ from those for low tack adhesive-coated or adhesiveless masking tapes. Soft, conformable, carrier-less tapes need only low application forces. For medical tapes, deapplication is more important than application. Such examples illustrate the variable application conditions for tapes as a function of their end-use.

1. Influence of the Product Surface on Tape Application

Application of PSPs generally consists of both delamination and lamination. The PSP is delaminated and builds up a new laminate with the adherend substrate. For labels, the surface quality of the face stock material does not influence the application procedure. Labels are applied by delaminating a multiweb composite. The adhesive coated surface (i.e., the PSA) and the release coated surface (i.e., the liner) are the work surfaces in this process. For tapes, delaminating consists of unwinding. The quality of the nonadhesive, coated surface of the carrier material is determinative for the delamination. This surface together with the adhesive-coated surface forms the working zone in delamination of the monoweb. After delaminating, both the label and the tape are laminated on a new substrate whose surface characteristics affect the bonding.

Solid-state components with various surface qualities are used for tapes. Tapes for rubber profiles in automobiles have to adhere on difficult surfaces like ethylene–propylene–diene multipolymer (EPDM), chloroprene rubber (CR), styrene butadiene rubber (SBR), according to VW specification TL 52018 [132]. EPDM foams are difficult to tape, they give only 2.4–3.6 N/25 mm peel resistance, compared with foam failure with PE or PUR foams [133]. Wire-wound tapes have to adhere to modified flame- and heat-resistant EPDM formulations [134]. EPDM roof systems use PSA tape for long time performance. Pipe insulation tapes should adhere also on oiled surfaces [135]. Pipe wrapping tapes have to adhere to metal surfaces, wood, rubber, or ceramics. Repair tapes have to adhere to metal surfaces, wood, rubber, or ceramics [90]. A double-layered pressure-sensitive tape intended for industrial and commercial use has to bond to glass, concrete, steel, ceramic drywall, wood, tarpaulins, and covers [136]. Butt-splicing tapes has to adhere on PP, PE, PA, and coated paper [90]. Medical tapes must adhere to skin [137]. Diaper closure tapes are used as refastenable closure systems for disposable diapers, incontinence garments, and similar items [138]. Such systems allow reliable closure and refastening from an embossed, corona-treated PE surface, as is typically used for diaper cover sheet. Fluorescent adhesive tapes for highlighting have to be removable from written surfaces. Such tapes are used on sheets bearing writing. These tapes are manufactured coating plastic films with fluorescent inks and adhesives with low adhesion [139].

Mounting tapes must adhere to glass, concrete, steel, ceramic, drywall, wood, tarpaulins, and covers [141]. Silicone mounting tapes adhere to a variety of surfaces such as metals, glass, paper, fabric, plastic, silicone rubber, silicone varnished glass cloth, and silicone/glass laminates [140]. Foam-like transfer tapes are used for structural bonding of aluminum or copper–titanium–zinc roof panels [131]. When tapes are delaminated from anodized aluminum, cohesive break may produce adhesive deposits [142]. Cables are coated with coextruded 28% vinyl acetate (VAc) [143], chlorinated PE, acrylonitrile–butadiene rubber, liquid chlorinated paraffin, and fire-retardant fillers (e.g., hydrated alumina, $CaCO_3$, zinc borate, and Sb_2O_3) have been compounded and formed into oil resistant tapes (0.7 mm thick) for electrical wires and cables [144]. Insulation tapes have to adhere to this material. It is evident that various application conditions require different adhesion performance characteristics. Table 11.4 lists some typical peel values for different tapes; Table 11.5 illustrates the broad range of shear resistance values for tapes with various applications.

TABLE 11.4
Peel Resistance Values for Selected Tapes

Tape Grade	Adhesive	Carrier	Peel Resistance	Reference
Packaging	HMPSA, SEBS	—	800 g/25 mm	[145]
Packaging	HMPSA, AC	—	570 g/10 mm	[146]
Packaging	HMPSA, SIS	—	720 g/82 mm	[147]
Packaging	AC, crosslinked	—	925 g/25 mm	[148]
Packaging	RR	PP	640 g/15 mm	[149]
Packaging	RR	—	280 g/10 mm	[150]
Packaging	AC, water-based	—	1450 g/25 mm	[151]
Packaging	AC, crosslinked	—	1200 g/25 mm	[152]
Packaging	AC, crosslinked	—	1000 g/25 mm	[153]
Removable	RR	—	40 g/10 mm	[154]
Removable	AC, EB-crosslinked	—	410 g/25 mm	[155]
Removable	—	—	200 g/25 mm	[156]
Removable	AC, crosslinked	PE	200 g/25 mm	[157]
Double-side coated	AC, water-based	PET	1300 g/20 mm	[158]
Temperature resistant	Silicone	Al	3500 g/25 mm	[159]
Temperature resistant	Silicone	PET	460 g/10 mm	[160]
Temperature resistant	Fluoropolymer	—	45 g/20 mm	[161]
Insulating	HMPSA, SBC	—	370 g/15 mm	[162]
Masking	AC, water-based	Foam	300 g/25 mm	[163]
Masking	AC	PET	120 g/20 mm	[164]
Sealing	AC, UV-cured	—	1640 g/10 mm	[165]
Medical	RR, solvent-based	PET	5 lbs/in.	[166]

2. Application Climate

The application temperature and atmospheric humidity during lamination and end-use of the tapes may vary widely [167,168,170]. Special conditions are given for medical, insulating, and sealing tapes. For instance, butt-splicing tapes have to resist up to 117°C [90]. Temperature-resistant silicone mounting tapes maintain their properties up to 288°C. They must exhibit good resistance to chemicals, as well as to fungus, moisture, and weathering. Insulation tapes have to resist temperatures between 5 and 60°C [171]. Various tapes used in automotive industry are often subjected to

TABLE 11.5
Shear Properties of Selected Tapes

Code	Adhesive	Manufacturing Method	Shear Resistance	Reference
Transfer tape	AC	Mechanical foaming of AC emulsion, impregnating, curing	4.5 kg/cm	[177]
Transfer tape	AC	AC oligomer, UV curing	>24 h	[178]
Insulating tape	Silicone	Coating	2.0 mm	[169]
Insulating tape	AC	Coating	48 h, 70°C	[179]
Packaging tape	RR, SB	Coating	>4 h	[180]
Packaging tape	AC	Coating	5 h	[156]
Packaging tape	NR	Coating	2.6 mm, 2 h	[159]

extremely high temperatures [172,173]. In a closed automobile in the sun, temperature can reach 100–120°C. Car undercarriage protection and insulation tapes have to resist 20 h at 150°C [174]. Their resistance to water or solvents is tested by immersion in the liquid for periods up to 1 week. Generally tapes should be applied between 20 and 30°C [143]. Products designed for short-term outdoor application should not be used for more than 3 days under such conditions. If the adhesive is to be used in a bathroom (e.g., for adhering soap dishes and utility racks to ceramic tiles), the adhesive have to adhere well to tile, have high cohesive strength, withstand temperatures in the 110–120°F range, and endure 100% relative humidity. Because of the extreme working conditions of several tapes, their test methods use elevated temperatures also (Table 11.6).

B. APPLICATION METHOD

Tapes are applied manually or with a special apparatus. One-side coated tapes are unwound; double-side coated tapes are delaminated. For transfer tapes, the adhesive is bonded to the sheet backing, which is a release liner, so that the exposed adhesive surface of this tape is placed in contact with the desired surface, the release liner is stripped away, and the newly exposed adhesive surface is bonded to a second substrate. Self-adhesive tear tape designed as an easy-opening device for flexible packaging can be used with a motorized applicator [175]. Sealing tapes (without carrier) based on tackified butyl rubber have been applied with extruder [176]. Butt-splicing tapes have high quick stick for high speed splicing applications [90]. According to Jahn [176], quick stick and high temperature shear are generally required for PSAs for tapes to allow their application. According to Ref. [43], packaging tapes need more shear resistance than labels. Low shear values are sufficient for office tapes, double-faced tapes, and foam mounting tapes. For closing PE bags or nonwoven envelopes with a surface tape, removability and reclosability is required [178,179]. Delamination conditions may also be various. For instance, radiation-cured automotive tapes have to present an initial breakaway peel, and then the force needed to continue the breaking of the bond; the initial continuing peel is measured (see also Chapter 7). Initial continuing peel is lower (by about 30%) than breakaway peel [180].

Labels are discrete elements during lamination. Their transformation from continuous web form to a discrete label is carried out before labeling. Tapes are generally web-like products

TABLE 11.6
Aging Conditions for the Test of Adhesive Properties of PSPs

| Product | Adhesive | Aging Conditions | | | Reference |
		Evaluation Criteria	Storage Temperature (°C)	Storage Time	
Tape	AC, UV-curable	Retention of adhesive properties	70	3 weeks	[175]
Tape	AC, water-based	Peel	40	60 days	[159]
Tape, cable	Polybutadiene, SBC	Peel, RB, creep	−5		[163]
			3		
			40		
			70	30 days	
Tape	AC, crosslinked	Retention of adhesive properties	80	—	[153]
Tape, masking	AC, water-based	Adhesion transfer	80	30 days	[164]
Sheet, removable	AC	Peel	60	100 h	[157]
Protective sheet	AC, crosslinked	Peel	60	3 days	[158]
Release liner	AC/SBR	Peel	70	28 days	[176]

during their application. Their transformation to finite webs is carried out during final lamination or after lamination. For instance, cuttability with hot knife and easy unwinding are required for bag closure tapes [140]. Easy-tear tapes or certain masking tapes need low mechanical resistance to be torn by hand. In contrast, transfer tapes are die cut for some applications [128].

Rolldown properties of tapes are very important. Such characteristics depend on the release properties of the back of the carrier. As discussed in Chapter 10, there are various possible manufacturing ways to improve these properties. For instance, biaxially oriented multilayered PP films have been manufactured for adhesive-coated tapes. Generally, to prepare an adhesive tape which can easily be drawn from a roll without requiring an additional coating on the reverse side, at least two different layers having different compositions are coextruded and the thin back layer (having a thickness ratio of one third) contains an antiadhesive component. Tapes with an olefinic carrier material are noisy during unwinding. According to Galli [181], modification of the adhesive by addition of a mineral oil reduces the unwinding noise level. Further improvement can be achieved by using a release coat or special treatment of the carrier back side. Polyvinyl carbamate or polyvinyl behenate can be used as release layer. It is supposed that the noise level is related to the mechanical properties and surface characteristics of the carrier material, to adhesive/abhesive properties, and to unrolling conditions. It has been demonstrated [182] that improving the surface tension (i.e., increasing it to more than 33 dyn/cm) reduces the noise level.

1. Application Equipment

Generally, packaging tapes are applied with the aid of machines. Special (medical, application, etc.) tapes are used manually. Foam-like transfer tapes are applied under pressure [141]. A hydraulic press is used for bonding of aluminum panels. Sealing tapes based on PVC are applied by hand or with a sealing machine [183]. Special application apparatus is suggested for folded tapes. Folding along the longitudinal axis is proposed to prevent the tapes from rewinding [184]. Thermal release tapes allow deapplication by using of heat [185] (see postapplication crosslinking).

C. Main Tapes

The range of tape-like products is wide. They can be classified according to their construction or their use (see also Chapter 2). Carrier-less tapes and paper, film, foam-based tapes, uncoated or one side- and double side-coated products are manufactured. Friction tapes, splicing tapes, and many other types are produced [49]. The classic type of tape used for fixing and fastening is known as packaging tape.

1. Packaging Tapes

The classic tape construction, a mechanically resistant nondeformable but flexible carrier material coated with high coating weight of an aggressive but inexpensive adhesive, is valid for packaging tapes without special requirements. The high processing speed for these tapes requires the decrease of the noise level. Low noise, color-printable, PP-based packaging tapes have been developed [86]. Packaging tapes are applied either by hand or by automatic packaging machines for the fastening of goods. Generally, packaging tapes are based on PVC or PP, and they may also have a reinforcing layer. Such products are some shade of brown [6]. Kraft paper, cloth, and oriented plastic carrier material are also used. Packaging tapes accounted for 66% of European tape consumption in 1987 [128]. Although their volume is increasing, their proportion of the global PSP production is strongly decreasing.

Special packaging tape constructions are also known. A packaging tape described in one patent [186] is composed of a flexible member and a central tear section. The central tear away portion is built up with two parallel lines of perforations extending along the entire length of the flexible part.

Adhesive tapes with the adhesive-coated area are narrower than the substrate and with one edge with an adhesive-free strip have an easy-untie property and are useful for bundling electronic parts, building materials, vegetables, and other items [187].

For packaging tapes on hot-melt pressure-sensitive adhesive (HMPSA) basis, adhesion on cardboard is of special importance [188]. The preferred adhesion value is 100 min. A flap test is also carried out; this is a combination of shear and peel resistance and tack. According to Ref. [189], for such tapes, one of the most important properties is open face aging.

Common HMPSA systems for tapes are based on SIS, a resin that is partially compatible with the midblock of the SBC and a plasticizer (oil). Generally, for such tapes, there is no need for a very light color, a color number of six for the resin is adequate [189]. The hot-melts for packaging tape are coated on prereleased biaxially oriented polypropylene (BOPP) $(21-22 \, g/cm^2)$. According to Ref. [188], for packaging tapes, a peel value of 18 N/25 mm (on steel), loop tack values of 20–30 N/25 mm, and rolling ball values below 5 cm are preferred. Typical SIS-based tape PSA performance characteristics are discussed in Ref. [190]. For an 1100/100/10/2 resin/SIS/oil/antioxidant composition, rolling ball tack values of 11 cm, polyken tack of 1400 g, loop tack of 16 N/25 mm, and peel value of 18 N/25 mm (on stainless steel) have been obtained. Packaging tapes need shear values higher than 680 min [42].

2. Tapes for Corrugated Board

Tapes for corrugated cardboard use PP as carrier material and have a silicone release coating [79]. Their application temperature is situated between -40 and $+100°F$ [191]. Such products must have an excellent tack and shear on cardboard. Although at normal temperatures rubber-resin adhesives have better shear resistance than waterborne acrylics, acrylic adhesives display superior performances at 120°F. There is a trend on the market to provide customers with printed packaging tapes. The text or graphics are imprinted on the nonadhesive side of the carrier material before the tape is made.

3. Box Closure-Seal Tapes

Tapes on cellulose hydrate or PVC as carrier material have been designed for box closures [192]. These are less heat and aging resistant and less tamper-proof than wet tapes [182]. According to UN Recommendation R 001 [193], packages of dangerous goods should be closed with tapes. Such tapes are so-called security tapes and have to ensure antitheft protection, drop destroying, and tear resistance. Reinforced wet glue tapes display 4% elongation, whereas nonreinforced plastic tapes 40–70% [194]. When compared with wet tapes, PSA tapes are more sensitive to dirt. Wet tapes do not need an unwinding force. These characteristics were the main advantages that allowed a fast exclusive application of wet tapes for security closures some years ago. Sealing tapes based on PVC carrier are printable. Sandwich printing (flexo printing on the top side, and gravure printing for the back side between the film and the adhesive) is also possible. Such tapes are applied by hand or by using a sealing machine. These products are used for closing, manufacturer identification, antitheft protection, copyright protection, quality assurance, advertising, information, classification, and administration [183].

4. Diaper Closure Tapes

Special pressure-sensitive closure tapes are used for diapers [195]. Adhesives for diaper tapes have to display high shear [196–199]. Such systems have to allow a reliable closure and refastenability. Tack values of 16–24 N/25 mm are acceptable. These tapes have to exhibit a maximum of the peel force at a peel rate between 10 and 400 cm/min and a log peel rate between 1.0 and 2.6 cm/min (see also Chapter 7). Tapes exhibiting these values have been found to be strongly preferred by consumers.

5. Bag Lip Tapes

Double-side-coated PE tape has been designed for PE bag manufacturers [130]. The tape is applied to the lip of the bag. Bag closure tape has to be cut with the same hot knife (hot wire) that is used to form the bag. It should be easily unwound for automatic applications [86]. Tapes for closing PE bags, like those used for nonwoven envelopes, must be reclosable. A clear PE carrier material is suggested [178]. A closure tape based on crosslinked acrylate terpolymer uses a reinforced web and a reinforcing filler. Such adhesive is useful for bonding the edges of a heat recoverable sheet to one another to secure closure [200].

6. Tear Tapes

Self-adhesive tear tapes are designed to be used as an easy-opening device for flexible packaging. They can be used to remove the top of a pack or to form a lid or in applications where a heavy duty opening system is required. These tapes provide tamper evidence indicating when the packaging is intact. They are effective on overwrap, shrink sleeves, flow packs, and blister packs for foodstuffs and pharmaceutical products. Some are provided with holograms, and some combine easy opening, tamper evidence, and inherent security using a motorized applicator [175]. An adhesive tape used for fastening printer paper together is perforated along the centerline in the long direction to allow easy tear [201]. Easy-tear, breakable tapes allow easy tearing or splitting of the tape during handling. Easy-tear breakable hank tapes are used in the same finished product as easy-tear paper tapes. The tape is hand applied and is used for hanking and spot tape when fast easy removal is required. There is a difference between the tear behavior of paper-based and film-based tapes. Although a film-based tear tape can be used in the same finished product as the paper tape, its physical properties, application method, and resultant assembly are different. Because of the higher tear resistance of film-based tear tapes, care must be taken to ensure a maximum of two laps [202]. Tapes for easy handling comprise continuous film-adhesive, layer-perforated film laminate [203].

7. Freezer Tapes

Like freezer labels, freezer tapes are used at low temperatures (see also Freezer Labels). They are applied as packaging tapes. They can be coated with an adhesive based on natural rubber, SBR, regenerated rubber, rosin ester, or polyether [204]. They are based on primed soft PVC and have to resist temperatures between -29 and $+113°C$.

8. Pharmaceutical Sealing Tapes (Antitheft Tapes)

Some years ago, wet gummed products were applied almost exclusively as sealing tapes for pharmaceutical use. Now PSA tapes are applied (see Antitheft labels) [194]. Such tapes are used as security closures for pharmaceuticals.

9. Repair Tapes

Repair tapes are flexible plastic tapes reinforced with nylon cord and have to adhere to metallic surfaces, wood, rubber, or ceramics. They are designed to offer puncture and tear resistance [90]. Such tapes are supplied as continuous web or die-cut finite elements. They have to be conformable and provide excellent bonding.

10. Insulating Tapes

There are insulating tapes used for a variety of applications, mainly for electrical or heat insulation or as a sealant. They must have excellent mechanical, electrical, and thermal properties. Wire-wound tapes, electrically conductive or electrically insulating tapes, and thermally insulating tapes are the most important representatives of this product class.

11. Wire-Wound Tapes (Electrical Insulation Tapes)

Electrical insulation tapes are used for taping generator motors and coils and transformers, where the tape serves as overwrap, layer insulation, or connection and lead-in tape [205]. The carrier is woven glass cloth impregnated with a high temperature-resistant polyester resin or woven polyester–glass cloth impregnated with polyester resin. The latter variant is used where conformability is required. Oriented PP film ($30-50$ μm) is used for tapes and oriented LDPE ($80-110$ μm) for insulation tapes [206]. Such nonflammable tapes can have an adhesive with $0-40\%$ polychloroprene rubber [207]. An insulating tape with an acetate carrier is used for wire winding by telephone manufacturers [208]. It resists up to $130°C$ [209]. It has to adhere to modified flame- and heat- resistant EPDM formulations [210]. High temperature-resistant insulating tapes on a PET basis can be used at $180°C$ and an electrical tension of 5000 V [209]. Cable tapes have to resist 1 month at $70°C$ [211]. Fixing, transfer, carpet, and electrical insulating tapes are coated with an acrylic PSA [212]. Certain electrical insulation tapes have to resist transformer oil; cross-linked acrylics are used for such products [213]. Electrical tapes must be approved by Underwriter Laboratories [214] and require elongation, shear, and peel resistance. Shear resistance should be measured after solvent exposure also. Electrical tapes have to possess high dielectric strength and good thermal dissipation properties. The edges of electrical insulating tapes used for taping must be straight and unbroken to provide distinct boundaries to the finished areas when the taping is complete [215]. Such tapes may be perforated.

12. Electrically Conductive Tapes

The principles of formulation of electrically conductive PSPs are discussed in Chapter 3, Chapter 5, Chapter 7, and Chapter 8. Generally, electrical conductivity is achieved using special filler. Such adhesives containing so-called extrinsic conductive polymers may exhibit anisotropic conductivity. Tapes with adequate electrical conductivity are made using a metal carrier (nickel foil) and an adhesive having nickel particles as filler ($3-20\%$) in an acrylic emulsion [216]. Electrically conductive adhesive tapes have been manufactured by coating an acrylic emulsion containing $3-20\%$ nickel particles onto a rough flexible material (e.g., Ni foil). Such products can contain silicium carbide as filler also [217]. Silver-coated glass powder (20%) or glass microbubbles (33%) as filler lead to a specific surface conductivity of $1.1 \times 10^{-11}-7.6 \times 10^{-5}$ G [218]. The adhesive may include a metallic network to ensure electrical conductivity [219]. Transparent electroconductive films have to be able for smooth writing demanded by pen input screens and touch panels. They have good transparency and surface conductivity [220].

13. Thermal Insulation Tapes

Double-side-coated foam tapes are designed for industrial gasketing application. They are based on closed-cell PE foams, coated with a high tack medium shear adhesive [221]. Pipe insulation tape is used for gas pipelines [222]. Pipe wrapping tapes are plastic tapes reinforced with nylon cord and have to adhere to metallic surfaces, wood, rubber, or ceramics. They are designed to offer puncture and tear resistance [90]. Pipe insulation tapes should also adhere to oiled surfaces [136]. An aluminum carrier and a tackified carboxylated butadiene rubber coated with 40 g/m^2 PSA are used for insulating tape in climatechnics [223]. Tapes applied as anticorrosion protection for steel pipes contain butyl rubber, crosslinked butyl rubber, regenerated rubber, tackifier (polybutene or resins), filler, and antioxidants [224]. Such adhesives are coated on a nonwoven carrier also. Air-seal tape is made with butyl rubber on ethylene–propylene foam. Insulating tapes for gas and oil pipelines are manufactured by coextrusion of butyl rubber and a PE carrier. There are also foamable insulation tapes. For these products, PUR foam is bonded *in situ* onto the (specially treated) carrier back [225].

Thermally conductive pressure-sensitive insulation tapes are manufactured as anticorrosion protection tapes also [226]. Pipe insulation tapes are wounded by overlapping (50%) using special machines [209]. Bilayered, heat-shrinkable insulating tapes for anticorrosion protection of petroleum and gas pipelines have been manufactured from photochemically cured LDPE with an EVAc copolymer as an adhesive sublayer [227].The liner dimensions of the two-layered insulating tape decrease by 10–50%, depending on the degree of curing (application temperature 180°C). The application conditions and physicomechanical properties of the coating were determined for heat-shrinkable tapes with 5% shrinkage, at a curing degree of 30%, and shrinkage force of 0.07 MPa.

Special insulating tapes have to adhere to refrigerated surfaces. A self-sticking tape for thermal insulation, used to secure a tight connection between heat exchange members (metal foils) and polyurethane foam-based thermal insulation materials, is prepared by applying an adhesive that is able to adhere to refrigerated surfaces on an olefin copolymer carrier material (containing release) that is bonded to the polyurethane [228]. Low temperature-curable silicone, PSAs have been formulated from gum-like silicones containing alkenyl groups, a tackifying silicone resin, a curing agent, and a platinum catalyst [229]. Such adhesives for heat-resistant aluminum tapes are coated with a thickness of 50 μm and cure in 5 min at 80°C.

14. Foam Tapes

Classic foam tapes are foam-like PSPs with a foamed plastic carrier that is coated with a PSA. Carrier-less foam-like tapes also exist. Such products are plastic foams impregnated with PSA or made from a foamed PSA. The foam carrier provides excellent conformability and stress distribution. In some applications, the PSA is coated first on a flexible PUR, acrylic, or other foam and then laminated onto the carrier surface. The opposite surface of the foam may also be provided with a PSA layer. Foam sealing tapes are made of a foam sealing tape layer and an interlayer strip rolled up together in a compressed form [230]. For such tapes, there are standard thicknesses, roll lengths, and ranges of width given by the supplier [231]. Special foam mounting tapes are used for bonding of soft printing plates for flexo printing [232]. Foam tapes can compensate for tolerances in the printing process and assure that dot gain is minimized [233].

Automotive insert tapes are used as sealants. Such products are tapes based on a foam carrier and have to exhibit aging stability, weatherabilty, sealing, and nonflammability. The tape roll is compressed before use (it is supplied in compressed one-fifth thickness). Double side-coated tapes with neoprene, PVC, PUR, or PE foam carrier have been developed [233]. When used as mounting tapes, such products have to satisfy requirements for stress relaxation, stress distribution, mechanical resistance, sealing and anticorrosion, and reinforcing and vibration/noise damping.

15. Mounting Tapes

Mounting tapes have been developed for a number of applications. They can be used for the temporary or permanent mounting of parts of an assembly (e.g., fixing films in cartridge, mounting of glass panes, windowpanes, etc.) [234]. Temperature-resistant silicone mounting tapes maintain their performance characteristics up to 288°C and adhere to a variety of surfaces including metals, glass, paper, fabric, plastic, silicone rubber, silicone varnished glass cloth, and silicone glass laminates. They must exhibit good electrical properties and good resistance to chemicals, as well as to fungus, moisture, and weathering [235].

Double-faced mounting tapes can be manufactured as double-side-coated carrier-based products or carrier-less, foam-like webs. Certain double-faced mounting tapes are built up with PET as the carrier material. They have a silicone adhesive on one side and a silicone release layer on the other. They are used in electronics [236].

Mounting tapes used in buildings must have the same expectancy as the building elements (i.e., 10 yr) [237]. Such mounting tapes can replace other classic (mechanical) methods of fixing [238].

For instance, a glass-mounting tape is specially designed to bond two sheets of glass around the edges for a composite glass with a strengthening layer. The old method used black butyl tape to create a layer of air between two sheets [239]. Such tapes give a long-lasting waterproof seal. Another double-coated adhesive tape for mounting provides a cleaner watertight system and cuts labor costs. One side of the tape is applied to the window frame, and the glass pane is pressed against the other side of the tape. The pane is thus held securely when laths are nailed to the surrounding frame [240]. In a first step of mounting, one-side coated PE pressure-sensitive tape is applied [241]. Pressure-sensitive sealant tapes are soft, and their PSAs are tackified or plasticized formulations. They can be filled and crosslinked. Their quick stick values are low. Their peel resistance value (on aluminum) is about 120–140 oz/in.

Special sealant tapes for gasket type application are made by extrusion onto release film. Such tapes must possess high tack and be resistant to gasoline.

Tissue tapes are applied to temporarily hold seat upholstery together. The double-faced tape holds leather in place prior to stitching and eliminates problems of leather creasing or slipping [240]. Pressure-sensitive tapes can be used to bond exterior trim to automobiles [242].

Certain mounting tapes have to be repositionable [243]. A special mounting tape for instrument housing assembly is designed like a tear type with a transfer adhesive [244]. The tape consists of a metallic foil strip of predetermined thickness and an acrylic transfer adhesive strip affixed to one side of the metallic foil and forming a tear band. Mounting tapes are used in fashion and textile industry also for attaching textile parts to one another. They are protected from telescoping during storage against with a woven on nonwoven supplementary layer [245]. Fasteners are used in attaching automobile seat covers [171]. Certain mounting operations require high peel removable adhesives. In such cases, postcrosslinking can be carried out. Postapplication crosslinking is proposed to achieve easier delamination. According to Charbonneau and Groff [246–248], the procedure allows delicate electronic components to be removed from the tape, even though they were nonremovable before crosslinking.

Adhesives with excellent water resistance (resistance to humidity and condensed water) are required for technical tapes and pressure-sensitive assembly parts in the automotive industry [233]. Shear values of about 600–1000 min (25 × 25 mm, 1000 g, on steel) and peel values of 25–35 N/25 mm (20 min, on steel) are required. High-shear, low-tack applications are suggested for certain mounting tapes and double-coated foam tapes [50]. For these mounting tapes, the target values of (180°, instantaneous) peel, loop tack, and 72°F shear are 34.5/6.5 N/25 mm and 400 min. High coating weight is used. For mounting tapes, a 88 μm thick adhesive layer has been proposed [50]. Peel was tested on steel and PC. A formulation of a PSA based on tackified SBS for double-faced carpet tape having 220 parts resin and 20 parts oil to 100 parts block copolymer (and a high coating weight of 40 g/m^2) exhibits relatively low adhesive characteristics (a peel strength of 16–21 N/25 mm), loop tack of 17–18 N/25 mm, rolling ball tack of 3–5 cm, and shear on steel of more than 100 h [189].

Foam-backed adhesive tapes are commonly used to adhere an article to a substrate. The foam carrier is pigmented with carbon black to camouflage the presence of the tape [249]. Foam tapes used as structural bonding elements exhibit the following advantages [141]: (i) they do not damage the substrate; (ii) no mechanical processing operations (drilling, screwing, riveting, etc.) are needed; (iii) they give better and more uniform stress distribution; (iv) they are lightweight; (v) they produce a water-tight seal; and (vi) they protect against corrosion. They might be used, for example, in the design of a thermal break. As defined in Ref. [141], a thermal break prevents differences in temperature within a structure, reducing stresses that arise from different rates of expansion. Self-sticking, double-faced foam tapes should also be fire and mildew resistant [45]. According to Ref. [152], there are two main categories of double-faced tapes: double-side coated tapes with carrier and double-side coated tapes without carrier (foam-like tapes).

A foam-like pressure-sensitive tape is manufactured according to Vesley [249] using glass microbubbles as filler. Dark glass microbubbles are embedded in a pigmented adhesive matrix.

Transparent microbubbles are included in the composition of a foam-like tape [250]. Pressure-sensitive tapes filled with glass microbubbles have a foam-like appearance and character and are useful for purposes previously requiring a foam-backed PSA tape. The average diameter of the microbubbles should be between 5 and 200 μm. The thickness of the pressure-sensitive layer should exceed three times the average diameter of the microbubbles, to enhance the flow of the bubbles in the matrix under applied pressure. This enables intimate contact to be built up with rough and uneven surfaces, while retaining the foam-like character. Optimum performance is attained if the thickness of the PSA layer exceeds seven times the average diameter of the bubbles.

Plotter films are used as chablones that are cut from the film material. The negative letter signs are applied on the substrate. Thus, spray lacquering of the chablones is eliminated. The decorative and informational elements are placed and temporarily fixed with application tapes. Such tapes can be considered as temporary mounting tapes also.

16. Transfer Tapes

Transfer tapes are tapes without carrier. They have a solid-state component only temporarily. The sheet backing is a release liner, and in use, the exposed adhesive surface of this tape is placed in contact with a desired substrate, the release liner stripped away, and the newly exposed adhesive surface bonded to the second surface. Their construction includes a release liner and the adhesive core. The adhesive core may be a continuous homogenous adhesive layer or a semicontinuous heterogeneous adhesive layer. The heterogeneity of the adhesive layer in the latter case is due to embedded solid-state, liquid, or gaseous particles (holes), that is, the adhesive layer can be a foam [141].

Double-faced tapes were used some years ago only for low-stress applications. Cohesive, high-resistance transfer tapes have been developed about a decade ago [251]. They can be used for structural bonding. Transfer tapes for structural and semistructural bonding have the following properties:

1. High temperature resistance (does not depend on the carrier)
2. Good environmental resistance
3. Conformability to the substrate shape
4. High transparency (for homogeneous tapes)
5. Low and constant adhesive thickness and variable adhesive thickness

According to the type of adhesive, carrier-less tapes may be classified as (i) tapes having an acrylic film or (ii) tapes having an adhesive foam core.

Traditional foam tapes display the disadvantage that the foam can split from the carrier (adhesive break) at too high peel force. Foam-like transfer tapes have the advantages of (i) softness, thickness, and elasticity of the foam which allow reliable bonding of uneven and textured surfaces; (ii) the ability to dampen sound and vibration; and (iii) resistance to temperatures of 90°C in the long term and 150°C in the short term.

The transfer process of the tape from the temporary liner can be regulated by means of the release, but it can also be controlled by using different adhesives or adhesives having different degrees of crosslinking. The radiation-curable compositions and curing procedure suggested in Refs. [246,247] can also be used for transfer tapes. The carrier can also be foam coated on either one side or both sides and having an overall thickness of 0.1–2.0 mm [250]. Both surfaces of the carrier may have low adhesion coatings, one of which is more effective than the other. When the tape is used as a transfer tape, when it is unwound, the adhesive layer remains wholly adhered to the higher adhesion surface from which it can be subsequently removed. To enhance immediate adhesion to rough and uneven surfaces, a resilient foam backing can be used. Glass microbubbles can be incorporated also in order to enhance immediate adhesion to rough and uneven surfaces.

For such tapes, both a breakaway cleavage peel value and a continuing cleavage peel value have been measured. The continuing cleavage peel value is about 50–60% of the initial value. The cohesive (tensile) strength of the crosslinked adhesive (transferred core) may attain 6000 kPa (depending on the crosslinking conditions).

Transfer tapes with contoured adhesive have been also proposed. According to Ref. [252], an uneven adhesive surface can be achieved by transferring adhesive during rewinding. This tape has a carrier with recesses on one side that are filled with adhesive. When the tape is unwound, the adhesive transfers from the recesses to the carrier. The use of EVAc copolymers for carrier-less, self-adhesive films (SAFs) (tapes) used to bond roofing insulation and laminate dissimilar materials has been proposed in Ref. [253]. Such formulations contain EVAc–polyolefin blends, rosin, waxes, and antioxidant. They are processed as hot-melt and cast as 1–4 mm films (application temperature 180°C). The adhesive strength of such tapes decreases by 10–45% in 1 yr.

17. Splicing Tapes

Splicing tapes have been used in paper, converting, and printing industries. Papermaking and printing technology require splicing the end of one roll of paper to the beginning of another, as well as splicing parts of the roll after defective material has been cut out. Such splices have to be made quickly and easily with an adhesive that rapidly attains maximum strength [254].

The adhesive and end-use requirements for splicing tapes are very severe. High tack and fast grabbing are required for high speed flying splicing. The tapes have to exhibit adhesion to various papers, liner board, foils, and films as well as high-shear resistance and temperature resistance while traveling through drying oven. The application time (i.e., bonding time) for splicing tapes is very short (less than 60 sec according to Ref. [255]). Reel changeover follows in less than 15 min. The joint length is less than 100 mm. The most difficult cases are characterized by high running speed, small reel diameter (for the next reel), maximum thickness of the web for the new, small diameter reel, and short distance between splicing point and the cutting line. High tack and fast grabbing are required for high speed flying splicing.

Butt-splicing tapes, built up as a single-coated PET, have to adhere on PP, PE, PA, and coated paper, resist up to 117°C, and possess high quick stick for high speed splicing applications [90]. For instance, the splice for flame-resistant tapes shall be constructed in such a fashion that the roll can be unwound at a rate of 7.3 m/min [256]. Splicing tapes for the paper manufacturing have to be recyclable [257] and resist temperatures up to 220–240°C. The paper used as carrier material has to be dispersible, and the adhesive must be water soluble, but resistant to organic solvents. No migration is allowed. The adhesive has to bond at a running speed of 160 m/min. Such formulations must adhere to paper substrate running with 1200 m/min [258]. The tear resistance of the tape has to be as high as that of the paper to be bonded, and high instantaneous peel and shear resistance are required. For splicing tapes, the corresponding tack/peel and shear values are 2.2/1.5/360 [50]. The shear value was measured at 250°F (50 μm adhesive thickness). The main types of splicing tapes are [257] one-side-coated tape, double-side-coated tape, and transfer tapes.

In one-sided splicing tapes, the siliconized paper liner remains on the adhesive and has to be recyclable also. The carrier is water dispersible or water soluble. Carrier-free splicing tapes have to be water soluble also. The liner for transfer tapes has to allow the detachment of the adhesive, that is, it has to display different degrees of release on its two sides. The requirement concerning the water solubility or dispersibility of the paper carrier used for splicing tapes is an old criterion [132,259]. Tack and heat stability together with water solubility or dispersibility are required for splicing tapes in papermaking and printing [260]. Such tapes are used for bioelectrodes also because of their low electrical impedance and conformability to skin. Water-soluble compositions used for splicing tapes contain polyvinyl pyrrolidone with polyols or polyalkylglycol ethers as plasticizer, vinyl ether copolymers, neutralized acrylic acid–alkoxyalkyl acrylate copolymers and water-soluble waxes [258] or acrylics [54]. Water solubility is given by a special composition containing vinyl carboxylic acid neutralized with an alkanolamine [260]. Other formulations contain acrylic acid polymerized in a water-soluble polyhydric alcohol and crosslinked with a multifunctional unsaturated monomer. Repulpable splicing tape, especially adapted for splicing carbonless paper, uses a water-dispersible PSA based on acrylate–acrylic acid copolymer, alkalies, and ethoxylated plasticizers. A PA–epichlorohydrin crosslinker may also be included [254] (see also Chapter 8).

Splicing tapes have to be tested for the following properties: tack and adhesion on paper (on silicone raw paper also), dynamic and static shear strength at normal and high temperatures, resistance to greasing and bleed through, water solubility at pH 3, 7, and 12 [54]. According to Ref. [258], water solubility between pH 3 and 9 should be given. As shown in Ref. [54], special acrylate-based, water-soluble adhesives for splicing tapes can change their adhesive properties as a function of the atmospheric humidity (see also Chapter 7).

18. Medical Tapes

Surgical, PSA sheet products include any product that has a flexible carrier material and a PSA [261]. Labels, tapes, adhesive bandages, adhesive plasters, adhesive surgical sheets, adhesive corn plaster, and adhesive absorbent dressings have been manufactured. Natural adhesives mixed with natural additives have been used [213]. Medical tapes have been developed for pharmaceutical companies, ostomy appliances, diagnostic apparatus, surgical grounding pads, transdermal drug delivery systems, and wound care products (see also Medical Labels and Chapter 9).

Physiologically compatible adhesives are based on polyurethanes, silicones, acrylics, polyvinyl ether, and styrene copolymers. These should be breathable, cohesive, conformable, and self-supporting compositions that cannot be degraded by body fluids [262]. They minimize skin irritation due to their ability to transmit air and moisture through the adhesive system. New ways are opened by use of antiviral agents as fillers for the carrier or the adhesive [263].

Air- and moisture-permeable nonwovens (PET nonwoven, embossed nonwoven) or air-permeable tissue (breathable face stock) is used for medical tapes coated with a nonsensitizing acrylic porous adhesive for proponged application of a medical device on the human skin. The air porosity rate has to fulfill a given value (50 sec/100 cc per square in.) [264]. Skin tolerance, no physiological effects, resistance to skin humidity, and sterilizing capability without color changes are required for medical tape adhesives [265]. Certain applications require water-soluble or crosslinked adhesives [137].

Surgical tapes are used for fastening cover materials during surgery. Cotton cloth and hydrophobic special textile materials coated with low energy polyfluorocarbon resins are used as carrier materials and coated with polyacrylate-based water-soluble PSAs. Such adhesive has to be insensitive to moisture and insoluble in cold water and must display adequate skin adhesion and skin tolerance. The water solubility of the product is reached above 65°C at a pH above 9 [54,265]. Reusable, hydrophobized textile materials and extensible, deformable paper are required for medical tapes [235,266]. PVC can be used for self-adhesive medical tape or as an adhesive coating also [267]. Preferred carrier materials are those that permit transpiration and perspiration or the passage of tissue or wound exudate therethrough [268]. They should have a moisture vapor transmission of at least 500 g/m^2 over 24 h at 38°C with a humidity differential of at least about 1000 g/m^2. Special fillers allow the production of breathable polyolefin films [269]. For medical tapes, the conformability of PSA, its initial skin adhesion value, the skin adhesion value after 24–48 h, and adhesive transfer [270] are measured as the main performance characteristics. A preferred skin adhesive will generally exhibit an initial peel value of 50–100 g and final peel value (after 48 h) of 150–300 g.

Rubber-based and acrylate homo- and copolymers have been used for medical tapes [270,271] (see also Chapter 8). For a composition containing polar comonomers with built-in (carboxyl or hydroxyl group-containing) crosslinking comonomers, a coating weight of $34–68 \text{ gm}^2$ has been suggested. Unfortunately, many of these compositions lose their adhesiveness in hot or humid environments and allow adhesive transfer during removal. The adhesives used for surgical tapes often exhibit a dynamic modulus too low for outstanding wear performance. A low storage and loss modulus result in a soft adhesive with adhesive transfer. A creep compliance of 1.2×10^{-5} cm^2/dyn is preferred for a skin adhesive [272] (see also Chapter 3, Section I). Irritation caused by removal of the tape was overcome by making the composition water-soluble [273]. So-called wet-stick adhesives (sterilized with ethylene oxide or gamma radiation) are used for surgical towels; they have to adhere to skin, wet textiles, and nonwovens [274].

19. Application Tape

Application tapes are special paper- or plastic-based mounting tapes. They are used together with plotter films as removable mounting aids. Application tapes are applied to ensure the transfer and temporary fixing of another permanent pressure-sensitive element (letters or written text) [275] (see also Chapter 8). Paper and plastic film-based application tapes are manufactured. Paper premasking/application tapes are made from a latex saturated paper that is coated with a PSA. In most cases, since low adhesion adhesives are utilized, the decal premask/application tape does not require a release coating. Adhesion levels of 130–300 g/15 mm (PSTC-1) are typical [276]. Since the adhesion to decal is usually a direct function of the area of the decal, commonly larger decals require lower adhesion products, whereas smaller decals, such as individual letters, require higher adhesion decal premask/application tapes. The premask/application tape has several functions. Applied to the decal, the tape protects the finished sign surface during the succeeding stages. It provides a means of maintaining absolute registration of the prespaced lettering, whether die cut or cut on computerized letter cutting equipment, and the application tape is essential in successfully applying the decal. The application tape provides the "body" with which the thin, stretchy expensive decal can be handled and protects the decal during the adhesion process when the decal is squeezed to increase its adhesion to the final surface and to eliminate trapped air.

The plastic-based tapes have PVC, PE, or PP carrier material that can be embossed to lower unwinding resistance. Plastic-based application tapes can also have paper release liners: these tapes are used for signs with small dimensions. Special application papers can be used as stencil paper also (see Plotter Film).

20. Protective Tapes

Protective tapes are used as temporary or permanent surface-protective or reinforcing elements. Most of them are really protective films with a tape-like geometry. There are various application domains of such tapes, including face or back-side protection.

21. Car Masking Tape

According to Ref. [140], the first masking tapes allowed the two-color painting of cars. These tapes are used during spray painting of vehicles. Masking tapes may have special carrier constructions to give them conformability and temperature resistance. Some masking tapes have to be flexible, stretchable, and contractible, capable of maintaining the contour and the curvature of the position in which they are applied [277]. The adhesive is covered with a release liner. According to Lipson [278], a masking tape carrier has a stiffened, longitudinal section extending from one edge, with an accordion-pleated structure, to conform to small radii. The nonadhesive face of the tape is heat reflective. Masking tape based on foam has been disclosed in Ref. [279]. Masking tapes on paper basis (90 g/m^2) can be coated with 30 g/m^2 HMPSA. EVAc copolymers are used for HMPSA for weatherstripping tapes. A coating weight of 60 g/m^2 is applied for a bonding surface, with 50 g/m^2 on the opposite side, to bond the wipe clean surface.

Masking tapes used for spray varnishing in the automotive industry have been partially replaced by adherent, nonpressure-sensitive cover films which allow a less expensive, full surface, and easily removed protection [280]. A PSA tape for protection of printed circuits, based on acrylics and terpene phenol resin with a $100-200$ g/m^2 siliconized crepe paper, is made by coating the adhesive mass first on a siliconized polyurethane release liner, then transferring it under pressure ($10-30$ N/cm^2) on the final carrier material [281]. Special office masking tapes are applied to cover written text passages; the covered surface portion should give a shadow-free copy [282]. Generally, masking tapes are rigorously tested for deposit-free removability at elevated temperatures also (see Table 11.7).

22. Mirror Tapes

Such tapes are large surface security films applied to the back of fragile substrates (see Protection Films). Security lamination films for glass is carried out with security films and with antisplitting films. According to DIN52290, an 1 m^2 glass panel has to resist three perpendicular consecutive impacts with a 4.1 kg steel ball [283].

23. Other Tapes

The number of other special tapes is large. Some of them have an ordinary construction, whereas others are more sophisticated. Certain products function mechanically to fix and fasten elements; others have dosage, insulation, or other special functions.

24. Test Tapes

Special or standard tapes are used to test the adhesivity of varnish or printing ink [284]. Tapes can be used in testing laminates also. To evaluate a laminate section by light microscopy, a microtome is used to produce perfect cuts. A rapid method for fixing (reinforcing) of the sample is to laminate it with a tape [285]. The surface treatment is tested according to ASTM 2141-68 by using a PSA tape [286–288].

IV. PROTECTIVE FILMS

The use of protective films is a relatively new field. At the end of the 1980s, protective films have been suggested for the posttreatment of metal coils as an alternative to oiling or wax coating [289]. The main criteria given by the German Association of Surface Coated Fine Coil (Fachverband oberflächenveredeltes Feinblech) for the choice of protective films are the nature, thickness, adhesion characteristics, forming, tear resistance, and light stability of the film, with the recommendation that "well-defined protective films should be applied outdoors for a given time only." First,

TABLE 11.7
Conditions for Removability Test of Masking PSPs

PSP	Storage Conditions		Reference
	Room Temperature	Elevated Temperature	
Masking tape	30 days	3 days, 80°C	[163]
Removable tapes or sheets	100 h	3 days, 60°C	[156]
Masking sheet	—	2 days, 60°C	[164]
Masking sheet, tape	—	3 days, 60°C	[157]

the protective papers were developed and later, the protective films. Protective masking papers can be classified as light weight, standard weight, medium weight, and heavy weight ones [290].

A. APPLICATION CONDITIONS

The application and deapplication are of equal importance for protective films because they are used for only temporary bonding. If the film is not detached completely, the different thermal dilatation of the metallic coil and plastic film may lead to stress cracking [289]. For film application, the chemical composition and processing quality of the surface play a special role.

1. Product Surface

As stated in Ref. [291], for protective films, the adhesion of the PSA should be tailored to the surface to be protected. Protective films are used on very different substrates depending on the nature of products to be protected. Generally, suppliers checklists concerning the application conditions of a protective film should contain the following data: UV resistance, desired adhesive strength, nature of the surface to be protected, thickness of the web to be protected, and future processing (forming) operations.

It should be taken into account that because of the higher elasticity, the processing of plastic surfaces for a given smoothness is more difficult [49]. The surface itself is a multilayered structure. There is a layer of dirt (3×10^{-6} mm), then an adsorbtion layer (about 3×10^{-7} mm) and a reaction layer (1×10^{-4} mm), followed by the material itself [292].

Various types of substrates have to be coated with protective films. Metallic or plastic surfaces are the most frequently encountered adherends. Various metals are used — unpainted and painted, powder- and thermo-painted, bright, glossy, and matte. Stainless steel, benzoin acrylate (BA), cold rolled, polished, or brushed and aluminum, mill finish, anodized, or brushed have been protected. Coil coating is the most important domain of metal coating. Here uncoated "pure" metal surfaces and precoated, varnished surfaces are laminated with protective films. Coated steel materials include plastisol-coated, epoxy-modified, polyester-coated, and hot dip galvanized steel [293]. Alkyde-, polyurethane-, silicone-modified, and fluorine-containing macromolecular compounds have been suggested also for coil coating [294]. Acrylics, melamine, PVC, lacquered, powder-coated, and thermo-lacquered surfaces may be used also as substrates. Coils are classified according to the metal, coating, and application. The main application fields are the automotive industry, construction, and household machines [295]. Coil-coating varnishes have to resist processing steps such as deep drawing, drilling, and cutting [296].

Quite different surfaces have to be protected within the same application field. As shown in Ref. [297], a variety of lacquers are used for lacquering of various plastic automotive components (depending on their elasticity). For instance, for low elasticity parts ($E < 3000$ N/mm^2), one-component acrylic-melamine, epoxide and two-component PUR lacquers are applied. For parts having medium elasticity ($E = 1800–3000$ N/m^2), one- or two-component PUR acrylic- or polyester-based varnishes are suggested. For elastic parts, two-component PUR lacquers have been proposed.

Customers refer to the typical stainless finishes as dull, bright, or annealed. Steel and copper surfaces may be chemically etched to produce fibrous oxide growth on the surface [49]. The level of roughness is achieved by transferring a pattern from the working (calendering) rolls to the material, which then goes to the tempering mill to be (eventually) tinned or coated with chromium or other metal.

A dull finish is developed by final rolling, followed by annealing and hot acid pickling. A low-gloss finish results that is suitable for functional parts, where ultimate lubricant retention is required or where severe forming with possible intermediate annealing occurs. Extensive polishing and buffing are required to acquire the desired finish quality.

A bright finish is achieved via rolling with high polish rolls, followed by conventional anneal-ing and electrolytic pickling. A relatively glossy gray surface results. This is a general-purpose finish used where a minimum of polishing and butting is employed to develop a bright luster. Typical uses include wheel covers, cookware, lids, etc.

For a bright annealed finish, the final rolling is done with highly polished rolls and annealing in a protective atmosphere. It is used for applications where only a color coating is used to achieve the desired luster. A hardened finish is practicable only to certain martensitic, hardenable grades of steel. It is achieved by heat treating, annealing, and pickling. Additionally, there are many degrees of finish within each category [298]. For packaging steel, the measured roughness is expressed as the RA value (in μm). Roughness may vary from mirror finish (RA less than 0.17 μm) to matte finish (RA 0.70–1.00 μm) [299]. Generally a steel surface is passivated. Passi-vation is a chemical or electrochemical process that increases the metal surface's resistance to oxi-dation and facilitates lacquering. Tin, chromium, and chromium oxide may be used for topcoating. Usually, coils are oiled in order to ensure a better handling and chemical protection. Plasticizers like dioctyl sebacate or acetyltributyl citrate are used as the "oil". Their coating weight depends on the coil quality and use. Tin-coated coils are coated with 0.5–11.2 mg/m^2 oil. Uncoated packa-ging steel (black plate) has 25 mg/m^2 oil. For lacquering or printing, 4.4 g/m^2 oil is used [299]; for overseas shipments, a larger quantity (10 mg/m^2) is used.

For stainless steel mill finish, polished, BA, protective films with rubber/resin adhesive are recommended. It would be difficult to achieve an adequate contact between a hard acrylic adhesive and this surface. It would make the edging difficult, the film could loosen. Furthermore, it is useful to have higher adhesive strength when forming special steel. As is known from the practice of dry lamination, rubber-based PSAs are used for Zellglas/PE lamination, where the mechanical properties of the components are quite different or where a soft bonding layer is necessary [300]. The ability to bond through oil is essential in many processing applications. For metal protection for applications where chemical and water resistance are required, immersion bath compositions based on thermoplasts, chlorinated paraffin, and esters of phtalic acid have also been suggested [301].

The surface of aluminum with a mill finish is softer. Acrylic adhesive-coated protective films may be used for their protection. Various surface qualities, mill finish, polished, matte, and anodized, are manufactured. With anodized aluminum, the kind of treatment and its age influence the adhesion. The surface composition of aluminum changes during storage [302]. The concen-tration of alkali and alkaline earth metals increases. An anodized aluminum surface can have a very different oxide morphology [303]. Generally, the surface of metals is rough and oxidized; the oxide layer contains water and the water contains salts [304]. For electronic parts, brushed aluminum is used [305].

All lacquered surfaces (unpolished, polished, and brightly polished) should be pretested before coating. The nature of the varnishes may be very different. Steel may be coated with epoxy-, polyester-, acrylic-, or melamine-based paints [306]. Flexible polyester is used for coil coating [307]. Polyester coil-coating systems may contain nonvolatile esteralcohols [308]. Silicone acrylates are also used for coil coating (surface panels for buildings). Epoxide and polyester, zinc, or acrylics and melamine are used for coating cans and coils [309]. Cellulose, PA, epoxy, vinyl, acryl, and polyester powders may be used for metal coating [310]. For coil coating, high solids paints, powder paints, and systems capable of being hardened by radiation have been developed [311]. Varnishes based on plasticized polymethyl methacrylate (PMMA) are also used.

Electrostatic powder spraying is also used for coil coating and automotive parts. According to Ref. [312], in Europe, the first powder coatings have been developed in the early 1960s. The first powder coatings based on PE were introduced in the U.S.A. in 1971 [313]. The mostly used powder lacquers are based on epoxy, epoxy–polyester, polyester, and polyurethane derivatives. These are crosslinked systems cured at (140–220°C) [314]. The epoxy–polyester systems are used particularly in the automotive industry where acrylate powder lacquers are also used [315]. Powder-molded compounds, where powder lacquering of the product occurs in the mold, are used for sanitary items and automotive parts [316].

It is well known that for certain acrylic coil-coating formulations, 1–5% of special PE dispersions are added in the recipe to impart better deformability and prevent blocking of the lacquer surface. Therefore, the application of a PE-based hot laminating film for such coils may cause a PE/PE adhesion, that is, a higher adhesion level. Such limitations exist for adhesive-coated protective films also. Rubber-resin adhesive-coated protective films cannot be used for copper or brass (bright or mill finish). Sulfur containing copper reacts with the rubber.

Other surfaces such as glass, plastic profiles, acrylic sheets, PC plates, textiles, carpets, linoleum, parquetry, marble, wood, and melaminated kitchen furniture also have to be protected. Most of them are manufactured in different qualities (grades). Furniture is lacquered with polyacrylates containing OH groups. Polyester acrylates in tripropylene glycol diacrylate are UV-cured on plastics [317]. Glass is covered with PVA/plasticizer [318]. PC plates are produced with bright, glossy, and matte surfaces; melamines may be bright or matte; and synthetic marble are produced with bright and matte surfaces. PVC plates (soft and hard PVC) are manufactured [319]. PVC plates may be coated to increase their abrasion resistance and gloss with a 0.003–0.004 in. layer of deck compound that displays adhesion other than common PVC [320].

Windows frames are produced at a speed of 3 m/min from high impact hard PVC [321]. Other polymers like, ABS, PP, PPE–HIPS blends, PMMA, etc., have been also tested. PVC window profiles were first marketed 40 yr ago [322]. Chlorinated PE or polyacrylester is used as impact modifier. SPVC, filler (TiO_2, $CaCO_3$), and chlorinated PVC are used as components in the formulation. After warm storage of the profiles, dimensional changes of 1.7% may occur. As is known, screw rotation speed and the temperature in extruder are also known to influence the shrinkage of the extruded product. Films to protect plastic window profiles must be aging and weathering resistant (months) and must have improved transparency to allow visual surface inspection. Generally, a 45 μm polyolefin film is used as carrier for these protective films [323].

For furniture, UV-curable unsaturated polyester coatings and UV-curable polyester coatings are used [324]. Decorative laminates are manufactured from paper impregnated with melamine resins [319]. Modified polyethyl methacrylate may be used as protective film for substrates such as PVC, ABS, and PS, wood, paper, and metal [325].

PMMA forming materials and semifinished products are produced from methyl methacrylate [324]. They are used for automotive, building, light technical, advertising, household wares, and hygienic products. Mobile phones, pocket PCs, and digital versatile discs are items where PMMA replaces the PC [326]. Semifinished products, cast and extruded PMMA plates, and extruded PC plates have been developed. The so-called acrylic glass is manufactured with a minimum thickness of 2–3 mm [327]. Cast PMMA plates possess a thickness of 2–100 mm, and extruded PMMA plates have a thickness of 1.5–6 mm. Extruded high impact (HI) PMMA plates have a gauge of 2–6 mm. Extruded PC plates have a gauge of 1–12 mm. According to Ref. [328], cast PMMA plates are manufactured with thicknesses of 2–25 mm and 30–100 mm; extruded plates with 1.5–6 mm; extruded HI PMMA with 2–6 mm; and extruded PC with 1–2 mm. Their abrasion resistance is improved using special silicone–alkoxy condensation products [329]. Their surface quality may be quite different. For example, PC plates can be patterned on both sides, with a pinspot pattern on one side and prism pattern on other side or both sides polished or patterned raindrop/haircell [330].

PC having a thickness of 0.25 mm is used for overlays for keyboards [327]. Plastic-coated glass surfaces are used as building panels. The coating is an ionomer. The same system is used for bottle manufacturing (see Label End-Use) [331].

As discussed earlier, there are both static (packaging) and dynamic (processing-related) uses for protective films. The requirements for these application fields are quite different.

2. Packaging Use of Protective Films

Certain protective films are used only as packaging materials. The protected item does not undergo working processes. Very often films are used for protection during transport, for example, when stacking profiles; there is no paper between stacked items. PC or PMMA sheets are provided with protective films on both sides to avoid damage to the mirror-polished surfaces during transport (and processing). Time of storage on the sheet or on the roll should not exceed 1–1.5 yr. In packaging uses, the self-deformation of the coated plastic material has to be taken into account as a dynamic force. Plastic development reduces this risk. In the last decade, the stability of plastic pipes at 80°C and 4 N/mm^2 force has increased in the main national norms from 170 to 1000 h [31]. Thermally-induced extension on plates and profiles may produce tensions and cause delamination of the protective film. PVC profiles display a coefficient of thermal expansion of 8×10^{-5} K^{-1}. Other protected films are processed together with the protected item and have to resist mechanical and thermal stresses during the technological cycle.

3. Processing Use

Processing (forming) of coated surfaces requires good adhesion between the protected surface and the protective film (Table 11.8). The film must be able to undergo the processing. The main work processes are bending, punching, drilling, and deep drawing. Cutting, drilling, shearing, routing, and heat bending are the most important working methods for protective films used for plastic plate protection. PC plates may undergo sawing, shearing, punching (sometimes punching tools are heated to 140–180°C), drilling, turning, trapping, milling, grinding, and polishing [330]. During these operations, PE protective films should not come into contact with oily or greasy substances. They should not be used at temperatures of more than 70–80°C. The film may be damaged by bending if the pressure of a tool is concentrated at a single point.

The thickness recommended for a protective film depends on thickness of the protected item (e.g., a thin stainless steel sheet of 3 mm needs a protective film at least 90 μm thick). For drilling

---

TABLE 11.8
Dependence of the Protective Film Grade Used on the Characteristics of the Protected Product

PSP Characteristics						
		Peel Resistance (N/10 mm)		Product to Be Protected		
		Dwell Time				
Thickness (μm)	Coating Weight (g/m^2)	10 min	3 month	Material	Roughness	Processing
50	0.7	0.3–1.0	0.15	SST, aluminum	High gloss and lacquered	Slight processing
70	5.0	0.8–1.8	0.25	SST, aluminum, plastics	Glossy and lacquered	Medium processing
75	3.0	>1.8	0.25	SST, aluminum, plastics	High gloss and lacquered	Medium processing
90	3.5	0.8–1.8	0.25	SST, aluminum, plastics	High gloss and lacquered	High degree of processing
100	2.0	0.8–1.8	0.25	SST, aluminum	High gloss and lacquered	High degree of processing

and punching aluminum or stainless steel sheets, it is important to know that only a very few liquid processing aids can be used because PE films are oil permeable. The penetration of liquid between the sheet and film may produce residues. In this case, 70–90 μm films should be used.

Protective films are also appropriate for deep drawing. It can happen that wrinkles will form on the edge (PVC would be better than PE). When the sheets are processed or embossed wrinkles can be pressed into the surface. The processing conditions are quite different for different plastics and plastic plates. For example, drilling speed during milling is the highest (2000 m/min) for PMMA when compared with PC (1000 m/min) or styrene-acrylonitrile (SAN) (200–300 m/min) [332]. The high processing speed must be considered with the low thermal conductivity of the material also taken into account. When compared with steel (47–58 W/mK), PC possesses a coefficient of thermal conductivity of only 0.21–0.23 W/mK. Vibration welding is also used for plastic plates [333].

In some cases, infrared radiant heating is used for warm-formed plastic plates. The halogen heating elements give their maximum radiation energy at 1.0 μm wavelength. However, at wavelengths of 0.5–1.5 μm, both the processed plastic plate and the protective plastic film on it may absorb radiation energy [334]. According to the nature of the protected surface, SAFs can be classified as SAF for hot laminating uncoated and coated lacquered metal surface; SAF for hot laminating plastic surfaces; SAF for hot laminating paper; SAF for mold release casting; SAF for back protection of glass; and SAF for special surfaces (fabric).

In choosing the SAF, the intrinsic adhesive properties of different raw material classes have to be taken into account also (Table 11.9).

It should be noted that coextruded SAFs have been used as hot laminating films for other applications for many years [341]. Peel strengths of 1–20 lb/in. have been achieved depending on the product surface and application conditions. For their application, a laminator roll with a temperature of 300–500°F, pressure of up to 400 pli, and line speeds 45–200 ft/min are suggested. Such products have a warranted shelf life of at least 6 months.

TABLE 11.9
Self-Adhesion of Elastomers/Plastomers Used for PSPs

Material	Adhesion		Reference
	Value	Method	
SIS	70 g	—	[335]
Ethylene-α-olefin copolymer, tackified with PIB, cold stretchable adhesive film	65 g	ASTM 3354-74	[336]
Ethylene–propylene block copolymer grafted with maleic anhydride	8.5 kg/25 mm	On SST	[337]
	3.8 kg/25 mm	On aluminum	
Ethylacrylate–ethylene–maleic anhydride copolymer, grafted with triethoxyvinylsilane	0.93 kg/cm	On glass	[338]
	0.87 kg/cm	On aluminum	
PIB, APP, polybutadiene, butyl rubber	65 g	ASTM 3354-74	[336]
Plasticized PVC	0.07–0.25 N/cm	On siliconized release liner	[339]
NR, synthetic rubber, filler	0.5 N/cm	—	[340]
PE, hot laminated	250 g/30 mm	On SST	—

4. Application/Deapplication Climate

PE films become brittle at low temperatures and it becomes very difficult to remove them. At temperatures below zero, the films should be warmed before removal. The recommended temperature for their delamination is 8–22°C. According to Ref. [342], protective films should be applied between +15 and +50°C. Table 11.10 summarizes the application conditions for the most common protective films. As Table 11.10 indicates, very different application temperatures and pressures are used for protective films laminated on metallic or plastic surfaces. The application climate is strongly influenced by the nature of the substrate, substrate manufacturing procedures and laminating equipment, and construction (grade) of the protective film.

a. Application Conditions According to Laminating Equipment

It is evident that laminating equipment determines the laminating temperature and pressure, which are the main laminating parameters [341]. The application temperature of the protective film depends on the manufacture technology used for the protected item. Extruded plastic plates have an elevated temperature after extrusion. This is used to improve the adhesion of the protective film. Such plastic plates are laminated with adhesiveless protective films at high temperatures. As an example, extruded double-layered PMMA and PC plates are laminated with an SAF (40 μm, EVAc-based) at 60–70°C. Cast items or plates are laminated with an SAF at room temperature, but they have to support the processing temperature (170°C, 30 min for PMMA, 160°C, 10 min for PC). In this case, a higher film thickness (80 μm) is suggested. Such high temperature extruded plates may be laminated with adhesive-coated protective films also. In this case, the adhesion of the film does not require the elevated laminating temperature. It is imposed by the manufacture procedure of the substrate to be protected. The laminating speed may vary with the equipment and the products to be protected. This influences the high temperature contact time of the film. For instance, the same coil coating production line may run at 20–120 m/min.

b. Application Conditions According to the Substrate Surface

Cast (chill roll) polyolefin films are used as self-adhesive protective films (without PSA). They have a thickness of 40–80 μm. They have to adhere to extruded and cast plastic plates. After cooling, the film has to be removed deposit-free and with low peel resistance from PC and AC plates. For cast plates, the application is carried out at room temperature; for extruded plates, at 60–70°C. In several cases, such films have to survive thermal forming. In this process, the protective film-coated plastic plate is heated for 5–15 min at 170–180°C.

The roughness of the surface (determined by manufacturing and treatment procedures) influences the type of the protective film also used. For coil coating, the gloss of the lacquer

TABLE 11.10
Application Conditions for Main Protective Film Grades

Application	Protective Film		Application Conditions	
	Adhesiveless	Adhesive-Coated	Temperature (°C)	Pressure
Metal coils	Hot laminating	—	220–250	High
	—	AC or RR	RT	Low
Plastic plates	Warm laminating	—	50–70	Medium
	—	AC or RR	RT-70	Low
Finished plastic products	—	AC or RR	RT	Low
Carpet	—	AC or RR	RT	Low

used is a decisive parameter. For instance, 30–40% gloss is recommended for hot laminating films. Generally, protective films with higher adhesion (peel) are required for complex processing (multiple deep drawing) and for matte surfaces. Protective films used for deep drawing have to function as a lubricating agent also.

5. Application Method

Generally, protective films are applied by lamination which involves the use of laminating equipment. The properties of the protective film depend on the chemical and physical properties of the adherend, cleanness, elongation of the film during lamination (i.e., laminating method and equipment), and surface temperature of the adherend [342].

6. Application Equipment

Large-volume objects and also profiles are applied with application lines. Such lines are supplied by specialized firms. Automatic lines with stacking and cutting devices are used. Generally, the lines are built with an expander (banana) roller. The expander roller ensures a bubble- and wrinkle-free lamination. The brake should not be too strong, just strong enough to keep the films in a stretched position, in order to avoid wrinkles. In some cases where the film has to be removed and reapplied, the relamination is carried out by hand. This is the case in small firms where short (100–200 m) rolls are used. In many cases (coil coating, plastic plate manufacturing, extrusion of plastic profiles), the lamination equipment is part of the production equipment for the web to be protected. Here, the laminating speed and temperature are given by the main equipment. For instance, PVC window frames are produced at a speed of 3 m/min, so the lamination speed of the protective film has to be the same.

B. Main Protective Films

Generally, the classification of protective films according to their end-use is related to the nature of the surface to be protected (Table 11.11). As discussed earlier, metals (coils) and plastics (profiles,

---

TABLE 11.11
Main Protective Films

Application	Product Buildup
Coil coating	Adhesiveless hot laminating film
	Adhesive-coated film
Plastic plate protection	Adhesiveless warm laminating film
	Adhesive-coated film
Automotive storage and transport	Adhesive-coated film
	Adhesiveless warm laminating film
Deep drawing	Adhesive-coated film
	Adhesiveless film
Carpet protection	Adhesive-coated film
Product security	Adhesive-coated film
Furniture protection	Adhesive-coated film
	Adhesiveless warm laminating film
Building protection	Adhesive-coated film

---

plates, films, and other plastic-coated materials) are the most important application domains. Independently, from their special use, the following general requirements exist for protective films: adequate elongation at break, opacity, coefficient of friction (COF) (film-to-film), release of backing, stability of adhesion, low unwinding resistance, high tensile strength, and no telescoping. In the choice film grades, their adhesive properties must be taken into account relative to the surface in their application domain (Table 11.12).

1. Protective Films for Coil Coating

Coil coating is web coating where calendered metal (steel and aluminum, copper, etc.) coils are coated with organic materials (lacquers, varnishes, or films). Liquid or dry coating materials can be used. Thin ($10-20\,\mu m$) layers and thicker layers ($60-100\,\mu m$ for dry lacquers) are applied at a machine speed of $15-18$ m/min. The protective film is laminated at the end of the varnishing process.

Because of outsourcing, the importance of coil coating has increased over recent years. This has led to the "finish first, fabricate later" principle, that is, the prefinishing of the metal plate surface before the coil is processed. Steel plates $0.1-0.8$ mm thick are used in automotive, building, furniture, packaging, and other industries [343].

Adhesive-coated and adhesive-free protective films are used for coil coating. The most known adhesive-free protective films in this field are hot laminating films (Table 11.13). LDPE may be sealed at $110°C$. If the temperature is increased, a seal strength of $1-2$ N/15 mm is achieved at $120°C$ [344]. According to Ref. [345], in thermal laminating, two thermoplastic substrates (with thermoplastic coatings) are combined with the application of heat and pressure. By the coating of coils the protective film is thermoplastic and it may or may not have an adhesive coating (or embedded PSA).

In many applications, the coil ($0.4-1.5$ mm) to be protected by a film is first lacquered. Chromated, primed ($5\,\mu m$) coils are lacquered ($20\,\mu m$) continuously at a running speed of $50-80$ m/min. The lamination temperature of the self-adhesive polyolefine protective film is $200-250°C$. After laminating, the film is cooled in a water bath. The debonding ($90°$) peel of such films is about $180-250$ g/30 cm (with tolerances of 50 g). The test conditions of such films are described in Ref. [346]. Generally, relatively thick films (120, 150, or $170\,\mu m$) are used. Such films are corona pretreated at $44-50$ dyn/cm (see also Chapter 8).

TABLE 11.12
Adhesive Characteristics of Protective Films

Product	Adherend Surface	Carrier Material	Thickness (μm)	Coating Weight (g/m²)	Peel Resistance
Hot laminated film	Glossy (40%) varnished metal	PE	120	—	185.0 g/30 mm
Automotive protective film	Lacquered surface	PE/PP	50	5–10	5.0–8.0 N/25 mm
Thermally formed protective film	PA	PE/PP	70	6–7	1.0 N/10 mm
Automotive technological protective film	PP, ABS, rough	PE/PP	100	14	1.5 N/10 mm
Carpet protective film	PP, PA, PET textured	PE	80	10–12	5.0 N/10 mm
Carpet underlay film	PA, PP, PET, EVAc	PE	25	—	5.0 N/10 mm
Plastic plate protective film	PC	EVAC	75–100	—	140–280 g/25 mm

TABLE 11.13
Manufacture and Application Characteristics of High Temperature Laminated Films

Polymer	BUR	MFI 190/2.16	Density (g/cm^3)	Thickness (μm)	Surface Tension (dyn/cm)	Tensile Strength (N/mm^2)	Peel	T (°C)
LDPE	1:1.8	0.6–0.8	0.923	50, 60	46–48	18/16	350 g/50 mm	
	1:2.4		0.02				Max. 750 g/50 mm	
							250 ± 50 g/30 mm	241
							250 g/30 mm	200
EVAc,	—	—	—	—	—	—	80–120 g/10 mm	60–70
EBA,								
PE + PB								

Note: T, application temperature.

The protection of aluminum, especially of anodized aluminum surfaces, is a complex domain of coil coating. Standard tests of the adhesion are generally made on stainless steel. Although the correlation between peel resistance on stainless steel and aluminum can be clear, the increase of the adhesion with the dwell time on stainless steel and aluminum may be quite different (Table 11.14). Other problems appear to be related to the use of protective films with "easy peel" and "cleavage peel" (see also Chapter 7). In this case, the dependence of the cleavage peel on the coating weight (Table 11.15), the interdependence between cleavage peel and standard peel (Table 11.16), and the interdependence between peel on stainless steel and aluminum must be cleared (Table 11.17).

2. Protective Films for Plastic Plates and Other Items

Various plastic plates and other items are laminated together with protective films, during and after their manufacture, to protect them during storage and processing. The main polymers used for plastic plates are PMMA, PC, PS, SAN copolymers, PVC, and PP [347]. Styrene-based safety

TABLE 11.14
Peel Buildup as a Function of the Substrate[a]

Peel Resistance on SS after Aging (N/10 mm)	Peel Resistance on Aluminum after Aging (N/10 mm)
0.05 Instantaneous peel resistance	0.5 Instantaneous peel resistance
0.10	0.7
0.20	1.0
0.30	1.2
0.40	1.2
0.60	1.5
0.80	2.0
1.00 Peel resistance after 7 days	3.0 Peel resistance after 7 days

Note: SS, stainless steel.

[a]Peel adhesion was tested at intervals of 1 day.

TABLE 11.15
Dependence of the Cleavage Peel on Coating Weight

Coating Weight (g/m²)	Cleavage Peel (N/10 mm)
4	0.06
5	0.07
6	0.09
7	0.10
8	0.11
9	0.11
10	0.12
12	0.12

glass for windows was also developed [348]. Butadiene–styrene copolymers for sheet extrusion replace PET [349]. A new range of compounds based on syndiotactic PS, a semicrystalline, metal-locene-catalyzed engineering polymer with a very high melting point, was developed. This material exhibits high continuous service temperature, hydrolysis resistance, dimensional stability, and good processing. The protective films are generally applied via in-plant lamination. A hot roll laminator nip or a hot roll press is used, and heat and pressure cause the film to conform and adhere to the surface. As can be seen from Table 11.18, the grade of the protective film used for a plastic plate depends on the plate material, manufacture process, and geometry.

Generally, adhesive-coated and self-adhesive protective films are used for the protection of plastic films and plates. Relatively hard water-based acrylic adhesive-coated protective films have been suggested as adhesive-coated protective films for plastic plates and films. Nontackified polyethylene (LDPE), PE EVAc blends, and polybutylene-tackified PE (coextrudate) have been proposed as SAFs for plastic plates. Nonpolar, common PE films are corona treated; blends with polar monomers can be corona treated also. Such film are generally laminated under pressure at elevated temperature. Generally, the adhesion of the corona-treated PE film to the plastic plate substrate increases with the application temperature.

For instance, such corona-treated LDPE (density of 0.923 g/cm³) film (50–60 μm) is manufac-tured via a blow procedure, with a BUR of 1:1.8–1:2.4 depending on its dimensions. It is pretreated for a surface tension of 46–48 dyn/cm. The adhesion of such films attains 350–750 g/50 mm.

The requirements concerning the dimensional stability and plasticity–elasticity balance of a protective film depend on the dimensional stability of the protected item. For instance, high molecular weight cast PMMA displays in a broad temperature range mainly thermoelastic beha-vior, therefore, formed PMMA items do not display plastic deformation, that is, dimensional

TABLE 11.16
Interdependence of Standard Peel and Cleavage Peel

Peel Resistance (N/10 mm)	Cleavage Peel (N/60 mm)
0.2	2.0
0.3	4.0
0.4	6.5
0.5	8.5
0.6	10.6
0.7	13.0

TABLE 11.17
Interdependence of Peel Adhesion on Stainless Steel and Aluminum

Peel on Stainless Steel (N/10 mm)	Peel on Aluminum (N/10 mm)	Cleavage Peel on Stainless Steel (N/10 mm)
0.05	No adhesion	No adhesion
0.10	0.05	0.008
0.20	0.12	0.015
0.30	0.17	0.020
0.40	0.25	0.025
0.60	0.40	0.050
0.80	0.55	0.065
1.00	0.67	0.080

Note: Instantaneous adhesion has been measured for different coating weight values.

changes [333]. The protective film used has to possess elastic properties also. In contrast, side low molecular weight extruded PMMA displays a narrow elastic domain during high temperature deformation. The transition from elastic to plastic deformation is not clear and forming leads to dimensional changes. It should be taken into account that when the PMMA or PC items (plates) are first heated a manufacture technology dependent shrinkage also occurs.

To allow easier application, the plates are given protective films of different colors on their front and back, and plates on different chemical bases (e.g., PMMA or PC) or produced with different manufacturing methods are laminated with protective film of different colors. If these plates are used outdoor under severe weathering conditions, the PE protective films have to be peeled within 4 weeks, in order to avoid the degradation (hardening) of the protective film and buildup of the adhesion.

Another problem is the hygroscopicity of plastic plates. They absorb humidity and reabsorb it after cooling under 100°C. The protective film seals the plate against water absorbtion also. Film-protected plates do not require elimination of water before use [350].

Cast PMMA has a broader processing temperature range than extruded PMMA, which has a more exactly controlled thickness tolerance and lower shrinkage than cast materials. Therefore, extruded PMMA is easier to coat with a protective film. The product range for extruded grades is broader. Many surface qualities are manufactured into the product. Generally, cast PMMA is

TABLE 11.18
Application and Processing Characteristics of Protective Films Used for Plastic Plates

Protective Film		Protected Plastic Plate		Laminating Temperature (°C)		Processing Conditions (°C/min)	Adhesive Properties (N/10 mm)
Grade	Thickness (μm)	Manufacture	Material	Laminating Cylinder	Plate		
Adhesive coated	50	Extrusion Casting	PMMA	70	70	150/15	0.5–0.8
Adhesiveless	40	Extrusion	PMMA PC	70	70	150/15 170/15	0.05–0.5
	80	Casting	PMMA	70	30	170/30	0.01–0.04

manufactured with gauges of 1.5–250 mm. Extruded PMMA can have a thickness of 1.5–18 mm. A quality required for food contact (XII.BGA Recommendation, and FDA Regulation 177.1010) is also available.

PMMA is used for illuminated advertisements, light domes, displays, pictures, lamps, vehicles, noise protection panels, solar panels, furniture, and sanitary items among other things. The plates are processed in various ways. Warm deep drawing is one of the main processing operations for PMMA plates. The maximal pressing force by heat pressing may attain 200 kN. Form stability after processing depends on the product type. Cast PMMA achieves its form stability at 70°C, extruded PMMA at 60°C, and PC at 110°C. PC needs a higher forming temperature (190–210°C); its thermal processing temperature range is narrow.

For extruded PC and PMMA plates, a 70/40 μm SAF with 18% EVAc is suggested. Such films are laminated at 60–70°C. For instance, a 40 μm coextrudate is having an 8 μm EVAc adhesive layer (based on a copolymer with 7–9% VAc). Extruded PC or cast PMMA are processed as sanitary materials. The carrier film for such applications has to be deep drawable and to support the processing conditions (160–170°C, 10–30 min) of the protected item. For such applications, self-adhesive or adhesive-coated films with a minimum thickness of 70–80 μm can be used.

Generally, thin plates (less than 6–10 mm) are protected with SAF (70 μm). LDPE/EVAc films (80/20) with an EVAc copolymer having 9–12% VAc have been suggested. For thicker plates, adhesive-coated protective films with a higher thickness (80–100 μm) have been proposed.

PC is manufactured as film (5–750 μm) and as plates. PC films are laminated with the protective film by using a laminating station (with rubber-coated cylinders) at 60°C and a lamination pressure of about 50 kPa/cm^2. Laminating pressure, cylinder hardness, cylinder cooling, temperature, running speed, and web tension influence the final adhesion level and film shrinkage. Extruded PC plates are laminated after processing and before cooling (at 60°C). Such plates are processed for 2–5 min at 180°C (e.g., for light domes, 2 min at 180°C). Protective films with a thickness of 25–100 μm (9–12% VAc) have been tested for PC plates. Passive protective films for PC plates have a thickness of 50 μm; processible protective films have a thickness of 70 μm. Such films are tested for adhesion buildup at normal and elevated temperature. Both the short-term and the long-term adhesion performances have to be tested. Instantaneous adhesion at room temperature and adhesion after storage (2 days) at room temperature, respective adhesion buildup after storage at 80°C (4 h), are evaluated. As is known, PC plates are hygroscopic and are generally dried before use. Drying is carried out at 130°C for 0.5–48 h depending on the plate thickness (0.75–12 mm). Masking should be applied at both 50 and 100°C. Therefore, short term adhesion buildup is tested after storage at 130°C. Shrinkage at 130°C is very important. Long-term adhesion buildup is tested after storage at room temperature and at 40°C for 2–8 weeks. The end adhesion should be situated at about 4 N/20 cm. If adhesiveless films are used, the strong influence of the laminating temperature should be taken into account (Table 11.19). If adhesive-coated films are applied, attention should be paid to the dependence of the peel buildup on the coating weight (Table 11.20).

Laminating can ensure sufficient instantaneous adhesion. If the product stiffens when it is cooled and during storage at room temperature, surface contact decreases and delaminating may occur. In contrast, for the same product, high temperature laminating (80°C) may lead to unacceptable adhesion buildup. For thick masking films used for PC or PMMA plates, shrinkage and lay flat are important quality criteria (Table 11.21).

3. Protective Films for Automotive Storage and Transport

Protective films for storage and transport have been developed to replace conventional systems used to protect the vehicle surfaces from scratches, iron dust, bird droppings, acid rain, etc. (see Table 2.13). The best known conventional protective systems are dispersions of low molecular,

TABLE 11.19
Dependence of Adhesion of Hot Laminated Film on Laminating Temperature[a]

Laminating Temperature (°C)	Peel Resistance (N/10 mm)
50	0.1
60	0.2
70	0.4
80	0.7
90	0.7
100	0.7

[a]The instantaneous peel resistance of a corona treated (52 dyn/cm), 60 μm polyethylene film on extruded PMMA was measured.

nonadhesive, or slightly adhesive polymeric products like waxes or acrylic compounds (see SAFs Based on Other Compounds, Chapter 8, Section III.k). Such products buildup a protective layer of about 15 μm thickness on the protected surface. These compounds are also applied in liquid state and are eliminated as a liquid system. Solvents are used for paraffin and alkalies (WB solutions) for acrylics. Such systems giving full surface protection are more effective than carrier-based films. Their disadvantages include working time and pollution.

Protective coatings generally have to fulfill the following requirements: chemical resistance (resistance to organic solvents, grease, oils); versatility (suitable for all types of varnishes); weathering resistance (outdoor exposure at least 6 months, to pass the Florida Test on cars, to resist harsh winter climates); minimum light transmission (lower than 0.04%); good environmental aspects (recycling, burning); and low costs compared with waxes (material, labor).

As for their adhesive properties, an initial peel of 4 N/25 mm, easy removability (hand peel off), and no visible changes to the protected surface are required. Table 11.22 lists the application requirements and test methods for automotive protective films. As can be seen from this table, these tests are related to the practical transport, storage, and end-use conditions of vehicles and include mostly empirically stated conditions.

The surface to be protected must be clean; dust, stains, and waterdrops should be removed. The film should be stretched tightly to minimize wrinkles and trapped air. It is desirable that the boundary line between the wrapped and unwrapped areas be round and vertical. The film should be put on

TABLE 11.20
Dependence of the Peel Buildup on Coating Weight[a]

	Peel Resistance (N/10 mm)	
Coating Weight (g/m²)	After 3 days	After 7 days
1.0	0.35	0.20
1.2	0.43	0.24
1.3	0.70	0.30
1.4	0.82	0.35
1.6	—	0.40

[a]A solvent-based acrylic adhesive was crosslinked with isocyanate curing agent (2.5%).

TABLE 11.21

End-Use Characteristics of Masking Films for Plastic Plates[a]

Product Code	Lay Flat (mm) 90°C	Lay Flat (mm) 120°C	Shrinkage (%)
1	1	1	1.1
2	1	1	1.0
3	1	3	1.0
4	1	2	1.1

[a] A 250 μm polyolefin film on PC was tested.

the substrate by slightly pressing it from the center area outwardly with a cloth and by hand or with a sponge roller. Deapplication should be carried out by hand peel-off. Quality problems that may appear include air marks, a visible borderline, and adhesive deposits. Generally, the chemical composition of the lacquers used strongly influences the applicability of such protective film. Soft lacquers, for example, lacquers based on low T_g amino-alkyl resins, are more sensitive than others. Highly crosslinked adhesives are suggested for automotive use (Table 11.23).

According to automotive suppliers, protective films for cars should meet special requirements concerning outward appearance, adhesive properties, stainability, and paint protection. Outward appearance is controlled by visual evaluation and concerns color, thickness, and light permeability. Adhesive properties are measured initially and after aging on the paint by overlapping. Stainability is tested on automotive paint, but component stainability should also be tested, which means that the film should be tested on various automotive parts (glass, sun roof, rubber seal, emblems, stainless steel moldings, headlight lens, etc.). Paint protection has to include resistance to acids, alkalies, rail dust, gasoline, and oil.

TABLE 11.22
Requirements for Automotive Protective Films

Application Criteria		Test Method	
Criterion	Parameter	Method	Conditions
Temperature resistance	High temperature	Storage stability	10–15 h/50°C, 10–15 h/80°C
	Low temperature		10–15 h/50°C
Chemical resistance	Battery acid	Contact	1 h, RT, pH 3
	Alkali		0.1 N NaOH, 80°C, 4 h,
	Motor oil		1 h, 70°C
	Calcium chloride		Saturated solution
	Washer fluid		4 h, 80°C
	Gasoline		10 min
Resistance to air pollutants	Metal dust (iron)	Contact	Iron dust, 200 mesh, 80°C
	Grain dust		80°C, 4 h
	Bird drop		80°C, 4 h
	Smoke		80°C, 4 h
Sealing quality	—	—	Water leakage test
Humidity resistance	—	Storage	5 days
Aging resistance	External exposure	Storage	1, 2, and 3 months, Florida test

TABLE 11.23
Crosslinking Agent Concentration as a Function of Product Application

Product	Curing Agent/Adhesive Ratio	
	In Adhesive	In Primer
Automotive protective tape	1/76	1/250
Thermoformed protective film	1/1000	—
Common protective film	1/40	1/250

Automotive undercarriage protective coatings and films have also been developed. Such films have to display resistance to abrasion, light, weather, aging, washing, and gasoline. They have to be heat deformable (150°C). Certain films are based on a soft PVC carrier material [351].

4. Deep Drawable Protective Film

Deep drawability is a general requirement for the protective films used as technological aids. Metals (coils) and plastics (films, plates, wovens, and nonwovens) are formed by deep drawing. Deep drawable films can be classified according to the surface protected into films for processing of metals and films and for processing of plastics (Figure 11.1).

Deep drawing of metals is more complex than that of plastics because metals are less ductile and require much higher forces for cold drawing. According to the depth of the drawing, standard (depth of about 25 mm) and special deep drawing (more than 300 mm) are carried out. Advanced deep drawing may be carried out in one step or in multiple (one to four) steps. The forces acting on the film increase with the number of steps.

Common deep drawable film is used to protect metal plates (steel, copper, etc. with a thickness of 0.7–0.8 mm [352]) which are deep drawn in one or more steps. This procedure is carried out at room temperature. Friction may cause slight increase of the temperature. The protective film coated on the substrate has to display the same deformability as the adherend, that is, a plastic deformation. The elastic forces in the film should be limited to prevent delamination. The film has to resist to the higher temperature produced by friction and must also play (partially) the role of lubricating agent.

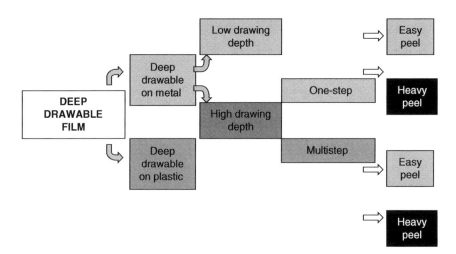

FIGURE 11.1 Classification of deep drawable protective films.

As discussed in Ref. [352], the friction (F) during processing is the sum of the solid-state friction between metal and metal (F_s), friction (F_L) in the lubricant layer at the boundary between the coil and the tool, and the hydrodynamic and hydrostatic friction (F_H) in the thick lubricant layer (Figure 11.2).

$$F = F_s + F_L + F_H \tag{11.1}$$

For a coil covered with protective film, the PE film participates in the friction between the solid-state components. In this process, a high contact pressure ($15-70 \text{ N/mm}^2$) is applied [352].

The time-dependent interlayer phenomena during deep drawing are influenced by the compression stress caused by friction (τ_F) which depends on the temperature (T), normal stress (σ_N), relative speed (v_{rel}), and degree of forming (ε^0) [353]:

$$\tau_F = f(T, \sigma_N, v_{rel}, \varepsilon^0) \tag{11.2}$$

The relative speed affects the temperature, and both depend on the drawing steps. The temperature influences the normal stress in the plastic component (protective film) also.

Plastics can be processed by deep drawing at room temperature or at elevated temperatures. Among the common polymers, PC, PVC, and ABS can be processed at room temperature via deep drawing [209]. As for PMMA, the deep drawing temperature of different PVC grades differ according to the polymerization procedure and film manufacturing process. Suspension-polymerized PVC has a lower and broader processing temperature range ($125-165°C$). High temperature deep drawing of PE can be carried out at a temperature as low as $30°C$. The deep drawing of PC is made at $170-180°C$. In such processes, heat is also evolved by friction between the deep drawing tool and the material processed. Therefore, water-based or organic lubricants are used. In the deep drawing of plastics, the different thermal expansion coefficients of different materials should be taken into account. For instance, the coefficient of thermal expansion for PC is higher ($65 \times 10^{-6} \text{ K}^{-1}$) than that of steel ($11 \times 10^{-6} \text{ K}^{-1}$) [330].

The common deep drawable protective films are built up on a clear, $50-70 \text{ μm}$ thin film of PVC or on polyolefin. No slip agents are allowed in their composition. However, protective film should function as a lubricating agent also [350]. The carrier film for deep drawable protective films should have good coatability, good slip, excellent deformability, and long-term deformability also. Coatability refers to one-side wettability and adhesive anchorage. Good slip means that the

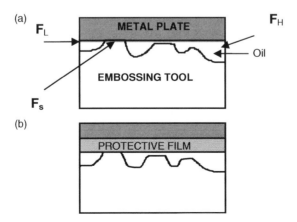

FIGURE 11.2 The role of the protective film as lubricating agent by deep drawing. (a) Processing without protective film; (b) processing with protective film.

bonded film should allow movement parallel to the drawing device. Deformability means that during embossing the film should follow the deformation of the metal plate. Metal plates are deep drawn up to 40–60 cm. Long-time deformability means that the film preserves its deformation after deep drawing and does not change form after the embossing device is removed. Such properties are given by soft PVC. Deformability is characteristic for LLDPE also, but its elastic forces do not permit its use alone. Excellent deep drawability is given by EVAc. According to Miura [354], the deep drawability of hydrocarbon-based films decreases as follows:

$$EVAc > PE > PP > PS \tag{11.3}$$

The processing temperature ($^\circ$C) of the previously mentioned polymers varies as follows:

$$PE\,(135) > PP\,(125) > PSt\,(105) > EVAc\,(90) \tag{11.4}$$

Unfortunately, the high blocking tendency allows the use of EVAc only as coextrudate. Special polymer compounds have been also suggested. For instance, acrylic copolymers have been suggested as additives for PP to improve its deep drawability [355]. Polyolefin-based deep drawable carrier film compositions contain an extensible component like EVAc, PP, or LLDPE and other components which ensure adequate mechanical resistance, friction behavior, and machinability.

Deep drawable protective films are generally coated with a medium coating weight of 5.5–8.5 g/m^2 of a water-based acrylic PSA or 4.5–5.5 g/m^2 of a solvent-based rubber-resin PSA. Because of the higher adhesion of the rubber-resin based formulation, such products are also release coated. For complex shapes and advanced deep drawing, rubber-based adhesives are preferred.

5. Protective Films for Carpets

Carpet-protecting films can be used for long fiber tufted carpets (high tack protection film), thermoformed embossed carpet (high tack deformable, deep-drawable protective film), and processing protection. Carpet films applied for protection during storage differ from automotive carpet protection films, which are undergoing the forming processes of automotive carpets.

For the protection of textile webs, different films are used. Their characteristics have to be tailored to the manufacturing technology and buildup of the carpet, that is, its adhesivity and its postprocessing. Buildup includes fiber nature, length, and construction. Tufted carpets generally have PA pile fibers and polyester-based fabric. Carpet processing depends on the end-use. Automotive carpets are generally molded. Common carpet-protecting films for normal use are based on SBAC adhesives coated on an 80 μm PE carrier material. Automotive carpet films protect automotive carpets during the manufacture and transport of the vehicle. They have to possess deep drawability and good adhesion. Special carpet-protective films must have antislip properties [356]; they have to fulfill the requirements of DIN 51130 [357]. Like other adhesives used for automobiles, the PSAs coated on such products have to meet the requirements for volatile organic compounds according to DIN 75201 (Foggingtest) and PV 3341 (VW) or PB VWL 709 (Daimler Chrysler AG) [358].

6. Mirror Tape

Mirror tape is a permanent security film used to avoid dangerous destroying (cracking) of large glass surfaces. This film is laminated as a reinforcing layer on the back of mirrors or on other impact-sensitive large surface items. The film has to have a high level of permanent adhesivity in order to avoid delamination of the glass by stress and dart drop resistance against glass splits.

7. Overlamination Protection Film

Self-wound overlamination protection films are used for overlapping and permanent protection of labels or other printed items. They have to display clarity, impact resistance, UV resistance, color stability, resistance to outdoor weathering, and printability. A matte finish, nonreflective character is also needed. PP and PET have been suggested as carrier material for such PSPs. BOPP (25–50 μm, standard and battery approved, normal and heat-resistant lacquered) and PET (25 μm) are generally used.

For screen printers and litho printers, thermal laminating was used as the finishing process. For overlaminating of screen-printed labels, a product buildup of 50/150, that is, 12.5 μm PET/75 μm heat-activated adhesive, has been recommended. This combination gives a greater gloss depth especially with embossed rough paper. UV-resistant thermolaminating films and inexpensive olefin copolymers have also been developed. Actually 10–15% of the printed products are overlaminated. Heat laminating has been (partially) replaced by pressure-sensitive protective laminating [359]. PET and BOPP are used for pressure-sensitive overlaminating. Thermal laminators vary in size and complexity. Tabletop models, machines with automatic feeders, and automatically synchronized cut-off have been developed.

8. Separating Films

Such products are used to protect web- or sheet-like metallic, plastic, or varnished items during storage. They must have a very low peel. Therefore, they are based on rubber-resin formulations with a very low tackifier level (less than 10%) and a high degree of crosslinking. Because of the polar surface of the protected items (e.g., anodized aluminum or stainless steel) and the high laminating pressure, peel buildup occurs rapidly.

9. Other Protective Films

Polarizing films are used for liquid crystal displays and have to offer superior transparency and humidity and heat resistance. The so-called E-masks are films protecting electronic components during the etching process.

V. FORMS

As discussed earlier, forms are multilayered laminates that have exactly controlled adhesion forces in the laminate layers and are discontinuous/finite elements on or in a web-like product, which allow detachment of one or many parts of the forms, making simultaneous multipurpose information transfer possible. Piggy back mailing labels are a typical example of forms; they are sandwich labels. Such products can contain two or more peel-off layers over a single backing paper. The complete label, which is die-cut and personalized in one pas, is affixed to the consignment, and two small numbered slips are peeled off and attached to control forms during the routing process [52]. Forms are applied by detaching or separating the parts of the form. This is carried out manually. Detachment and refixing of parts of forms (e.g., cards) are carried out by hand, but certain parts of a form (e.g., main web) can be processed automatically also.

VI. PLOTTER FILMS

Plotter films are web-like PSPs that serve as raw material for the manufacture of information-carrying items. Special equipments are used to cut letters, signs, and repromasks from plotter films [360]. The plotter film is used as carrier material for self-adhesive signs, letters, typefaces, or decorative elements. It is a bulk or surface colored (coated) film with excellent dimensional stability. That means minimal deformation during processing, storage, and application. Plotter film has to display very good die cuttability, sharp contours, no plasticity or elasticity during cutting, and excellent lay flat and stiffness. It must also be weather resistant. Such carrier materials display elongation values of about 300% MD and 300% CD and modulus values of about $7-15\,N/mm^2$. High quality products exhibit elongation values of 24–200% (DIN 53455) [361]. The first U.S. made pressure-sensitive PVC films were tested in Europe as plotter film in 1946 [362]. The first usable PVC films were manufactured in the 1950s. In 1984, computerized cutting plotters have been launched.

PVC (cast and calendered) and polyolefin films are manufactured as carriers for plotter film. Plasticized PVC and vinyl chloride copolymers have been developed for plotter films. Plasticized PVC films can be used outdoors for 3–5 yr. Films based on vinyl chloride copolymers can withstand 5–7 yr of external use [359]. The finished product must exhibit excellent chemical and environmental resistance. Resistance to water, acid, alkalies, surface active agents, ethylene glycol, motor oil, methanol, gasoline, and certain other substances is required.

Such films are processed using special plotters to cut them. A procedure for the application of plotter films is described in Ref. [361]. Letter cutting plotters have been developed [359]. Such cutting machines can use reels with a width of up to 1300 mm [362] and may run at speeds of up to 400–600 mm/s [362,363] and tolerances of 0.1 mm. The pressure of the cutter is adjustable for different film thicknesses. Rollplotter can process up to $1,320 \times 25,000$ mm web with a minimum sign dimension of 3 mm [364]. Generally, the working speed and acceleration are given as performance characteristics [365].

Plotter films are used together with a handling or fixing aid, that is, the application tape (see Application Tape). Such plotter films have to be applied (laminated) at temperatures 4–50°C [361]. Wet- and dry-application methods are used. In the wet technique, warm water should be used for calendered film and cold water for the cast film. Their end-use temperature is in the range of −56 to +107°C.

There are special application papers (or films) that can be used like a plotter film to produce logos or decorative elements. For such products, the continuous, web-like part of the paper or film is used after the places to be lacquered or printed are cut off. Such stencil films or papers generally have a release liner because of their improved adhesivity. This is necessary to avoid ink penetration at the edges of the carrier [366].

VII. OTHER PSPs

A number of pressure-sensitive imaging materials have been developed. Their number is growing with the increase in the number of new carrier materials, printing techniques, and packaging methods. In 1974, PSA stamps have been introduced on a trial basis [367]. They cost five times as much as conventional stamps. Pressure-sensitive envelopes have been developed [368]. Message reply pads use low peel PSAs [369]. Cast vinyl marking films are used for producing labels, emblems, stripes, and other decorative markings for trucks, automobiles, and other equipment. They are durable, conformable, and dimensionally stable. Decal films have been developed for marking functional machine, building parts, etc. [370]. Two-way decal films are used for push/pull glass doors. Security decals, and association membership decals, etc. have been produced [371]. Collections are manufactured [372].

PSAs can be applied as laminating adhesives also. Decorative panels bearing PSA have been manufactured by pressing substrates (e.g., veneer) precoated with metallic alkoxides or chelates onto decorative sheets coated with PSA containing reactive groups (e.g., *N*-methylol acrylamide) [373]. Films based on cellulose have been applied by hot-pressing on paper to protect it [374]. Electrically conductive PSAs are used for biomedical electrodes [375]. Laminating films for book binding are used as reinforcement, repair, binding, and lamination [376].

Decor films are used in household and technical domains [243,377–380]. They are manufactured as adhesive-coated or self-adhesive films. Self-adhesive decoration films contain a high level of plasticizers, therefore, it is difficult to print on them. For PVC decor films, water-based gravure printing is suggested [381]. Such products can have a printed release liner too.

Self-adhesive wall cover consists of a layer of fabric with a visible surface, a barrier of paper that has one surface fixed to another fabric layer, a PSA coated on the barrier paper, and a release paper [382]. Low styrene content (15%) styrene–ethylene–butylene–styrene (SEBS) block copolymers have been proposed for adhesives for wall covering. Metallized film used for electromagnetic protection can be applied by using PSAs. Such products are suggested for low volume items with simple shapes and for optically low quality applications [383].

Self-sticking carpet tiles have a pressure-sensitive layer [384]. The system comprises tiles having adhesive and nonsticky material on the back. Self-adhesive wall carpets have been also manufactured. PSAs or PSPs may be used as formulation or constructional components in other products also. PSAs can be used for attachment of plastic cards (see also Forms) [385]. Diapers include standing gather, frontal tape, and foamed waist elastics [386]. Postexpansion adhesive sheets are used to cover odd-shaped spaces between parts [387]. A cleaning device based on PSPs has also been developed [388]. Scent labels were produced placing original perfume oils in a special two layer label structure [389].

Pressure-sensitive aerosol adhesive is used to bond paper, cardboard, plastics, films, foil, felt, cloth, metal, glass, or wood [390]. The PSA market for do-it-yourself applications has a higher growth rate than the average for PSAs [391]. Sprayable HMPSAs are suitable for bonding laminating panels, thin plastics, fibrous paneling, and foam materials [392]. It is evident that new application fields and requirements, new raw materials, and advances in manufacturing technology will force an increase in the number of special PSPs.

REFERENCES

1. I. Benedek and L.J. Heymans, *Pressure-Sensitive Adhesives Technology*, Marcel Dekker, Inc., New York, Basel, Hong Kong, 1997, chap. 5.
2. I. Benedek, *Developments in Pressure-Sensitive Products*, Marcel Dekker, New York, 1999, chap. 8.
3. I. Benedek, *Pressure Sensitive Adhesives and Applications*, Marcel Dekker, New York, 2004, chap. 5.
4. M. Fairley, *Labels, Label.*, (7/8), 272, 1995.
5. *Adhes. Age*, (7), 34, 1983.
6. R. Jordan, *Adhäsion*, (1/2), 22, 1987.
7. *Finat Label. News*, (3), 12, 1996.
8. J.M. Casey, *Tappi J.*, (6), 42, 1998.
9. M.H. Mishne, *Screen Printing*, (1), 60, 1986.
10. K. Fust, *Coating*, (2), 65, 1988.
11. K. Heger, *Coating*, (5), 180, 1987.
12. *Druckwelt*, (14/15), 6, 1988.
13. *Kaut. Gummi, Kunstst.*, 42(5), 420, 1988.
14. *Adhäsion*, (11), 9, 1988.
15. *Der Siebdruck*, (8), 75, 1986.
16. Ball Bros Co., U.S. Patent 3,352,708/2, *Coating*, (7), 210, 1969.
17. Owens Illinois Glass Company, OH, USA, U.S. Patent 31,47,135, *Adhäsion*, (2), 80, 1966.
18. *Adhäsion*, (5), 28, 1982.
19. Ball Brothers Co., Muncie, Ind., USA, DBP 1,284,585, 1977.
20. *Coating*, (9), 269, 1972.
21. C.S. Maxwell, *Tappi J.*, (8), 1464, 1970.
22. L. Ridgeway and L. Pyle, *Tappi J.*, (12), 129, 1968.
23. *Papier Kunstst. Verarb.*, (1), 24, 1996.
24. *Etiketten-Labels*, (5), 22, 1995.
25. *Coating*, (7), 243, 1987.

26. *Coating*, (5), 151, 1969.
27. B.H. Gregory, *Extrusion Coasting Advances Resins, Processing, Applications, Markets*, in Proceedings of Polyethylene '93, The Global Challenge for Polyethylene in Film, Lamination, Extrusion, Coating Markets, October 4, 1993, Maack Business Services, Zurich, Switzerland.
28. A. Ridgeway and L.K. Mengenhagen, TAPPI Proceedings, EAA/Polybutylene Blend for Packaging Applications, in Proceedings of Polymers, Laminations & Coatings Conference, 1992, p. 52.
29. *Coating*, (12), 450, 1986.
30. G. Bulian, *Extrusion Coating and Adhesive Laminating: Two Techniques for the Converter*, in Proceedings of Polyethylene '93, Maack Business Service Conference, Zürich, Switzerland, 4/6, 1993.
31. C. Gondro, *Kunststoffe*, (10), 1082, 1990.
32. U. Reichert, *Kunststoffe*, 83(10), 737, 1993.
33. *Verpackungs Rundsch.*, (12), 1394, 1985.
34. C. Stöver, *Kunststoffe*, 84(10), 1426, 1994.
35. H. Maack, *Coextruded and Coinjected Packaging Developments and Recycling Aspects*, in Proceedings of Maack Business Service, Speciality Plastics Conference, Zurich, Switzerland, 1989.
36. *Adhes. Age*, (11), 10, 1986.
37. *Brit. Plast.*, (10), 64, 2001.
38. Nichimen Corporation, Exceed (Biaxially oriented EVOH film), Technical Data, 1995.
39. J.A. Fries, *New Developments in Hot Melt PSA and Heat Seals*, in Proceedings of the Tappi Hot Melt Symposium, June 7–10, 1987, Monterey, CA, USA.
40. *Coating*, (6), 214, 1996.
41. G. Bonneau and M. Baumassy, *New Tackifying Dispersions for Water Based PSA for Labels*, in Proceedings of the 19th Munich Adhesive and Finishing Seminar, Munich, Germany, 1994, p. 82.
42. P. Dunckley, *Adhäsion*, (11), 19, 1989.
43. T. Kishi, Japanese Patent, 6,369,879, (Sekisui Chem. Co. Ltd.), 1988, *CAS, Adhesives*, 19, 6, 1988.
44. Raflatac Adhesives for Extreme Conditions, Raflatac, Data Sheet, 1992.
45. *Kunststoff J.*, (9), 61, 1986.
46. *Brit. Plast.*, (10), 42, 2001.
47. *Labels, Label.*, (3/4), 11, 1994.
48. A. Dobmann and J. Planje, *Papier Kunstst. Verarb.*, (1), 37, 1986.
49. *Adhes. Age*, (12), 38, 1987.
50. D.G. Pierson and J.J. Wilczynski, *Adhes. Age*, (8), 52, 1990.
51. *Adhes. Age*, (3), 10, 1974.
52. J. Müller and D. Eisele, ATP News-Letter, 3, 4, 2004, ATP Adhesive Systems GmbH, Wuppertal, Germany.
53. A. Prittie, *Finat News*, (3), 35, 1988.
54. Z. Czech, *Eur. Adhes., Seal.*, (6), 4, 1995.
55. W. Grebe, *Papier Kunstst. Verarb.*, (2), 49, 1985.
56. *Verpackungs Rundsch.*, (9), 994, 1983.
57. T.G. Wood, *Adhes. Age*, (7), 19, 1987.
58. H.J. Teichmann, *Papier Kunstst. Verarb.*, (11), 10, 1994.
59. A.W. Norman, *Adhes. Age*, (4), 35, 1974.
60. *Labels, Label. Int.*, (5/6), 24, 1997.
61. Lasertechnik in der Papierverarbeitung, *Papier Kunstst. Verarb.*, (11), 57, 1994.
62. *Etiketten-Labels*, (3), 9, 1995.
63. *Labels, Label. Int.*, (5/6), 20, 1997.
64. P. Thorne, *Finat News*, (3), 47, 1988.
65. A. Linder, *Hauptmarkttendenzen im Einsatz von Selbstklebenden Etiketten*, in Proceedings of 19th Munich Adhesive and Finishing Seminar, Munich, Germany, 1994, p. 118.
66. *Adhäsion*, (5), 36, 1988.
67. *Kunststoff J.*, (9), 82, 1986.
68. *Adhes. Age*, (1), 6, 1985.
69. F.T.R. Mecker, *Low Molecular Weight Isoprene Based Polymers-Modifiers for Hot Melts*, in Proceedings of TAPPI Hot Melt Symposium 1984, in *Coating*, (11), 310, 1984.

70. *Coating*, (3), 65, 1974.
71. I. Benedek, *Pressure-Sensitive Adhesives and Applications*, Marcel Dekker, New York, 2004, chap. 10.
72. J. Lechat, European Tape and Label Conference, Exxon, Brussel, Belgium, April 1989, p. 67.
73. *Adhäsion*, (11), 9, 1988.
74. *Verpackungs Rundsch.*, (2), 154, 1986.
75. *Etiketten-Labels*, (5), 9, 1995.
76. *Papier und Kunststoff Verarbeiter*, (12), 44, 1986.
77. H. Mueller and J. Tuerk, EP 0,118,726 (BASF A.G., Ludwigshafen, Germany), 1984.
78. *Papier Kunstst. Verarb.*, (11), 57, 1994.
79. *Druckwelt*, (14/15), 27, 1985.
80. *Tappi J.*, (6), 159, 1988.
81. *Etiketten-Labels*, (5), 100, 1995.
82. Packlabel News, Packlabel Europe 97, Ausgabe 2.
83. I. Benedek, *Adhäsion*, (3), 22, 1987.
84. *Etiketten-Labels*, (5), 24, 1995.
85. *Etiketten-Labels*, (5), 24, 1995.
86. *Neue Verpackung*, (3), 10, 1995.
87. *Labels, Label., Int.*, (5/6), 18, 1997.
88. *Avery Rating Plates and Type Plates Last a Fairly Long Time*, Avery Etikettier-Logistik GmbH, Eching b. München, Germany, Technical Booklet, 1995.
89. *Package Print, Design Int.*, (1/2), 8, 1997.
90. *Adhes. Age*, (11), 10, 1986.
91. P. Broschk, *Adhäsion*, (6), 14, 1985.
92. J. Schaetti, *Aufbügelbare chemisch reinigungsbeständige Etiketten*, in Proceedings of 9th Munich Adhesive and Finishing Seminar, Munich, Germany, 1984.
93. D. Lacave, *Labels, Label.*, (3/4), 54, 1994.
94. *Labels, Label.*, (3), 82, 1994.
95. LOS Lager Organisations System GmbH, Mannheim, Germany, Gebrauchsmuster, G 8,706,322.0, 1987.
96. D. Allen Jr. and E. Flam, U.S. Patent, 4,650,817 (Bard Inc., Murray Hill, NJ, USA), 1987.
97. Morstik 103 Adhesive, Morton Thiokol Inc., Adhesives and Coatings, MTD-MS 103-12, 1988.
98. R.L. Sun and J.F. Kenney, U.S. Patent, 4,762,688 (Johnson and Johnson Products Inc.), 1988, in *CAS, Hot Melt Adhesives*, 26, 1, 1988.
99. S.E. Krampe and C.L. Moore, EP 0,202,831 A2, Minnesota Mining and Manuf. Co., St. Paul, MN, USA, 1986.
100. U.S. Patent, 3,661,874, in F.D. Blake, EP 0,141,504 A1, Minnesota Mining and Manufacture, Co., St. Paul, MN, USA, 1985.
101. U.S. Patent, Blake, 3.865770, in F.C. Larimore and R.A. Sinclair, EP 0,197,662 A1, Minnesota Mining and Manuf. Co., St. Paul, MN, USA, 1986.
102. *Adhes. Age*, (4), 6, 1983.
103. DDR Patent 64,111, *Coating*, (1), 24, 1969.
104. J.R. Pennace, C. Ciuchta, D. Constantin, and T. Loftus, WO 8,703,477 A, 1987.
105. *Neue Verpackung*, (1), 156, 1991.
106. Z. Czech, *Eur. Adhes., Seal.*, (6), 4, 1995.
107. U.S. Patent 3,321,451, in S.E. Krampe and C.L. Moore, EP 0,202,831 A2, Minnesota Mining and Manuf. Co., St. Paul, MN, USA, 1986.
108. *Pack Report*, (19), 28, 1986.
109. *Seifen Öle Fette Wachse*, 112(17), 613, 1986.
110. *Labels, Label.*, (3/4), 44, 1994.
111. *Verpackungs-Berater*, (3), 18, 1997.
112. Tagsa, *Newsletter*, 1995, p. 1.
113. D. Boettger, *Labels Label.*, (3/4), 58, 1994.
114. S.W. Thompson, Malvern, PA, USA, U.S. Patent 4,674,771, 1987.
115. J. Nentwig, *Converter, Flessibili-Carta-Cartone*, (1), 66, 1991.

116. Oji Paper Co., Japanese Patent, 07,325,391, 1995, *Adhes. Age*, (5), 12, 1996.
117. *Labels, Label.*, (3/4), 50, 1994.
118. *Labels, Label.*, (3/4), 34, 1994.
119. *Adhes. Age*, (8), 8, 1983.
120. K. Unbehaun, *Deutscher Drucker*, (6), 43, 1988.
121. Interpack-Scotch 3560-O, 3M Presse information, 04.02, Düsseldorf, Germany.
122. J. Dostal, *Verpackungs-Rundsch.*, (5), 74, 2004.
123. *Allg. Papier Rundsch.*, 22, 787, 1986.
124. *Brit. Plast.*, (10), 47, 2001.
125. R. Davis, *Fassson Facts Int.*, (1), 2, 1969.
126. *Etiketten-Labels*, (5), 91, 1995.
127. *Coating*, (7), 232, 1989.
128. *Adhäsion*, (1/2), 78, 1987.
129. H. Roder, *Adhäsion*, (6), 16, 1983.
130. W.E. Broxtermann, *Adhes. Age*, (2), 20, 1988.
131. K.F. Schroeder, *Adhäsion*, (5), 161, 1971.
132. W. Möhren, Doppelseitiges Haftklebeband zur Klebung von Gummiprofilen im Fahrzeugbau, *Adhäsion*, (6), 14, 1985.
133. M. Haufe, ATP News-Letter, 3, 9 (2004), ATP Adhesive Systems GmbH, Wuppertal, Germany.
134. H. Kato, H. Adachi, and H. Fujita, *Rubber Chem. Technol.*, 56, 287, 1983.
135. E.L. Scheinbart and J.E. Callan, *Adhes. Age*, (3), 17, 1973.
136. *Adhes. Age*, (3), 8, 1987.
137. E.H. Andrews, T.A. Khan, and K.A. Majid, *J. Mater. Sci.*, 20, 3621, 1985.
138. J.A. Miller and E.A. Jakusch (Minnesota Mining and Manufacture Co., St. Paul, MN, USA), EP 0,306,232B1, 1993.
139. G. Sala, EP 273,997, (International Carbon Solvent S.p.A.), 1988, *CAS, Adhesives*, 24, 5, 1988.
140. *Adhes. Age*, (9), 8, 1986.
141. F. Altenfeld and D. Breker, Semi Structural Bonding with high Performance Pressure Sensitive Tapes, 3M Deutschland, Neuss, 2nd ed., February 1993, p. 278.
142. TESA, Technologie des Klebens, Beiersdorf AG, Hamburg, Germany, 1997, p. 55.
143. M. Toshio, Japanese Patent 6,263,865, (Furukawa Electric Co. Ltd.), 1988, *CAS, Adhesives*, 24, 6, 1988.
144. *Coating*, (11), 424, 1985.
145. K. Mizui, PCT, WO 8,8 02,767 (Mitsui Petrochemical Ind. Ltd.) 1988, *CAS, Adhesives*, 22, 4, 1988.
146. R.M. Enanoza, EP 259,968, (Minnesota Mining and Manufacture Co.), 1986, *CAS, Adhesives*, 22, 3, 1988.
147. *Coating*, (1), 8, 1986.
148. T. Sugiyama, N. Miyaji, I. Yoshihide, and T. Tange, Japanese Patent 62,199,672 (Nippon Carbide Ind. Co. Inc.), 1987, *CAS, Adhesives*, 14, 3, (1988).
149. T. Hiroyoshi, H. Kuribayashi, and E. Usuda, P 254,002, (Sumitomo Chem Co. Ltd.), 1988, *CAS, Adhesives*, 14, 3, 1988.
150. C. Cervellati, G. Capaldi, and J. Lutz Erich, EP 273,585, (Exxon Chemical Patents Inc.), 1988.
151. R.G. Czerepinski and R. Gundermann, U.S. Patent 4,713,412 (Dow. Chem. Co.), 1987.
152. T. Tsubakimoto, K. Minami, A. Baba, and M. Yoshida, Japanese Patent 6,335,676 (Nippon Shokubai Kagaku Kogyo Co., Ltd.), 1988, *CAS, Adhesives*, 16, 1, (1988).
153. I. Yorinobu, W. Yasuhhisa, and T. Hiroshi, Japanese Patent 6,386,777 (Japan Synthetic Rubber Co. Ltd.), 1988, *CAS, Adhesives*, 24, 3, 1988.
154. C.H. Hill, N.C. Memmo, W.L. Phalen Jr., and R.R. Suchanec, EP 259,697 (Hercules Inc.), 1988, *CAS, Adhesives*, 14, 5, 1988.
155. D. Akihito, Ota Tomohisa, and U. Toshishige, Japanese Patent 6,368,683 (Hitachi Chem. Co. Ltd.), 1988, *CAS, Crosslinking Reactions*, 14, 7, (1988).
156. K. Akasha, S. Sanuki, and T. Matsuyama, Japanese Patent 6,224,336,669, Nippon Synth. Chem. Ind. Co. Ltd., 1987.
157. S. Shinji and Y. Yoshiuki, Japanese Patent 62,243,670, Nippon Synth. Chem. Co. Ind. Ltd., 1987, *CAS, Adhesives*, 13, 4, 1988.

158. Y. Moroishi, T. Sugii, and K. Noda, Japanese Patent 6,381,183, Nitto Electric Ind. Co. Ltd., 1988, *CAS, Adhesives*, 22, 6, 1988.

159. H. Yaguchi, H. Fukuda, T. Masayuki, and T. Ohashi, Japanese Patent, 6,327,583, Bridgestone Corp., 1988, *CAS, Adhesives*, 19, 6, 1988.

160. B. Copley and K. Melancon, U.S. Patent 878,816, Minnesota Mining and Manuf. Co., 1986, *CAS, Siloxanes & Silicones*, 14, 2, 1988.

161. A. Saburo, Japanese Patent, 63,117,085, Central Glass Co. Ltd., 1988.

162. L.E. Grunewald and D.J. Classen, EP 256,662, Minnesota Mining and Manufacture Co., St. Paul, MN, USA, 1986.

163. S. Koriki, R. Mokino, K. Ito, and T. Unno, EP 239,348, Nagoya Oil Chem. Co. Ltd., 1987, *CAS, Colloids*, 4, 5, 1988.

164. H. Kenjiro and S. Kotaro, Japanese Patent 63,118,383, Nitto Electric Ind. Co. Ltd., 1988, *CAS, Adhesives*, 24, 5, 1988.

165. Japanese Patent, 63,178,091, Minnesota Mining and Manuf. Co. Ltd., St. Paul, 1988, *CAS, Adhesives*, 25, 7, 1988.

166. Morstik 103 Adhesive, Morton Thiokol Inc., Adhesives and Coatings, MTD-MS103–12/1988.

167. A. Nagasuka and M. Kobari, Japanese Patent 6,389,585, Sekisui Chem. Co., Ltd., 1988, in *CAS, Adhesives*, 19, 6, 1988.

168. J.P. Kealy and R.E. Zenk, Canadian Patent, 1,224,678, Minnesota Mining and Manufacture Co., St. Paul, MN, USA, 1987.

169. M.A. Johnson, *J. Plast. Film Sheeting*, (1), 50, 1988.

170. Ashland Chemicals, Aroset, EPO 32216-APS-102, Data Sheet, 1995.

171. A.B. Wechsung, *Coating*, (9), 268, 1972.

172. G.W.H. Lehmann and H.A.J. Curts, Hamburg, Germany, Beiersdorf AG, U.S. Patent 4,038,454, 1977.

173. Y. Sasaaki, D.L. Holguin, and R. Van Ham, EP 0,252, 717, Avery International Co., Pasadena, CA, USA, 1988.

174. *Coating*, (1), 12, 1984.

175. *Openlines*, (8), Spring, 93, P.P. Payne Ltd., England.

176. R.G. Jahn, *Adhes. Age*, (12), 35, 1977.

177. *Adhes. Age*, (3), 6, 1985.

178. E.B. Richmann, S. Nemth, S.W. Tomlinson, and D.W. Wilson, U.S. Patent 624,870, Avery Internat. Co., San Marino, CA, USA, 1975.

179. D.K. Fisher and B.J. Briddell, EP 0,426,198 A2, Adco Product Inc., Michigan Center, MI, USA, 1991.

180. *Adhes. Age*, (12), 8, 1986.

181. G. Galli, EP 0,191,191 A1, Manuli Autoadesivi Spa., Cologno Monzese, Italy, 1986.

182. Italian Patent 21,842 A, 1982, in G. Galli, EP 0,191,191 A1, Manuli Autoadesivi Spa., Cologno Monzese, Italy, 1986.

183. Siegel Band, Kalle Folien, Hoechst, Mi 1984, 38T 5.84 LVI, Hoechst AG, Wiesbaden, Germany.

184. Better Packages Inc., U.S. Patent 3,510,037, *Coating*, (5), 130, 1971.

185. Nitto Denko Products, Catalog Code 02002, Tokyo, Japan, p. 9.

186. R.E. Nelson, U.S. Patent 4,647,485, Vega, Texas, USA, 1987.

187. M. Satsuma, Japanese Patent 6,348,381, Nitto Electric Ind. Co., 1988, *CAS, Adhesives*, 14, 4, 1988.

188. C. Donker, R. Luth, and K. van Rijn, *Hercules MBG 208 Hydrocarbon Resin: A New Resin for Hot Melt Pressure Sensitive (HMPSA) Tapes*, in Proceedings of 19th Munich Adhesive and Finishing Seminar, Munich, Germany, 1994, p. 64.

189. L. Jacob, *New Development of Tackifiers for SBS Copolymers*, in 19th Munich Adhesive and Finishing Seminar, 1994, Munich, Germany, p. 107.

190. M.E. Ahner and M.L. Evans, *Light Colour Aromatic Containing Resins*, in Proceedings of TAPPI Hot Melt Symposium '85, June 16–19, 1985, Hilton Head, SC, USA, *Coating*, (1), 6, 1986.

191. K. Nakamura and Y. Yoshiuki, Japanese Patent 6,386,786, Nitto Electric Industrial Co. Ltd., 1988, *CAS, Adhesives*, 21, 5, 1988.

192. B. Hanka, *Adhäsion*, (10), 342, 1971.

193. *Papier Kunstst. Verarb.*, (1), 20, 1996.

194. *Allg. Papier Rundsch.*, (18), 572, 1987.
195. C. Parodi, S. Giordano, A. Riva, and L. Vitalini, *Styrene–Butadiene Block Copolymers in Hot Melt Adhesives for Sanitary Application*, in Proceedings of 19th Munich Adhesive and Finishing Seminar, 1994, Munich, Germany, p. 119.
196. W.E. Lenney, Canadian Patent, 1,225,176, Air Products and Chemical Inc., USA, 1987.
197. U.S. Patent 3,257,478, in W.E. Lenney, Canadian Patent 1,225,176, Air Products and Chemical Inc., USA, 1987.
198. U.S. Patent 3,697,618, in W.E. Lenney Canadian Patent 1,225,176, Air Products and Chemical Inc., USA, 1987.
199. U.S. Patent 3,971,766, in W.E. Lenney Canadian Patent 1,225,176, Air Products and Chemical Inc., USA, 1987.
200. T.J. Bonk, T.I. Cheng, P.M. Olson and D.E. Weiss, PCT, WO 87/00189/15.01.1987.
201. H. Nakahata, Japanese Patent 07,316,510, 1995, *Adhes. Age*, (5), 11, 1996.
202. Packard Electric, Engineering Specification, ES-M-2359.
203. N. Yamazaki, Japanese Patent 6,386,782, Matsushita Electric Ind. Co., Ltd., 1988, *CAS, Adhesives*, 12, 5, 1988.
204. H.K. Porter Co. Inc., Pittsburgh, PA, U.S. Patent 3,149,997, *Adhäsion*, (3), 79, 1966.
205. *Adhes. Age*, (12), 352, 1984.
206. B. Kunze, S. Sommer, and G. Düsdorf, *Kunststoffe*, 84(10), 1337, 1994.
207. Johns Manville Corp., U.S. Patent 3,356,635, *Coating*, (7), 210, 1969.
208. *Coating*, (5), 151, 1969.
209. P. Kriston, *Müanyag Fóliák*, Müszaki Könyvkiadó, Budapest, 1976, p. 45.
210. H. Kato, H. Adachi, and H. Fujita, *Rubber Chem. Technol.*, 56, 287, 1983.
211. T. Kishi (Sekisui Chem.Co.Ltd), Japanese Patent 6369879/29.03.1988, in *CAS, Adhesives*, 19, 6, 1988.
212. J. Andres, *Allg. Papier Rundsch.*, (16), 444, 1986.
213. Minnesota Mining and Manuf. Co., St. Paul, MN, USA, U.S. Patent 3,307,690, *Coating*, (4), 114, 1969.
214. *Adhes. Age*, (8), 62, 1986.
215. Packard Electric, Engineering Specification, ES-M-1881.
216. R. Shibata and H. Mixagawa, Japanese Patent 6,386,785, Hitachi Condenser Co. Ltd., 1988, *CAS, Adhesives*, 21, 5, 1988.
217. Allmänna Svenska Elekriska, AB,DBP 1,276,771.
218. M. Haufe, ATP News-Letter, 5, 6, 2004, ATP Adhesive Systems GmbH, Wuppertal, Germany.
219. T.J. Kilduff and A.M. Biggar, U.S. Patent 3,355,545, in *Coating*, (7), 210, 1969.
220. Nitto Denko Products, Catalog Code 02,002, Tokyo, Japan, p. 6.
221. *Adhes. Age*, (1), 6, 1985.
222. *Coating*, (3), 210, 1985.
223. A. Dobmann and A.G. Viehofer, in Proceedings of 19th Munich Adhesive and Finishing Seminar, Munich, Germany, 1994, p. 168.
224. P. Penczek and B. Kujawa-Penczek, *Coating*, (6), 232, 1991.
225. Minnesota Mining and Manuf. Co., St. Paul., MN, USA, DBP 1,486,514.
226. Y. Torigoe U.S. Patent 4,645,697 Dainichi Nippon Cable Ltd., Amagasaki, Japan, 1987, *Adhes. Age*, (5), 26, 1987.
227. V.M. Ryabov, O.I. Chernikov, and M.F. Nosova, *Plast. Massy*, (7), 58, 1988.
228. G. Camerini, EP 248,771, Coverplast Italiana SpA, 1986.
229. Y. Hamada and O. Takuman, EP 253,601, Toray Silicone Co., Ltd., 1988.
230. R. Tschudin Mahrer U.S. Patent 4,564,550, Irbit Research Consulting AG, Fribourg, Switzerland, 1986.
231. *Adhes. Age*, (8), 62, 1986.
232. B. Ellegard, *Etiketten Labels*, (3), 47, 1995.
233. *Labels, Label.*, (2), 22, 1997.
234. Henkel & Cie. GmbH, Düsseldorf, Germany, DBP 1,179,660, *Adhäsion*, (2), 84, 1966.
235. *Adhäsion*, (6), 14, 1985.
236. Johnson and Johnson, U.S. Patent 3,403,018, *Coating*, (1), 24, 1969.
237. *Adhäsion*, (11), 37, 1994.

238. *Kaut. Gummi, Kunstst.*, 37 (1), 17, 1984.

239. *Eur. Adhes. J.*, (6), 23, 1995.

240. *Adhes. Age*, (12), 30, 1987.

241. W. Endlich, Klebende Dichtstoffe in der industriellen Fertigung, T.3, Seminar an der TA Wuppertal, *Adhäsion*, (3), 26, 1982.

242. M. Gerace, *Adhes. Age*, (8), 1983.

243. J.W. Otter and G.R. Watts, U.S. Patent 5,346,766, Avery International Corp., Pasadena, CA, USA, 1994.

244. G.D. Bennett (Simmonds Precision, New York), U.S. Patent 4,395,250, 1987, in *Adhes. Age*, (5), 28, 1988.

245. Kendall Co., Canadian Patent 853,145, 1987, in *Coating*, (9), 33, 1988.

246. R.R. Charbonneau and G.L. Groff, EP 0,106,559, B1, Minnesota Mining and Manuf. Co., St. Paul, MN, USA, 1984.

247. U.S. Patent 2,925,174, in R.R. Charbonneau and G.L. Groff, EP 0,106,559 B1, Minnesota Mining and Manuf. Co., St. Paul, MN, USA, 1984.

248. U.S. Patent 4,286,047, in R.R. Charbonneau and G.L. Groff, EP 0,106,559 B1, Minnesota Mining and Manuf. Co., St. Paul, MN, USA, 1984

249. G.F. Vesley, A.H. Paulson, and E.C. Barber, EP 0,202,938, A2, 1986.

250. U.S. Patent 4,223,067, in R.R. Charbonneau and G.L. Groff, EP 0,106,559 B1, Minnesota Mining and Manuf. Co., St. Paul, MN, USA, 1984.

251. G. Bennett, P.L. Geib, J. Klingen, and T. Neeb, *Adhäsion*, (7–8), 19, 1996.

252. D.C. Stillwater, D.C. Kostenmaki, and M.H. Mazurek, U.S. Patent 5,344,681, Minnesota Mining and Manuf. Co., MN, USA, 1994.

253. J. Suchy, J. Hezina, and J. Matejka, Czech Patent 247,802, 1987, *CAS, Adhesives*, 19, 4, 1988.

254. F.D. Blake, EP 0,141,504, A1, Minnesota Mining and Manuf. Co., St. Paul, MN, USA, 1985.

255. *Coating*, (9), 33, 1988.

256. Packard Electric, Engineering Specification, ES-M-2147.

257. Z. Czech, *Adhäsion*, (11), 26, 1994.

258. U.S. Patent 3,096,202, in P. Gleichenhagen and I. Wesselkamp, EP 0,058, 382 B1, Beiersdorf AG, Hamburg, Germany, 1982.

259. *Adhäsion*, (5), 161, 1971.

260. U.S. Patent Blake 3.865770, in F.C. Larimore and R.A. Sinclair, EP 0,197,662 A1, Minnesota Mining and Manuf. Co., St. Paul, MN, USA, 1986.

261. T. Huang Haddock, EP 0,130,080 B1, Johnson & Johnson Products Inc., New Brunswick, NJ, 1985.

262. D. Allen Jr. and E. Flam, U.S. Patent 4,650,817, Bard Inc., Murray Hill, NJ, 1987, *Adhes. Age*, (5), 24, 1987.

263. *Modern Plast. Int.*, Show Daily, 23 & 24 Oct. 2004, p. 17.

264. *Adhes. Age*, (4), 6, 1983.

265. Z. Czech and D. Sander, DE 44, 005, Lohmann GmbH, Neuwied, Germany, 1994.

266. Johnson & Johnson, U.S. Patent 3,483,018, *Coating*, (1), 23, 1971.

267. M. von Bittera, D. Schäpel, U. von Gizycki, and R. Rupp, EP 0,147,588 B1, Bayer AG, Leverkusen, Germany, 1985.

268. US Patent 3,321,451, in S.E. Krampe and C.L. Moore, EP 0,202,831 A2, Minnesota Mining and Manuf. Co., St. Paul, MN, USA, 1986.

269. *Caoutchoucs Plastiques*, (9), 64, 2004.

270. R.L. Sun and J.F. Kennedy, U.S. Patent 4,762,888, Johnson and Johnson Products, Inc., USA, 1988, *CAS, Hot Melt Adhesives*, 26, 1, 1988.

271. US Patent 1,760,820, 3,189,581, 3,371,071, 3,509,111, and 3,475,363, in T. Huang Haddock, EP 0,130,080 B1, Johnson & Johnson Products Inc., New Brunswick, NJ, 1985.

272. J.N. Kellen and C.W. Taylor, EP 0246 A2, Minnesota Mining and Manuf. Co., St. Paul, MN, USA, 1987.

273. Martens et al., U.S. Patent 4,181,752, in J.N. Kellen and C.W. Taylor, EP 0,246A2, Minnesota Mining and Manuf. Co., St. Paul, MN, USA, 1987.

274. *Adhäsion*, (7/8), 124, 1991.

275. *Verpackungs Rundsch.*, (9), 994, 1983.

276. *Kimberly Clark Corporation, Technical Paper*, TP-C-10−104/11/89, Neenah, WI, USA.

277. M.J. Huber, U.S. Patent 546,692, Quality Manufacturing Inc., 1995, *Adhes. Age*, (5), 12, 1996.

278. R.B. Lipson, U.S. Patent 5,468,533, Kwik Paint Products, *Adhes. Age*, (5), 12, 1996.

279. H. Meichner, Gebrauchsmuster, GM 8,607,368, Rehau, Germany, 1986.

280. I. Benedek, E. Frank, and G. Nicolaus, DE 4,433,626 A1, Poli-Film Verwaltungs GmbH, Wipperfürth, Germany, 1994.

281. T. Moldvai and N. Piatkowski, Romanian Patent 93,124, Inst. Cerc. Pielarie, Incaltaminte, Bucharest, 1987.

282. *Adhäsion*, (11), 481, 1967.

283. C. Cordorch, *VDI Nachrichten*, (33), 14, 1994.

284. K. Taubert, *Adhäsion*, (10), 379, 1970.

285. *Papier Kunstst Verarb.*, (19), 32, 1990.

286. G. Menges, W. Michaeli, R. Ludwig, and K. Scholl, *Kunststoffe*, 80, (11), 1245, 1990.

287. E. Höflin and H. Breu, *Adhäsion*, 10(6), 252, 1966.

288. *Allg. Papier Rundsch.*, 29, 798, 1988.

289. *Coating*, (8), 98, 1987.

290. Kimberly Clark Corporation, Technical Paper, TP-C-10−107/11/89, Neenah, WI, USA.

291. Schutzfolien, Tl-2.2−21, d/November 1979, Teil 3, b.5, BASF AG, Ludwigshafen, Germany.

292. G. Fauner, *Adhäsion*, (4), 27, 1985.

293. *Eur. Adhes. J.*, (6), 22, 1995.

294. H. Bucholz, *Oberflächentechnik*, (10), 484, 1973.

295. *Coating*, (3), 66, 1984.

296. G. Allwyn, *American Paint J. Convention Daily Number*, (15), 33, 1969, in *Coating*, (12), 368, 1970.

297. *Kaut. Gummi, Kunstst.*, 40(10), 981, 1987.

298. Product Application Guide, Poli Film-America, Cary, IL, USA, 1994.

299. Hoogovens Packaging Steel, Ijmujden, The Netherlands, Technical Booklet, 1997.

300. W.E. Havercroft, *Paper, Film Foil Conv.*, (10), 52, 1973.

301. J.W. Kühr, *Farbe u. Lack*, (5), 475, 1970.

302. *Coating*, (2), 78, 1984.

303. W. Brockman, O. Diedrich Hennemann, H. Kollek, and C. Marx, *Adhäsion*, (9), 24, 1986.

304. W.S.D. Bascom and R.L. Patrick, *Adhes. Age*, (19), 25, 1974.

305. *Coating*, (10), 264, 1979.

306. V. Antonioli, *La Revista del Colore*, (58), 79, 1973.

307. Dynamit Nobel A.G, Dutch Patent 72.07134, *Coating*, (5), 122, 1974.

308. *Adhäsion*, (1/2), 28, 1983.

309. G.F. Bond and J.N. Ralston, *Ind. Finish. Surf. Coat.*, (305), 4, 1973.

310. C.W. Uezelmeier, *SPE-J.*, (5), 69, 1970.

311. G. Kuehl, *Metalloberflaeche*, 42(3), 139, 1988, in *CAS, Crosslinking Reactions*, 12, 11, 1988.

312. K.H. Seifert, *Coating*, (2), 41, 1972.

313. W. Brushwell, *Farbe u. Lack*, (1), 34, 1974.

314. *Scope*, (3), 92, 1988.

315. *Coating*, (3), 66, 1984.

316. *Coating*, (12), 450, 1986.

317. *Farbe u. Lack*, 92(4), 296, 1986.

318. Ball Brothers, Muncie, Ind., USA, DBP 128,585, *Coating*, (7), 274, 1969.

319. *Eur. Plast. News*, (1), 28, 1996.

320. M. Saada and M. El-Moualled, *Chimie et Ind.*, (15), 1917, 1970.

321. E. Röhrl and L. Schwiegk, *Plastverarb.*, 42(10), 39, 1991.

322. J.P. Weibel, W. Körmer, and S. Caporusso, *Kunststoffe*, 83(7), 541, 1993.

323. Products and more, *Plastverarb*, K2004, Oct. 20−27, 2004, p. 17.

324. W. Kremer, *Farbe u. Lack*, 94(3), 205, 1988, in *CAS, Crosslinking Reactions*, 12, 11, 1988.

325. *Kunststoff Information*, (1020), 2, 1991.

326. *Plastics Information Eur.*, 25(22), 5, 2001.

327. P. Dippel, *Coating*, (4), 44, 1988.

328. *Plastverarb.*, (9), 9, 1994.

329. *Coating*, (7), 246, 1993.
330. *Polycarbonate*, Technical Booklet, 15/3/82, ERTA Tielt, Belgium.
331. *Coating*, (5), 170, 1996.
332. M. Heinze, *Kunststoffe*, 83(8), 630, 1993.
333. A.K. Schlarb and G.W. Ehrenstein, *Kunststoffe*, 83(8), 597, 1993.
334. W. Daum, *Kunststoffe*, 84(10), 1433, 1994.
335. D.J.P. Harrison, J.F. Johnson, and W. Ross Yates, *Polymer Eng. Sci.*, 22(14), 865, 1982.
336. A. Haas, U.S. Patent 4,624,991, 1986, Soc. Chim. de Charbonnage-CdF, *Adhes. Age*, (5), 26, 1987.
337. H. Inada, T. Tomomori, and K. Kawakura, Japanese Patent 62,270,642, Tokuyama Soda Co., Ltd., 1987, *CAS, Adhesives*, 13, 1, 1988.
338. O. Tadayuki, M. Kentaro, F. Kazunori, and K. Toshio, EP 249,442, Sumitomo Chem. Co., Ltd., 1987.
339. Klebebandträger und Abdeckfolie, Metallisierungs und Dekorationsfolie, Datenblatt, Hoechst Folien, Ausgabe 07/92.
340. British Patent 1,081,291, Bakelite Xylonite Ltd., *Coating*, (4), 114, 1969.
341. DAF, Dow Adhesive Films, Dow Chemical Co., Horgen, Switzerland, 1983.
342. Protective by Nitto, Nitto, Deutschland, 04.11.1986, Data Sheets.
343. A.S. Jandel, *VDI*, (37), 29, 1996.
344. J.F. Kuik, *Papier Kunstst. Verarb.*, (10), 26, 1990.
345. R.M. Podhajny, *Converting and Packaging*, (3), 21, 1986.
346. *Adhérence des Films Pelables*, ECCA,T 17/1985.
347. *Prodoc.*, (4), 14, 1988.
348. *Br. Plast.*, (10), 48, 2001.
349. *Br. Plast. Rub.*, (10), 56, 1998.
350. Plexiglas GS, XT, Ke. Nr.212–1, Technische Information, Produktbeschreibung, November, 1993, Röhm AG, Darmstadt, Germany.
351. D. Symietz, *Kunststoffe*, (11), 11, 1994.
352. E. Doege and U. Hesberg, *Blech, Rohre, Profile*, 39(1), 25, 1992.
353. M. Herrmann, *Blech, Rohre, Profile*, 40(2), 164, 1993.
354. T. Miura, *Giesserei*, 62(17), 437, 1975.
355. *Kunststoff Information*, (1169), 5, 1994.
356. A. Hollman, ATP News-Letter, 1, 6, 2004, ATP Adhesive Systems GmbH, Wuppertal, Germany.
357. D. Eisele, ATP News-Letter, 2, 4, 2004, ATP Adhesive Systems GmbH, Wuppertal, Germany.
358. D. Eisele, ATP News-Letter, 2, 5, 2004, ATP Adhesive Systems GmbH, Wuppertal, Germany.
359. *Druckwelt*, (5), 51, 1988.
360. R. Uhlemayr, Gebrauchsmuster, PS 3,508,114, O.T. Drescher GmbH, Rutesheim, Germany, 1986.
361. Selbstklebefolien maßgeschnitten, Booklet, Grafityp, Houthalen, Belgium, October 1992.
362. *Grafityp Newsletter*, (11), 10, 1997.
363. Perigraf, Computer und Informationssysteme GmbH, Eibau, Germany, 2002.
364. Graphtec, RFC3100–60, Tischschneideplotter, Multiplot Grafiksysteme, Bad Emstal, Germany, 1997.
365. Cal Comp, Data Sheet, Sign Europe, 1997, Düsseldorf, Germany.
366. Regulus Schablonen Papiere, Application tape, Schutzfolie, Klaus Koenig KG, Germany, 1997, Booklet.
367. J.W. Prane, *Adhes. Age*, (1), 44, 1989.
368. *Papier Kunstst. Verarb.*, (2), 44, 1996.
369. Emulsion Adhesive E 959, Sealock, Andover, Hampshire, UK, 1995.
370. *Adhes. Age*, (9), 62, 1986.
371. *Labels, Label.*, (3/4), 10, 1994.
372. *Papier Kunstst. Verarb.*, (9), 57, 1968.
373. Y. Ohata, K. Awano, M. Atsuji, and T. Hattori, Japanese Patent, 63,89,345, Toa Gohsei Chemical Industry Co., Ltd., Aica Kogyo Co., Ltd., 1988, *CAS, Adhesives*, 21, 10, 1988.
374. DBP 1777807, Rasvino Predpriatie za organiziona Technika, Silistra, Bulgary, *Coating*, (7), 210, 1969.
375. Z. Czech and H.D. Sander, DE 4,219,368, Lohmann GmbH, Neuwied, 1992.
376. *Coating*, (1), 29, 1978.
377. *Adhäsion*, (6), 270, 1965.

378. G. Fuchs, *Adhäsion*, (3), 24, 1982.
379. *Coating*, (12), 342, 1984.
380. *Allg. Papier Rundsch.*, (6), 159, 1986.
381. *Allg. Papier Rundsch.*, (18), 686, 1986.
382. I.P. Rothernberg, U.S. Patent 4,650,704, Stik-Trim Industries Inc., New York, 1987.
383. J. Weiss, *Metall*, (5), 367, 1994.
384. W.C. Zybco, W. Wald, and T.W. McCure, U.S. Patent 4,680,209, Burlington Industries Inc., Greensboro, NC, USA, 1987, *Adhes. Age*, (2), 58, 1988.
385. J. Weidauer, *Eur. Adhes., Seal.*, (5), 26, 1995.
386. F.S. Thomas, Los Angeles, CA, U.S. Patent 3148/398, *Adhäsion*, (2), 78, 1966.
387. Nitto Denko Products, Catalog Code 02002, p. 16, Tokyo, Japan.
388. *Adhes. Age*, (3), 10, 1987.
389. *Pack Report*, (3), 26, 1997.
390. *Der Siebdruck*, (8), 7, 1988.
391. *Kunststoff J.*, (9), 15, 1986.
392. *Eur. Adhes., Seal.*, (6), 16, 1996.

Abbreviations and Acronyms

I. COMPOUNDS

AA	acrylic acid
AB	α-allyl benzoin
ABP	4-acryloiloxy benzophenone
ABS	acrylonitrile–butadiene–styrene copolymer
AC	acrylic
ACDB	4-acrylamidocabonyldioxy benzophenone
AlAcAc	aluminium(III) acetylacetonate
AN	acrylnitrile
APAO	amorphous polyalphaolefine
APO	amorphous polyolefin
APP	atactic polypropylene
B	butadiene
BA	benzoin acrylate
BBBF	butanediol-1,4-*bis*-benzophenoxy formiate
BBPF	butanediol-1,4-*bis*-propylene imine formiate
BMN-s-T	2,4-*bis*-trichloromethyl-6[1-(4-methoxynaphtyl)-*s*-triazine
BMP-s-T	2,4-*bis*-trichloromethyl-6(3,4,5-trimethoxyphenyl)-*s*-triazine
BN-s-T	2,4-*bis*-trichloromethyl-6(1-naphtyl)-*s*-triazine
BOPP	biaxially oriented polypropylene
BPAA	*N,N'-bis*-propylene adipic acid amide
BPAD	*N,N'-bis*-propylene phenyl phosphonic acid diamide
BPFA	*N,N'-bis*-propyleneoctafluoro adipic acid amide
BPIA	*N,N'-bis*-propylene isophtalic acid amide
BPO	benzoyl peroxide
BR	butyl rubber
i-BT	tetraisobutyl titanate
n-BT	tetranormalbutyl titanate
BuAc	butyl acrylate
BVCN	4-benzophenyl vinyl carbonate
CIPCB	4-chloro-4'-propylene imine carbonyl benzophenone
CoACA	cobalt(II) acetylacetonate
CrACA	chromium(III) acetylacetonate
CSBR	carboxylated butadiene rubber
Cymel 1123	highly alkylated benzoguanamine resin
Cymel 1170	highly *n*-butylated glycoluril resin
Cymel 303	highly methylated melamine resin
Cymel 370	partially methylated melamine resin
DBP	dibutyl phtalate
Desmodur TT	uretdione of 2,4-diisocyanate toluene
DMAEMA	dimethylamino ethyl methacrylate
DMPA	5-dimethyl maleininimidyl-*N,N'-bis*-propylene isophtalic acid amide
Dynomin UB-24-BX	partially *n*-butylated urea resin

EAA	ethylene-acrylic acid
EBA	ethylene-butyl acrylate
EHA	ethylhexyl acrylate
2-EHT	tetra-2-ethylhexyl titanate
EMAA	ethylene-maleic anhydride
EPDM	ethylene–propylene–diene multipolymer
EPR	ethylene propylene rubber
EPVC	emulsion PVC
ET	tetraethyl titanate
EVAc	ethylene-vinyl acetate copolymer
FeAcAc	iron(III) acetylacetonate
HDPE	high density polyethylene
HIPS	high impact polystyrene
HMDI	hexamethylene diisocyanate
HMHDPE	high molecular weight high density polyethylene
I	Isolene (plasticizer)
IBMA	N-(isobutoxymethylene)acrylamide
IPDI	isophorone diisocyanate
IPPU	isophorone dipropylene urea
KT	tetracresol titanate
LDPE	low-density polyethylene
LLDPE	linear low-density polyethylene
Lupranate A270	4,4′-diisocyanatediphenyl methane
MAA	methacrylic acid
MAGME	methyl acrylamidoglycolate methyl ether
MDI	methylene diisocyanate
MDPE	medium-density polyethylene
MeP	methyl pentene
MMA	methyl methacrylate
MnAcAc	manganese(III) acetylacetonate
MOST	2,4-*bis*-(trichloromethyl)-6-*p*-methoxystyryl-*s*-triazine
Neocryl CX-100	*tris*-methylolpropane-*tris*-N-methylaziridinyl) propionate
NiAcAc	nickel(II) acetylacetonate
NMA	N-methylol acrylamide
NR	natural rubber
NRL	natural rubber latex
NVP	N-vinylpyrrolidone
OPP	oriented polypropylene
OPS	oriented polystyrene
PA	polyamide
PAC	phenyl-(1-acryloyloxy)-cyclohexyl ketone
PB	polybutylene
PC	polycarbonate
PCA	4-propylene imine carbonyl benzophenone
PCF	2-propylene imine carbonyl-9-fluorenone
PE	polyethylene
PEG	poly(ethylene glycol)
PET	polyethylene terephtalate

PIB	polyisobutylene
PP	polypropylene
PS	polystyrene or pressure-sensitive
i-PT	tetraisopropyl titanate
n-PT	tetranormalpropyl titanate
PUR	polyurethane
PVA	polyvinyl alcohol
PVC	polyvinyl chloride
PVE	polyvinyl ether
PVP	polyvinyl pyrrolidone
R	Regalite (tackifier) or gas constant
SAN	styrene–acrylnitrile
SBC	styrene block copolymers
SBR	styrene–butadiene–rubber
SBS	styrene–butadiene–styrene
SEP	styrene–ethylene–propylene
SIS	styrene–isoprene–styrene
SPVC	suspension PVC
T	tackifier or temperature
TBPO	*tris*-benzophenyloxy phosphine oxide
TEC	triethyl citrate
TiAcAc	titanium(IV) acetylacetonate (TiACA-75)
TMDI	isomeric mixture (1:1) of 2,2,4- and 3,4,4-trimethyl-1,6-hexamethylene diisocyanate
TPAT	2,4,6-*tris*-propylene propionic acid amide)-1,3,5-triazine
TPU	thermoplastic polyurethane
VAc	vinyl acetate
p-VB	*p*-vinyl benzophenone
VC	vinyl chloride
VLDPE	very low-density polyethylene
VP	vinyl pyrrolidone
VTAS	vinyltrimethylaziridinyl silane
XeF	xenone fluoride
XL-353	2,4-*bis*-trichlorometly-6(4-methoxy-phenyl)-*s*-triazine
XR 5551	polycarbodiimide, Permutex XR-5551
XR 5580	polycarbodiimide, Permutex XR-5589
ZLI 3331	2-hydroxy-1-[4-(2-acryloyloxyethoxy)phenyl]-2-methyl-1-propanone
ZnAcAc	zinc(II) acetylacetonate
ZrAcAc	zirconium(IV) acetylacetonate

II. TERMS

A	area or a constant
a_T	temperature-dependent shift factor
A_0	reference area
AFERA	Association des Fabricants Européens de Rubans Autoadhesifs
ASTM	American Society for the Testing of Materials
B	constant
BGA	Bundes Gesundheitsamt
BWB	German military procurement office

BWB-TL	German military norm
BUR	blow up ratio
C	constant or cuttability or cohesion
C_w	coating weight
CC	carbon–carbon
cf	cohesive failure
CI	cuttability index
CD	cross-direction
CH	carbon-hydrogen
CLC	carcass-like crosslinker
COF	coefficient of friction
DDI	dart drop impact
DIN	German standard
DMA	Dynamic Mechanical Analysis
DPI	dots per inch
DSC	Differential Scanning Calorimetry
E	modulus
EB	electron beam
EDP	electronic data processing
EN	European norms
EPSMA	European Pressure Sensitive Manufacturers Association
eV	electron volt
EV	viscosity test method
FDA	food and drug administration
FFP	film forming polymer
FINAT	Federation Internationale des Fabricants Transformateurs d'Adhésifs et Thermocollants sur Papier et Autres Supports
FTIR	Fourier-transformed Infrared Spectroscopy
FTM	FINAT Test Method
G'	storage modulus
G''	loss modulus
GID	gear-in-die
HC	hydrocarbon
HD	high density
HI	high impact
HM	hot-melt
HMA	hot-melt adhesive
HMPSA	hot-melt pressure-sensitive adhesive
HS	hot shear
HSG	hot shear gradient
IPN	inter-penetrating network
IR	infrared
ISO	International Standards Organization
kp	kilopond
LLC	ladder-like crosslinker
LT	loop tack
MD	machine direction
MFI	melt flow index
MFT	minimum film forming temperature
M_n	number average molecular weight
MP	melting point

mPa sec	milliPascal second
MPR	melt-processed rubber
MW	molecular weight
M_w	weight average molecular weight
MWD	molecular weight distribution
NMR	nuclear magnetic resonance
P	pressure or plasticizer
PALS	Positron Annihilation Life-Time Spectroscopy
PEG	poly(ethylene glycol)
PEGDA	poly(ethylene glycol) diacrylate
PEGMMA	poly(ethylene glycol) monomethacrylate
PN	plasticity number (Williams)
Ppt	parts per thousand
PS	pressure-sensitive or polystyrene
PSA	pressure-sensitive adhesive
PSP	pressure-sensitive product
PSTC	Pressure-Sensitive Tape Council
Pts	parts
PTS	Papiertechnische Stiftung
Q	heat
R	universal gas constant
RB	rolling ball
RBT	rolling ball tack
R&B	ring and ball softening point
RCT	rolling cylinder tack
RF	radiofrequency
RH	relative humidity
RR	rubber–resin
RT	room temperature
RVT	known Brookfield viscosity method
SAF	self-adhesive film
SAFT	shear adhesion failure temperature
SB	solvent-based
SF	solvent-free
SH	shear
SL	solventless
SP	softening point
T	temperature or tackifier
T_g	glass transition temperature
TD	transverse direction
TLMI	Tag and Label Manufacturers Institute
TMA	thermomechanical analysis
TPE	thermoplastic elastomer
UV	ultraviolet
VIP	variable image printing
W	work
W_a	work of adhesion
WB	water-based
WBA	wet bonding adhesion
WLF	William–Landel–Ferry
ZN	Ziegler–Natta

Index